ESSENTIAL
CIRCUITS
REFERENCE
GUIDE

The McGraw-Hill
Engineering Reference Guide Series

This series makes available to professionals and students a wide variety of engineering information and data available in McGraw-Hill's library of highly acclaimed books and publications. The books in the series are drawn directly from this vast resource of titles. Each one is either a condensation of a single title or a collection of sections culled from several titles. The Project Editors responsible for the books in the series are highly respected professionals in the engineering areas covered. Each Editor selected only the most relevant and current information available in the McGraw-Hill library, adding further details and commentary where necessary.

Hicks · CIVIL ENGINEERING CALCULATIONS REFERENCE GUIDE

Hicks · MACHINE DESIGN CALCULATIONS REFERENCE GUIDE

Hicks · PLUMBING DESIGN AND INSTALLATION REFERENCE GUIDE

Hicks · POWER GENERATION CALCULATIONS REFERENCE GUIDE

Hicks · POWER PLANT EVALUATION AND DESIGN REFERENCE GUIDE

Higgins · PRACTICAL CONSTRUCTION EQUIPMENT MAINTENANCE
      REFERENCE GUIDE

Johnson & Jasik · ANTENNA APPLICATIONS REFERENCE GUIDE

Markus and Weston · ESSENTIAL CIRCUITS REFERENCE GUIDE

Merritt · CIVIL ENGINEERING REFERENCE GUIDE

Perry · BUILDING SYSTEMS REFERENCE GUIDE

Rosaler and Rice · INDUSTRIAL MAINTENANCE REFERENCE GUIDE

Rosaler and Rice · PLANT EQUIPMENT REFERENCE GUIDE

Woodson · HUMAN FACTORS REFERENCE GUIDE FOR ELECTRONICS AND
      COMPUTER PROFESSIONALS

Woodson · HUMAN FACTORS REFERENCE GUIDE FOR PROCESS PLANTS

ESSENTIAL CIRCUITS REFERENCE GUIDE

JOHN MARKUS (deceased)

Consultant, McGraw-Hill Book Company
Senior Member, Institute of Electrical and Electronics Engineers

CHARLES WESTON

Technical Editor
Byte Magazine
Peterborough, N.H.

McGRAW-HILL BOOK COMPANY

New York St. Louis San Francisco Auckland Bogotá
Caracas Colorado Springs Hamburg Lisbon
London Madrid Mexico Milan Montreal
New Delhi Oklahoma City Panama Paris
San Juan São Paulo Singapore
Sydney Tokyo Toronto

Library of Congress Cataloging-in-Publication Data

Markus, John, 1911–
    Essential circuits reference guide.

    (McGraw-Hill engineering reference guide series)
    Includes index.
    1. Electric circuits—Handbooks, manuals, etc.
2. Electronic circuits—Handbooks, manuals, etc.
I. Weston, Charles (Charles D.)   II. Title.   III. Series.
TK454.M29   1988   621.319′2   88–569
ISBN   0-07-040462-3

1234567890   HAL/HAL   8921098

ISBN 0-07-040462-3

The material in this volume has been published previously in the following
books by John Markus (ed.): *Modern Electronic Circuits Reference Manual*,
Copyright © 1980 by McGraw-Hill, Inc., *Electronic Circuits Manual*, Copyright © 1971 by McGraw-Hill, Inc., and *Guidebook of Electronic Circuits*,
Copyright © 1974 by McGraw-Hill, Inc. All rights reserved. Printed and
bound by Arcata graphics/Halliday.

Contents

Preface   vii
Abbreviations Used   ix
Semiconductor Symbols Used   xiii
Addresses of Sources Used   xv

1. Antenna Circuits   1
2. Audio Control Circuits   5
3. Automotive Circuits   11
4. Battery-Charging Circuits   17
5. Bridge Circuits   25
6. Capacitance Control Circuits   31
7. Chopper Circuits   36
8. Code Circuits   41
9. Comparator Circuits   49
10. Contact Bounce Suppression Circuits   58
11. Converter Circuits—Analog-to-Digital   62
12. Converter Circuits—DC-to-DC   72
13. Converter Circuits—General   78
14. Converter Circuits—Digital-to-Analog   89
15. Data Transmission Circuits   96
16. Fiber-Optic Circuits   106
17. Filter Circuits—Passive   111
18. Filter Circuits—Active   116
19. Frequency Divider Circuits   143
20. Function Generator Circuits   146
21. Gain Control Circuits   153
22. Infrared Circuits   159
23. Instrumentation Circuits   166
24. Lighting Control Circuits   173
25. Measuring Circuits—Audio   180
26. Measuring Circuits—Capacitance   183
27. Measuring Circuits—Current   192
28. Measuring Circuits—Frequency   199
29. Measuring Circuits—General   208
30. Measuring Circuits—Power   215

31. Measuring Circuits—Resistance   219
32. Measuring Circuits—Temperature   225
33. Metal Detector Circuits   230
34. Microprocessor Circuits   234
35. Modem Circuits   237
36. Modulator Circuits   243
37. Modulator/Demodulator Circuits   251
38. Motor Control Circuits   258
39. Multiplexer Circuits   267
40. Multiplier Circuits   273
41. Multivibrator Circuits   279
42. Music Circuits   285
43. Noise Circuits   294
44. Operational Amplifier Circuits   299
45. Optoelectronic Circuits   315
46. Oscillator Circuits—Audio Frequency   320
47. Oscillator Circuits—Radio Frequency   330
48. Phase Control Circuits   343
49. Photoelectric Circuits   347
50. Power Control Circuits   351
51. Power Supply Circuits   358
52. Protection Circuits   367
53. Pulse Generator Circuits   373
54. Receiver Circuits   387
55. Regulated Power Supply Circuits   398
56. Regulator Circuits   422
57. Servo Circuits   453
58. Switching Regulator Circuits   460
59. Telephone Circuits   479
60. Voltage-Level Detector Circuits   490
61. Voltage Reference Circuits   503
62. Zero-Voltage Detector Circuits   509

Index   515

Preface

Five hundred pages of electronic circuits are arranged here in 62 logical sections for convenient browsing and reference by electronics engineers, technicians, students, microprocessor enthusiasts, amateur radio fans, and experimenters. Each circuit has type numbers or values of all significant components, an identifying title, a concise description, performance data, and suggestions for other applications. At the end of each description is a citation giving the title of the original article or book, its author, and the exact location of the circuit in the original source.

The circuits for this book were taken from the *Modern Electronics Circuits Reference Manual*, the *Guidebook of Electronic Circuits*, and the *Electronic Circuits Manual*. Although this reference guide only includes a small fraction of these encyclopedic works, we hope the user will find this selection of "classic" circuits of interest.

Engineering libraries, particularly in foreign countries, have found these circuit abstracts to be a welcome substitute for the original sources when facing limitations on budgets, shelving, or search manpower. As further evidence of their usefulness in other countries, some of the books have been translated into Greek, Spanish, or Japanese.

To find a desired circuit quickly, start with the alphabetically arranged table of contents at the front of the book. Note the sections most likely to contain the desired type of circuit, and look in these first. Remember that most applications use combinations of basic circuits, so a desired circuit could be in any of several different sections. Scope notes following section titles define the basic circuits covered and sometimes suggest other sections for browsing.

If a quick scan does not locate the exact circuit desired, use the index at the back of the book. Here the circuits are indexed in depth under the different names by which they may be known. Hundreds of cross-references in the index aid searching. The author index will often help find related circuits after one potentially useful circuit is found, because authors tend to specialize in certain circuits.

Values of important components are given for every circuit because these help in reading the circuit and redesigning it for other requirements. The development of a circuit for a new application is speeded when design work can be started with a working circuit, instead of starting from scratch. Research and experimentation are thereby cut to a minimum, so even a single use of this circuit-retrieval book could pay for its initial cost many times over. Drafting errors on diagrams are minimized because any corrections pointed out in subsequently published errata notices have been made; this alone can save many frustrating hours of troubleshooting.

Most circuit elements are still available from the original manufacturers. In some cases, new types with better stability and operating characteristics are direct replacements for these older circuit elements.

This book is organized to provide a maximum of circuit information per page, with minimum repetition. The section title at the top of each page and the original title in the citation should therefore be considered along with the abstract when evaluating a circuit.

Abbreviations are used extensively to conserve space. Their meanings are given after this preface. Abbreviations on diagrams and in original article titles were unchanged and may differ slightly, but their meanings can be deduced by context.

Mailing addresses of all cited original sources are given at the front of the book, for convenience in writing for back issues or copies of articles when the source is not available at a local library. These sources will often prove useful for construction details, performance graphs, and calibration procedures.

To the original publications cited and their engineering authors and editors should go major credit for making possible this contribution to electronic circuit design. The diagrams have been reproduced directly from the original source articles, by permission of the publisher in each case.

CHARLES D. WESTON
Technical Editor

Abbreviations Used

A	ampere	CRO	cathode-ray oscilloscope	F	farad
AC	alternating current			°F	degree Fahrenheit
AC/DC	AC or DC	CROM	control and read-only memory	FET	field-effect transistor
A/D	analog-to-digital			FIFO	first-in first-out
ADC	analog-to-digital converter	CRT	cathode-ray tube	FM	frequency modulation
		CT	center tap		
A/D, D/A	analog-to-digital, or digital-to-analog	CW	continuous wave	4PDT	four-pole double-throw
		D/A	digital-to-analog		
ADP	automatic data processing	DAC	digital-to-analog converter	4PST	four-pole single-throw
				FS	full scale
AF	audio frequency	dB	decibel	FSK	frequency-shift keying
AFC	automatic frequency control	dBC	C-scale sound level in decibels		
				ft	foot
AFSK	audio frequency-shift keying	dBm	decibels above 1 mW	ft/min	foot per minute
		dBV	decibels above 1 V	ft/s	foot per second
AFT	automatic fine tuning	DC	direct current	ft²	square foot
		DC/DC	DC to DC	F/V	frequency-to-voltage
AGC	automatic gain control	DCTL	direct-coupled transistor logic	F/V, V/F	frequency-to-voltage, or voltage-to-frequency
Ah	ampere-hour	diac	diode AC switch		
ALU	arithmetic-logic unit	DIP	dual in-line package	G	giga- (10⁹)
AM	amplitude modulation	DMA	direct memory access	GHz	gigahertz
				G-M tube	Geiger-Mueller tube
AM/FM	AM or FM	DMM	digital multimeter	h	hour
AND	type of logic circuit	DPDT	double-pole double-throw	H	henry
AVC	automatic volume control			HF	high frequency
		DPM	digital panel meter	HFO	high-frequency oscillator
b	bit	DPST	double-pole single-throw		
BCD	binary-coded decimal			hp	horsepower
BFO	beat-frequency oscillator	DSB	double sideband	Hz	hertz
		DTL	diode-transistor logic	IC	integrated circuit
b/s	bit per second	DTL/TTL	DTL or TTL	IF	intermediate frequency
C	capacitance; capacitor	DUT	device under test		
		DVM	digital voltmeter	IGFET	insulated-gate FET
°C	degree Celsius; degree Centigrade	DX	distance reception; distant	IMD	intermodulation distortion
CATV	cable television	EAROM	electrically alterable ROM	IMPATT	impact avalanche transit time
CB	citizens band				
CCD	charge-coupled device	EBCDIC	extended binary-coded decimal interchange code	in	inch
				in/s	inch per second
CCTV	closed-circuit television			in²	square inch
		ECG	electrocardiograph	I/O	input/output
cm	centimeter	ECL	emitter-coupled logic	IR	infrared
CML	current-mode logic	EDP	electronic data processing	JFET	junction FET
CMOS	complementary MOS			k	kilo- (10³)
CMR	common-mode rejection	EKG	electrocardiograph	K	kilohm (,000 ohms); kelvin
		EMF	electromotive force		
CMRR	common-mode rejection ratio	EMI	electromagnetic interference	kA	kiloampere
cm²	square centimeter	EPROM	erasable PROM	kb	kilobit
coax	coaxial cable	ERP	effective radiated power	keV	kiloelectronvolt
COHO	coherent oscillator			kH	kilohenry
COR	carrier-operated relay	ETV	educational television	kHz	kilohertz
				km	kilometer
COS/MOS	complementary-symmetry MOS (same as CMOS)	eV	electronvolt	kV	kilovolt
		EVR	electronic video recording	kVA	kilovoltampere
				kW	kilowatt
CPU	central processing unit	EXCLUSIVE-OR	type of logic circuit	kWh	kilowatthour
		EXCLUSIVE-NOR	type of logic circuit	L	inductance; inductor
CR	cathode ray			LASCR	light-activated SCR

LASCS	light-activated SCS	NMOS	N-channel MOS	QRP	low-power amateur radio
LC	inductance-capacitance	NOR	type of logic circuit	R	resistance; resistor
LCD	liquid crystal display	NPN	negative-positive-negative	RAM	random-access memory
LDR	light-dependent resistor	NPNP	negative-positive-negative-positive	RC	resistance-capacitance
LED	light-emitting diode	NRZ	nonreturn-to-zero	RF	radio frequency
LF	low frequency	NRZI	nonreturn-to-zero-inverted	RFI	radio-frequency interference
LIFO	last-in first-out	ns	nanosecond	RGB	red/green/blue
lm	lumen	NTSC	National Television System Committee	RIAA	Recording Industry Association of America
LO	local oscillator				
logamp	logarithmic amplifier				
LP	long play	nV	nanovolt	RLC	resistance-inductance-capacitance
LSB	least significant bit	nW	nanowatt		
LSI	large-scale integration	OEM	original equipment manufacturer	RMS	root-mean-square
m	meter; milli- (10^{-3})	opamp	operational amplifier	ROM	read-only memory
M	mega- (10^6); meter (instrument); motor	OR	type of logic circuit	rpm	revolution per minute
		p	pico- (10^{-12})	RTL	resistor-transistor logic
mA	milliampere	P	peak; positive		
Mb	megabit	pA	picoampere	RTTY	radioteletype
MF	medium frequency	PA	public address	RZ	return-to-zero
mH	millihenry	PAL	phase-alternation line	s	second
MHD	magnetohydro-dynamics	PAM	pulse-amplitude modulation	SAR	successive-approximation register
MHz	megahertz				
mi	mile	PC	printed circuit		
mike	microphone	PCM	pulse-code modulation	SAW	surface acoustic wave
min	minute				
mm	millimeter	PDM	pulse-duration modulation	SCA	Subsidiary Communications Authorization
modem	modulator-demodulator				
		PEP	peak envelope power		
mono	monostable	pF	picofarad	scope	oscilloscope
MOS	metal-oxide semiconductor	PF	power factor	SCR	silicon controlled rectifier
		phono	phonograph		
MOSFET	metal-oxide semiconductor FET	PIN	positive-intrinsic-negative	SCS	silicon controlled switch
		PIV	peak inverse voltage	S-meter	signal-strength meter
MOST	metal-oxide semiconductor transistor	PLL	phase-locked loop		
		PM	permanent magnet; phase modulation	S/N	signal-to-noise
MPU	microprocessing unit			SNR	signal-to-noise ratio
ms	millisecond	PMOS	P-channel MOS	SPDT	single-pole double-throw
MSB	most significant bit	PN	positive-negative		
MSI	medium-scale integration	PNP	positive-negative-positive	SPST	single-pole single-throw
m^2	square meter	PNPN	positive-negative-positive-negative	SSB	single sideband
μ	micro- (10^{-6})			SSI	small-scale integration
μA	microampere	pot	potentiometer		
μF	microfarad	P-P	peak-to-peak	SSTV	slow-scan television
μH	microhenry	PPI	plan-position indicator	SW	shortwave
μm	micrometer			SWL	shortwave listener
μP	microprocessor	PPM	parts per million; pulse-position modulation	SWR	standing-wave ratio
μs	microsecond			sync	synchronizing
μV	microvolt	preamp	preamplifier	T	tera- (10^{12})
μW	microwatt	PRF	pulse repetition frequency	TC	temperature coefficient
mV	millivolt				
MVBR	multivibrator	PROM	programmable ROM	THD	total harmonic distortion
mW	milliwatt	PRR	pulse repetition rate		
n	nano- (10^{-9})	ps	picosecond	TR	transmit-receive
N	negative	PSK	phase-shift keying	TRF	tuned radio frequency
nA	nanoampere	PTT	push to talk		
NAB	National Association of Broadcasters	PUT	programmable UJT	triac	triode AC semiconductor switch
		pW	picowatt		
NAND	type of logic circuit	PWM	pulse-width modulation	TTL	transistor-transistor logic
nF	nanofarad				
nH	nanohenry	Q	quality factor		

TTY	teletypewriter	V	volt	VSWR	voltage standing-wave ratio
TV	television	VA	voltampere	VTR	videotape recording
TVI	television interference	VAC	volts AC	VTVM	vacuum-tube voltmeter
TVT	television typewriter	VCO	voltage-controlled oscillator	VU	volume unit
TWX	teletypewriter exchange service	VDC	volts DC	VVC	voltage-variable capacitor
UART	universal asynchronous receiver-transmitter	V/F	voltage-to-frequency	VXO	variable-frequency crystal oscillator
		VFO	variable-frequency oscillator		
		VHF	very high frequency	W	watt
UHF	ultrahigh frequency	VLF	very low frequency	Wh	watthour
UJT	unijunction transistor	VMOS	vertical metal-oxide semiconductor	WPM	words per minute
UPC	universal product code	VOM	volt-ohm-milliammeter	WRMS	watts RMS
				Ws	wattsecond
UPS	uninterruptible power system	VOX	voice-operated transmission	Z	impedance
		VRMS	volts RMS		

Abbreviations on Diagrams. Some foreign publications, including *Wireless World*, shorten the abbreviations for units of measure on diagrams. Thus, μ after a capacitor value represents μF, n is nF, and p is pF. With resistor values, k is thousand ohms, M is megohms, and absence of a unit of measure is ohms. For a decimal value, the letter for the unit of measure is sometimes placed at the location of the decimal point. Thus, 3k3 is 3.3 kilohms or 3,300 ohms, 2M2 is 2.2 megohms, 4μ7 is 4.7 μF, 0μ1 is 0.1 μF, and 4n7 is 4.7 nF.

Semiconductor Symbols Used

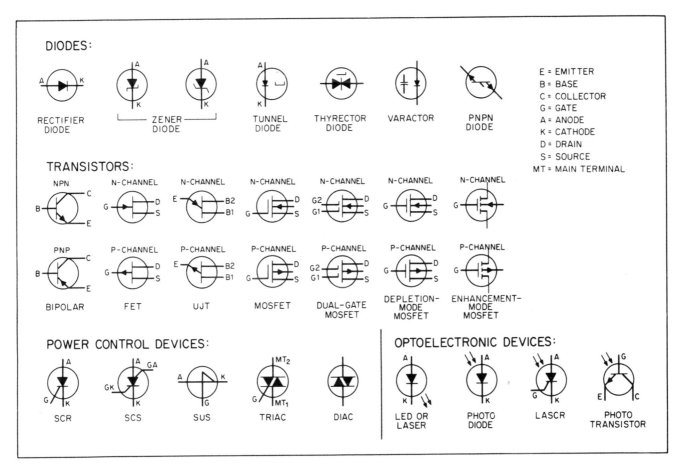

DIODES:

RECTIFIER DIODE | ZENER DIODE | TUNNEL DIODE | THYRECTOR DIODE | VARACTOR | PNPN DIODE

E = EMITTER
B = BASE
C = COLLECTOR
G = GATE
A = ANODE
K = CATHODE
D = DRAIN
S = SOURCE
MT = MAIN TERMINAL

TRANSISTORS:

NPN — BIPOLAR | PNP — BIPOLAR | N-CHANNEL / P-CHANNEL — FET | N-CHANNEL / P-CHANNEL — UJT | N-CHANNEL / P-CHANNEL — MOSFET | N-CHANNEL / P-CHANNEL — DUAL-GATE MOSFET | N-CHANNEL / P-CHANNEL — DEPLETION-MODE MOSFET | N-CHANNEL / P-CHANNEL — ENHANCEMENT-MODE MOSFET

POWER CONTROL DEVICES:

SCR | SCS | SUS | TRIAC | DIAC

OPTOELECTRONIC DEVICES:

LED OR LASER | PHOTO DIODE | LASCR | PHOTO TRANSISTOR

The commonest forms of the basic semiconductor symbols are shown here. Leads are identified where appropriate, for convenient reference. Minor variations in symbols, particularly those from foreign sources, can be recognized by comparing with these symbols while noting positions and directions of solid arrows with respect to other symbol elements.

Omission of the circle around a symbol has no significance. Arrows are sometimes drawn open instead of solid. Thicker lines and open rectangles in some symbols on diagrams have no significance. Orientation of symbols is unimportant; artists choose the position that is most convenient for making connections to other parts of the circuit. Arrow lines outside optoelectronic symbols indicate the direction of light rays.

On some European diagrams, the position of the letter k gives the location of the decimal point for a resistor value in kilohms. Thus, 2k2 is 2.2K or 2,200 ohms. Similarly, a resistance of 1R5 is 1.5 ohms, 1M2 is 1.2 megohms, and 3n3 is 3.3 nanofarads.

Substitutions can often be made for semiconductor and IC types specified on diagrams. Newer components, not available when the original source article was published, may actually improve the performance of a particular circuit. Electrical char-

acteristics, terminal connections, and such critical ratings as voltage, current, frequency, and duty cycle, must of course be taken into account if experimenting without referring to substitution guides.

Semiconductor, integrated-circuit, and tube substitution guides can usually be purchased at electronic parts supply stores.

Not all circuits give power connections and pin locations for ICs, but this information can be obtained from manufacturer data sheets. Alternatively, browsing through other circuits may turn up another circuit on which the desired connections are shown for the same IC.

When looking down at the top of an actual IC, numbering normally starts with 1 for the first pin *counterclockwise* from the notched or otherwise marked end and continues sequentially. The highest number is therefore next to the notch on the other side of the IC, as illustrated in the sketches below. (*Actual positions* of pins are rarely shown on schematic diagrams.)

Addresses of Sources Used

In the citation at the end of each abstract, the title of a magazine is set in italics. The title of a book or report is placed in quotes. Each source title is followed by the name of the publisher of the original material, plus city and state. Complete mailing addresses of all sources are given below, for the convenience of readers who want to write to the original publisher of a particular circuit. When writing, give the complete citation, exactly as in the abstract.

Books can be ordered from their publishers, after first writing for prices of the books desired. Some electronics manufacturers also publish books and large reports for which charges are made. Many of the books cited as sources in this volume are also sold by bookstores and by electronics supply firms. Locations of these firms can be found in the YELLOW PAGES of telephone directories under headings such as "Electronic Equipment and Supplies" or "Television and Radio Supplies and Parts."

Only a few magazines have back issues on hand for sale, but most magazines will make copies of a specific article at a fixed charge per page or per article. When you write to a magazine publisher for prices of back issues or copies, give the *complete* citation, *exactly* as in the abstract. Include a stamped self-addressed envelope to make a reply more convenient.

If certain magazines consistently publish the types of circuits in which you are interested, use the addresses below to write for subscription rates.

American Microsystems, Inc., 3800 Homestead Rd., Santa Clara, CA 95051

Audio, 401 North Broad St., Philadelphia, PA 19108

BYTE, 70 Main St., Peterborough, NH 03458

Computer Design, 11 Goldsmith St., Littleton, MA 01460

CQ, 14 Vanderventer Ave., Port Washington, L.I., NY 11050

Delco Electronics, 700 East Firmin, Kokomo, IN 46901

Dialight Corp., 203 Harrison Place, Brooklyn, NY 11237

EDN, 221 Columbus Ave., Boston, MA 02116

Electronics, 1221 Avenue of the Americas, New York, NY 10020

Electronic Servicing, 9221 Quivira Rd., P.O. Box 12901, Overland Park, KS 66212

Exar Integrated Systems, Inc., 750 Palomar Ave., Sunnyvale, CA 94086

Ham Radio, Greenville, NH 03048

Harris Semiconductor, Department 53-35, P.O. Box 883, Melbourne, FL 32901

Hewlett-Packard, 1501 Page Mill Rd., Palo Alto, CA 94304

Howard W. Sams & Co. Inc., 4300 West 62nd St., Indianapolis, IN 46206

IEEE Publications, 345 East 47th St., New York, NY 10017

Instruments & Control Systems, Chilton Way, Radnor, PA 19089

Kilobaud, Peterborough, NH 03458

McGraw-Hill Book Co., 1221 Avenue of the Americas, New York, NY 10020

Modern Electronics, 14 Vanderventer Ave., Port Washington, NY 11050

Motorola Semiconductor Products Inc., Box 20912, Phoenix, AZ 85036

Mullard Limited, Mullard House, Torrington Place, London WC1E 7HD, England

National Semiconductor Corp., 2900 Semiconductor Dr., Santa Clara, CA 95051

Optical Electronics Inc., P.O. Box 11140, Tucson, AZ 85734

Popular Science, 380 Madison Ave., New York, NY 10017

Precision Monolithics Inc., 1500 Space Park Dr., Santa Clara, CA 95050

QST, American Radio Relay League, 225 Main St., Newington, CT 06111

Radio Shack, 1100 One Tandy Center, Fort Worth, TX 76102

Raytheon Semiconductor, 350 Ellis St., Mountain View, CA 94042

RCA Solid State Division, Box 3200, Somerville, NJ 08876

Howard W. Sams & Co. Inc., 4300 West 62nd St., Indianapolis, IN 46206

73 Magazine, Peterborough, NH 03458

Siemens Corp., Components Group, 186 Wood Ave. South, Iselin, NJ 08830

Signetics Corp., 811 East Arques Ave., Sunnyvale, CA 94086

Siliconix Inc., 2201 Laurelwood Rd., Santa Clara, CA 95054

Sprague Electric Co., 479 Marshall St., North Adams, MA 01247

Teledyne Philbrick, Allied Drive at Route 128, Dedham, MA 02026

Teledyne Semiconductor, 1300 Terra Bella Ave., Mountain View, CA 94040

Texas Instruments Inc., P.O. Box 5012, Dallas, TX 75222

TRW Power Semiconductors, 14520 Aviation Blvd., Lawndale, CA 90260

Unitrode Corp., 580 Pleasant St., Watertown, MA 02172

Wireless World, Dorset House, Stamford St., London SE1 9LU, England

SECTION 1
Antenna Circuits

Includes VSWR bridges, modulation monitors, field-strength metering circuits, attenuators, and preamplifiers. See other chapters on noise circuits, amplifier circuits, and receiver circuits.

FAR-FIELD TRANSMITTER—Provides far-field signal source for tuning Yagi and other beam antennas used on amateur radio frequencies. Q1 is Pierce oscillator operating in fundamental mode of 7.06-MHz crystal to permit field-strength measurements at 14.12, 21.18, and 28.24 MHz for 20-, 15-, and 10-meter bands. Antenna uses two 5-foot lengths of wire connected as dipole. T1 is Amidon core T50-2 with 22 turns on primary and 20 turns center-tapped on secondary. T2 is same core with 22-turn primary and 5-turn secondary.—G. Hinkle, Closed Loop Antenna Tuning, *73 Magazine,* May 1976, p 32–33.

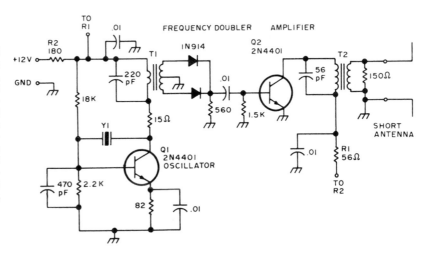

VSWR METER—Simple, easily transported VSWR meter consists of high-gain amplifier, narrow-bandwidth (100-Hz) selective amplifier tuned to 1000 Hz, and variable-gain output amplifier driving low-cost VU meter. Ideal for nulling-type VSWR measurements. Draws only about 6 mA from 9-V transistor battery. Closing S1 increases gain about 100 times for low-level readings. R1 sets U1B to 1000 Hz, while R2 sets reference on VU meter.—J. Reisert, Matching Techniques for VHF/UHF Antennas, *Ham Radio,* July 1976, p 50–56.

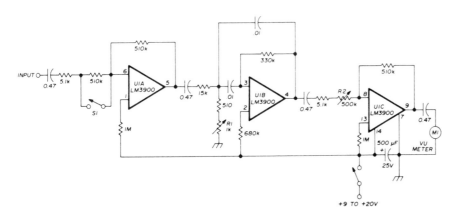

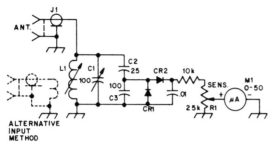

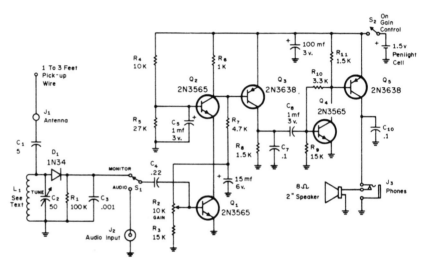

FIELD-STRENGTH METER—Useful for antenna experiments and adjustments in amateur bands from 160 to 10 meters. Increasing size of pickup antenna increases sensitivity. Far-field measurements are made with alternate input circuit, in which reference dipole or quarter-wave wire cut for frequency of interest is connected to input link. Diodes are 1N34A germanium or equivalent. M1 is 50 μA. Table gives values of tuned-circuit components for six amateur bands.—D. DeMaw, A Simple Field-Strength Meter and How to Calibrate It, *QST*, Aug. 1975, p 21–23.

Band	160 M	80 M	40 M	20 M	15 M	10 M
L1 (μH)	100 (Nom.)	25 (Nom.)	10 (Nom.)	2.2 (Nom.)	1.3 (Nom.)	0.5 (Nom.)
C2 (pF)	25	25	15	15	10	10
C3 (pF)	100	100	68	68	47	47
Miller Coil	4409	4407	4406	4404	4403	4303

MODULATION MONITOR—Provides off-the-air monitoring of RF signals up to 200 MHz by rectified detection of AM signals and by slope detection of FM signals. Can also be used as signal tracer, audio amplifier, or hidden-transmitter locator. High-gain audio amplifier has low-noise cascode input stage and output stage driving headphone or loudspeaker. S_1 selects RF signals detected by D_1 or AF applied to J_2. L_1 is 4 turns No. 18 for monitoring 75-150 MHz. Will also monitor VHF transmissions from pilot to ground stations while in commercial aircraft, using 24-inch wire antenna near window and earphone. Passive-type receiver is safe in aircraft because it has no oscillators that could interfere with navigation equipment. —W. F. Splichal, Jr., Sensitive Modulation Monitor, *CQ*, Jan. 1973, p 59–61.

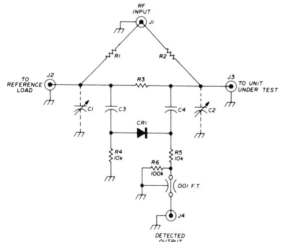

C1, C2	small capacitive tab required for balance
C3, C4	0.001 μF (small disc ceramic or chip capacitor)
CR1	1N82A or equivalent germanium diode
J1, J4	UG-290A/U BNC connector
J2, J3	UG-58/U type-N connector
R1, R2	47 to 55 ohms, matched
R3	51 ohms, ¼-watt carbon composition
R4, R5	10k ohms, ¼-watt carbon composition
R6	100k ohms, ¼-watt carbon composition

VSWR BRIDGE—Works well through 450 MHz for measuring and matching VHF and UHF antennas. If identical load impedances are placed at J2 and J3, signals at opposite ends of R3 are equal and in phase and there is no output at J4. If impedances are different, output proportional to difference appears at J4. Impedance values can be from 25 to 100 ohms, although circuit is designed for optimum performance at 50 ohms.—J. Reisert, Matching Techniques for VHF/UHF Antennas, *Ham Radio*, July 1976, p 50–56.

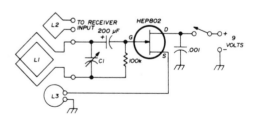

Q MULTIPLIER FOR LOOP—Improves performance of loop antenna on 40, 80, and 160 meters. Feedback control is obtained with adjustable single-turn loop L3 coupled to L1, and receiver input is taken from L2. L3 is rotated within field of L1 to adjust amount of regeneration, optimize circuit Q, and make directional null more pronounced. Article gives loop construction details. Ground lower end of 100K resistor to provide ground return for FET.—K. Cornell, Loop Antenna Receiving Aid, *Ham Radio*, May 1975, p 66–70.

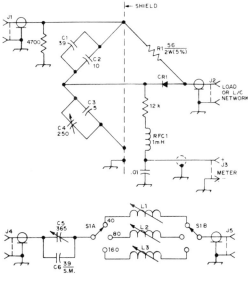

C1, C2 — 39- and 10-pF silver micas in parallel.
C3 — 5-pF silver mica.
C4 — 250-pF straight-line-wavelength variable (Hammarlund MC-250M).
C5 — 365-pF miniature variable (Archer-Allied 695-1000).
CR1 — Germanium diode.
J1, J2, J4, J5 — Coaxial receptacle.
J3 — Phono jack.
L1 — 15 turns No. 24 enamel close-wound on Miller 66A022-6 form (purple slug).
L2 — 30 turns like L1.
L3 — 63 turns like L1, but scramble-wound.
S1 — 2-pole 3-position wafer switch.

RF BRIDGE FOR COAX—Simplifies adjustment of vertical antenna for 40, 80, and 160 meters. S1 in add-on LC unit switches coil for desired band. Values of C1-C4 and standard resistor R1 give range of 10 to 150 ohms for measurement of radiation resistance. Meter can be from 50 to 200 μA full scale if 500 mW of power is available as signal source. For shorter-wavelength bands, change resistance in parallel with J1 to 5600 ohms and omit C6. L1 for 10 meters should then have 3½ turns No. 18 spaced to occupy ¼ inch on Miller 4200 coil form. L2 (15 meters) is 6 turns No. 16 enamel closewound on similar form. L3 (20 meters) is 11 turns No. 14 enamel on Miller 66A022-6 form.—J. Sevick, Simple RF Bridges, *QST*, April 1975, p 11–16 and 41.

5-STEP ATTENUATOR—Applications include comparing performance of various receiving antennas and measuring gain of preamp used ahead of receiver. Dashed lines represent required shield partitions. All resistors are ¼-W composition with 5% tolerance.—D. DeMaw, What Does My S-Meter Tell Me?, *QST*, June 1977, p 40–42.

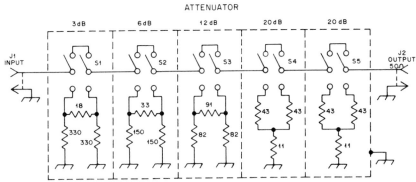

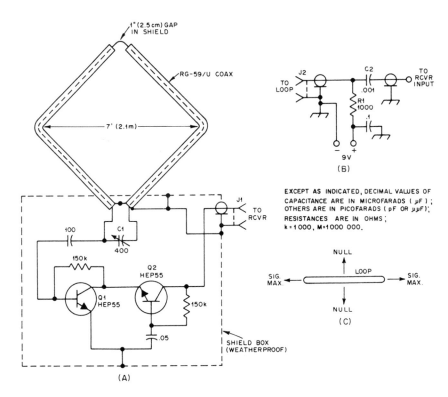

160-METER LOOP-PREAMP—Shielded 5-foot square loop and single preamp pull signals out of noise when propagation conditions make other antennas unsatisfactory. Operating voltage is supplied through coax feeder. R1 isolates signal energy from ground, and C2 keeps DC voltage out of receiver input. Nulls are off broad side of loop.—B. Boothe, Weak-Signal Reception on 160—Some Antenna Notes, *QST*, June 1977, p 35–39.

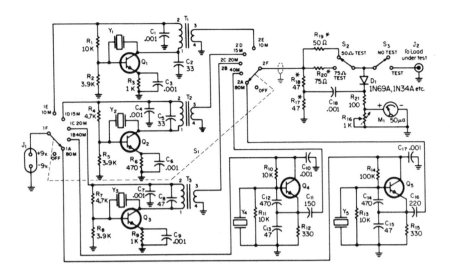

SELF-EXCITED SWR BRIDGE—Portable bridge has built-in signal sources for each band from 80 through 10 meters, for tuning antenna on tower before transmission line is connected. Oscillators are crystal controlled at desired antenna tune-up frequencies. Separate oscillators for each band simplify switching problems, so only supply voltage from J_1 and oscillator outputs to meter circuit need be switched. Current drain from 9-V battery is maximum of 12 mA. R_{17} and R_{18} should be closely matched, while R_{19} and R_{20} should have 5% tolerance.—T. P. Hulick, An S.W.R. Bridge with a Built-In 80 Through 10 Meter Signal Source, *CQ*, June 1971, p 64–66, 68, and 99.

Q_1-Q_5—RCA 40245.
S_1—2 pole 6· position subminiature rotary switch. (Centerlab PA-2005).
S_2—S.p.d.t. slide switch.
S_3—S.p.s.t. slide switch.
T_1—Pri.: 11 t. #36 e. Sec.: 3 t. #36 e. on Indiana General CF-101 Q2 toroid.
T_2—Pri.: 16 t. #36 e. Sec.: 4 t. #36 e. Same core as T_1.

T_3—Pri.: 20 t. #36 e. Sec.: 5 t. #36 e. Same core as T_1.
Y_1, Y_2, Y_3—Overtone crystals for 10, 15 and 20 meter bands respectively. HC-6U holders.
Y_4, Y_5—40 and 80 meter crystals respectively in HC-6U holders.

Audio Control Circuits

Includes amplifiers, squelch circuits, level clippers, compressors, and equalizers.

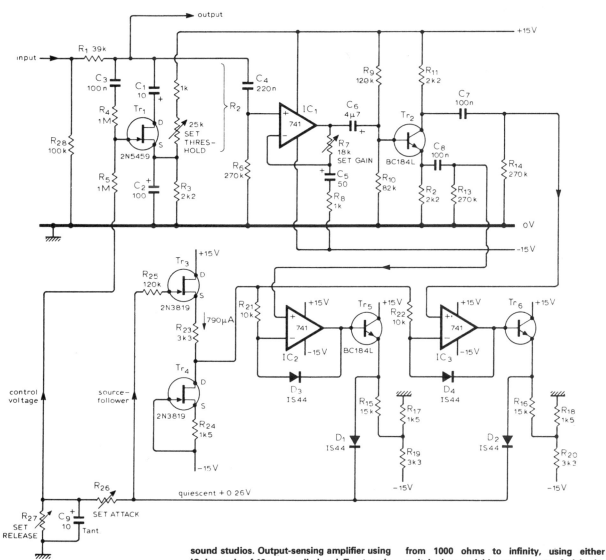

COMPRESSOR/LIMITER—High-fidelity circuit uses voltage-controlled attenuator to increase attenuation of input signal in response to voltage of control loop. Designed for use in modern sound studios. Output-sensing amplifier using IC_1 has gain of 19 over audio band. Tr_2 stage is phase-splitter driving precision rectifiers IC_2 and IC_3. Final part of circuit defines attenuation time constants; R_{26} sets attack time and R_{27} decay time. R_{26} can range from 0 to 1 megohm and R_{27} from 1000 ohms to infinity, using either switched or variable components. Article describes circuit operation and adjustment in detail. Tr_6 is BC184L or equivalent.—D. R. Self, High-Quality Compressor/Limiter, *Wireless World,* Dec. 1975, p 587–590.

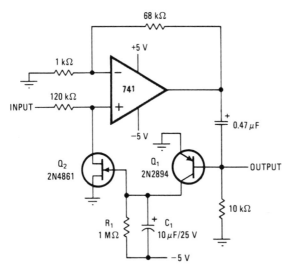

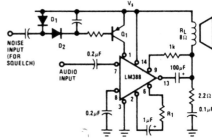

60-dB RANGE FOR AUDIO—JFET acts as voltage-controlled resistor in peak-detecting control loop of 741 opamp. Input range is 20 mV to 20 V, with response time of 1–2 ms and delay of 0.4 s. Output is about 1.4 V P-P over entire 60-dB range.—N. Heckt, Automatic Gain Control Has 60-Decibel Range, *Electronics*, March 31, 1977, p 107.

AUDIO-OPERATED RELAY—Addition of two general-purpose transistors to 555 timer gives audio-triggered relay that can be used for automatic recording of output of channel-monitoring radio receiver or data from any audio link. Adjustable time delay R keeps control circuit actuated up to 5 s (determined by R and C) to avoid cycling relay during pauses in speech or dropouts in data. Q1 is NPN, and Q2 is PNP. Attack time equals very short pull-in time of 5-V reed relay K. Adjust 10K input pot just below point at which K pulls in when there is no audio input.—R. Taggart, Sound Operated Relay, *73 Magazine*, Oct. 1977, p 114–115.

DIFFERENTIAL-AMPLIFIER CLIPPER—Provides gain as well as precise symmetrical clipping for improving intelligibility of speech fed into radio transmitter. Circuit reduces dynamic range of energy peaks to bring them closer to average energy level. When inserted in series with microphone, use of clipper gives at least 6-dB increase in effective power. Signals are passed up to certain amplitude but limited above this level.—B. Kirkwood, Principles of Speech Processing, *Ham Radio*, Feb. 1975, p 28–34.

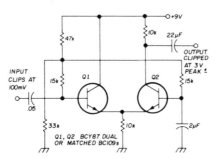

SQUELCHABLE AMPLIFIER—Circuit designed for portable FM scanners and two-way walkie-talkie radios can be turned off by noise or by control signal to minimize battery drain. When squelched, LM388 opamp-transistor-diode array draws only 0.8 mA from 7.5-V supply. Diodes rectify noise from limiter or discriminator of receiver, producing direct current that turns on Q1 and thereby clamps opamp off. Voltage gain is 20 to 200, depending on value used for R1. Power output without squelch is about 0.5 W for 8-ohm loudspeaker.—"Audio Handbook," National Semiconductor, Santa Clara, CA, 1977, p 4-37–4-41.

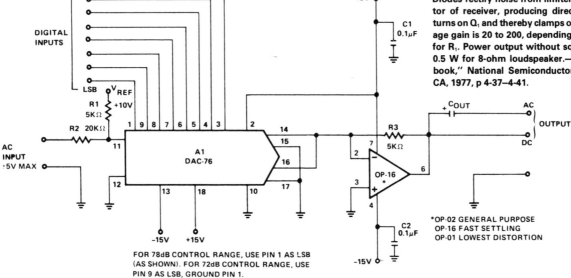

FOR 78dB CONTROL RANGE, USE PIN 1 AS LSB (AS SHOWN). FOR 72dB CONTROL RANGE, USE PIN 9 AS LSB, GROUND PIN 1.

*OP-02 GENERAL PURPOSE
OP-16 FAST SETTLING
OP-01 LOWEST DISTORTION

TWO-QUADRANT EXPONENTIAL CONTROL—Decibel-weighted control characteristic of Precision Monolithics DAC-76 D/A converter matches natural loudness sensitivity of human ear, to provide much greater useful dynamic range for controlling audio level. Control range can be either 72 or 78 dB, depending on pin connections used. 8-bit word control input can be interfaced with standard TTL-compatible microprocessor outputs. To avoid annoying output transients during large or rapid gain changes, use clickless attenuator/amplifier (also given in application note).—W. Jung and W. Ritmanich, "Audio Applications for the DAC-76 Companding D/A Converter," Precision Monolithics, Santa Clara, CA, 1977, AN-28, p 2.

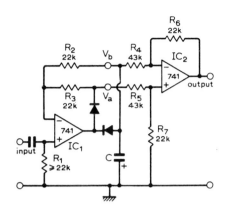

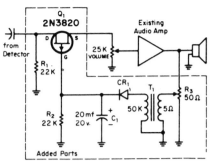

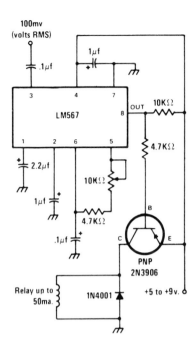

CLAMPING WITH OPAMPS—Circuit is used after stage of AC amplification to clamp minimum level of signal voltage to 0 V for signals having amplitudes between 10 mV and 10 V. With 250-μF electrolytic for C, sinusoidal waveforms between 3 and 10,000 Hz are clamped with little distortion. Overall gain is unity.—C. B. Mussell, D.C. Level Clamp, *Wireless World*, Feb. 1975, p 93.

AF COMPRESSOR—Developed for use in communication receiver where signals vary so greatly that even modern AVC systems cannot level all signals. Circuit is AVC that sets maximum audio level which will not be exceeded. Uses one FET as series attenuator controlled by DC voltage derived from audio output. R_3 permits adjustment of compression level.—C. E. Richmond, A Receiver Audio Compressor, *CQ*, June 1970, p 35 and 86.

TONE-DRIVEN RELAY—LM567 tone decoder will respond to frequency between 700 and 1500 Hz, determined by setting of 10K pot. When input of 100 mVRMS at preset frequency arrives, output of IC goes low and energizes relay through transistor. Tone can be obtained from audio oscillator or telephone Touch-Tone pad. Relay contacts can be used to turn desired device on or off.—J. A. Sandler, 9 Easy to Build Projects under $9, *Modern Electronics*, July 1978, p 53–56.

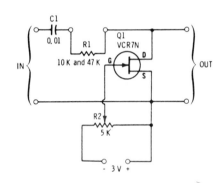

VOLTAGE-CONTROLLED ATTENUATOR—Used to control low-level audio signals with variable DC voltage of ±3 V. Control pot can be remotely located. Highest possible output is equal to input level, occurring when gate bias is set close to pinchoff value. Output is minimum when gate bias is zero.—E. M. Noll, "FET Principles, Experiments, and Projects," Howard W. Sams, Indianapolis, IN, 2nd Ed., 1975, p 258–260.

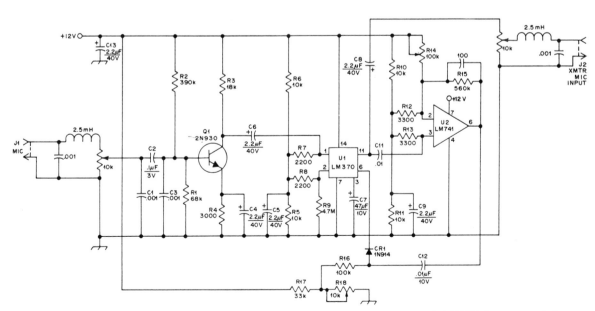

EQUALIZER—Designed for use between input jack and microphone of amateur transmitter, to keep bandpass response between limits of about 200 and 3100 Hz. Circuit also provides measure of volume compression, improving transmitter efficiency. Construction and adjustment details stress importance of eliminating ground loops and RF feedback. U1 is voltage-controlled amplifier in feedback loop, with 741 opamp U2 as compression detector. U2 is biased so output is almost at ground, and no feedback voltage is applied until input to U2 exceeds 0.9 V. U1 thus operates in linear mode at maximum gain until output voltage exceeds 0.9 V, when voltage is applied to U1 and gain of IC is reduced.—R. Tauber, The Equalizer, *QST*, March 1977, p 18–20.

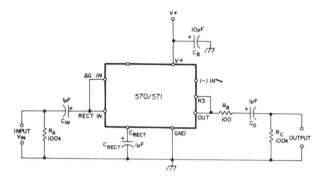

EXPANDER—Uses Signetics dual-channel compandor IC; 571 has lower inherent distortion and higher supply voltage range (6–24 V) than 571 (6–18 V). Values shown are for 15-V supply with either IC. Gain through expander is 1.43 V_{IN}, where V_{IN} is average input voltage. Unity gain occurs at RMS input level of 0.775 V, or 0 dBm in 600-ohm systems.—W. G. Jung, Gain Control IC for Audio Signal Processing, *Ham Radio*, July 1977, p 47–53.

SWITCHING-CLICK SUPPRESSOR—Correction network shown can be inserted in audio channel of mixing console without producing transients or level changes. Although Baxandall network is shown, switching technique is applicable to other filters. With S_1 normally open and S_2 normally closed, circuit operation is normal. If switch positions are simultaneously reversed, response remains flat regardless of positions of bass and treble pots, center-frequency gain remains unchanged, and phase shift is unchanged. There is then no transient interruption of AF signals. Switching clicks cannot occur because there is no direct current in the network.—J. S. Wilson, Click-Free Switching for Audio Filters, *Wireless World*, Jan. 1975, p 12.

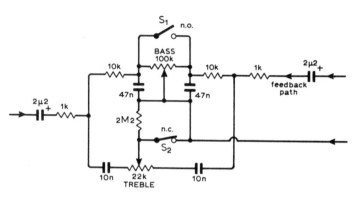

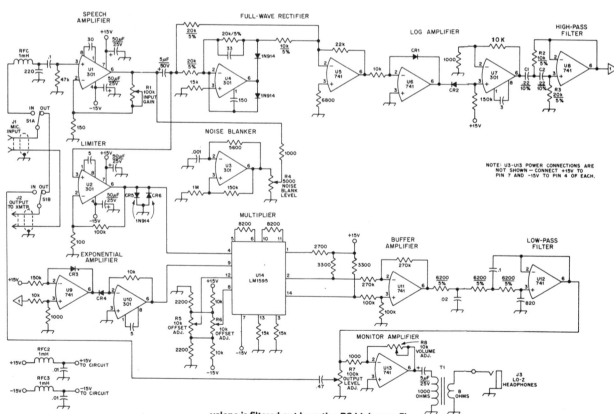

SPEECH PROCESSOR—Can improve signal strength and intelligibility of voice signals up to 6 dB without unpleasant changes in fidelity. Used between microphone and input of AM or SSB transmitter. Based on separation of signal envelope from constant-amplitude carrier that together make up voice signal. After logamp U6 separates components of speech waveform, en- velope is filtered out by active RC high-pass fil- ter U8 having 50-Hz cutoff, with exactly unity gain above cutoff. Filtered signal goes to ex- ponential amplifier U9-U10 and is then multi- plied by correct sign information in U14. Sign information is obtained by hard-limiting input voice signal with diode clipper CR5-CR6. Re- sulting square-wave output is multiplied by sig- nal from UJT in U14. Processed signal goes to transmitter input through low-pass filter U12 having sharp cutoff above 3 kHz to eliminate unwanted high-frequency energy. CR1-CR4 are 1N914 or other matched silicon diodes. T1 is 250-mW audio transformer. Article gives con- struction and adjustment details.—J. E. Kauf- mann and G. E. Kopec, A Homomorphic Speech Compressor, *QST*, March 1976, p 33–37.

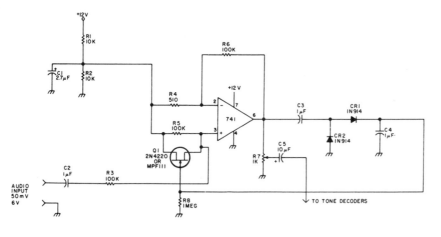

CONSTANT AF LEVEL—Provides constant output level even though input may vary between 50 mV and 6 V, for distribution to tone decoders of autocall system used to monitor simplex or repeater channel to which amateur radio receiver is tuned. Positive terminal of electrolytic C3 must go to pin 6 of 741.—C. W. Andreasen, Autocall '76, *73 Magazine*, June 1976, p 52–54.

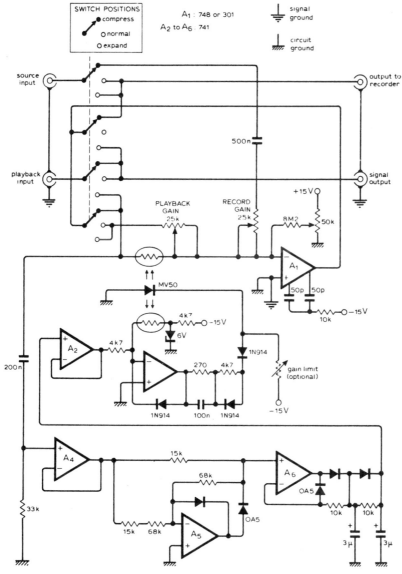

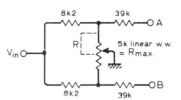

PANNING MIXER POT—Circuit gives best possible approach to sine law so $A^2 = B^2$ is constant for all positions of wiper. Calculated error is less than 1 dB over full range of wiper. Use wirewound pot to minimize crosstalk.—J. Dawson, Single Gang Pan-Pot, *Wireless World*, Feb. 1976, p 78.

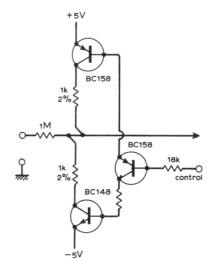

COMPANDER WITH 100-dB RANGE—Simple square-law circuit preserves dynamic range of virtually any input signal when recorded by ordinary tape recorder. Suitable for speech signals as well as for recording or playback in noisy environments. Opamp A$_1$ should have separately decoupled supply. Switching provides compression during recording and expansion during playback. Tracking of photocells is es-

sential for accurate power-law compansion. LED can be glued with clear epoxy to matched photocells. Use silicon signal diodes such as 1N914, 1N4148, or 1S44. Inexpensive photocells such as Vactec VT-833 gave suitably low distortion. Article gives performance characteristics and operating details.—J. Vanderkooy, Wideband Compander Design, *Wireless World*, July 1976, p 45–49.

AUDIO SWITCHING GATE—Can be used with programmed channel selectors, as required in music synthesis for controlling audio signals by means of TTL levels. DC offset at output is negligible when gate is off, simplifying design of subsequent stages. Use logic 1 (+5 V) to open gate, and logic 0 (0 V) to close it.—L. Cook, Analogue Gate with No Offset, *Wireless World*, Feb. 1975, p 93.

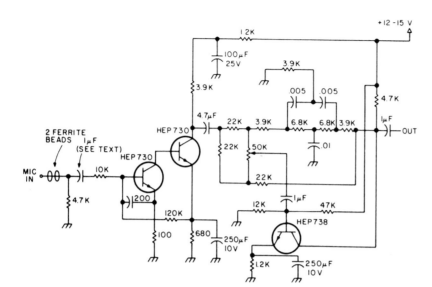

PREEMPHASIS AT 1500 Hz—Single-transistor peaking filter is combined with low-noise RF-protected preamp stage to improve speech intelligibility for any type of modulation. Effectiveness is most noticeable with deep bass voice, where soft peak around 1500–2000 Hz improves speech intelligibility. Can also serve as audio-type CW or SSB filter.—You Can Sound Better with Speech Pre-Emphasis, *73 Magazine,* Feb. 1977, p 42–43.

SECTION 3
Automotive Circuits

Includes capacitor-discharge, optoelectronic, and other types of electronic ignition, tachometers, dwell meters, idiot-light buzzer, audible turn signals, headlight reminders, mileage computer, cold-weather starting aids, wiper controls, oil-pressure and oil-level gages, solid-state regulators for alternators, overspeed warnings, battery-voltage monitor, and trailer-light interface. For auto theft devices, see Burglar Alarm chapter.

SOLID-STATE AUTO REGULATOR—Replaces and outperforms electromechanical charging-voltage regulator in autos using alternator systems. Prolongs battery life by preventing undercharging or overcharging of 12-V lead-acid battery. Uses LM723 connected as switching regulator for controlling alternator field current. R2 is adjusted to maintain 13.8-V fully charged voltage for standard auto battery. Article gives construction details and tells how to use external relay to maintain alternator charge-indicator function in cars having idiot light rather than charge-discharge ammeter. Q1 is 2N2063A (SK3009) 10-A PNP transistor.—W. J. Prudhomme, Build Your Own Car Regulator, *73 Magazine,* March 1977, p 160–162.

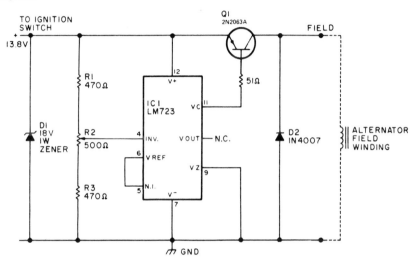

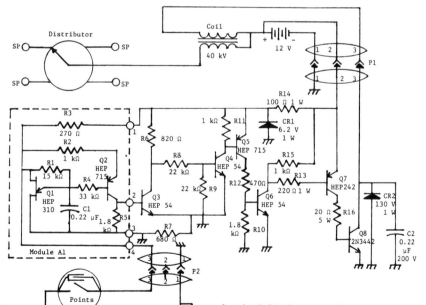

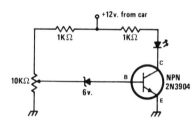

BATTERY MONITOR—Basic circuit energizes LED when battery voltage drops to level set by 10K pot. Any number of additional circuits can be added, for reading battery voltage in 1-V steps or even steps as small as 0.1 V. Circuit supplements idiot light that replaces ammeter in most modern cars. LED type is not critical.—J. Sandler, 9 Projects under $9, *Modern Electronics,* Sept. 1978, p 35–39.

COLD-WEATHER IGNITION—Multispark electronic ignition improves cold-weather starting ability of engines in arctic environment by providing more than one spark per combustion cycle. Circuit uses UJT triangle-wave generator Q1, emitter-follower isolator Q2, wave-shaping Schmitt trigger Q3-Q4, three stages of square-wave amplification Q5-Q7, and output switch-ing circuit Q8, all operating from 12-V negative-ground supply. 6.2-V zener provides regulated voltage for UJT and Schmitt trigger. Initial 20,-000- to 40,000-V ignition spark produced by opening of breaker points is followed by continuous series of sparks at rate of about 200 per second as long as points stay open.—D. E. Stinchcomb, Multi-Spark Electronic Ignition for Engine Starting in Arctic Environment, *Proceedings of the IEEE 1975 Region Six Conference,* May 1975, p 224–225.

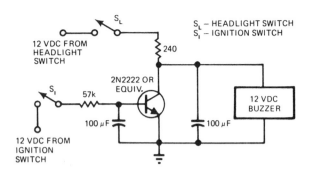

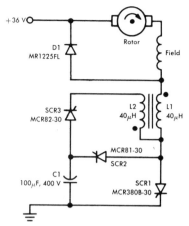

HEADLIGHTS-ON ALARM—Designed for cars in which headlight switch is nongrounding type, providing 12 V when closed. When both light and ignition switches are closed, transistor is saturated and there is no voltage drop across it to drive buzzer. If ignition switch is open while lights are on, transistor bias is removed so transistor is effectively open and full 12 V is applied to buzzer through 240-ohm resistor until lights are turned off.—R. E. Hartzell, Jr., Detector Warns You When Headlights Are Left On, *EDN Magazine,* Nov. 20, 1975, p 160.

ELECTRIC-VEHICLE CONTROL—SCR1 is used in combination with Jones chopper to provide smooth acceleration of golf cart or other electric vehicle operating from 36-V on-board storage battery. Normal running current of 2-hp 36-V series-wound DC motor is 60 A, with up to 300 A required for starting vehicle up hill. Chopper and its control maintain high average motor current while limiting peak current by increasing chopping frequency from normal 125 Hz to as high as 500 Hz when high torque is required.—T. Malarkey, You Need Precision SCR Chopper Control, *New Motorola Semiconductors for Industry,* Motorola, Phoenix, AZ, Vol. 2, No. 1, 1975.

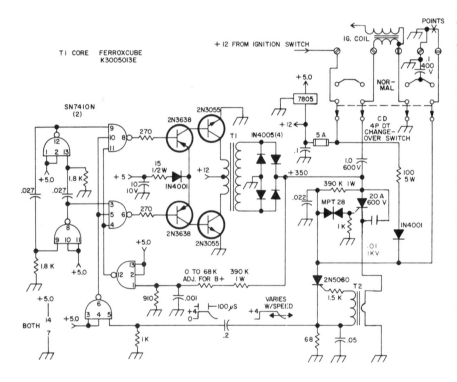

CD IGNITION—Uses master oscillator—power amplifier type of DC/DC converter in which two sections of triple 3-input NAND gate serve as 10-kHz square-wave MVBR feeding class B PNP/NPN power amplifier through two-gate driver. Remaining two gates are used as logic inverters. Secondary of T1 has 15.24 meters of No. 26 in six bank windings, with 20 turns No. 14 added and center-tapped for primary. T2 is unshielded iron-core RF choke, 30–100 μH, with several turns wound over it for secondary. When main 20-A SCR fires, T2 develops oscillation burst for firing sensitive gate-latching SCR. Storage capacitor energy is then dumped into ignition coil primary through power SCR.—K. W. Robbins, CD Ignition System, *73 Magazine,* May 1974, p 17 and 19.

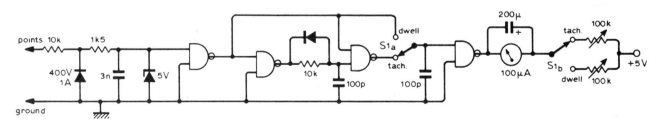

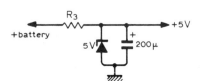

TACH/DWELL METER—Built around SN7402 NOR-gate IC. Requires no internal battery; required 5 V is obtained by using 50 ohms for R_3 in zener circuit shown if car battery is 6 V, and 300 ohms if 12 V. Article gives calibration procedure for engines having 4, 6, and 8 cylinders; select maximum rpm to be indicated, multiply by number of cylinders, then divide by 120 to get frequency in Hz.—N. Parron, Tach-Dwell Meter, *Wireless World,* Sept. 1975, p 413.

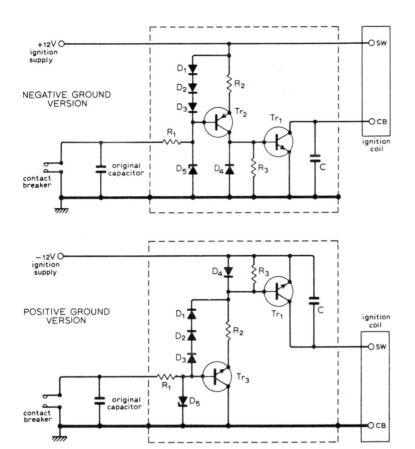

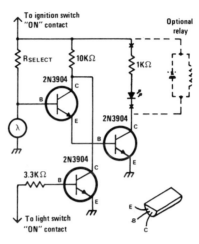

HEADLIGHT REMINDER—Photocell energizes circuit at twilight to remind motorist that lights should be turned on. Indicator can be LED connected as shown or relay turning on buzzer for more positive signal. Circuit can be made automatic by connecting relay contacts in parallel with light switch, provided delay circuit is added to prevent oncoming headlights from killing circuit. Mount photocell in location where it is unaffected by other lights inside or outside car.—J. Sandler, 9 Projects under $9, *Modern Electronics,* Sept. 1978, p 35–39.

TRANSISTORIZED BREAKER POINTS—Uses Texas Instruments BUY23/23A high-voltage transistors that can easily withstand voltages up to about 300 V existing across breaker points of distributor in modern car. Circuit serves as electronic switch that isolates points from heavy interrupt current and high-voltage backswing of ignition coil, thereby almost completely eliminating wear on points. Values are: Tr$_2$ 2N3789; Tr$_3$ (for positive ground version) 2N3055; D$_1$-D$_4$ 1N4001; D$_5$ 18-V 400-mW zener; R$_1$ 56 ohms; R$_2$ 1.2 ohms; R$_3$ 10 ohms; C 600 VDC same size as points capacitor. Article covers installation procedure.—G. F. Nudd, Transistor-Aided Ignition, *Wireless World*, April 1975, p 191.

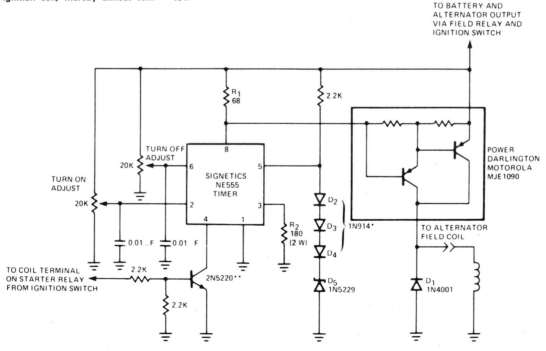

VOLTAGE REGULATOR—Timer and power Darlington form simple automobile voltage regulator. When battery voltage drops below 14.4 V, timer is turned on and Darlington pair conducts. Separate adjustments are provided for preset turn-on and turnoff voltages.—"Signetics Analog Data Manual," Signetics, Sunnyvale, CA, 1977, p 731.

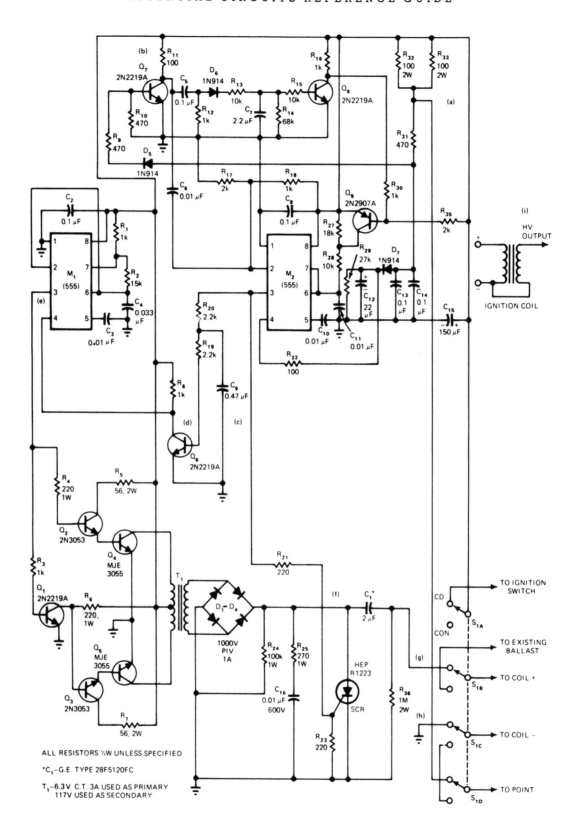

LOW-EMISSION CD—Solid-state capacitor-discharge ignition system improves combustion efficiency by increasing spark duration. For 8-cylinder engine, normal CD system range of 180 to 300 μs is increased to 600 μs below 4000 rpm. Oscillation discharge across ignition coil primary lasts for two cycles here, but above 4000 rpm the discharge lasts for one cycle or 300 μs because at higher speeds the power cycle has shorter times. Circuit uses 555 timer M_1 as 2-kHz oscillator, with Q_1-Q_3 providing drive to Q_4-Q_5 and T_1 for converting battery voltage to about 400 VDC at output of bridge rectifier. When distributor points open, Q_7 turns on and triggers M_2 connected as mono that provides gate drive pulses for SCR. Article describes operation of circuit in detail and gives waveforms at points a-i.—C. C. Lo, CD Ignition System Produces Low Engine Emissions, *EDN Magazine,* May 20, 1976, p 94, 96, and 98.

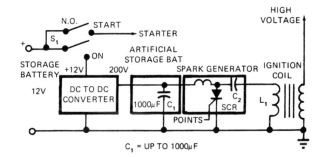

CAPACITOR SERVES AS IGNITION BATTERY— Developed for use with capacitor-discharge ignition systems to provide independent voltage source for ignition when starting car in very cold weather. Before attempting to start car, S_1 is set to ON position for energizing DC-to-DC converter for charging C_1 with DC voltage between 200 and 400 V. Starter is now engaged. If voltage of storage battery drops as starter slowly turns engine over, C_1 still represents equivalent of fully charged 12-V storage battery that is capable of driving ignition system for almost a minute.—W. Stalzer, Capacitor Provides Artificial Battery for Ignition Systems, *EDN Magazine*, Nov. 15, 1972, p 48.

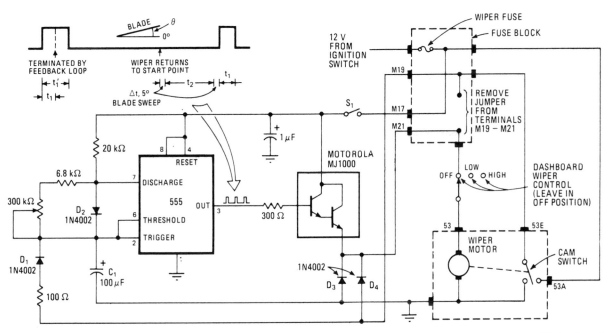

WIPER-DELAY CONTROL— 555 timer provides selectable delay time between sweeps of wiper blades driven by motor in negative-ground system. Article also gives circuit modification for positive-ground autos. Delay time can be varied between 0 and 22 s. Timer uses feedback signal from cam-operated switch of motor to synchronize delay time with position of wiper blades.— J. Okolowicz, Synchronous Timing Loop Controls Windshield Wiper Delay, *Electronics*, Nov. 24, 1977, p 115 and 117.

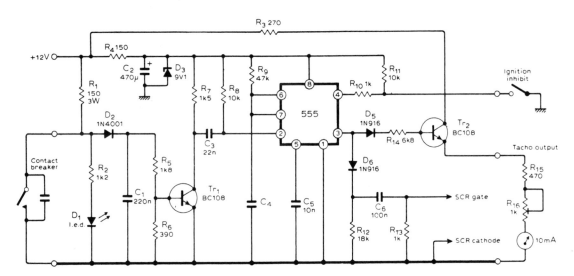

RPM-LIMIT ALARM— Used with capacitor-discharge ignition system to provide tachometer output along with engine speed control signal. When breaker contacts open, C_1 charges and turns Tr_1 on, triggering 555 timer used in mono MVBR mode. Resulting positive pulse from 555 fires control SCR through D_6 and C_6. When contacts close, D_2 isolates C_1 to reduce effect of contact bounce. With values shown, for speed limit between 8000 and 9000 rpm, use 0.068 μF for C_4 with four-cylinder engine, 0.047 μF for six cylinders, and 0.033 μF for eight cylinders. LED across breaker contacts can be used for setting static timing.—K. Wevill, Trigger Circuit for C.D.I. Systems, *Wireless World*, Jan. 1978, p 58.

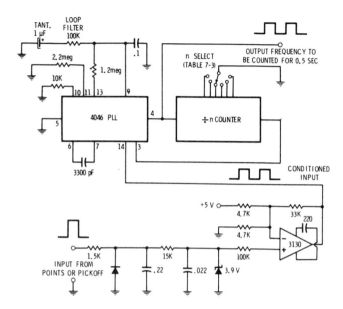

DIGITAL TACHOMETER—Pulses from auto engine points or other pickoff are filtered before feed to 3130 CMOS opamp used as comparator to complete conditioning of input. Pulses are then fed through 4046 PLL to divide-by-N counter that is set for number of cylinders in engine (60 for four cylinders, 45 for six, and 30 for eight). Output frequency is then counted for 0.5 s to get engine or shaft speed in rpm.—D. Lancaster, "CMOS Cookbook," Howard W. Sams, Indianapolis, IN, 1977, p 366–367.

REGULATOR FOR ALTERNATOR—Simple and effective solid-state replacement for auto voltage regulator can be used with alternator in almost any negative-ground system. Circuit acts as switch supplying either full or no voltage to field winding of alternator. When battery is below 13 V, zener D1 does not conduct, Q1 is off, Q2 is on, and full battery voltage is applied to alternator field so it puts out full voltage to battery for charging. When battery reaches 13.6 V, Q1 turns on, Q2 turns off, alternator output is reduced to zero, and battery gets no charging

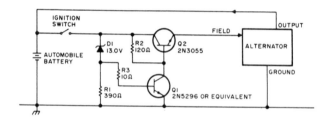

current. Circuit can also be used with wind-driven alternator systems.—P. S. Smith, $22 for a

Regulator? Never!, *73 Magazine,* Holiday issue 1976, p 103.

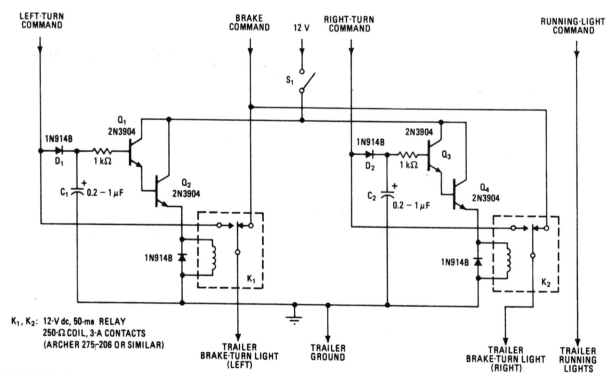

K_1, K_2: 12-V dc, 50-ma RELAY
250-Ω COIL, 3-A CONTACTS
(ARCHER 275-206 OR SIMILAR)

AUTO-TRAILER INTERFACE FOR LIGHTS—Low-cost transistors and two relays combine brake-light and turn-indicator signals on common bus to ensure that trailer lights respond to both commands. C_1 and C_2 charge to peak amplitude of turn signal, which flashes about 2 times per second. Values are selected to hold relay closed between flash intervals; if capacitance is too large, brake signal cannot immediately activate trailer lights after turn signal is canceled. Developed for new cars in which separate turn and brake signals are required for safety.—M. E. Gilmore and C. W. Snipes, Darlington-Switched Relays Link Car and Trailer Signal Lights, *Electronics,* Aug. 18, 1977, p 116.

SECTION 4
Battery-Charging Circuits

Includes constant-voltage, constant-current, and trickle chargers operating from AC line, solar cells, or auto battery. Some circuits have automatic charge-rate control, automatic start-up, automatic shutoff, and low-charge indicator.

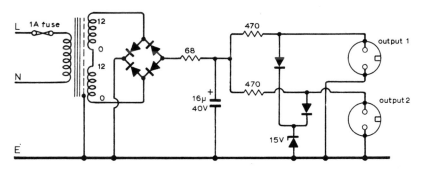

9.6 V AT 20 mA—Developed to charge 200-mAh nickel-cadmium batteries for two transceivers simultaneously. Batteries will be fully charged in 14 hours, using correct 20-mA charging rate. Zener diode ensures that voltage cannot exceed safe value if battery is accidentally disconnected while under charge. Diode types are not critical.—D. A. Tong, A Pocket V.H.F. Transceiver, *Wireless World,* Aug. 1974, p 293–298.

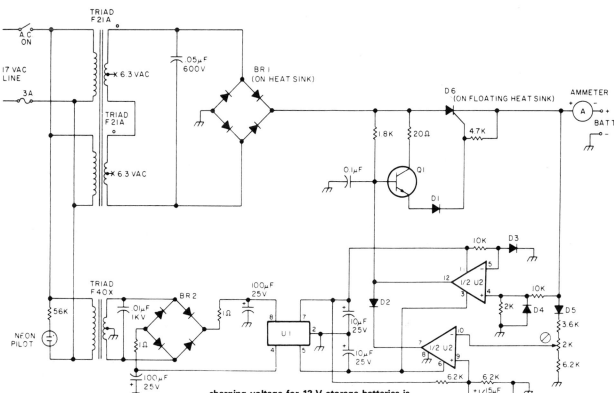

ADJUSTABLE FINISH-CHARGE—Uses National LM319D dual comparator U2 to sense end-of-charge battery voltage and provide protection against shorted or reversed charger leads. Final charging voltage for 12-V storage batteries is adjustable with 2K trimpot. Separate ±15 V supply using Raytheon RC4195NB regulator U1 is provided for U2. D1-D5 are 1N4002 or HEP-R0051. D6 is 2N682 or HEP-R1471. BR1 is Motorola MDA980-2 or HEP-R0876 12-A bridge. BR2 is Varo VE27 1-A bridge. Q1 is 2N3641 or HEP-S0015.—H. Olson, Battery Chargers Exposed, *73 Magazine,* Nov. 1976, p 98–100 and 102–104.

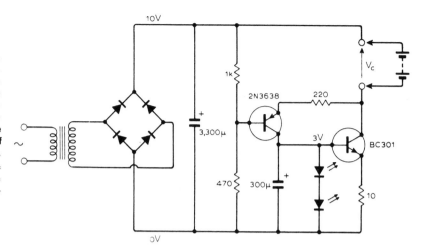

NICKEL-CADMIUM CELL CHARGER—Charges four size D cells in series at constant current, with automatic voltage limiting. BC301 transistor acts as current source, with base voltage stabilized at about 3 V by two LEDs that also serve to indicate charge condition. Other transistor provides voltage limiting when voltage across cells approaches that of 1K branch of voltage divider. Values shown give 260-mA charge initially, dropping to 200 mA when V_c reaches 5 V and decreasing almost to 0 when V_c reaches 6.5 V.—N. H. Sabah, Battery Charger, *Wireless World,* Nov. 1975, p 520.

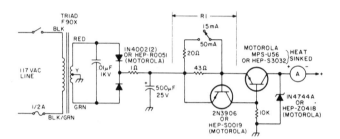

12-V FOR NICADS—Produces constant current with simple transistor circuit, adjustable to 15 or 50 mA with switch and R1. Zener limits voltage at end of charge. Developed for charging 10-cell pack having nominal 12.5 V, as used in many transceivers.—H. Olson, Battery Chargers Exposed, *73 Magazine,* Nov. 1976, p 98–100 and 102–104.

CHARGING SILVER-ZINC CELLS—Used for initial charging and subsequent rechargings of sealed dry-charged lightweight cells developed for use in missiles, torpedoes, and space applications. Article covers procedure for filling cell with potassium hydroxide electrolyte before placing in use (cells are dry-charged at factory and have shelf life of 5 or more years in that condition). Charge current should be 7 to 10% of rated cell discharge capacity; thus, for Yardney HR-5 cell with rated discharge of 5 A, charge at 350 to 500 mA. Stop charging when cell voltage

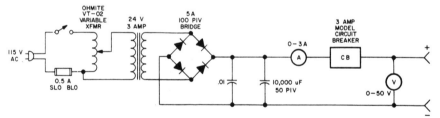

reaches 2.05 V. If used only for battery charging, large filter capacitor can be omitted.—S. Kelly, Will Silver-Zinc Replace the Nicad?, *73 Magazine,* Holiday issue 1976, p 204–205.

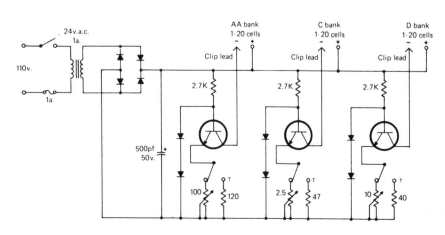

BULK NICAD CHARGER—Can handle up to 20 AA cells, 20 C cells, and 20 D cells simultaneously, with charging rate determined separately for each type. Single transformer and full-wave rectifier feed about 24 VDC to three separate regulators. AA-cell regulator uses 100-ohm resistor to vary charge rate from 6 mA to above 45 mA. C-cell charge-rate range is 24 to 125 mA, and D-cell range is 60 to 150 mA. Batteries of each type should be about same state of discharge. Batteries are recharged in series to avoid need for separate regulator with each cell. Trickle-charge switches cut charge rates to about 2% of rated normal charge (5 mA for 500-mAh AA cells). Transistors are 2N4896 or equivalent. Use heatsinks. All diodes are 1N4002.—J. J. Schultz, A Bulk Ni-Cad Recharger, *CQ,* Dec. 1977, p 35–36 and 111.

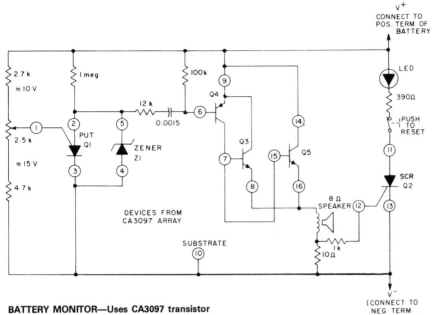

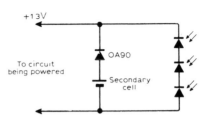

SOLAR-POWER BACKUP—If solar-cell voltage drops 0.2 V below battery voltage, circuit is powered by storage cell feeding through forward-biased OA90 or equivalent germanium diode. When solar-cell voltage exceeds that of battery, battery is charged by approximately constant reverse leakage current through diode. Battery can be manganese-alkaline type or zinc-silver oxide watch-type cell.—M. Hadley, Automatic Micropower Battery Charger, *Wireless World,* Dec. 1977, p 80.

BATTERY MONITOR—Uses CA3097 transistor array to provide active elements required for driving indicators serving as aural and visual warnings of low charge on nicad battery. LED remains on until circuit is reset with pushbutton switch.—"Circuit Ideas for RCA Linear ICs," RCA Solid State Division, Somerville, NJ, 1977, p 9.

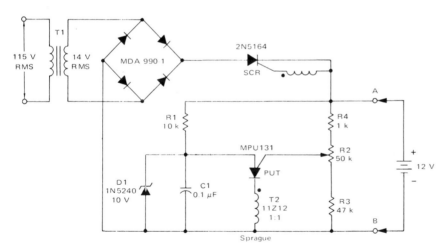

12 V AT 8 A—Charging circuit for lead-acid storage batteries is not damaged by short-circuits or by connecting with wrong battery polarity. Battery provides current for charging C1 in PUT relaxation oscillator. When PUT is fired by C1, SCR is turned on and applies charging current to battery. Battery voltage increases slightly during charge, increasing peak point voltage of PUT and making C1 charge to slightly higher voltage. When C1 voltage reaches that of zener D1, oscillator stops and charging ceases. R2 sets maximum battery voltage between 10 and 14 V during charge.—R. J. Haver and B. C. Shiner, "Theory, Characteristics and Applications of the Programmable Unijunction Transistor," Motorola, Phoenix, AZ, 1974, AN-527, p 10.

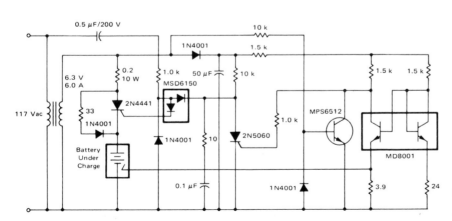

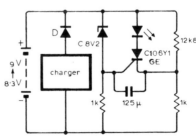

THIRD ELECTRODE SENSES FULL CHARGE—Circuit is suitable only for special nickel-cadmium batteries in which third electrode has been incorporated for use as end-of-charge indicator. Voltage change at third electrode is sufficient to provide reliable shutoff signal for charger under all conditions of temperature and cell variations.—D. A. Zinder, "Fast Charging Systems for Ni-Cd Batteries," Motorola, Phoenix, AZ, 1974, AN-447, p 7.

LED VOLTAGE INDICATOR—Circuit shown uses LED to indicate, by lighting up, that battery has been charged to desired level of 9 V. Circuit can be modified for other charging voltages. Silicon switching transistor can be used in place of more costly thyristor.—P. R. Chetty, Low Battery Voltage Indication, *Wireless World,* April 1975, p 175.

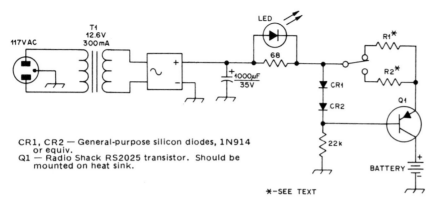

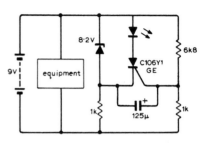

CR1, CR2 — General-purpose silicon diodes, 1N914 or equiv.
Q1 — Radio Shack RS2025 transistor. Should be mounted on heat sink.

✳—SEE TEXT

NICAD CHARGER—Switch gives choice of two constant-current charge rates. With 10 ohms for R1, rate is 60 mA, while 200 ohms for R2 gives 3 mA. Silicon diodes CR1 and CR2 have combined voltage drop of 1.2 V and emitter-base junction of Q1 has 0.6-V drop, for net drop of 0.6 V across R1 or R2. Dividing 0.6 by desired charge rate in amperes gives resistance value.—M. Alterman, A Constant-Current Charger for Nicad Batteries, *QST*, March 1977, p 49.

LED INDICATES LOW VOLTAGE—LED lights when output of 9-V rechargeable battery drops below minimum acceptable value of 8.3 V, to indicate need for recharging. Can also be used with transistor radio battery to indicate need for replacement. Zener is BZY85 C8V2 rated at 400 mW, with avalanche point at 7.7 V because of low current drawn by circuit. LED can be Hewlett-Packard 5082-4440.—P. C. Parsonage, Low-Battery Voltage Indicator, *Wireless World*, Jan. 1973, p 31.

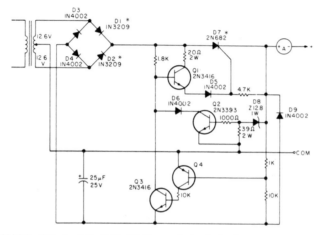

12-V AUTOMATIC—Circuit of Heathkit GP-21 automatic charger is self-controlling (Q1 and Q2) and provides protection against shorted or reversed battery leads (Q3 and Q4). Zener D8 is not standard value, so may be obtainable only in Heathkits. D1, D2, and D7 should all be on one heatsink.—H. Olson, Battery Chargers Exposed, *73 Magazine*, Nov. 1976, p 98–100 and 102–104.

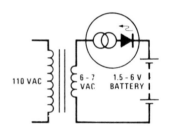

LED TRICKLE CHARGER—Constant-current characteristic of National NSL4944 LED is used to advantage in simple half-wave charger for batteries up to 6 V.—"Linear Applications, Vol. 2," National Semiconductor, Santa Clara, CA, 1976, AN-153, p 2.

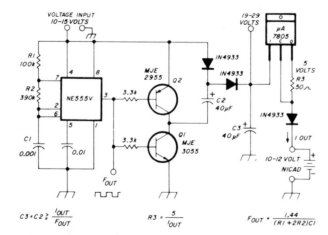

$$C3 \times C2 \geq \frac{I_{OUT}}{F_{OUT}}$$

$$R3 = \frac{5}{I_{OUT}}$$

$$F_{OUT} = \frac{1.44}{(R1 + 2R2)C1}$$

NICAD CHARGER FOR AUTO—Voltage doubler provides at least 20 V from 12-V auto battery, for constant-current charging of 12-V nicads, using NE555 timer and two power transistors. Doubled voltage drives source current into three-terminal current regulator. Switching frequency of NE555 as MVBR is 1.4 kHz. Charging current is set at 50 mA for charging ten 500-mAh nicads.—G. Hinkle, Constant-Current Battery Charger for Portable Operation, *Ham Radio*, April 1978, p 34–36.

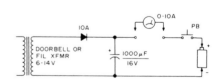

NICAD ZAPPER—Simple circuit often restores dead or defective nicad battery by applying DC overvoltage at current up to 10 A for about 3 s. Longer treatment may overheat battery and make it explode.—Circuits, *73 Magazine*, July 1977, p 35.

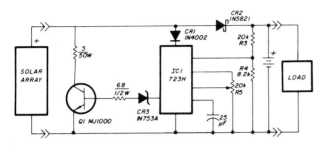

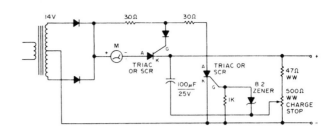

SOLAR-POWER OVERCHARGE PROTECTION—Voltage regulator is connected across solar-cell array as shown to prevent damage to storage battery by overcharging. Series diode prevents array from discharging battery during hours of darkness. Regulator does not draw power from battery, except for very low current used for voltage sampling. Battery can be lead-calcium, gelled-electrolyte, or telephone-type wet cells. For repeater application described, two Globe Union GC12200 40-Ah gelled-electrolyte batteries were used to provide transmit current of 1.07 A and idle current of 12 mA.—T. Handel and P. Beauchamp, Solar-Powered Repeater Design, *Ham Radio*, Dec. 1978, p 28–33.

AUTOMATIC SHUTOFF—Prevents overcharging and dryout of battery under charge by shutting off automatically when battery reaches full-charge voltage. Accepts wide range of batteries. Choose rectifying diodes and triacs or SCRs to handle maximum charging current desired. For initial adjustment, connect fully charged battery and adjust charge-stop pot until ammeter just drops to zero.—Circuits, *73 Magazine*, July 1977, p 34.

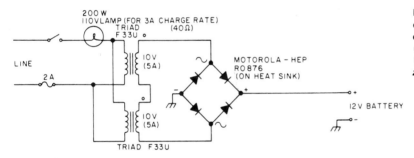

BASIC 12-V CHARGER—Uses 200-W lamp as current-limiting resistor in transformer primary circuit. Serves in place of older types of chargers using copper-oxide or tungar-bulb rectifiers.—H. Olson, Battery Chargers Exposed, *73 Magazine*, Nov. 1976, p 98–100 and 102–104.

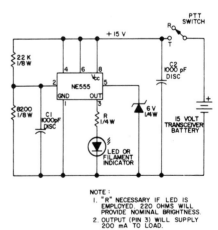

NOTE:
1. "R" NECESSARY IF LED IS EMPLOYED. 220 OHMS WILL PROVIDE NOMINAL BRIGHTNESS.
2. OUTPUT (PIN 3) WILL SUPPLY 200 mA TO LOAD.

NICAD MONITOR—Uses two comparators, flip-flop, and power stage all in single NE555 IC. When battery voltage drops below 12-V threshold set by R1 and R2 for 15-V transceiver battery, one comparator sets flip-flop and makes output at pin 3 go high. IC then supplies up to 200 mA to LED or other indicator. For other battery voltage value, set firing point to about three-fourths of fully charged voltage. Since battery voltage will show biggest drop when transmitting, connect monitor across transmit supply only so as to minimize battery drain.—A. Woerner, Ni-Cad Lifesaver, *73 Magazine*, Nov. 1973, p 35–36.

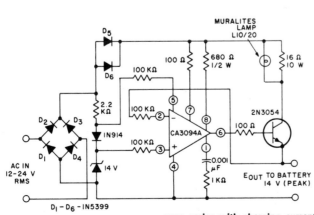

14-V MAXIMUM—Circuit accurately limits peak output voltage to 14 V, as established by zener connected between terminals 3 and 4 of CA3094A programmable opamp. Lamp brightness varies with charging current. Reference voltage supply does not drain battery when power supply is disconnected.—"Circuit Ideas for RCA Linear ICs," RCA Solid State Division, Somerville, NJ, 1977, p 19.

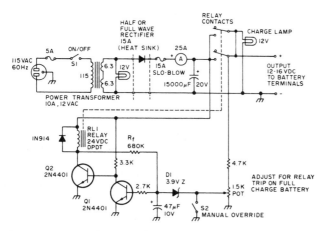

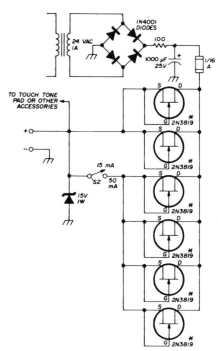

AUTOMATIC SHUTOFF—Charger automatically turns itself off when 12-V auto storage battery is fully charged. Setting of 1.5K pot determines battery voltage at which zener D1 conducts, turning on Q2 and pulling in relay that disconnects charger. If battery voltage drops below threshold, relay automatically connects charger again. S2 is closed to bypass automatic control when charger itself is to be used as power supply.—G. Hinkle, The Smart Charger, *73 Magazine,* Holiday issue 1976, p 110–111.

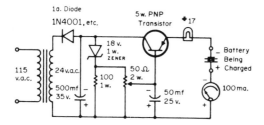

NICAD CHARGER—Pot is adjusted to provide 10% above rated voltage (normal full-charge voltage) while keeping charging current below 25% of maximum. For 10-V 1-Ah battery, set voltage at 11 V and current below 250 mA.—G. E. Zook, F.M., *CQ,* Feb. 1973, p 35–37.

NICAD CHARGER—Developed for recharging small nickel-cadmium batteries used in hand-held FM transceivers. Field-effect transistors serve as constant-current sources when gate is shorted to source. Practically any N-channel JFET having drain-to-source current of 8–15 mA will work. FETs shown were measured individually and grouped to give desired choice of 15- or 50-mA charging currents.—G. K. Shubert, FET-Controlled Charger for Small Nicad Batteries, *Ham Radio,* Aug. 1975, p 46–47.

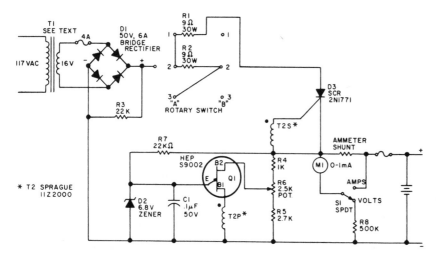

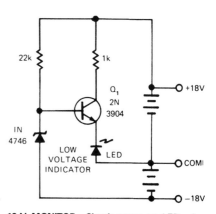

UJT CHARGER FOR 12 V—Keeps 12-V auto storage battery fully charged, for immediate standby use when AC power fails. Power transformer secondary can be 14 to 24 V, rated at about 3 A. Two-gang rotary switch gives choice of three charging rates. Pulse transformer T2 is small audio transformer rewound to have 1:1 turns ratio and about 20 ohms resistance, or can be regular SCR trigger transformer. UJT relaxation oscillator stops when upper voltage limit for battery is reached, as set by pot R6. If oscillator fails to start, reverse one of pulse transformer windings.—F. J. Piraino, Failsafe Super Charger, *73 Magazine,* Holiday issue 1976, p 49.

18-V MONITOR—Circuit turns on LED when ±18 V battery pack discharges to predetermined low level, while drawing less than 1 mA when LED is off. Zener is reverse-biased for normal operating range of battery. When lower limit is reached, zener loses control and Q₁ becomes forward-biased, turning on LED or other signal device to indicate need for replacement or recharging.—W. Denison and Y. Rich, Battery Monitor Is Efficient, yet Simple, *EDN Magazine,* Oct. 5, 1974, p 76.

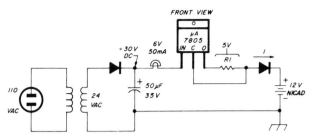

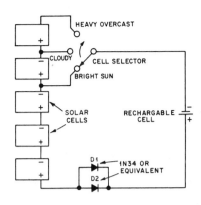

CONSTANT-CURRENT NICAD CHARGER— Constant current is obtained from voltage regulator by floating common line and connecting R1 from output to common terminal. Regulator then tries to furnish fixed voltage across R1. Input voltage must be greater than full-charge battery voltage plus 5 V (for 5-V regulator) plus 2 V (overhead voltage). Changing R1 varies charging current. If R1 is 50 ohms and V is 5 V, constant current is 50 mA through nicad being charged.—G. Hinkle, Constant-Current Battery Charger for Portable Operation, *Ham Radio*, April 1978, p 34—36.

SOLAR-ENERGY CHARGER—Single solar cell on bright day delivers 0.5 V at 50 mA, so three cells are used in bright sun to recharge secondary cell. Switch permits use of additional solar cells on cloudy days. Solar cells can be Radio Shack 276-128.—J. Rice, Charging Batteries with Solar Energy, *QST*, Sept. 1978, p 37.

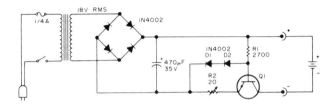

NICAD CHARGER—Regulated charger circuit will handle variable load from 1 to 18 nicad cells. Current-limiting action holds charging current within 1 to 2 mA of optimum value (about one-tenth of rated ampere-hour capacity) from 0 to 24 V. Q1 should have power rating equal to twice supply voltage multiplied by current-limit value. If charging 450-mAh penlight cells, charge current is 45 mA and transistor should be 2 W.—A. G. Evans, Regulated Nicad Charger, *73 Magazine*, June 1977, p 117.

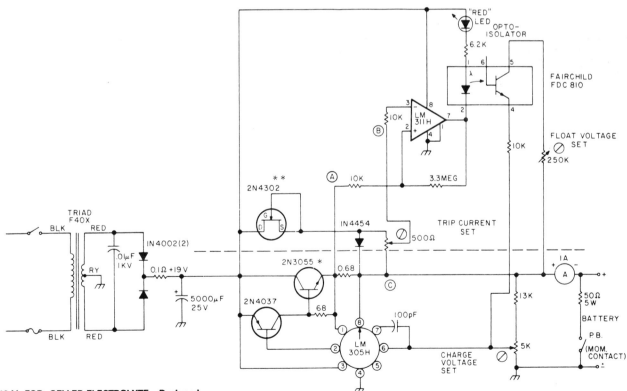

12-V FOR GELLED-ELECTROLYTE—Designed to charge 12-V 3-Ah gelled-electrolyte battery such as Elpower EP1230A at maximum of 0.45 A until battery reaches 14 V, then at constant voltage until charge current drops to 0.04 A. Charger is then automatically switched to float status that maintains 2.2 V per cell or 13.2 V for battery. Circuit is constant-voltage regulator with current limiting as designed around National LM305H, with PNP/NPN transistor pair to increase current capability. Circuit above dashed line is added to standard regulator to meet special charging requirement. Article covers operation and use of circuit in detail.—H. Olson, Battery Chargers Exposed, *73 Magazine*, Nov. 1976, p 98—100 and 102—104.

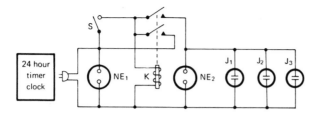

NICAD CHARGE CONTROL—Prevents double-charging if someone forgets to turn off 24-h time clock after recommended 16-h charge period. Nicad devices with built-in chargers are plugged into jacks J_1-J_3, and timer dial is advanced until clock switch is triggered. Neon lamp NE_1 should now come on. Momentary pushbutton switch S is pushed to energize relay K and start charge. When timer goes off, K releases to end charge.—M. Katz, Battery Charge Monitor, *CQ,* July 1976, p 27.

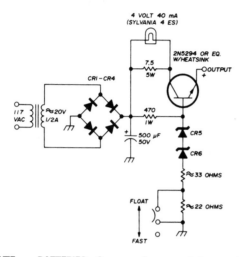

GELLED-ELECTROLYTE   BATTERIES—Constant-voltage charger for Globe-Union 12-V gelled-electrolyte storage batteries can provide either fast or float charging. Constant voltage is maintained by series power transistor and series-connected zeners. Output voltage is 13.8 V for float charging and 14.4 V for fast charging.—E. Noll, Storage-Battery QRP Power, *Ham Radio,* Oct. 1974, p 56–61.

SECTION 5
Bridge Circuits

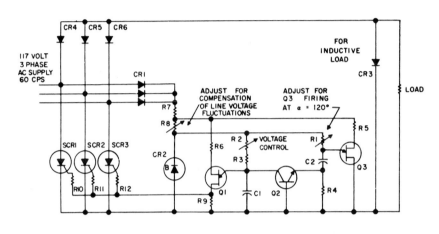

THREE-PHASE SCR CONTROL—Simple three-transistor scr firing circuit provides stepless control of d-c output voltage of three-phase bridge between 25% and 100% of maximum output voltage, or from 40 to 150 V. With 93-V output for line voltage of 130 V a-c, drop to 100 V a-c reduces d-c output only 1 V.—"SCR Manual," 4th Edition, General Electric, 1967, p 202—204.

CR1 — (3) G E 1N1695
CR2 — 20 VOLT, 1 WATT ZENER DIODE, 1N1527
CR3, CR4, CR5, CR6 — AS REQUIRED FOR LOAD (e.g. 1N2156)
SCR1, SCR2, SCR3 — AS REQUIRED FOR LOAD (e.g. C35B)
Q1, Q3 — G E 2N2646   Q2 — G E 2N2923
R1 — 10K POT
R2 — 20K POT
R3 — 470 Ω

R4 — 100 Ω
R5, R6 — 390 Ω
R7 — 3.3 K, 2 W
R8 — 500 Ω POT, 2 W
R9 — 100 Ω
R10, R11, R12 — 25 Ω
C1 — 0.5 MFD
C2 — 1.0 MFD

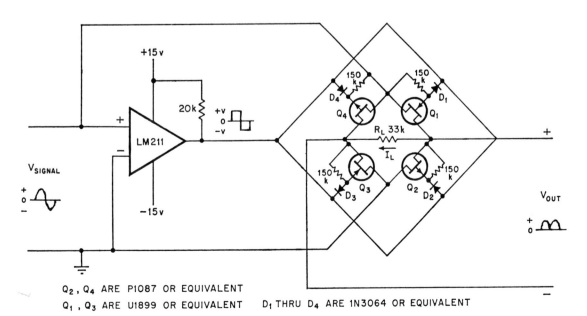

Q_2, Q_4 ARE P1087 OR EQUIVALENT
Q_1, Q_3 ARE U1899 OR EQUIVALENT D_1 THRU D_4 ARE 1N3064 OR EQUIVALENT

COMPARATOR-CONTROLLED FET BRIDGE—Diodes have secondary role of simply switching fet's. With zero offset voltage for fet's, output closely follows or inverts half of sine-wave input signal, to give full-wave rectification with no distortion.—L. Accardi, Diode-Switched FET's Rectify the Full Wave, *Electronics*, Aug. 3, 1970, p 76.

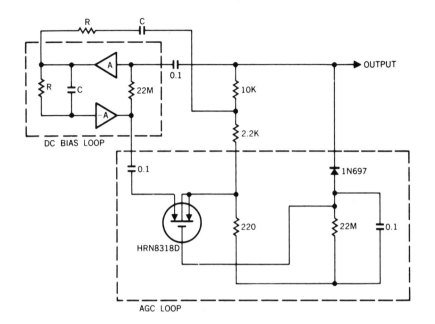

WIEN-BRIDGE OSCILLATOR—Mosfet is used as linear resistance whose value depends on level of oscillator output signal. Increase in output level gives bias voltage for mosfet by detection of peak negative value, for controlling gain in oscillator automatically.—C. R. Perkins, "Application of MOSFET Devices to Electronic Circuits," Hughes, Newport Beach, Cal., 1968, p 29.

LAMPS REGULATE 1-V A-C—Inexpensive voltage-regulating bridge depends on ballast action of 2-V lamps operated at about 0.6 V to give 1-V a-c output that varies only 0.25% for line voltage change of 105 to 125 V. Regulation is independent of frequency over range of 25 to 800 Hz. Thermal inertia makes correction slow, so large input change may require almost 1 s for full correction, but this is usually not serious drawback. Bulbs should be soldered into circuit.—D. Kelly, Small Lamp Bridge Regulates Line Voltage, *Electronics*, March 20, 1967, p 89—90.

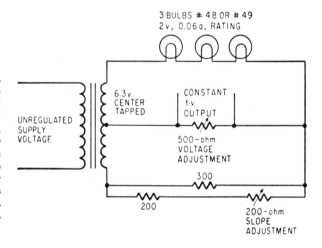

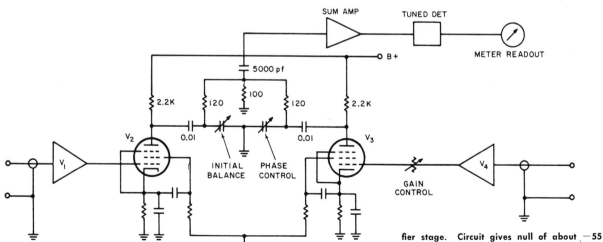

MEASURING PHASE AT 100 MHZ—Broad null of conventional balanced bridge is sharpened in Teltronics PD-200 phase detector by using two similar channels exactly 180 deg out of phase. Pentode types for V2 and V3 are not critical, and component values for tube circuits are those of conventional pentode amplifier stage. Circuit gives null of about —55 dB, permitting phase measurement accuracy to within 0.1 deg in frequency range of 15 to 100 MHz.—R. O. Goodwin, Unbalanced Bridge Simplifies Phase Measurements, *Electronic Design*, March 15, 1965, p 52—53.

10-V BIPOLAR FROM BRIDGE—Grounding common point of two zeners connected across output of power supply eliminates need for two secondary windings on power transformer and provides protection against short-circuits. If unsymmetrical voltages are required, use zeners with different ratings.—S. Ritterman, Single Transformer Provides Positive and Negative Voltages, *Electronic Design*, Jan. 18, 1970, p 86.

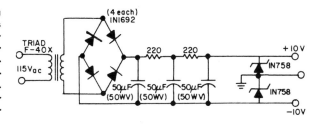

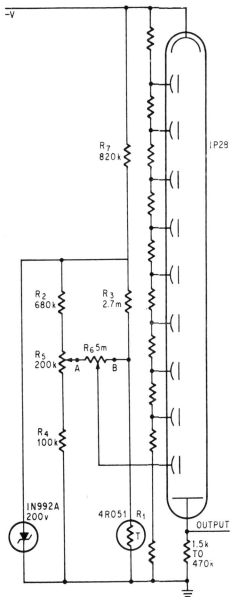

TUBE WARMUP CONTROL—Thermistor mounted on socket of photomultiplier forms one leg of bridge used to compensate for decrease in gain of tube as ambient temperature increases during warm-up period. Control circuit acts on voltage of last dynode. Output amplitude is constant within 2% from 24 to 40 C with compensation, as compared to 8% variation without compensation. Article gives calibration procedure.—A. E. Martens, Photomultiplier's Gain is Temperature Compensated, *Electronics*, April 14, 1969, p 97–98.

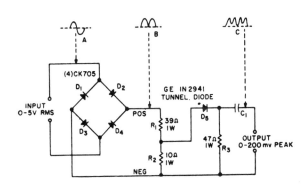

QUADRUPLER—Tunnel diode doubles double-frequency output of bridge rectifier. Adjust input voltage to give best output waveform on cro. Circuit operates up to several MHz. —R. P. Turner, Tunnel Diode Doubles Doubled Frequency, "400 Ideas for Design Selected from Electronic Design," Hayden Book Co., N.Y., 1964, p 170.

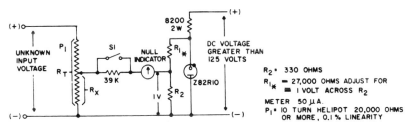

POTENTIOMETER BRIDGE—Uses precision neon voltage regulator. Potentiometer divides unknown voltage down and compares it to 1-V reference across R2. As balance is approached, S1 is closed to increase sensitivity of null meter.—E. Bauman, "Applications of Neon Lamps and Gas Discharge Tubes," Signalite, Neptune, N.J., p 89.

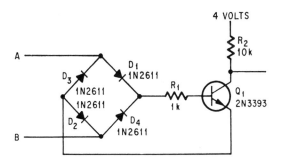

EXCLUSIVE-OR—Simple circuit gives output of 1 when input A or input B is 1, but no output (0) when both inputs are 1.—R. C. Hoyler, Bridge and Transistor are Exclusive-or Gate, *Electronics*, Aug. 19, 1968, p 88.

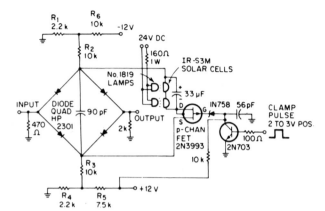

SOLAR-CELL ISOLATION—Cells provide floating low-impedance bias source supplying 1.6 V to turn on fast-response hot-carrier-diode bridge. Bridge input will accept signal voltages up to 4 V p-p. On-off ratios of 60 dB are possible up to 20 MHz. Response is down 5 dB at 40 MHz. Corners of diode bridge are switched simultaneously by low-impedance source, to minimize transients in output. Gating period can range from several ns to several hours.—J. J. Contus, A Broadband Low-Noise Gate Using Hot-Carrier Diodes, *EEE*, March 1969, p 122 and 124.

LOGARITHMIC NULL VOLTMETER—May be used as indicator for Wheatstone bridge, comparison bridge, or differential voltmeter. Will also serve as solid-state galvanometer for laboratory use. Model 9156 IC operates as active attenuator for 1,000-V full-scale setting. Precision resistors are not required because absolute measurements are seldom made off null. Meter should be 10—0—10 V. Model 2245 IC provides four-decade bipolar log function.—A Logarithmic Null-Voltmeter Design, Optical Electronics, Tucson, Ariz., No. 10084.

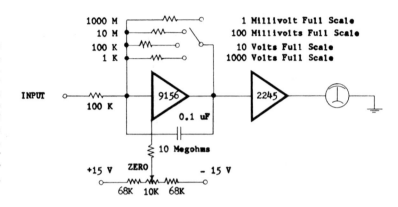

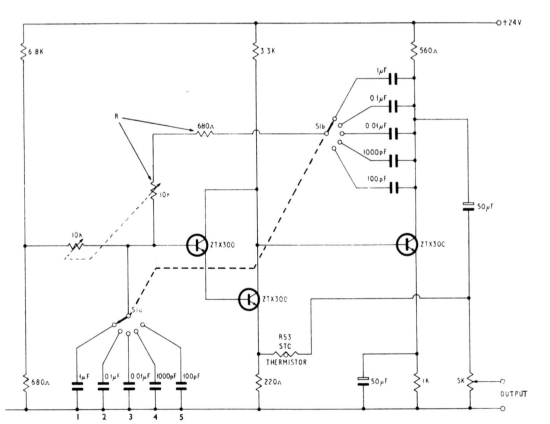

10 HZ—5 MHZ WIEN-BRIDGE OSCILLATOR— Output is 1 V rms, with less than 0.2% total harmonic distortion at 1 kHz. Switch gives five ranges: 15–200 Hz; 150–2,000 Hz; 1.5–20 kHz; 15–200 kHz; 150–2,000 kHz.—"E-Line Transistor Applications," Ferranti Ltd., Oldham, Lancs., England, 1969, p 35.

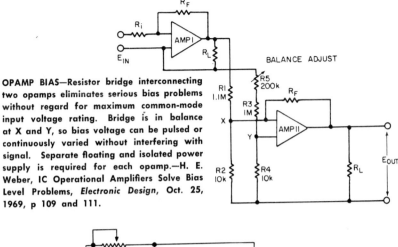

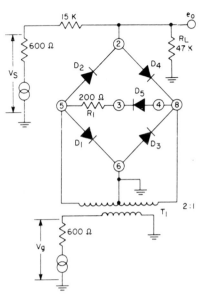

OPAMP BIAS—Resistor bridge interconnecting two opamps eliminates serious bias problems without regard for maximum common-mode input voltage rating. Bridge is in balance at X and Y, so bias voltage can be pulsed or continuously varied without interfering with signal. Separate floating and isolated power supply is required for each opamp.—H. E. Weber, IC Operational Amplifiers Solve Bias Level Problems, *Electronic Design*, Oct. 25, 1969, p 109 and 111.

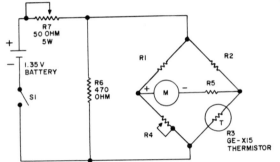

T₁—TECHNITROL No. 851166 OR EQUIV.

THERMISTOR THERMOMETER—Will indicate on meter the temperature inside deep freeze, solution temperature in darkroom, or any other temperature measurable with thermistor that can be connected into bridge-meter circuit shown. For range of 32 to 122 F, R1 and R2 are 1,000, R4 is 5,000, and R5 is 9,500 ohms. For low-temperature range of −40 to +32 F, R1 and R2 are 7,300, R4 is 50,000, and R5 is 4,850 ohms, with battery changed to 1.5-V mercury cell. Both ranges use 50-μA d-c meter (GE type DW-91 having 1,500 ohms resistance). Resistor values in bridge are critical. Calibrate with crushed ice in water and with calibrating resistors supplied with thermistor.—"Hobby Manual," General Electric, Owensboro, Ky., 1965, p 150.

SHUNT GATE—Uses CA3019 diode IC array with diode bridge shunting load resistance and balancing out gating signal to provide pedestal-free output. When gating voltage Vg is of sufficient amplitude, bridge conducts for half of each gating cycle and prevents input signal Vs from reaching output. Gating voltage should be 0.8 to 1.2 V rms at 1 to 100 kHz, and input signal 0 to 1 V rms from d-c to 500 kHz.—"Linear Integrated Circuits," RCA, Harrison, N.J., IC-41, p 303.

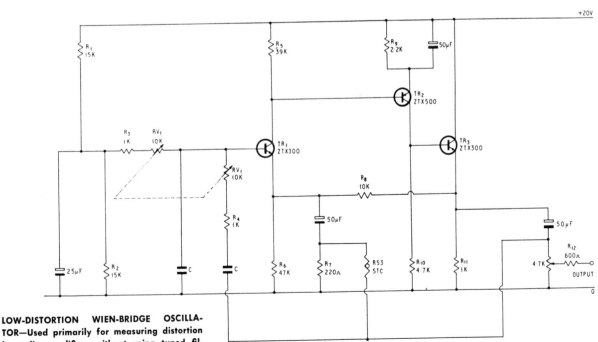

LOW-DISTORTION WIEN-BRIDGE OSCILLATOR—Used primarily for measuring distortion in audio amplifiers without using tuned filters. Delivers pure sine wave with very low distortion. Values of R3 and R4 determine frequency ratio covered on each range; 1K gives maximum to minimum frequency ratio of 11:1, for small overlap on decade ranges. Output is 1 V rms into 600 ohms over frequency range of 10 Hz to 100 kHz.—"E-Line Transistor Applications," Ferranti Ltd., Oldham, Lancs., England, 1969, p 36.

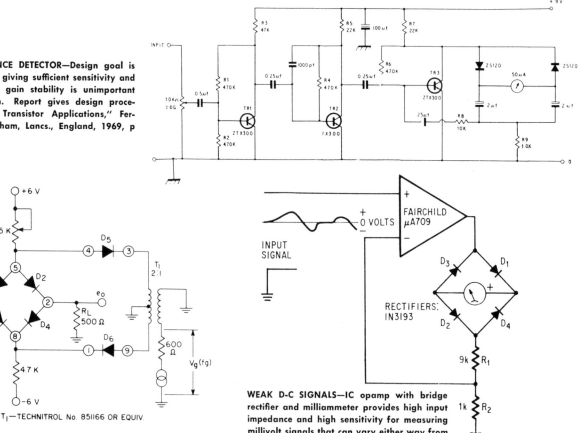

BRIDGE BALANCE DETECTOR—Design goal is simplest circuit giving sufficient sensitivity and linearity, since gain stability is unimportant for application. Report gives design procedure.—"E-Line Transistor Applications," Ferranti Ltd., Oldham, Lancs., England, 1969, p 38.

T_1—TECHNITROL No. 851166 OR EQUIV.

SERIES GATE—Uses CA3019 four-diode IC array in diode-quad bridge that balances out undesired gating signal at output and reduces pedestal to extent that bridge is balanced. Gating voltage should be 1 to 3 V rms at 1 to 500 kHz.—"Linear Integrated Circuits," RCA, Harrison, N.J., IC-41, p 301.

WEAK D-C SIGNALS—IC opamp with bridge rectifier and milliammeter provides high input impedance and high sensitivity for measuring millivolt signals that can vary either way from zero. Meter reads upward regardless of input polarity. Full-scale sensitivity is 1 V. R1 insures that meter overload will not exceed 5%. For measuring a-c voltages, meter indicates average value of rectified waveform; for rms readings, reduce R2 by 11%.—J. P. Budlong, Bridge and Amplifier Monitor D-c Level, *Electronics*, Sept. 2, 1968, p 71–72.

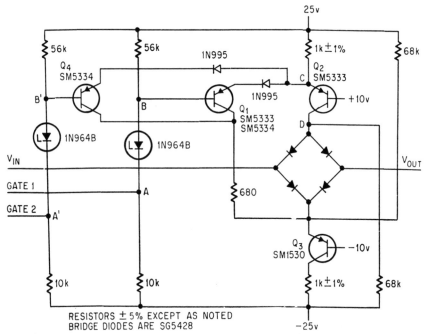

DIAMOND GATE—Used in Paramatrix system for preprocessing analog information, chiefly photos and graphs, while converting to digital signals under computer control. System can enlarge or shrink picture, move it around, rotate it, correct blurs, and fill in gaps. Cuts computer requirements by factor of 10. Can also be applied to automatic drafting. Article presents theory and many block diagrams along with examples of results achieved. Circuit shown is used in interpolator, and requires input signals at both gates before diode bridge transmits analog voltage.—W. J. Poppelbaum, M. Faiman, and E. Carr, Paramatrix Puts Digital Computer In Analog Picture, And Vice Versa, *Electronics*, Sept. 4, 1967, p 99–108.

RESISTORS ± 5% EXCEPT AS NOTED
BRIDGE DIODES ARE SG5428

SECTION 6
Capacitance Control Circuits

NO-PUSH ELEVATOR BUTTON—Finger near button, even if gloved, actuates appropriate relay in modern elevator control system and turns on lamp behind button. Screen behind button picks up hum signal from building ground, capacitively coupled through body of person. This signal is passed to MEM511 mosfet which in turn makes GE C106 scr turn on No. 1829 lamp and relay.—F. G. Geil, *MOS FET Takes the Push Out of Elevator Push Button, Electronics,* Oct. 30, 1967, p 70–71.

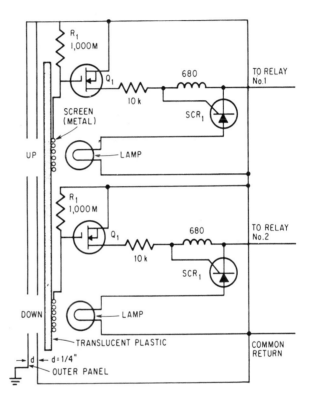

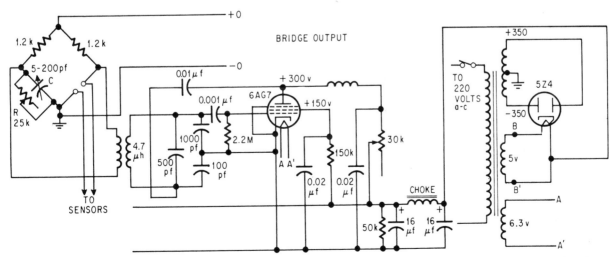

LIQUID LEVEL IN TANK—Movement of liquid in glass gage tube outside tank changes capacitance between liquid and metal-band sensor around glass. Sensor is connected in one arm of bridge, which is first balanced by adjusting C and R with liquid at top of the two sensor electrode rings. Bridge receives 10-V excitation from 3.2-MHz oscillator using 6AG7 tube. Article gives detection sensitivity for ten different liquids.—P. K. Mital, *Capacitance Sensor Monitors Stored Liquid Levels, Electronics,* Oct. 30, 1967, p 71–72.

31

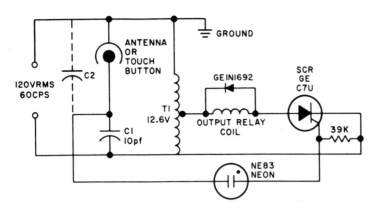

T1 :— 117/12.6 VOLT AUTO TRANSFORMER (OR FILAMENT TRANSFORMER)

PROXIMITY SWITCH—Can be used for elevator call buttons, supermarket and other door controls, burglar alarms, and other applications where load is to be energized by momentary touch of finger on button. C2 represents capacitance between touch button and ground that is provided by body of person. Size of touch button or sensing plate depends on distance to object being sensed; if distance is small, as with touch control, sensing plate need be no larger than a penny. For latching action, drive scr anode circuit only with d-c.— D. R. Grafham, Using Low Current SCR's, General Electric, Syracuse, N.Y., No. 200.19, 1967, p 30.

BODY-CAPACITANCE CONTROL—Bistable neon mvbr is triggered by momentary finger contact with ON touch point, to pull in relay. Relay releases when other point is touched. Chief requirement is sufficient capacitance between circuit and ground. With a-c/d-c supply, ground side of power line should go to circuit ground. With battery supply, larger touch points are needed and circuit ground should be metal chassis.—W. G. Miller, "Using and Understanding Miniature Neon Lamps," H. W. Sams & Co., Indianapolis, Ind., 1969, p 74.

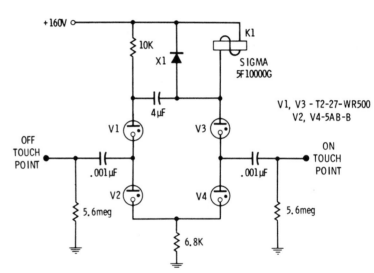

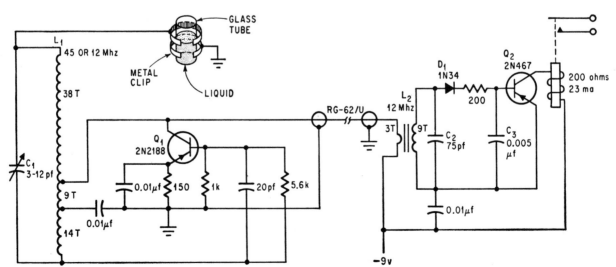

LIQUID-SENSING CAPACITOR—Conventional Hartley oscillator resonates at 45 MHz when glass tube is empty. When liquid reaches critical level inside metal bands surrounding tube, increased dielectric constant of liquid column changes capacitance enough to reduce frequency to 12 MHz. Circuit will trigger reliably with differential capacitance of 0.1 pF between sensing point and ground.—J. K. Marsh, Two-Frequency Oscillator Detects Level of Liquid, *Electronics*, March 20, 1967, p 90.

PROXIMITY SWITCH—Load is energized as long as button of sensor plate capacitor C2 is touched. When button is released, load is de-energized. Latching action can be obtained by replacing CR1 with connection shown as dashed line. Reset will then require auxiliary contact in series with SCR1. Used for door safety controls and floor selector buttons in elevators, supermarket door control, safe monitor in banks, flow switches, and conveyor counting systems.—"SCR Manual," 4th Edition, General Electric, 1967, p 169.

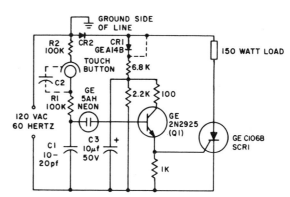

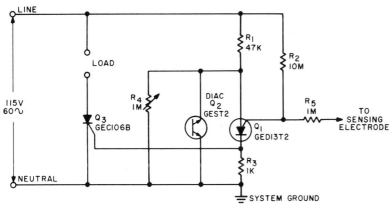

100-W A-C PROXIMITY SWITCH—High sensitivity of D13T2 programmable ujt makes triggering of circuit possible with only about 10 pF capacitance between sensing electrode and ground. This means that finger does not have to touch sensing button. Circuit is non-latching, and will therefore open when body or other capacitance is removed. R4 adjusts sensitivity. Applications include counting freshly painted objects moving past on conveyor line.—E. K. Howell, Small Scale Integration in Low Cost Control Circuits, General Electric, Syracuse, N.Y., No. 671.9, 1968, p 19.

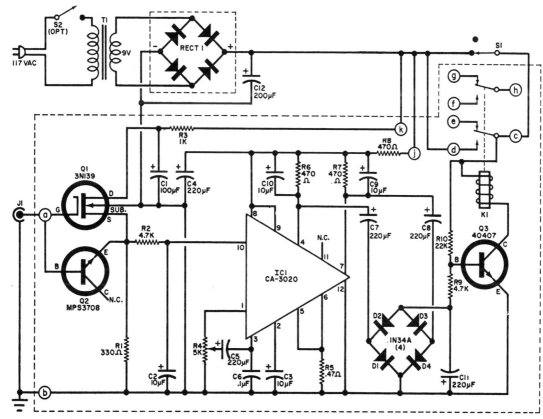

INTRUSION DETECTOR—Responds to change in capacitance between sensor antenna (J1) and ground when person is within several feet of antenna, to energize relay that trips local or remote alarm, turns on lights, summons guard, or applies output of auto ignition coil to intruder. Antenna may be decorative metal art object. For fail-safe operation, Q3 may be biased to energize relay under normal conditions. Presence of intruder or failure of power will then release relay and set off alarm. For protection against burglars in stores, antenna can be safe or cash register.—L. E. Garner, Jr., The Amazing "People Detector," *Popular Electronics*, June 1968, p 27—32 and 93.

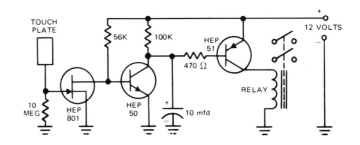

TOUCH SWITCH—Can be used to turn on or off variety of circuits, depending on type of relay used. Touching metal plate, which should be kept close to gate of fet, operates relay.—"Tips on Using FET's," Motorola, Phoenix, Ariz., HMA-33, 1969.

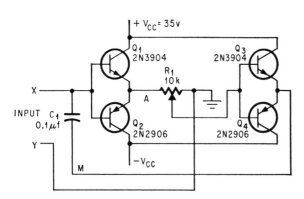

1,000:1 VARIABLE CAPACITOR—Potentiometer R1, buffered on both sides by complementary emitter-followers, reduces effective value of fixed capacitor C1 over much greater range than with conventional variable capacitors. Useful for capacitive tuning of tank circuits over wide frequency ranges.—J. Gaon, Feedback Turns Fixed Capacitor into Variable Capacitance, *Electronics*, Nov. 28, 1966, p 80–81.

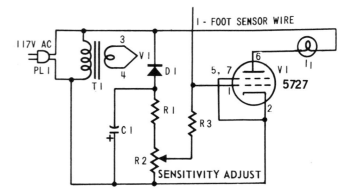

PROXIMITY DETECTOR—Turns on light when person approaches sensor wire. R2 should be well insulated because it is connected to one side of power line when at lower end of range. Values are: I1 10-W lamp; T1 Stancor P-6134; D1 125-V 65-mA selenium rectifier; C1 10 μF; R1 39K; R2 15K pot; R3 3.9 meg.—R. M. Brown, "104 Simple One-Tube Projects," Tab Books, Blue Ridge Summit, Pa., 1969, p 144.

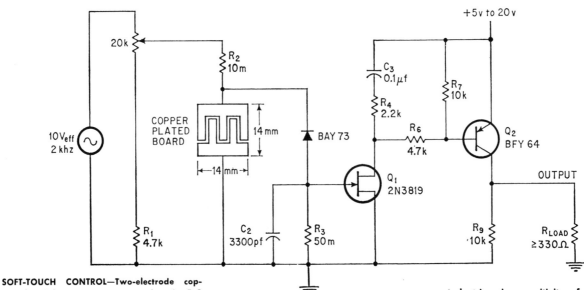

SOFT-TOUCH CONTROL—Two-electrode copper-plated button acts as capacitor in R-C reactive bleeder circuit connected to 2-kHz 10-V source. With button not touched, fet Q1 and transistor Q2 are off and output is at ground or logic zero. When button is touched, its capacitance increases about five times, triggering both transistors on and giving positive output voltage or logic one. Pot controls triggering sensitivity of circuit and triggering level. May be used for elevators, typewriters, and computers.—F. Minder, Touch-Activated Switch Built With Copper-Plated Board, *Electronics*, Sept. 1, 1969, p 79.

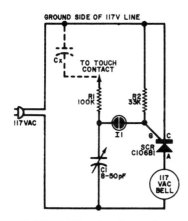

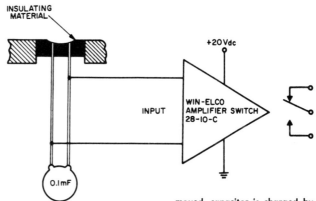

TOUCH ALARM—When intruder touches metal plate or other metal object connected to free end of R1, scr is triggered on and alarm is energized. May be connected to doorknob, metal screen door, foil strip on window sill, or other metal object that is not too large. Object must not be grounded through building structure, because alarm depends on change in capacitance between R1 and ground. Alarm stops when touch contact is broken.—R. F. Graf, Build Low-Cost Touch Alarm, *Popular Electronics*, Feb. 1969, p 92—93.

TOUCH SWITCH—Finger in depression of switch plate discharges capacitor and energizes double Darlington amplifier driving reed switch. Relay remains energized as long as finger makes contact. When finger is re-

moved, capacitor is charged by IC and relay stays energized for about 10 s until charging current drops below threshold current of amplifier.—J. H. Still, Time Delay Touch Switch Uses Body Stray Voltage, *Electronic Design*, July 19, 1967, p 106.

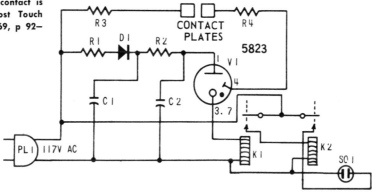

TOUCH CONTROL—Lamp, appliance, tv, or stereo set can be turned on simply by touching two small pieces of tin foil or metal with finger to bridge gap between them. Uses grid-glow tube. Relays are 2.5K. When V1 pulls in K1, its contacts pull in K2 to apply

power to lamp or other appliance plugged into socket. Once fired, V1 stays on and relays remain energized until power from a-c line is interrupted. Values are: D1 125-V 65-mA selenium rectifier; C1 0.47 μF; C2 2μF; R1 27; R2 110K; R3 1.1 meg; R4 110K. Warn-

ing: if R3 is shorted or much smaller resistor used, circuit can be dangerous for one position of plug because one contact plate goes to a-c line.—R. M. Brown, "104 Simple One-Tube Projects," Tab Books, Blue Ridge Summit, Pa., 1969, p 140.

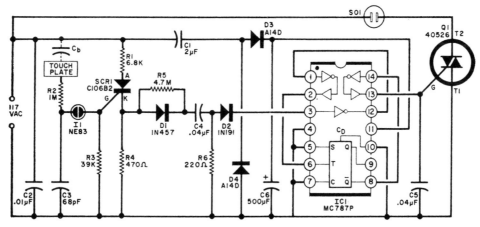

TOUCH SWITCH—Body capacitance is sufficient to trigger scr through neon and energize load through IC and triac Q1, which

will handle up to 150 W. Alternate touches of metal plate will turn load on and off. Reversing line cord plug at wall outlet may

improve operation.—J. Bechtold, TC Switch For Remote Control, *Popular Electronics*, April 1970, p 52—55.

SECTION 7
Chopper Circuits

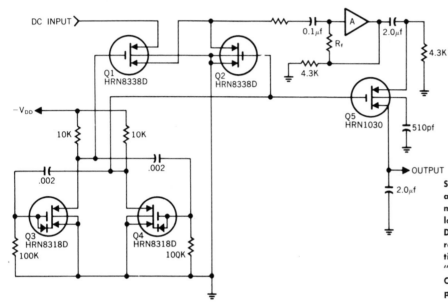

SERIES-SHUNT CHOPPER—Series switch **Q1** and shunt switch **Q2** are driven by astable mvbr Q3-Q4 for detection of signal levels as low as 30 μV and amplification by opamp. Demodulator Q5 is simple series switch. Arrangement gives stable and accurate detection of very small d-c signals.—C. R. Perkins, "Application of MOSFET Devices to Electronic Circuits," Hughes, Newport Beach, Cal., 1968, p 22.

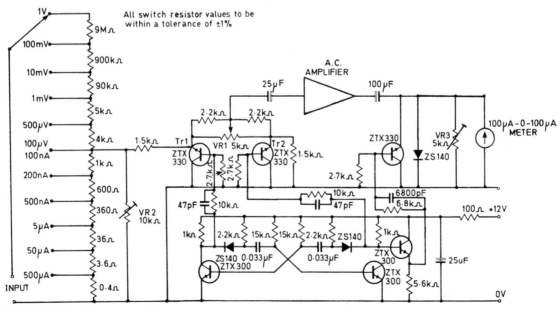

BALANCED CHOPPER—Used here as combination microvoltmeter and microammeter. A-c amplifier can be discrete components or IC having power gain of 86 dB, bandwidth of 20 Hz to 150 kHz at 3 dB down, and p-p noise less than 10 μV referred to input. Chopping frequency is 1.5 kHz. Input impedance on voltage ranges is 10 meg per V. Requires regulated power supply providing 12 V at about 35 mA. Accuracy depends on initial adjustment procedure, as given in report.—E-Line Transistor Applications, Ferranti Ltd., Oldham, Lancs., England, 1969, p 46.

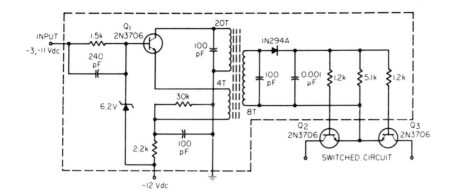

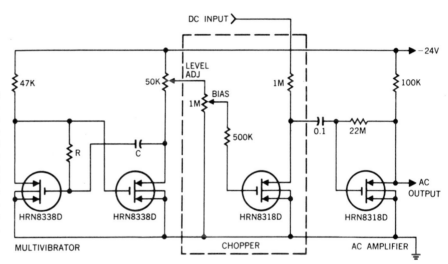

CHOPPER-DRIVER—Switches at rates from d-c to 10 kHz, with output completely isolated from input control signal just as in conventional relay. Q1 is 10-MHz common-base oscillator receiving bias from input logic signals. Rectified and filtered transformer secondary voltage controls bipolar switch Q2-Q3 having ON resistance of about 20 ohms and OFF leakage below 1 μA. Transformer is wound on Arnold A4-134P toroidal core.—J. E. Frecker, Solid-State Relay, *EEE*, June 1967, p 136.

D-C MODULATOR—Useful for modulating or chopping low-level d-c signals for further amplification or for detection of d-c level. Left-hand pair of mosfets forms astable mvbr whose period of oscillation is determined by values of R and C used. Mvbr output drives simple shunt mosfet chopper switch, while mosfet at right provides signal gain and transforms high impedance of chopper to low impedance for a-c output.—C. R. Perkins, "Application of MOSFET Devices to Electronic Circuits," Hughes, Newport Beach, Cal., 1968, p 21.

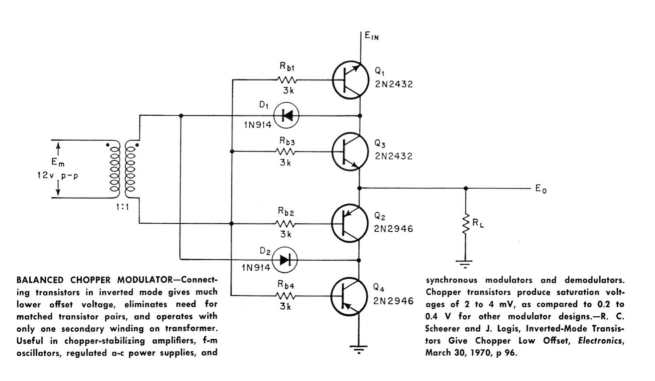

BALANCED CHOPPER MODULATOR—Connecting transistors in inverted mode gives much lower offset voltage, eliminates need for matched transistor pairs, and operates with only one secondary winding on transformer. Useful in chopper-stabilizing amplifiers, f-m oscillators, regulated a-c power supplies, and synchronous modulators and demodulators. Chopper transistors produce saturation voltages of 2 to 4 mV, as compared to 0.2 to 0.4 V for other modulator designs.—R. C. Scheerer and J. Logis, Inverted-Mode Transistors Give Chopper Low Offset, *Electronics*, March 30, 1970, p 96.

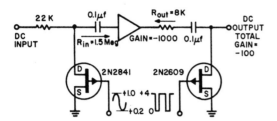

FET CHOPPERS GIVE GAIN OF 100—Input fet chopper is driven by sine wave, and output by 4-V p-p square wave. Input current is less than 20 pA, input impedance is 33K, and offset is less than 4.5 μV at room temperature. Any IC amplifier having gain of 1,000 can be used.—DC Amplifier With FET Chopper, *Electronic Design*, April 13, 1964, p 66.

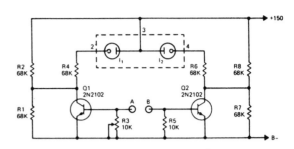

LAMP DRIVER FOR CHOPPER—Used with full-wave laboratory-type photoelectric chopper amplifier. Lamps are neons.—The Dual Photocell, Hewlett Packard, No. 924.

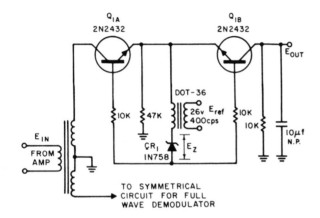

AMPLIFYING LOW-LEVEL D-C—Chopper-amplifier converts d-c signal to a-c at 400 Hz, passes signal through a-c narrow-band amplifier, and synchronously demodulates to give amplified d-c signal. Zener in chopper switching leg suppresses transients and minimizes drift and offset errors.—T. B. Hooker, Zener Diode Aids Chopper in Demodulator Application, *Electronic Design*, April 12, 1965, p 56—57.

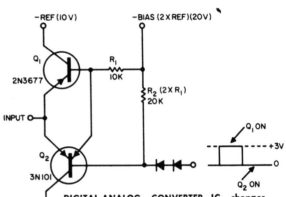

DIGITAL-ANALOG CONVERTER—IC chopper Q2 acts as spdt switch operating from single-polarity drive between logic levels of 0 and +3 V, to give digital-to-analog conversion without complex drive circuit or multiplicity of supply voltages. Input terminal is connected to leg in ladder network, for switching between ground and minus reference voltage.—Integrated Chopper Forms Simple Digital-to-Analog Converter, "Microelectronic Design," Hayden Book Co., N.Y., 1966, p 135.

FULL-WAVE PHOTOELECTRIC—Uses pair of dual photocells in push-pull for both modulator and demodulator, to minimize noise for laboratory applications.—The Dual Photocell, Hewlett Packard, No. 924.

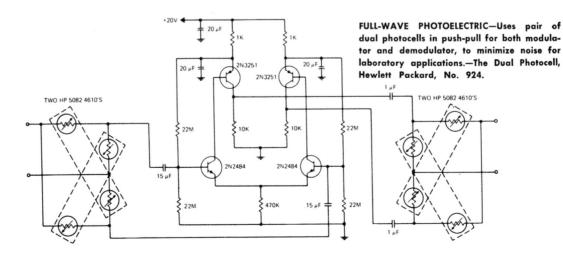

FAST SWITCH—Chops analog signals by using output of pulse-saving network (C1-D1) to switch fet on and off, without introducing common-mode noise. Pulse transformer can maintain 6-V on or off signals for over 100 ms. Transformer is Pulse Engineering Co. model 2228.—C. A. Walton, Pulse-Saving Network Permits Signal Switching, *Electronics*, Sept. 18, 1967, p 108–109.

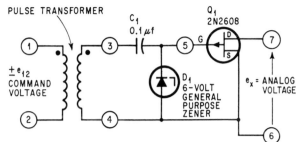

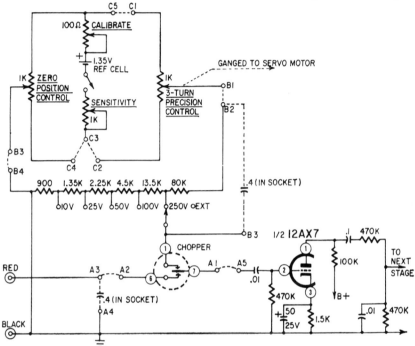

SERVO-DRIVEN PEN—Uses potentiometric metering system, in which input voltage is continuously compared with reference voltage from mercury cell by means of 3-turn slide-wire pot ganged to servo motor. Voltage difference is chopped, and resulting a-c voltage is amplified. Second pot, at left, is used to set marking pen at any desired point on strip chart paper for zero input voltage. Four stages of amplification round off corners of square-wave chopper output so drive for servo is practically sine wave.—E. Leslie, Low-Cost Strip-Chart Recorders, *Radio-Electronics*, June 1965, p 58–59.

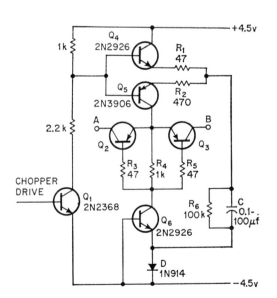

TRANSFORMERLESS WIDEBAND CHOPPER—Provides high isolation and reliable operation over wide range of mark-space ratios and repetition frequencies. During positive half-cycles of square-wave chopper drive, signals are transferred from A to B, but blocked during negative half-cycles.—H. Riddle, Transformer is Eliminated in Transistor Chopper, *Electronics*, Dec. 22, 1969, p 77.

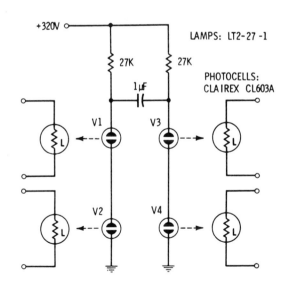

20-HZ PHOTOCHOPPER—Four neons in astable mvbr act on four photocells to provide synchronous chopping at 20 Hz. Developed for use in voltmeters and chopper-stabilized power supplies.—W. G. Miller, "Using and Understanding Miniature Neon Lamps," H. W. Sams & Co., Indianapolis, Ind., 1969, p 68.

NONINVERTING CHOPPER OPAMP—Noninverting mode of operation gives greatly improved accuracy when handling signals from high-impedance sources. Applications include microvolt measurements in biological and research applications. Maximum current drift of opamp is 10 pA per degree C and common-mode rejection ratio is 300,000. Report also covers differential and off-ground applications, and use with bridge signals.—P. Zicko, Designing With Chopper Stabilized Operational Amplifiers, Analog Devices, Cambridge, Mass., 1970.

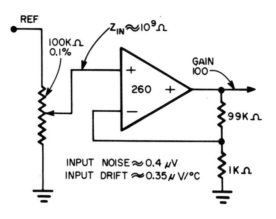

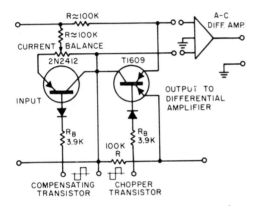

NOISE CANCELLER—Arrangement provides almost complete cancelling of switching spikes and voltage offsets. Dual emitters of chopper transistor convert these noise sources into common-mode noise, which is rejected by differential amplifier.—S. W. Holcomb, F. Opp, and J. A. Walston, Low-Noise Chopper, Electronic Design, Jan. 6, 1964, p 78.

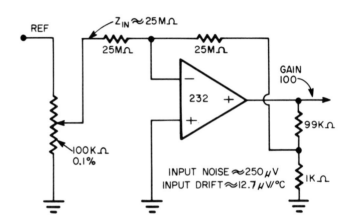

INVERTING-ONLY CHOPPER OPAMP—Used for high-accuracy low-level measurements where low noise and low drift are major requirements, as when measuring signals from 100K precision pot. Common-mode rejection ratio is 300,000.—P. Zicko, Designing With Chopper Stabilized Operational Amplifiers, Analog Devices, Cambridge, Mass., 1970.

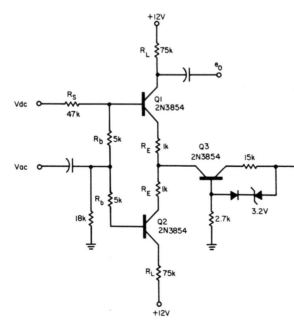

TRANSFORMERLESS CHOPPER—Uses differential amplifier Q1-Q2 and constant-current source Q3. Differential input stage has good temperature and drift stabilization. For optimum conversion of very small d-c signals, differential amplifier should be operated at currents of about 10 μA.—A. C. Caggiano, Transformerless Chopper Circuit Built With a Differential Amplifier, Electronic Design, Oct. 11, 1967, p 106.

SECTION 8
Code Circuits

Covers Morse-code circuits as used in amateur, maritime, and other CW communication applications for keyers, monitors, code generators and regenerators, decoders, practice oscillators, CW filters, and call-letter generators. For circuits capable of handling CW along with other types of modulation and for circuits handling other types of codes, see also Filter, IF Amplifier, Keyboard, Memory, Microprocessor, Single-Sideband, Receiver, Transceiver, and Transmitter chapters.

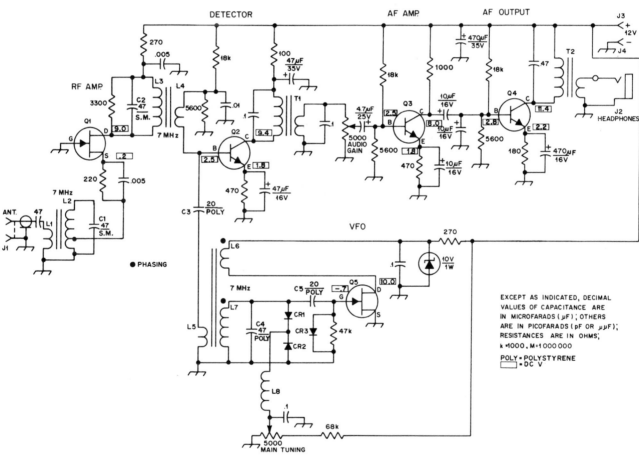

CR1 - CR3, incl. — High-speed switching diode (Radio Shack type 276-1620).
J1 — RCA-type phono jack.
J2 — 1/4-inch phone jack.
J3, J4 — Binding post.
L1 — 3 turns insulated hookup wire wound over (ground) end of L2.
L2 — Radio Shack type 273-101 rf choke. Tap at 4 turns above ground end.

L3 — Radio Shack type 273-101 rf choke.
L4 — 4 turns insulated hookup wire wound over cold end of L3.
L5 — 5 turns insulated hookup wire wound over ground end of L7.
L6 — 4 turns insulated hookup wire wound adjacent to high end of L7.
L7 — Radio Shack type 273-101 rf choke with six of the original turns removed.

L8 — Radio Shack type 273-102 rf choke.
Q1, Q5 — JFET (Radio Shack type RS-2035).
Q2 - Q4, incl. — Transistor (Radio Shack type 276-1617).
T1 — Audio transformer (Radio Shack type 273-1378).
T2 — Audio transformer (Radio Shack type 273-1380).

40-METER DIRECT-CONVERSION—Simple, foolproof circuit design uses discrete components mounted on printed-circuit board shaped to fit in oval herring can. Single 7-MHz RF stage and voltage-tuned VFO feed product detector Q2 that drives 2-stage AF amplifier having peak response at about 650 Hz for most comfortable CW listening. VFO uses Armstrong or tickler-feedback circuit, with CR1 and CR2 connected as voltage-variable-capacitance diodes. Zener regulator powers VFO circuit for good frequency stability. Receiver will tune any 100-kHz segment of 40-meter band.—J. Rusgrove, The Herring-Aid Five, *QST*, July 1976, p 20–23.

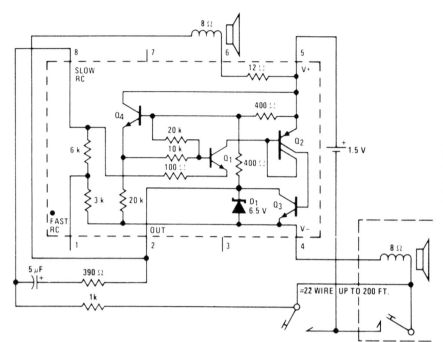

MORSE-CODE SET—National LM3909 flasher IC is connected as tone oscillator that simultaneously drives loudspeakers at both sending and receiving ends of wire line used for Morse-code communication system. Single alkaline penlight cell lasts 3 months to 1 year depending on usage. Three-wire system using parallel telegraph keys eliminates need for send-receive switch. Tone frequency is about 400 Hz.—"Linear Applications, Vol. 2," National Semiconductor, Santa Clara, CA, 1976, AN-154, p 5–6.

CQ ON TAPE—Frequently used code message such as amateur radio CQ call is recorded by keying audio oscillator with desired message and picking up oscillator output with microphone of endless-loop cassette or other tape recorder. Rewound recording is played back through single-transistor stage connected as shown for driving keying relay of transmitter. Circuit requires shielding.—Circuits, *73 Magazine*, July 1977, p 34.

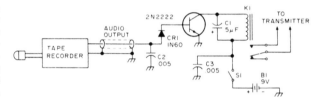

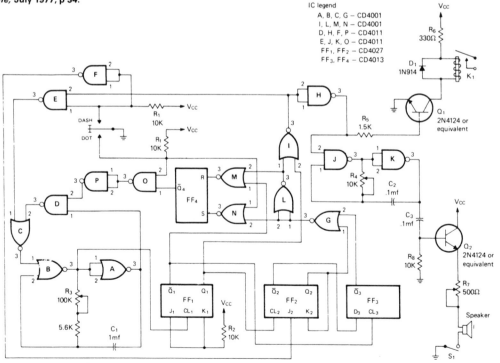

KEYER WITH MEMORY—Includes sidetone oscillator and dash-dot memory along with variable speed, automatic spacing, and self-completing dots and dashes. If dot paddle is pressed and released while keyer is generating dash, dot is generated with correct spacing after dash is completed. Gates A, B, and C form gated MVBR. Gates D, E, O, and P serve to complete characters. JK flip-flops FF_1 and FF_2, D flip-flop FF_3, and gates F, G, and L provide character-shaping required for dash-dot memory using gates M, N, and RS flip-flop FF_4. Gates J and K generate audio sidetone. K_1 is B & F Enterprises ERA-21061 SPST reed relay. Supply can be 9-V battery.—T. R. Crawford, A Low-Power Cosmos Electronic Keyer in Two Versions, *CQ*, Nov. 1975, p 17–24.

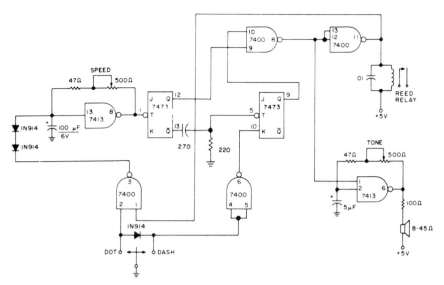

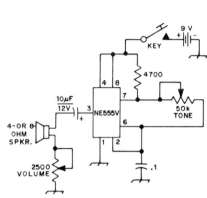

KEYER—Uses gating and flip-flop functions to generate dots and dashes under control of gated clock. SN7413 Schmitt trigger is connected as relaxation oscillator. Circuit provides minimum spacing between dots and dashes regardless of paddle movements.—A. D. Helfrick, A Simple IC Keyer, *73 Magazine*, Dec. 1973, p 37–38.

TIMER FOR CODE PRACTICE—Signetics NE555V timer operating on 9-V supply serves as AF oscillator providing adequate volume for classroom instruction. Output tone can be varied from several hundred to several thousand hertz.—J. Burney, Code Practice Oscillator, *QST*, July 1974, p 37.

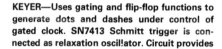

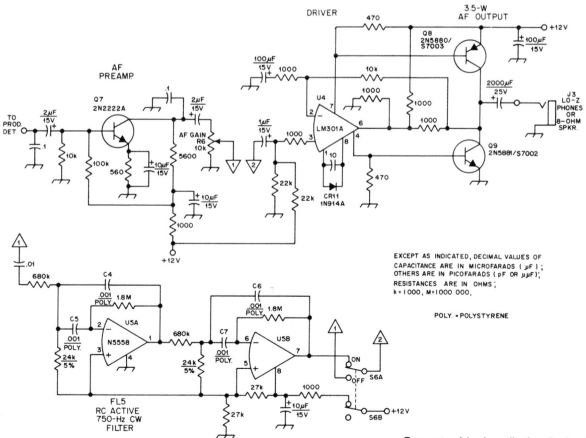

EXCEPT AS INDICATED, DECIMAL VALUES OF
CAPACITANCE ARE IN MICROFARADS (μF);
OTHERS ARE IN PICOFARADS (pF OR μμF);
RESISTANCES ARE IN OHMS ;
k = 1000, M = 1000 000.

POLY. = POLYSTYRENE

3.5 W FOR CW—Discrete devices minimize distortion and eliminate fuzziness while listening to low-level CW signals in communication receiver covering 1.8–2 MHz. RC active bandpass filter peaked at 800 Hz improves S/N ratio for weak signals. Adjust BFO of receiver to 800 Hz. Two-part article gives all other circuits of receiver.—D. DeMaw, His Eminence—the Receiver, *QST*, Part 2—July 1976, p 14–17 (Part 1—June 1976, p 27–30).

SENSOR KEYER—Skin resistance of about 10K creates dashes when finger touches grid pattern on left side of paddle and dots when other finger touches pattern on other side. Transistors act as solid-state switches. Developed for use with Heathkit CW keyer HD-10. Supply is 10 V, obtained from 10-V zener connected through appropriate dropping resistor to higher-voltage source. Article covers construction of paddle by etching printed-wiring board.—T. Urbizu, Try a Sensor Keyer, *73 Magazine*, Jan. 1978, p 184–185.

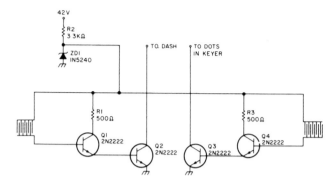

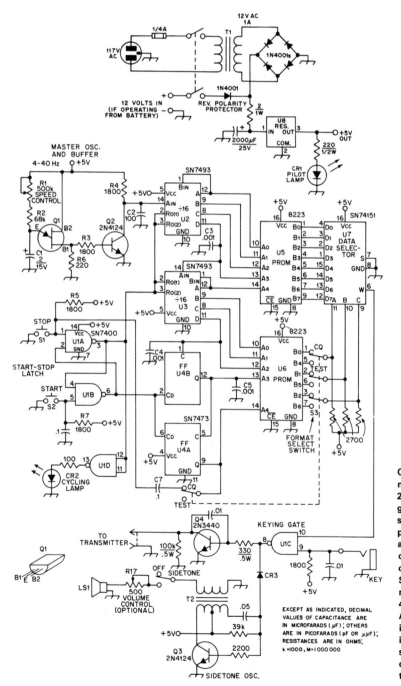

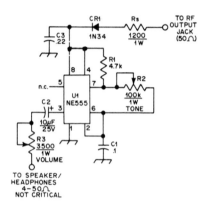

AF OSCILLATOR MONITORS CW—Can be added to any transceiver not already having built-in sidetone oscillator, to hear keying of transmitter. RF input from transmitter is rectified by CR1 to provide about 6-VDC supply. Keying of carrier on and off turns NE555 AF oscillator on and off correspondingly.—J. Arnold, A CW Monitor for the Swan 270, *QST*, Aug. 1976, p 44.

CQ CALL SYNTHESIZER—Uses only two Signetics 8223 256-bit PROMs for storing up to 2048 bits of code information, for automatic generation of Morse-code CQ calls, test messages, and other frequently used messages. Repeated words are stored in only one location and selected as needed, to quadruple capacity of memory. PROMs can be programmed in field or custom-programmed by manufacturer. Speed and timing of code characters are determined by UJT oscillator Q1, variable from about 4 to 40 Hz or 5 to 50 WPM. CR1 and CR2 are Archer (Radio Shack) 276-042 or equivalent. CR3 is 1N34A, 1N270, or equivalent germanium. Q1 is Motorola MU4891 or equivalent. Article describes circuit operation and programming in detail.—J. Pollock, A Digital Morse Code Synthesizer, *QST*, Feb. 1976, p 37–41.

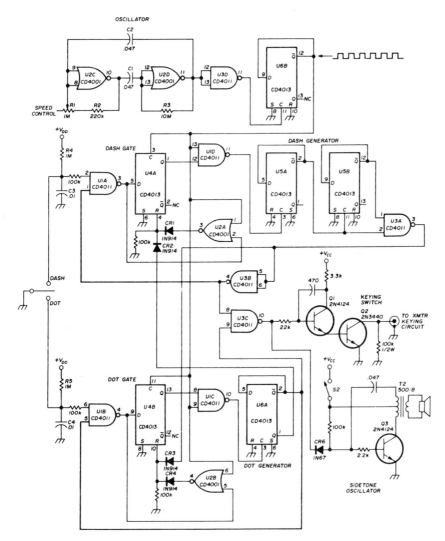

CMOS KEYER—Draws only 0.4 mA on standby and 2 mA with key down if supply is 10 V. Will work properly with 4 to 15 V. Features include self-completing dots, dashes, and spaces, along with sidetone generator and built-in transmitter keying circuit. Ratio of dashes to dots is 3:1, and space has same duration as dot. Time base of keyer is generated by NOR gates U2C and U2D connected as class A MVBR. Frequency of oscillator is inversely linear with setting of R1. Inverter U3D buffers oscillator and squares its output. Flip-flop U6B divides frequency by 2 and provides clock source with perfect 50% duty cycle. Once enabled, gates ensure completion along with following space. Article gives power supply circuit operating from AC line and 12-V battery.—J. W. Pollock, COSMOS IC Electronic Keyer, *Ham Radio*, June 1974, p 6–10.

SIDETONE MONITOR—Mostek MK5086N IC is used with crystal in range from 2 to 3.5 MHz as signal generator driving FET audio amplifier. Switch S1 gives choice of four AF tones, determined by dividing crystal frequency in hertz by 5120 for T1, 4672 for T2, 4234 for T3, and 3776 for T4. Can also be used as code practice set and as audio signal generator.—J. Garrett, A Sidetone Monitor-Oscillator-Audio Generator, *QST*, June 1978, p 43.

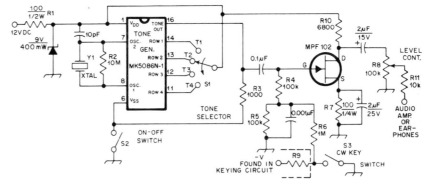

REGENERATED CW—Audio oscillator whose frequency can be varied is keyed in accordance with incoming CW signal, to give clean locally generated audio signal without background noise and interference. NE567 phase-locked loop serves as tunable audio filter and LED switch driver for activating NE555 variable-frequency tone oscillator. LED serves as visual tuning aid to indicate that PLL is locked on to incoming signal.—Regenerated CW, *73 Magazine*, Dec. 1977, p 152–153.

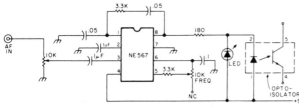

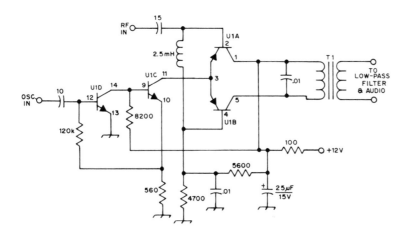

PRODUCT DETECTOR—Designed for use in 40-meter CW direct-conversion receiver, in which oscillator input is from 3.5–4 MHz VFO. U1 is RCA CA3046 transistor quad. Circuit provides bias stabilization for constant-current transistor and some amplification of AF output. T1 is audio transformer.—A. Phares, The CA3046 IC in a Direct-Conversion Receiver, *QST*, Nov. 1973, p 45.

TONE DECODER—Decodes audio output of amateur radio receiver. Resulting audio tone burst corresponds to CW signal being received, with tone frequency varying with receiver tuning. Center frequency of NE567 phase-locked loop is adjusted with R1. Audio is translated into digital format of 1s and 0s, with tones for 0s. Output can be fed into computer for automatic translation of Morse code and printout as text.—W. A. Hickey, The Computer Versus Hand Sent Morse Code, *BYTE*, Oct. 1976, p 12–14 and 106.

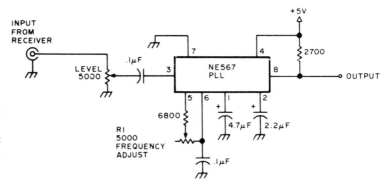

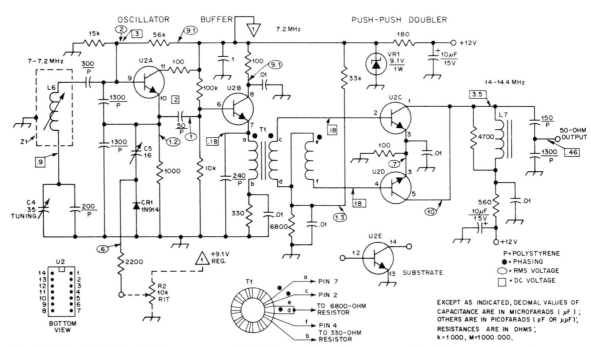

C4 — 35-pF air variable (Millen 26035 or Hammarlund HF-35).
C5 — 16-pF air trimmer, pc-board mount Johnson 187-0109-005.
CR1 — Silicon high-speed switching diode, 1N914 or equiv.
L6 — Slug-tuned inductor, 3.6 to 8.5 μH (Mil-

ler 42A686CBI or equivalent). Use shield can (35-mm film canister or Miller S-33).
L7 — Toroidal inductor, 0.9 μH. Use 12 turns of No. 24 enam. wire on a T50-6 toroid core. (See *QST* ads for toroid suppliers — Amidon, G. R. Whitehouse and Palomar Engrs.).

R2 — Optional circuit (see text). 10,000-ohm linear-taper composition control.
T1 — Trifilar-wound trans. 2 μH, 20 turns, twisted six turns per inch. No. 28 enam. wire on a T50-2 toroid core (see text).
U2 — RCA CA3045 array IC.
VR1 — Zener diode, 9.1 V, 1 W.

BFO FOR 20 METERS—Uses CA3045 transistor array, with U2A as series-tuned Clapp oscillator covering 7–7.2 MHz. Tuned emitter-follower U2B provides push-pull drive at 7 MHz to bases of push-push doubler U2C-U2D. Output of BFO is applied to product detector rather than to mixer of receiver. Audio signal from detector is frequency difference between BFO and incoming signal, typically 700 Hz for CW reception. Article covers construction and adjustment.—D. DeMaw, Understanding Linear ICs, *QST*, Feb. 1977, p 19–23.

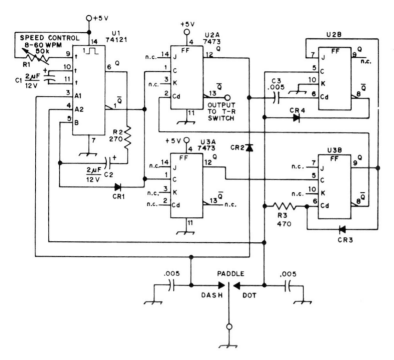

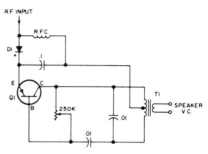

KEYER WITH DOT MEMORY—Features include self-completing characters, exact timing of characters, and dot memory. Timing circuit uses 74121 mono MVBR U1, serving dot generator and output stage U2A, dot memory U2B, and dash generator U3A-U3B. U2 and U3 are 7473 dual JK flip-flops. Length of timing pulse is determined by R1-C1, with R1 controlling speed of keyer. Pulse-width stability at all speeds is better than 5% between first and all following pulses. Dot memory U2B allows keying of dot at any time, even if dash has not yet been completed. Dot is held in memory and keyed out automatically after dash. Diodes are 1N914.—J. H. Fox, An Integrated Keyer/TR Switch, *QST*, Jan. 1975, p 15–20.

CODE MONITOR—Works with any transmitter, regardless of type of keying. Use any good PNP transistor. With NPN transistor, reverse connections to diode. Frequency of tone gets higher as resistance of 250K pot is reduced. Monitor is turned off at minimum resistance. Enough RF to operate monitor can be obtained simply by connecting it to chassis of receiver or transmitter.—J. Smith, Yet Another Code Monitor, *73 Magazine*, Sept. 1971, p 58.

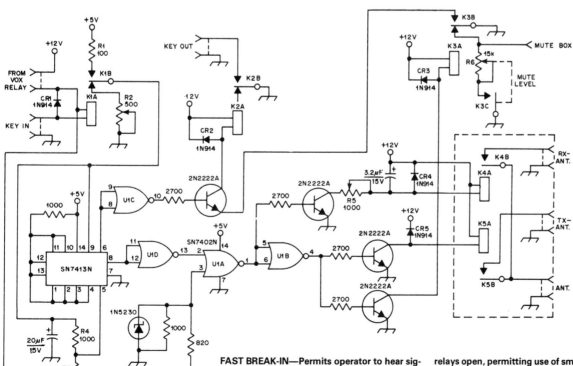

FAST BREAK-IN—Permits operator to hear signals even between dots while calling DX station, so call can be stopped if DX station answers someone else. Timing circuit ensures that transmitter is not producing power when relays open, permitting use of small high-speed relays. K1-K4 are common reed relays. K5 should have contacts rated for 300 VAC at 500 mA.—A. Pluess, A Fast QSK System Using Reed Relays, *QST*, Dec. 1976, p 11–12.

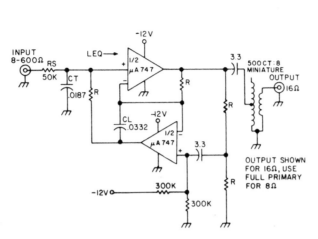

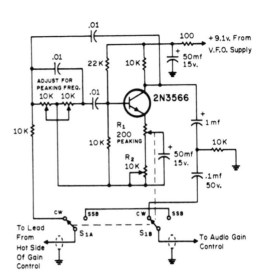

CW FILTER—Simple single-section parallel-tuned active filter uses negative-impedance converter or gyrator to replace hard-to-get inductor of passive code filter. Capacitor CL is gyrated from 0.0332 μF to effective inductance of 1.87 H. Filter has 6-dB gain at resonance and essentially zero output impedance. Bandpass is 85 Hz centered at about 865 Hz. Uses single −12 V supply. Resistors R are 7.5K, matched to about 2%.—N. Sipkes, Build This CW Filter, *73 Magazine,* June 1977, p 55.

REGENERATIVE CW FILTER—Can be added between product detector and volume control of SSB receiver or transceiver that does not have CW filter. Just before oscillation occurs, gain becomes extremely high with very narrow bandpass. Regeneration and bandpass can be adjusted as required. Filter typically has 40-Hz bandwidth centered on 800 Hz.—R. A. Yoemans, Further Enhancing the Yaesu FTDX-560 Transceiver, *CQ,* July 1972, p 16–18 and 20–22.

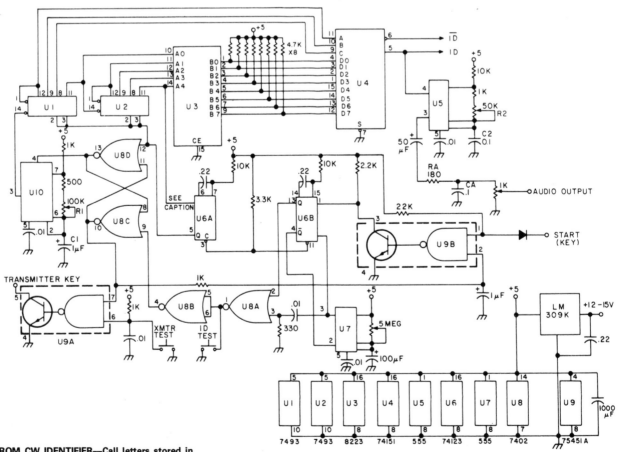

PROM CW IDENTIFIER—Call letters stored in 8223 PROM U3 drive 74151 multiplexer/data selector U4 for keying NE555 audio oscillator U5 which feeds transmitter mike input through RA. Timed holdoff keeps identifier from being re-keyed within specified time period, with reidentifying at end of period. Article covers operation of circuit and gives construction and programming details.—W. Hosking, ID with a PROM, *73 Magazine,* Nov. 1976, p 90–92.

SECTION 9
Comparator Circuits

Used to compare two values of voltage, frequency, phase, or digital inputs and provide logic output for driving variety of control circuits and indicating devices. See also Logic, Logic Probe, and Voltage-Level Detector chapters.

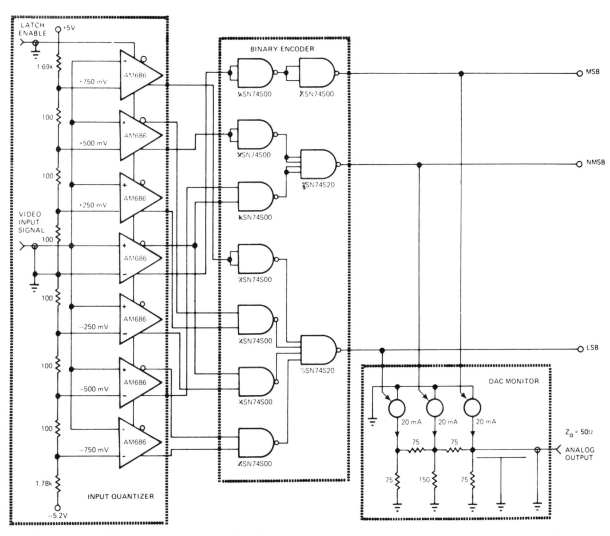

LATCH COMPARATORS FORM 3-BIT A/D CONVERTER—Seven Advanced Micro Devices AM686 comparators are arranged for direct parallel conversion of rapidly changing input signals, without prior sample-and-hold conditioning. Comparators feed Schottky TTL binary encoder logic for encoding to 3-bit offset binary. Quantization process is monitored by D/A converter. Article describes operation in detail and gives performance graphs which show freedom from output glitching at conversion speeds under 12 ns.—S. Dendinger, Try the Sampling Comparator in Your Next A/D Interface Design, *EDN Magazine,* Sept. 20, 1976, p 91–95.

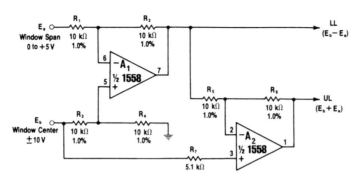

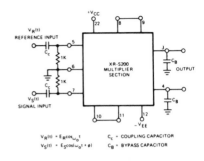

VARIABLE WINDOW—Single comparator can be programmed for wide variety of applications. One reference input voltage positions center of window, and other sets width of window. Sum or difference of reference voltages must not exceed ±10 V; if larger voltages must be handled, add voltage divider to scale them down into comparison range. A_1 is subtractor, generating voltage $E_b - E_a$ for use as lower limit voltage. Lower limit is added to $2E_b$ at A_2 to derive upper limit voltage $E_b + E_a$.—W. G. Jung, "IC Op-Amp Cookbook," Howard W. Sams, Indianapolis, IN, 1974, p 232–233.

PHASE COMPARATOR—High-level reference or carrier signal and low-level reference signal are applied to multiplier inputs of Exar XR-S200 PLL IC. If both inputs are same frequency, DC output is proportional to phase angle between inputs. For low-level inputs, conversion gain is proportional to input signal amplitude. For high-level inputs (V_S above 40 mVRMS), conversion gain is constant at about 2 V/rad.— "Phase-Locked Loop Data Book," Exar Integrated Systems, Sunnyvale, CA, 1978, p 9–16.

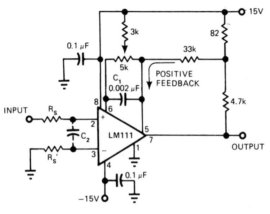

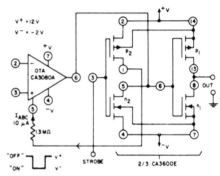

SUPPRESSING OSCILLATION—Use of positive feedback to pin 5 of comparator gives sharp and clean output transitions even with slow triangle-wave inputs, with no possibility of comparator bursting into oscillation near crossing point. Input resistors should not be wirewound. Circuit will handle triangle-wave inputs up to several hundred kilohertz.—P. Lefferts, Overcome Comparator Oscillation Through Use of Careful Design, *EDN Magazine,* May 20, 1978, p 123–124.

STROBED COMPARATOR—Combination of CA3080A opamp and two CMOS transistor pairs from CA3600E array gives programmable micropower comparator having quiescent power drain of about 10 μW. When comparator is strobed on, opamp becomes active and circuit draws 420 μW while responding to differential-input signal in about 8 μs. Common-mode input range is −1 V to +10.5 V. Voltage gain of comparator is typically 130 dB.—"Linear Integrated Circuits and MOS/FET's," RCA Solid State Division, Somerville, NJ, 1977, p 279.

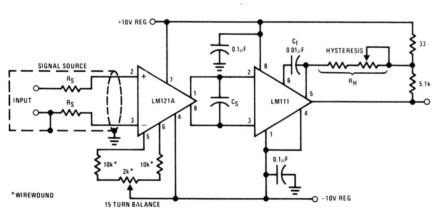

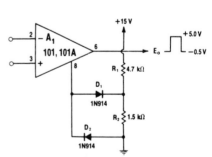

MICROVOLT COMPARATOR—Combination of National LM121A preamp and LM111 comparator serves for comparing DC signal levels that are only within microvolts of each other. With bias network shown, preamp has open-loop temperature-stable voltage gain close to 100. Separation of preamp from comparator chip minimizes effects of temperature variations. Circuit hysteresis is 5 μV, which under certain conditions can be trimmed to 1 μV.—"Linear Applications, Vol. 2," National Semiconductor, Santa Clara, CA, 1976, LB-32.

5-V CLAMPED COMPARATOR—R_1 and R_2 provide +3.8 V bias for D_1, clamping positive output of comparator opamp to +5 V. D_2 limits negative output swing to −0.5 V. Open-loop circuit means that output voltage will vary in proportion to load current.—W. G. Jung, "IC Op-Amp Cookbook," Howard W. Sams, Indianapolis, IN, 1974, p 226–228.

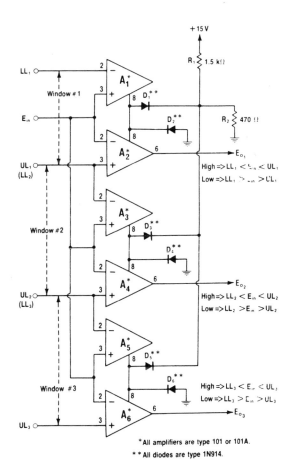

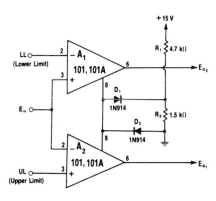

*All amplifiers are type 101 or 101A.

**All diodes are type 1N914.

STAIRCASE WINDOW COMPARATOR—Cascading of 101-type window comparators for sequential operation indicates which of three windows input voltage is in. Input voltage is applied in parallel to all comparators. Output goes high only for comparator whose range includes voltage value of input. Lamp or other indicator can be added to each output line to give visual indication of voltage range.—W. G. Jung, "IC Op-Amp Cookbook," Howard W. Sams, Indianapolis, IN, 1974, p 233–234.

INTERNALLY GATED WINDOW COMPARATOR—Operation is based on fact that source and sink currents available at pin 8 of 101 opamp are unequal, with negative-going drive being larger. Voltage at pin 8 is low if either comparison input (A_1 or A_2) so dictates. Both A_1 and A_2 must have high outputs for pin 8 to be high. Outputs of A_1 and A_2 thus follow pin 8 since opamps have unity gain. D_1 and D_2 form clamp network. Either output of A_1 or A_2 can be used. Outputs go to +5 V only when input voltage is in window established by upper and lower voltage limits.—W. G. Jung, "IC Op-Amp Cookbook," Howard W. Sams, Indianapolis, IN, 1974, p 231–232.

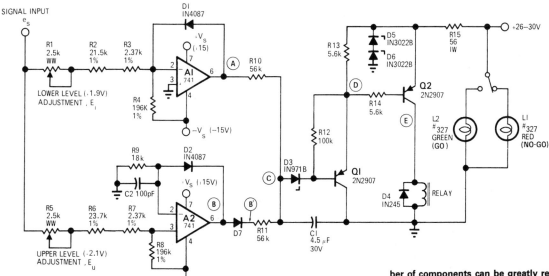

1.9–2.1 V WINDOW COMPARATOR—When positive input voltage is between levels set by R1 and R5, relay is actuated and green indicator lamp is turned on. Red lamp is on for voltages outside limits of window. Article gives design equations and traces operation of circuit. Number of components can be greatly reduced by changing opamps to LM111 comparators.—J. C. Nirschl, 'Window' Comparator Indicates System Status, *EDN/EEE Magazine*, June 15, 1971, p 49–50.

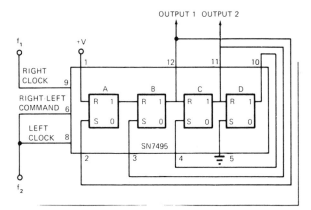

FREQUENCY/PHASE UP TO 25 MHz—Universal shift register such as 5495/7495 is connected to compare both frequency and phase of two carrier signals anywhere in range from DC to 25 MHz. When f_1 is greater than f_2, output is 1; when f_1 is less than f_2, output is 0. For $f_1 = f_2$, output is square wave whose duty cycle varies linearly with phase difference between f_1 and f_2. Comparisons are almost instantaneous, requiring at most two carrier cycles.—J. Breese, Single IC Compares Frequencies and Phase, *EDN Magazine*, Sept. 15, 1972, p 44.

VARIABLE BIPOLAR CLAMPING—Precision comparator provides independent regulation of both output voltage limits without connection to comparison inputs. A_2 and A_3 are complementary precision rectifiers having independent positive and negative reference voltages, with both rectifiers operating in closed loop through A_1. A_2 senses positive peak of E_o and maintains it equal to $+V_{clamp}$ by adjusting voltage applied to D_1. A_3 and D_3 perform similar function on negative peaks. Feedback network around output stage of A_1 regulates output voltage independently of inputs to A_1.—W. G. Jung, "IC Op-Amp Cookbook," Howard W. Sams, Indianapolis, IN, 1974, p 228–229.

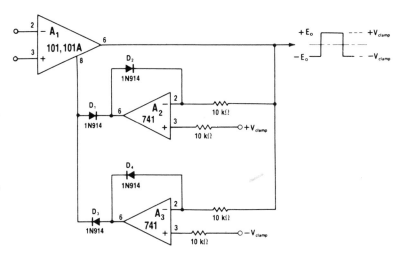

LEVEL-CROSSING DISPLAY—Uses Monsanto MV5491 dual red/green LED, with 220 ohms in upper lead to +5 V supply and 100 ohms in lower +5 V lead because red and green LEDs in parallel back-to-back have different voltage requirements. Circuit requires SN75451 driver ICs and one section of SN7404 hex inverter, with LM311 comparator. All operate from single +5 V source. Provides indicator change from red to green with input change of only a few millivolts.—K. Powell, Novel Indicator Circuit, *Ham Radio*, April 1977, p 60–63.

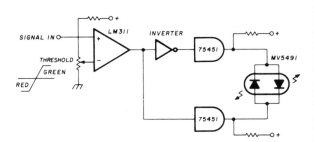

FREQUENCY COMPARATOR—Can be used with wide range of clock frequencies up to 5.3 MHz to provide output frequency that is equal to absolute difference between input frequencies f_1 and f_2. Article traces operation of circuit and gives design equations.—P. B. Morin, Frequency Comparator Provides Difference Frequency, *EEE Magazine*, April 1971, p 65–66.

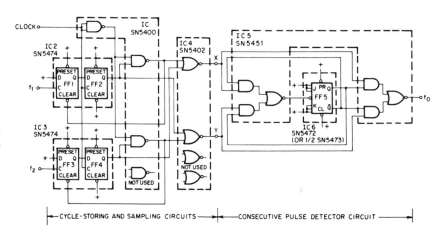

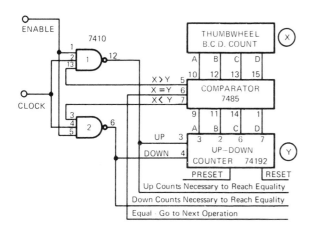

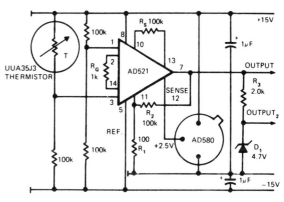

4-BIT BCD COMPARATOR—Provides less than, equal to, or greater than comparison between setting of BCD thumbwheel switch at X and BCD input digit at Y (Y is count preset into 74192 up/down counter). If equality does not exist, circuit will count up or down until it reaches equality, and thereby calculate difference between BCD values. Separate register can be used to store up or down counts required to reach equality.—R. A. Scher, Digital Comparator Is Self-Adjusting, *EDN Magazine,* Sept. 1, 1972, p 51.

INDEPENDENT SIGNALS—Single AD521 instrumentation amplifier compares two independent signal levels from sources having no common reference point. When one differential signal is applied to usual input of opamp and other to reference input, output is proportional to difference. Positive feedback provides small amount of hysteresis, to eliminate ambiguity and reduce noise susceptibility. Stable threshold of about 25 mV is derived from AD580 low-voltage reference circuit. Reference voltage is 2.5 V, but values used for R_S and R_G are in ratio of 1:100 so comparator output switches when normal input is about 1/100 of reference input. Output is negative when normal input is zero, and switches positive when input exceeds threshold. Output swings ±12 V as inputs go through critical ratio. R_3 and D_1 provide TTL-compatible second output.—A. P. Brokaw, You Can Compare Two Independent Signal Levels with Only One IC, *EDN Magazine,* April 5, 1975, p 107–108.

VOLTAGE COMPARATOR—Motorola MC1539 opamp provides excellent temperature characteristics and very high slewing rate for comparator applications. Zener connected to pin 5 limits positive-going waveform at output to about 2 V below zener voltage. Silicon diode connected to output limits negative excursion of output to give protection for logic circuit being driven. Parallel RC network in output provides impedance matching and minimizes output current overload problems.—E. Renschler, "The MC1539 Operational Amplifier and Its Applications," Motorola, Phoenix, AZ, 1974, AN-439, p 18.

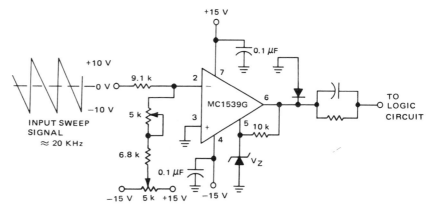

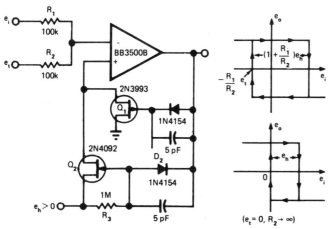

VOLTAGE-CONTROLLED HYSTERESIS—Precise, independent control of comparator trip point and hysteresis is achieved by switching hysteresis control signal e_h to comparator input with Q_1 and Q_2 when opamp changes state. Circuit avoids hysteresis feedback error while achieving inherent 0.01% trip-point accuracy of comparator. Control voltage e_t determines first trip point. When opamp output is negative, Q_2 is held off and Q_1 is on for connecting noninverting input to ground. Output switching occurs when input signal e_i drives input of inverting amplifier to zero.—J. Graeme, Comparator Has Precise, Voltage-Controlled Hysteresis, *EDN Magazine,* Aug. 20, 1975, p 78 and 80.

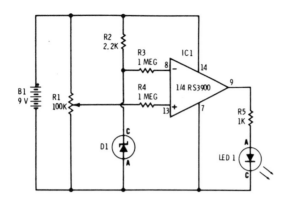

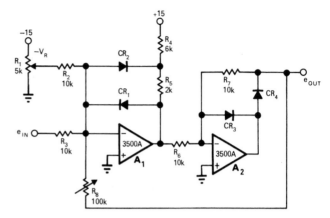

ZENER REFERENCE—One section of RS3900 quad opamp is connected as comparator using zener D1 for reference voltage. When voltage applied to pin 13 by R1 exceeds breakdown voltage of zener D1, comparator amplifies difference voltage to produce output voltage high enough to turn on LED. Can be used for classroom demonstration of comparator action. Zener breakdown should be under 9 V. LED can be Radio Shack 276-041.—F. M. Mims, "Semiconductor Projects, Vol. 2," Radio Shack, Fort Worth, TX, 1976, p 35–42.

INDEPENDENT HYSTERESIS ADJUSTMENT—Trip point and hysteresis of comparator opamp A_1 can be adjusted independently, with trip point being determined by setting of R_1 or programmed by DC voltage applied to R_2. Opamp A_2 provides polarity inversion and rectification of A_1 output. Hysteresis control R_8 is in feedback path from A_2 back to A_1. Amount of hysteresis is determined by ratio of R_3 to R_8. With values shown, circuit output levels are 0 and 5 V.—G. Tobey, Comparator with Noninteracting Adjustments, *EDN/EEE Magazine*, Oct. 1, 1971, p 43.

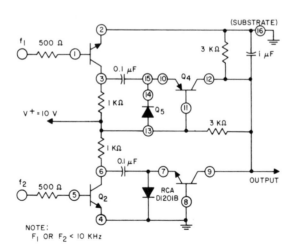

FREQUENCY COMPARATOR—Circuit using CA3096 transistor array plus one discrete diode develops DC output voltage that is proportional to difference between frequencies of input signals f_1 and f_2. Maximum input frequency is 10 kHz.—"Circuit Ideas for RCA Linear ICs," RCA Solid State Division, Somerville, NJ, 1977, p 17.

SLEW RATE—Circuit measures slew rate of input signal with Am685 comparator in circuit having delay-line length under 10 ns. When slew rate exceeds predetermined limit set by R_6, comparator changes state and latches, turning on LED. Pushing reset switch restores normal operation. Based on comparison of input signal with time-delayed counterpart. Derivative of input signal, equal to its instantaneous slew rate, is measured accurately for swings of 6 V P-P as found in most 50-ohm video signals. Action is fast enough to detect glitches.—R. C. Culter, Slew-Rate Limit Detector Is Simple, yet Versatile, *EDN Magazine*, Aug. 20, 1977, p 140–141.

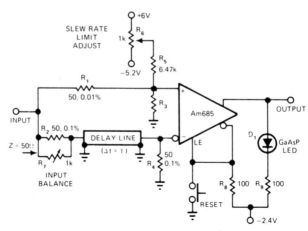

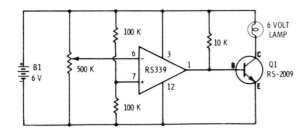

COMPARATOR DRIVES LAMP—Classroom demonstration circuit for comparator action uses transistor to amplify output of one section of RS339 quad comparator, to boost output current enough for driving 60-mA lamp. Lamp comes on when voltage at movable arm of 500K pot is greater than half of supply voltage.—F. M. Mims, "Integrated Circuit Projects, Vol. 6," Radio Shack, Fort Worth, TX, 1977, p 33–41.

STROBED MICROPOWER—Uses CA3080A variable opamp and CA3600E CMOS transistor array. Quiescent power drain from ±12 V supply is only 10 μW, increasing to 420 μW when comparator is strobed on to make CA3080A active.—"Circuit Ideas for RCA Linear ICs," RCA Solid State Division, Somerville, NJ, 1977, p 16.

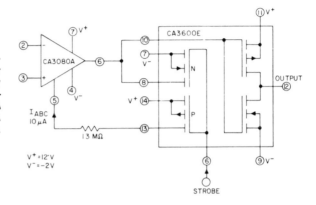

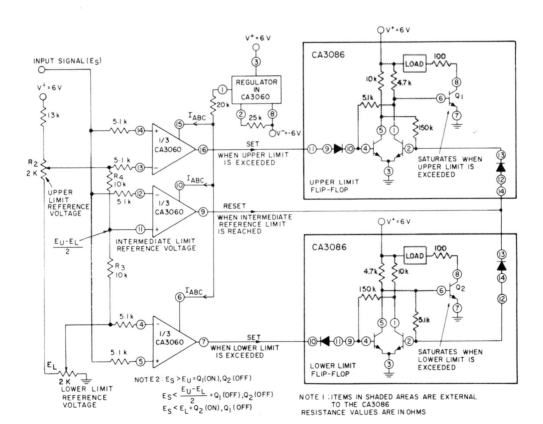

THREE-LEVEL COMPARATOR—All three sections of CA3060 three-opamp array are used with CA3086 transistor arrays to provide three adjustable limits for comparator. If upper or lower limit is exceeded, appropriate output is activated until input signal returns to preselected intermediate limit. Suitable for many types of industrial control applications.—"Circuit Ideas for RCA Linear ICs," RCA Solid State Division, Somerville, NJ, 1977, p 17.

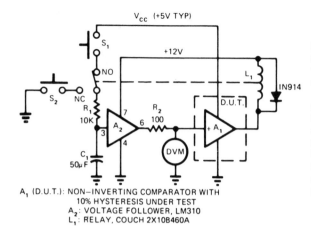

MEASURING THRESHOLDS—Upper and lower thresholds of noninverting comparator under test (A_1) are read on DVM at end of capacitor charge and discharge cycles initiated by S_1 and S_2. With C_1 discharged, relay L_1 is energized. Closing S_1 allows C_1 to charge toward V_{CC}. When upper threshold is reached, relay drops out and meter is read. Closing S_2 starts discharge cycle which stops at lower threshold. Reverse relay connections when testing inverting comparator.—E. S. Papanicolaou, Comparator Is Part of Its Own Measuring System, *EDN Magazine,* Aug. 5, 1974, p 76.

A_1 (D.U.T.): NON-INVERTING COMPARATOR WITH 10% HYSTERESIS UNDER TEST
A_2: VOLTAGE FOLLOWER, LM310
L_1: RELAY, COUCH 2X10B460A

COMPARATOR DRIVES LED—Simple classroom demonstrator of comparator action uses one section of RS339 quad comparator. Reference voltage applied to positive input of comparator is half of supply voltage. R1 serves as voltage divider applying variable voltage to inverting input. When voltage applied to pin 6 by R1 exceeds reference voltage on pin 7, comparator switches on and LED lights. R4 is chosen for use with Radio Shack 276-041 red LED.—F. M. Mims, "Integrated Circuit Projects, Vol. 6," Radio Shack, Fort Worth TX, 1977, p 33–41.

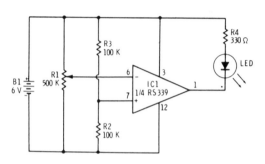

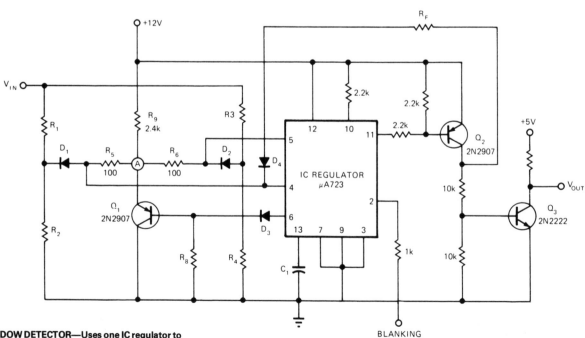

WINDOW DETECTOR—Uses one IC regulator to compare output voltages of two separate voltage dividers with fixed reference voltage. Resulting absolute error signal is amplified and converted to TTL-compatible logic signal. Voltage divider for lower limit of window detector is R_1-R_2 and for upper limit is R_3-R_4. Article covers circuit operation in detail.—N. Pritchard, Window Detector Uses One IC Regulator, *EDN Magazine,* May 20, 1973, p 81 and 83.

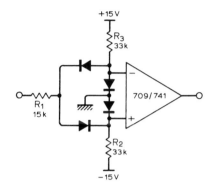

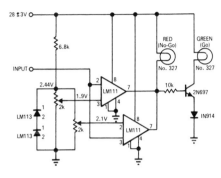

DUAL LIMITS—Opamp used without frequency compensation gives positive output only when input voltage exceeds 8.5 V in either polarity. Resistors in supply leads determine limit points. For inverted output, reverse inputs to opamp. Diodes are 1N914.—K. Pickard, Dual Limit Comparator Using Single Op-Amp, *Wireless World,* Dec. 1974, p 504.

VOLTAGE-WINDOW COMPARATOR—Use of LM111 opamps minimizes number of components required to turn on green indicator lamp when input voltage is between predetermined limits set by 2K pots. Similar circuit using 741 opamps requires total of 31 components. Improved circuit draws only 120 nA from voltage level being monitored, and operates within 0.3% threshold level stability using single unregulated supply varying ±3 V from 28 V.—D. Priebe, Comparators Compared, *EDN/EEE Magazine,* Oct. 1, 1971, p 61.

SECTION 10
Contact Bounce Suppression Circuits

Used to solve bounce problems of switch and relay contacts during closing or opening.

BOUNCELESS SQUARE OUTPUT—NE555 timer eliminates need for gates to suppress contact bounce. Timer can provide pulse at least 5 ms long (much longer if desired) and can remain on as long as trigger input (key pulse) is low (grounded). Timer triggers on negative-going edge of low-going pulse, such as key down to ground. Common negative is isolated from ground. V_{CC} can be 5 to 15 VDC. Timer output can be connected directly to exciter keying input for negative grid keying. Because of square-wave output on make or break (100 ns each), circuits must be added in exciter or between keying transistors to provide at least 5-ms rise and fall times for Morse or RTTY keying.—B. Conklin, Improving Transmitter Keying, *Ham Radio*, June 1976, p 44–47.

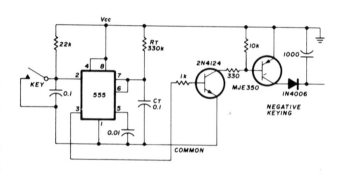

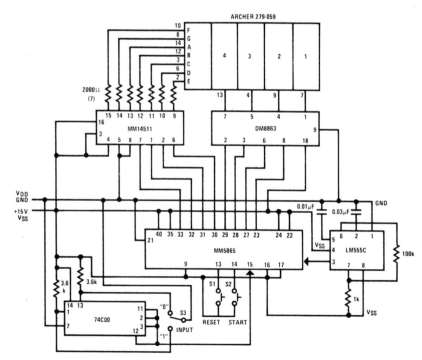

SWITCH-CLOSURE COUNTER—National MM5865 universal timer and counter chip is used with 74C00 debouncer and LM555C timer to drive digital display that counts closures of manual switch S3. Reset transition restores display to 0000. BCD segment outputs of MM5865 feed LED 4-digit display through MM14511 interface, while digit enable outputs go to display through DM8863 driver.—"MOS/LSI Databook," National Semiconductor, Santa Clara, CA, 1977, p 2-23–2-32.

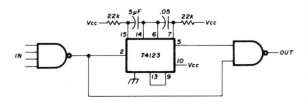

DELAYED START—Keyed output of RTTY terminal equipment or other keys and relays is delayed by 74123 dual mono for at least 5 ms while contact bounce settles down. Can be used for calculator keyboards, flip-flop testers, and other applications in which final clean pulse length is not highly important.—B. Conklin, Improving Transmitter Keying, *Ham Radio*, June 1976, p 44–47.

KEYBOARD BOUNCE ELIMINATOR—Dual 9602 mono MVBR is used with Harris HD-0165 keyboard encoder to generate delayed strobe pulse St', with delay set at about 10 ms by first mono. Pulse width is determined by second mono and should be set to meet system requirements. Circuit eliminates effects of arcing or switch bounce and provides proper encoding under two-key rollover conditions.—"Linear & Data Acquisition Products," Harris Semiconductor, Melbourne, FL, Vol. 1, 1977, p 6-4.

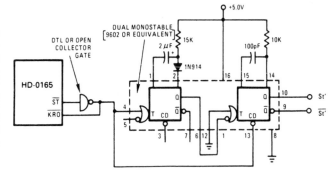

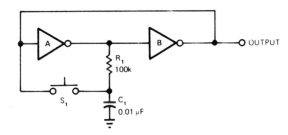

GATES FORM SWITCH—Each time pushbutton switch is closed momentarily, voltage on C_1 makes inverter A change state, with positive feedback from inverter B, to give alternate ON and OFF action. R_1 delays charging and discharging of C_1, making circuit essentially immune to contact bounce. Switch works equally well with either CMOS or TTL gates. Values of R_1 and C_1 are not critical.—T. Tyler, Inverters Provide "ON-OFF" from Momentary Switch, *EDN Magazine*, June 20, 1976, p 126.

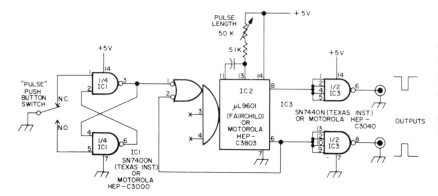

ONE PULSE PER PUSH—Circuit generates only one rectangular pulse for each actuation of pushbutton switch, even if contacts bounce. TTL gates IC1 are wired as RS flip-flop (latch) that triggers mono MVBR IC2 having fixed-duration positive and negative output pulses. Output drives are increased by TTL inverting buffer gates.—H. Olson, Further Adventures of the Bounceless Switch, *73 Magazine,* Feb. 1975, p 111–114.

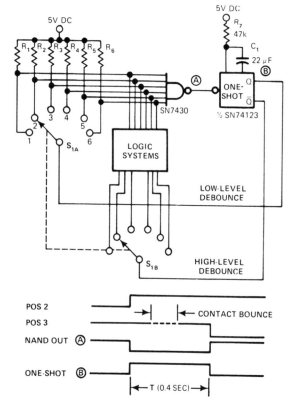

ROTARY SWITCH DEBOUNCE—Outputs from mono (one-shot) provide common returns for rotary switch. Multi-input NAND gate, tied to normally high signals from one deck of rotary switch, instantly detects opening of one contact and triggers mono. Mono then simulates open contact for interval determined by values used for R_7 and C_1; for values shown, delay is 400 ms.—E. S. Peltzman, Circuit Eliminates Rotary-Switch Bounce Problems, *EDN Magazine,* April 20, 1978, p 132.

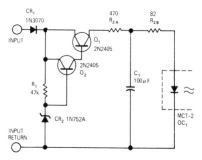

BOUNCELESS ISOLATOR—Integrating filter C_1-R_2 eliminates effects of contact bounce that may be superimposed on digital input signal feeding optoisolator. Photodiode in optoisolator drives Schmitt trigger that makes output to TTL circuits change state when LED is turned on by input signal.—C. E. Mitchell, Optical Coupler and Level Shifter, *EDN/EEE Magazine,* Feb. 1, 1972, p 55.

DEBOUNCING WITH COUNTER—Circuit uses CMOS counter/decoder with any inexpensive 200-Hz or higher clock such as CMOS two-gate oscillator or 555 timer. Signal to be debounced is fed directly to reset input of counter, with no preconditioning. When contact is made by switch, counter unclears and starts counting up. Each bounce of contact resets counter, so it cycles between states 0 and 1 until contacts settle. Counter then delivers clean nonoverlapping pulses to remaining output lines, any of which may be used as conditioned output signal. When counter reaches state 7, it inhibits itself to prevent repeated pulsing of output lines. When switch is opened, cycling action is repeated during bounces, with output never going higher than state 1. After bouncing, counter is held in clear state ready for next closing.—L. T. Hauck, Solve Contact Bounce Problems Without a One-Shot, *EDN Magazine*, Sept. 5, 1975, p 80 and 82.

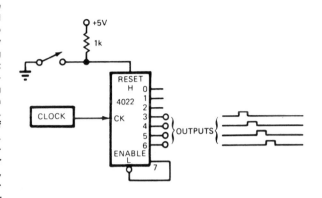

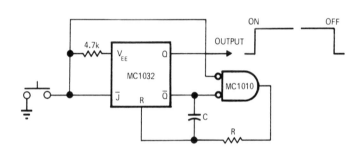

BOUNCELESS MAKE/BREAK—Circuit eliminates switch bounce problems during closing as well as opening. When switch is closed, Q output of flip-flop goes to logic 1 for delay period determined by RC time constant. Releasing switch operates NAND gate, making its output go to logic 1. This charges C through R until reset level is reached. Flip-flop then resets, changing Q output to logic 0. Values for R and C are chosen according to bounce duration of switch used. For typical 1-A SPST switch, 10,000 ohms and 0.47 μF were used.—L. F. Walsh and T. W. Hill, Make-and-Break Bounceless Switching, *EDN/EEE Magazine*, July 15, 1971, p 49.

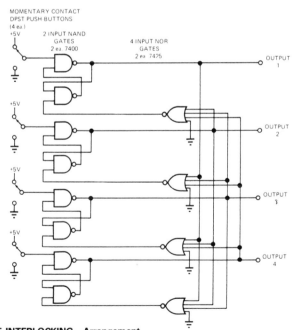

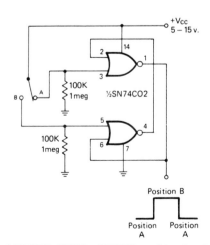

BOUNCE-FREE INTERLOCKING—Arrangement provides low-cost equivalent of mechanically interlocked switch assembly, while providing TTL compatibility and freedom from switch bounce. Momentary pressing of any pushbutton restores its associated RS flip-flop to normal and makes output of that channel high. Arrangement uses cross-coupled two-input NAND gates for each flip-flop, connected so each actuation produces an output and resets all other flip-flops. If two or more buttons are pushed simultaneously, all their channels will go high, but only last one released will stay on. Any number of channels may be added.—B. Brandstedt, Digital Interlocking Switch Is Inexpensive to Build, *EDN Magazine*, Dec. 15, 1972, p 42.

LATCHING GATES—SN74C02 quad two-input NOR gate forms latching circuit in which first noise pulse produced by switch latches circuit, making it immune to contact bounce.—I. Math, Bounceless Switch, *CQ*, July 1976, p 50.

SECTION 11
Converter Circuits—Analog-to-Digital

Includes circuits for converting DC, audio, and video analog inputs to linearly related binary, BCD, or Gray-code digital outputs. Some circuits have autoranging or some type of input compression, input multiplexing, and input buffering.

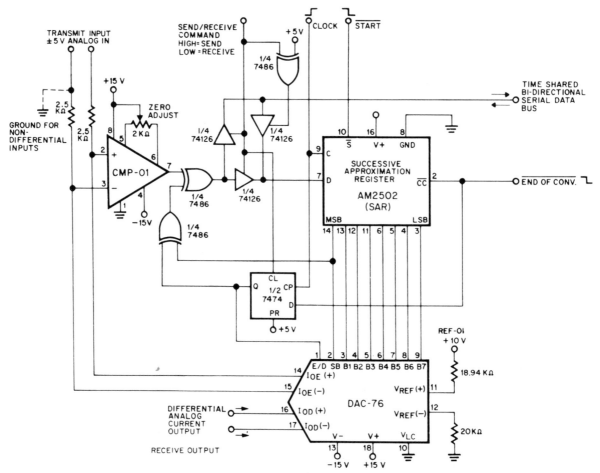

RECEIVE OUTPUT

SERIAL DATA OUTPUT—Precision Monolithics ICs form transceiving converter suitable for use in control systems incorporating 8-bit microprocessors. Output conforms with Bell-System μ-255 logarithmic law for PCM transmission.

Applications include servocontrols, stress and vibration analysis, digital recording, and speech synthesis. Start must be held low for one clock cycle to begin send or receive cycle. Conversion is completed in nine clock cycles, and output is

available for one full clock cycle. Other half of system is identical.—"COMDAC Companding D/A Converter," Precision Monolithics, Santa Clara, CA, 1977, DAC-76, p 12.

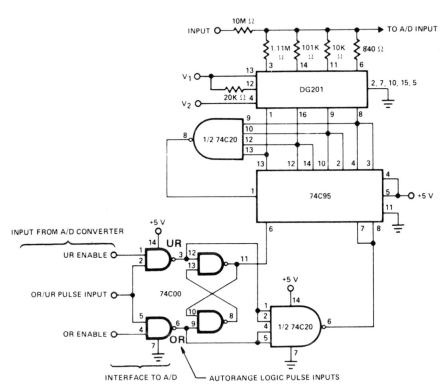

SELF-CONTROLLED AUTORANGING—DG201 quad analog switch inserts one of four attenuator resistors in input circuit of Siliconix LD130 or comparable A/D converter under control of autoranging pulse output derived from converter. Control logic includes 74C00 quad two-input NAND gate with two sections connected as flip-flop, 74C95 4-bit right-shift left-shift register, and 74C20 dual four-input NAND gate.—"Analog Switches and Their Applications," Siliconix, Santa Clara, CA, 1976, p 6-28–6-29.

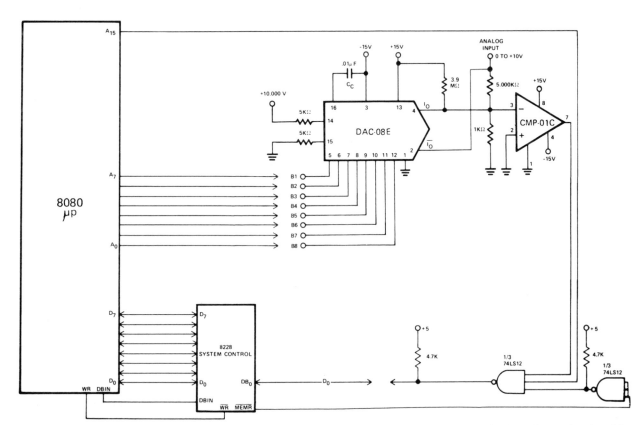

SOFTWARE CONTROL—Innovative software for Intel 8080A microprocessor eliminates need for peripheral isolation devices when using Precision Monolithics DAC-08E D/A converter and CMP-01C comparator for 8-bit A/D conversion. Technique can easily be expanded to 10-bit or 12-bit conversions and adapted to other microprocessors. Logic of microprocessor replaces conventional successive-approximation register. 8 lowest-order address bits control data bit input to DAC, using software given in article.—W. Ritmanich and W. Freeman, "Software Controlled Analog to Digital Conversion Using DAC-08 and the 8080A Microprocessor," Precision Monolithics, Santa Clara, CA, 1977, AN-22, p 3.

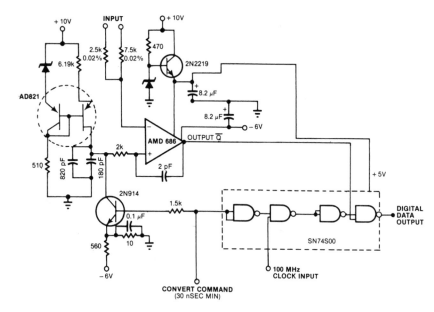

10-BIT ACCURACY—Single-slope A/D converter gives high-speed conversion of DC input voltage to digital data output. For 0–10 V input, 1024 pulses of 100-MHz clock appear at full scale and 512 at half scale. When command pulse is applied, 2N914 transistor resets 1000-pF capacitor (820 and 180 in parallel) to 0 V. Capacitor begins to charge linearly on falling edge of command pulse, to 2.5 V. 10-μs ramp is applied to AMD686 for comparison with unknown voltage. Output of opamp is pulse whose width is proportional to input voltage and can therefore be used to gate 100-MHz clock.—J. Williams, Low-Cost, Linear A/D Conversion Uses Single-Slope Techniques, *EDN Magazine*, Aug. 5, 1978, p 101–104.

VIDEO COMPRESSOR—Nonlinear function amplifier IC-2 compresses video input signals as required to compensate for inefficient quantization where there are too many levels for small signals and too few levels for large signals. Designed to feed 6-bit analog-to-digital converter, IC-1 attenuates input −20 dB and shifts level. Output of IC-2 is amplified by IC-3 to voltage range comparable to that of input signal. IC-4 acts as temperature compensator and output level shifter. R_7 nulls small output offsets.—J. B. Frost, Non-Linear Function Amplifier, *EEE Magazine*, March 1971, p 78.

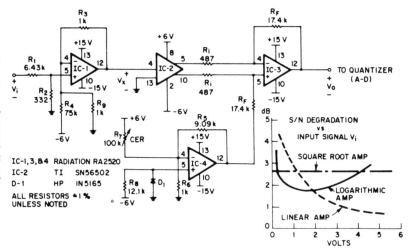

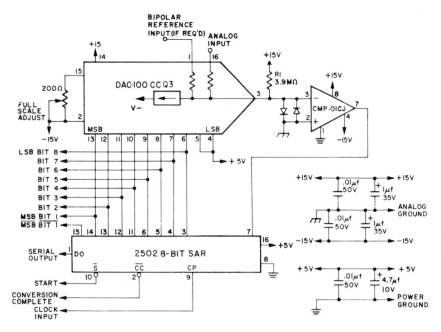

8-BIT SUCCESSIVE APPROXIMATION—Uses Precision Monolithics DAC-100 CCQ3 D/A converter and CMP-01CJ fast precision comparator in combination with Advanced Micro Devices AM2502PC or equivalent successive approximation register to compare analog input with series of trial conversions. Clamp diodes minimize settling time and prevent large inputs from damaging DAC output. Digital output is available in serial nonreturn-to-zero format at data output DO shortly after each positive-going clock transition.—D. Soderquist, "A Low Cost, Easy-to-Build Successive Approximation Analog-to-Digital Converter," Precision Monolithics, Santa Clara, CA, 1976, AN-11, p 3.

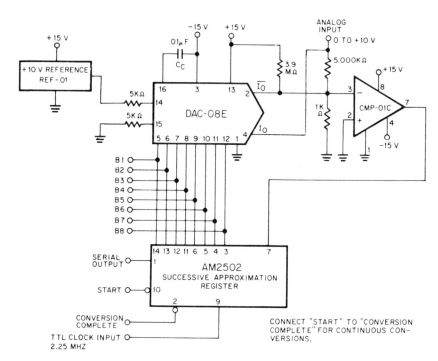

4-μs CONVERSION TIME—Provides conversion of analog input to 8-bit digital output by successive approximation, with conversion time of 4 μs. Advanced Micro Devices AM2502 successive-approximation register contains logic for Precision Monolithics DAC-08E and CMP-01C comparator.—D. Soderquist and J. Schoeff, "Low Cost, High Speed Analog-to-Digital Conversion with the DAC-08," Precision Monolithics, Santa Clara, CA, 1977, AN-16, p 3.

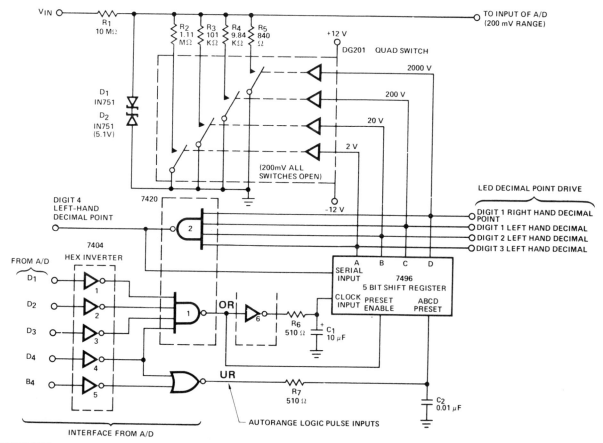

AUTORANGING—Digitally controlled attenuator uses DG201 quad analog switch as input ladder attenuator switches for A/D converter. Switches are controlled by digital logic that detects overrange and underrange information from A/D converter and closes appropriate attenuator path. Circuit is suitable for Siliconix LD110/111 or LD111/114 A/D converter.—"Analog Switches and Their Applications," Siliconix, Santa Clara, CA, 1976, p 6-28.

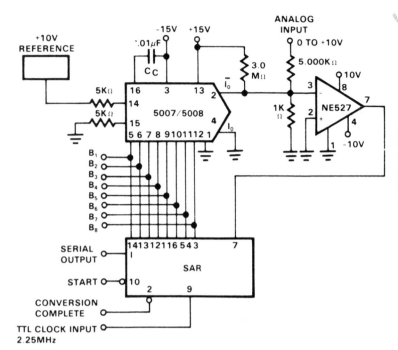

0–10 V ANALOG INPUT—Used to provide digital input to computer for processing and storage of analog signals. Requires only three ICs in addition to external +10 V reference and 2.25-MHz TTL clock. Successive-approximation register (SAR) can be Motorola MC1408 or equivalent. For continuous conversions, connect pins 10 and 2 of SAR.—"Signetics Analog Data Manual," Signetics, Sunnyvale, CA, 1977, p 677–685.

HIGH-IMPEDANCE BUFFER—Two sections of Motorola MC3403 quad opamp serve as voltage followers for differential inputs of third section connected as buffer for MC1505 A/D converter. Dual transistor Q1, connected as dual diode, provides 0.6-V offset at inputs of voltage followers, to obtain temperature tracking and predictable performance at low bias currents of opamp.—D. Aldridge and S. Kelley, "Input Buffer Circuits for the MC1505 Dual Ramp A-to-D Converter Subsystem," Motorola, Phoenix, AZ, 1976, EB-24A.

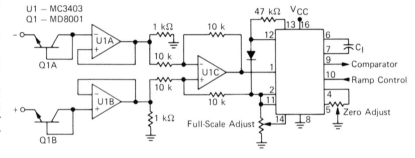

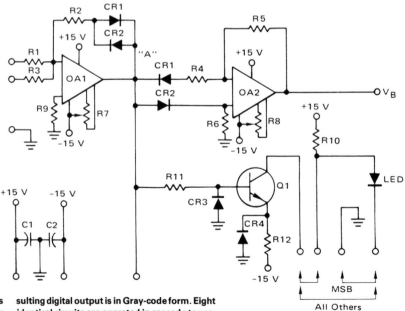

R1, 3	10 k 0.1%
R2, 4, 5, 6	20 k 0.1%
R7, 8	10 k Pot
R9	3.9 k
R10, 12	1.2 k
R11	10 k
OA1, 2	MC1456C
CR1	MSD6100
CR2	MSD6150
CR3, 4	1N914
Q1	MPS6415
C1, 2	0.1
LED	MLED 630

CYCLIC CONVERTER—Unknown voltage is successively compared to reference voltage for determining each digital bit. After determining bit, voltage difference between unknown and reference is operated on, then sent to successive stages to determine less significant bit. Resulting digital output is in Gray-code form. Eight identical circuits are operated in cascade to provide 8-bit A/D converter having accuracy within 1 LSB and full-scale range of 0-8 V. Circuit requires only two MC1456CG opamps per stage, with MPS6514 transistor as comparator. Switching diode CR1 is MSD6100, and CR2 is MSD6150. Other dioodes are 1N914.—J. Barnes, "Analog-to-Digital Cyclic Converter," Motorola, Phoenix, AZ, 1974, AN-557, p 7.

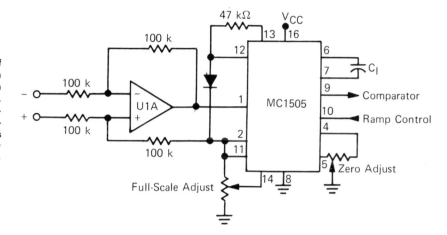

DIFFERENTIAL OPAMP AS BUFFER—Section of Motorola MC3403 quad opamp, operating from single supply, serves as low-cost unity-gain buffer for MC1505 dual-ramp A/D converter. Opamp is used as differential amplifier referenced to MC1505 reference voltage of 1.25 V.— D. Aldridge and S. Kelley, "Input Buffer Circuits for the MC1505 Dual Ramp A-to-D Converter Subsystem," Motorola, Phoenix, AZ, 1976, EB-24A.

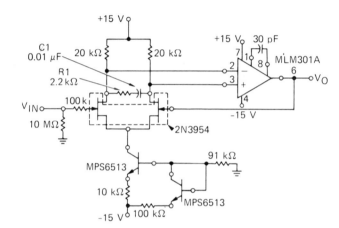

FET-INPUT BUFFER—Used ahead of Motorola MC1505 A/D converter to provide input impedance of 10 megohms. FETs are connected as differential amplifier having common source leads returned to constant-current generator built from bipolar transistor, with similar transistor providing temperature compensation. Temperature drift of amplifier is well under 1 mV from 0 to 50°C.—D. Aldridge and S. Kelley, "Input Buffer Circuits for the MC1505 Dual Ramp A-to-D Converter Subsystem," Motorola, Phoenix, AZ, 1976, EB-24A.

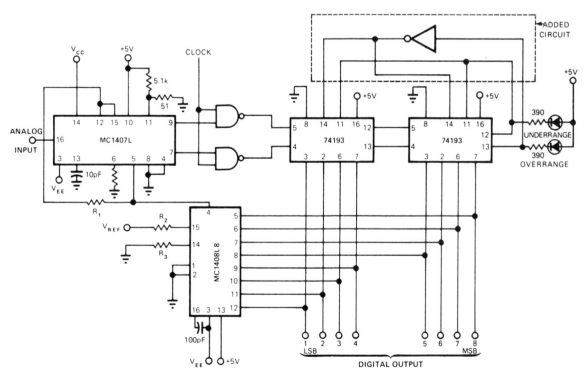

TRACKING A/D CONVERTER—Addition of one gate to tracking or servo-type A/D converter, as shown in dashed box, overcomes instability problems otherwise occurring when input voltages are less than zero or greater than full scale.

With 8-bit converter shown, count of 11111111 when counting up makes carry output and load inputs go low, holding counter in this state so subsequent up clocks are ignored. When count is all 0s, borrow output goes low and clear input

goes high, so counter is free to count up only.— A. Helfrick, Tracking A/D Converters Need Another Look, *EDN Magazine*, June 20, 1975, p 118 and 120.

REPETITIVE-MODE OPERATION—Quicker conversion is obtained in Teledyne Philbrick 4109 or 4111 A/D converter by restarting converter within a few microseconds after status signal, using sure-start circuit shown. Reset pulse is fed to converter when status signal is held at low DC level. When status command is high, oscillator A-B is disabled. If reset pulse is not obeyed and status signal remains low, oscillator starts up until conversion does occur.—R. W. Jacobs, "Repetitive Mode Operation for Models 4109/4111 Integrating A/D Converters," Teledyne Philbrick, Dedham, MA, 1977, AN-28.

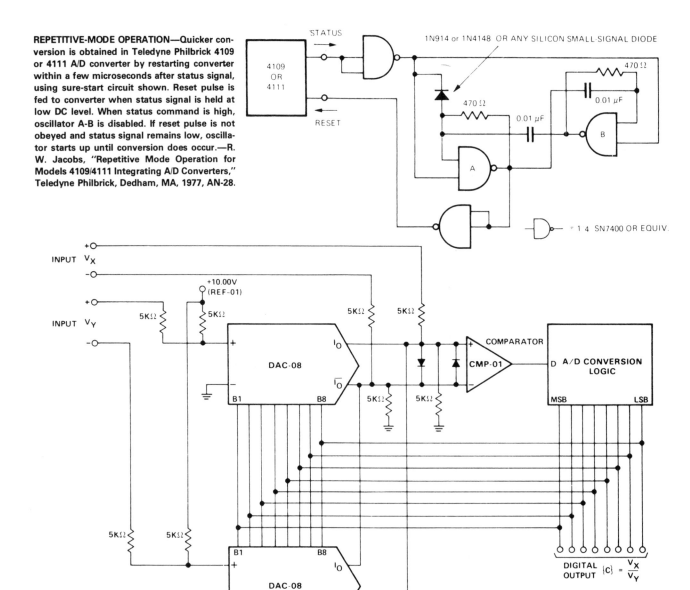

FOUR-QUADRANT RATIOMETRIC—Uses Precision Monolithics DAC-08 D/A converters and CMP-01 comparator to drive successive-approximation conversion logic using REF-01 +10 V reference and 2502-type successive-approximation register. Imputs V_X are connected conventionally, and inputs V_Y are connected in multiplying fashion. I_{REF} for both DACs is modulated between 1 and 3 mA. Resulting output currents are differentially transformed into voltages by 5K resistors at comparator inputs and compared with V_X differential input. When conversion process is complete (comparator inputs differentially nulled to less than ½ LSB), digital output corresponds to quotient V_X/V_Y. Diodes are 1N4148.—J. Schoeff and D. Soderquist, "Differential and Multiplying Digital to Analog Converter Applications," Precision Monolithics, Santa Clara, CA, 1976, AN-19, p 5.

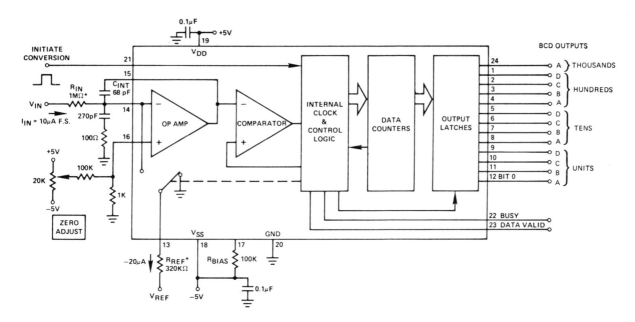

BCD OUTPUT—Latched nonmultiplexed parallel BCD outputs from Teledyne 8750 3½-digit CMOS analog-to-digital converter are suitable for liquid crystal and gas-discharge displays. 2-mA drain on ±5 V supply permits battery operation. Features include high linearity, noise immunity, and 3½-digit resolution within 0.025% error. Circuit is based on switching number of current pulses needed to bring analog current to zero at input of opamp, then determining digital equivalent by counting these pulses. Values shown are for full-scale voltage input of 10 V and voltage reference of −6.4 V.—CMOS A-D Converter Provides BCD Output, *Computer Design*, Nov. 1977, p 156 and 158.

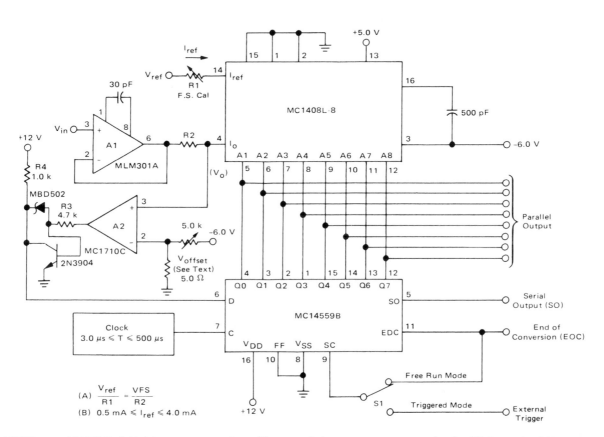

$$(A) \quad \frac{V_{ref}}{R1} = \frac{VFS}{R2}$$

$$(B) \quad 0.5 \text{ mA} \leq I_{ref} \leq 4.0 \text{ mA}$$

HIGH-SPEED SUCCESSIVE-APPROXIMATION—Total conversion time for 8-bit system is about 4.5 μs. Clock rate is up to 2 MHz. Serial output is used for transmission to one or more other locations.—T. Henry, "Successive Approximation A/D Conversion," Motorola, Phoenix, AZ, 1974, AN-716, p 5.

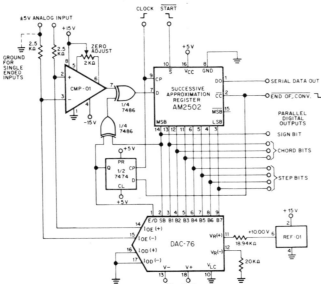

COMPRESSING A/D CONVERSION—Step size increases as output changes from zero scale to full scale, in contrast to conventional linear converter in which step size is constant percentage of full scale. Uses Precision Monolithics DAC-76 D/A converter in combination with CMP-01 comparator, any standard EXCLUSIVE-OR gate, and successive-approximation register for conversion logic. Encoding sequence begins with sign-bit comparison and decision. Bits are converted with successive-removal technique, starting with decision at code 011 1111 and turning off bits sequentially until all decisions have been made. Conversion is completed in nine clock cycles.—"COMDAC Companding D/A Converter," Precision Monolithics, Santa Clara, CA, 1977, DAC-76, p 12.

CMOS-COMPATIBLE SUCCESSIVE-APPROXIMATION—Converts analog input to 8-bit digital output by using MC14559 CMOS successive-approximation register with Precision Monolithics DAC-100 D/A converter and CMP-01 comparator. Conversion sequence is initiated by applying positive pulse, with width greater than one clock cycle, to START CONVERSION input. Analog input is then compared successively to ½ scale, ¼ scale, and remaining binarily decreasing bit weights until it has been resolved within ½ LSB. END OF CONVERSION then changes to logic 1 and parallel answer is present in negative-true binary-coded format at register outputs.—D. Soderquist, "Interfacing Precision Monolithics Digital-to-Analog Converters with CMOS Logic," Precision Monolithics, Santa Clara, CA, 1975, AN-14, p 4.

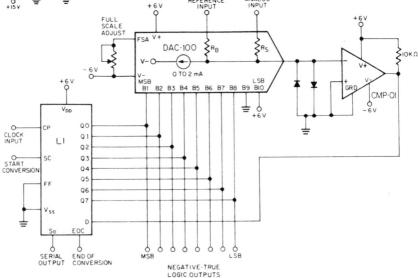

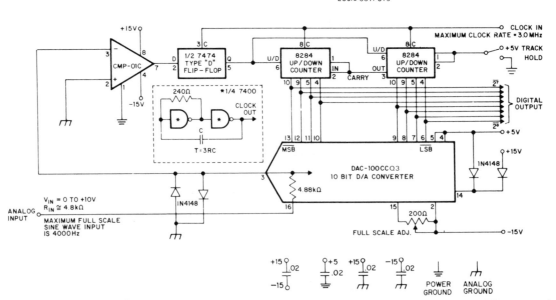

8-BIT TRACKING—Uses Precision Monolithics DAC-100 CCQ3 D/A converter and CMP-01CJ fast precision comparator to make digital data continuously available at output while tracking analog input. Diode clamps hold DAC output near zero despite input and turn-on transients. Unused least significant digital inputs of 10-bit DAC are turned off by connecting to +5 V as shown. Simple clock circuit shown in dashed box is stable over wide range of temperatures and supply voltages. D/A converter is used in feedback configuration to obtain A/D operation.—"A Low Cost, High-Performance Tracking A/D Converter," Precision Monolithics, Santa Clara, CA, 1977, AN-6, p 2.

MOSFET-INPUT BUFFER—Uses Motorola MC14007 dual complementary pair plus inverter, with two of MOSFETs connected as differential amplifier for buffering opamp and third serving as current source for differential amplifier. Arrangement gives high input impedance required in some applications of MC1505 A/D converter for which buffer was designed. 1-megohm pot controls gate voltage for current source. Temperature drift is well under 2 mV over range of 0–50°C. Pin 14 of MC14007 should be tied to +5 V.—D. Aldridge and S. Kelley, "Input Buffer Circuits for the MC1505 Dual Ramp A-to-D Converter Subsystem," Motorola, Phoenix, AZ, 1976, EB-24A.

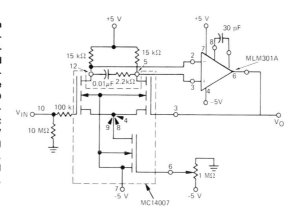

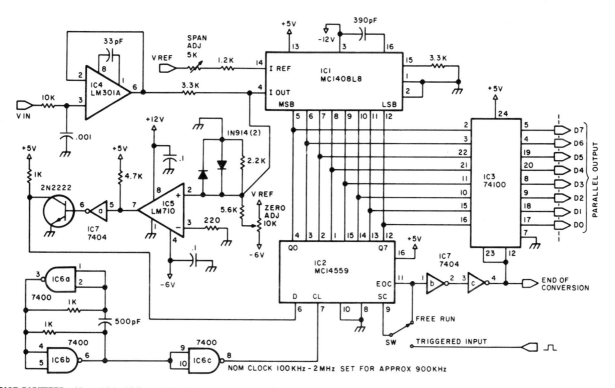

VOICE DIGITIZER—Uses 8-bit ADC capable of sampling AF input signal 100,000 times per second when using 900-kHz clock. 100-kHz clock gives 9000 samples per second, about minimum for human voice. Digital output is stored in computer memory for later conversion back to analog form for such applications as synthesis of speech from phonemes and providing voice answers to queries. Requires about 10,000 bytes in memory for 1 s of voice data. Pin 7 of IC4 is +12 V, and pin 4 is −6 V. For IC6 and IC7, pin 14 is +5 V and pin 7 is ground. 8080 assembler programs are given for input and output of memory.—S. Ciarcia, Talk to Me! Add a Voice to Your Computer for $35, *BYTE*, June 1978, p 142–151.

Converter Circuits—DC-to-DC

Use inverters typically operating from DC supplies in range of 2–15 V to generate AC voltage at frequency typically in range of 16–25 kHz, for step-up by voltage-doubling rectifier or transformer-rectifier combination to give desired new positive or negative DC supply voltage that can be as high as 10 kV.

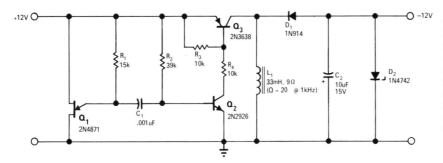

+12 V TO −12 V—Transformerless inverting DC-to-DC converter has above 55% efficiency and can withstand output shorts lasting up to several minutes. UJT Q_1 and base-emitter diode of transistor Q_2 form free-running MVBR whose 25-kHz output is amplified by Q_2 to drive switching-mode converter Q_3-L_1-D_1-C_2. Zener D_2 regulates output for variations in input voltage or output loads up to 40 mA.—G. Bank, Transformerless Converter Supplies Inverted Output, *EDN/EEE Magazine,* July 1, 1971, p 48.

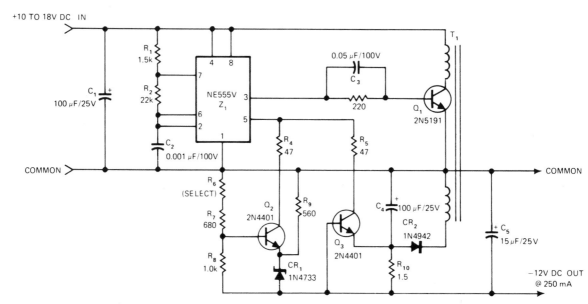

+12 V TO −12 V—Transforms unregulated +12 VDC to current-limited regulated −12 VDC. Front end of 555 is connected in astable configuration, with R_2 selected to give about 25 kHz at pin 3. Control of modulation input to pin 5 gives voltage regulation and current limiting. Circuit tolerates continuous operation under short-circuit conditions. With 10-V nominal output, line regulation is within ±0.05% for input and output voltage ranges of 0.3 to 10 V. Load regulation is 0.2% for loads from 10 μA to 10 mA when load impedance is 10 ohms.—R. Dow, Build a Short-Circuit-Proof +12V Inverter with One IC, *EDN Magazine,* Sept. 5, 1977, p 177–178.

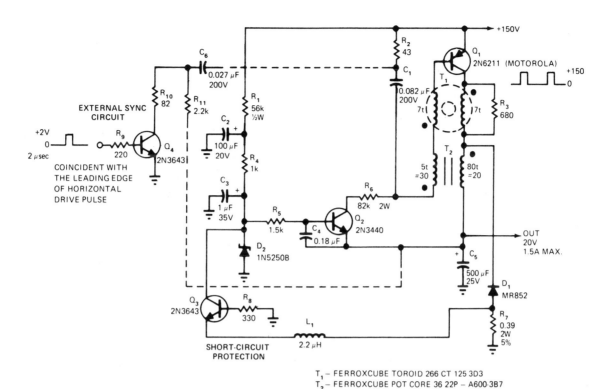

T₁ — FERROXCUBE TOROID 266 CT 125 3D3
T₂ — FERROXCUBE POT CORE 36 22P — A600-3B7

2 V TO 20 AND 150 V—Use of 7-turn toroidal transformer in self-excited ringing-choke blocking-oscillator circuit improves efficiency of converter circuit by providing fast switching time. Circuit is practical only when input and output voltages differ significantly. Blocking oscillator is formed by Q_1, T_2, C_1, R_2, and base-bias network R_6-Q_2. Q_4 makes possible external synchronization, permitting use in television systems for triggering regulator with leading edge of horizontal drive pulse. This ensures completion of cycle within blanking interval.—N. Tkacenko, Transformer Increases DC-DC Converter Efficiency to 80%, *EDN Magazine,* May 5, 1976, p 110 and 112.

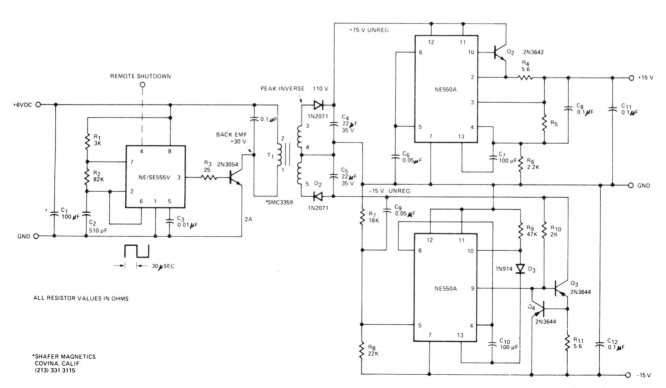

+6 V TO ±15 V—Combination of 555 timer and two NE550A precision adjustable regulators gives 0.1% line and load regulation. Timer operates as oscillator driving step-up transformer which feeds full-wave rectifier.—"Signetics Analog Data Manual," Signetics, Sunnyvale, CA, 1977, p 726–727.

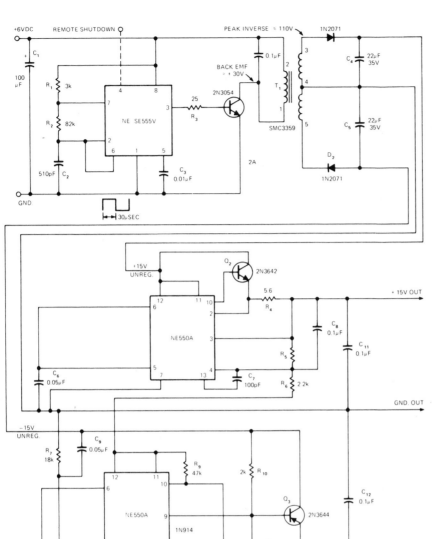

6 V TO ±15 V—Combination of 555 timer and two NE550 voltage regulators provides voltage multiplication along with regulation of independent DC outputs. Selected oscillator frequency of 17 kHz optimizes performance of transformer. Can be used to power opamps from either TTL supplies or 6-V batteries. Line and load regulation are 0.1%, while power efficiency at full load of 100 mA is better than 75%.—R. Solomon and R. Broadway, DC-to-DC Converter Uses IC Timer, *EDN Magazine,* Sept. 5, 1973, p 87, 89, and 91.

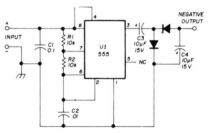

+12 V TO −8 V—Developed for use with mobile equipment when DC voltage is required with opposite polarity to that of auto battery. U1 is 555 timer operated as free-running square-wave oscillator. Frequency is determined by R1, R2, and C2; with values shown, it is about 6 kHz. C1 reduces 6-kHz signal radiated back through input lines. For 12-V input, typical outputs are −8.4 V at 10 mA, −7.9 V at 20 mA, and −5.7 V at 50 mA. All diodes are 1N914, 1N4148, or equivalent.—G. A. Graham, Low-Power DC-DC Converter, *Ham Radio,* March 1975, p 54—56.

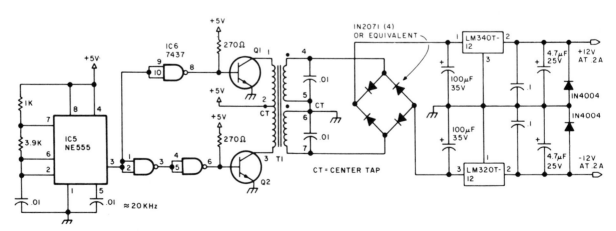

±12 V FROM +5 V—NE555 timer connected as 20-kHz oscillator drives pair of D44H4 transistors through 7437 quad two-input NAND buffer to produce full 200 mA of regulated output for each polarity. Circuit uses push-pull inverter technique to generate AC for driving transformer constructed by rewinding 88-mH toroid to have 40 turns No. 20 center-tapped for primary and 350 turns No. 26 center-tapped for secondary.—S. Ciarcia, Build a 5 W DC to DC Converter, *BYTE,* Oct. 1978, p 22, 24, 26, 28, and 30—31.

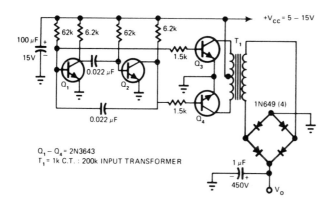

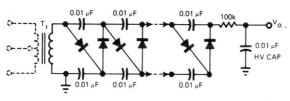

- ADD ADDITIONAL STAGES AS REQUIRED
- DIODES ARE 1N649 OR EQUIVALENT
- CAPACITORS ARE CERAMIC DISC 1 kV

5 V TO 400 V—Astable MVBR operating at 2.174 kHz for values shown drives push-pull transistor pair feeding primary of audio input transformer T_1. Secondary voltage is rectified by diode bridge to provide DC output voltage ranging from 100 to 400 V depending on load resistance and exact value of supply voltage V_{cc}. Bridge rectifier can be replaced by 40-stage multiplier as shown in lower diagram, to give 10-kVDC output.—A. M. Hudor, Jr., Power Converter Uses Low-Cost Audio Transformer, *EDN Magazine,* April 20, 1977, p 139.

280 V TO 600 V—Cascode push-pull transistor switch conversion circuit uses low-voltage transistors and provides automatic equalization of transistor storage time. Drive-signal input to cascode push-pull switch is symmetrical 50-kHz 15 V P-P square wave from 50-ohm source. Q1 and Q2 each see only half of DC source voltage because C1 and C2, in series across 280-V input, charge to 140 V each. Circuit is adaptable to wide range of output voltages and currents because identical units can be connected in series or parallel to obtain desired rating.—L. G. Wright and W. E. Milberger, HV Building Block Uses Series Transistor Switches, *EDN Magazine,* Feb. 15, 1971, p 39–40.

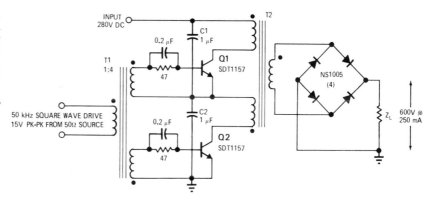

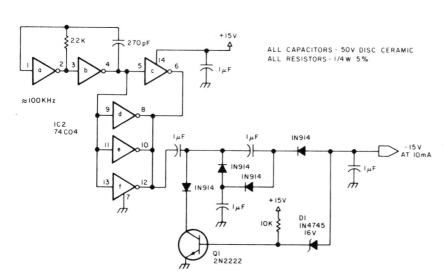

−15 V FROM +15 V—First two sections of 74C04 hex inverter form 100-kHz oscillator, with other sections connected to provide inversion of standard microprocessor source voltage as required for some interfaces and some D/A converters. Shunt regulator formed by D1 and Q1 maintains output voltage relatively constant. Changing zener D1 to 13 V makes output −12 V.—S. Ciarcia, Build a 5 W DC to DC Converter, *BYTE,* Oct. 1978, p 22, 24, 26, 28, and 30–31.

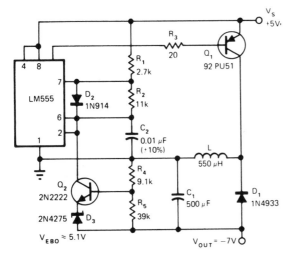

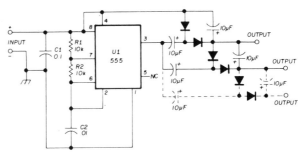

+5 V TO −7 V—Uses LM555 timer as variable-duty-cycle pulse generator controlling transistor switch Q_1 which in turn drives flyback circuit. Regulator Q_2-D_3-R_4-R_5 varies duty cycle according to load, and flyback circuit L-D_1-C_1 develops negative output voltage. When Q_1 is on, current flows through L to ground. When Q_1 turns off, polarity across L reverses, diode becomes forward-biased, and negative voltage appears across C_1 and load. When Q_1 turns on again, voltage across L reverses for start of new cycle. Circuit eliminates separate transformer supply for negative supply of microprocessor. Efficiency is about 60%, load regulation 1.3%, and supply rejection 30 dB. Article gives design equations.—P. Brown, Jr., Converter Generates Negative μP Bias Voltage from +5V, *EDN Magazine*, Aug. 5, 1977, p 42, 44, and 46.

DC MULTIPLIERS—Voltage output from positive voltage booster (+12 VDC to +20 VDC) is increased by using diode-capacitor voltage-doubler sections as shown. Diodes are 1N914, 1N4148, or equivalent. Doubling is achieved at expense of available current. Same technique may be used to increase output of DC/DC converter having negative output voltage.—G. A. Graham, Low-Power DC-DC Converter, *Ham Radio*, March 1975, p 54–56.

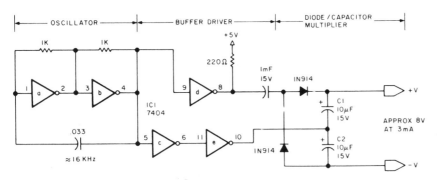

+8 V FROM +5 V—Oscillator operating at about 16 kHz steps up 5-V supply voltage of microprocessor to 8 V for driving special interface circuits. Sections c, d, and e of 7404 hex inverter form buffer and driver for voltage-doubling rectifier.—S. Ciarcia, Build a 5 W DC to DC Converter, *BYTE*, Oct. 1978, p 22, 24, 26, 28, and 30–31.

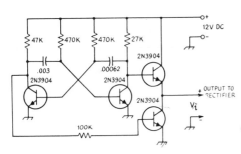

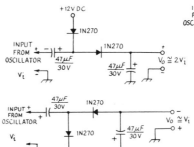

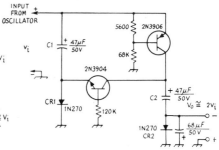

POLARITY REVERSER—Simple RC oscillator operating at about 1200 Hz can be used with choice of rectifier circuits to provide negative or positive voltages equal to or higher than DC supply, without use of transformer. Output transistors connect load alternately to positive supply and to ground for high operating efficiency. Two-diode voltage doubler with connection to 12-V supply gives positive output. Other diode rectifier circuit doubles oscillator output and gives negative supply. Negative doubler uses switching transistors. All three rectifier circuits provide common ground from supply to output.—J. M. Pike, Negative and High Voltages from a Positive Supply, *QST*, Jan. 1974, p 23–25.

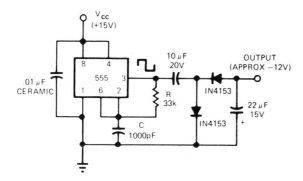

+15 V TO −12 V—Simple transformerless power converter uses 555 timer in self-triggered mode as square-wave generator, followed by voltage-doubling rectifier. Values shown for R and C give frequency of about 20 kHz, which permits good filtering with relatively small capacitors. Maximum load current is about 80 mA.—M. Strange, IC Timer Makes Transformerless Power Converter, *EDN Magazine*, Dec. 20, 1973, p 81.

±15 V FROM 12 V—Steps up output of 12-V battery to voltages required by PLL such as NE561. Uses 900-Hz sine-wave oscillator and LM380N AF amplifier to drive voice-coil side of standard 500-ohm to 3.2-ohm output transformer having bridge rectifier across center-tapped primary. With 10-mA loads, maximum ripple is 15 mV P-P. With receiver quiet, 900-Hz hum is audible, but is normally lost under background noise. Oscillator choke (about 700 mH) is 800 turns of No. 44 magnet wire in Ferroxcube 3C pot core.—R. Megirian, Build a Noise-Free Power Supply, *73 Magazine*, Dec. 1977, p 208–209.

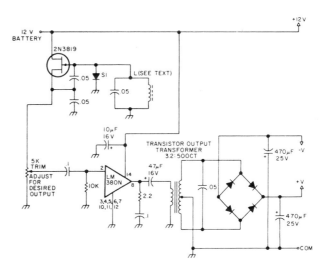

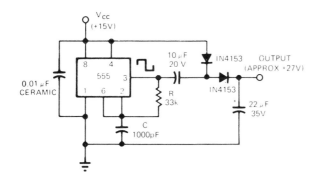

+15 V TO +27 V—Uses 555 timer in self-triggered mode as square-wave generator operating at about 20 kHz, followed by voltage-doubling rectifier. Provides approximate doubling of voltage without use of transformer. Maximum load current is about 80 mA.—M. Strange, IC Timer Makes Transformerless Power Converter, *EDN Magazine*, Dec. 20, 1973, p 81.

SECTION **13**

Converter Circuits—General

Includes V/F, V/I, V/pulse width, V/time, F/V, 7-segment/BCD, BCD/7-segment, Gray/BCD, Gray/binary, binary/BCD, time/V, pulse height/time, I/V, and other converter circuits for changing one parameter linearly to another. See also other Converter chapters.

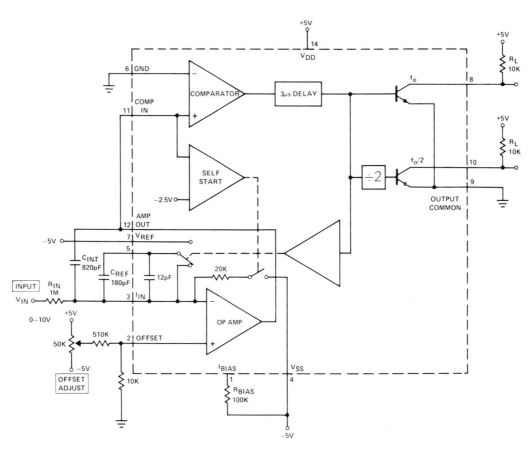

10 Hz TO 10 kHz V/F—External circuit shown for Teledyne 9400 voltage-to-frequency converter provides means for trimming zero location and full-scale frequency value of output. For 10-kHz full-scale value, set V_{IN} to 10 mV and trim with 50K offset adjust pot to get 10-Hz output, then set V_{IN} to 10.000 V and trim either R_{IN}, V_{REF}, or C_{REF} to obtain 10-kHz output.—M. O. Paiva, "Applications of the 9400 Voltage to Frequency Frequency to Voltage Converter," Teledyne Semiconductor, Mountain View, CA, 1978, AN-10, p 3–5.

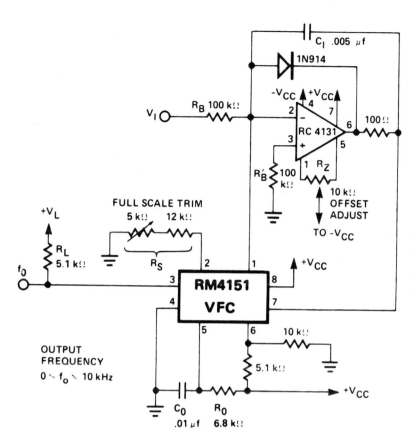

V/F CONVERTER WITH 0.05% LINEARITY—Raytheon RM4151 converter is used with integrator opamp to give highly linear conversion of inputs up to −10 VDC to proportional frequency of square-wave output. With maximum input of −10 V, adjust 5K full-scale trimpot for maximum output frequency of 10 kHz. Set offset adjust pot to give 10-Hz output for input of −10 mV. To operate from single positive supply, change opamp to RC3403A.—"Linear Integrated Circuit Data Book," Raytheon Semiconductor Division, Mountain View, CA, 1978, p 7-38.

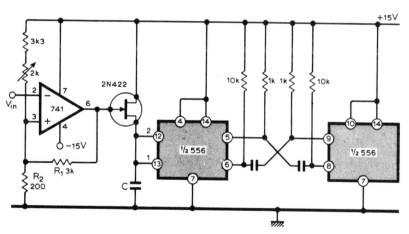

0.1 Hz–100 kHz V/F—Uses NE556 timer in dual mode in combination with opamp and FET for linear voltage-to-frequency conversion with output range from 0.1 Hz to 100 kHz. Operating frequency is 0.91/2RC where R is resistance of FET.—K. Kraus, Linear V-F Converter, *Wireless World*, May 1977, p 80.

BCD	GRAY	X-3 GRAY
0000	0000	0010
0001	0001	0110
0010	0011	0111
0011	0010	0101
0100	0110	0100
0101	0111	1100
0110	0101	1101
0111	0100	1111
1000	1100	1110
1001	1101	1010

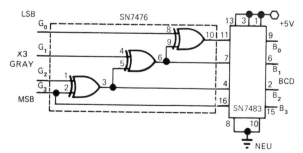

EXCESS-THREE GRAY CODE TO BCD—Developed for use with shaft encoder providing excess-three Gray-code output. Requires only two TTL ICs, connected as shown. To convert regular Gray code to BCD, omit SN7483 4-bit adder. Tabulation shows how circuit accomplishes conversion for both types of Gray codes.—D. M. Risch, Two ICs Convert Excess-Three Gray Code to BCD, *EDN Magazine*, Nov. 1, 1972, p 44.

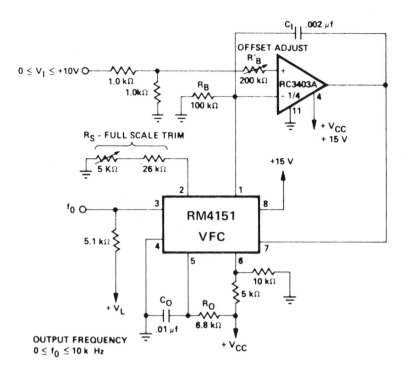

HIGH-PRECISION V/F CONVERTER—Active integrator using one section of RC3403A quad opamp improves linearity, frequency offset, and response time of Raytheon RM4151 converter operating from single supply. Opamp develops null voltage.—"Linear Integrated Circuit Data Book," Raytheon Semiconductor Division, Mountain View, CA, 1978, p 7-38.

DC VOLTAGE TO TIME—Opamp connected as integrator feeds opamp comparator to produce output pulse whose width is proportional to magnitude of DC input voltage. Circuit shown is for positive inputs only; for both positive and negative inputs, article tells how to add another comparator. Circuit can then be used to generate start and stop pulses applied to digital timer of digital voltmeter.—G. B. Clayton, Experiments with Operational Amplifiers, *Wireless World*, Sept. 1973, p 447–448.

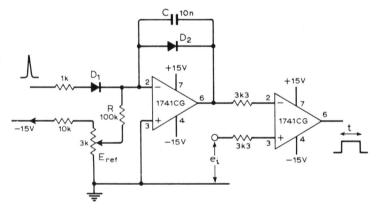

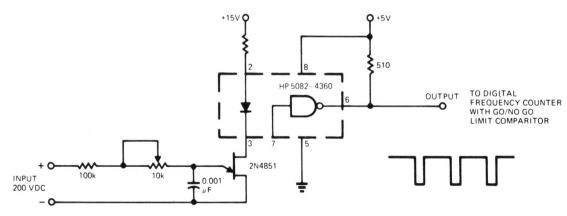

VOLTAGE-TO-FREQUENCY GO/NO-GO—Single UJT is used as V/F converter to provide completely isolated inputs and outputs for high-voltage go/no-go test monitor. When voltage exceeds predetermined limit, output to digital frequency counter exceeds corresponding frequency limit. Output can be fed directly into digital frequency-limit detector that provides go/no-go indication.—T. H. Li, VFC Used in Isolated GO/NO GO Voltage Monitor, *EDN Magazine*, July 5, 1974, p 75.

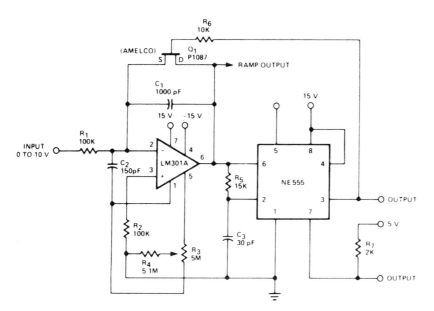

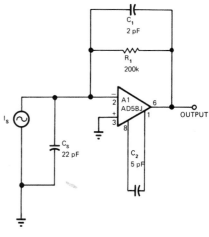

VOLTAGE TO FREQUENCY—Input voltage range of 0 to −10 VDC is converted by opamp and timer to proportional frequency with good linearity. Circuit is TTL-compatible. Accuracy is 0.2%.—"Signetics Analog Data Manual," Signetics, Sunnyvale, CA, 1977, p 727–729.

CURRENT TO VOLTAGE—Developed for use with current-output transducers such as silicon photocells. For widest frequency response, circuit values may need some adjusting for source current and capacitance. C_1, across feedback resistor of opamp, eliminates ringing around 500 kHz. If input coupling capacitor is added to reduce DC gain, circuit can be used with inductive source such as magnetic tape head.—R. S. Burwen, Current-to-Voltage Converter for Transducer Use, *EDN Magazine*, Dec. 15, 1972, p 40.

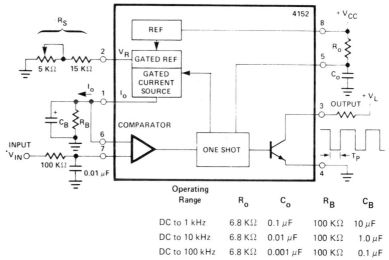

0.01–10 V to 1, 10, OR 100 kHz—Simple voltage-to-frequency converter uses Raytheon 4152 operating from single 15-V supply to convert analog input voltage to proportional frequency of square-wave output. Maximum output frequency depends on values used for resistors and capacitors, as given in table. Suitable for applications where input dynamic range is limited and does not go to zero.—"Linear Integrated Circuit Data Book," Raytheon Semiconductor Division, Mountain View, CA, 1978, p 7-45–7-46.

Operating Range	R_o	C_o	R_B	C_B
DC to 1 kHz	6.8 KΩ	0.1 μF	100 KΩ	10 μF
DC to 10 kHz	6.8 KΩ	0.01 μF	100 KΩ	1.0 μF
DC to 100 kHz	6.8 KΩ	0.001 μF	100 KΩ	0.1 μF

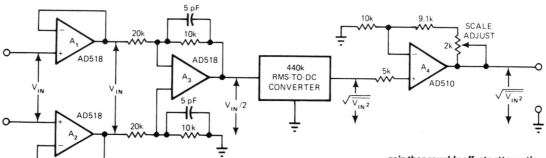

20-kHz SQUARE-WAVE TO DC—Provides accuracy within 0.1% for square-wave inputs of 3 to 7 VRMS in frequency range of 5 to 20 kHz when duty cycle is 50%. Opamps A_1 and A_2 are connected as differential input voltage-followers to provide high input impedance. A_3 converts to single-ended output as required by Model 440 converter IC. A_4 provides adjustable gain than roughly offsets attenuation of A_3, with 2K pot being adjusted to provide desired ratio of DC output voltage to RMS value of input.—J. Renken, Differential High-Z RMS-DC Converter Has 0.1% Accuracy, *EDN Magazine*, May 5, 1977, p 114 and 116.

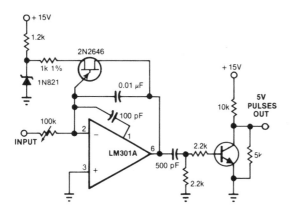

SINGLE-SLOPE V/F CONVERSION—UJT forms reference that determines reset point of LM301A integrator for converting analog input voltage to proportional frequency. Output of integrator ramps negative until UJT switches and drives output positive at high slew rate. Positive edge of integrator output is differentiated by RC network and level-shifted by NPN bipolar transistor to provide logic-compatible pulse.—J. Williams, Low-Cost, Linear A/D Conversion Uses Single-Slope Techniques, *EDN Magazine*, Aug. 5, 1978, p 101–104.

GRAY TO BINARY—Converts first 4 bits of Gray-code word to binary output. Uses two MC7496 shift registers and logic elements to transfer data serially from input register through MC1812 EXCLUSIVE-OR IC to output register. One requirement is that strobe on pin 8 of input register must complete its function before clock appears on pin 1 of register. When this and other timing conditions are satisfied, converter will work at speeds up to about 10 megabits per second.—J. Barnes, "Analog-to-Digital Cyclic Converter," Motorola, Phoenix, AZ, 1974, AN-557, p 9.

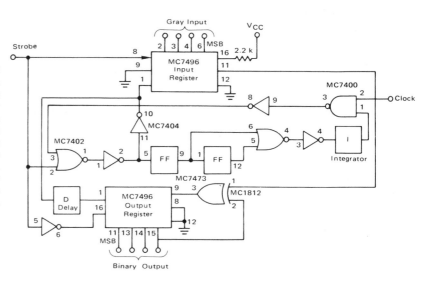

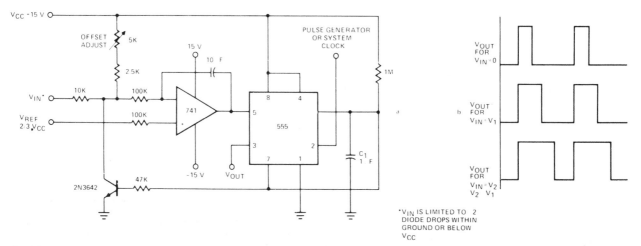

VOLTAGE TO PULSE WIDTH—Opamp and timer together convert input voltage level to width of output pulse with accuracy better than 1%. Output is at same frequency as input.— "Signetics Analog Data Manual," Signetics, Sunnyvale, CA, 1977, p 726–727.

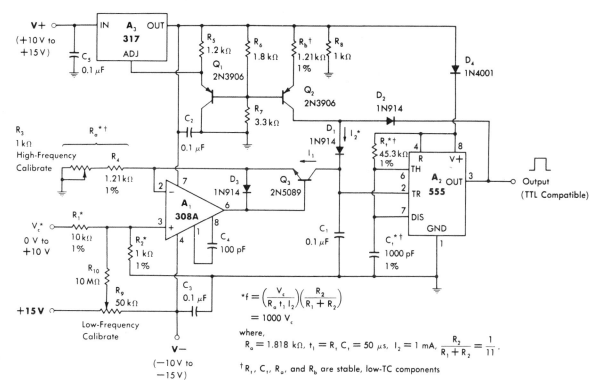

$$*f = \left(\frac{V_c}{R_a t_1 I_2}\right)\left(\frac{R_2}{R_1 + R_2}\right)$$
$$= 1000 \, V_c$$

where,

$R_a = 1.818$ kΩ, $t_1 = R_t C_t = 50$ μs, $I_2 = 1$ mA, $\dfrac{R_2}{R_1 + R_2} = \dfrac{1}{11}$.

†R_t, C_t, R_a, and R_b are stable, low-TC components

POSITIVE-INPUT V/F—Input voltages from 0 to 10 V are divided by R_1 and R_2 for application to noninverting input of current source A_1. 555 timer A_2 provides functions of precision mono MVBR and level sensor. Regulator A_3 acts as gated current source and provides stabilized voltage output for 555 and 308A.—W. G. Jung, "IC Timer Cookbook," Howard W. Sams, Indianapolis, IN, 1977, p 184–192.

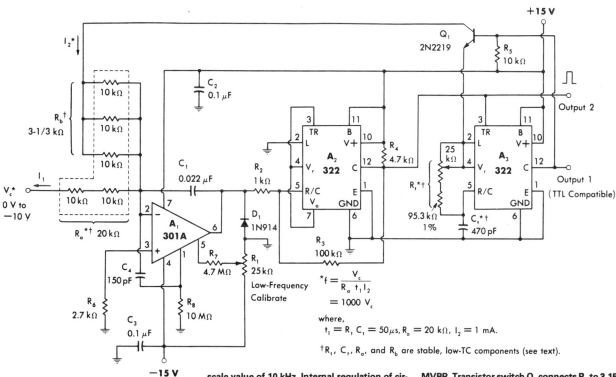

$$*f = \frac{V_c}{R_a t_1 I_2}$$
$$= 1000 \, V_c$$

where,

$t_1 = R_t C_t = 50$ μs, $R_a = 20$ kΩ, $I_2 = 1$ mA.

†R_t, C_t, R_a, and R_b are stable, low-TC components (see text).

−10 V GIVES 10 kHz—Control voltage input in range of 0 to −10 V is converted linearly to frequency of digital output pulse train having full-scale value of 10 kHz. Internal regulation of circuit makes operation essentially independent of ±15 V supply level. A_1 is opamp integrator, A_2 is comparator, and A_3 is precision mono MVBR. Transistor switch Q_1 connects R_b to 3.15-V reference voltage during t_1 timing period of A_3.—W. G. Jung, "IC Timer Cookbook," Howard W. Sams, Indianapolis, IN, 1977, p 184–192.

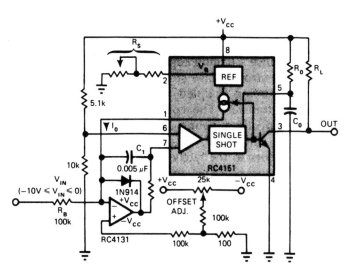

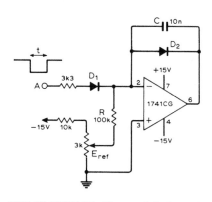

V/F AND F/V—Although based on Raytheon 4151 IC voltage-to-frequency converter, circuit is readily adapted to other modern V/F converters now costing under $10 each. With values shown, input of 0 to −10 VDC provides proportional frequency change from 0 to 10 kHz at out- put. Design equations are given. Article also covers F/V operation of same IC for demodulating FSK data.—T. Cate, IC V/F Converters Read- ily Handle Other Functions Such as F/V, A/D, *EDN Magazine,* Jan. 5, 1977, p 82–86.

TIME TO VOLTAGE—Time period of negative gating signal determines amplitude of linear output ramp generated by integrator opamp. Amplitude of ramp, proportional to input time, is observed on calibrated screen of oscillo- scope.—C. B. Clayton, Experiments with Op- erational Amplifiers, *Wireless World,* Sept. 1973, p 447–448.

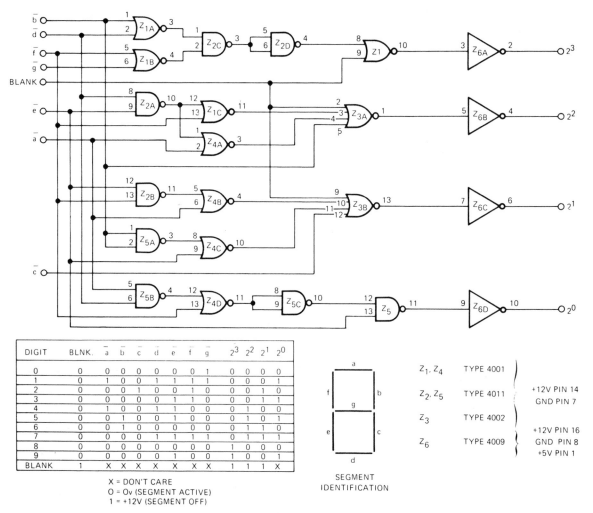

DIGIT	BLNK.	$\bar{a}$	$\bar{b}$	$\bar{c}$	$\bar{d}$	$\bar{e}$	$\bar{f}$	$\bar{g}$	2^3	2^2	2^1	2^0
0	0	0	0	0	0	0	0	1	0	0	0	0
1	0	1	0	0	1	1	1	1	0	0	0	1
2	0	0	0	1	0	0	1	0	0	0	1	0
3	0	0	0	0	0	1	1	0	0	0	1	1
4	0	1	0	0	1	1	0	0	0	1	0	0
5	0	0	1	0	0	1	0	0	0	1	0	1
6	0	0	1	0	0	0	0	0	0	1	1	0
7	0	0	0	0	1	1	1	1	0	1	1	1
8	0	0	0	0	0	0	0	0	1	0	0	0
9	0	0	0	0	0	1	0	0	1	0	0	1
BLANK	1	X	X	X	X	X	X	X	1	1	1	X

X = DON'T CARE
O = 0v (SEGMENT ACTIVE)
1 = +12V (SEGMENT OFF)

SEGMENT
IDENTIFICATION

Z_1, Z_4	TYPE 4001
Z_2, Z_5	TYPE 4011
Z_3	TYPE 4002
Z_6	TYPE 4009

+12V PIN 14
GND PIN 7

+12V PIN 16
GND PIN 8
+5V PIN 1

7-SEGMENT TO BCD—Uses six CMOS pack- ages to convert 7-segment display to corre- sponding four-line positive-logic BCD code for digits 0-9. Added feature is blank input which, when high, forces blank code (1110 or 1111) into readout, for use in suppressing leading zeros with some types of data storage. Use 4010 in place of 4009 for Z_6 when negative-logic BCD output is required.—R. Sturla, Real-Time 7-Seg- ment to BCD Converter, *EDN Magazine,* June 20, 1973, p 89.

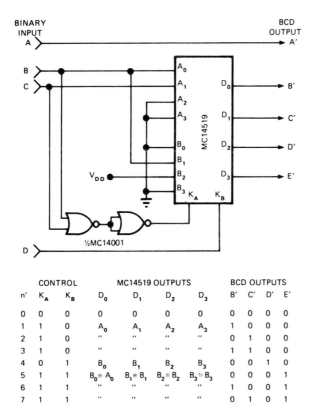

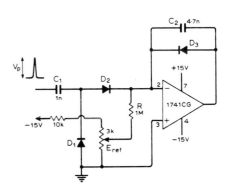

	CONTROL		MC14519 OUTPUTS				BCD OUTPUTS			
n'	K_A	K_B	D_0	D_1	D_2	D_3	B'	C'	D'	E'
0	0	0	0	0	0	0	0	0	0	0
1	1	0	A_0	A_1	A_2	A_3	1	0	0	0
2	1	0	"	"	"	"	0	1	0	0
3	1	0	"	"	"	"	1	1	0	0
4	0	1	B_0	B_1	B_2	B_3	0	0	1	0
5	1	1	$B_0 \otimes A_0$	$B_1 \otimes B_1$	$B_2 \otimes B_2$	$B_3 \otimes B_3$	0	0	0	1
6	1	1	"	"	"	"	1	0	0	1
7	1	1	"	"	"	"	0	1	0	1

4-BIT BINARY TO 5-BIT BCD—Converts binary number within machine to BCD value from 0 to 15, for driving visual displays. Requires only quad two-channel data selector with EXCLUSIVE-NOR function, available in IC packages.

Article gives truth tables and traces operation step by step.—J. Barnes and J. Tonn, Binary-to-BCD Converter Implements Simple Algorithm, *EDN Magazine*, Jan. 5, 1975, p 56, 57, and 59.

PULSE HEIGHT TO TIME—Simple opamp circuit produces time interval proportional to height of positive input pulse. Opamp is connected as integrator whose output is held at about zero by negative feedback through D_3. Positive input pulse charges C_1 and C_2, amplifier output steps down, and D_3 is reverse-biased. Time for output to charge back up to zero, as observed on oscilloscope, is then directly proportional to input pulse height. Article gives design equations.—G. B. Clayton, Experiments with Operational Amplifiers, *Wireless World*, Sept. 1973, p 447–448.

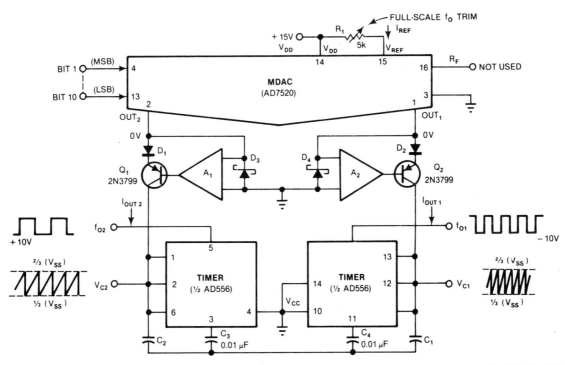

DIGITAL TO FREQUENCY—Combination of multiplying DAC and 556 dual timer provides complementary output frequencies under control of digital input. Opamp and diode types are not critical. Output frequency of each timer depends on supply voltages, capacitor values, and setting of R_1.—J. Wilson and J. Whitmore,

MDAC's Open Up a New World of Digital-Control Applications, *EDN Magazine*, Sept. 20, 1978, p 97–105.

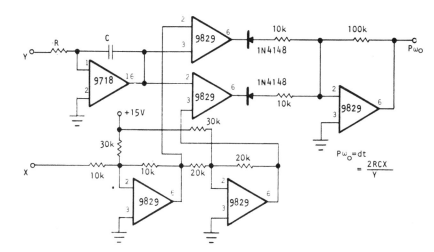

VOLTAGE TO PULSE DURATION—Optical Electronics 9829 opamps are used as fast comparators and 9718 FET opamp as fast integrator to give high precision at high speed for converting analog voltage to pulse duration for such applications as A/D conversion, delta code generation, motor speed control, and pulse-duration modulation. Output pulse durations can be as short as 1 μs. Conversion linearity is better than 0.1%. Minimum pulse duration is 100 ns, and maximum dynamic range is 40 dB. Reference voltages are determined by X input; if X is 3 V, reference voltages differ by 6 V. Two 9829 opamps present reference voltages to two comparator opamps. Fifth 9829 sums comparator outputs and gives positive output.—"Voltage to Pulse Width Converter," Optical Electronics, Tucson, AZ, Application Tip 10230.

$P\omega_o = dt$

$= \dfrac{2RCX}{Y}$

VOLTAGE TO CURRENT—Circuit is capable of supplying constant alternating current up to 1 A to variable load. Actual value of load current is determined by input voltage, values of R_1-R_3, and value of R_5. Input of 250 mV gives 0.5 A through load (RMS values) with less than 0.5% total harmonic distortion. Applications include control of electromagnet current.—"Audio Handbook," National Semiconductor, Santa Clara, CA, 1977, p 4-21–4-28.

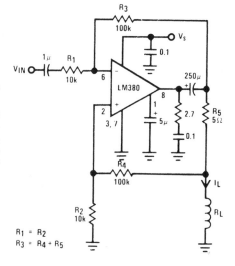

$R_1 = R_2$

$R_3 = R_4 + R_5$

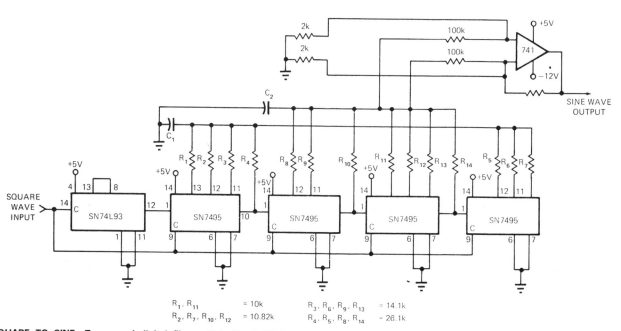

SINE WAVE OUTPUT

R_1, R_{11}	= 10k
R_2, R_7, R_{10}, R_{12}	= 10.82k
R_3, R_6, R_9, R_{13}	= 14.1k
R_4, R_5, R_8, R_{14}	= 26.1k

SQUARE TO SINE—Transversal digital filter suppresses harmonics present on input square wave, to give pure sine wave. Resistors weight data as it passes through 16-bit shift register, so sine wave is sampled at 16 times its frequency and theoretically has no harmonics below the 16th. Simple RC filter removes remaining harmonics. Input is clock whose repetition rate is 16 times desired frequency. SN74L93 4-bit ripple counter divides this down to provide square wave of desired frequency. Square wave is sampled 16 times per cycle and shifted down SN7495 16-bit shift register. C_1 and C_2 are selected to eliminate higher harmonics. Sine-wave output has harmonic distortion of less than −50 dB.—L. J. Mandell, Sine-Wave Synthesizer Has Low Harmonic Distortion, *EDN Magazine,* Aug. 15, 1972, p 52.

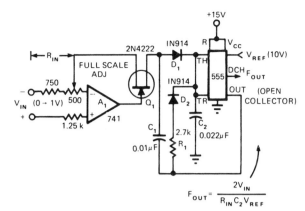

$$F_{OUT} = \frac{2V_{IN}}{R_{IN}C_2V_{REF}}$$

V/F GIVES 10 to 10,000 Hz—Current proportional to input voltage is balanced via periodic charging of C_1 to precisely repeatable voltage by opamp A_1 and FET Q_1. With values shown, nominal scale factor is 10 kHz/V. Input of 0 to 1 V gives output of 10 to 10,000 Hz with better than 0.05% linearity. Article gives operating details and design equations.—W. S. Woodward, Simple 10 kHz V/F Features Differential Inputs, *EDN Magazine*, Oct. 20, 1974, p 86.

0–10 VDC TO 0–10 kHz—Single-supply voltage-to-frequency converter produces square-wave output at frequency varying linearly with input voltage. Linearity error is typically only 1%. For values shown, response time for step change of input from 0 to +10 V is 135 ms. Uses Raytheon 4151 converter. Supply can be 15 V.—"Linear Integrated Circuit Data Book," Raytheon Semiconductor Division, Mountain View, CA, 1978, p 7-38.

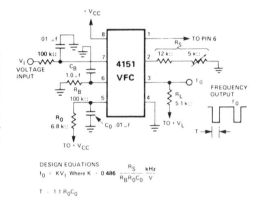

DESIGN EQUATIONS

$$f_0 = KV_I \text{ Where } K = 0.486 \frac{R_S}{R_B R_0 C_0} \frac{kHz}{V}$$

$$T = 1.1 R_0 C_0$$

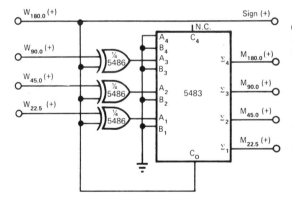

0–360° to 0–180°—Used for converting angular information in 360° wrap-around code to ±180° sign-plus-magnitude code. For values under 180°, converter outputs and inputs are identical. For larger input angles, output code is complement of input plus one. Used for interfacing shaft encoders and synchro-to-digital converters to digital display. Article gives truth table showing which lines are high and which are low at input and at output for angular increments of 22.5°.—J. N. Phillips, Convert Wrap-Around Code to Sign-Plus-Magnitude, *EDN Magazine*, Jan. 5, 1973, p 103.

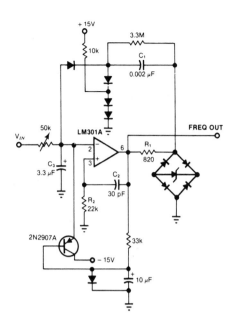

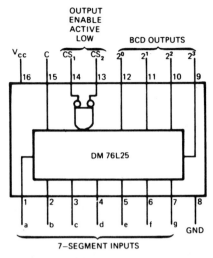

CHARGE-DISPENSING V/F CONVERSION— Output state of opamp switches C_1 between reference voltage provided by diode bridge and its inverting input. Network R_2-C_2 reinforces direction of opamp output change. Circuit can deliver 0–10 kHz output with 0.01% linearity for 0–10 V input. —J. Williams, Low-Cost, Linear A/D Conversion Uses Single-Slope Techniques, *EDN Magazine*, Aug. 5, 1978, p 101–104.

BCD FROM 7-SEGMENT DISPLAY—Single National DM76L25 read-only memory provides conversion from 7-segment outputs of MOS chip driving display to BCD inputs for data processing. Typical power dissipation is 75 mW. Access time is 70 ns when using 5-V supply. Article gives truth table for all standard and special characters of 7-segment display.—U. Priel, 7-Segment-to-BCD Converter: The Last Word?, *EDN Magazine*, Aug. 20, 1974, p 94–95.

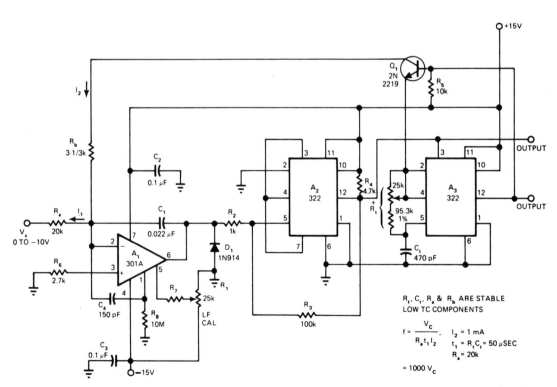

VOLTAGE-TO-FREQUENCY USING IC TIMERS—Two 322 IC timers and single 301A opamp provide all functions required for charge-balancing type of voltage-to-frequency converter, including integrator, level sensor or comparator, precision mono, and gated current source. Circuit accepts control voltage inputs of 0 to −10 V, corresponding to output pulse stream range of 0 to 10 kHz. Article describes operation in detail. R_4 should be 4.7 megohms. Output pulses of comparator A_2 trigger mono A_3, which generates pulse having duration t_1 that saturates Q_1, to force reference current I_2 into summing point of opamp integrator.—W. G. Jung, Take a Fresh Look at New IC Timer Applications, *EDN Magazine*, March 20, 1977, p 127–135.

SECTION 14
Converter Circuits—Digital-to-Analog

Includes circuits for converting variety of digital inputs to linearly related analog output voltage or current, providing analog sum of two digital inputs, or converting stored digital speech back to analog form.

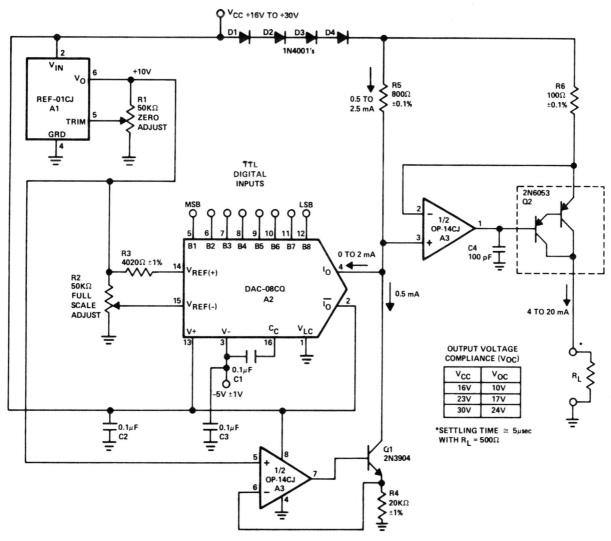

OUTPUT VOLTAGE COMPLIANCE (V_{OC})	
V_{CC}	V_{OC}
16V	10V
23V	17V
30V	24V

*SETTLING TIME ≅ 5μsec WITH R_L = 500Ω

8-BIT BINARY TO PROCESS CURRENT—Uses only three Precision Monolithics ICs operating from −5 V and +23 V supplies to convert 8-bit binary digital input to process current in range of 4-20 mA. Fixed current of 0.5 mA is added to DAC output current varying between 0 and 2 mA, with resulting total current multiplied by factor of 8 to give up to 20 mA through 500-ohm load.—D. Soderquist, "3 IC 8 Bit Binary Digital to Process Current Converter with 4-20 mA Output," Precision Monolithics, Santa Clara, CA, 1977, AN-21.

HIGH-SPEED OUTPUT OPAMP—Precision Monolithics OP-17F opamp optimizes DAC-08E D/A converter for highest speed in converting DAC output current to output voltage up to 10 V under control of digital input. Settling time is 380 ns.—G. Erdi, "The OP-17, OP-16, OP-15 as Output Amplifiers for High Speed D/A Converters," Precision Monolithics, Santa Clara, CA, 1977, AN-24, p 2.

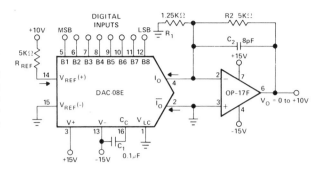

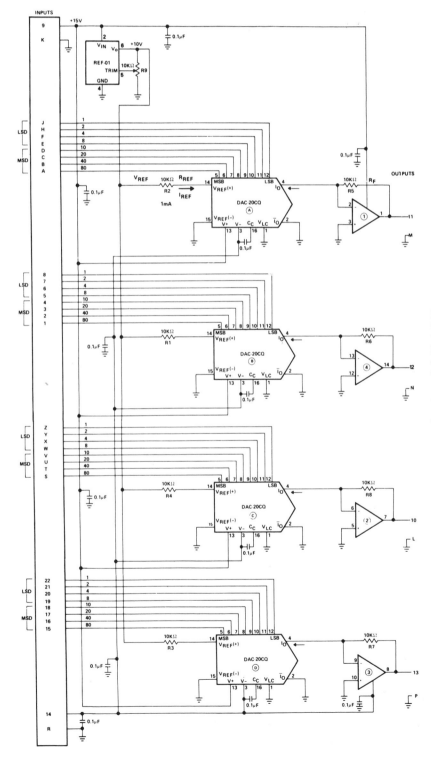

FOUR-CHANNEL BCD—Uses four Precision Monolithics DAC-20CQ 2-digit BCD D/A converter, OP-11FY precision quad opamp, and REF-01HJ +10 V voltage reference to convert 2-digit BCD input coding to proportional analog 0 to +10 V output for each of four channels. Same configuration will handle binary inputs, as covered in application note. For output range of 0 to +5 V, change voltage reference to REF-02.—D. Soderquist, "Low Cost Four Channel DAC Gives BCD or Binary Coding," Precision Monolithics, Santa Clara, CA, 1977, AN-26, p 3.

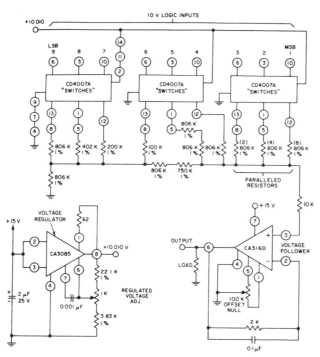

9-BIT USING DIGITAL SWITCHES—Combination of CD4007A multiple-switch CMOS ICs, ladder network of discrete metal-oxide film resistors, CA3160 voltage-follower opamp, and CA3085 voltage regulator gives digital-to-analog converter that is readily interfaced with 10-V logic levels of CMOS input. Required resistor accuracy, ranging from $\pm0.1\%$ for bit 2 to $\pm1\%$ for bits 6-9, is achieved by using series and parallel combinations of 806K resistors.—"Linear Integrated Circuits and MOS/FET's," RCA Solid State Division, Somerville, NJ, 1977, p 267–268.

6 BITS TO ANALOG—Uses Motorola MC1723G voltage regulator to provide reference voltage and opamp for MC1406L 6-bit D/A converter. Output current can be up to 150 mA. Full-scale output is about 10 V, but can be boosted as high as 32 V by increasing value of R_2 and increasing +15 V supply proportionately to maximum of 35 V.—D. Aldridge and K. Huehne, 6-Bit D/A Converter Uses Inexpensive Components, *EDN Magazine,* Dec. 15, 1972, p 40–41.

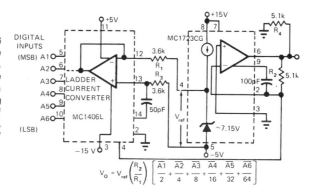

$$V_O = V_{ref}\left(\frac{R_2}{R_1}\right)\left[\frac{\overline{A1}}{2} + \frac{\overline{A2}}{4} + \frac{\overline{A3}}{8} + \frac{\overline{A4}}{16} + \frac{\overline{A5}}{32} + \frac{\overline{A6}}{64}\right]$$

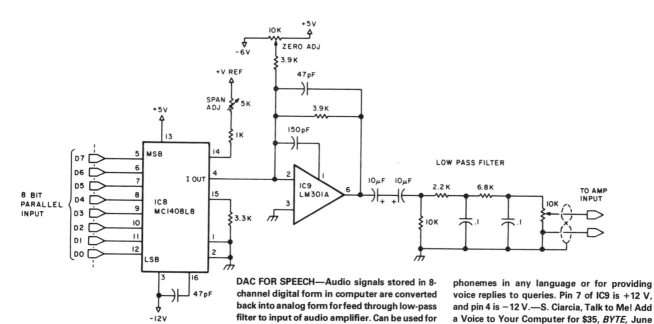

DAC FOR SPEECH—Audio signals stored in 8-channel digital form in computer are converted back into analog form for feed through low-pass filter to input of audio amplifier. Can be used for computer-controlled synthesis of speech from phonemes in any language or for providing voice replies to queries. Pin 7 of IC9 is +12 V, and pin 4 is −12 V.—S. Ciarcia, Talk to Me! Add a Voice to Your Computer for $35, *BYTE,* June 1978, p 142–151.

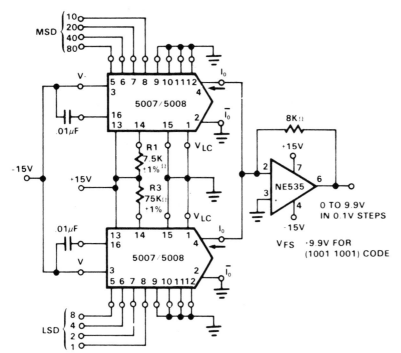

2-DIGIT BCD INPUT—Each Signetics 5007/5008 multiplying D/A converter serves one digit of input voltage to give output current that is product of digital input number and input reference current. Opamp combines currents and converts them to analog output voltage proportional to digital input value.—"Signetics Analog Data Manual," Signetics, Sunnyvale, CA, 1977, p 677–685.

ANALOG SUM OF DIGITAL NUMBERS—Two Precision Monolithics DAC-100 D/A converters and OP-01 opamp combine conversion with adding to give high-precision DC output voltage. 200-ohm pots are adjusted initially to give exactly desired output for input of all 0s.—"8 & 10 Bit Digital-to-Analog Converter," Precision Monolithics, Santa Clara, CA, 1977, DAC-100, p 5.

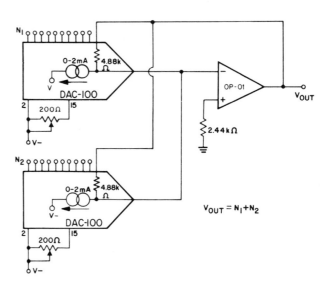

SIMPLE DAC—Transistors are either saturated or cut off by outputs of clock-controlled SN7490 BCD counter. Portions of emitter voltages of the four transistors are added in ratios 1:2:4:8 by 741 summing opamp to obtain analog output. Article tells how two such circuits can be combined for use in two-digit DVM.—D. James, Simple Digital to Analogue Converter, *Wireless World,* June 1974, p 197.

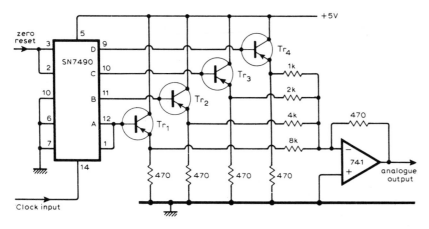

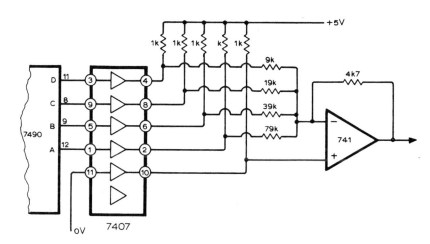

TEMPERATURE COMPENSATION—Use of 7407 hex buffer following SN7490 of D/A converter permits satisfactory performance over reasonably wide temperature range even when driving several TTL stages. Noninverting input of 741 opamp is connected to output of unused buffer at logic 0. Circuit is modification of D/A converter developed by D. James for use in simple two-digit DVM.—R. J. Chance, Improved Simple D. to A. Converter, *Wireless World*, Dec. 1974, p 503.

2-DIGIT BCD—Output current of Precision Monolithics DAC-100 D/A converter can be adjusted to exactly desired value with 200-ohm pot for each DAC; adjustment is made with input of all 0s. Circuit can be expanded to 3 digits by adding third DAC and adding 99 to current divider.—"8 & 10 Bit Digital-to-Analog Converter," Precision Monolithics, Santa Clara, CA, 1977, DAC-100, p 5.

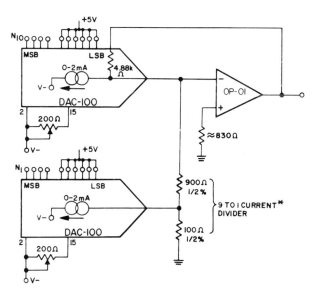

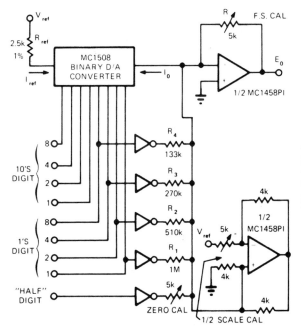

2½-DIGIT INPUT FOR 199 COUNT—Addition of ½-digit circuit to basic 2-digit BCD DAC increases count from 99 to 199. Circuit sequences to 99 while ½-digit section of MC14009 hex two-input NOR gate has low output, and goes through steps 100 to 199 while ½-digit output is high. Reference voltage is 5.0 V. Calibration procedure is given.—T. Henry, Binary D/A Converters Can Provide BCD-Coded Conversion, *EDN Magazine*, Aug. 5, 1973, p 70–73.

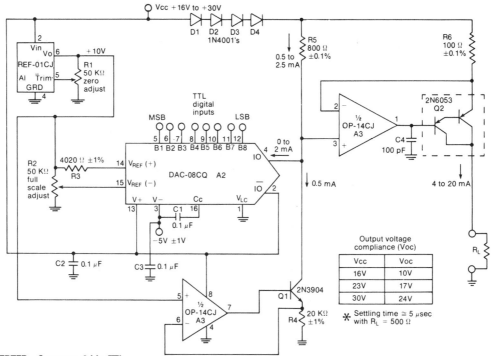

CURRENT CONVERTER—Converts 8-bit TTL digital inputs to process current in range of 4 to 20 mA, for microprocessor control of industrial operations. Fixed 0.5-mA current is added to DAC output current varying between 0 and 2.0 mA and multiplied by factor of 8 to produce final output current of 4–20 mA. To calibrate, connect ammeter between output and ground, then apply +23 V ± 7 V and −5 V ± 1 V to converter. Make digital inputs all 0s (less than +0.8 V). Adjust R1 until output current is 4.0 mA. Change digital inputs to all 1s (greater than +2.0 V), and adjust R2 until output current is 20 mA.—D. Soderquist, Build Your Own 4-20 mA Digital to Analog Converter, *Instruments & Control Systems,* March 1977, p 57–58.

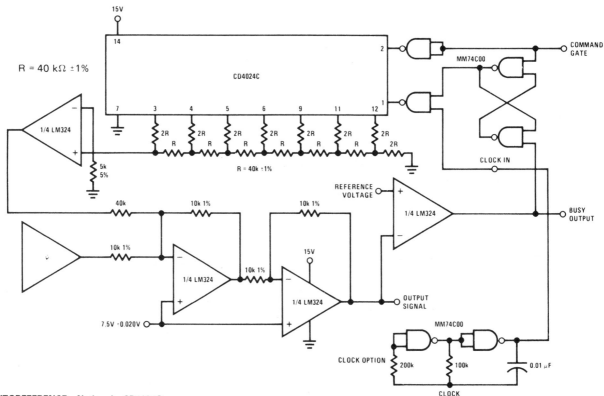

AUTOREFERENCE—National CD4024C converter is used with logic and summer elements to eliminate virtually all offset errors induced by time and temperature changes in process control system fed by transducer. Best suited for applications having short repeated duty cycles, each containing reference point. Examples include weighing scale in which transducer is load cell, pressure control systems, fuel pumps, and sphygmomanometers. Circuit eliminates warm-up errors.—"Pressure Transducer Handbook," National Semiconductor, Santa Clara, CA, 1977, p 7-4–7-8.

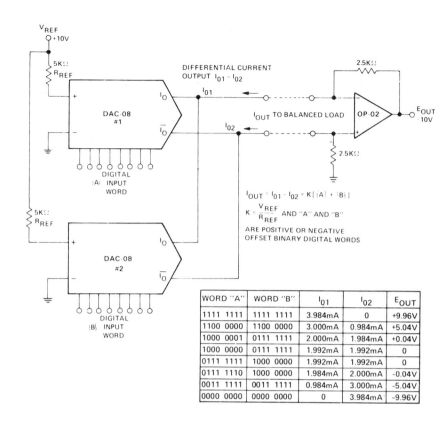

WORD "A"	WORD "B"	I_{01}	I_{02}	E_{OUT}
1111 1111	1111 1111	3.984mA	0	+9.96V
1100 0000	1100 0000	3.000mA	0.984mA	+5.04V
1000 0001	0111 1111	2.000mA	1.984mA	+0.04V
1000 0000	0111 1111	1.992mA	1.992mA	0
0111 1111	1000 0000	1.992mA	1.992mA	0
0111 1110	1000 0000	1.984mA	2.000mA	-0.04V
0011 1111	0011 1111	0.984mA	3.000mA	-5.04V
0000 0000	0000 0000	0	3.984mA	-9.96V

FOUR-QUADRANT ALGEBRAIC—Two Precision Monolithics DAC-08 D/A converters perform fast algebraic summation of two digital input words and feed OP-02 opamp that provides direct analog output which is algebraic sum of words A and B in all four quadrants.—J. Schoeff and D. Soderquist, "Differential and Multiplying Digital to Analog Converter Applications," Precision Monolithics, Santa Clara, CA, 1976, AN-19, p 7.

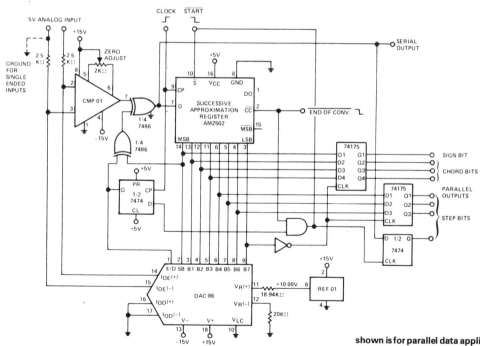

BELL-SYSTEM μ-255 COMPANDING LAW— Precision Monolithics DAC-86 is used in circuit that provides 15-segment linear approximation by using 3 bits to select one of eight binarily related chords, then using 4 bits to select one of sixteen linearly related steps within each chord. Sign bit determines signal polarity, and encode/decode select bit determines operation. Circuit shown is for parallel data applications. For serial data, omit inverter, two 74175 chips, and half of 7474. Power supplies should be well bypassed.—"COMDAC Companding D/A Converter," Precision Monolithics, Santa Clara, CA, 1977, DAC-86, p 6.

SECTION 15
Data Transmission Circuits

Includes line driver, line receiver, modem, bit-rate generator, coder-decoder, FSK demodulator, signal conditioner, optoisolator, PDM telemetry, active bandpass filter, and other circuits used for transmitting digital data and digital speech over twisted-pair, coaxial, or balanced line.

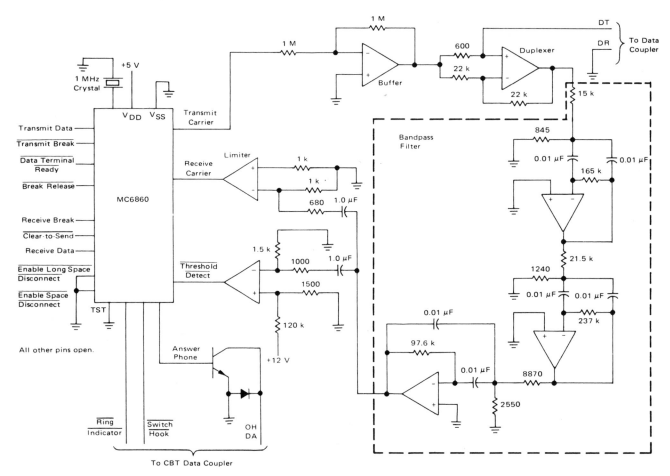

ANSWER MODEM—Transmits on upper channel (mark 2225 Hz and space 2025 Hz) and receives on lower channel (mark 1270 Hz and space 1070 Hz). Buffer and duplexer provide modem interface to transmission network. Bandpass filter allows only desired receive signals to be seen by limiter and demodulator. Motorola MC6860 modem IC contains modulator, demodulator, and supervisory control functions.—G. Nash, "Low-Speed Modem Fundamentals," Motorola, Phoenix, AZ, 1974, AN-731, p 6.

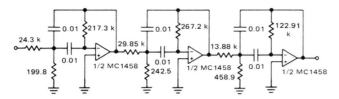

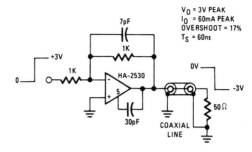

BANDPASS ORIGINATE FILTER—Provides gain of over 15 dB between 1975 and 2275 Hz, to accept 2025–2225 Hz signals of low-speed modem system using Motorola MC6860 IC.—J. M. DeLaune, "Low-Speed Modem System Design Using the MC6860," Motorola, Phoenix, AZ, 1975, AN-747, p 13.

FAST-SETTLING COAX DRIVER—Suitable for use as radar pulse driver, video sync driver, or pulse-amplitude-modulation line driver. Uses Harris HA-2530/2535 wideband amplifier having high slew rate. Usable bandwidth is about 100 kHz when connected for noninverting operation as shown. Driver output is 60 mA into 60-ohm load. 5% settling time is 60 ns.—"Linear & Data Acquisition Products," Harris Semiconductor, Melbourne, FL, Vol. 1, 1977, p 7-54 (Application Note 516).

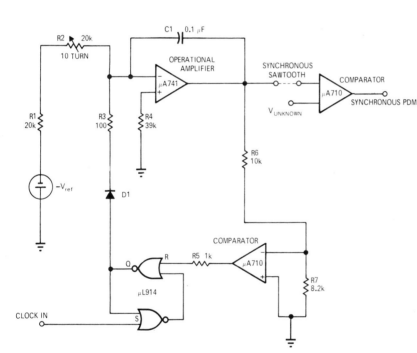

SYNCHRONOUS SAWTOOTH FOR PDM TELEMETRY—Circuit generates highly linear ramp that is reset to zero by each clock pulse. When ramp exceeds analog value of unknown input voltage, pulse is terminated. R1, R2, and C1 form integrating network around opamp. Varying R2 changes slope of ramp output.—J. Springer, Build a Sawtooth Generator with Three ICs, *EDN Magazine,* Nov. 15, 1970, p 49.

INTERFACES FOR 100-OHM LINE—Permits transferring data signals from SA900/901 diskette storage drive to location of MC6800 microprocessor up to maximum of 20 feet away through 100-ohm coax. Data line drivers used are capable of sinking 100-mA in logic true state with maximum voltage of 0.3 V with respect to logic ground. When line driver is in logic false state, driver transistor is cut off and voltage at output of driver is at least 3 V with respect to logic ground.—"Microprocessor Applications Manual" (Motorola Series in Solid-State Electronics), McGraw-Hill, New York, NY, 1975, p 5-211–5-212.

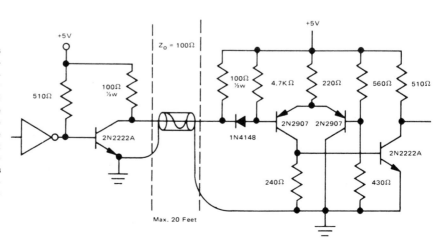

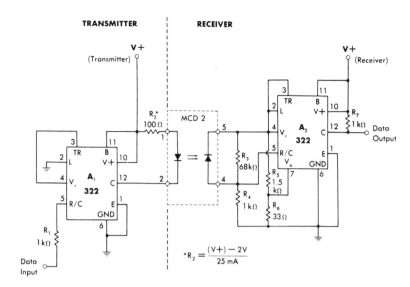

OPTICALLY COUPLED DATA LINK—322 comparator at transmitter end of link drives LED of MCD 2 optoisolator which accepts TTL input. Receiver is similar comparator having additional biasing to match photodiode output of optoisolator. Complete system is noninverting, with delay of about 2 μs. Receiver can have any supply within 4.5–40 V range of 322. Transmitter should be matched to its supply voltage by selecting R₂ according to equation shown.—W. G. Jung, "IC Timer Cookbook," Howard W. Sams, Indianapolis, IN, 1977, p 156–158.

$$*R_2 = \frac{(V+) - 2V}{25\ mA}$$

RECEIVE FILTER—Used as prefilter having controlled group-delay distortion, ahead of receiving modem in data transmission system. Values shown are for 950–1400 Hz answer filter. For 1900–2350 Hz originate filter, change critical values to those given in parentheses.—D. Lancaster, "TV Typewriter Cookbook," Howard W. Sams, Indianapolis, IN, 1976, p 180–182.

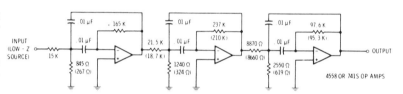

950-1400 Hz • NORMAL VALUES • ANSWER FILTER
1900-2350 Hz • (PARENTHETICAL VALUES) ORIGINATE FILTER

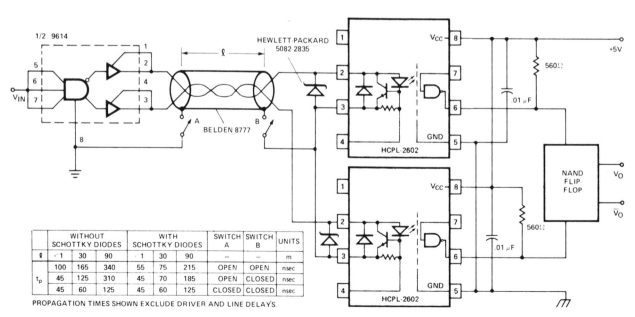

	WITHOUT SCHOTTKY DIODES			WITH SCHOTTKY DIODES			SWITCH A	SWITCH B	UNITS
ℓ	<1	30	90	<1	30	90	—	—	m
t_p	100	165	340	55	75	215	OPEN	OPEN	nsec
	45	125	310	45	70	185	OPEN	CLOSED	nsec
	45	60	125	45	60	125	CLOSED	CLOSED	nsec

PROPAGATION TIMES SHOWN EXCLUDE DRIVER AND LINE DELAYS.

POLARITY-REVERSING SPLIT-PHASE DRIVE— Half of 9614 polarity-reversing line driver feeds pair of Hewlett-Packard HCPL-2602 optically coupled line receivers through coax cable. Cable-grounding switches A and B change performance. Closing only switch B enhances common-mode rejection but reduces propagation delay slightly. Closing both switches optimizes data rate. Schottky diodes at receiver inputs improve data rate. NAND flip-flop at output greatly improves system noise rejection in split-phase termination of line.—"Optoelectronics Designer's Catalog 1977," Hewlett-Packard, Palo Alto, CA, 1977, p 158–159.

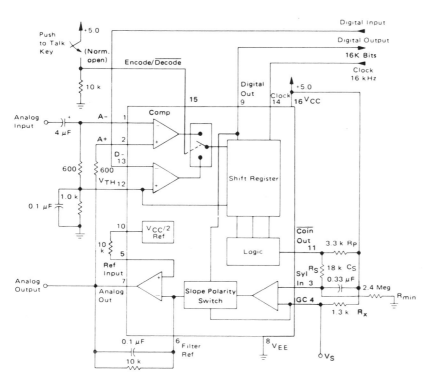

CVSD ENCODER FOR SECURE RADIO—Motorola MC3417 continuously variable slope delta modulator-demodulator IC is used as 16-kHz simplex voice coder-decoder for systems requiring digital communication of analog signals. Clock rate used depends on bandwidth required and can be 9.6 kHz or less for voice-only systems. Analog output uses single-pole integration network formed with 0.1 μF and 10K. Report covers circuit operation in detail for various applications.—"Continuously Variable Slope Delta Modulator/Demodulator," Motorola, Phoenix, AZ, 1978, DS 9488.

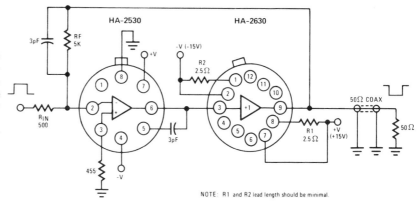

5-MHz COAX LINE DRIVER—Combination of Harris HA-2530 wideband inverting amplifier and HA-2630 unity-gain current amplifier provides 20-dB gain with extremely high slew rate and full power bandwidth even under heavy output loading conditions.—"Linear & Data Acquisition Products," Harris Semiconductor, Melbourne, FL, Vol. 1, 1977, p 2-47–2-50.

NOTE: R1 and R2 lead length should be minimal.

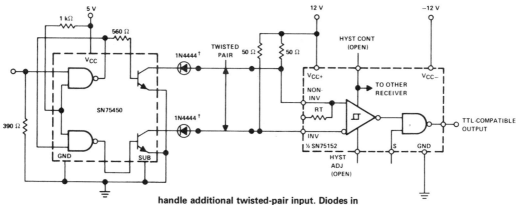

BALANCED-LINE TRANSMISSION—Transmits data at rates up to 0.5 MHz over twisted pair to Texas Instruments SN75152 dual-line receiver. Other section of receiver is identical and can handle additional twisted-pair input. Diodes in lines are required only for negative common-mode protection at driver outputs. System has high common-mode voltage capability. SN75450 is dual peripheral driver for high-current switching at high speeds.—"The Linear and Interface Circuits Data Book for Design Engineers," Texas Instruments, Dallas, TX, 1973, p 8-78.

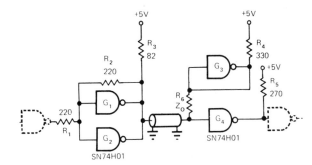

COAX DRIVER AND RECEIVER—Uses two TTL gates of SN74H01 package to form either driver or receiver for transmitting data over RG59 or RG174 coax at rates exceeding 10 megabits per second, with distance increasing from 400 meters at 10 Mb/s to over 1000 meters at 100 kb/s for RG59 and lesser distances for RG174. Can also be used for twisted-pair lines but at lower data rates. Bias gate G_3 exhibits low output impedance, for terminating channel load resistor R_6.—R. W. Stewart, Two TTL Gates Drive Very Long Coax Lines, *EDN Magazine*, Oct. 1, 1972, p 49.

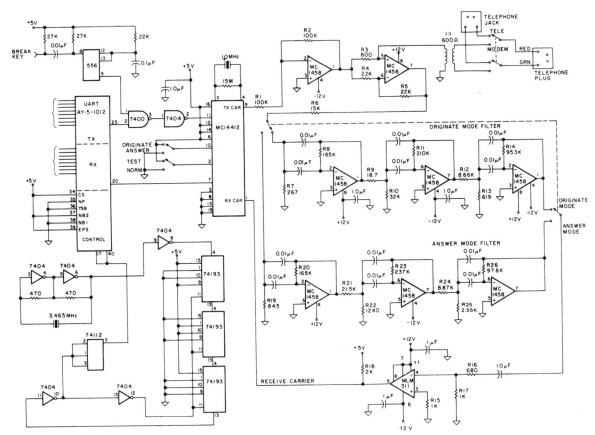

MODEM—Developed as part of TV terminal for microprocessor, to permit communication over telephone line with time-sharing computer system. Uses Motorola MC14412 modem chip for full-duplex FSK modulation having originate frequencies of 1270 Hz for mark and 1070 Hz for space, with answer frequencies of 2225 Hz for mark and 2025 Hz for space. AY-5-1012 UART serves as parallel interface to microprocessor. Article covers operation, construction, testing, connection to telephone lines, and use of modem.—R. Lange, Build the $35 Modem, *Kilobaud*, Nov. 1977, p 94–96.

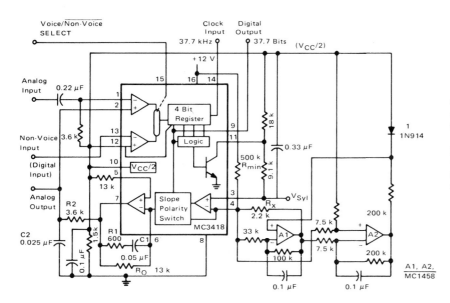

TELEPHONE-QUALITY CODER-DECODER—Uses Motorola MC3418 continuously variable slope delta modulator-demodulator IC to give over 50 dB of dynamic range for 1-kHz test at 37.7K bit rate. At this rate, 40 voice channels can be multiplexed on standard 1.544-megabit telephone carrier facility. IC includes active companding control and double integration for improved performance in encoding and decoding digital speech. Opamp types are not critical.—"Continuously Variable Slope Delta Modulator/Demodulator," Motorola, Phoenix, AZ, 1978, DS 9488.

REZEROING AMPLIFIER—Used where input signal has unknown and variable DC offset, as in telemetry applications. Rezero command line is enabled while ground reference signal is applied to input, making C1 charge to level proportional to DC offset of system. When rezero line is deactivated, amplifier becomes conventional inverter, subtracting system offset and giving true ground-referenced output. For 10-V full-scale system requiring 0.1% (10-mV) accuracy, amplifier needs rezeroing reference every 100 ms.—"Linear Applications, Vol. 1," National Semiconductor, Santa Clara, CA, 1973, AN-63, p 1–12.

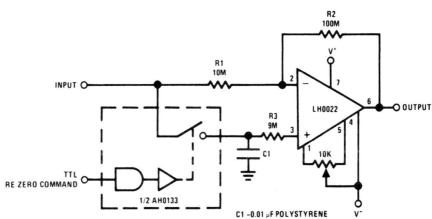

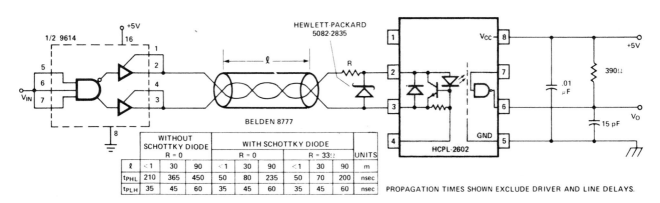

	WITHOUT SCHOTTKY DIODE			WITH SCHOTTKY DIODE						
	R = 0			R = 0			R = 33Ω			UNITS
ℓ	<1	30	90	<1	30	90	<1	30	90	m
t_{PHL}	210	365	450	50	80	235	50	70	200	nsec
t_{PLH}	35	45	60	35	45	60	35	45	60	nsec

PROPAGATION TIMES SHOWN EXCLUDE DRIVER AND LINE DELAYS.

POLARITY-REVERSING DRIVE—Half of 9614 polarity-reversing line driver feeds Hewlett-Packard HCPL-2602 optically coupled line receiver through shielded, twisted-pair, or coax cable. Data rate is improved considerably by using Schottky diode at input of receiver. Best data rates are achieved when t_{PHL} (propagation delay time to low output level) and t_{PLH} (propagation delay time to high output level) are closest to being equal.—"Optoelectronics Designer's Catalog 1977," Hewlett-Packard, Palo Alto, CA, 1977, p 158–159.

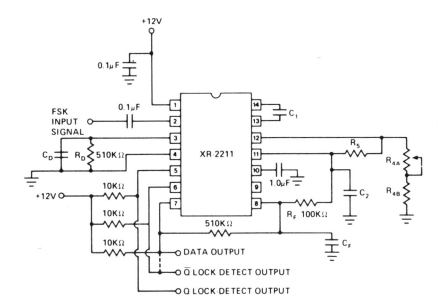

FSK DEMODULATOR WITH CARRIER DETECT— Exar XR-2211 FSK demodulator operating with PLL provides choice of outputs when carrier is present; pin 5 goes low and pin 6 goes high when carrier is detected. With pins 6 and 7 connected, output from these pins provides data when FSK is applied but is low when no carrier is present. Circuit performance is independent of input signal strength over range of 2 mV to 3 VRMS. Center frequency is $1/C_1R_4$ Hz, with values in farads and ohms. Choose frequency to fall midway between mark and space frequencies. Used in transmitting digital data over telecommunication links.—"Phase-Locked Loop Data Book," Exar Integrated Systems, Sunnyvale, CA, 1978, p 57–61.

FSK DETECTOR—Exar XR-S200 PLL IC is connected as modem suitable for Bell 103 or 202 data sets operating at data transmission rates up to 1800 bauds. Input frequency shift corresponding to data bit reverses polarity of DC output voltage of multiplier. DC level is changed to binary output pulse by gain block connected as voltage comparator.—"Phase-Locked Loop Data Book," Exar Integrated Systems, Sunnyvale, CA, 1978, p 9–16.

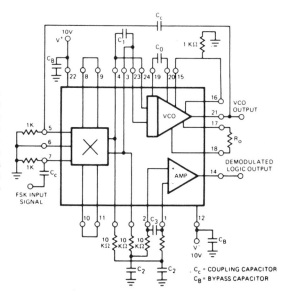

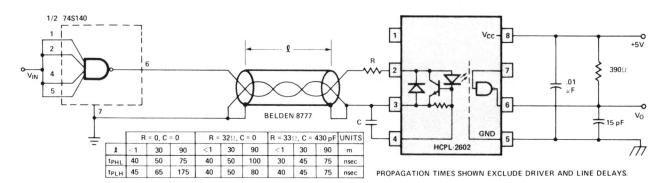

ℓ	R = 0, C = 0			R = 32Ω, C = 0			R = 33Ω, C = 430 pF			UNITS
ℓ	<1	30	90	<1	30	90	<1	30	90	m
t_{PHL}	40	50	75	40	50	100	30	45	75	nsec
t_{PLH}	45	65	175	40	50	80	40	45	75	nsec

PROPAGATION TIMES SHOWN EXCLUDE DRIVER AND LINE DELAYS.

POLARITY-NONREVERSING DRIVE—Hewlett-Packard HCPL-2602 optically coupled line receiver handles high data rates from shielded, twisted-pair, or coax cable fed by 74S140 line driver. Reflections due to active termination do not affect performance. Peaking capacitor C and series resistor R can be added to achieve highest possible data rate. C should be as large as possible without preventing regulator in line receiver from turning off during negative excursions of input signal. Highest data rates are achieved by equalizing t_{PHL} (propagation delay time to low output level) and t_{PLH} (propagation delay time to high output level).—"Optoelectronics Designer's Catalog 1977," Hewlett-Packard, Palo Alto, CA, 1977, p 158–159.

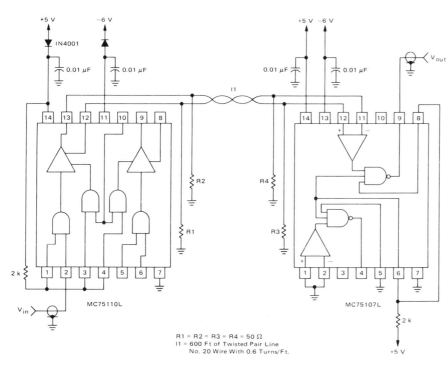

DIFFERENTIAL LINE DRIVER—Uses Motorola MC75110L line driver and MC75107L receiver with twisted-pair transmission line having attenuation of 1.6 dB per 100 feet at 10 MHz. Clock rate is 18.5 MHz. With push-pull driver shown, single pulse corresponds to transmission of 1 followed by series of 0s; one line is then at ground and the other at −300 mV. Arrangement is suitable for party-line or bus applications.— T. Hopkins, "Line Driver and Receiver Considerations," Motorola, Phoenix, AZ, 1978, AN-708A, p 11.

R1 = R2 = R3 = R4 = 50 Ω
I1 = 600 Ft of Twisted Pair Line
No. 20 Wire With 0.6 Turns/Ft.

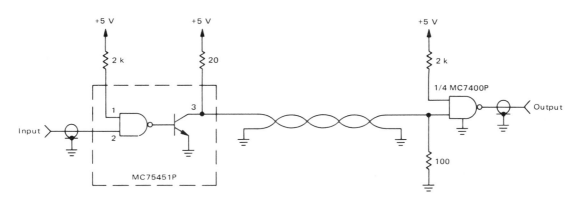

SINGLE-ENDED LINE DRIVER—Supplies 4.2-V input pulse to twisted-pair transmission line for point-to-point system. Requires only single +5 V supply.—T. Hopkins, "Line Driver and Receiver Considerations," Motorola, Phoenix, AZ, 1978, AN-708A, p 14.

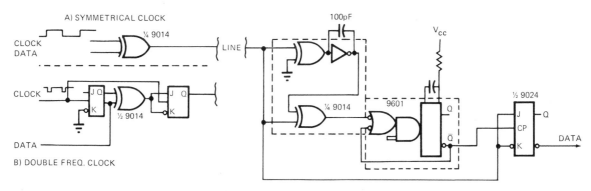

EXCLUSIVE-OR GATES—Use of retriggerable mono with EXCLUSIVE-OR gates simplifies design of both transmitter and receiver for handling binary phase-modulated digital data over single line. With 50% duty-cycle clock at transmitter, clock and data signals are applied to inputs of 9014 to generate output signal for line. At receiver, clock and data stream are regenerated by 9601 adjusted to 75% of data-bit time and connected in nonretriggerable mode. One EXCLUSIVE-OR gate and an EXCLUSIVE-NOR gate connected as inverting delay element will trigger 9601. System remains synchronized as long as pulse width of mono is between 50% and 100% of data-bit time.—P. Alfke, Exclusive-OR Gates Simplify Modem Designs, *EDN Magazine*, Sept. 15, 1972, p 43.

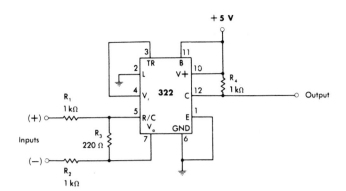

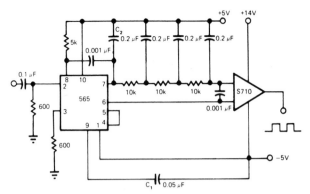

DIFFERENTIAL LINE RECEIVER—Responds to balanced-input drive signals fed to both comparator inputs of 322. Output is undisturbed even with up to 1 V of common-mode noise on input lines. TTL-compatible output is in phase with positive input. Overall delay is about 1 μs.—W. G. Jung, "IC Timer Cookbook," Howard W. Sams, Indianapolis, IN, 1977, p 153–155.

ANALOG PLL IN FSK DEMODULATOR—Developed for frequency-shift keying used in data transmission over wires, in which inputs vary carrier between two preset frequencies corresponding to low and high states of binary input signal. Circuit uses elaborate filter to separate modulated signal from carrier signal passed by PLL. 565 PLL provides reference for S710 comparator. Article gives design equations.—E. Murthi, Monolithic Phase-Locked Loops—Analogs Do All the Work of Digitals, and Much More, *EDN Magazine*, Sept. 5, 1977, p 59–64.

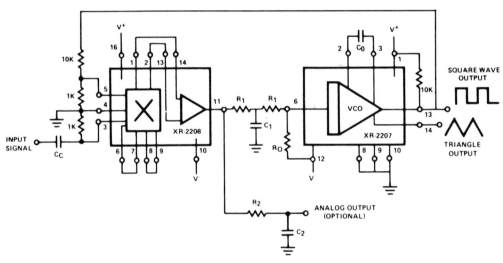

PLL FOR 0.01 Hz TO 100 kHz—Highly stable and precise phase-locked loop system using Exar XR-2207 VCO and XR-2208 operational multiplier is suitable for wide range of applications in data transmission and signal conditioning. Supply voltage range is ±6 V to ±13 V. For 10-kHz center frequency, R_O is 10K and C_O is 0.01 μF. R_1 and C_1, which determine tracking range and low-pass filter characteristics, are 45K and 0.032 μF.—"Phase-Locked Loop Data Book," Exar Integrated Systems, Sunnyvale, CA, 1978, p 62–64.

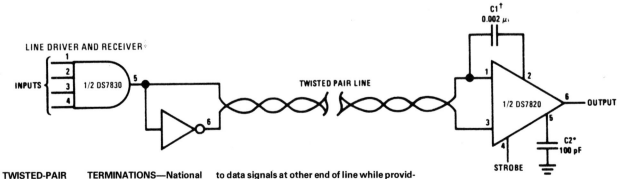

TWISTED-PAIR TERMINATIONS—National DS7830 line driver applies digital data to twisted-pair transmission line in high-noise environment, and DS7820 line receiver responds to data signals at other end of line while providing immunity to noise spikes. Exact value of C1 depends on line length. Supply voltage is 4.5 to 5 V for both receiver and driver. C2 is optional and controls response time.—"Interface Integrated Circuits," National Semiconductor, Santa Clara, CA, 1975, p 8-1–8-16.

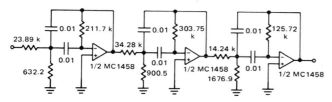

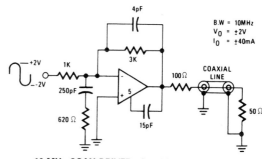

All capacitors are in μF.

BANDPASS ANSWER FILTER—Provides gain of 15 dB over bandwidth of 1020 to 1320 Hz for low-speed modem system using Motorola MC6860 IC. Attenuation is 35 dB at 2225 Hz, as required for answer-only modem system. Equations for values of filter components are given.—J. M. DeLaune, "Low-Speed Modem System Design Using the MC6860," Motorola, Phoenix, AZ, 1975, AN-747, p 10.

10-MHz COAX DRIVER—Provides high output current to coaxial line over bandwidth limited only by single-pole response of feedback components. Response is flat with no peaking and distortion is low. Uses Harris HA-2530/2535 wideband amplifier having high slew rate.— "Linear & Data Acquisition Products," Harris Semiconductor, Melbourne, FL, Vol. 1, 1977, p 7-54 (Application Note 516).

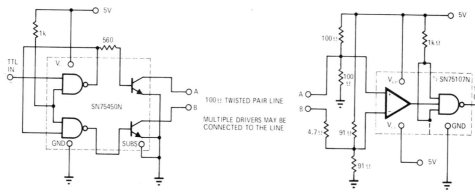

WIRED-OR TERMINALS—Arrangement permits connecting several IC line drivers in parallel for feeding single 100-ohm twisted-pair data line. With wired-OR transmitting capability, TTL output of receiver at right is logic 1 only if all paralleled drivers are transmitting logic 1. If any one or all of drivers transmit logic 0, output of receiver is logic 0.—D. Pippenger, Termination Is the Key to Wired-OR Capability, *EDN/EEE Magazine*, Dec. 15, 1971, p 17.

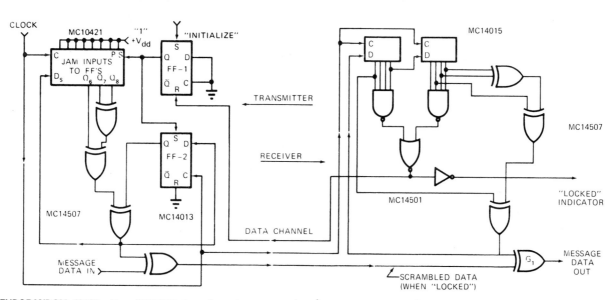

PSEUDORANDOM CMOS—Uses MC14021 8-bit shift register in conjunction with MC14507 EXCLUSIVE-OR gates to generate pseudorandom digital code. To develop code pattern, 1st, 6th, 7th, and 8th bits are sent through EXCLUSIVE-OR gates and fed back to shift-register input. Output can be used as random test signal or for protecting messages sent over public channels or stored in public files. Digital message is scrambled by mixing it with output of code generator in EXCLUSIVE-OR gate. Functionally identical 255-bit random generator is used at receiver to unscramble data. Decoding circuit must have access to sending clock and means for synchronizing so as to put both registers into all-1 state. Register in receiver goes through all its states within 255 clock pulses; when it reaches all-1 state, signal is fed back to sender for releasing FF-1 so scrambling can commence. Article traces operation in detail.—J. Halligan, Pseudo-Random Number Generator Uses CMOS Logic, *EDN Magazine*, Aug. 15, 1972, p 42–43.

SECTION 16
Fiber-Optic Circuits

Includes LED modulators and photodiode or phototransistor receivers for single- or multiple-fiber data links handling audio, data, and teleprinter signals. Circuits are also given for infrared receivers and transmitters, high-voltage isolator links, laser-diode modulator, Manchester-code demodulator, and fiber light-transmission checker.

DATA COUPLER WITH ISOLATION—Length of fiber or polystyrene rod determines amount of voltage isolation provided between digital or analog signal input and Fairchild FPT 100 photodetector driving Optical Electronics 9720 opamp having 100-mA output for driving cables, relays, or loudspeakers. LED can be Monsanto MV50 handling up to 200 mA. Output of opamp is zero for no light. Pulse-duration modulation should be used for transmission of analog data.—"High Voltage Optically Isolated Data Coupler," Optical Electronics, Tucson, AZ, Application Tip 10266.

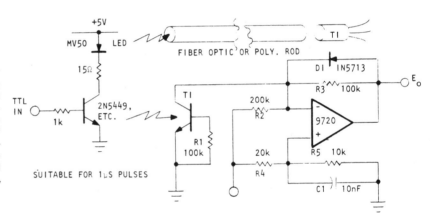

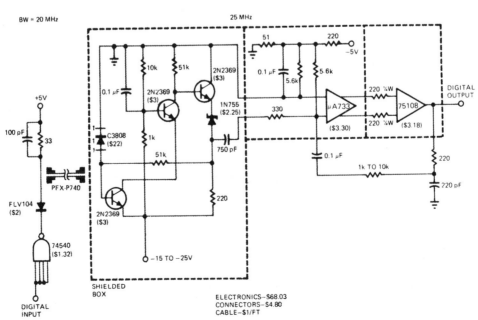

ELECTRONICS—$68.03
CONNECTORS—$4.80
CABLE—$1/FT

24-MEGABIT DATA LINK—High data-rate capability for square-wave pulses is achieved by increasing complexity of receiver feeding digital output to microprocessor from remote teleprinter. Preamp design compensates for noise over limited frequency range, giving uniform S/N ratio to about 20 MHz. With demonstration setup, visible-spectrum LED and photodetector shown performed acceptably with 40-foot cable.—O. E. Marvel and J. C. Freeborn, A Little Hands-On Experience Illuminates Fiber-Optic Links, *EDN Magazine*, Nov. 5, 1977, p 71–75.

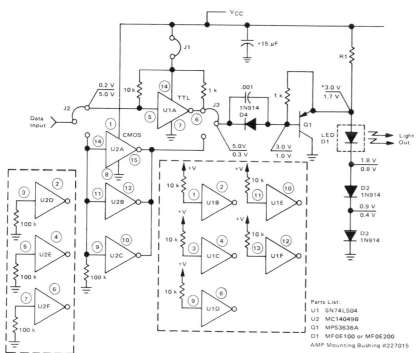

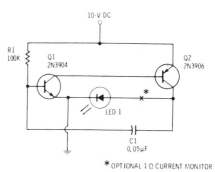

HIGH-CURRENT INFRARED LED PULSER—Circuit operates as regenerative amplifier for delivering 10-μs pulses with amplitude of 1.1 A and repetition rate of 1.4 kHz to infrared LED. Suitable for infrared beacon in fiber-optic communication and optical radar applications. Drain is 100 mA from 10-V supply. Use gallium arsenide LED such as SSL-55C or TIL32 for high-output infrared transmitter. Q2 can be changed to germanium transistor such as 2N1305 to give peak current of 2 A at pulse width of 15 μs and 750-Hz repetition rate.—F. M. Mims, "Electronic Circuitbook 5: LED Projects," Howard W. Sams, Indianapolis, IN, 1976, p 33–35.

FIBER-OPTIC TRANSMITTER—Will handle NRZ data rates to 10 megabits or square waves to 5 MHz. Input is TTL- or CMOS-compatible depending on circuit selected. Transmitter draws only 150 mA from 5-V supply for TTL or from 5-15 V supply for CMOS. Choose R1 to give LED drive current for proper operation of system. For TTL operation, jumpers J1, J2, and J3 are connected as shown. For CMOS operation, remove J1 and transfer J2 and J3 to alternate positions for connecting to U2. Choice of LED depends on system length and desired data rate. Power supply can be HP6218A or equivalent. DC voltages shown are for TTL interface, with upper value for LED on at 50 mA and lower value for LED off.—"Basic Experimental Fiber Optic Systems," Motorola, Phoenix, AZ, 1978.

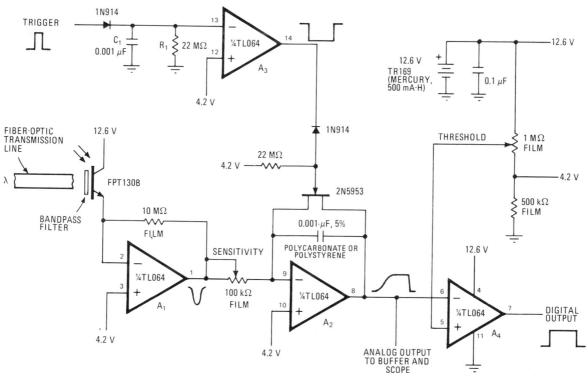

TL064: TEXAS INSTRUMENTS QUAD BI-FET AMPLIFIER OR EQUIVALENT

LIGHT TRANSMISSION CHECKER—Phototransistor and quad opamp serve as total-energy detector of pulsed-light signals propagated through fiber-optic cable of communication system. Can be used for checking and comparing condition of long fibers if light intensity of source is held constant. Will also detect changes in light intensity and changes in pulse width. Circuit gives linear response to light levels from 100 to 10,000 ergs/cm^2 if minimum pulse width is at least 10 μs. A$_2$ acts as RC integrator, giving voltage proportional to total light energy received. A$_4$ provides comparison to fixed level.—E. W. Rummel, Low-Level-Light Detector Checks Optical Cables Fast, *Electronics*, April 27, 1978, p 148 and 150.

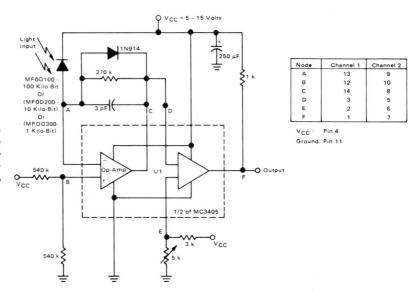

Node	Channel 1	Channel 2
A	13	9
B	12	10
C	14	8
D	3	5
E	2	6
F	1	7

V_{CC}: Pin 4
Ground: Pin 11

1/10/100-KILOBIT FIBER-OPTIC RECEIVER—Choice of input device determines operating speed of receiver. MC3405 contains two opamps and two comparators, permitting use as two-channel receiver. Table gives pin connections for each channel.—"Basic Experimental Fiber Optic Systems," Motorola, Phoenix, AZ, 1978.

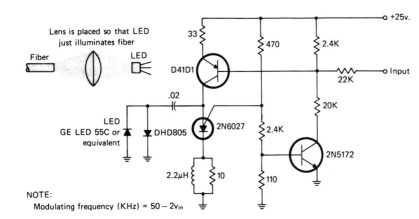

NOTE:
Modulating frequency (KHz) = 50 − 2v_{in}

50-kHz FM OPTICAL TRANSMITTER—Uses pulse-rate modulation system with center frequency of 50 kHz. Audio fed into transmitter varies pulse rate, for driving LED coupled to optical fiber. Phototransistor at other end of fiber receives and demodulates light signal for reconstruction of audio.—I. Math, Math's Notes, *CQ*, July 1977, p 67–68 and 90.

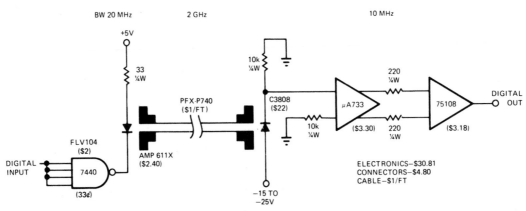

ELECTRONICS—$30.81
CONNECTORS—$4.80
CABLE—$1/FT

10-MEGABIT LINK—Transmitter and receiver for fiber-optic data link between teleprinter and microprocessor utilize wide bandwidth of cable for transmitting data at 10-megabit rate. Receiver input requires C3808 PIN photodiode.—O. E. Marvel and J. C. Freeborn, A Little Hands-On Experience Illuminates Fiber-Optic Links, *EDN Magazine*, Nov. 5, 1977, p 71–75.

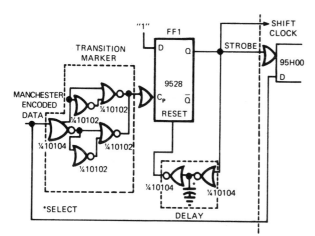

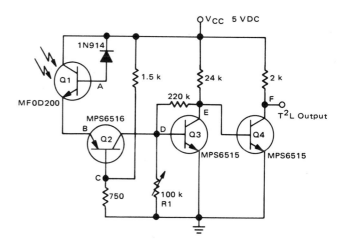

1-GHz MANCHESTER DECODER—Use of ECL flip-flop with toggle rates above 1 GHz makes decoding of bit rates approaching gigabit speeds feasible. Article gives step-by-step design procedure for 48-Mb telemetry application using PCM over single optical-fiber cable.—B. R. Jarrett, Could You Design a High-Speed Manchester-Code Demodulator?, *EDN Magazine*, Aug. 20, 1974, p 75–80.

20-KILOBIT FIBER-OPTIC RECEIVER—Phototransistor driving three-transistor amplifier provides TTL output for data rates up to 20 kilobits.—"Basic Experimental Fiber Optic Systems," Motorola, Phoenix, AZ, 1978.

IR DETECTOR—Photodiode transforms light-signal output of fiber-optic cable to electric signal. Spectral response of detector closely matches that of IR-emitting diode at other end of cable, for maximum system efficiency. Rise and fall times of detector can be less than 35 ns when properly biased and loaded by receiver circuit. Developed by Augat, Inc., Attleboro, MA, as part of fiber-optic evaluation kit for TTL applications.—Fiber-Optic Kit Allows Engineering Evaluation of Complete Interconnection System, *Computer Design*, Nov. 1977, p 27 and 30.

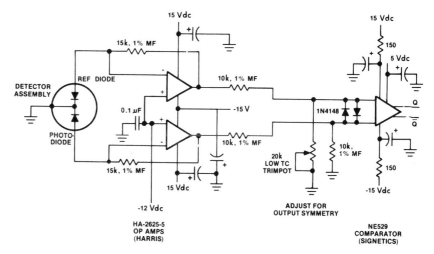

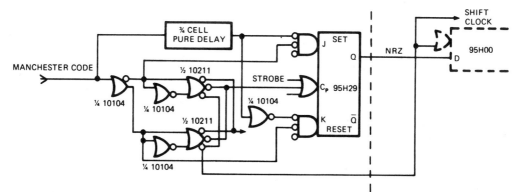

MANCHESTER-CODE DEMODULATOR—Digital approach using ECL provides maximum speed, is self-synchronizing for alternate bit-pairs, and has minimum complexity. Developed for optically coupled 25-channel PCM telemetry system used over single optical-fiber channel. Undesired transitions in input data are masked by creating strobe. Approach recognizes distinction between identical sequences that would give some output except for time-of-occurrence restriction. Article gives step-by-step design procedure, waveforms, and excitation table.—B. R. Jarrett, Could You Design a High-Speed Manchester-Code Demodulator?, *EDN Magazine*, Aug. 20, 1974, p 75–80.

1-MHz LED PULSE MODULATION—Circuit provides required low driving-point impedance for fast turn-on of gallium arsenide phosphide LED used as source for high-speed pulse modulation of fiber-optic or other light beam. Q₁ supplies DC level and modulation information to emitter-follower output stage Q₃. Output current is sensed and limited to about 30 mA by Q₂. Turn-on time for full brightness is 12 ns.—G. Schmidt, LED Modulator, *EDN/EEE Magazine*, June 15, 1971, p 57.

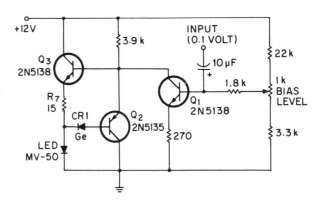

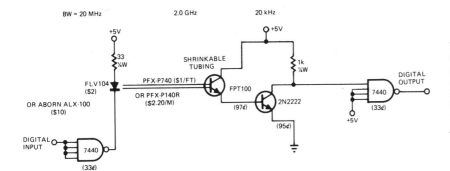

TTY LINK FOR MICROPROCESSOR—Demonstration circuit illustrates use of fiber-optic cable with low-cost components for relatively narrow-band application, to provide feed from remote teleprinter to microprocessor.—O. E. Marvel and J. C. Freeborn, A Little Hands-On Experience Illuminates Fiber-Optic Links, *EDN Magazine*, Nov. 5, 1977, p 71–75.

SECTION 17
Filter Circuits—Passive

Includes low-pass, high-pass, and bandpass AF filters for improving reception of voice, CW, SSB, and RTTY signals, along with higher-frequency circuits for suppressing broadcast-band interference in communication receivers and minimizing other types of interference.

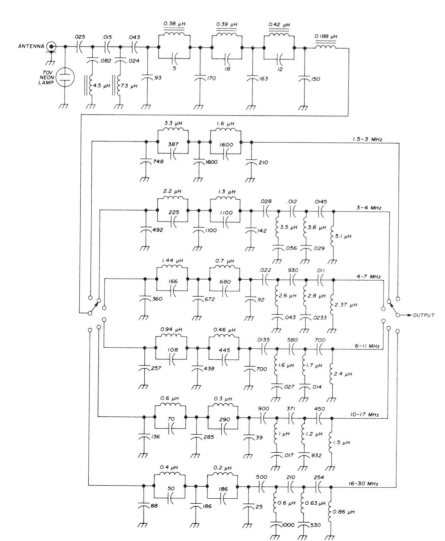

ELLIPTIC HIGH-PASS/LOW-PASS—Covers 1.45 to 32 MHz in six steps, for use at front end of high-frequency communication receiver to suppress unwanted broadcast signals. Low-pass filter section, acting with one of six following low-pass sections, gives over 90-dB image suppression. Special Bessel-Cauer elliptic filter having Chebyshev response in passband provides required 50-ohm impedance matching so filters can be cascaded.—U. L. Rohde, Optimum Design for High-Frequency Communications Receivers, *Ham Radio*, Oct. 1976, p 10–25.

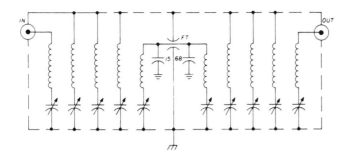

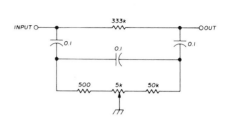

50.5-MHz BANDPASS—Provides 60% bandwidth with only 4-dB insertion loss. Each coil is about 2.2 μH, and trimmer capacitors are 1.5–7 pF. Sweep signal generator and 5-in CRO are essential for alignment.—P. H. Sellers, 50-MHz Bandpass Filter, *Ham Radio,* Aug. 1976, p 70–71.

60-Hz TUNABLE NOTCH—Can be used to minimize hum pickup from AC line. Circuit tunes from 40 to 120 Hz with single pot. Article gives design equations. With unmeasured ceramic disk capacitors and 5% resistors, notch depth at 60 Hz was 44.5 dB. By selecting capacitors with equal values and replacing 333K with 500K trimpot, careful adjustment increases notch depth to 57 dB.—C. Hall, Tunable RC Notch Filter, *Ham Radio,* Sept. 1975, p 16–20.

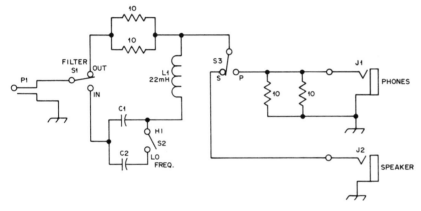

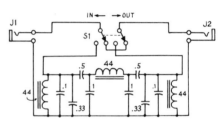

CW FILTER FOR INTERFERENCE—Audio bandpass filter, designed for connection between loudspeaker jack of receiver and external loudspeaker or phone, has half-power bandwidth of about 70 Hz but rolls off gradually without causing ringing. Series LC combination, connected in hot line to loudspeaker, looks like 5-ohm resistance at resonance, cutting signal amplitude about in half. At lower frequencies filter looks like large capacitive reactance and at higher frequencies it resembles large inductive reactance, both causing high attenuation. Filter thus discriminates against all except switch-selected resonant frequency, either 760 or 1070 Hz. Choose best frequency for particular receiving situation by trial.—F. Noble, A Passive CW Filter to Improve Selectivity, *QST,* Nov. 1977, p 34–35.

BANDPASS FOR CW—Provides bandwidth of about 400 Hz (3 dB down) centered on 875 Hz, for improving reception of CW signals with amateur receiver. Uses three 44-mH toroids.—D. C. Rife, Low-Loss Passive Bandpass CW Filters, *QST,* Sept. 1971, p 42–44.

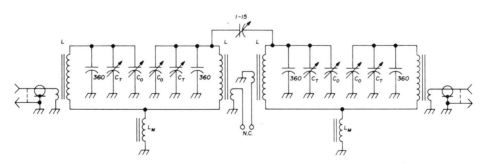

| L | 33 turns no. 20 enamelled on Amidon T-106-2 toroid cores (approximately 14 μH). Links are each 2 turns |
| L_m | 6 turns no. 20 enamelled on Amidon T-30-2 toroid core |

| C_o | 4-section air variable, 10-160 pF per section |
| C_T | 35-pF trimmer capacitors |

160-METER BANDPASS—Four-resonator filter is tunable from 1.8 to 2 MHz and has insertion loss of 5 dB. 3-dB bandwidth is 30 kHz, and 6–60 dB shape factor is 4.78. Stopband attenuation is over 120 dB. Key to high performance is use of high-Q toroid cores. Article covers theory, construction, and adjustment.—W. Hayward, Bandpass Filters for Receiver Preselectors, *Ham Radio,* Feb. 1975, p 18–27.

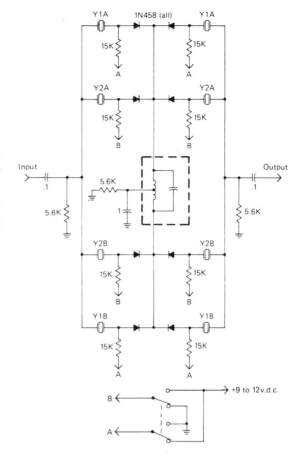

DIODE-SWITCHED FOUR-CRYSTAL IF FILTER— Application of 9–12 VDC to control points A or B gives choice of two different selectivities for IF amplifier in amateur communication receiver. For 500-Hz bandwidth at 455 kHz, frequencies of crystals in use should be 300 Hz apart for CW, 1.8 kHz apart for 2.7-kHz SSB bandwidth, and 1.25 kHz apart for 2.1-kHz SSB bandwidth. Article gives design graphs.—J. J. Schultz, Economical Diode-Switched Crystal Filters, *CQ,* July 1978, p 33–35 and 91.

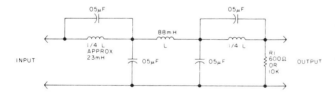

2125-Hz LOW-PASS—Used with AFSK keyer to convert 2125-Hz square wave to sine wave by removing third and fifth harmonics. All three coils are toroids, with its two windings in series for 88 mH and in parallel for 23 mH.—L. J. Fox, Dodge That Hurricane!, *73 Magazine,* Jan. 1978, p 62–69.

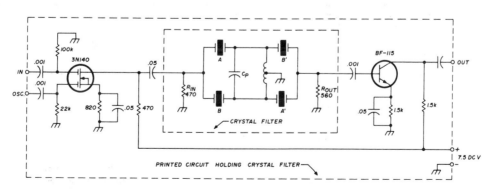

FOUR-CRYSTAL FILTER—Uses two matched sets of crystals, with each pair having maximum frequency difference of 25 Hz. Transistors serve as input and output isolating stages. Each matched pair, such as A-A', should be from same manufacturer and have same nominal parallel capacitance for circuit, same activity, and same resonant frequency within 25 Hz. Article gives detailed instructions for grinding crystal to increase resonant frequency when necessary for matching. Use frequency counter for checking frequency. Values given in circuit are for 5.645-MHz crystal filter with −6 dB bandpass of 1.82 kHz and insertion loss of about 5 dB. Crystals used are 5.644410 MHz and 5.644416 MHz for A and A', and 5.645627 MHz and 5.645641 MHz for B and B'. Coil has 7 + 7 turns No. 28 enamel bifilar wound on 10.7-MHz IF transformer having 2.4-mm slug diameter. C_p is 39 to 47 pF.—J. Perolo, Practical Considerations in Crystal-Filter Design, *Ham Radio,* Nov. 1976, p 34–38.

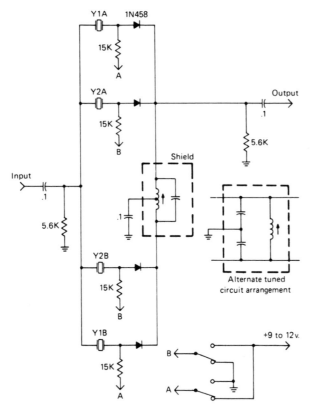

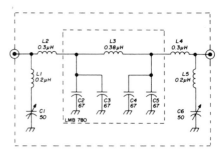

C1,C6	50 pF APC or MAPC variable
C2,C3	67 pF ±5%, 7500 working volts dc (Centralab type 850S ceramic capacitor, 6 for $1.00 from John Meshna, P.O. Box 62, E. Lynn, Massachusetts 01904)
C4,C5	
L1,L5	0.2 μH, 3 turns no. 16 or no. 14 enamelled, ½ inch (13mm) ID, spaced 1/8 inch (3mm) per turn
L2,L4	0.3 μH, 5½ turns no. 14 enamelled, ½ inch (13mm) ID, spaced 1/8 inch (3mm) per turn
L3	0.38 μH, 7 turns no. 14 enamelled, ½ inch (13mm) ID, spaced 1/8 inch (3mm) per turn

DIODE-SWITCHED CRYSTALS—1N458 diodes switch crystals in pairs to provide two different degrees of selectivity for 455-kHz IF filter. For 500-Hz bandwidth in amateur communication receiver, spacing between crystal frequencies should be 300 Hz, which is obtained with 455.150 kHz for Y1A and 454.850 kHz for Y1B. Provides adequate CW selectivity for transceiver having good SSB filter.—J. J. Schultz, Economical Diode-Switched Crystal Filters, *CQ*, July 1978, p 33–35 and 91.

LOW-PASS WITH 42.5-MHz CUTOFF—Designed for insertion in antenna coax of amateur radio station up to 1 kW, to cure TVI problems. Provides 60-dB attenuation on channel 2. Filter uses m-derived terminating half-sections at each end, with two constant-K midsections. End sections are tuned either to channel 2 (55 MHz) or channel 3 (61 MHz). Article covers construction and tune-up.—N. Johnson, High-Frequency Lowpass Filter, *Ham Radio*, March 1975, p 24–27.

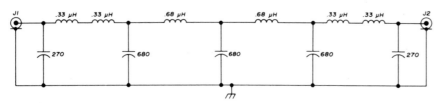

2.955-MHz HIGH-PASS—Used in offset frequency-measuring system for amateur-band signals. Nine-section Chebyshev high-pass filter with 1-dB passband ripple attenuates undesired 2.045–2.245 MHz image 16 dB while selecting desired 2.955–3.155 MHz signal. Filter has sharper cutoff characteristic, for given number of sections, than Butterworth or image parameter designs.—J. Walker, Accurate Frequency Measurement of Received Signals, *Ham Radio*, Oct. 1973, p 38–55.

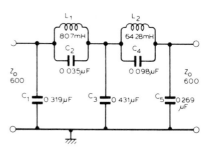

L_1	253 turns 28 swg 35mm VINKOR LA1211
L_1-C_2	Tune to 3kHz
L_2	226 turns 28 swg 35mm VINKOR LA1211
L_2-C_4	Tune to 2kHz

1-kHz FIFTH-ORDER LOW-PASS—Used with 1-kHz signal generator to remove unwanted harmonics, leaving pure sine wave as required for measuring distortion in modern audio amplifiers. Attenuation peaks are carefully positioned to coincide with second and third harmonics, giving 65-dB attenuation of these harmonics and at least 50-dB attenuation of higher harmonics.—J. A. Hardcastle, 1 kHz Source Cleaning Filter, *Wireless World*, Oct. 1978, p 59.

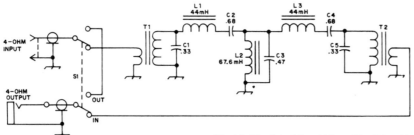

VOICE BANDPASS—Used between 8-ohm output of communication receiver and 8-ohm loudspeaker or low-impedance phones, to suppress Continuous Random Unwanted Disturbances on voice transmissions. Passband is 355 to 2530 Hz at 3-dB points. L1 and L3 are 44-mH toroids. L2 is 88-mH toroid with 94 turns removed. T1 and T2 are 88-mH toroids with 100 turns No. 28 enamel wound over original winding of each for primary.—R. M. Myers, The SSB Crud-O-Ject, *QST*, May 1974, p 23–25 and 56.

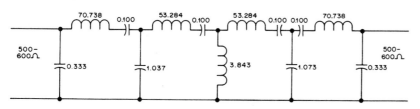

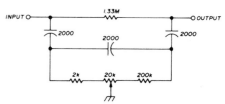

225-Hz BANDPASS RTTY—Used ahead of limiter in 170-Hz-shift RTTY receiving converter. Chebyshev mesh configuration with 0.1-dB ripple uses inductor to ground for sharpening lower skirt, with capacitive coupling for sharpening upper skirt, to give good symmetry for response curve. Capacitors should be high-Q types, well matched. Take turns off inductors as required to move passband higher if initially low in frequency. Insertion loss is 6.6 dB and 3-dB bandwidth is 225 Hz, which makes mark and space tones only 1.5 dB down.—A. J. Klappenberger, A High-Performance RTTY Band-Pass Filter, *QST*, Jan. 1978, p 33.

693–2079 Hz TUNABLE NOTCH—Requires only one tuning pot to cover entire frequency range. Developed for use in tunable narrow-band audio amplifier. Article gives design equations. Depth of notch is greater than 50 dB. Doubling capacitor values changes tuning range to 355–1028 Hz, while cutting values in half gives range of 1340–4110 Hz.—C. Hall, Tunable RC Notch Filter, *Ham Radio*, Sept. 1975, p 16–20.

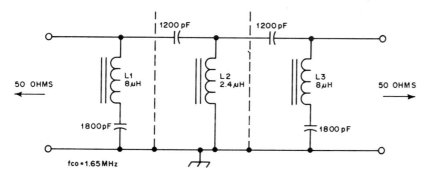

AM BROADCAST REJECTION—Seven-element m-derived high-pass filter provides 30-dB rejection at AM broadcast-band frequencies while passing signals in 160-meter band. Midsection m-derived branch of circuit was eliminated to simplify construction, but can be added and tuned to particular broadcast station that presents difficult interference problem. L1 and L3 are 40 turns No. 30 enamel wound on T50-2 powdered-iron toroid. L2 has 22 turns No. 30 on T50-2 core.—D. DeMaw, Low-Noise Receiving Antennas, *QST*, Dec. 1977, p 36–39.

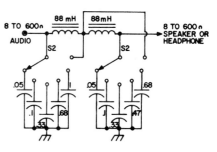

SWITCHABLE AF FILTER—Provides wide range of switch-selected capacitor values for varying cutoff frequencies, to permit use of filter for either phone or CW reception. On CW, circuit improves reception by eliminating higher frequencies that are largely interference.—J. J. Schultz, The Quiet Maker, *73 Magazine*, March 1974, p 81–84.

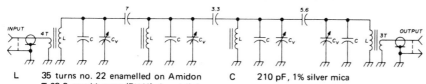

L — 35 turns no. 22 enamelled on Amidon T-68-2 toroid cores (7 μH). Input link is 4 turns, output link is 3 turns

C — 210 pF, 1% silver mica

C_V — 60-pF mica compression trimmers

80-METER BANDPASS—Four-resonator filter for use in 80-meter amateur band has 100-kHz bandwidth, 4.4-dB insertion loss, and 6–60 dB shape factor of 5.16. Filter was designed and aligned at 3.75 MHz; realignment at 3.6 and 3.9 MHz yielded similar results. Article covers theory, construction, and adjustment.—W. Hayward, Bandpass Filters for Receiver Preselectors, *Ham Radio*, Feb. 1975, p 18–27.

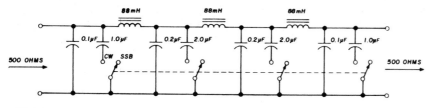

LOW-PASS PI-SECTION AF—Four-pole double-throw switch gives choice of 650-Hz cutoff for CW or 2000 Hz for SSB. Filter capacitors are matched. Response decreases continuously beyond cutoff frequency, with no loss of attenuation.—E. Noll, Circuits and Techniques, *Ham Radio*, April 1976, p 40–43.

Filter Circuits—Active

Includes low-pass, high-pass, bandpass, notch, state-variable (2 to 4 functions), tracking, and equalizing filters covering from 1 Hz to limits of audio spectrum, along with gyrator, Q multiplier and variable-Q circuits, crossover networks for loudspeakers, and RF circuits providing frequency emphasis.

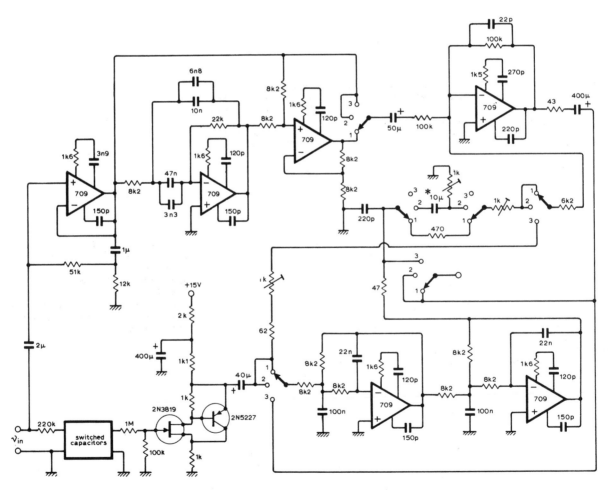

✱ non-polarized polycarbonate

TRACKING LINE-FREQUENCY FILTER—Improvements in commutating RC network filter extend dynamic range without sacrificing signal bandwidth, for reducing interference at fundamental of power-line frequency and harmonics up to fifth. Although values in circuit are for British 50-Hz mains frequency, circuit can readily be adapted for 60-Hz rejection. Operation involves commutating 16 capacitors electronically at 16 times line frequency. Article gives one method of doing this, by driving two 8-way multiplexers alternately. Each multiplexer has eight MOSFETs, each switched on in turn by consecutive input clock pulses. Circuit details, design equations, and performance graphs are given. Three-position switch gives choice of filter characteristics.—K. F. Knott and L. Unsworth, Mains Rejection Tracking Filter, *Wireless World,* Oct. 1974, p 375–379.

10-kHz VARIABLE-Q—Second-order state-variable filter having center frequency of 10 kHz uses all four sections of OP-11FY quad opamp. Center frequency can be tuned by varying 1.6K feedback resistors or by changing 0.01-μF feedback capacitors. Value of feedback resistor for opamp D determines Q of filter, for adjusting circuit bandwidth or damping. For higher-frequency operation, use high-speed opamps such as OP-15 or OP-16.—D. Van Dalsen, Need an Active Filter? Try These Design Aids, *EDN Magazine,* Nov. 5, 1978, p 105–110.

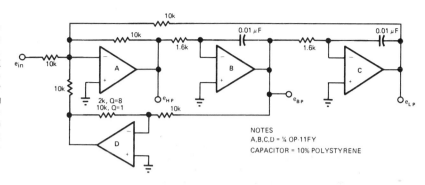

NOTES
A,B,C,D = ¼ OP-11FY
CAPACITOR = 10% POLYSTYRENE

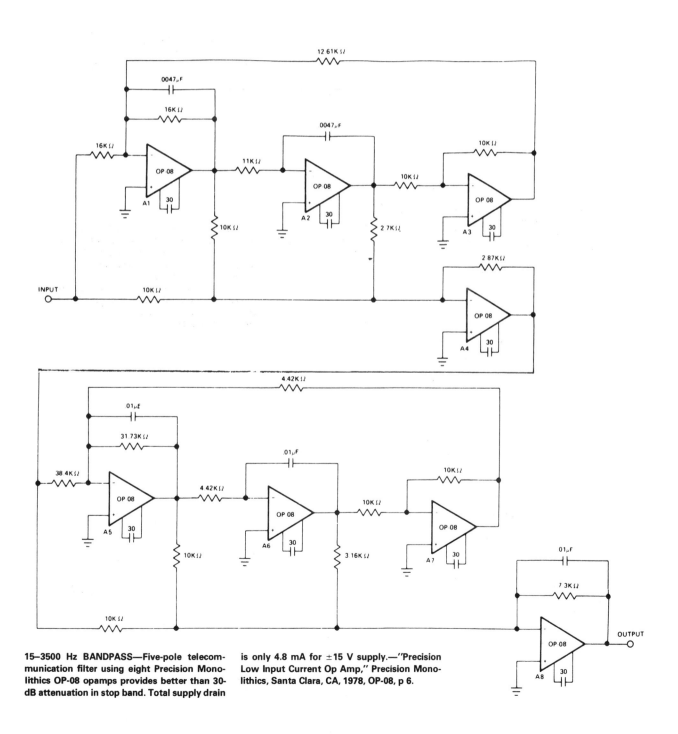

15–3500 Hz BANDPASS—Five-pole telecommunication filter using eight Precision Monolithics OP-08 opamps provides better than 30-dB attenuation in stop band. Total supply drain is only 4.8 mA for ±15 V supply.—"Precision Low Input Current Op Amp," Precision Monolithics, Santa Clara, CA, 1978, OP-08, p 6.

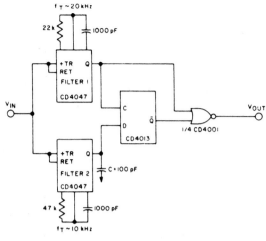

10–20 kHz BANDPASS—Two CD4047A low-pass filters, one connected for 10-kHz cutoff and other for 20-kHz cutoff, drive CD4013A flip-flop. If output of filter 2 is delayed by C, flip-flop clocks high only when input pulse frequency exceeds 10-kHz cutoff of filter 2. Waveforms show performance when input signal is swept through passband.—"CQS/MOS Integrated Circuits," RCA Solid State Division, Somerville, NJ, 1977, p 619.

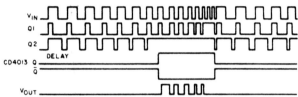

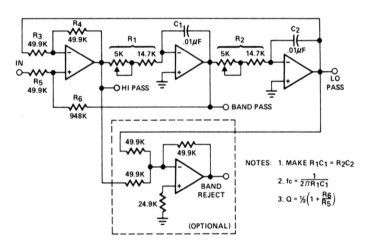

1-kHz STATE-VARIABLE WITH Q OF 10—Use of all four sections of Harris HA-4602/4605 quad opamp gives four types of 1-kHz second-order filtering simultaneously. Pot adjustments permit matching of various RC products allowing for noninteractive adjustment of Q and center frequency.—"Linear & Data Acquisition Products," Harris Semiconductor, Melbourne, FL, Vol. 1, 1977, p 2-84.

NOTES: 1. MAKE $R_1C_1 = R_2C_2$

2. $f_c = \dfrac{1}{2\pi R_1 C_1}$

3. $Q = \frac{1}{2}\left(1 + \dfrac{R_6}{R_5}\right)$

5-kHz SERIES-SWITCHED BANDPASS—N-path filter having N of 4, Q of 500, and voltage gain of 2 uses DG509 four-channel CMOS differential multiplexer having necessary pairs of analog switches and decode logic. Dual D flip-flop generates 2-bit binary sequence from 20-kHz clock signal. Bandwidth is about 10 Hz for 3 dB down, centered on 5 kHz.—"Analog Switches and Their Applications," Siliconix, Santa Clara, CA, 1976, p 5-15–5-17.

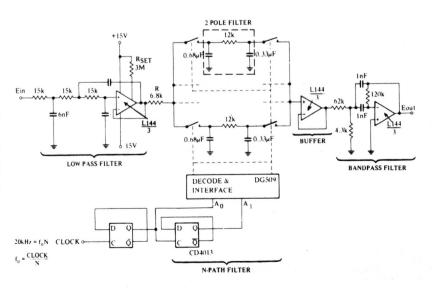

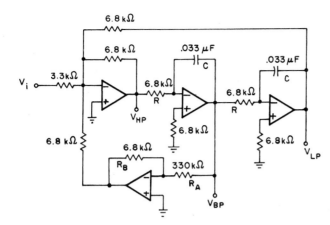

700-Hz STATE-VARIABLE—Provides voltage gain at center frequency of 100 (40 dB) and Q of 50. Used when simultaneous low-pass, high-pass, and bandpass output responses are required. Cutoff frequency of low-pass and high-pass responses is equal to center frequency of bandpass response. Opamps can be 741. Based on use of 5% resistors.—H. M. Berlin, "Design of Active Filters, with Experiments," Howard W. Sams, Indianapolis, IN, 1977, p 184–187.

1.5-kHz NOTCH—Unity-gain state-variable filter consists of low-pass and high-pass sections combined with two-input summing amplifier to give notch response for suppression of 1.5-kHz signals. Opamps can be 741.—H. M. Berlin, "Design of Active Filters, with Experiments," Howard W. Sams, Indianapolis, IN, 1977, p 186–189.

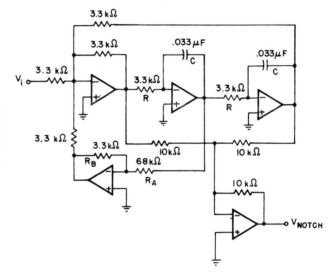

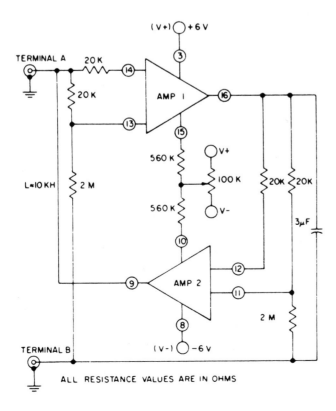

10-kH GYRATOR—Active filter circuit uses two sections of CA3060 three-opamp array as gyrator that makes 3-μF capacitor function as floating 10-kilohenry inductor across terminals A and B. Q of inductor is 13. 100K pot tunes inductor by changing gyration resistance.—"Linear Integrated Circuits and MOS/FET's," RCA Solid State Division, Somerville, NJ, 1977, p 152.

60-Hz NOTCH FILTER—Design is based on passband gain of 3 and Q of 6. Resistors can be 5%. Opamps can be 741. Notch response is obtained by subtracting output signal of bandpass filter from its input signal with R_6.—H. M. Berlin, "Design of Active Filters, with Experiments," Howard W. Sams, Indianapolis, IN, 1977, p 155.

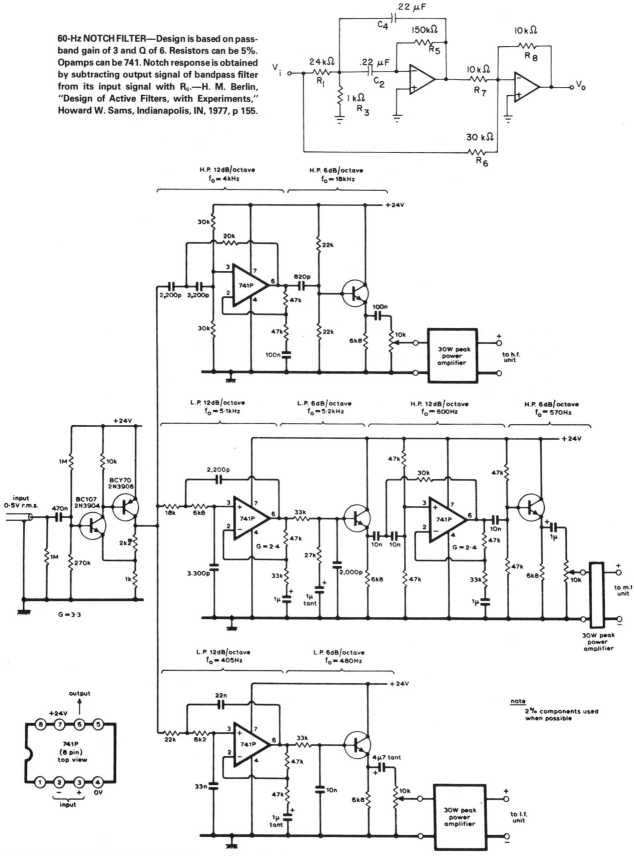

THREE-LOUDSPEAKER CROSSOVERS—Active filter network splits AF input into three frequency bands each feeding separate 30-W power amplifier. Design allows adjustment of any part of frequency characteristic to any desired level and gives choice of slopes in any part of frequency band. Article gives design equations and construction details. NPN transistors can be BC107 or 2N3904; PNP transistors can be BCY70, BCY71, BCY72, or 2N3906. Article also gives circuit of suitable 30-W amplifier.—D. C. Read, Active Filter Crossover Networks, *Wireless World*, Dec. 1973, p 574–576.

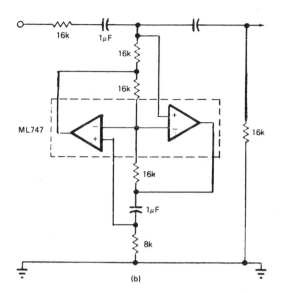

10-Hz HIGH-PASS—Equiterminated Butterworth high-pass ladder filter has corner frequency of 10 Hz and output impedance level of 16K. Opamps are matched pair in single ML747 package. Article covers design procedure based on use of generalized impedance converters and gives frequency response curve.—L. T. Burton and D. Treleaven, Active Filter Design Using Generalized Impedance Converters, *EDN Magazine,* Feb. 5, 1973, p 68–75.

10-kHz VOLTAGE-TUNED—High-Q circuit using Optical Electronics 9831 opamp has sharp resonance, as required for analysis of spectrum of incoming signal. Reverse-biased silicon junctions serve as voltage variable capacitors for sweeping center frequency over 3:1 range. Values shown for three resistors in twin-T network give center frequency of 10 kHz.—"Voltage Tuned High-Q Filter," Optical Electronics, Tucson, AZ, Application Tip 10207.

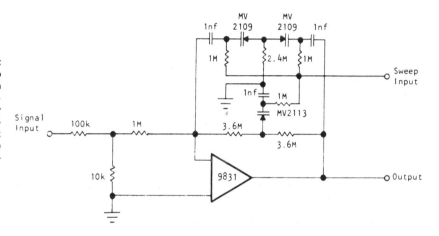

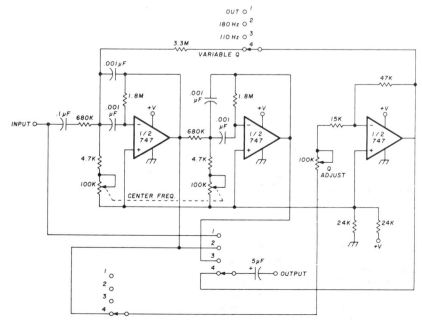

ACTIVE CW FILTER—Modifications made on MFJ Enterprises CWF-2 active audio filter permit maximum flexibility. Circuit provides fixed bandwidth of 180 or 110 Hz centered on 750 Hz, or optional variable bandwidth for which center frequency can be adjusted in range of 280 to 1590 Hz.—H. M. Berlin, Increased Flexibility for the MFJ Enterprises CW Filters, *Ham Radio,* Dec. 1976, p 58–60.

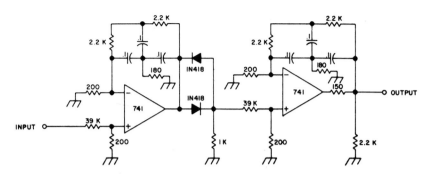

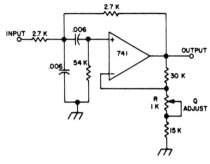

TWO-STAGE CW—Uses diode threshold detector between stages to prevent weak undesired signals from passing through until CW signal of desired frequency is present, so as to provide quiet tuning between signals. Bandwidth of filter is sharp (16 Hz), and keyed waveform is good. Gain is near unity, and frequency and Q are both fixed.—A. F. Stahler, An Experimental Comparison of CW Audio Filters, *73 Magazine*, July 1973, p 65–70.

VARIABLE Q FOR CW—Fixed-frequency active filter gives slowly rising and falling keyed waveform with good slope considering narrowness of bandwidth, which is 75 Hz at 3 dB down. Adjusting Q with 1K pot changes bandwidth.—A. F. Stahler, An Experimental Comparison of CW Audio Filters, *73 Magazine*, July 1973, p 65–70.

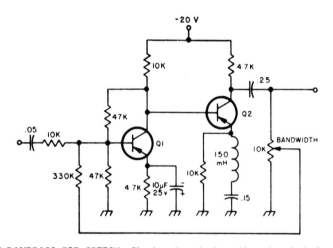

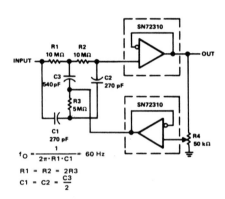

NARROW BANDPASS FOR SPEECH—Simple audio filter provides about 20-dB gain at bandwidth of 80 Hz. Bandwidth can be narrowed to limits of unintelligibility by adjusting 10K pot. Input is plugged into phone jack of receiver, and headphones are connected to output. Transistors are SK3004 or equivalent.—Circuits, *73 Magazine*, Jan. 1974, p 125.

60-Hz ADJUSTABLE-Q NOTCH—Connection shown for two SN72310 voltage-follower opamps provides attenuation of 60-Hz power-line frequency. Setting of R4 determines Q of filter.—"The Linear and Interface Circuits Data Book for Design Engineers," Texas Instruments, Dallas, TX, 1973, p 4-39.

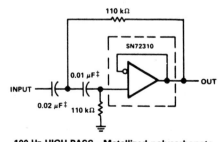

100-Hz HIGH-PASS—Metallized polycarbonate capacitors are required for good temperature stability in high-pass active filter using voltage-follower opamp. Cutoff frequency is 100 Hz.—"The Linear and Interface Circuits Data Book for Design Engineers," Texas Instruments, Dallas, TX, 1973, p 4-39.

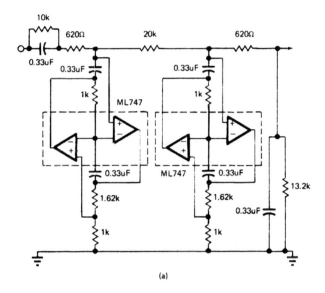

(a)

480-kHz LOW-PASS—Butterworth low-pass active filter uses pair of dual opamps with external resistors and capacitors to give corner frequency of 480 kHz and output impedance level of 1K. Article presents design procedure in detail and gives frequency response curve.—L. T. Burton and D. Treleaven, Active Filter Design Using Generalized Impedance Converters, *EDN Magazine*, Feb. 5, 1973, p 68–75.

300–3000 Hz WIDEBAND—Used in voice communication systems where signals below 300 Hz and above 3000 Hz must be rejected. Second-order Butterworth stopband responses are achieved by combining low-pass and high-pass sections of equal-component voltage-controlled voltage-source filters. Overall passband gain is 8 dB. Opamps can be 741.—H. M. Berlin, "Design of Active Filters, with Experiments," Howard W. Sams, Indianapolis, IN, 1977, p 148–151.

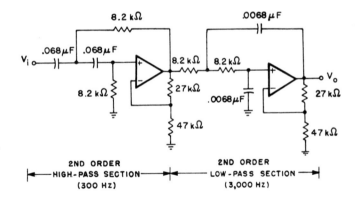

18 dB PER OCTAVE EMPHASIS—Circuit shown is result of design procedure given in article for active filter that provides frequency emphasis at rate of 18 dB per octave between 5 and 15 kHz. Emphasis does not exceed 40 dB at 20 kHz. Design equations include parameters for closed-loop gain of opamp. Scale factor is applied to input and feedback networks individually after design, to give reasonable component values.—B. Brandstedt, Tailor the Response of Your Active Filters, *EDN Magazine*, March 5, 1973, p 68–72.

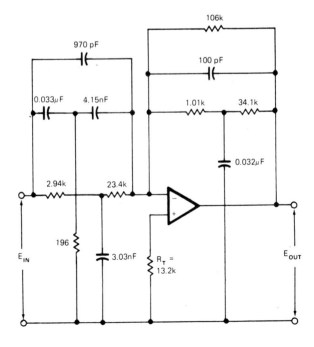

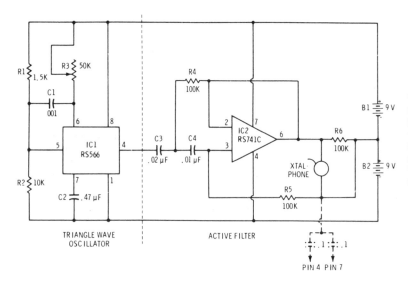

TRIANGLE WAVE
OSCILLATOR

ACTIVE FILTER

PIN 4   PIN 7

150-Hz HIGH-PASS—Circuit includes variable high-frequency source supplying triangle-wave input to filter for demonstrating high-pass action. If long supply leads cause oscillation, connect 0.1-μF capacitors between ground and supply pins 4 and 7 as shown.—F. M. Mims, "Integrated Circuit Projects, Vol. 4," Radio Shack, Fort Worth, TX, 1977, 2nd Ed., p 87–94.

750-Hz SIXTH-ORDER BANDPASS—Provides passband gain of 6 (15.6 dB) and Q of 8.53 by cascading three identical second-order filter sections. Each section uses multiple feedback.—H. M. Berlin, "Design of Active Filters, with Experiments," Howard W. Sams, Indianapolis, IN, 1977, p 147–148.

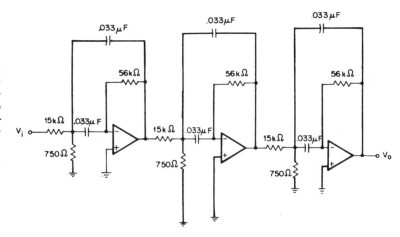

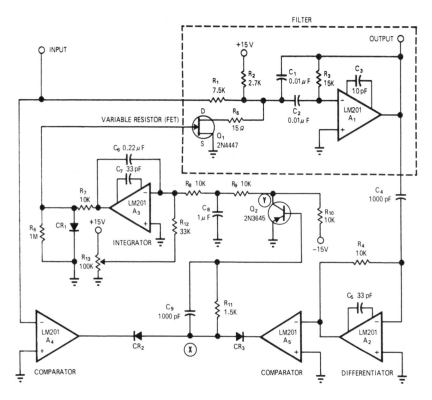

2–20 kHz SELF-TUNING BANDPASS—Center frequency of filter is automatically adjusted to track signal frequency, for optimum noise rejection when input frequency varies over wide range as it does with many types of vibrating transducers. Requires no reference frequency and no internal oscillator or synchronizing circuits. Frequency range can be extended in decade steps by capacitor switching. When filter is not tuned to input frequency, phase shift is not 180° and phase detector applies error signal to gate of FET to control its drain-source resistance. Phase detector A_4-A_5-CR_2-CR_3 and FET form part of negative-feedback loop around filter, so error in phase changes resistance of FET and thereby retunes filter. Article gives design equations, operational details, and waveforms at various points in circuit.—G. J. Deboo and R. C. Hedlund, Automatically Tuned Filter Uses IC Operational Amplifiers, *EDN/EEE Magazine*, Feb. 1, 1972, p 38–41.

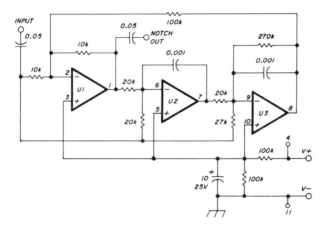

3-kHz NOTCH—Uses three sections of National LM324 quad opamp to provide fixed center frequency of 3 kHz for notch. Single supply can be 5–25 V.—P. A. Lovelock, Discrete Operational Amplifier Active Filters, *Ham Radio*, Feb. 1978, p 70–73.

STATE-VARIABLE DESIGN—Universal filter network using three opamps can provide low-pass, high-pass, or bandpass audio response for CW and SSB reception. Filter uses one summing block U1, two identical integrators, U2 and U3, and one damping network. Cutoff frequencies are same as center frequency for bandpass response. Article gives design equations and graph for choosing values to give optimum performance for type of response desired. For unity-gain second-order Butterworth filter with low-pass or high-pass cutoff of 700 Hz, R is 6800 ohms and C is 0.033 μF. Q must be fixed at 0.707 and R_A must equal $1.12 \times R_B$. Thus, if R_B is 2700 ohms, R_A should be 3000.—H. M. Berlin, The State-Variable Filter, *QST*, April 1978, p 14–16.

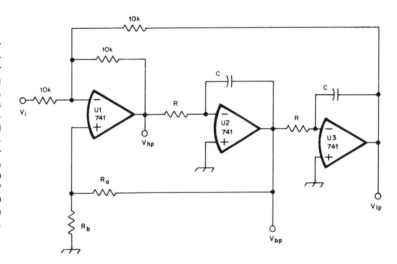

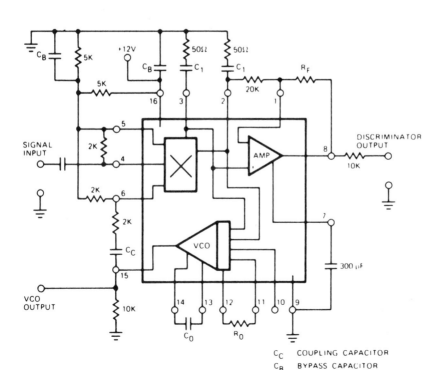

3:1 TRACKING FILTER—Connection shown for Exar XR-215 PLL IC tracks input signal over 3:1 frequency range centered on free-running frequency of VCO. Tracking range is maximum when pin 10 is open. R_0 is typically between 1K and 4K. C_1 is between 30 and 300 times C_0 where timing capacitor C_0 depends on center frequency. System can also be operated as linear discriminator or analog frequency meter covering same 3:1 change of input frequency. R_F can be 36K.—"Phase-Locked Loop Data Book," Exar Integrated Systems, Sunnyvale, CA, 1978, p 21–28.

C_C COUPLING CAPACITOR
C_B BYPASS CAPACITOR

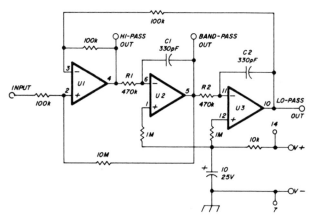

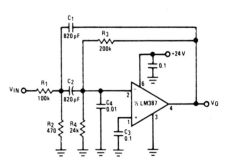

1-kHz THREE-FUNCTION—Three-function fixed-frequency active filter uses three sections of RCA CA3401E, Motorola MC3301P, or National LM3900N quad opamp. Circuit uses high- value series resistors for noninverting inputs to limit bias current to between 10 and 100 μA.— P. A. Lovelock, Discrete Operational Amplifier Active Filters, *Ham Radio*, Feb. 1978, p 70–73.

20-kHz BANDPASS—Provides bandwidth of 2000 Hz and midband gain of 1 for applications requiring narrow-bandwidth bandpass active filter. Design procedure is given.—"Audio Handbook," National Semiconductor, Santa Clara, CA, 1977, p 2-52–2-53.

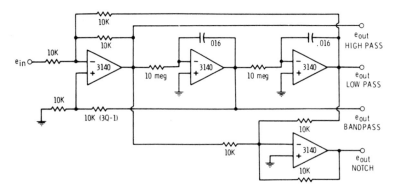

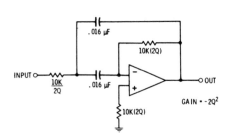

1-Hz STATE-VARIABLE FILTER—Universal filter has simultaneous low-pass, bandpass, high-pass, and notch outputs all with cutoff frequency of 1 Hz. To scale circuit up to 1-kHz cut- off, replace 10-megohm resistors with 10K.—D. Lancaster, "CMOS Cookbook," Howard W. Sams, Indianapolis, IN, 1977, p 343–344.

1-kHz MULTIPLE-FEEDBACK BANDPASS—Single 741 or equivalent opamp is suitable for applications where bandwidth is less than 100%. Gain is fixed at $-2Q^2$, where Q is reciprocal of damping d and ranges from less than 1 to above 100. Q is changed by varying ratio of input and feedback resistors while keeping their product constant. For Q of 3, feedback resistor should be 36 times value of input resistor.—D. Lancaster, "Active-Filter Cookbook," Howard W. Sams, Indianapolis, IN, 1975, p 150–154.

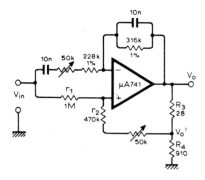

50-Hz WIEN-BRIDGE NOTCH FILTER—Uses opamp in circuit having essentially zero output impedance, making additional buffer amplifier unnecessary. Article gives design theory and covers many other types of notch filters.—Y. Nezer, Active Notch Filters, *Wireless World*, July 1975, p 307–311.

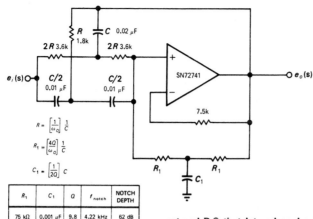

$$R = \left[\frac{1}{\omega_o}\right]\frac{1}{C}$$

$$R_1 = \left[\frac{4Q}{\omega_o}\right]\frac{1}{C}$$

$$C_1 = \left[\frac{1}{2Q}\right]C$$

R_1	C_1	Q	f_{notch}	NOTCH DEPTH
75 kΩ	0.001 μF	9.8	4.22 kHz	62 dB
150 kΩ	660 pF	18.4	4.22 kHz	62.7 dB
220 kΩ	360 pF	25.0	4.22 kHz	62 dB

4.22-kHz NOTCH—Circuit consists of positive unity-gain opamp, RC twin-T network, and T network R_1C_1 that determines circuit Q. Variable Q feature is controlled by single passive RC network. Center frequency of notch filter is about 4.22 kHz. Table gives values of R_1 and C_1 for three different values of Q.—H. T. Russell, Notch Filter Has Passive Q Control, *EDN/EEE Magazine*, July 1, 1971, p 43 and 45.

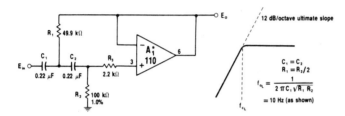

10-Hz HIGH-PASS UNITY-GAIN—Low cutoff frequency is 10 Hz in active filter using opamp as voltage-controlled voltage source. Alternative opamps can be 1556 and 8007.—W. G. Jung, "IC Op-Amp Cookbook," Howard W. Sams, Indianapolis, IN, 1974, p 331–333.

$$C_1 = C_2$$
$$R_1 = R_2/2$$
$$f_{cL} = \frac{1}{2\pi C_1\sqrt{R_1\,R_2}}$$
$$= 10\text{ Hz (as shown)}$$

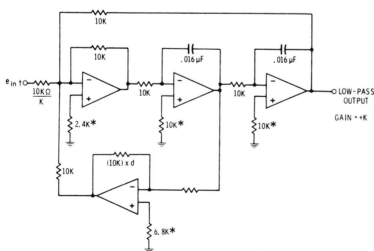

† must return to ground via low-impedance dc path.
*optional offset compensation, may be replaced with short in noncritical circuits.

1-kHz VARIABLE-GAIN STATE-VARIABLE—Damping signal is inverted with fourth opamp to make gain and damping as well as frequency independently adjustable. Damping is in range of 0–2, with critical value of 1.414 giving flattest response. For high pass, take output from first opamp; for bandpass, take output from second opamp. Value of input resistor is 10K (10,000 ohms) when gain K is 1.—D. Lancaster, "Active-Filter Cookbook," Howard W. Sams, Indianapolis, IN, 1975, p 135–136.

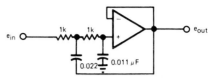

10-kHz LOW-PASS SALLEN-KEY—Article gives design equations from which values of components were obtained. Critical damping (Q = 0.71) is provided. Frequency can be tuned over range of two decades by changing resistor values simultaneously. Opamp can be one section of OP-11FY. For equivalent high-pass filter, transpose positions of resistors and capacitors. Gain is unity.—D. Van Dalsen, Need an Active Filter? Try These Design Aids, *EDN Magazine*, Nov. 5, 1978, p 105–110.

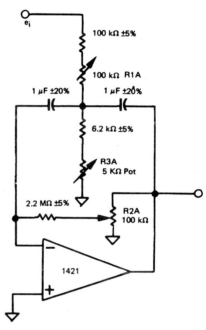

1 Hz WITH 0.1-Hz BANDWIDTH—Three pots provides easy trimming to precise values desired. Use R2A to trim bandwidth to exactly 0.100 Hz. Use R1A to trim gain to exactly 10.00. Finally, trim center frequency to exactly 1.000Hz. Adjustments are almost perfectly noninteracting if made in sequence given.—R. A. Pease, "Band-Pass Active Filter with Easy Trim for Center Frequency," Teledyne Philbrick, Dedham, MA, 1972, Applications Bulletin 4.

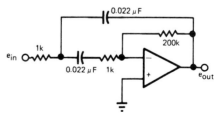

100-Hz BANDPASS SALLEN-KEY—Uses one section of OP-11FY quad opamp or equivalent in circuit having Q of 4.7 and providing closed-loop gain of 200 or 46 dB. Opamp selected should have open-loop gain of 5 to 10 times required gain at resonance. Adjust resistor values to tune center frequency.—D. Van Dalsen, Need an Active Filter? Try These Design Aids, *EDN Magazine*, Nov. 5, 1978, p 105–110.

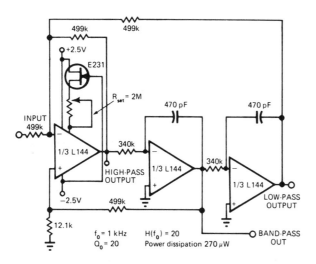

1-kHz STATE-VARIABLE—Low-power filter uses three opamps to provide simultaneous high-, low-, and bandpass outputs. Article presents complete design procedure for keeping current drain at minimum while providing desired gain-bandwidth product of 240 kHz.—L. Schaeffer, Op-Amp Active Filters—Simple to Design Once You Know the Game, *EDN Magazine*, April 20, 1976, p 79–84.

$f_0 = 1$ kHz $\quad H(f_0) = 20$
$Q_0 = 20$ $\quad$ Power dissipation 270 μW

1 Hz–500 kHz VOLTAGE-TUNED BANDPASS—Coupling FET opamps with analog multiplier gives simple two-pole bandpass filter that can be tuned by external voltage of 0–10 VDC to give center frequency anywhere in range from 1 Hz to 500 kHz with components shown. Article gives design equations.—T. Cate, Voltage Tune Your Bandpass Filters with Multipliers, *EDN Magazine*, March 1, 1971, p 45–47.

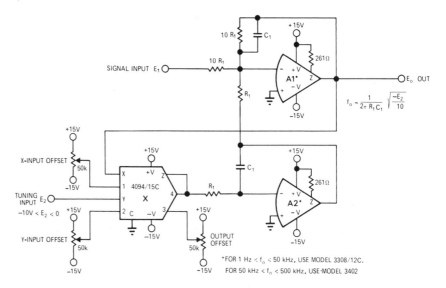

$$f_0 = \frac{1}{2\pi R_1 C_1}\sqrt{\frac{-E_2}{10}}$$

*FOR 1 Hz < f_0 < 50 kHz, USE MODEL 3308/12C.
FOR 50 kHz < f_0 < 500 kHz, USE MODEL 3402

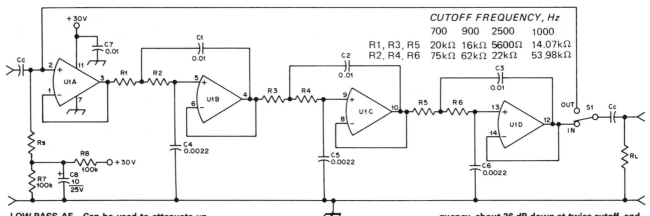

CUTOFF FREQUENCY, Hz			
700	900	2500	1000
R1, R3, R5 20kΩ	16kΩ	5600Ω	14.07kΩ
R2, R4, R6 75kΩ	62kΩ	22kΩ	53.98kΩ

LOW-PASS AF—Can be used to attenuate undesired high-frequency audio response in superhet or direct-conversion receivers having inadequate IF selectivity, to improve CW or SSB reception. Resistor values determine cutoff frequency; 700 and 900 Hz are for CW and 2500 Hz for SSB. Insert filter at point having low audio level. Filter has input buffer, three cascaded active low-pass filter stages, and IN/OUT switch. Overall gain is unity. U1 is Fairchild μA4136, Raytheon RC4136, or equivalent quad opamp. Overall response is 1.5 dB down at cutoff frequency, about 36 dB down at twice cutoff, and about 60 dB down at three times cutoff. R7 and R8 provide pseudoground of half supply voltage, to eliminate need for negative supply. Will operate with supply from 6 to 36 V, drawing about 7 mA.—T. Berg, Active Low-Pass Filters for CW or SSB, *QST*, Aug. 1977, p 40–41.

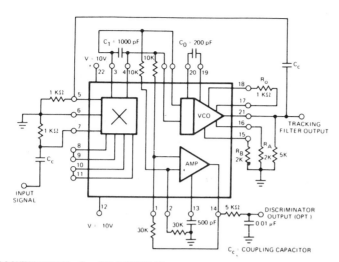

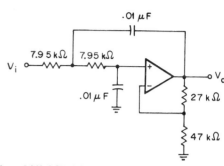

1-MHz TRACKING FILTER—Exar XR-S200 PLL IC is connected to function as frequency filter when phase-locked loop locks on input signal, to produce filtered version of input signal frequency at VCO output. Because circuit can track input over 3:1 range of frequencies around free-running frequency of VCO, it is known as tracking filter. Optional wideband discriminator output is also provided.—"Phase-Locked Loop Data Book," Exar Integrated Systems, Sunnyvale, CA, 1978, p 9–16.

2-kHz LOW-PASS—Voltage-controlled voltage-source filter uses equal-value input resistors and equal-value capacitors, simplifying selection of components. Equation for cutoff frequency then simplifies to f = 1/6.28RC or 1/(6.28)(7950)(0.01)(10⁻⁶). Opamp can be 741.—H. M. Berlin, "Design of Active Filters, with Experiments," Howard W. Sams, Indianapolis, IN, 1977, p 85–86.

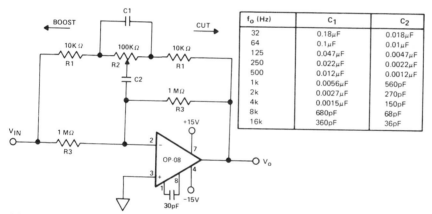

f₀ (Hz)	C₁	C₂
32	0.18μF	0.018μF
64	0.1μF	0.01μF
125	0.047μF	0.0047μF
250	0.022μF	0.0022μF
500	0.012μF	0.0012μF
1k	0.0056μF	560pF
2k	0.0027μF	270pF
4k	0.0015μF	150pF
8k	680pF	68pF
16k	360pF	36pF

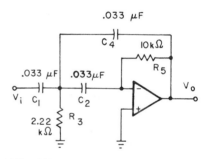

OCTAVE AUDIO EQUALIZER—R2 provides up to 12-dB boost or cut at center frequency determined by values of C1 and C2 as given in table. Uses Precision Monolithics OP-08 opamp. Low input bias current of opamp permits scaling resistors up by factor of 10, to reduce values of C1 and C2 at low-frequency end. Same circuit is used for all 10 sections of equalizer, which together draw only 6 mA maximum from supply.—"Precision Low Input Current Op Amp," Precision Monolithics, Santa Clara, CA, 1978, OP-08, p 7.

1-kHz HIGH-PASS UNITY-GAIN—Passband gain of 741 or equivalent opamp circuit is set by ratio of C₄ to C₁ rather than by resistors. Values shown give unity gain for passband above 1-Hz cutoff. Circuit uses multiple feedback.—H. M. Berlin, "Design of Active Filters, with Experiments," Howard W. Sams, Indianapolis, IN, 1977, p 100–102.

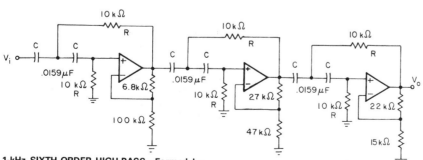

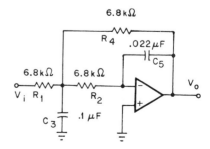

1-kHz SIXTH-ORDER HIGH-PASS—Formed by cascading three different second-order sections. Passband gain is 12.5 dB. Opamps can be 741 or equivalent. Used when high rejection is needed for signals just below passband, in application where such rejection justifies cost of extra filter sections.—H. M. Berlin, "Design of Active Filters, with Experiments," Howard W. Sams, Indianapolis, IN, 1977, p 122–125.

500-Hz LOW-PASS UNITY-GAIN—Multiple-feedback filter using 741 or equivalent opamp has unity gain in passband below 500-Hz cutoff. Resistors can be 5% tolerance.—H. M. Berlin, "Design of Active Filters, with Experiments," Howard W. Sams, Indianapolis, IN, 1977, p 99–100.

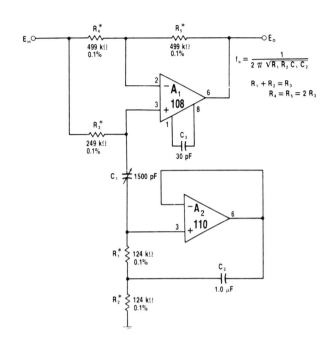

$$f_o = \frac{1}{2\pi\sqrt{R_1 R_2 C_1 C_2}}$$

$$R_1 + R_2 = R_3$$
$$R_4 = R_5 = 2R_3$$

AF NOTCH—Notch frequency is easily tuned at frequencies below 1 kHz with single capacitor C_1 or by replacing R_1 and R_2 with 249K pot. For higher frequencies, use 118 opamp for A_1 and 5K for R_3 while lowering other resistances in proportion to R_3. Indicated resistance tolerances are necessary for optimum notch depth.—W. G. Jung, "IC Op-Amp Cookbook," Howard W. Sams, Indianapolis, IN, 1974, p 340–341.

1.4-kHz TWIN-T BANDPASS—Combination of passive twin-T bandpass filter and 741 opamp gives simple audio filter for amplifying narrow frequency band (about 300 Hz wide) centered on 1.4 kHz. Filter can be tuned to other frequencies by replacing R1 and R2 with 10K pots. Frequency is equal to 1/6.28RC where R is value in ohms of R1 and R2 and C is capacitance in farads of C1 and C2. R3 is half of R1.—F. M. Mims, "Integrated Circuit Projects, Vol. 2," Radio Shack, Fort Worth, TX, 1977, 2nd Ed., p 71–80.

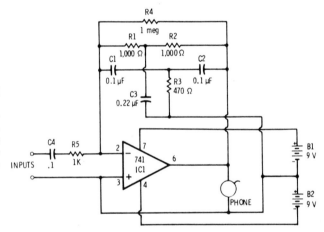

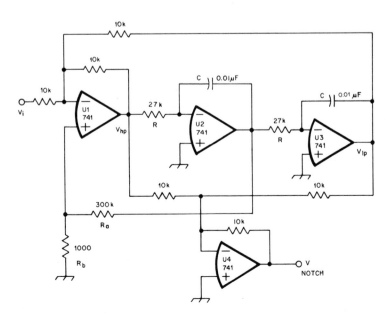

600-Hz NOTCH—With values obtained from design equations and graph in article, state-variable or universal filter provides Q of 100 with four opamps. Notch filter is achieved by adding low-pass and high-pass outputs equally, for feed to dual-input summing amplifier.—H. M. Berlin, The State-Variable Filter, *QST*, April 1978, p 14–16.

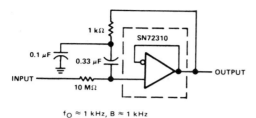

$f_O \approx 1 \text{ kHz}, B \approx 1 \text{ kHz}$

1-kHz BANDPASS—Simple circuit using voltage-follower opamp provides bandpass of 1 kHz centered on 1 kHz, to give output range of 500–1500 Hz.—"The Linear and Interface Circuits Data Book for Design Engineers," Texas Instruments, Dallas, TX, 1973, p 4-39.

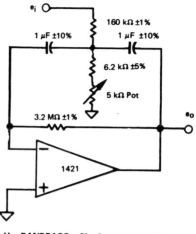

1-Hz BANDPASS—Single pot provides easy trimming to exact center frequency desired without change in bandwidth or gain. Q is 10. Design equations are given.—R. A. Pease, "Band-Pass Active Filter with Easy Trim for Center Frequency," Teledyne Philbrick, Dedham, MA, 1972, Applications Bulletin 4.

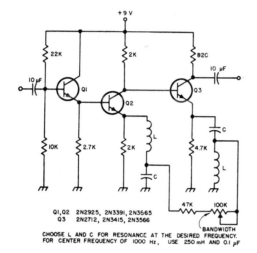

Q1,Q2   2N2925, 2N3391, 2N3565
Q3   2N2712, 2N3415, 2N3566

CHOOSE L AND C FOR RESONANCE AT THE DESIRED FREQUENCY.
FOR CENTER FREQUENCY OF 1000 Hz, USE 250 mH AND 0.1 μF

1-kHz BANDPASS—Three-stage audio filter uses two series resonant circuits to give very narrow audio passband. Amount of feedback determines Q and bandwidth.—Circuits, *73 Magazine,* March 1974, p 89.

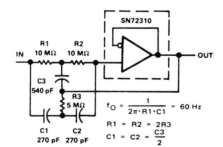

$$f_O = \frac{1}{2\pi \cdot R1 \cdot C1} = 60 \text{ Hz}$$

$$R1 = R2 = 2R3$$

$$C1 = C2 = \frac{C3}{2}$$

60-Hz HIGH-Q NOTCH—Input network for SN72310 voltage-follower opamp provides attenuation of 60-Hz power-line frequency. Use high-quality capacitors for maximum Q.—"The Linear and Interface Circuits Data Book for Design Engineers," Texas Instruments, Dallas, TX, 1973, p 4-39.

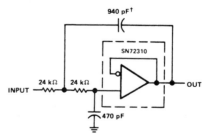

LOW-PASS WITH 10-kHz CUTOFF—Simple circuit uses only one Texas Instruments SN72310 voltage-follower opamp. For good temperature stability, use silvered mica capacitors.—"The Linear and Interface Circuits Data Book for Design Engineers," Texas Instruments, Dallas, TX, 1973, p 4-39.

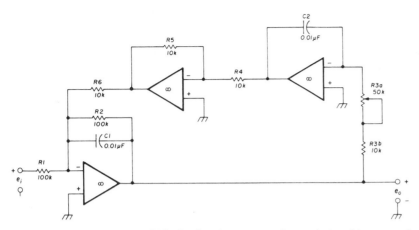

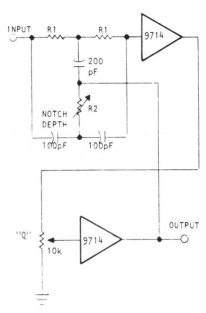

1-kHz CASCADED-OPAMP BANDPASS—Bandwidth of only 71 Hz is achieved by using four identical three-opamp filters in series, for increased selectivity in communication or RTTY receiver. Values for R1 and R2 are made variable for first filter stage, but pot is used for R3 in all four stages so they can be tuned to same center frequency. Batteries are used for supply, as filter draws only 17 mA.—F. M. Griffee, RC Active Filters Using Op Amps, *Ham Radio,* Oct. 1976, p 54–58.

19-kHz NOTCH—Used in commercial FM transmitters to eliminate 19-kHz program material from stereo encoder. Uses Optical Electronics 9714 opamp in circuit that gives unity passband gain below center frequency, 0.7 gain above center frequency, and less than 0.001 gain at notch frequency. Provides adjustments for notch rejection level and Q. R1 is 84K, and R2 is 36K in series with 10K pot.—"Precision Notch Filter," Optical Electronics, Tucson, AZ, Application Tip 10255.

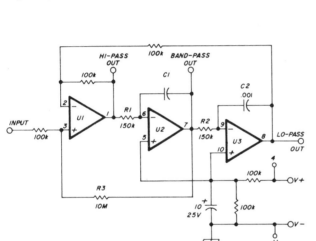

1-kHz THREE-FUNCTION—Uses National LM324 quad opamp, with appropriate biasing for single supply of +5 to +25 VDC. Values of R1 and R2 establish f_c at 1000 Hz, while R3 gives Q of 50. Values of R1 and R2 for other bandpass center and cutoff frequencies f_c can be calculated from $R = 15 \times 10^7/f_c$. Fourth opamp may be used as output amplifier or for summing high-pass and low-pass outputs. C1 is same as C2.—P. A. Lovelock, Discrete Operational Amplifier Active Filters, *Ham Radio,* Feb. 1978, p 70–73.

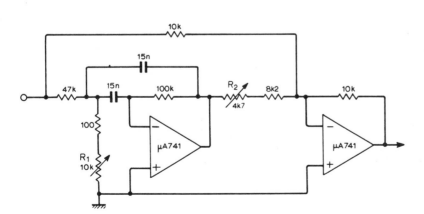

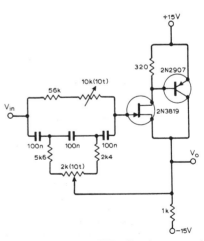

TUNABLE NOTCH—Opamp circuit requires only one pot (R_1) to vary notch frequency. R_2 is used to set noise rejection to maximum. With values shown, filter tunes from 170 Hz to 3 kHz, with 3-dB bandwidth of 230 Hz and notch rejection better than 40 dB over entire range. Circuit can be voltage-tuned by replacing R_1 with FET operated as voltage-variable resistor.—R. J. Harris, Simple Tunable Notch Filter, *Wireless World,* May 1973, p 253.

TUNABLE NOTCH FILTER—Simple pot-tuned active notch filter has tuning range of 200 Hz in audio band and 3-dB rejection bandwidth of 10 Hz, as required for tuning out whistle or power-line hum that is interfering with radio program. Article gives design theory for many other types of notch filters.—Y. Nezer, Active Notch Filters, *Wireless World,* July 1975, p 307–311.

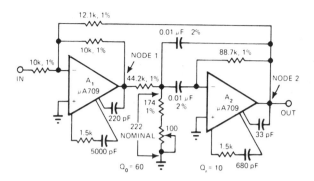

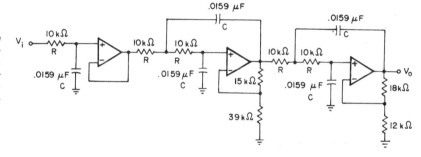

Q MULTIPLIER—Article gives design procedure and equations for utilizing Q multiplication to simplify circuit for active bandpass filter. With values shown, center frequency is 3.6 kHz and Q of 10 is multiplied by gain of 6 to give effective Q of 60.—A. B. Williams, Q-Multiplier Techniques Increase Filter Selectivity, *EDN Magazine,* Oct. 5, 1975, p 74 and 76.

1-kHz FIFTH-ORDER LOW-PASS—Uses single first-order section and two different second-order sections to give passband gain of 10.3 dB. Opamps can be 741 or equivalent.—H. M. Berlin, "Design of Active Filters, with Experiments," Howard W. Sams, Indianapolis, IN, 1977, p 119–122.

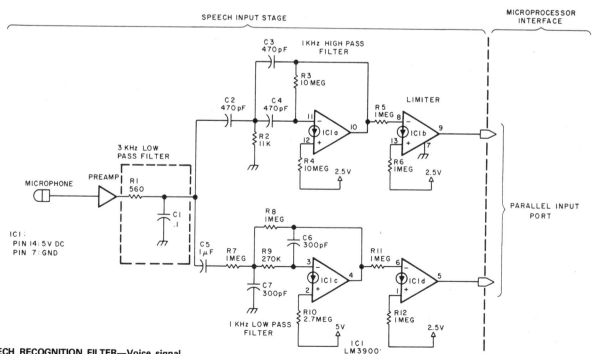

SPEECH RECOGNITION FILTER—Voice signal picked up by microphone is preamplified and sent through 3-kHz low-pass passive filter C1-R1 to 1-kHz high-pass active filter and 1-kHz low-pass active filter using sections of LM3900 quad opamp. Diode symbols on opamps indicate use of current mirrors for noninverting inputs. Outputs are sampled about 60 times per second to implement speech recognition algorithm of computer, which counts number of high-pass and low-pass zero crossings per second and compares results with series of word models in memory to determine most likely match.—J. R. Boddie, Speech Recognition for a Personal Computer System, *BYTE,* July 1977, p 64–68 and 70–71.

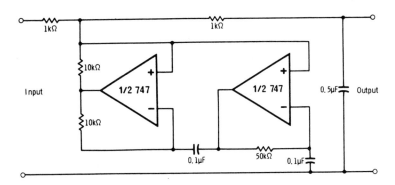

320-Hz LOW-PASS—Frequency-dependent negative-resistance circuit uses 747 dual opamp. Signal source used as input should have low resistance, and load should have high resistance. Voltage-follower stages can be used to isolate both input and output of filter.—R. Melen and H. Garland, "Understanding IC Operational Amplifiers," Howard W. Sams, Indianapolis, IN, 2nd Ed., 1978, p 104–105.

VOLTAGE-TUNED STATE-VARIABLE—Provides choice of high-pass, bandpass, and low-pass outputs, each with cutoff frequency variable between 1 and 6 kHz by varying control voltage between −10 V and +15 V. Output load resistor sets voltage gain between input and output. Gain-control input varies gain from maximum set by load resistor down to zero. Input signals must be limited to 100 mV because input circuit is differential amplifier operating without feedback.—D. Lancaster, "Active-Filter Cookbook," Howard W. Sams, Indianapolis, IN, 1975, p 203–205.

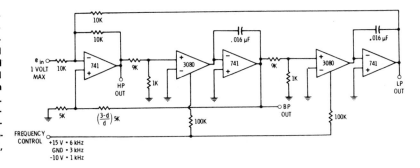

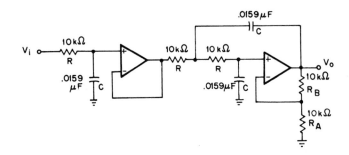

1-kHz THIRD-ORDER LOW-PASS—Circuit using 741 or equivalent opamp consists of unity-gain first-order section followed by equal-component voltage-controlled voltage-source second-order section.—H. M. Berlin, "Design of Active Filters, with Experiments," Howard W. Sams, Indianapolis, IN, 1977, p 113–114.

1-kHz BIQUAD BANDPASS—Three 741 opamps are connected to give two integrators and inverter. Overall gain is −Q, determined by value of input resistor used. Circuit is tuned by varying capacitors in steps. Absolute bandwidth remains constant as frequency changes. Chief applications are in telephone systems, where identical absolute-bandwidth channels are required.—D. Lancaster, "Active-Filter Cookbook," Howard W. Sams, Indianapolis, IN, 1975, p 159–164.

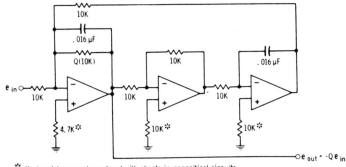

✳offset resistors may be replaced with shorts in noncritical circuits.

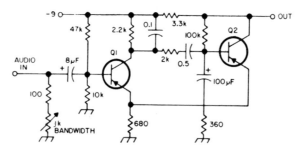

VARIABLE-BANDWIDTH AF—Audio filter using 1000-Hz Wien bridge provides bandwidths from 70 to 600 Hz. Transistors can be SK3004, GE-2, or HEP-254.—Circuits, *73 Magazine,* Jan. 1974, p 124.

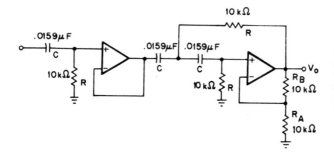

1-kHz THIRD-ORDER HIGH-PASS—Passband gain is 6 dB for Butterworth filter above 1-kHz cutoff. Damping factor is 1.000 for both sections, each using 741 or equivalent opamp.—H. M. Berlin, "Design of Active Filters, with Experiments," Howard W. Sams, Indianapolis, IN, 1977, p 115–116.

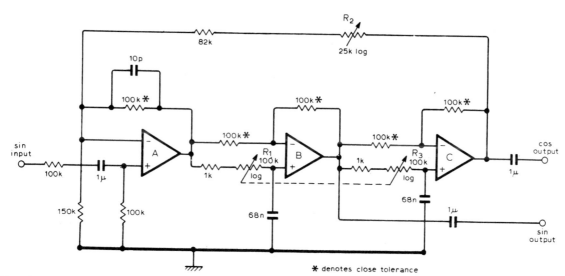

20–2000 Hz VARIABLE BANDPASS—High-Q active bandpass filter can be adjusted over wide frequency range (100:1) while maintaining Q essentially constant over 100. Two-phase output is available. Opamps can be 741 or equivalent. Cascaded all-pass networks B and C each have 0 to 180° phase variation and unity gain at all frequencies. These are driven by opamp A whose feedback signal is sum of input and output of all-pass networks. R_2 adjusts Q, and ganged log pots change center frequency.—J. M. Worley, Variable Band-Pass Filter, *Wireless World,* April 1977, p 61.

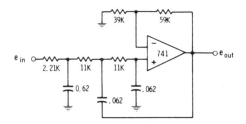

250-Hz THIRD-ORDER LOW-PASS—Values shown place cutoff at 250 Hz, with 1-dB dip in response curve. Input must be returned to ground with low-impedance DC path.—D. Lancaster, "Active-Filter Cookbook," Howard W. Sams, Indianapolis, IN, 1975, p 146.

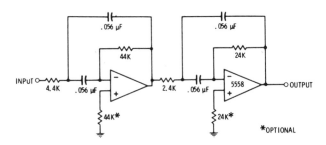

200–400 Hz PASSBAND—Design is based on use of 3.2 for value of Q, to hold passband dip at 1 dB for two-pole filter. Multiple feedback is used for each pole. First opamp can be 741 or equivalent. Center frequency is 283 Hz.—D. Lancaster, "Active-Filter Cookbook," Howard W. Sams, Indianapolis, IN, 1975, p 166.

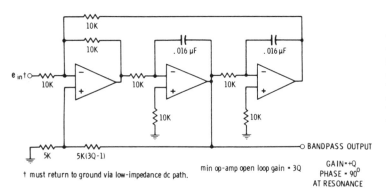

† must return to ground via low-impedance dc path.

min op-amp open loop gain = 3Q

GAIN = +Q
PHASE = 90°
AT RESONANCE

1-kHz STATE-VARIABLE BANDPASS—With three 741 opamps or equivalent, circuit gain is Q (reciprocal of damping). Frequency is changed by changing 10K coupling resistors between opamps while keeping their values equal. Increasing resistors 10 times increases frequency 10 times. High-pass output is obtained from first opamp and low pass from second opamp.—D. Lancaster, "Active-Filter Cookbook," Howard W. Sams, Indianapolis, IN, 1975, p 156–159.

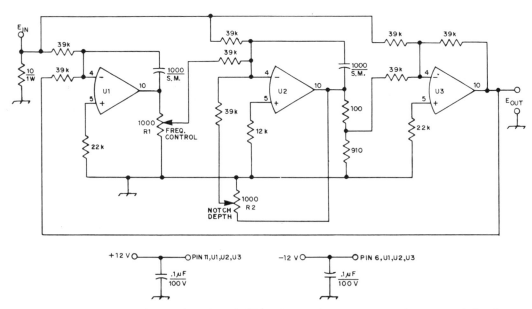

AF NOTCH—Center frequency of notch can be varied with single control R1; upper limit is about 4 kHz. Circuit Q and notch depth are constant over range. R2 is adjusted initially for best notch depth. All opamps are 741 14-pin DIP, such as Motorola MC1741L. U1 and U2 are integrators with DC gain of about 2500, and U3 is summing device. Notch depth is at least 50 dB. Input to filter is taken from loudspeaker or headphone jack of receiver. High-impedance headset may be connected directly across output, or buffer stage can be added to drive lower-impedance loudspeaker or headset. Use with AGC off.—A. Taflove, An Analog-Computer-Type Active Filter, *QST*, May 1975, p 26–27.

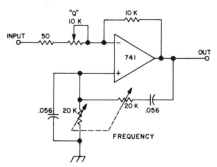

VARIABLE Q AND FREQUENCY—Bandwidth can be made extremely sharp (less than 9 Hz) or very broad (greater than 300 Hz). Adjusting Q to change bandwidth also changes gain of filter. Center frequency of filter is independently adjustable.—A. F. Stahler, An Experimental Comparison of CW Audio Filters, *73 Magazine*, July 1973, p 65–70.

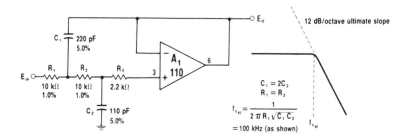

100-kHz LOW-PASS UNITY-GAIN—Opamp serves as active element in voltage-controlled voltage-source second-order filter. Other opamps having required high input resistance, low input current, and high speed are 1556 and 8007.—W. G. Jung, "IC Op-Amp Cookbook," Howard W. Sams, Indianapolis, IN, 1974, p 331–333.

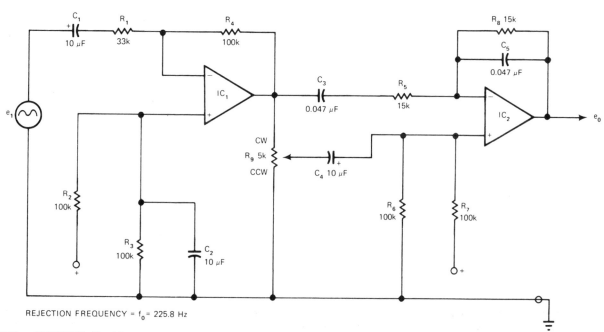

REJECTION FREQUENCY = f_0 = 225.8 Hz

225.8-Hz REJECTION—Provides extremely sharp adjustable-depth notch with only two low-gain opamps. Suitable for single-ended supplies. Article gives equation for transfer function.—R. Carter, Sharp Null Filter Utilizes Minimum Component Count, *EDN Magazine*, Sept. 20, 1976, p 110.

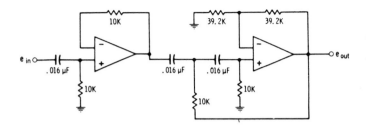

1-kHz HIGH-PASS FLATTEST-RESPONSE—Values are chosen for flattest possible response obtainable with third-order configuration of two 741 opamps. Tripling capacitance values cuts cutoff frequency by one-third and vice versa. Component tolerance can be 10%. Gain is 2.—D. Lancaster, "Active-Filter Cookbook," Howard W. Sams, Indianapolis, IN, 1975, p 186.

2.4-kHz LOW-PASS/HIGH-PASS—Three 741 opamps are connected to provide separate low-pass and high-pass outputs simultaneously for complex synthesis problem requiring state-variable filter. Gain is 1.—D. Lancaster, "Active-Filter Cookbook," Howard W. Sams, Indianapolis, IN, 1975, p 192–193.

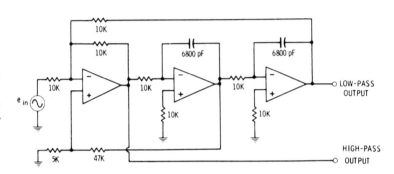

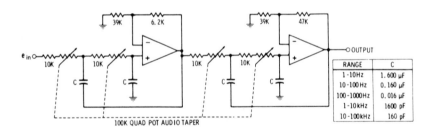

TUNABLE FOURTH-ORDER LOW-PASS—Use of four ganged pots permits varying cutoff frequency over 10:1 range. Table gives ranges obtainable with five different values for C. Opamps can be 741 or equivalent. Tracking of 5% for pots calls for expensive components, but ordinary snap-together pots may prove satisfactory if tuning range is restricted to 3:1 or less and more capacitor switching is used.—D. Lancaster, "Active-Filter Cookbook," Howard W. Sams, Indianapolis, IN, 1975, p 195–197.

RANGE	C
1-10 Hz	1.600 µF
10-100 Hz	0.160 µF
100-1000 Hz	0.016 µF
1-10 kHz	1600 pF
10-100 kHz	160 pF

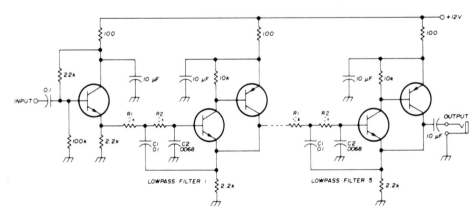

AF LOW-PASS FOR CW—Design using 10% tolerance components gives sufficiently wide bandwidth while maintaining steep skirt response for CW reception in direct-conversion communication receiver. Filter has five identical three-transistor sections, each peaked at cutoff frequency. Q of each section is about 1.9, which gives 6-dB bandwidth of about 200 Hz. With center frequency at 540 Hz, attenuation is 75 dB at 1200 Hz. Net gain of system is 28 dB at resonance. NPN transistors are 2N3565, 2N3904, or similar; PNP transistors are 2N3638, 2N3906, or similar.—W. Howard, Simple Active Filters for Direct-Conversion Receivers, *Ham Radio*, April 1974, p 12–15.

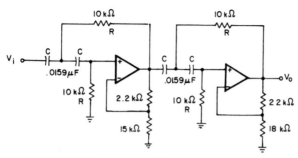

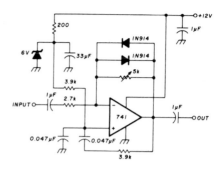

1-kHz FOURTH-ORDER HIGH-PASS—First section is second-order high-pass filter having gain of 1.2 dB, and second section has gain of 7 dB. Opamps are 741 or equivalent.—H. M. Berlin, "Design of Active Filters, with Experiments," Howard W. Sams, Indianapolis, IN, 1977, p 116–117.

800-Hz BANDPASS—Active filter has 800-Hz center frequency for optimum CW reception. Bandwidth is adjustable. Back-to-back diodes provide noise-limiting capability.—U. L. Rohde, IF Amplifier Design, Ham Radio, March 1977, p 10–21.

1.8–1.9 MHz BANDPASS—Butterworth bandpass filter, suitable for use with broadband preamp, helps reject out-of-band signals. Filter also protects preamp from signals across response range from broadcast band through VHF. C1 and C2 are mica trimmers. L1 and L2 have 30 turns No. 22 enamel on Amidon T68-6 toroid cores to give 5.1 μH.—D. DeMaw, Beat the Noise with a "Scoop Loop," QST, July 1977, p 30–34.

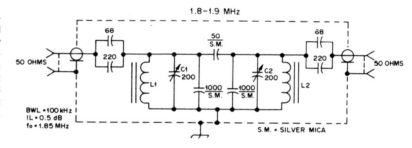

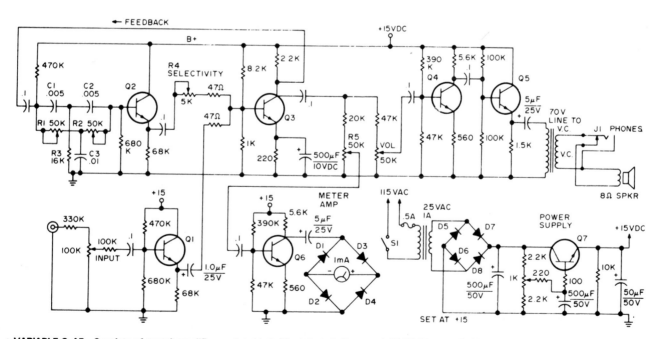

VARIABLE-Q AF—Consists of tuned amplifier having inverse feedback, connected so bandwidth at −6 dB is variable from 50 to 400 Hz for center frequency of 1 kHz. Improves selectivity of amateur receivers. Audio from receiver is applied to inverter Q3 through Q1. Part of inverted signal is fed back to twin-T network R1-R2-R3-C1-C2-C3 which has high impedance to ground except at its resonant frequency. Unattenuated signal goes through Q2 for adding to uninverted output at base of Q3. Degree of cancellation by two out-of-phase signals feeding Q3 is controlled by R4 to adjust selectivity. Filtered output is boosted by Q4 and Q5. Article covers construction, calibration, and operation. Q1-Q6 are GE-20, and Q7 is GE-14 or GE-28.—C. Townsend, A Variable Q Audio Filter, 73 Magazine, Feb. 1974, p 54–56.

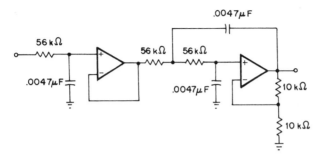

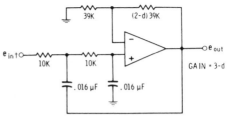

600-Hz THIRD-ORDER LOW-PASS—Butterworth filter using 741 or equivalent opamp provides gain of 6 dB in passband below 600-Hz cutoff. All components can be 5% tolerance.—H. M. Berlin, "Design of Active Filters, with Experiments," Howard W. Sams, Indianapolis, IN, 1977, p 114–116.

SECOND-ORDER 1-kHz LOW-PASS—Circuit using 741 opamp has equal-value series input resistors and high-pass capacitors. Cutoff frequency can be increased by changing 10K resistors to higher values while keeping their values identical. 10:1 resistance change provides 10:1 frequency change. Damping d is adjustable; critical value of 1.414 gives maximum flatness of response without overshoot. Interchange 10K resistors and 0.016-μF capacitors to convert circuit to 1-kHz high-pass filter.—D. Lancaster, "Active-Filter Cookbook," Howard W. Sams, Indianapolis, IN, 1975, p 127–129.

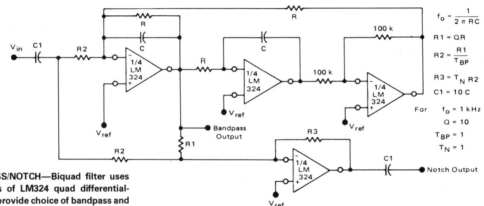

1-kHz BANDPASS/NOTCH—Biquad filter uses all four sections of LM324 quad differential-input opamp to provide choice of bandpass and notch outputs. Supply voltage range can be 3–32 V, with reference voltage equal to half of supply value used. For center frequency of 1 kHz, R is 160K, C is 0.001 μF, and R1-R3 are 1.6 meg-ohms. Coupling capacitors C1 can be 10 times value used for C.—"Quad Low Power Operational Amplifiers," Motorola, Phoenix, AZ, 1978, DS 9339 R1.

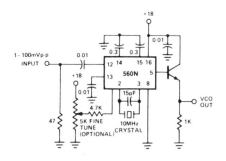

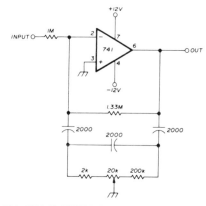

10-MHz TRACKING—Connection shown for 560N PLL tracking filter uses crystal to keep free-running frequency at desired value for signals near 10 MHz. Lock range varies with input amplitude, from about 0.3 kHz for 1 mV P-P input to about 3 kHz for 100 mV P-P.—"Signetics Analog Data Manual," Signetics, Sunnyvale, CA, 1977, p 850–851.

700–2000 Hz TUNABLE BANDPASS—Uses RC notch circuit as feedback element for active-filter opamp. With tuning pot set for center frequency of 1000 Hz, 3-dB bandwidth is 23 Hz and 10-dB bandwidth is 68 Hz. At 1000 Hz, voltage gain is 36 dB. High-frequency rolloff is good, being 43 dB down at 2000 Hz, so circuit converts 1000-Hz square wave into sine wave. Article gives design equations.—C. Hall, Tunable RC Notch Filter, *Ham Radio*, Sept. 1975, p 16–20.

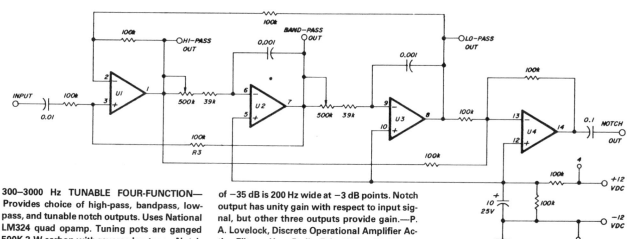

300–3000 Hz TUNABLE FOUR-FUNCTION—Provides choice of high-pass, bandpass, low-pass, and tunable notch outputs. Uses National LM324 quad opamp. Tuning pots are ganged 500K 2-W carbon with reverse log taper. Notch of −35 dB is 200 Hz wide at −3 dB points. Notch output has unity gain with respect to input signal, but other three outputs provide gain.—P. A. Lovelock, Discrete Operational Amplifier Active Filters, *Ham Radio*, Feb. 1978, p 70–73.

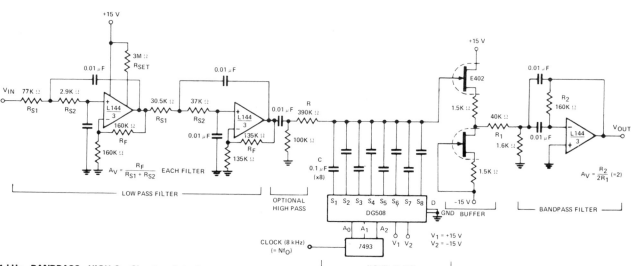

1-kHz BANDPASS HIGH-Q—Shunt-switched bandpass filter with Q of 1000 and voltage gain of about 7 uses DG508 CMOS multiplexer containing required analog switches, interface circuits, and decode logic for 8-path filter. Bandwidth for 3 dB down is 1 Hz centered on 1 kHz, with asymptotic slope of 6 dB per octave. Clock controls shunt-switched filter action.—"Analog Switches and Their Applications," Siliconix, Santa Clara, CA, 1976, p 5-12–5-14.

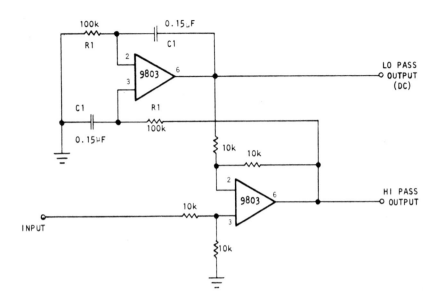

DC LEVEL SHIFTER FOR AF—Circuit using Optical Electronics 9803 opamps separates AF input signal into two outputs. Low-pass output contains DC to 10 Hz, and high-pass output has frequency content above 10 Hz to upper frequency limit approaching 10 MHz for opamp used. Dynamic output impedance of both outputs is less than 1 ohm. Both outputs have DC continuity. DC output of high-pass terminal is equal to offset voltage of integrator. DC output of low-pass terminal equals DC input plus offset voltages of both opamps.—"Automatic DC Level Shifter," Optical Electronics, Tucson, AZ, Application Tip 10226.

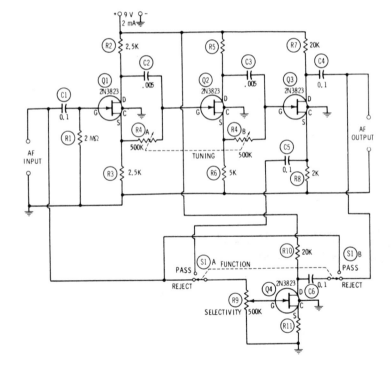

PASS/REJECT TUNABLE NOTCH—Full rotation of ganged tuning control R4 tunes circuit from 100 Hz to 10 kHz, with position of switch S1 determining whether circuit passes or rejects frequency to which it is tuned. Maximum selectivity, corresponding to maximum height of pass curve or depth of reject curve and minimum width of either curve, is obtained when R9 is set for maximum gain in FET Q4. If R9 is advanced far enough with switch set to pass, circuit will oscillate and give sine-wave output at tuned frequency.—R. P. Turner, "FET Circuits," Howard W. Sams, Indianapolis, IN, 1977, 2nd Ed., p 71–73.

SECTION 19
Frequency Divider Circuits

Includes programmable counters and divide circuits of various modulos.

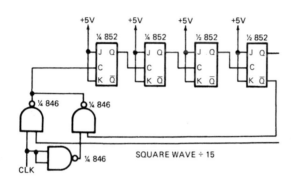

DIVIDE BY 15—Input clock is alternately inverted and noninverted by gates operating in conjunction with 4 bits of storage using 852 JK flip-flops, to give square-wave output at 1/15 of clock frequency.—C. W. Hardy, Reader Responds to Odd Modulo Divider in July 1st EDN, *EDN Magazine,* Oct. 1, 1972, p 50.

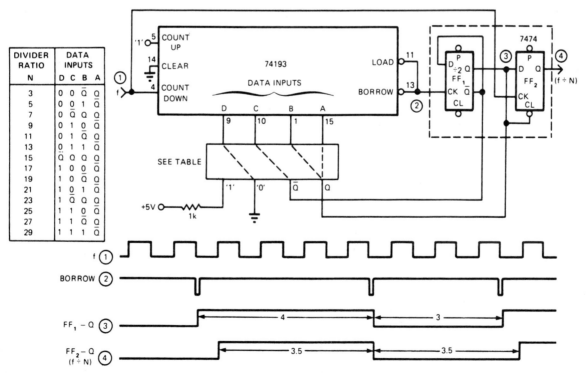

DIVIDER RATIO	DATA INPUTS			
N	D	C	B	A
3	0	0	$\bar{Q}$	Q
5	0	0	1	$\bar{Q}$
7	0	$\bar{Q}$	Q	$\bar{Q}$
9	0	1	0	$\bar{Q}$
11	0	1	$\bar{Q}$	Q
13	0	1	1	$\bar{Q}$
15	$\bar{Q}$	Q	Q	$\bar{Q}$
17	1	0	0	$\bar{Q}$
19	1	0	$\bar{Q}$	Q
21	1	0	1	$\bar{Q}$
23	1	$\bar{Q}$	Q	$\bar{Q}$
25	1	1	0	$\bar{Q}$
27	1	1	$\bar{Q}$	Q
29	1	1	1	$\bar{Q}$

3 TO 29 ODD-MODULO—Basic divider using 74193 4-bit up/down counter and single 7474 dual D flip-flop provides any odd number of divider ratios from 3 to 29 by changing feedback connections as shown in table, all with symmetrical output waveforms. Based on writing any odd number N as $N = M + (M + 1)$, where M is integer. Circuit forces counter to divide alternately by M and M + 1. Connection shown is for divide-by-7.—V. R. Godbole, Simplify Design of Fixed Odd-Modulo Dividers, *EDN Magazine,* June 5, 1975, p 77–78.

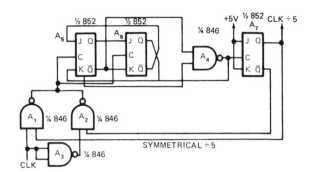

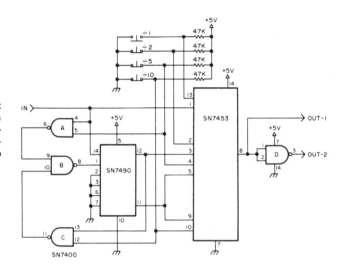

DIVIDE BY 5—Requires only two digital chip types. Input clock is alternately inverted and noninverted for clocking divide-by-3 counter, to give effect of dividing by 2½ which toggles A_3 to give symmetrical divide-by-5 output with 50% duty cycle for pulses. Article gives timing diagram and traces operation of circuit.—C. W. Hardy, Reader Responds to Odd Modulo Divider in July 1st EDN, *EDN Magazine*, Oct. 1, 1972, p 50.

SQUARE-WAVE DIVIDER—Divides input square wave by 1, 2, 5, or 10 depending on which switch is open. Signal at OUT-1 is inverted with respect to input, and OUT-2 is noninverted.—Circuits, *73 Magazine*, June 1977, p 49.

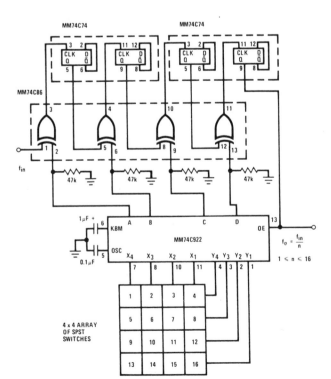

PROGRAMMABLE COUNTER—Input frequency can be divided by any number between 1 and 16 by pressing appropriate key on keyboard connected to National MM74C922 16-key encoder. Output frequency is symmetrical for odd and even divisors. Can be used for simple frequency synthesis or as keyboard-controlled CRO trigger. Operates over standard CMOS supply range of 3–15 V. Typical upper frequency limit is 1 MHz with 10-V supply. Circuit uses two MM74C74 dual D flip-flops and MM74C86 EX-CLUSIVE-OR package.—"CMOS Databook," National Semiconductor, Santa Clara, CA, 1977, p 5-50–5-51.

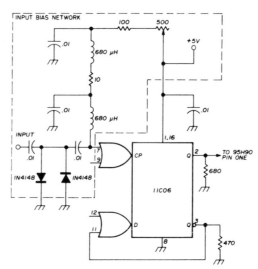

UHF PRESCALER—Uses Fairchild 11C06 700-MHz D flip-flop as divide-by-20 UHF prescaler with toggle rates in excess of 550 MHz from 0 to 75°C. Amplifier may be used in place of input bias network shown. Developed for use with 95H90 decade divider. Unused CP and D inputs are tied to ground.—D. Schmieskors, 1200-MHz Frequency Scalers, *Ham Radio,* Feb. 1975, p 38–40.

SECTION 20
Function Generator Circuits

Used for generating various combinations of sine, square, and triangle waveforms, usually with manual or external variations of frequency in AF or RF ranges by DC control voltage. Also includes circuits for generating cubic, quadratic, hyperbolic, trigonometric, ramp, and other mathematical waveforms, as well as circuits for converting one of these waveforms to one or more others. See also Multivibrator, Oscillator, Pulse Generator, Signal Generator, and Sweep chapters.

FSK SINE-SQUARE-TRIANGLE GENERATOR— Exar XR-2206 modulator-demodulator (modem) is connected as function generator providing high-purity sinusoidal output along with triangle and square outputs, for FSK applications. Circuit has excellent frequency stability along with TTL and CMOS compatibility. Total harmonic distortion in 3 V P-P sine output is about 2.5% untrimmed, but can be trimmed to 0.5%. High-level data input signal selects frequency of $1/R_6C_3$ Hz, while low-level input selects $1/R_7C_3$ Hz. For optimum stability, R_6 and R_7 should be in range of 10K to 100K. Adjust R_8 and R_9 for minimum distortion.—"Phase-Locked Loop Data Book," Exar Integrated Systems, Sunnyvale, CA, 1978, p 57–61.

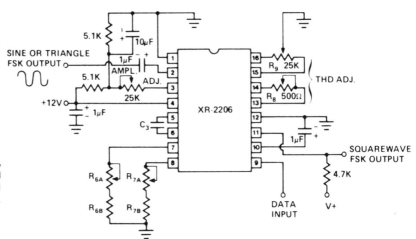

HYPERBOLIC A/X FUNCTION—Uses Precision Monolithics DAC-20EX D/A converter with OP-17G opamp to generate extended-range hyperbolic functions of the type A/X, where A is analog constant and X represents decimally expressed digital divisor. R5 provides simultaneous adjustment of scale factor and output amplifier offset voltage. Same circuit serves for −A/X function if DAC reference amplifier and output opamp terminals are reversed.—W. Ritmanich, B. Blair, and B. Debowey, "Digital-to-Analog Converter Generates Hyperbolic Functions," Precision Monolithics, Santa Clara, CA, 1977, AN-23, p 2.

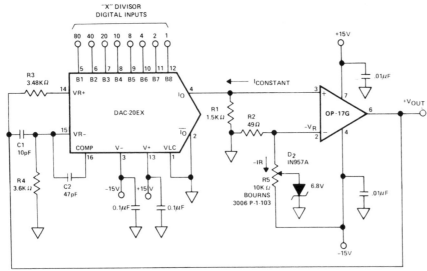

R_1, R_2, R_3 = 1% METAL FILM
ALL CAPS = CERAMIC DISC

GROUND PLANE

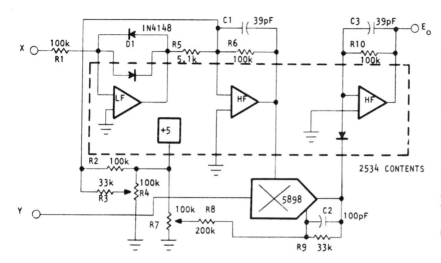

VOLTAGE-CONTROLLED NONLINEAR—Circuit produces function $E_0 = X^{Y/2}$, where X is input voltage in range of +10 mV to +10 V and Y is analog programming voltage in range of −0.4 V to −10 V. Uses Optical Electronics 2534 temperature-compensated log feedback elements, +5 V reference, two high-frequency opamps, and one low-frequency opamp. 2534 produces log conversion of input signal. 5898 multiplier serves to vary scale factor of log signal. With offsets used as shown, +10 V input will always produce +10 V output regardless of Y input. To set up, adjust R4 until output does not change with Y for +10 V input, then adjust R7 for +10 V output with +10 V input.—"Voltage-Controlled Non-Linear Function Generator," Optical Electronics, Tucson, AZ, Application Tip 10263.

0.5 Hz TO 1 MHz SINE-SQUARE-TRIANGLE— Uses Exar XR-2206 IC function generator in simple circuit that operates from dual supply ranging from ±6 V to ±12 V. With 1-μF capacitor for C, 2-megohm frequency control covers range of 0.5–1000 Hz. Range is 5–10,000 Hz with 0.1 μF, 50 Hz to 100 kHz with 0.01 μF, and 500 Hz to 1 MHz with 0.001 μF. Designed for experiments with active filters.—H. M. Berlin, "Design of Active Filters, with Experiments," Howard W. Sams, Indianapolis, IN, 1977, p 9–10.

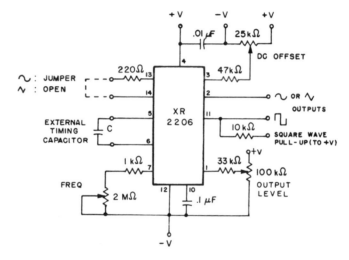

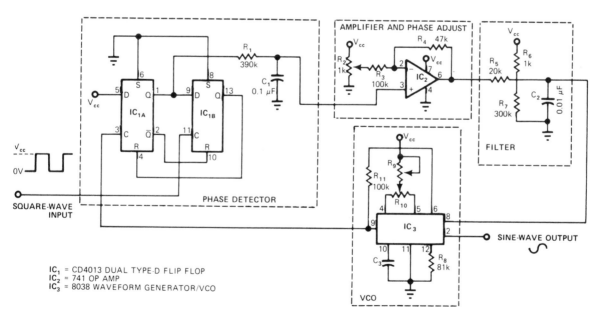

IC$_1$ = CD4013 DUAL TYPE-D FLIP FLOP
IC$_2$ = 741 OP AMP
IC$_3$ = 8038 WAVEFORM GENERATOR/VCO

SQUARE TO SINE WITH PLL—8038 waveform generator simultaneously generates synthesized sine wave and square wave. Square-wave output closes phase-locked loop through 741 opamp IC$_2$ and dual flip-flop IC$_1$, while sine-wave output functions as converted output. Center frequency is $0.15/R_9C_3$. R_{10} should be at least 10 times smaller than R_9. If center frequency is 400 Hz, capture range is half that or ±100 Hz. When input is applied, phase comparator generates voltage related to frequency and phase difference of input and free-running signals. IC$_2$ amplifies and offsets phase-difference signal. Sine output has less than 1% distortion, DC component of 0.5 V_{CC}, and minimum amplitude of 0.2 V_{CC} P-P.—L. S. Kasevich, PLL Converts Square Wave into Sine Wave, *EDN Magazine*, June 20, 1978, p 128.

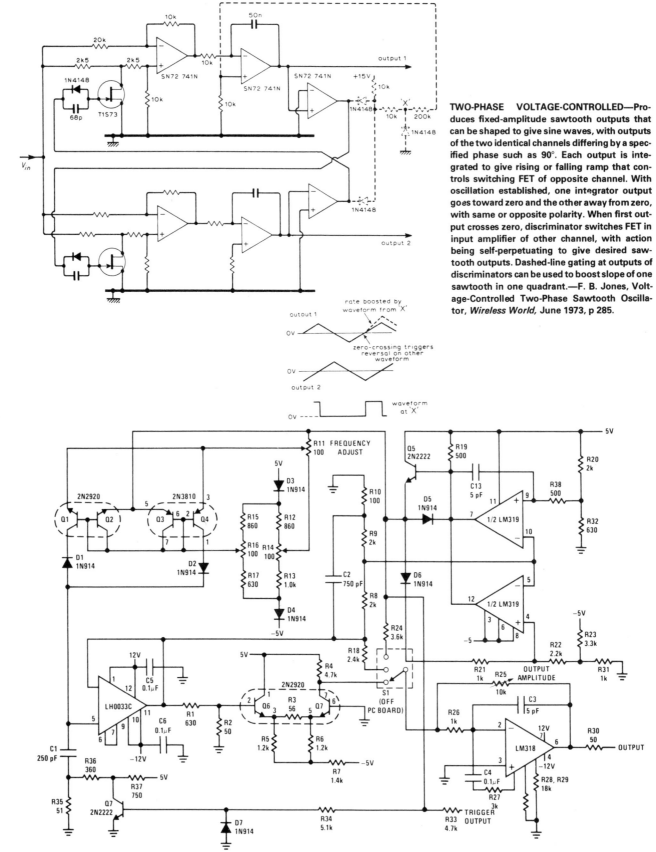

TWO-PHASE VOLTAGE-CONTROLLED—Produces fixed-amplitude sawtooth outputs that can be shaped to give sine waves, with outputs of the two identical channels differing by a specified phase such as 90°. Each output is integrated to give rising or falling ramp that controls switching FET of opposite channel. With oscillation established, one integrator output goes toward zero and the other away from zero, with same or opposite polarity. When first output crosses zero, discriminator switches FET in input amplifier of other channel, with action being self-perpetuating to give desired sawtooth outputs. Dashed-line gating at outputs of discriminators can be used to boost slope of one sawtooth in one quadrant.—F. B. Jones, Voltage-Controlled Two-Phase Sawtooth Oscillator, *Wireless World*, June 1973, p 285.

10 Hz TO 2 MHz—Triangle wave is generated by switching current-source transistors to charge and discharge timing capacitor. Precision dual comparator sets peak-to-peak amplitude. Sine converter requires close amplitude control to give low-distortion output from triangle input. Square-wave output is obtained at emitter of Q5, for driving current switches Q1-Q4 and LM318 output amplifier. Scaling permits adjusting all three waveforms to ±10 V. Waveforms are symmetrical up to 1 MHz, and output is usable to about 2 MHz.—R. C. Dobkin, "Wide Range Function Generator," National Semiconductor, Santa Clara, CA, 1974, AN-115.

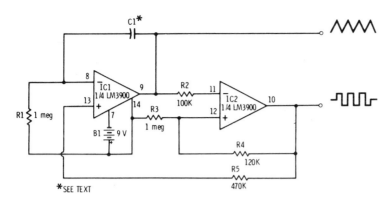

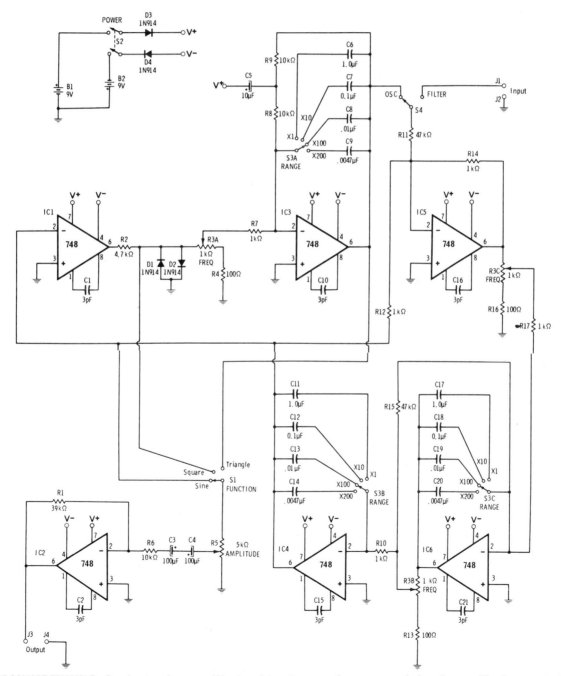

SQUARE-TRIANGLE AF—Two sections of LM3900 quad opamp are connected to generate dual-polarity triangle- and square-wave AF outputs while operating from single supply, by using current mirror circuit at noninverting input. Value used for C1 determines frequency and pulse width; frequency ranges from 0.5 Hz with 1 μF to 3800 Hz with 0.0001 μF and 21 kHz with C1 omitted. Pulse-width range is 35 μs without C1 to 1.6 s with 1 μF.—F. M. Mims, "Integrated Circuit Projects, Vol. 5," Radio Shack, Fort Worth, TX, 1977, 2nd Ed., p 57–63.

AF SINE-SQUARE-TRIANGLE—Can be tuned over entire audio spectrum in four ranges for generation of low-distortion waves for laboratory use. IC1 converts sine wave to square wave. IC3 acts as integrator converting square-wave output of IC1 to triangle wave. IC4-IC6 form state-variable filter for removing sine-wave component from triangle wave. IC2 is simple inverting amplifier for output.—R. Melen and H. Garland, "Understanding IC Operational Amplifiers," Howard W. Sams, Indianapolis, IN, 2nd Ed., 1978, p 130–134.

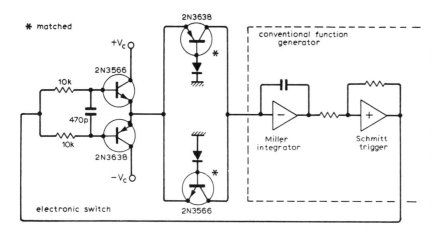

1000:1 FREQUENCY SWEEP—Permits varying output frequency of function generator over wide frequency range by using pot to vary control voltage V_c. Network consisting of two transistors and two diodes replaces usual charging resistor of Miller integrator in function generator, and has output current varying exponentially with input voltage. Electronic switch using pair of transistors is controlled by Schmitt trigger of function generator, which connects $+V_c$ and $-V_c$ alternately to charging circuit. If frequency pot is mechanically connected to strip-chart recorder, Bode plots of audio equipment can be made over entire audio range.—P. D. Hiscocks, Function Generator Mod. for Wide Sweep Range, *Wireless World,* Aug. 1973, p 374.

CURRENT-CONTROLLED SQUARE-TRIANGLE GENERATOR—CA3080 opamp is connected as current-controlled integrator of both polarities for use in current-controlled triangle oscillator. Frequency depends on values of C and opamp bias current and can be anywhere in audio range of 20 Hz to 20 kHz. Square-wave output is obtained by using LM301A opamp as Schmitt trigger.—S. Franco, Current-Controlled Triangular/Square-Wave Generator, *EDN Magazine,* Sept. 5, 1973, p 91.

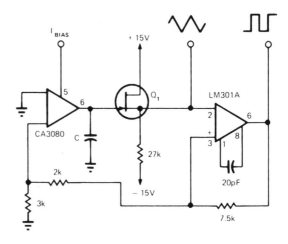

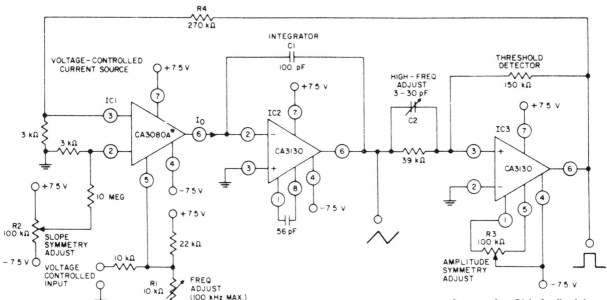

SINGLE CONTROL FOR 1,000,000:1 FREQUENCY RANGE—Uses two RCA CA3130 opamps and CA3080A operational transconductance amplifier to generate square and triangle outputs that can be swept over range of 0.1 Hz to 100 kHz with single 100K pot R1. Alternate voltage-control input is available for remote adjustment of sweep frequency. IC1 is operated as voltage-controlled current source whose output current is applied directly to integrating capacitor C1 in feedback loop of integrator IC2. R2 adjusts symmetry of triangle output. IC3 is used as controlled switch to set excursion limits of triangle output when square wave is desired.—"Linear Integrated Circuits and MOS/FET's," RCA Solid State Division, Somerville, NJ, 1977, p 236–244.

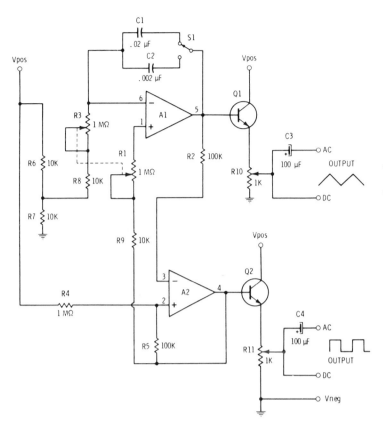

VARIABLE SQUARE-TRIANGLE—Dual pot R1-R3 varies frequency over range of 15–500 Hz when C1 is in circuit and 150–4800 Hz when C2 is in circuit. Each output has amplitude control. Opamps are Motorola MC3401P or National LM3900, and transistors are 2N2924 or equivalent NPN. Supply can be 12 VDC.—C. D. Rakes, "Integrated Circuit Projects," Howard W. Sams, Indianapolis, IN, 1975, p 19–20.

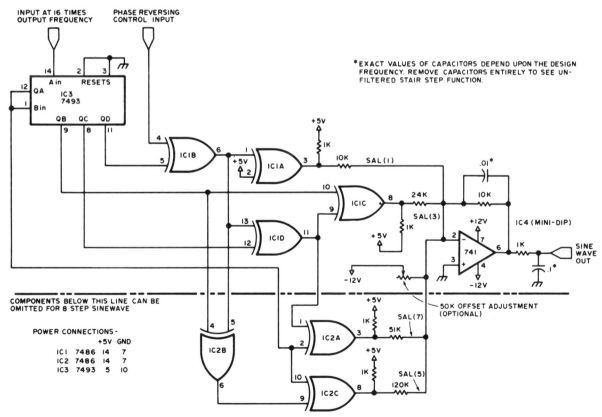

DIGITAL FOURIER—Sine-wave generator produces Walsh-function approximation of sine function. Frequency of sine wave is set by square-wave input to pin 14 of 7493. Filter components of opamp help smooth staircase waveform generated by summing Walsh-function components as weighted by resistors. Circuit is converter consisting of digital expander that expands input square wave into variety of digital waveforms and analog combiner that adds these waveforms to produce periodic analog output. Negative signs of Walsh harmonics are handled with digital inverter, and magnitudes are handled by choice of resistor value in summing junction. Signs and magnitudes are under microprocessor control. Net output is stairstep approximation to desired output, which can be smoothed by low-pass filter.—B. F. Jacoby, Walsh Functions: A Digital Fourier Series, *BYTE,* Sept. 1977, p 190–198.

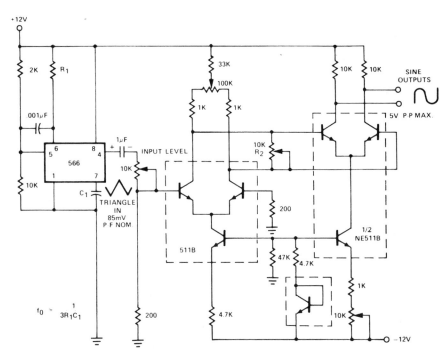

TRIANGLE-TO-SINE CONVERTER—Nonlinear emitter-base junction characteristic of 511B transistor array is used for shaping triangle output of 566 function generator to give sine output having less than 2% distortion. Amplitude of triangle is critical and must be carefully adjusted for minimum distortion of sine wave by varying values of R_1, R_2, and input level pot while monitoring output with Hewlett-Packard 333A distortion analyzer.—"Signetics Analog Data Manual," Signetics, Sunnyvale, CA, 1977, p 851–853.

PULSE/SAWTOOTH GENERATOR—Pulse output is obtained from Exar XR-2006C function-generator IC when pin 9 is shorted to square-wave output at pin 11. Pulse duty cycle, along with rise and fall times of ramp from pin 2, is determined by values of R1 and R2. Both can be adjusted from 1 to 99% by proper selection of resistor values as given in formulas alongside diagram.—E. Noll, VHF/UHF Single-Frequency Conversion, *Ham Radio*, April 1975, p 62–67.

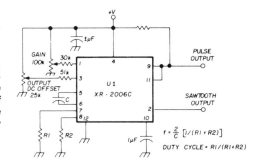

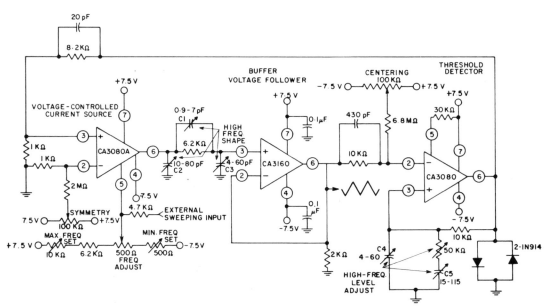

SINGLE FREQUENCY CONTROL—Adjustment range of over 1,000,000 to 1 for frequency is achieved by using CA3080A as programmable current source, CA3160 opamp as voltage follower, and CA3080 variable opamp as high-speed capacitor. Variable capacitors C1-C3 shape triangle waveform between 500 kHz and 1 MHz. C4 and C5 with 50K trimmer in series with C5 maintain constant amplitude within 10% up to 1 MHz.—"Circuit Ideas for RCA Linear ICs," RCA Solid State Division, Somerville, NJ, 1977, p 6.

SECTION 21
Gain Control Circuits

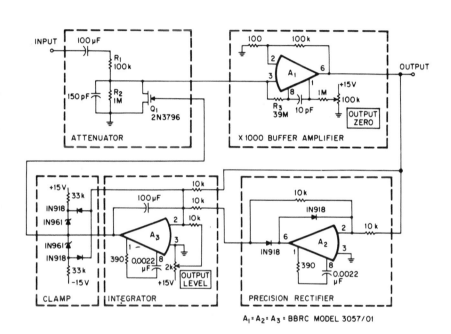

1,000:1 GAIN RANGE—Insulated-gate fet increases controlled gain range to about 60 dB. Opamp A1 is connected noninverting to prevent loading of variable attenuator R1-R2-Q1. Output is full-wave rectified by A2 and fed to A3 along with reference from 2K output-level pot. A3 integrates sum and applies it to Q1 to complete feedback loop. Clamp circuit around A3 prevents saturation by zero or overload input signals.—E. Guenther, MOS-FET Provides 60-dB Dynamic Range Low-Frequency AGC Circuit, *EEE*, Nov. 1969, p 107.

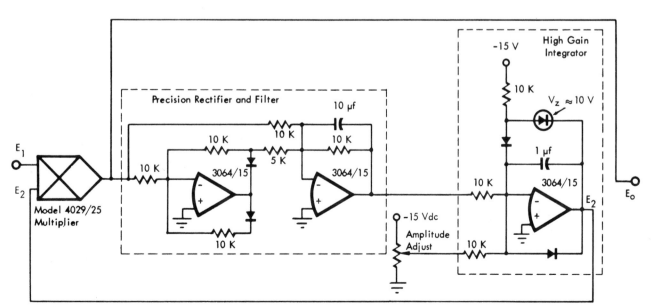

MULTIPLIER MODULE FOR AGC—Burr-Brown 4029/25 quarter-square multiplier-divider is combined with amplifier module to give stable, reliable, and accurate agc loop for such military applications as stabilizing oscillator signal amplitude or holding amplitude constant while phase angle is varied by filtering. Designed for frequency range of 10–500 Hz. Amplitude of output signal is converted to d-c and compared with reference voltage to generate error signal for integrator whose output is multiplied by input signal to vary gain. Rate of restabilization of loop after sudden change in input level depends on time constant of integrator and rectifier-filter. —Quarter-Square Multiplier/Dividers, Burr-Brown Research, Tucson, Ariz., PDS-201C, 1969.

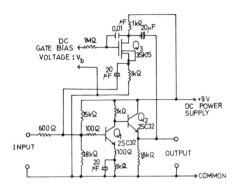

MOS TRANSISTOR AS VARIABLE RESISTOR—Provides resistance range of 0.8 to 50,000 ohms depending on strength of applied signal. When used in feedback path of agc amplifier (Q3), gives agc range of 3:1. Input signal applied to base of Q1 appears at emitter of Q2 and is fed back through mos transistor Q3 to base of Q1.—H. Ikeda, MOST Variable Resistor and Its Application to an AGC Amplifier, *IEEE Journal of Solid-State Circuits*, Feb. 1970, p 43—45.

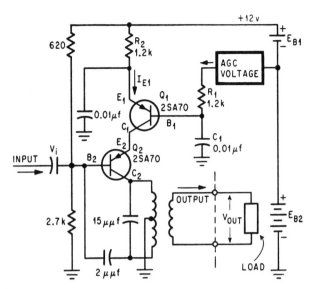

50-DB AGC—Used in Sony portable tv sets to handle large input signals without distortion or overload. Two transistors give far greater control range than conventional single-stage agc. Q2 amplifies, while Q1 serves are variable impedance in emitter circuit of Q2 and controls bias of Q2.—Two-Stage Gain Control for Portable TV Set, *Electronics*, Nov. 14, 1966, p 171—172.

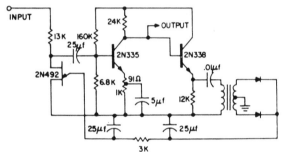

LIMITED AGC FOR OSCILLATOR—Ujt compensates oscillator whose output varies inversely with frequency, to keep output essentially constant at 7-V p-p over frequency range of 2.8 to 3.1 kHz. Operation is based on negative resistance characteristic of ujt.—R. S. Hughes, Automatic Gain Control Circuit Uses Unijunction Transistor, "400 Ideas for Design Selected from Electronic Design," Hayden Book Co., N.Y., 1964, p 111.

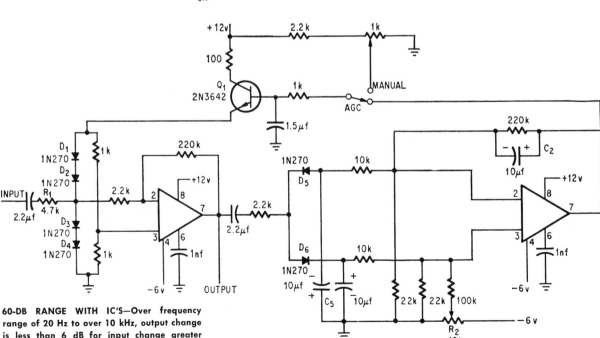

60-DB RANGE WITH IC'S—Over frequency range of 20 Hz to over 10 kHz, output change is less than 6 dB for input change greater than 60 dB. Signal from voltage divider R1-D1-D2-D3-D4 is amplified by first IC opamp having gain of 100, rectified by D5-D6, then fed back to input through second opamp and buffer Q1. For manual control, switch inserts potentiometer in feedback loop. Input signal should be several volts, for maximum agc range.—W. H. Ellis, Jr., AGC Circuit Possesses 60-Decibel Gain, *Electronics*, Dec. 12, 1966, p 107—108.

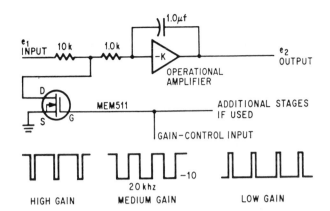

HIGH GAIN   MEDIUM GAIN   LOW GAIN

20 khz

CHOPPER-CONTROLLED GAIN—Gain of any opamp may be varied over 32-dB range by varying ratio of OFF and ON times of 20-kHz pulse-width-modulated source applied to single fet. Resulting chopping of input to opamp, at significantly higher rate than upper cutoff frequency of amplifier, attenuates input signal during pulse ON time while giving normal gain during OFF time.—D. E. Lancaster, Operational Amplifier Gain Varied by FET Chopper, *Electronics*, Oct. 2, 1967, p 98.

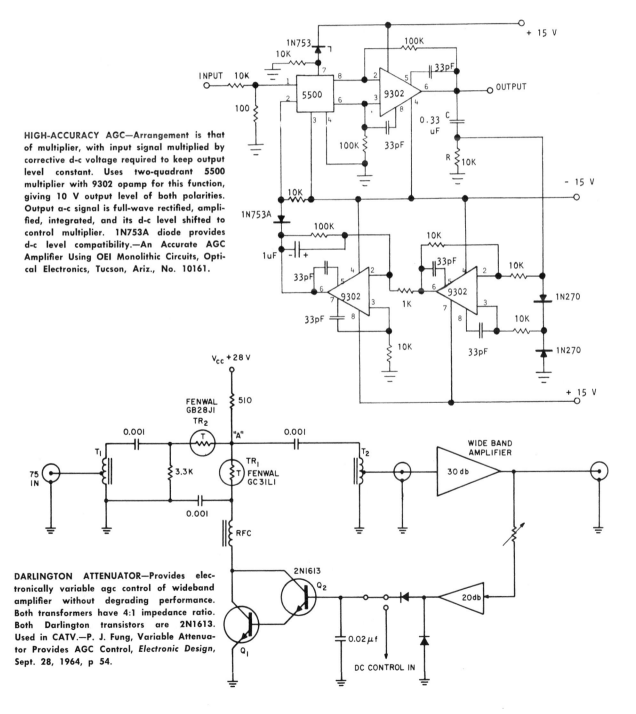

HIGH-ACCURACY AGC—Arrangement is that of multiplier, with input signal multiplied by corrective d-c voltage required to keep output level constant. Uses two-quadrant 5500 multiplier with 9302 opamp for this function, giving 10 V output level of both polarities. Output a-c signal is full-wave rectified, amplified, integrated, and its d-c level shifted to control multiplier. 1N753A diode provides d-c level compatibility.—An Accurate AGC Amplifier Using OEI Monolithic Circuits, Optical Electronics, Tucson, Ariz., No. 10161.

DARLINGTON ATTENUATOR—Provides electronically variable agc control of wideband amplifier without degrading performance. Both transformers have 4:1 impedance ratio. Both Darlington transistors are 2N1613. Used in CATV.—P. J. Fung, Variable Attenuator Provides AGC Control, *Electronic Design*, Sept. 28, 1964, p 54.

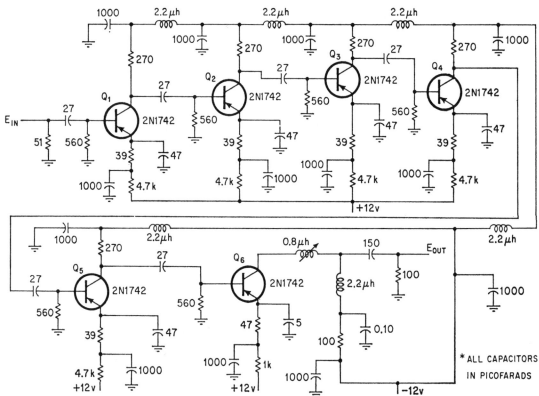

CONSTANT-OUTPUT LIMITER—Provides constant output essentially independent of changes in input amplitude, without introducing phase shift. Limiting begins at input level of 9 mV, and circuit holds output level constant within 0.06 dB even though input level increases 50 dB. Output stage Q6 has resonant circuit tuned to 80 MHz and drives 50-ohm or 100-ohm cable. Upper cutoff frequency of 100 MHz is limited only by transistors used.—R. J. Turner, Transistors Improve High-Frequency Limiter, *Electronics*, Oct. 13, 1969, p 94.

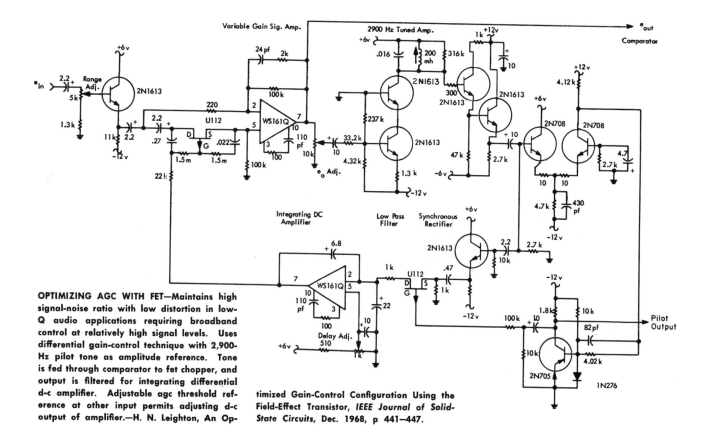

OPTIMIZING AGC WITH FET—Maintains high signal-noise ratio with low distortion in low-Q audio applications requiring broadband control at relatively high signal levels. Uses differential gain-control technique with 2,900-Hz pilot tone as amplitude reference. Tone is fed through comparator to fet chopper, and output is filtered for integrating differential d-c amplifier. Adjustable agc threshold reference at other input permits adjusting d-c output of amplifier.—H. N. Leighton, An Op-timized Gain-Control Configuration Using the Field-Effect Transistor, *IEEE Journal of Solid-State Circuits*, Dec. 1968, p 441–447.

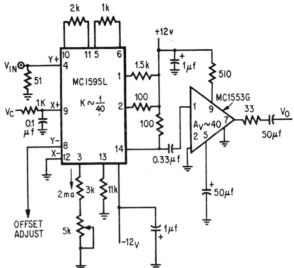

AGC WITH IC MULTIPLIER—Circuit requires 0-1 V d-c control voltage to provide linear gain control of 1-V p-p signal at 200 kHz. Peak-to-peak value of output is thus exactly the same as value of control voltage. Gain is recovered with IC video amplifier having gain of 40.—E. Renschler and D. Weiss, Try the Monolithic Multiplier as a Versatile A-C Design Tool, *Electronics*, June 8, 1970, p 100—105.

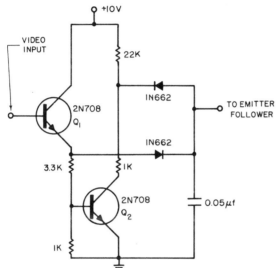

BOXCAR FOR PULSE AGC—Provides control over gain of video amplifier when both pulse width and frequency are variable. Used to keep amplitude of r-f burst constant despite variations in signal amplitude, frequency, or pulse width. Boxcar circuit is sample-and-hold integrator. Output of standard emitter-follower is desired agc voltage to be applied to controlled amplifier. With values shown, circuit will handle inputs of 2 to 8 V at 1 to 50 kHz and pulse widths of 5 to 500 μs.—G. P. Klein, Boxcar Circuit Provides Pulse Amplifier AGC, *Electronic Design*, Sept. 13, 1965, p 83—84.

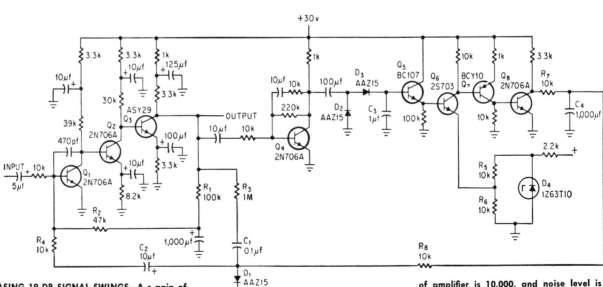

ERASING 19-DB SIGNAL SWINGS—A-c gain of 10—50,000 Hz amplifier is automatically controlled by error signal that changes dynamic resistance of diode D1 in feedback loop. Used in system for determining density of plasma in microwave cavity. Maximum gain of amplifier is 10,000, and noise level is 20 μV.—C. A. J. van der Geer, Amplifier Erases Swing of 19-DB in Input Signals, *Electronics*, May 29, 1967, p 85—86.

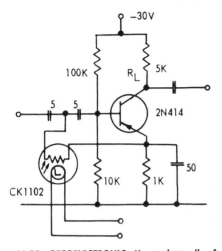

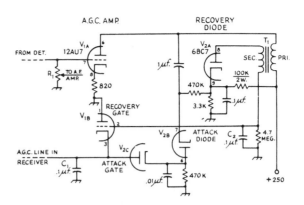

10-DB OPTOELECTRONIC—Uses photocell of Raysistor to shunt input of transistor, with one end of Raysistor tied to emitter to protect low-frequency response. Requires d-c control voltage of 0 to 1 V.—Raysistor Optoelectronic Devices, Raytheon, Quincy, Mass., 1967, p 12.

AGC FOR SSB—Provides fast attack and slow decay, for improved reception with ssb receiver. R1 is normal volume control in receiver and T1 is 1:3 step-up audio transformer such as Stancor A-53. Uses audio drive from output of detector in receiver.

Requires receiver having good skirt selectivity, with no d-c return from agc line to ground.—G. W. Luick, Improved A. G. C. For S. S. B. Reception, "Single Sideband for the Radio Amateur," ARRL, Newington, Conn., 1965, p 180.

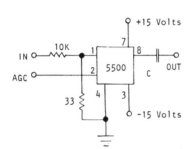

VOLTAGE-CONTROLLED GAIN—Analog multiplier makes agc voltage act on input signal up to 100 kHz, to give over 40 dB of agc range with low distortion. If signal level is under 10 mV at input, resistive divider is not needed and 5500 will then provide useful voltage gain.—Applying the Model 5500 Monolithic Analog Multiplier, Optical Electronics, Tucson, Ariz., No. 10138.

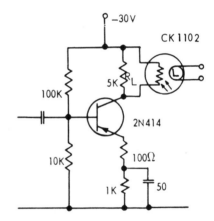

15-DB OPTOELECTRONIC—Uses Raysistor in collector circuit, so collector load decreases and reduces gain when input increases. Changing load resistance RL to 15K increases control range to over 30 dB for d-c control voltage of 0 to 1 V.—Raysistor Optoelectronic Devices, Raytheon, Quincy, Mass., 1967, p 12.

SECTION 22
Infrared Circuits

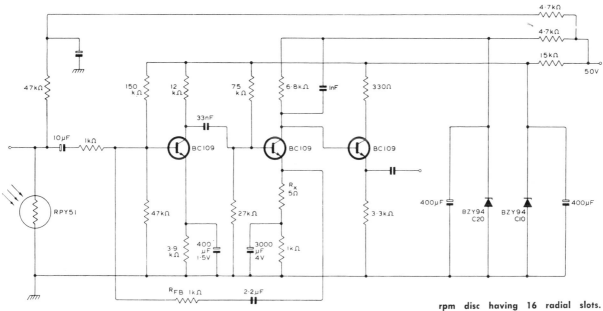

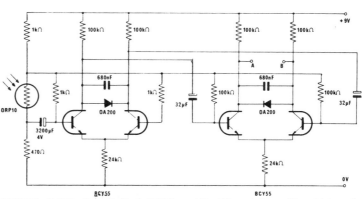

IC TEMPERATURE DISTRIBUTION—Infrared microscope uses Mullard RPY51 indium antimonide detector cooled by liquid nitrogen for measuring temperature of extremely small area, such as portion of integrated-circuit chip. Motor-driven chopper converts radiation from target to about 800 Hz with 3,000- rpm disc having 16 radial slots. Circuit features low noise and high gain stability. Article gives design equations and other circuits required for complete thermal microscope system.—T. J. Jarratt, Infrared Microscope for Temperature Measurement of Small Areas, *Mullard Technical Communications*, May 1968, p 101—107.

IR INTRUSION ALARM—Uses Mullard ORP10 uncooled indium antimonide photocell in unchopped system to detect objects with temperatures above 10 deg C moving across field of view against steady background at room temperature. Detector is a-c coupled to low- drift differential amplifier driving alarm circuit. Responds to minimum angular speed of about 1 milliradian per second by intruder.— R. A. Lockett, Passive Intruder Alarm Using Infrared Detector, *Mullard Technical Communications*, May 1968, p 88—90.

INTRUSION ALARM—When R2 is adjusted just below maximum ambient illumination, slight increase in light falling on photocell makes scr oscillate and produce loud clicking sounds in speaker. Frequency of clicks increases with illumination. Will respond to flashlight of intruder or to turning on of room lights in vacant office or home. Uses GE-X6 cadmium sulfide photoconductor. For infrared detection or fire alarm, use cadmium selenide cell.—"Hobby Manual," General Electric, Owensboro, Ky., 1965, p 124.

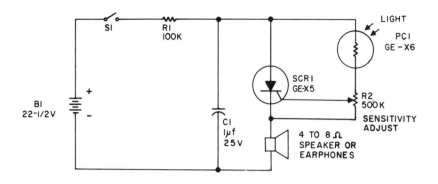

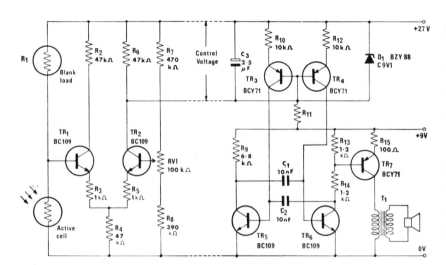

LOCATING FIRE IN SMOKE—Uses uncooled lead sulfide cell with lens to pick up source of fire or heat when visibility is completely obscured by smoke, dust, or fumes. Cell is Mullard developmental type 119CPY. Additional cell R1 is blacked out and serves as temperature-compensating load for active cell, which drives adjustable-gain preamp that controls base currents of square-wave relaxation oscillator TR5-TR6. Emitter-follower TR7 provides necessary current gain for driving speaker which gives low note of about 100 Hz for no fire and increasing pitch up to 10 kHz for increasing infrared radiation from fire. Unit is portable and battery-operated, to permit use by firemen in aiming hose.—P. R. D. Coleby, Simple Heat Locator Using Infrared Detector, *Mullard Technical Communications*, May 1968, p 86—88.

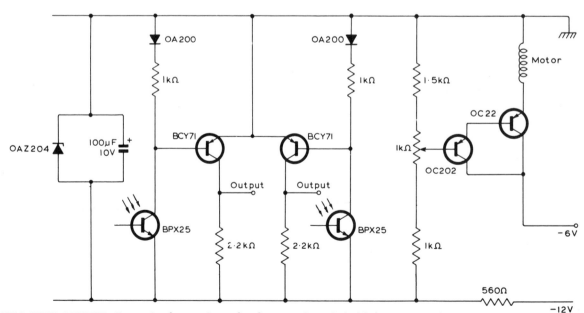

IR CAMERA PULSE AMPLIFIER—Uses pair of BPX25 silicon phototransistors receiving light pulses through rotating 30-hole Nipkow scanning disc on which infrared image is focused. One phototransistor receives sync pulses from uniformly spaced holes on disc, and other receives pulses from scanning spiral of holes. Resulting electrical pulses are squared, amplified, and applied to time bases of camera feeding standard oscilloscope. Motor drive circuit at right, for disc, applies constant voltage to motor, with 1K pot providing 10% range of speed control. Used for examining patients for cancerous growths, locating hot spots on circuit boards, and detecting flaws in glass furnace walls.—M. H. Jervis, Closed Circuit Infrared Television System, Mullard Ltd., London, TP950, 1967.

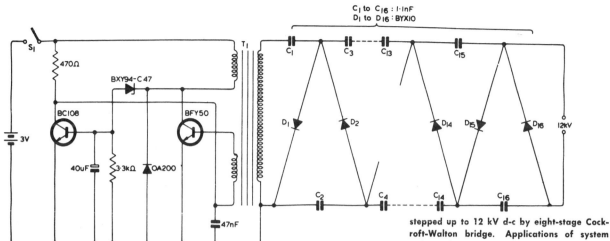

12 KV D-C—Used with Mullard 6929-1 image converter tube for converting near-infrared radiation to visible yellow-green emitted radiation. Uses simple blocking oscillator operating from 3-V battery to drive high-ratio transformer giving 1.5-kV pulses. These are stepped up to 12 kV d-c by eight-stage Cockroft-Walton bridge. Applications of system include observation of behavior of people in the dark, animals at night, scatter from laser beams, and hot-spots on defective high-voltage insulators.—Infrared Image Converter, Mullard Ltd., London, TP898, 1967.

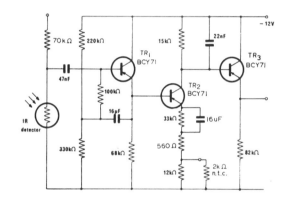

RADIATION THERMOMETER—Covers 100 to 500 C in five ranges, using linear amplification. Detector is Mullard 61SV lead sulfide cell having high infrared sensitivity at room temperature, for measuring radiant energy from surface of target. Energy from target is conducted to chopper wheel by light pipe, for a-c input to head amplifier that provides necessary transformation from high to low impedance and compensates for decrease in cell sensitivity with increasing ambient temperature.—P. R. D. Coleby, Radiation Thermometer for Temperatures Greater than 100 deg C, Mullard Technical Communications, May 1968, p 91—98.

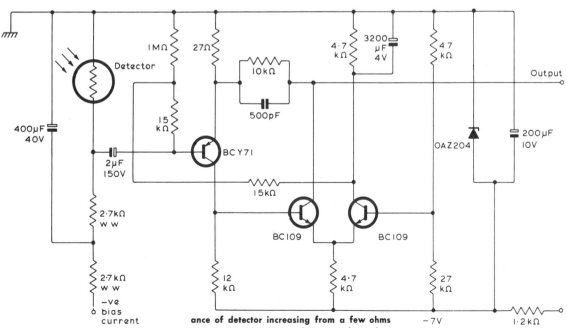

INDIUM ANTIMONIDE DETECTOR PREAMP—Will detect temperature changes from room temperature down to 77 deg K, with impedance of detector increasing from a few ohms at room temperature to about 200 ohms when back of detector is cooled by liquid nitrogen drip-feed system. Used in industrial and medical thermal scanning systems employing 30-hole Nipkow scanning disc.—M. H. Jervis, Closed Circuit Infrared Television System, Mullard Ltd., London, TP950, 1967.

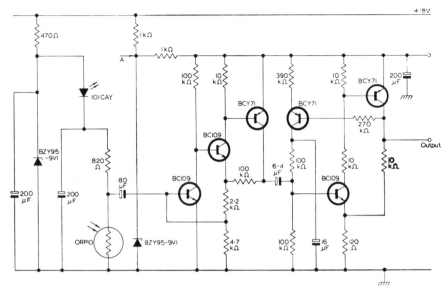

IR CLINICAL THERMOMETER—Covers range of 30 to 40 C, for measuring temperature of external auditory canal. Closely approximates oral measurements. Uses Mullard ORP10 uncooled indium antimonide cell for which radiation is interrupted at about 600 Hz by motor-driven chopper. Resulting a-c signal is amplified by flat-response wideband amplifier, mixed with signal corresponding to room temperature, then fed to phase-sensitive rectifier that derives its reference signal from interruption of secondary beam of radiation (between 101CAY emitter and BPX25 detector) by same chopper blade. Direct current from rectifier drives indicating meter. Article gives all circuits.—R. A. Lockett, Room Temperature Radiation Thermometer, *Mullard Technical Communications*, May 1968, p 98–101.

CODED-LIGHT DETECTOR—Tuned amplifier following phototransistor has twin-T network that greatly attenuates all signals more than 1 Hz off from 2.7 kHz. Only light within this narrow passband will turn on TR3 and energize load. One application is in photoelectric safety devices, where stray light might negate interruption of safety beam.—Applications of Silicon Planar Phototransistor BPX25, Philips, Pub. Dept., Elcoma Div., Eindhoven, The Netherlands, No. 316, 1967.

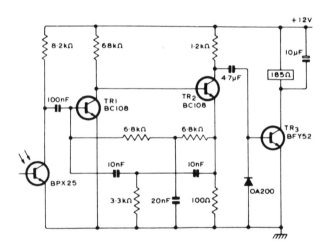

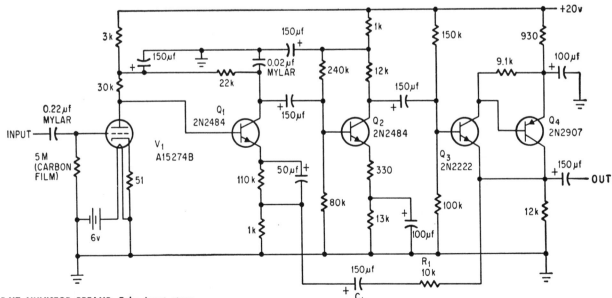

15-HZ NUVISTOR PREAMP—Tube input stage has much lower internal noise than bipolar transistor, and provides sufficient amplification to overcome noise of following transistor stages. Used in narrow-band signal processor for infrared radiometer. Gain overall is 40 dB, with less than 0.1 dB variation from −40 C to 60 C.—G. C. Kuipers, Front-End Nuvistor Lowers Transistor Amplifier Noise, *Electronics*, Oct. 31, 1966, p 71–72.

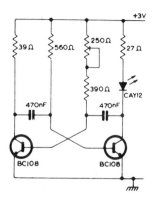

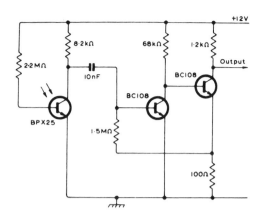

CODED INFRARED SOURCE—Used when un-modulated visible or infrared beam may be subject to interference from nearby stray light sources. Gallium arsenide light-emitting diode is energized by mvbr whose frequency is adjusted to match that of coded infrared source. Although mvbr modulator gives square-wave output, typically at 2.7 kHz, coded-beam detector responds to fundamental frequency.—Applications of Silicon Planar Phototransistor BPX25, Philips, Pub. Dept., Elcoma Div., Eindhoven, The Netherlands, No. 316, 1967.

INFRARED DETECTOR—Uses BPX25 phototransistor which has good sensitivity to near infrared wavelengths. With two amplifier stages, peak input of 1 lux will produce peak output of 400 mV. Response at 4 kHz is 3 dB below that at 1 kHz. Provides compact communication system giving secrecy and freedom from interference at ranges up to 100 ft when using either f:2 collimating lenses or parabolical reflectors at light source and at phototransistor. For near infrared, ordinary glass lenses are satisfactory.—Applications of Silicon Planar Phototransistor BPX25, Philips, Pub. Dept., Elcoma Div., Eindhoven, Netherlands, No. 316, 1967.

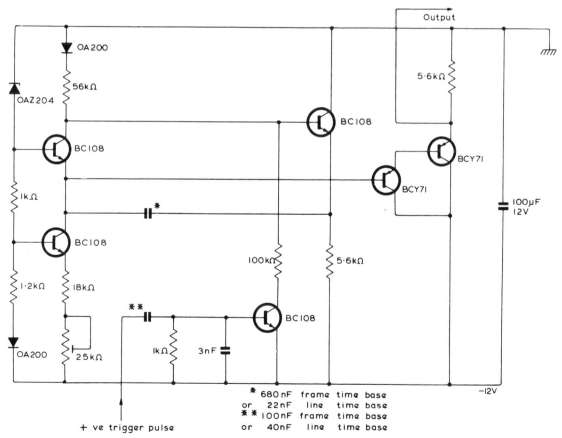

* 680 nF frame time base
or 22 nF line time base
** 100 nF frame time base
or 40 nF line time base

+ ve trigger pulse

IR CAMERA TIME BASES—Circuits for X and Y inputs of display cro for infrared Nipkow-disc scanning camera are identical sawtooth generators except for timing capacitor changes indicated below diagram. Trigger pulses are obtained from pulse amplifier fed by phototransistor. Used in medical and industrial infrared applications.—M. H. Jervis, Closed Circuit Infrared Television System, Mullard Ltd., London, TP950, 1967.

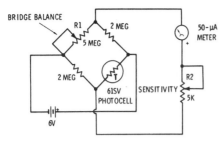

NONCONTACTING TEMPERATURE SENSOR—
Simple resistance bridge having Amperex
61SV infrared photocell in one leg permits
measuring temperature of furnace, flat iron,
or other heated objects from distance of
several feet. When bridge is balanced, meter
will read upscale when cell resistance is decreased by exposure to infrared radiation.
—J. P. Shields, "Novel Electronic Circuits," H.
W. Sams & Co., Indianapolis, Ind., 1968, p 11.

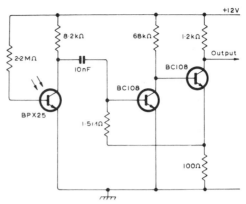

IR COMMUNICATION DETECTOR—Uses BPX25
phototransistor followed by two amplifying
stages to pick up modulated infrared radiation in communication system. Range is over
100 ft when using modulated gallium arsenide diode as light source.—Silicon Planar
Phototransistors, Mullard Ltd., London, TP1000,
1968.

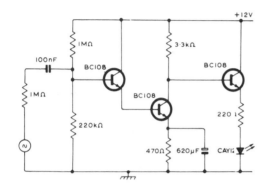

IR COMMUNICATION SOURCE—Gallium arsenide light-emitting diode is fed by microphone
or other signal source driving three-stage
amplifier. Response at 80 kHz is 3 dB below that at 1 kHz. Range is up to about 100
ft when using BPX25 phototransistor at receiving end.—Silicon Planar Phototransistors,
Mullard Ltd., London, TP1000, 1968.

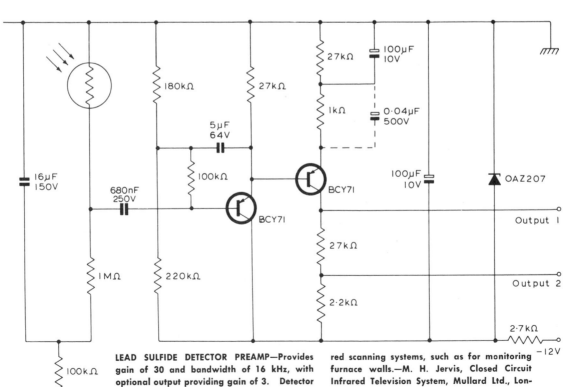

LEAD SULFIDE DETECTOR PREAMP—Provides
gain of 30 and bandwidth of 16 kHz, with
optional output providing gain of 3. Detector
requires no cooling, but minimum detectable
temperature is 120 C. Used in industrial infrared scanning systems, such as for monitoring
furnace walls.—M. H. Jervis, Closed Circuit
Infrared Television System, Mullard Ltd., London, TR950, 1967.

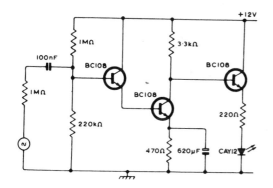

INFRARED MODULATOR—Serves as transmitter for short-range communication system using modulated infrared radiation. Light source at output of audio amplifier is CAY12 gallium arsenide diode. Can be used well above audio range, because response at 80 kHz is only 3 dB below that at 1 kHz. Peak diode current at 10 mA is obtained with input of 150 mV peak from microphone or other source.—Applications of Silicon Planar Phototransistor BPX25, Philips, Pub. Dept., Elcoma Div., Eindhoven, The Netherlands, No. 316, 1967.

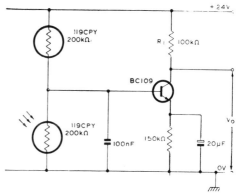

FIRE ALARM—Infrared flame detector produces voltage for actuating alarm as soon as flame occurs in field of view of Mullard 61SV lead sulfide detector which contains two 119CPY cells. Upper cell is covered, and serves to provide tracking with temperature changes. Circuit feeds bistable circuit (shown in article) that drives audible alarm such as Mallory Sonalert.—P. R. D. Coleby, Fire Alarm System Using Infrared Flame Detector, *Mullard Technical Communications*, May 1968, p 82—86.

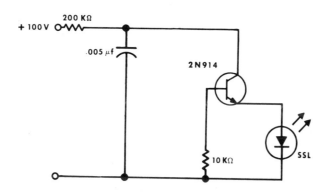

SSL OSCILLATOR—Gallium arsenide solid-state lamp in avalanche-transistor oscillator produces IR pulse durations in nanosecond range, with infrared power output of a few mW and pulse repetition rates of several kHz.—L. M. Hertz, Solid State Lamps—Part II, General Electric, Cleveland, Ohio, No. 3-0121, 1970, p 7.

NOTE: To locate additional circuits in the category of this chapter, use the index at the back of this book. Check also the author's "Sourcebook of Electronic Circuits," published by McGraw-Hill in 1968.

Instrumentation Circuits

Includes DC, AF, and wideband RF amplifiers with such special features as automatic nulling and automatic calibration, for use with resistance-bridge, photocell, strain-gage, and other input transducers. Applications include measurement of ionization, radiation, small currents, liquid flow and level, light level, pH, power, torque, weight, and wind velocity. Metal detectors and proximity detectors are also covered. See also chapters covering measurement of Capacitance, Frequency, Resistance, and Temperature.

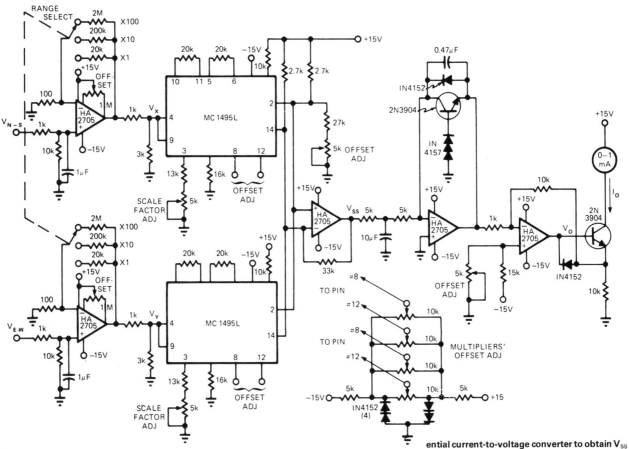

WIND SPEED—Developed to give magnitude of wind velocity over wide range of values when its two measured vectors are expressed as voltages. Output is in logarithmic form for easy adaptation to data processors. N-S and E-W vector voltages from strain-gage sensors are converted to normalized values V_x and V_y which are squared by MC1495L four-quadrant transconductance multipliers. Output currents are then summed, and HA2705 opamp is used as differential current-to-voltage converter to obtain V_{ss} as sum-of-squares of V_x and V_y. Range covered is 1–100 mph. Article covers operation of circuit in detail.—J. A. Connelly and M. B. Lundberg, Analog Multipliers Determine True Wind Speed, *EDN Magazine*, April 20, 1974, p 69–72.

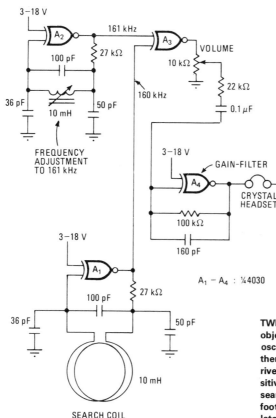

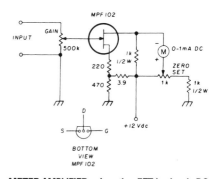

METER AMPLIFIER—Junction FET in simple DC amplifier circuit converts 0–1 mA DC milliammeter to 0–100 μA DC microammeter. Adjust zero-set control for zero meter current with no input, then apply input signal and adjust gain to desired value.—N. J. Foot, Electronic Meter Amplifier, *Ham Radio,* Dec. 1976, p 38–39.

TWIN-OSCILLATOR METAL DETECTOR—Metal object near search coil changes frequency of oscillator A$_1$ which is initially tuned to 160 kHz, thereby changing frequency of 1-kHz output derived by mixing with 161-kHz output of A$_2$. Sensitivity, determined largely by dimensions of search coil, is sufficient to detect coins about 1 foot away.—M. E. Anglin, C-MOS Twin Oscillator Forms Micropower Metal Detector, *Electronics,* Dec. 22, 1977, p 78.

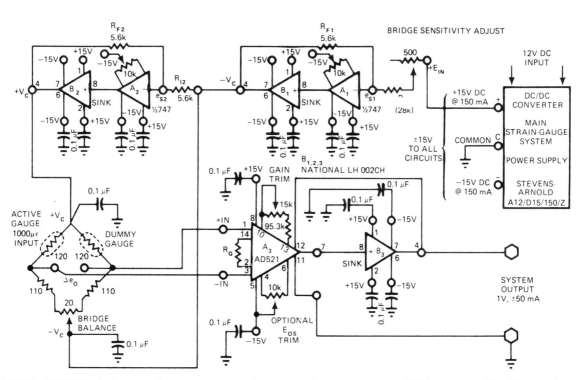

STRAIN-GAGE AMPLIFIER—Optimum performance is achieved in fully portable system by utilizing combination of 747 opamps for A$_1$ and A$_2$ with National LH002CH opamp for B$_1$-B$_3$ and special AD521K instrumentation amplifier for output stage. Bypass capacitors suppress undesirable high-frequency signals. Stevens-Arnold DC/DC converter operating from 12-V storage battery provides required regulated ±15 VDC for system while giving excellent power isolation.—D. Sheehan, Strain-Gauge Transducer System Uses Off-the-Shelf Components, *EDN Magazine,* Nov. 5, 1977, p 79–81.

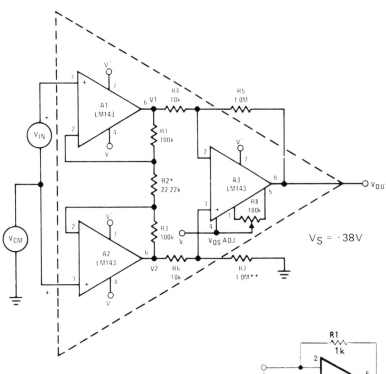

±34 V COMMON-MODE RANGE—Interconnections shown for three LM143 high-voltage opamps give equivalent of single differential-input opamp having wide common-mode range, high input impedance, and gain of 1000. Adjust R2 to trim gain. Adjust R7 for best common-mode rejection. With 10K load, frequency response is down 3 dB at 8.9 kHz.—"Linear Applications, Vol. 2," National Semiconductor, Santa Clara, CA, 1976, AN-127, p 2–3.

$V_S = \cdot 38V$

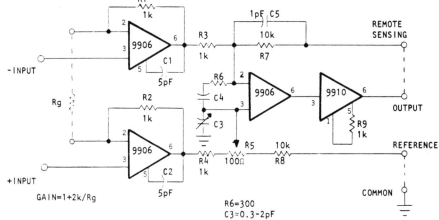

DIFFERENTIAL-INPUT AMPLIFIER—Provides gain up to 1000, depending on value of Rg, for video signals in radar, medical ultrasound, laser communication, and laser rangefinder applications. Uses three Optical Electronics 9906 wideband opamps and 9910 current booster for cable drive. Bandwidth is above 10 MHz for gains of 0.1 to 100, decreasing to 5 MHz at gain of 1000. Miller compensation of input amplifiers minimizes noise level and gives input impedance of 5 megohms and 5 pF.—"Wide Band Instrumentation Amplifier," Optical Electronics, Tucson, AZ, Application Tip 10276.

GAIN=1+2k/Rg

R6=300
C3≈0.3–2pF
C4=.033μF

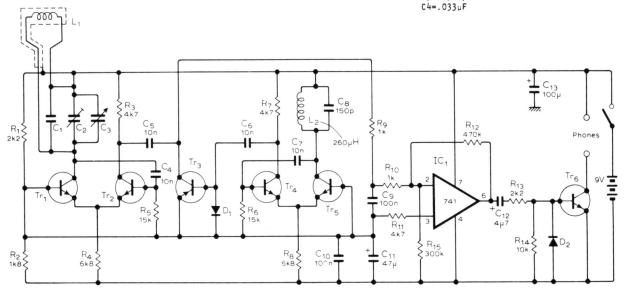

METAL DETECTOR—Will detect small coin up to about 5 inches underground and larger metal objects at much greater depths. Frequency of search oscillator Tr₁-Tr₂ depends on values used for three paralleled capacitors, search coil, and metal objects in vicinity of coil. Mixer Tr₃ feeds difference between search oscillator and reference oscillator Tr₄-Tr₅ to opamp and Tr₆ for driving phones or loudspeaker. Article gives construction and adjustment details, including dimensions for search coil. Reference oscillator is set to 625 kHz. C₁ is 560 pF, C₂ 150 pF, and C₃ 10 pF variable. C₂ is used for coarse tuning, and C₃ for fine adjustment to get beat note. Diodes are 1N4148. Tr₃ is BC308, BCY72, or equivalent, and other transistors are BF238, BC108, or equivalent.—D. E. Waddington, Metal Detector, *Wireless World*, April 1977, p 45–48.

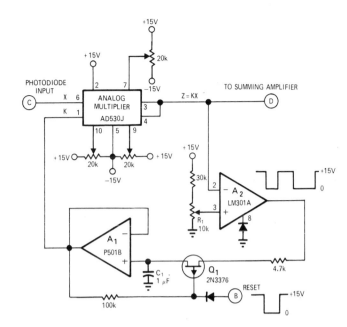

AUTOMATIC CALIBRATOR—Automatic scaling circuit permits frequent and fast recalibration for precision optical measurements, to compensate for variations in light intensity due to thermal cycling of lamp filament, dirty optics, and gain variations between photodetectors and between amplifiers. With reset pulse at point B, comparator A_2 compares output of multiplier to preset reference voltage on R_1. If A_2 input voltage is greater than reference applied to pin 3 by R_2, output switches to zero and remains there until C_1 has discharged enough to lower output of A_1 and output of multiplier below reference on pin 3. If input at pin 2 of A_2 is less than reference on pin 3, A_2 will switch to 15 V and output of multiplier will be adjusted upward until voltage on pin 2 of A_2 again exceeds that on pin 3. Output of A_2 is thus continually switching between 15 V and 0 V during reset or scaling. After reset pulse is removed, scale factor K is maintained constant by multiplier during measuring.—R. E. Keil, Automatic Scaling Circuit for Optical Measurements, *EDN/EEE Magazine*, Nov. 15, 1971, p 49–50.

HIGH GAIN FOR WEAK SIGNALS—National LM121 differential amplifier is operated open-loop as input stage for input signals up to ±10 mV. Input voltage is converted to differential output current for driving opamp acting as current-to-voltage converter with single-ended output. R4 is adjusted to set gain at 1000. Null pot R3 serves for offset adjustment.—"Linear Applications, Vol. 2," National Semiconductor, Santa Clara, CA, 1976, AN-79, p 7–8.

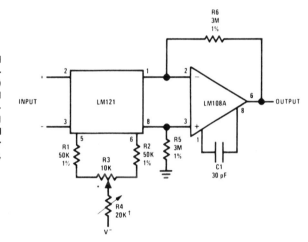

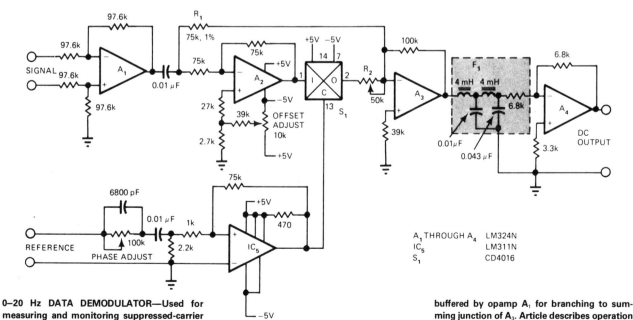

0–20 Hz DATA DEMODULATOR—Used for measuring and monitoring suppressed-carrier signal modulation from aircraft control systems. Provides data frequency response within 0.1 dB from DC to 20 Hz, with linearity better than 0.1%. In-phase reference voltage applied to comparator IC_5 controls gating of CD4016 MOS switch S_1. Suppressed-carrier signal is buffered by opamp A_1 for branching to summing junction of A_3. Article describes operation of circuit.—J. A. Tabb and M. L. Roginsky, Instrumentation Signal Demodulator Uses Low-Power IC's, *EDN Magazine*, Jan. 20, 1976, p 80.

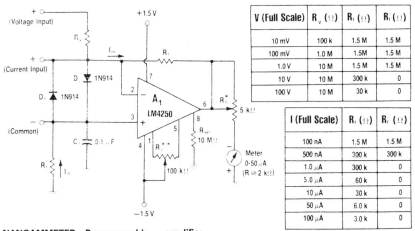

V (Full Scale)	R_v (Ω)	R_i (Ω)	R_i (Ω)
10 mV	100 k	1.5 M	1.5 M
100 mV	1.0 M	1.5M	1.5 M
1.0 V	10 M	1.5 M	1.5 M
10 V	10 M	300 k	0
100 V	10 M	30 k	0

I (Full Scale)	R_i (Ω)	R_i (Ω)
100 nA	1.5 M	1.5 M
500 nA	300 k	300 k
1.0 μA	300 k	0
5.0 μA	60 k	0
10 μA	30 k	0
50 μA	6.0 k	0
100 μA	3.0 k	0

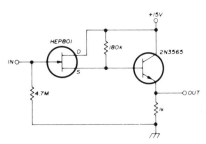

NANOAMMETER—Programmable amplifier operating from ±1.5 V supply such as D cells is used as current-to-voltage converter. Offset null of A_1 is used to minimize input offset voltage error. If programmed for low bias current, amplifier can convert currents as small as 100 nA with less than 1% error. Resistor values for variety of current and voltage ranges are given in tables. Adjust R_1 to calibrate meter, and adjust R_2 to null input offset voltage on lowest range. Not suitable for higher current ranges because power drain is excessive above 100 μA.—W. G. Jung, "IC Op-Amp Cookbook," Howard W. Sams, Indianapolis, IN, 1974, p 414–417.

FET-BIPOLAR DARLINGTON—Can be used as meter interface amplifier, impedance transformer, coax driver, or relay actuator. Combination of unipolar and bipolar transistors gives desirable amplifying features of each solid-state device.—I. M. Gottlieb, A New Look at Solid-State Amplifiers, *Ham Radio*, Feb. 1976, p 16–19.

DIGITAL pH METER—3130 CMOS opamp gives required high input impedance for pH probe at low cost. Output of probe, ranging from positive generated DC voltage for low pH to 0 V for pH 7 and negative voltages for high pH values, is amplified in circuit that provides gain adjustment to correct for temperature of solution being measured. For analog reading, output of opamp can be fed directly to center-scale milliammeter through 100K calibrating pot. For digital display giving reading of 7.00 for 0-V output, pH output is converted to calibrated current for summing with stable offset current equal to 700 counts. This is fed to current-to-frequency converter driving suitable digital display. Standard pH buffer solutions are used for calibration.—D. Lancaster, "CMOS Cookbook," Howard W. Sams, Indianapolis, IN, 1977, p 347–349.

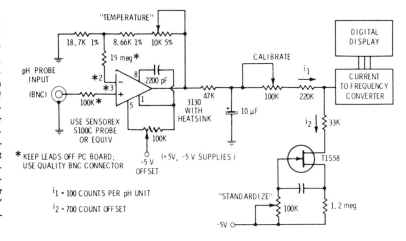

i_1 = 100 COUNTS PER pH UNIT
i_2 = 700 COUNT OFFSET

* KEEP LEADS OFF PC BOARD; USE QUALITY BNC CONNECTOR

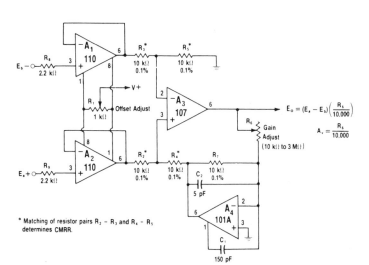

* Matching of resistor pairs R_2 – R_3 and R_4 – R_5 determines CMRR.

$$E_o = (E_a - E_b)\left(\frac{R_6}{10,000}\right)$$

$$A_v = \frac{R_6}{10,000}$$

DIFFERENTIAL-INPUT VARIABLE-GAIN—Gain of A_3 is varied by modifying feedback returned to R_4. A_4 serves as active attenuator in feedback path, presenting constant zero-impedance source to R_4 as required for maintaining good balance and high CMRR. With values shown, gain can be varied from unity to 300.—W. G. Jung, "IC Op-Amp Cookbook," Howard W. Sams, Indianapolis, IN, 1974, p 238–239.

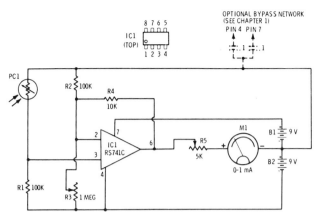

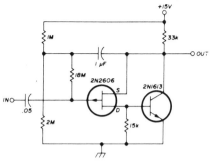

PHOTOCELL BRIDGE—Radio Shack 276-116 cadmium sulfide photocell is connected in Wheatstone bridge circuit. When bridge is balanced, RS741C opamp connected to opposite corners of bridge receives no voltage and meter reads zero. Light on photocell unbalances bridge and gives meter deflection. Can be used as high-sensitivity light meter. Adjust R3 until meter reads zero with photocell covered while R5 is at maximum resistance, adjust R5 until needle moves away from zero, rezero with R3, and repeat procedure until meter can no longer be brought to zero. Sensitivity is now maximum, and uncovered photocell will detect flame from candle at 20 feet.—F. M. Mims, "Integrated Circuit Projects, Vol. 4," Radio Shack, Fort Worth, TX, 1977, 2nd Ed., p 29–35.

HIGH INPUT Z—Suitable for use as active probe for CRO, as electrometer, and for instrumentation applications. Combination of unipolar and bipolar transistors gives desirable amplifying features of each solid-state device.—I. M. Gottlieb, A New Look at Solid-State Amplifiers, *Ham Radio*, Feb. 1976, p 16–19.

SIX-RANGE LIGHT METER—Switching of feedback resistors for opamp driven by Radio Shack 276-115 selenium solar cell gives multirange linear light meter. With 1000-megohm resistor for highest sensitivity, star Sirius will produce photocurrent of about 25 pA when solar cell is shielded from ambient light with length of cardboard tubing. Supplies are 9 V, and meter is 0–1 mA.—F. M. Mims, "Integrated Circuit Projects, Vol. 4," Radio Shack, Fort Worth, TX, 1977, 2nd Ed., p 45–53.

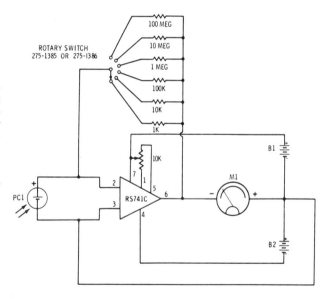

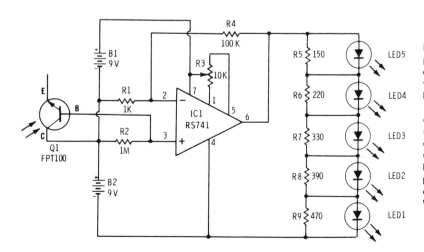

LIGHT METER WITH LED READOUT—Light on phototransistor Q1 (Radio Shack 276-130) produces voltage change across R2 for amplification by opamp whose output drives array of five LEDs forming bar graph voltage indicator. Adjust R3 initially for highest sensitivity by turning off room lights and rotating until LED 1 just stops glowing. Now, as light is gradually increased on sensor, LEDs come on one by one in upward sequence and stay on until all five are lit. Solar cells or selenium cells can be used in place of phototransistor.—F. M. Mims, "Optoelectronic Projects, Vol. 1," Radio Shack, Fort Worth, TX, 1977, 2nd Ed., p 85–93.

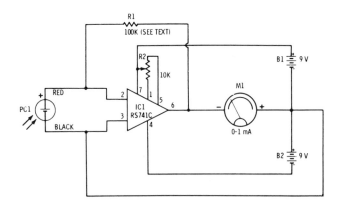

LINEAR LIGHT METER—Uses Radio Shack 276-115 selenium solar cell or equivalent photocell with high-gain RS741C opamp to drive meter. Sensitivity is sufficient to detect individual stars at night without magnifying lens if photocell is shielded from ambient light with length of cardboard tubing. Increasing value of R1 increases gain and sensitivity of circuit. R2 sets meter needle to zero when sensor is dark.—F. M. Mims, "Integrated Circuit Projects, Vol. 4," Radio Shack, Fort Worth, TX, 1977, 2nd Ed., p 45–53.

HODOSCOPE AMPLIFIER—Charge amplifier using Teledyne Philbrick 102601 opamp was developed for use with each Geiger counter of 132-counter array for ionization hodoscope used in tracing paths of cosmic rays. Charge-sensitive stage A_1 converts input charge pulse to voltage pulse significantly larger than noise of second stage. With 616-pF load capacitor, output is 12 V for input of 10 mV. Cost of charge amplifier is about $50.—H. C. Carpenter, Low Cost Charge Amplifier, *EDN Magazine,* May 20, 1973, p 83 and 85.

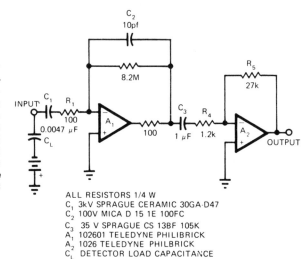

ALL RESISTORS 1/4 W
C_1 3kV SPRAGUE CERAMIC 30GA-D47
C_2 100V MICA D 15 1E 100FC
C_3 35 V SPRAGUE CS 13BF 105K
A_1 102601 TELEDYNE PHILBRICK
A_2 1026 TELEDYNE PHILBRICK
C_L DETECTOR LOAD CAPACITANCE

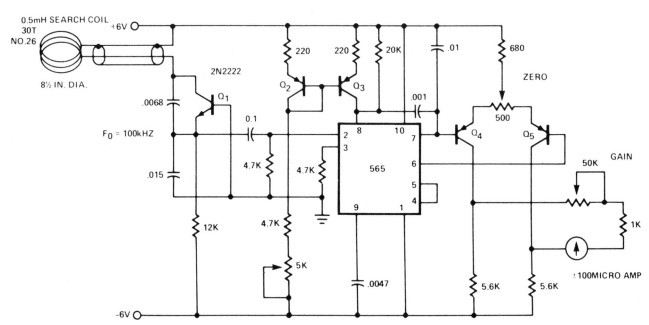

PLL DETECTOR FOR ALL METALS—Frequency change produced in Colpitts oscillator by metal object near tank coil is indicated by 565 PLL connected as frequency meter. Oscillator frequency increases when search coil is brought near nonferrous metal object. Oscillator frequency decreases, as indicated by lower meter reading, when coil is brought near ferrous object.—"Signetics Analog Data Manual," Signetics, Sunnyvale, CA, 1977, p 856–858.

SECTION 24

Lighting Control Circuits

Covers methods of triggering triacs and silicon controlled rectifiers for turning on, dimming, and otherwise regulating lamp loads in response to photoelectric, acoustic, logic, or manual control at input. Starting circuits for fluorescent lamps are also given.

AC CONTROL WITH TRIAC—Decoder outputs of microprocessor feed 7476 JK flip-flop that drives optocoupler which triggers triac for ON/ OFF control of lamp or other AC load. LED and cadmium sulfide photocell are mounted in light shield. When light from LED is on photocell, cell resistance drops and allows control voltage of correct direction and amplitude to trigger gate of triac, turning it on. When light disappears, triac remains on until voltage falls near zero in AC cycle.—R. Wright, Utilize ASCII Control Codes!, *Kilobaud,* Oct. 1977, p 80–83.

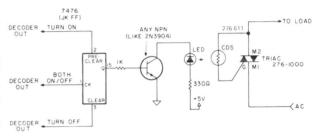

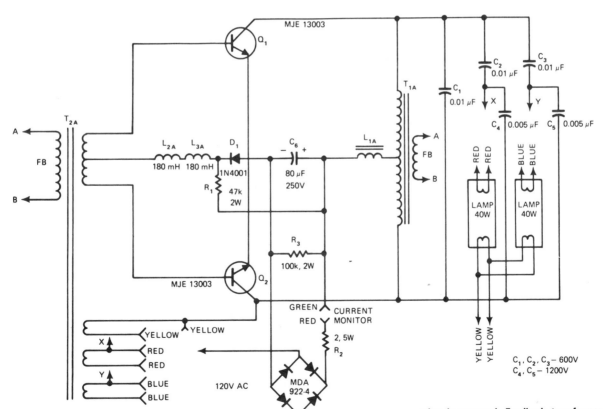

40-W RAPID-START BALLAST—AC line voltage is rectified by diode bridge and filtered by C_6-L_{1A}. Transistors Q_1-Q_2 with center-tapped tank coil T_{1A} and C_1-C_3 make up power stage of 20-kHz oscillator that develops 600 V P sine wave across T_{1A}. When fluorescent lamps ionize, current to each is limited to about 0.4 A. Lamps operate independently, so one stays on when other is removed. Feedback transformer T_{2A} supplies base drive for transistors and filament power for lamps. Article gives transformer and choke winding data.—R. J. Haver, The Verdict Is In: Solid-State Fluorescent Ballasts Are Here, *EDN Magazine,* Nov. 5, 1976, p 65–69.

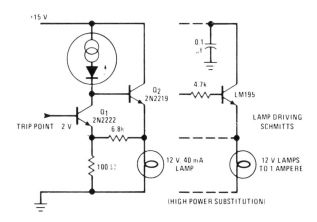

ACTIVE LOAD—National NSL4944 constant-current LED serves as current source for collector resistor of Schmitt trigger to provide up to 12-V output at 40 mA for lamp load. When lamp and Q_2 are off, most of LED current flows through 100-ohm resistor to determine circuit trip point of 2 V. When control signal saturates Q_1, Q_2 provides about 1 V for lamp to give some preheating and reduce starting current surge. When control is above trip point, Q_2 turns on and energizes lamp.—"Linear Applications, Vol. 2," National Semiconductor, Santa Clara, CA, 1976, AN-153, p 3.

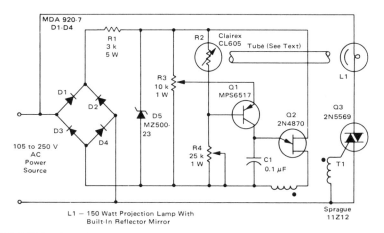

L1 — 150 Watt Projection Lamp With Built-In Reflector Mirror

PROJECTION-LAMP VOLTAGE REGULATOR—Circuit will regulate RMS output voltage across lamp to 100 V ± 2% for input voltages between 105 and 250 VAC. Light output of 150-W projection lamp is sensed indirectly for use as feedback to firing circuit Q1-Q2 that controls conduction angle of triac Q3. Light pipe, painted black, is used to pick up red glow from back of reflector inside lamp, which has relatively large mass and hence has relatively no 60-Hz modulation.—"Circuit Applications for the Triac," Motorola, Phoenix, AZ, 1971, AN-466, p 12.

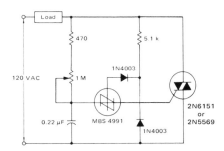

800-W TRIAC DIMMER—Simple circuit uses Motorola MBS-4991 silicon bilateral switch to provide phase control of triac. 1-megohm pot varies conduction angle of triac from 0° to about 170°, to give better than 97% of full power to load at maximum setting. Conduction angle is the same for both half-cycles at any given setting of pot.—"Circuit Applications for the Triac," Motorola, Phoenix, AZ, 1971, AN-466, p 5.

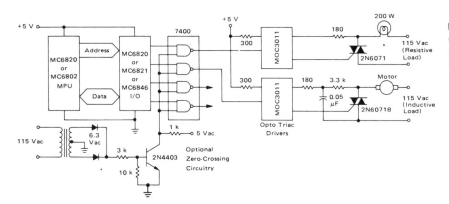

INTERFACE FOR AC LOAD CONTROL—Standard 7400 series gates provide input to Motorola MOC3011 optoisolators for control of triac handling resistive or inductive AC load. Gates are driven by MC6800-type peripheral interface adapters. If second input of two-input gate is tied to simple transistor timing circuit as shown, triac is energized only at zero crossings of AC line voltage. This extends life of incandescent lamps, reduces surge-current effect on triac, and reduces EMI generated by load switching.—P. O'Neil, "Applications of the MOC3011 Triac Driver," Motorola, Phoenix, AZ, 1978, AN-780, p 6.

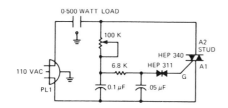

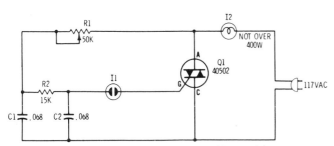

500-W MOOD-LIGHTING CONTROL—Pot provides phase control of triac to vary current through load up to 500 W, for changing brightness of lamps or speed of power tool plugged into output.—"Home Handyman's Construction Projects," Motorola Semiconductors, Phoenix, AZ, HMA-37, 1972.

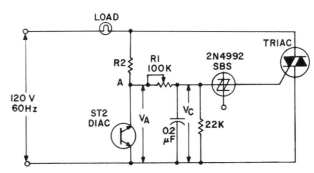

400-W TRIAC DIMMER—Uses inexpensive NE-2 or NE-83 neon lamp in place of trigger diode. As R1 is advanced, lamp turns on at about medium brilliance because neon must conduct before it can trip triac. R1 can then be backed down to give soft glow from lamp.—H. Friedman, "99 Electronic Projects," Howard W. Sams, Indianapolis, IN, 1971, p 61–62.

LINE COMPENSATION—Compensating circuit prevents lamp load from turning off if line voltage drops momentarily when dimmer control R1 is set for low light level.—R. W. Fox, "Solid State Incandescent Lighting Control," General Electric, Semiconductor Products Dept., Auburn, NY, No. 200.53, 1970, p 9.

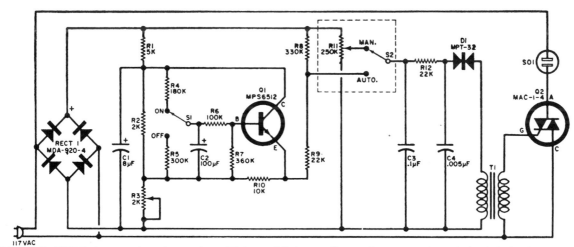

DIMMER WITH BRIGHTENING—With values shown, provides 1-min dimming cycle for room lights at start of home slide or movie show, and 20-s brightening cycle at end of show. Q2 is rated 5 A, so will control over 500 W. S2 gives choice of manual or automatic dimming. T1 is 1:1 pulse transformer (Sprague 11Z12). R3 sets interval, and S1 determines whether it is for dimming or brightening lights.—I. Gorgenyi, Build a "MALF", Popular Electronics, Sept. 1967, p 67–69 and 102.

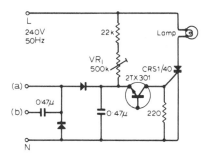

FIRING THYRISTOR—Uses npn silicon planar transistor (2TX301) connected to behave as zener diode for firing thyristor in simple half-wave lamp dimmer. Can be controlled manually with VR1, by d-c input of 0–10 V at (a), or by a-c input of 0–10 V p-p at (b). —J. A. H. Edwards, Negative Resistance of Transistor Junction, *Wireless World*, Jan. 1970, p 12.

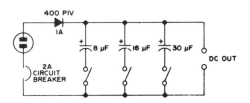

SIMPLE VOLTAGE REDUCER—Combination of three switched electrolytics and silicon rectifier gives seven different d-c output voltages for controlling lamps or other resistive loads of up to 50 W. Input at left is normal 120-V a-c line voltage. Theoretical maximum d-c output with all capacitors in use is 169 V, and without any capacitors is 60 V pulsating d-c.—N. Johnson, Simple Diode Controller, 73, March 1972, p 113–114.

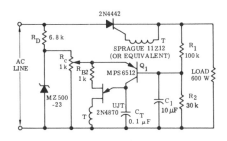

600-W HALF-WAVE AVERAGE-VOLTAGE FEEDBACK—Control responds to average load voltage. Pulse transformer is required for triggering of scr by ujt. Network R1-R2-C1 averages load voltage for comparison by Q1 with set point of Rc.—D. A. Zinder, "Unijunction Trigger Circuits for Gated Thyristors," Motorola Semiconductors, Phoenix, AZ, AN-413, 1972.

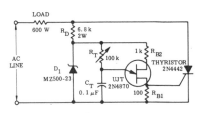

600-W HALF-WAVE—Basic ujt-triggered scr circuit is designed as two-terminal control replacing switch. If full-wave power is desired at upper limit of control, switch can be added that will short out thyristor (scr) when RT is set for maximum power. For inductive loads, two-position switch must be used to transfer load from scr to direct line.—D. A. Zinder, "Unijunction Trigger Circuits for Gated Thyristors," Motorola Semiconductors, Phoenix, AZ, AN-413, 1972.

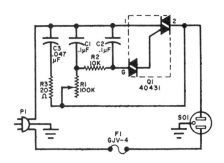

VARIABLE A-C VOLTAGE—Use of triac permits lowering a-c voltage of appliances up to 360 W for speed control, heat control, or dimming of lights.—R. C. Arp, Jr., Build a Solid-State Variable Transformer, *Popular Electronics*, Oct. 1969, p 42–44 and 117.

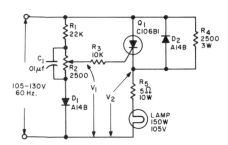

LAMP REGULATOR—Designed for use with specific lamps, for which circuit measures lamp power by measuring filament resistance. During negative half-cycle lamp receives full current through D2. At beginning of positive half-cycle, R4 provides bias current to develop voltage V2 that is dependent on lamp resistance. This voltage is compared with reference V1 by scr, to give phase-controlled triggering when V1 is more positive than V2. Possible drawback is 60-Hz component in lamp current, noticeable as flicker with peripheral vision.—R. W. Fox, "Solid State Incandescent Lighting Control," General Electric, Semiconductor Products Dept., Auburn, NY, No. 200.53, 1970, p 14–15.

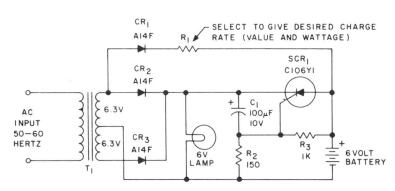

6-V STANDBY LIGHT—In event of a-c line failure, circuit switches automatically to 6-V lamp load. Turns off automatically when line voltage is restored, and converts to battery-charging mode that automatically keeps 6-V storage battery fully charged. Lamp rating should take into account battery capacity and maximum time of power failure.—"SCR Manual," General Electric, Syracuse, NY, 1972, 5th Ed., p 225–226.

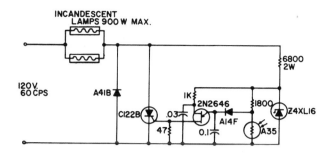

LIMITED-RANGE 900-W—For lamp loads in which power does not have to be reduced below half-power level. Rectifier supplies one half-cycle uncontrolled, and scr provides regulation of other half-cycle by phase control. Photocell used for automatic control is operated in low-resistance condition to achieve fast response time and eliminate hunting. Uses ramp-and-pedestal system.—R. W. Fox, "Solid State Incandescent Lighting Control," General Electric, Semiconductor Products Dept., Auburn, NY, No. 200.53 1970, p 9–10.

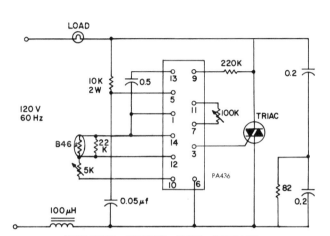

HIGH-GAIN CONTROL WITH FEEDBACK—Uses GE PA436 IC phase control, connected in simple light feedback arrangement with level set. Report gives optional soft-on circuit modification. Choose triac to match lamp load.—R. W. Fox, "Solid State Incandescent Lighting Control," General Electric, Semiconductor Products Dept., Auburn, NY, No. 200.53, 1970, p 14.

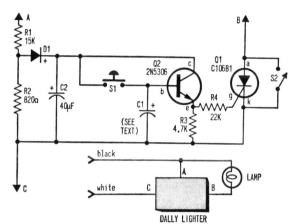

DELAYED TURN-OFF—Turns off garage, driveway, hall, or other light at any desired predetermined interval after pushbutton switch S1 is pressed, from 30 s to 15 min. S2 is conventional manual on-off switch used before. With 10 μF for C1, delay is 30 s; 100 μF gives 5 min and 300 μF 15 min. Use 12-V d-c electrolytic for C1. D1 is 200-V 50-mA or better silicon rectifier. B and C are house wiring leads to former switch and A goes to black wire.—R. Michaels, Dally Lighter, *Elementary Electronics*, May-June 1969, p 57–60.

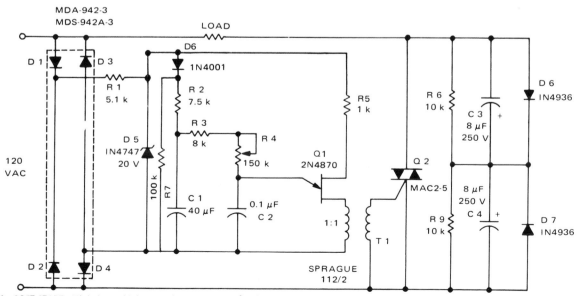

500-W SOFT-START—Minimizes high inrush current through very low resistance of cold filament, for protection of both lamp and dimmer circuit. Circuit is conservatively rated, because evaluation tests indicated triac was within ratings even for 1,000-W lamp load. Report describes operation of circuit in detail. —R. J. Haver and D. A. Zinder, "Conventional and Soft-Start Dimming of Incandescent Lights," Motorola Semiconductors, Phoenix, AZ, AN-436, 1972.

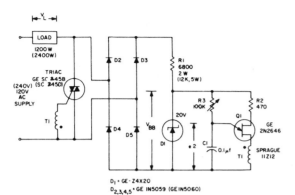

1.2-KW UJT-TRIAC—Uses zener to clamp control circuit voltage at fixed level of 20 V. R3 provides manual control of triac triggering over time-constant range of 0.3 to 8 ms.—"SCR Manual," General Electric, Syracuse, NY, 1972, 5th Ed., p 254–255.

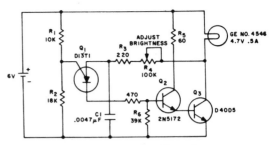

LOW-LOSS CONTROL—Provides almost continuous adjustment of light output between 0 and 100%, by changing average value of d-c supply voltage on 4.7-V tungsten lamp at high switching frequeny. Uses put Q1 as oscilator whose frequency is determined by R3, R4, and C1. Each time Q1 fires, Q2 drives Q3 into saturation and applies battery voltage to lamp.—"SCR Manual," General Electric, Syracuse, NY, 1972, 5th Ed., p 443–444.

TIME-DEPENDENT DIMMER—Designed for theater and other applications in which lamp load is to be turned on or off slowly. Uses full-wave ramp-and-pedestal control circuit. At maximum setting of R3, full transition from on to off takes about 20 min. Choose triac TR1 and fuse F1 to match load. (Use GE X12 for 500 W.)—R. W. Fox, "Solid State Incandescent Lighting Control," General Electric, Semiconductor Products Dept., Auburn, NY, No. 200.53, 1970, p 10–11.

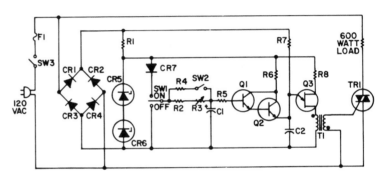

CR1 THRU CR4 : G-E (1N5059)RECTIFIER DIODE	R1 : 3.3 K OHM, 2 WATT RESISTOR
CR5, CR6 : G-E Z4XL7.5 ZENER DIODE	R2,R4: 4.7K OHM, 1/2 WATT RESISTOR
CR7 : G-E (1N5059) RECTIFIER DIODE	R3 : 5 MEGOHM, 1 WATT POTENTIOMETER
C1 : 100 μf, 15 WVDC ELECTROLYTIC CAPACITOR (G-E QT1-22)	R5,R7: 1 MEGOHM, 1/2 WATT RESISTOR
C2 : 0.1μf, 15 WVDC CAPACITOR	R6 : 2.2 K OHM, 1/2 WATT RESISTOR
Q1, Q2: G-E (2N5172)n-p-n TRANSISTOR	R8 : 470 OHM, 1/2 WATT RESISTOR
Q3 : G-E 2N2647 UNIJUNCTION TRANSISTOR	SW1 : SPDT SWITCH
	SW2 : SPST SWITCH
	T1 : SPRAGUE 11Z12 PULSE TRANSFORMER

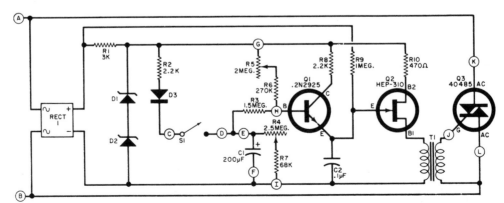

TIMER-DIMMER—Can be set to dim room lighting to any preset lower level, even full off, over period of 10 minutes so dimming is almost unnoticeable. Ideal for bachelor apartment. Closing S1 starts dimming cycle. R4 controls dimming time and R5 final level of light. D1 and D2 are GE Z4XL12B or similar 12-V 1-W zeners and D3 is HEP154. Use VS-248 bridge rectifier assembly. T1 is Sprague 11Z12 1:1 pulse transformer. Control is connected in series with lamp load being dimmed, via terminals A and B.—R. J. Bik, Build the Dynadim, *Popular Electronics*, Sept. 1968, p 71–74.

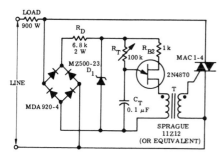

900-W FULL-WAVE—Basic ujt-triggered triac combined with bridge rectifier provides wide range of control with RT for resistive load. Pulse transformer T isolates triac gate from steady-state ujt current.—D. A. Zinder, "Unijunction Trigger Circuits for Gated Thyristors," Motorola Semiconductors, Phoenix, AZ, AN-413, 1972.

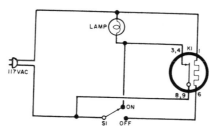

TURN-OFF DELAY—After lamp is switched off, light remains on long enough to get into bed or leave room. Delay is 30 s with Amperite 115C30T thermal relay, 60 s with 115C60T, and 120 s with 115C120T. Can usually be mounted in base of lamp.—J. Small, Mannerly Table Lamp, *Popular Electronics*, Nov. 1968, p 79 and 97.

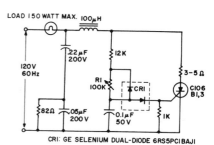

CR1: GE SELENIUM DUAL-DIODE 6RS5PC1BAJI

LOW-COST 150-W—Dual-diode CR1 eliminates more expensive switching device usually used in phase controls. Provides half-wave control, with scr triggered through diode when line voltage goes positive.—R. W. Fox, "Solid State Incandescent Lighting Control," General Electric, Semiconductor Products Dept., Auburn, NY, No. 200.53, 1970, p 9–10.

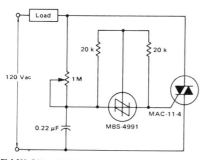

FLASH-ON SUPPRESSION—Shunting of sbs with two 20K resistors minimizes hysteresis effect wherein triac increases lamp brilliance suddenly as 1-meg control is turned up.—L. J. Newmire, "Theory, Characteristics and Applications of Silicon Unilateral and Bilateral Switches," Motorola Semiconductors, Phoenix, AZ, AN-526, 1970.

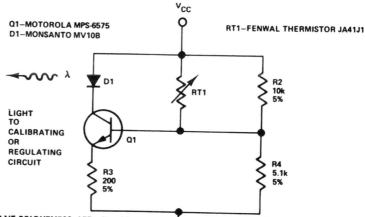

Q1—MOTOROLA MPS-6575
D1—MONSANTO MV10B
RT1—FENWAL THERMISTOR JA41J1

CONSTANT-BRIGHTNESS LED—Provides accurate and reliable brightness reference for industrial equipment, using thermistor R21 to compensate for temperature coefficient of led D1. Brightness is constant within 3% over temperature range of 10–50 C, with no need to compensate for aging or blackening as with incandescent lamps. Requires +12 V regulated d-c supply.—W. Otsuka, "Constant Brightness Light Source," *Monsanto GaAsLite Tips*, Vol. 1, 1970.

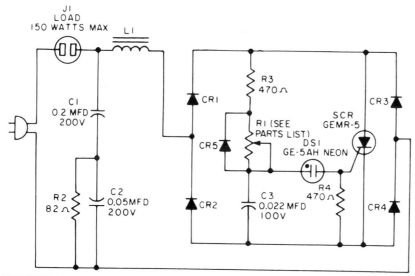

HIGH-INTENSITY LAMP—Designed for use with high-intensity study lamp having built-in transformer. Uses phase-control circuit acting on both halves of a-c cycle. Will also give 3:1 control of speed for shaded-pole fan motor. Diodes are GE-504A. R1 is 100K for fan and 250K for lamp. L1 is 65 turns No. 18 on ¼" ferrite rod.—"Electronics Experimenters Circuit Manual," General Electric, Owensboro, KY, 1971, 3rd Ed., p 183–186.

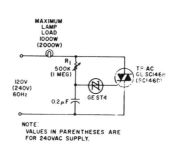

NOTE:
VALUES IN PARENTHESES ARE FOR 240VAC SUPPLY.

FULL-WAVE ASBS-TRIAC—Use of ST4 asymmetrical silicon bilateral switch eliminates snap-on problem of diac-triac phase control, in which load current suddenly snaps from 0 to intermediate point at which smooth control starts. Operation of circuit is covered in detail. Waveform asymmetry of load current gives d-c components, precluding use with loads having transformers or fluorescent-lamp ballasts.—"SCR Manual," General Electric, Syracuse, NY, 1972, 5th Ed., p 253.

SECTION 25
Measuring Circuits—Audio

Includes level detectors, S-meters, and distortion measuring circuits. See other sections on audio control circuits, audio amplifiers, and audio-frequency oscillators.

AF VOLTMETER—Although not calibrated on absolute basis, either 3 dB or 10 dB of attenuation can be switched in with S1 for measuring purposes. Internal adjustments are made easily by tacking 51-ohm resistor temporarily across input, then driving input with step attenuator fed with audio power at −10 dB by generator having 50-ohm pad in its output. CR1-CR4 are 1N914.—W. Hayward, Defining and Measuring Receiver Dynamic Range, *QST*, July 1975, p 15–21 and 43.

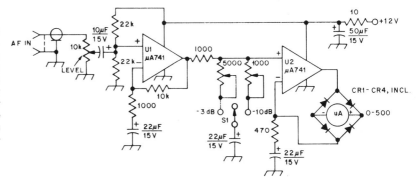

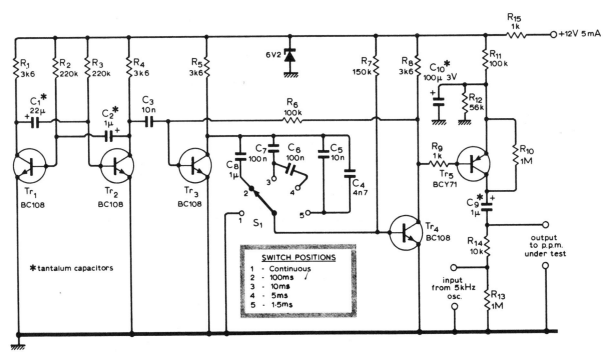

PEAK PROGRAM METER TESTER—Used with 5-kHz audio oscillator to produce tone bursts of 1.5, 5, 10, and 100 ms, as required for checking response of program meter to tone bursts. Transistors Tr_3 and Tr_4 form mono with switched timing capacitors. Article covers calibration and use.—E. T. Garthwaite, Tone Burst Generator for Testing P.P.Ms, *Wireless World*, Aug. 1976, p 53.

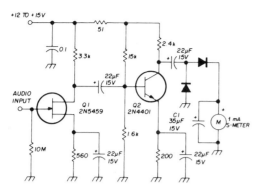

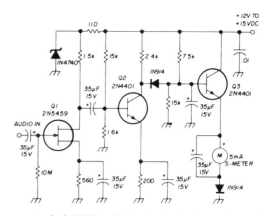

1-mA S-METER—Amplifier designed for 1-mA meter movement consists of two-stage voltage amplifier driving meter rectifier. FET input provides high impedance to detected audio and minimizes loading and distortion problems. Q2 is common-emitter voltage amplifier with simple positive-pulse rectifier for meter. C1 filters rectified audio signal. AF input for S-9 reading is 25—30 mV P-P and for full scale is 50—60 mV P-P. Frequency response is 500 Hz to 10 kHz.— M. A. Chapman, Solid-State S-Meters, *Ham Radio*, March 1975, p 20—23.

5-mA S-METER—Circuit designed for 5-mA meter movement uses two-stage voltage amplifier Q1-Q2 with emitter-follower output Q3 serving as impedance-matching stage. AF input for S-9 reading is 25—30 mV P-P and for full scale is 50—60 mV P-P. Frequency response is 500 Hz to 10 kHz.—M. A. Chapman, Solid-State S-Meters, *Ham Radio*, March 1975, p 20—23.

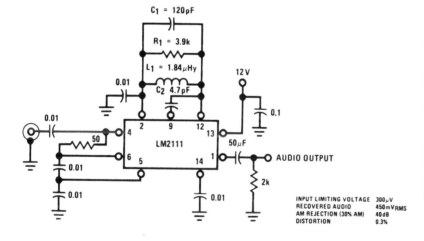

INPUT LIMITING VOLTAGE	300μV
RECOVERED AUDIO	450mVRMS
AM REJECTION (30% AM)	40 dB
DISTORTION	0.3%

AUDIO-FREQUENCY METER—Covers 0—100 kHz in four ranges. Meter reading is independent of signal amplitude from 1.7 VRMS upward and independent of waveform over wide range. Linear response means only one point need be calibrated in each frequency range. Circuit uses two overdriven FET amplifier stages in cascade. Square-wave output of last stage is rectified by X1 and X2. Deflection of meter depends only on number of pulses per second passing through meter so is proportional to pulse frequency. Battery drain is 1.4 mA.—R. P. Turner, "FET Circuits," Howard W. Sams, Indianapolis, IN, 1977, 2nd Ed., p 129—131.

LEVEL DETECTOR—Circuit lights green LED if signal at output of audio preamp exceeds 1 V peak for predetermined period. Red LED comes on when tone-control stage at output of preamp is on verge of clipping. VU meter driver circuit is also provided. Entire circuit must be duplicated for other stereo channel. Article describes circuit operation in detail and gives all associated circuits used in high-performance audio preamp. D_1 is 1N914; red LED is TIL209 or equivalent; green LED is TIL211.—D. Self, Advanced Preamplifier Design, *Wireless World*, Nov. 1976, p 41—46.

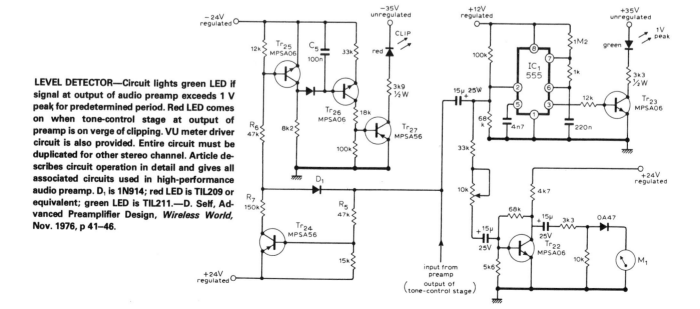

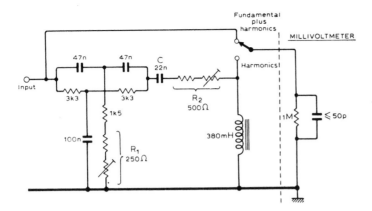

MEASURING AF DISTORTION—Passive high pass 1-kHz filter is used with audio millivoltmeter to improve accuracy of distortion measurements for low-impedance sources at 1 kHz. Filter removes low-frequency noise from input signal and compensates for loss of harmonic frequency. Applications include setting bias and recording levels of tape recorder. Adjust R_1 for best null, then adjust R_2 and value of C to equalize responses at harmonics.—J. B. Cole, Passive Network to Measure Distortion, *Wireless World,* Jan. 1978, p 60.

SECTION 26
Measuring Circuits—Capacitance

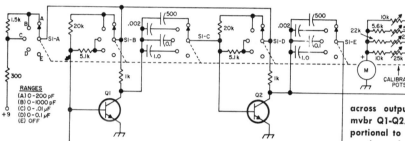

CAPACITANCE METER—Measures capacitors accurately from 4 pF to 0.1 μF in four ranges, and values down to 2 pF can be esti- mated. Uses 0–50 μA meter. Transistors can be 2N2926, SK3011, or HEP54. Capacitor under test is connected in series with meter across output of square-wave free-running mvbr Q1-Q2. Meter deflection is directly pro- portional to capacitance value, so calibration requires only precision capacitor for full-scale value of each range.—J. Fisk, "Useful Tran- sistor Circuits," 73 Inc., Petersborough, NH, 1967, p 30A.

CAPACITOR LEAKAGE—Requires disconnect- ing one capacitor lead from its circuit, before connecting test leads across capacitor and throwing S1. Repeated blinking of neon NE51 indicates small amount of leakage. Continu- ous glow means capacitor is shorted. Single blink means capacitor is good. Uses voltage- doubling circuit with SK3016 rectifiers to apply about 250 V to capacitor being checked, so should be used only for capaci- tors rated at this voltage or higher.—R. M. Brown and T. Kneitel, "101 Easy Test Instru- ment Projects," Howard W. Sams, Indianapo- lis, IN, 1968, p 75–76.

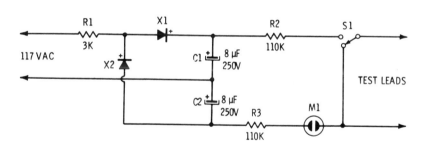

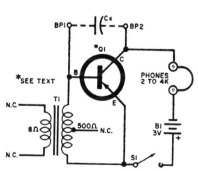

AUDIO-TONE CAPACITOR TESTER—Capacitor under test provides feedback in simple audio oscillator using general-purpose pnp power transistor such as 2N554, for checking 0.002 to 0.1 μF capacitors of any type for opens, shorts, and approximate value as determined by tone frequency heard in headphones. T1 is miniature output transformer used as choke. Higher pitch indicates smaller value.—K. Scharf, Reader's Circuits, *Popular Electronics*, April 1968, p 84–85.

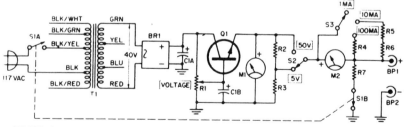

BP1,BP2—Insulated binding post
BR1—Full-wave bridge rectifier: 1 A, 200 PIV (Motorola HEP-176)
C1A,C1B—20/20 μf, 150 V dual electrolytic capacitor
M1—0-50 V DC voltmeter (Emico Model RF-2¼C, Allied 52 C 6097)
M2—0-1 ma DC milliammeter (Emico Model RF-2¼C, Allied 52 C 8012)
Q1—2N2405 transistor (RCA)
Resistors: ½ watt, 10% unless otherwise indi- cated
R1—50,000 ohm linear-taper potentiometer

R2—4,700 ohms
R3—470 ohms
R4—10 ohms, 5%
R5—39 ohms, 5%
R6—56 ohms, 5%
R7—22 ohms, 1 watt (see text)
S1—DPDT toggle or slide switch
S2—SPDT toggle or slide switch
S3—SR triple-throw rotary switch (see text)
T1—Low-voltage rectifier transformer; secon- daries: 10-20 V center tapped and 40 V center tapped @ 35 ma (Allied 54 C 4731)
Misc.—7 x 5 x 3-in. Minibox, terminal strips

LEAKAGE IN ELECTROLYTICS—Designed pri- marily for checking electrolytics rated up to 50 V d-c, as used in solid-state circuits where excess leakage can be serious. Leakage cur- rent limits for capacitors from 1 to 3,000 μF range from 0.31 mA to 10 mA (given in table in article), but 100-mA range is provided for initial use to protect meter if capacitor is shorted. Capacitor under test is connected be- tween BP1 and BP2, S2 is set for voltage range, and R1 adjusted to exact test voltage desired as indicated on M1. M2 reads leak- age current after capacitor charges.—V. Kell, Electrolytic Leakage Checker, *Electronics Illus- trated*, Nov. 1969, p 82–85.

CAPACITANCE METER—Range selector S1 provides four ranges, by switching different frequency-determining network into mvbr Q1-Q2. Output of mvbr is coupled through capacitor under test (at jacks J1-J2) to indicating meter having linear 0–100 scale for reading capacitance directly from a few pF to 0.1 μF. Zener D2 provides regulation for life of battery. Based on principle that square-wave alternating current passed by capacitor is directly proportional to capacitance value.—S. Sula, Build a Capacitance Meter, *Popular Electronics*, Oct. 1969, p 66–69 and 110.

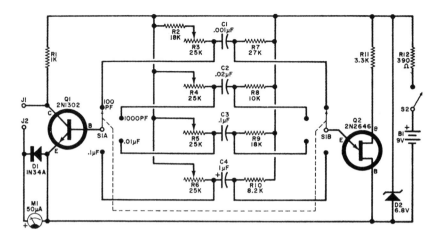

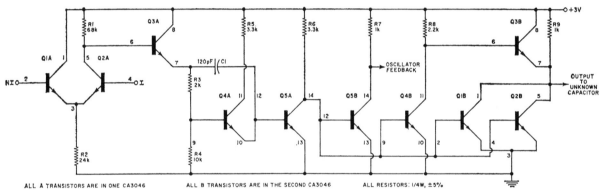

ALL A TRANSISTORS ARE IN ONE CA3046   ALL B TRANSISTORS ARE IN THE SECOND CA3046   ALL RESISTORS: 1/4W, ±5%

CAPACITANCE-METER OSCILLATOR—Used with switched frequency-determining R-C networks connected to terminal NI to generate seven different stable frequencies in range from 23 Hz to 460 kHz, as required for obtaining seven different capacitance ranges in meter. Article gives all other circuits and calibration procedure. Uses two RCA IC transistor arrays.—H. A. Wittlinger, IC Capacitance Meter, *Electronics World*, Sept. 1970, p 44–47 and 84.

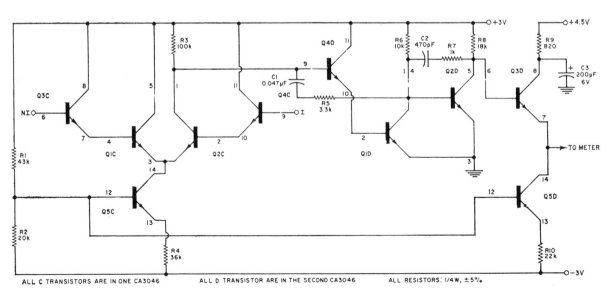

ALL C TRANSISTORS ARE IN ONE CA3046   ALL D TRANSISTOR ARE IN THE SECOND CA3046   ALL RESISTORS: 1/4W, ±5%

CAPACITANCE-METER AMPLIFIER—Uses two IC transistor arrays to form differential-input amplifier, two-stage d-c amplifier, and emitter-follower output for driving 1-mA meter of capacitance meter.—H. A. Wittlinger, IC Capacitance Meter, *Electronics World*, Sept. 1970, p 44–47 and 84.

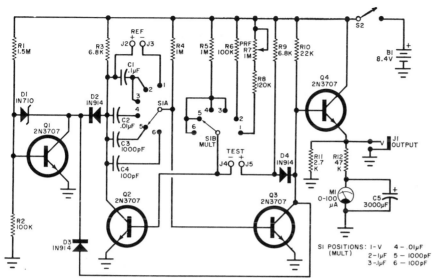

15 PF TO 10 μF—Test circuit is basically free-running mvbr which compares capacitances. Unkown capacitor connected to J4 and J5 becomes one of mvbr cross-coupling capacitors, while other is known precision value selected by S1A. At switch position 1, external reference capacitor is connected to J2 and J3 to determine pulse width, and R7 is used to adjust pulse repetition rate. Article gives construction and calibration details. For pulse generator applications, output pulses are available from J1.—D. Hileman, Direct-Reading Capacitance Meter, *Popular Electronics*, Feb. 1973, p 65–68.

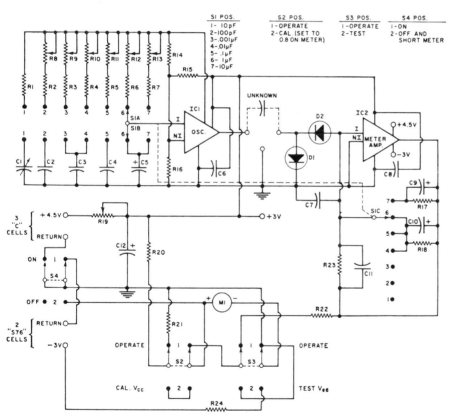

VOM CAPACITANCE METER—Accuracy is satisfactory with 5,000-ohm-per-volt meter set to 5-V a-c range, for measuring electrolytics up to 1,000 μF. Cs is 5% tantalum capacitor, at least 10 V, used as standard; use value equal to, or decade multiple of, desired capacitance value at midscale of ohms scale. Do not test capacitors rated under 10 V if using 6-V transformer tap.—H. Schoenbach, Direct-Reading Capacitance Meter For Electrolytics, *Ham Radio*, Oct. 1971, p 14–15.

R1, R2, R3, R4, R5, R6—5100 ohm, ¼ W res. ±5%
R7, R15, R16—10,000 ohm, ¼ W res. ±5%
R8, R9, R10, R11, R12—5000 ohm trimmer (Mallory MTC4-53L4)
R13—10,000 ohm trimmer (Mallory MTC4-14L4)
R14—15,000 ohm, ¼ W res. ±5%
R17—2000 ohm, ¼ W res. ±5%
R18—11,000 ohm, ¼ W res. ±5%
R19—500 ohm linear-taper pot (Mallory U-2)
R20—3600 ohm, ¼ W res. ±5%
R21—100 ohm, ¼ W res. ±10%
R22—910 ohm, ¼ W res. ±5%
R23—100,000 ohm, ¼ W res. ±5%
R24—5600 ohm. ¼ W res. ±5%
C1—7-100 pF trimmer capacitor (Elmenco 423)
C2—0.0027 μF polystyrene capacitor
C3—0.027 μF, 80 V capacitor
C4—0.27 μF, 80 V capacitor

C5—2.7 μF, 15 V tantalum capacitor
C6, C8—0.1 μF, 10 V disc ceramic capacitor
C7, C11—0.1 μF, 20 V disc ceramic capacitor
C9—150 μF, 6 V tantalum capacitor
C10—6.8 μF, 6 V tantalum capacitor
C12—1000 μF, 6 V elec. capacitor
D1, D2—1N914 diode
M1—0-1 mA meter
S1—3-pole, 11-pos. (7 pos. used) non-shorting rotary sw. (Centralab PA-1009)
S2—D.p.d.t. momenetary push-button sw. ("Cal")
S3—D.p.d.t. momentary push-button sw. ("Test")
S4—D.p.d.t. slide sw. ("On-Off")
IC1, IC2—Four RCA CA3046 integrated circuits
Three 1.5-volt "C" cells
Two 1.5-volt silver-oxide cells (Eveready S76 or Mallory MS76)

7-RANGE CAPACITANCE METER—Uses transistor arrays to form oscillator and meter amplifier, circuits for which are shown separately in article and in this chapter. Unknown capacitor is placed in series with square-wave oscillator, switching diode network D1-D2, and 1-mA indicating meter. Article covers construction and calibration.—H. A. Wittlinger, IC Capacitance Meter, *Electronics World*, Sept. 1970, p 44–47 and 84.

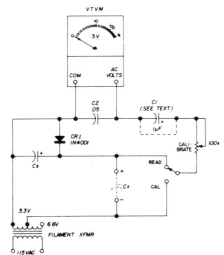

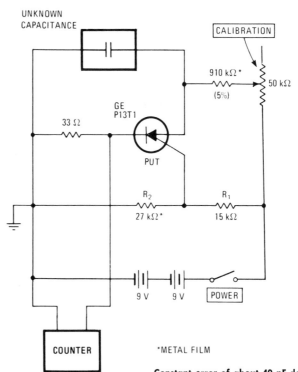

VTVM CAPACITANCE METER—Measures up to 1,000 μF for electrolytics, with value read directly on resistance scale. Use 3-V a-c volts range, and set calibration pot for 3 V full-scale. C1 is blocking capacitor; not needed if already in vtvm. Cs is 5% tantalum capacitor, at least 10 V, used as standard; its value should be equal to, or decade multiple of, desired capacitance value at midscale of ohms scale. If 100 μF is used, for vtvm reading 10 at midscale, range will be 10 to 1,000 μF for easy reading.—H. Schoenbach, Direct-Reading Capacitance Meter For Electrolytics, Ham Radio, Oct. 1971, p 14–15.

COUNTER AS CAPACITANCE METER—Relaxation oscillator converts any time-measuring counter into direct-reading capacitance meter in which seconds indicate μF, milliseconds nF, and microseconds pF values. Accuracy is better than 1% from 5,000 pF to over 10 μF.

Constant error of about 40 pF degrades accuracy of absolute-value measurements below 5,000 pF, but circuit can still be used for matching capacitors as small as 100 pF. Battery drain is only 0.5 mA.—M. J. Salvati, Oscillator Converts Counter to Capacitance Meter, Electronics, July 5, 1973, p 109.

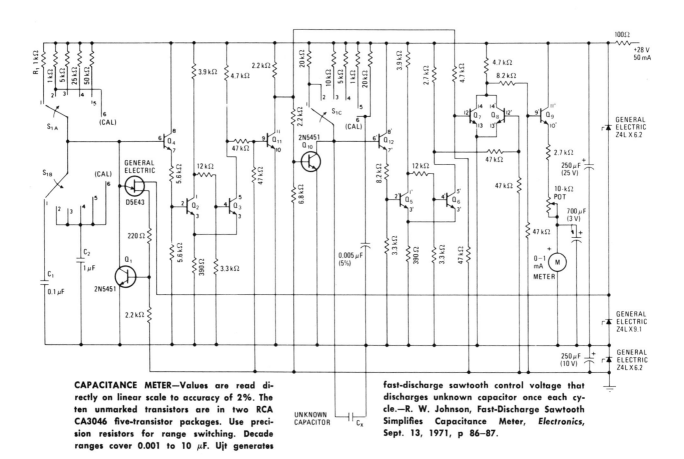

CAPACITANCE METER—Values are read directly on linear scale to accuracy of 2%. The ten unmarked transistors are in two RCA CA3046 five-transistor packages. Use precision resistors for range switching. Decade ranges cover 0.001 to 10 μF. Ujt generates fast-discharge sawtooth control voltage that discharges unknown capacitor once each cycle.—R. W. Johnson, Fast-Discharge Sawtooth Simplifies Capacitance Meter, Electronics, Sept. 13, 1971, p 86–87.

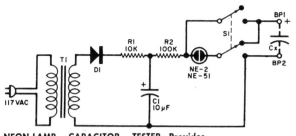

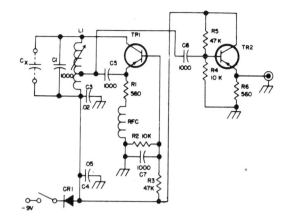

NEON-LAMP CAPACITOR TESTER—Provides visual indication for capacitors rated over 150 V, including electrolytics. With S1 as shown, neon flashes once with brightness and duration proportional to capacitor value for nonelectrolytics. If bulb stays dark, capacitor is open; flicker means intermittent, and dim continuous light means leakage. Use other position of S1 for electrolytics, observing polarity. Circuit now forms relaxation oscillator; the larger the capacitor, the lower the flash rate. If lamp stays on, capacitor is either open or so low in value that flashing is faster than eye can follow. No light means capacitor is leaky or shorted. T1 is 1:1 isolation transformer and D1 general-purpose silicon diode.—E. Richardson, Reader's Circuit, *Popular Electronics*, April 1968, p 84–86.

C BY FREQUENCY SHIFT—Simple 4,000-kHz Hartley oscillator with buffer amplifier is fed into antenna terminal of any amateur or other communication receiver covering 80-meter band, and oscillator is tuned for zero beat with receiver set at this frequency (by adjusting L1, which is Miller 4400 ⅜ inch slug-tuned form having 12 close-wound turns No. 26 tapped 3 turns from bottom). Unknown capacitor Cx is then connected, receiver retuned for zero beat at lower frequency, frequency shift is noted, and capacitance value is then read directly from computer-generated tabulation in article; range covered is from 2.574 pF for 5-kHz deviation to 313.829 pF for 499-kHz deviation. Accuracy is comparable to that of precision lab bridge. Article also gives equations. Transistors are HEP 1 or 2N964. RFC is 2.5 mH National R100S. CR1 is 1N34.—H. Lukoff, Capacitance Measurement by Frequency Shift, *73*, Feb. 1973, p 108–113.

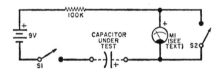

ELECTROLYTIC LEAKAGE TEST—Use 50-μA microammeter. Close S2 before closing S1, to prevent charging current of capacitor from damaging meter. Wait at least 3 minutes before opening S2 to read leakage, which should be less than 1 μA for high-quality capacitor. Switch to lower meter range if available. Large-value low-leakage low-voltage electrolytic connected across battery of transistor radio will give 30% longer play time from set of batteries by eliminating audio distortion, motorboating, and poor oscillator operation.—I. M. Gottlieb, Extending Battery Life, *Popular Electronics*, Feb. 1968, p 66.

ELECTROLYTICS TESTER—Provides mondestructive test of miniature high-capacitance low-voltage electrolytic capacitors over range of 0.001 to 1,000 μF. Uses type 6977 triode indicator tube having phosphorescent coating on anode to give pale blue glow when grid voltage is zero and tube conducts. Tube should glow brightly in proportion to charging current of capacitor under test, and fade slowly as full charge is reached. S2 is then operated to float capacitor and observe its ability to hold charge. Choose value of R1 to produce 40-V full-scale reading on meter.—D. F. Fleshren, Electrolytic-Capacitor Tester, *Electronics World*, Oct. 1969, p 82–83.

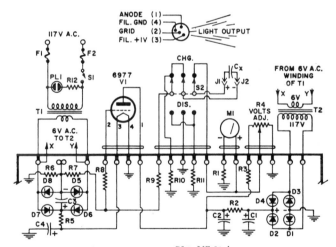

R1—Multiplier res. for meter (optional, see text)
R2—150 ohm, 2 W res.
R3—1000 ohm, 2 W res.
R4—1000 ohm, 2 W linear-taper pot
R5—100 ohm, ½ W res.
R6—150 ohm, ½ W res.
R7—10,000 ohm, ½ W res.
R8—120,000 ohm, ½ W res.
R9—15,000 ohm, ½ W res.
R10, R11—680,000 ohm, ½ W res. ±5%
R12—22,000 ohm, ½ W res.
C1, C2—10 μF, 150 elec. capacitor
C3, C4—100 μF, 20 V elec. capacitor
S1—S.p.s.t. toggle switch
S2—3-pole, d.t., center-off, lever switch

PL1—NE-51 lamp
F1, F2—½ A, plug-mounted 3AG fuse
J1, J2—Banana terminal post
M1—0-40 V d. c. full-scale meter
D1, D2, D3, D4—1N2070 diode or equiv.
D5, D6, D7, D8—1N1693 diode or equiv.
T1, T2—117 V a. c./6 V a. c. at 1 A trans.
V1—6977 indicator tube (Sylvania, Amperex, or Tung-Sol)
1—3" x 5" x 7" circuit box

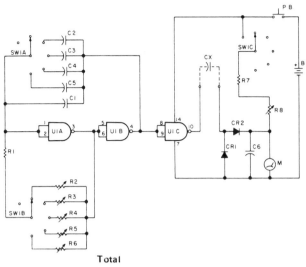

5 pF TO 1 μF—Consists of an oscillator using two gates from CD4011 quad NAND gate, separated from diode rectifier by another NAND gate. Increasing oscillator frequency gives more pulses per second and higher integrated meter reading. Each meter range is linear, so value of 5-pF capacitor can be read on lowest range. Diodes are 1N34 or equivalent. R1 is 12K, and R2-R6 are 50K trimpots set to values shown in table. R7 is 5K, and R8 is 10K trimpot. B1 is 9-V transistor battery. Article covers construction and calibration with known capacitors.—E. Landefeld, Build a Simple Capacitance Meter, *73 Magazine,* Jan. 1978, p 164–165.

Range	Total R	C	Frequency
0-100 pF	15k	5 pF	1100 kHz
0-1000 pF	31k	100 pF	112 kHz
.01 uF	36k	1500 pF	11.2 kHz
.1 uF	45k	.012 uF	1.170 kHz
1 uF	45k	.1 uF	109 Hz

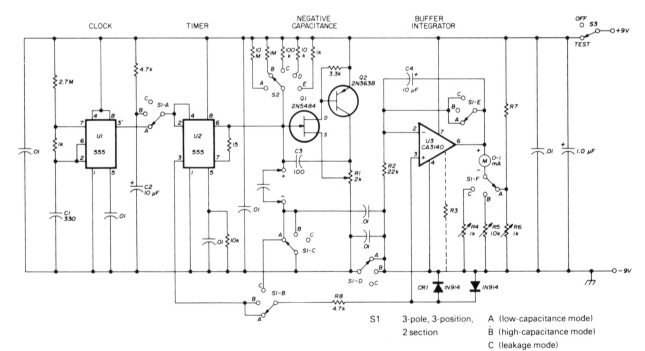

S1	3-pole, 3-position, 2 section	A (low-capacitance mode) B (high-capacitance mode) C (leakage mode)

S2	1-pole, 5-position	mode A μF	mode B μF	mode C
	A	0.0001	0.25	
	B	0.001	2.5	
	C	0.01	25.0	
	D	0.1	250.0	
	E	1.0	2500.0	leakage

S3	SPST (test)

CAPACITOR TESTER—Portable instrument measures capacitance values to 2500 μF and leakage current with up to 8 V applied. Timer U1 operates as clock providing about 350 negative-going pulses per second to trigger timer U2 and unclamp test capacitor so it charges through switch-selected resistor to half of supply voltage. U2 then resets, discharging capacitor through pin 7. During charge, pin 3 of U2 is high (about 8 V) and duration of high state is directly proportional to capacitance. Resulting rectangular waveform is applied to unity-gain buffer opamp U3 that feeds meter through calibrating trimpot R6. Meter deflection is proportional to average value of rectangular output waveform and is therefore proportional to capacitance. Table gives switch functions. Mode B uses larger clock timing capacitor to permit measuring larger capacitance values, for total of 10 ranges. Article covers construction, calibration, and use.—P. H. Mathieson, Wide-Range Capacitance Meter, *Ham Radio,* Feb. 1978, p 51–53.

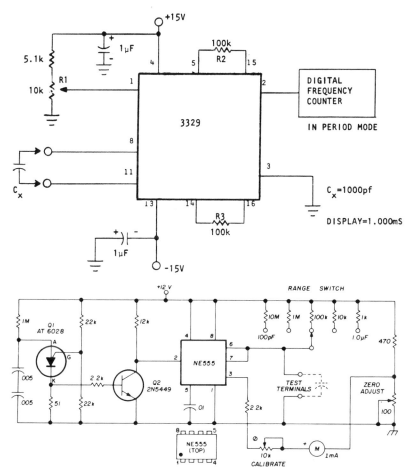

DIGITAL WITH 100:1 RANGE—Frequency counter operated in period mode serves as readout for Optical Electronics 3329 voltage-to-frequency converter. Unknown capacitance is connected as external timing capacitance for IC, so output period of IC is directly proportional to unknown capacitance. To calibrate, connect known C and adjust R1 for correct reading on digital frequency counter. With values shown, 1 nF gives period of 1 ms.—"Low Cost Capacitance Measurement," Optical Electronics, Tucson, AZ, Application Tip 10262.

DIRECT-READING FIVE-RANGE—Covers 1 pF to 1 μF in five ranges, using easily available components. Trigger source is free-running pulse generator using programmable UJT Q1 and inverter-amplifier Q2 to produce narrow −12 V output pulse at constant frequency of about 500 Hz. For trigger pulse, NE555V timer connected as mono MVBR initiates output pulse whose width increases with value of capacitor under test. Meter reads average value of pulse waveform and may be calibrated directly to read capacitance. Range resistors should be 5% or better. 10K trimpot in series with meter serves for initial calibration. Zero-adjustment pot is needed only for lower ranges. Use zener-regulated supply to provide 12 V at up to 50 mA. Full type number of Q1 is A7T6028; 2N6027, 2N6028, 2N6118, and HEP S9001 are similar. Single 0.0025-μF capacitor can be substituted for two 0.005-μF units in series.—C. Hall, Direct-Reading Capacitance Meter, *Ham Radio*, April 1975, p 32—35.

CHECKING BY SUBSTITUTION—Uses 1-MHz crystal oscillator with fixed-tuned tank circuit L1-C2 link-coupled to resonant measuring circuit consisting of L4, C4, C5, and unknown capacitance. Simple RF voltmeter is connected across measuring circuit as resonance indicator. C4 and C5 have calibrated dials reading directly in picofarads. L1 is Miller 20A224RBI slug-tuned unit adjusted to 250 μH. L4 is Miller 41A685CBI adjusted to 60 μH. Links L2 and L3 are 2 turns each. To use, close S1, set C5 to maximum, and adjust C4 for peak deflection of M1. Connect unknown capacitance to XX with shortest possible leads, retune C5 to resonance, then subtract this capacitance reading of C5 from maximum reading to get value of unknown capacitor.—R. P. Turner, "FET Circuits," Howard W. Sams, Indianapolis, IN, 1977, 2nd Ed., p 140—142.

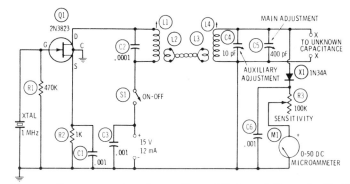

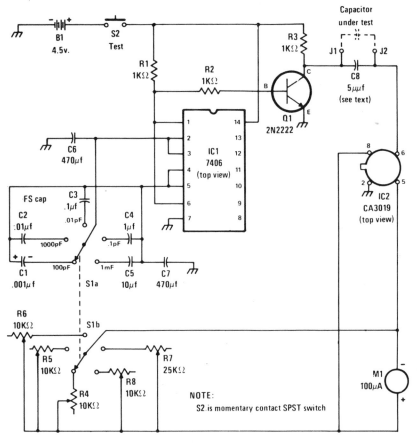

FIVE RANGES UP TO 1 μF—Direct-reading meter gives capacitance values in five ranges, all using same 0–100 scale on 100-μA meter. Operates from three penlight cells. To calibrate, connect known capacitor to jack, close S2, and adjust trimmer pot for each range in turn to give correct indication of capacitor value on meter.—C. Green, Build This Easy Capacitor Meter, *Modern Electronics*, Aug. 1978, p 78–79.

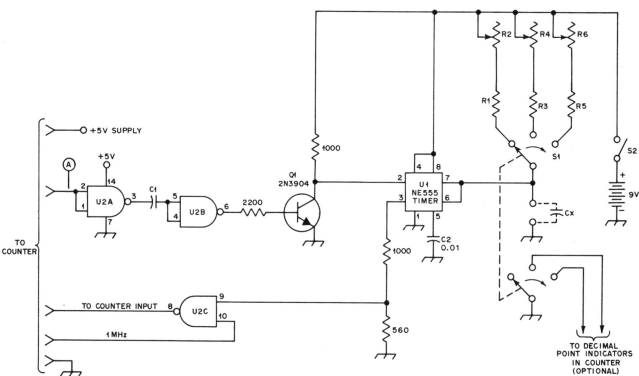

ADAPTER FOR COUNTER—Converts counter into digital capacitance meter for measuring values down to around 5 pF with better than 1% accuracy. Three ranges give full-scale values of 99,999 pF, 0.99999 μF, and 9.9999 μF. Positive-going count-enable command from frequency counter, applied to point A of gate U2A, re-moves short-circuit from unknown capacitor C_x and enables gate U2C. Capacitor charges ex-ponentially through R1 and R2 (range 1) to volt-age at which threshold comparator at U1 makes flip-flop change state, shorting C_x and disabling gate U2C. During charge time, 1-MHz pulses are applied to counter input. Counter reading then corresponds to capacitor value. C1 is 18 pF, R1 is 860K, R2 is 100K, R3 is 86K, R4 is 10K, R5 is 8.6K, R6 is 1K, and U2 is 7400 quad NAND gate.—R. F. Kramer, Using a Frequency Counter as a Capacitance Meter, *QST*, Aug. 1977, p 19–22.

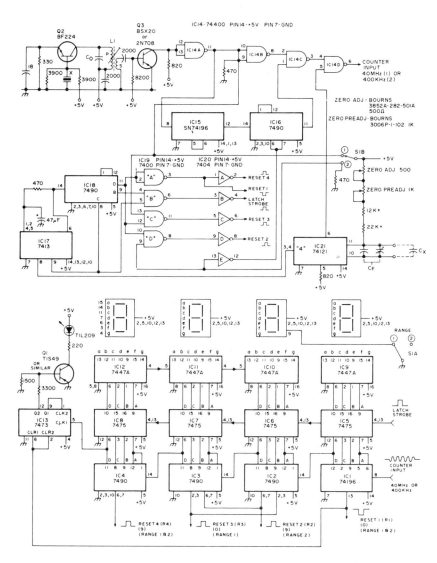

1 pF TO 1 μF—Presents instantly in digital form the value of unknown capacitor, in ranges of 1–9999 pF and 1–999.9 nF. Four digits are displayed, with leading-zero suppression and overflow indicator. Accuracy is better than 0.1% of full range ± 1 digit for higher values in both ranges. Mono MVBR IC21 produces pulse whose length is directly proportional to value of C_X plus about 980-pF total in C_F. This pulse enables gate IC14D whose output goes to counter. Oscillator Q2, buffer Q3, dividers IC15 and IC16, and gates IC14 together give 40-MHz (range 1) or 400-kHz (range 2) pulses that are counted while IC21 holds IC14D open. Article covers construction in detail.—I. M. Chladek, Build This Digital Capacity Meter, *73 Magazine*, Jan. 1976, p 70–78.

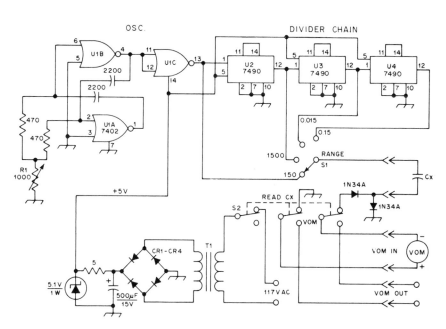

C WITH VOM—TTL-derived square-wave generator U1 charges unknown capacitor C_X to about 3.5 V at 285 kHz when using 150-μA scale of Heath MM-1 volt-ohm-milliammeter, to give 150-pF full-scale range. Larger values of capacitance are read by decreasing frequency with 7490 decade dividers. Use Mallory PTC401 for CR1-CR4. T1 is 6.3-VAC filament transformer. S2 restores normal VOM functions. Article gives design equations.—K. H. Cavcey, Read Capacitance with Your VOM, *QST*, Dec. 1975, p 36–37.

SECTION 27
Measuring Circuits—Current

GEIGER COUNTER—Used for locating uranium ore, but will also respond to other radioactivity such as from luminous-dial wristwatch. Normally gives faint clicking sound, which becomes faster and louder in vicinity of radioactive material. Can be mounted in aluminum box since metal does not block radiation being measured. Protect Geiger tube with foam rubber.—R. M. Brown and T. Kneitel, "49 Easy Transistor Projects," Howard W. Sams, Indianapolis, IN, 1972, p 47–48.

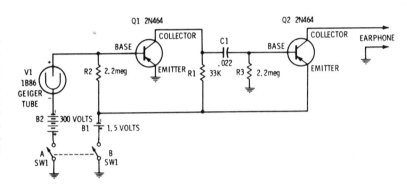

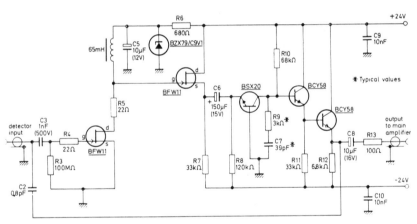

PREAMP FOR SOLID-STATE DETECTOR—Designed for use with silicon surface-barrier radiation detectors, which have 10 times better resolution than scintillation counter, even though requiring much higher gain. Uses low-noise fet's in charge-sensitive input section.—"Field Effect Transistors," Mullard, London, 1972, TP1318, p 75–78.

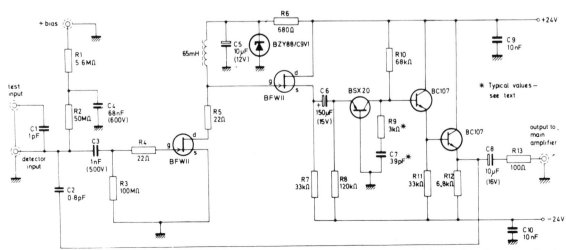

HIGH-GAIN PREAMP—Uses low-noise Mullard BFW11 fet's in charge-sensitive section. Used to convert into voltage the charge released by radiation detector connected to input. Test input is used to set up minimum preamp rise and fall times with minimum overshoot.— "Field-Effect Transistors in a Pre-Amplifier for Use with Solid-State Radiation Detectors," Mullard, London, 1969, TP1106, p 2.

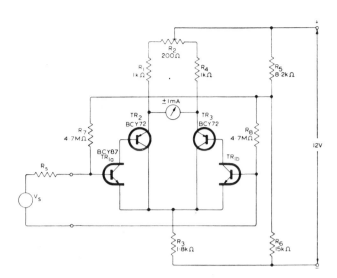

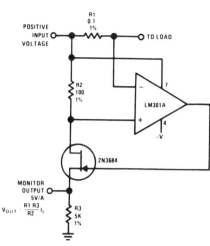

GALVANOMETER AMPLIFIER—Provides deflection from zero to full-scale value of 1 mA on rugged meter for input current change of 25 nA. Biasing is obtained by passing half the difference of input transistor base currents through signal source. Input current for zero output is typically between —10 and 10 nA.—"Circuits Using Low-Drift Transistor Pairs," Mullard, London, 1968, TP 994, p 16—17.

SUPPLY CURRENT MONITOR—R1 senses current flow of power supply. Combination of opamp and jfet buffer converts resulting voltage drop across R1 to output monitor voltage which accurately reflects current output of power supply.—"FET Circuit Applications," National Semiconductor, Santa Clara, CA, 1970, AN-32, p 11.

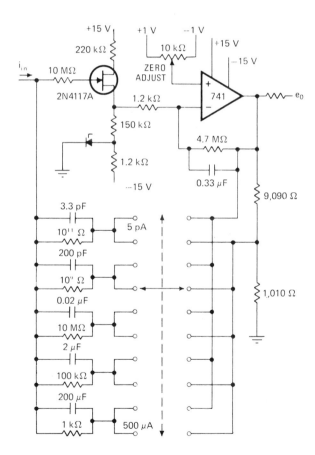

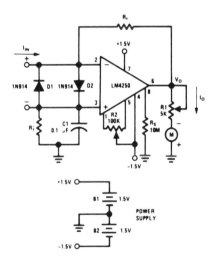

I FULL SCALE	R_f [Ω]	R'_f [Ω]
100 nA	1.5M	1.5M
500 nA	300k	300k
1 µA	300k	0
5 µA	60k	0
10 µA	30k	0
50 µA	6k	0
100 µA	3k	0

WIDE-RANGE ELECTROMETER—Measures low-frequency currents in ten ranges having full-scale values from 5 pA to 500 µA. Chief uses are in helium leak detectors, mass spectrometers, photomultipliers, and vacuum gages.—R. G. Weinberger, Solve Low-Current Measuring Woes by Designing Your Own Electrometer, *Electronics*, Aug. 30, 1971, p 58—62.

NANOAMMETER—Meter amplifier provides full-scale current ranges from 100 µA down to 100 nA by changing values of two resistors as shown in table. Circuit is built around programmable opamp, connected as differential current-to-voltage converter with input protection, zeroing and full-scale adjustments, and input resistor balancing for minimum offset voltage. Pair of D cells will serve minimum of 1 year even though operated continuously, corresponding to shelf life, so on-off switch is not needed.—M. K. Vander Kooi and G. Cleveland, "Micropower Circuits Using the LM4250 Programmable Op Amp," National Semiconductor, Santa Clara, CA, 1972, AN-71, p 5.

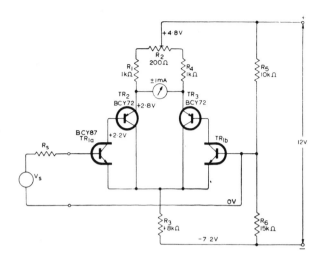

GALVANOMETER AMPLIFIER—Used in place of galvanometer to amplify small currents sufficiently for driving more rugged meter. Uses long-tailed pair directly coupled to two emitter-followers driving meter. Input current for zero output is typically 65 nA, and for full-scale deflection is 90 nA. Biasing is obtained by passing all base current of one input transistor through signal source.—"Circuits Using Low-Drift Transistor Pairs," Mullard, London, 1968, TP994, p 15–16.

COMPENSATED MOSFET ELECTROMETER—Copper-wire resistor Rcu in parallel with adjustable zero-temperature-coefficient resistor RT are adjusted to offset positive temperature coefficient of circuit. RU and VT1 in parallel compensate for decrease in battery voltage during discharge. Neon NB protects input against overvoltage (Czech mosfet has 100-V breakdown).—M. Pacak, Simple MOSFET Electrometer Circuits, *Electronic Engineering*, Sept. 1969, p 24–27.

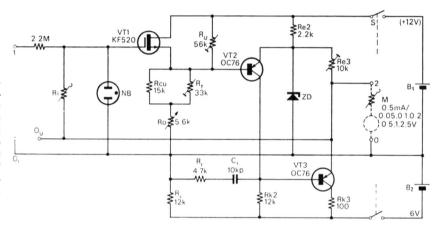

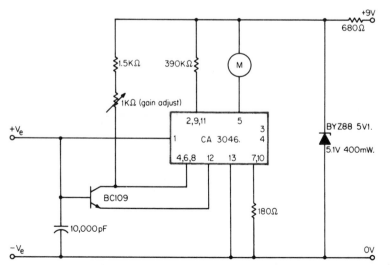

METER AMPLIFIER—Input of 1 µA gives full-scale deflection of 200-µA moving-coil meter. Accuracy and linearity are within 1% over temperature of 0 to 50 C and supply of 8 to 12 V. Warm-up drift is negligible.—H. MacDonald, Current Amplifier Circuit, *The Electronic Engineer*, Aug. 1971, p 70.

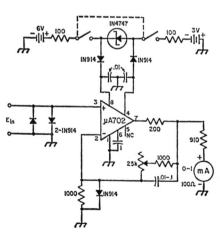

MICROAMMETER—Inputs as low as 40 mV will give full-scale reading. 25K control adjusts d-c gain of opamp from about 2 to 25. Uses 20-V zener for transient suppression.—J. M. Pike, The Operational Amplifier, *QST*, Sept. 1970, p 54–57.

CURRENT TO VOLTAGE—Values shown give bandwidth of 600 kHz and scale factor of 0.1 V per μA. Increasing R1 to 10 meg and reducing C2 to 1 pF, with C1 omitted, gives 25-kHz bandwidth and 10 V per μA. Useful for measuring small currents.—"AD513, AD516 I.C. Fet Input Operational Amplifiers," Analog Devices, Inc., Norwood, MA, Technical Bulletin, Aug. 1971.

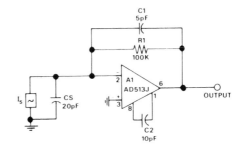

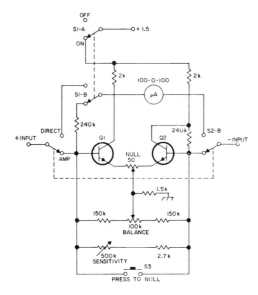

MICROAMMETER AMPLIFIER—Sensitivity can be adjusted with 500K pot so full-scale deflection on each side of zero is anywhere between 2 and 100 μA. Battery drain is only 1.5 mA. Uses differential amplifier with degenerative biasing and collector meter feed to give satisfactory compromise between sensitivity and stability. Transistors are 2N930, GE-10, or HEP50.—J. Fisk, "Useful Transistor Circuits," 73 Inc., Petersborough, NH, 1967, p 31A.

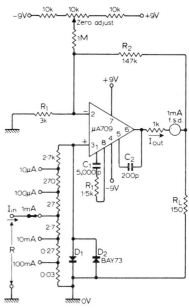

D-C MICROAMMETER—Gives accuracy of 1% at ambient temperature if meter is good quality and resistor values are accurate. Voltage drop is low.—D. C. Microammeter, *Wireless World*, Jan. 1970, p 16.

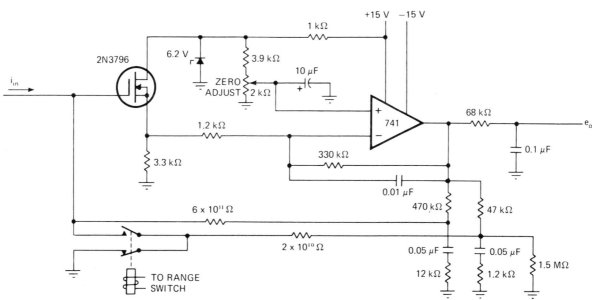

FAST-RESPONSE ELECTROMETER—Frequency-compensated feedback loop gives response time of 7 ms for ranges of 1 pA and 300 pA full-scale. Use of reed relay as range switch provides necessary isolation from ground.—R. G. Weinberger, Solve Low-Current Measuring Woes by Designing Your Own Electrometer, *Electronics*, Aug. 30, 1971, p 58–62.

I FULL SCALE	R_A [Ω]	R_B [Ω]	R_f [Ω]
1 mA	3.0	3k	300k
10 mA	.3	3k	300k
100 mA	.3	30k	300k
1A	.03	30k	300k
10A	.03	30k	30k

D-C AMMETER—Meter amplifier using programmable opamp provides full-scale ranges from 10 A down to 1 mA by changing resistor values as in table. Uses inverting amplifier configuration. Current drain from pair of flashlight D cells is so low that no on-off switch is needed.—M. K. Vander Kooi and G. Cleveland, "Micropower Circuits Using the LM4250 Programmable Op Amp," National Semiconductor, Santa Clara, CA, 1972, AN-71, p 5—6.

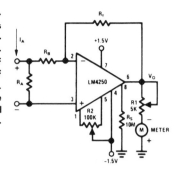

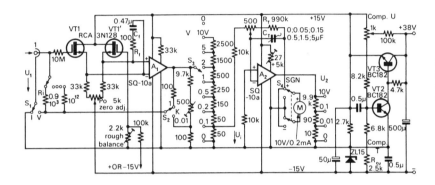

7-RANGE ELECTROMETER—Includes compensation for temperature and supply voltage. Opamp A1 supplies up to 10 V to range switch, and feedback-stabilized opamp A2 with gain of 100 boosts signal to 10 V full-scale on all ranges. Article covers construction and use.—M. Pacak, Simple MOSFET Electrometer Circuits, *Electronic Engineering*, Sept. 1969, p 24—27.

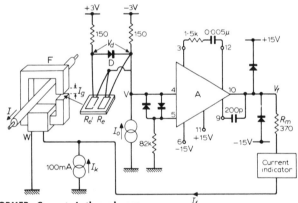

D-C TRANSFORMER—Current I through conductor is measured by means of magnetoresistance and Hall effects, using magnetoresistance Re in air gap of Ferroxcube core FX1795 having 500 turns of No. 24 s.w.g. wire. A is MC1709CP Motorola IC having gain of about 45,000. Indicator is 50-μA meter. Adjust offset current Io for zero meter deflection when I is zero.—B. E. Jones, Magnetoresistance and Its Application, *Wireless World*, Jan. 1970, p 17—19.

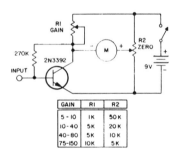

GAIN	RI	R2
5 - 10	1K	50 K
10 - 40	5K	20 K
40 - 80	5K	10 K
75 - 150	10K	5 K

METER AMPLIFIER—Resistor values used in bridge-type amplifier determine gain of circuit. With values for highest gain, 1-mA meter will read 10-μA full-scale. Requires high-quality linear meter. To calibrate, apply known desired full-scale current value to input and alternately adjust gain and zero controls until meter reads full scale.—J. Fisk, Transistor Meter Amplifiers, 73, Jan. 1966, p 44—45.

SIMPLE MOSFET ELECTROMETER—Uses feedback with medium loop gain of about 200. Requires far less space than comparable tube circuit.—M. Pacak, Simple MOSFET Electrometer Circuits, *Electronic Engineering*, Sept. 1969, p 24—27.

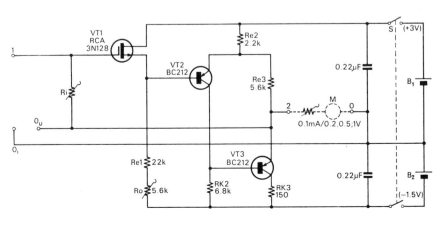

CURRENT-VOLTAGE CONVERTER—Current to be measured is injected directly into summing point of opamp connected in inverting configuration. Current is thus forced to flow through feedback resistor, for conversion into voltage with scaling factor of Rf volts per ampere. Input of 1 μA thus gives reading of 1 V on meter. Output voltage offset with zero input current is typically 0.2 V.—G. B. Clayton, Resistive Feedback Circuits, *Wireless World*, Aug. 1972, p 391–393.

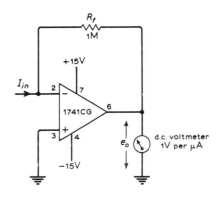

GEIGER COUNTER—Used in prospecting for uranium. Mount fragile Geiger tube V1 carefully in protective aluminum box, since gamma rays from uranium will pass through aluminum. Plug high-impedance headphones into jack M3, and check out circuit by holding radium-dial clock or watch near V1. Click rate of normal background radiation will then increase to that heard when V1 is close to samples of desired uranium ore.—R. M. Brown and T. Kneitel, "101 Easy Test Instrument Projects," Howard W. Sams, Indianapolis, IN, 1968, p 104–105.

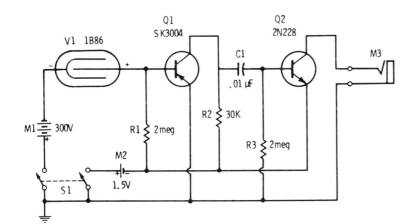

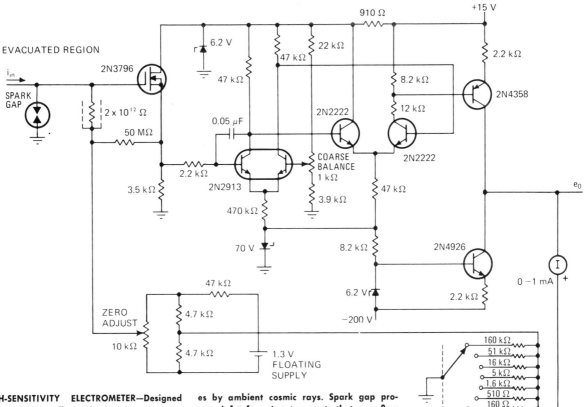

HIGH-SENSITIVITY ELECTROMETER—Designed for use as ion collector in mass spectrometer requiring noise level below 5×10^{-16} A. Input through 2N3796 must be in vacuum chamber to prevent production of noise pulses by ambient cosmic rays. Spark gap protects igfet from input currents that may flow when no power is applied. Measures maximum input current of 9×10^{-11} A in seven ranges covering 3½ decades.—R. G. Weinberger, Solve Low-Current Measuring Woes by Designing Your Own Electrometer, *Electronics*, Aug. 30, 1971, p 58–62.

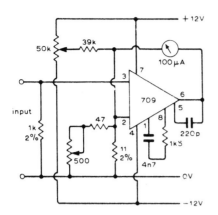

MICROAMMETER—Opamp increases sensitivity of moving-coil meter enough to give full-scale deflection for 1-μA input. Accuracy and linearity are within 1% for ambient range of 0—50 C. Range of 500—0—500 pA can be obtained by adjusting 50K pot. 500-ohm pot adjusts amplifier gain—H. MacDonald, Opamp Meter Amplifier, *Wireless World*, Sept. 1972, p 443.

5-A A-C AMMETER—Current passing through 0.75-A r-f thermocouple (such as BC-442 antenna current indicator taken from surplus airplane command set) heats dissimilar metal junction, producing small current which is indicated on meter. R1 is made from 31-inch length of tv twin-lead shorted at one end. Article gives calibration table.—N. Johnson, Low-Cost A.C. Ammeter, *Popular Electronics*, May 1969, p 50—52.

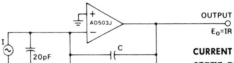

CURRENT TO VOLTAGE—Use of fet-input opamp permits measuring small currents such as are generated by photomultipliers and photodiodes. Bandwidth drops from 220 kHz to 12 kHz as R is increased from 100K to 10 meg and C is decreased from 100 pF to 1 pF. —"AD503, AD506 I.C. Fet Input Operational Amplifier," Analog Devices, Inc., Norwood, MA, Technical Bulletin, Aug. 1971.

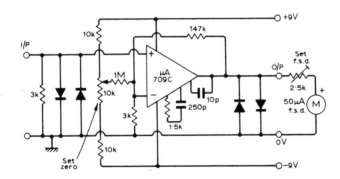

DIRECT-CURRENT AMPLIFIER—Use with 50-μA meter for measuring 1-μA d-c full-scale. Diodes are 1N914.—T. D. Towers, Elements of Linear Microcircuits, *Wireless World*, Feb. 1971, p 76—80.

SECTION 28
Measuring Circuits—Frequency

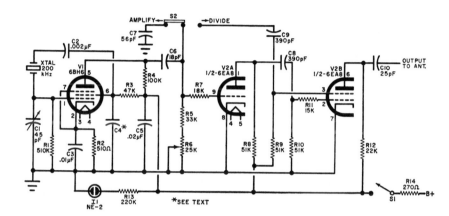

25-KHZ HAM-BAND CALIBRATOR—Absence of tuned circuits gives harmonic-rich output of V1 for feeding to frequency divider V2 controlled by R6. Designed to work with vacuum-tube receivers, from which between 150 and 200 V d-c for plate supply is obtained. C4 is between 50 and 150 pF, depending on crystal used. Article covers construction, calibration, and operation.—N. Johnson, Advanced Ham Frequency Calibrator, *Popular Electronics*, Jan. 1969, p 57–58.

50-KHZ MARKERS—Frequency standard for ham transmitter uses 100-kHz crystal oscillator Q1 to stabilize output of 50-kHz mvbr Q3-Q4. Q2 is pulse amplifier. Harmonics of mvbr serve to spot band edges in station receiver.—S. C. Creason, A Novice Frequency Standard, *QST*, Jan. 1967, p 22–23.

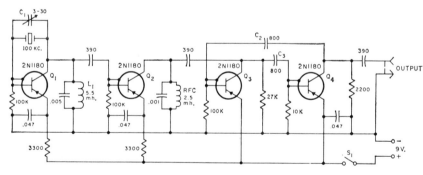

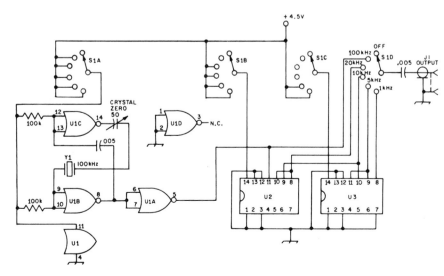

1-KHZ MARKERS—Uses low-cost Fairchild IC's with 100-kHz crystal mvbr to give strong marker signals at 100, 20, 10, 5, and 1 kHz intervals well into h-f range. U2 and U3 are MS19350 decade counters, and U1 is quad 2-input NOR gate such as International Rectifier 1C724-C. Power supply can be three penlight cells in series.—A. C. Beresford, An Inexpensive Secondary Frequency Standard, *QST*, May 1972, p 37–39.

1-MHZ REFERENCE—Provides square-wave output with amplitude of 5 V and about 1:1 on-off ratio, with long-term frequency accuracy of 1 part in 10^8. Article includes temperature control circuit for crystal oven. Output of oscillator Tr1 drives two-transistor shaping circuit.—L. O. Nelson-Jones, Crystal Oven and Frequency Standard, *Wireless World*, June 1970, p 269–273.

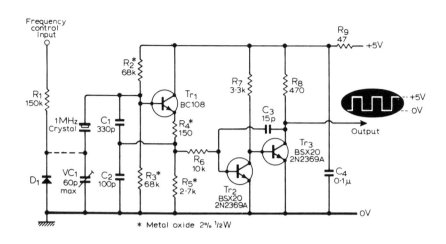

10-KHZ PIPS TO 30 MHZ—provides switch-selectable crystal-controlled marker pips at every 1,000, 500, 200, 100, 50, 20, and 10 kHz up to frequency limit of most commercial short-wave receivers, to simplify location of desired specific short-wave frequency. Closing S4 tone-modulates marker for easier spotting in crowded band. Use of IC flip-flops as frequency dividers makes one crystal do work of four. Different frequencies of square waves in circuit can be used for checking scope sweeps and testing audio amplifiers for ringing. XTAL1 is 100 kHz and XTAL2 is 1,000 kHz.—A. A. Mangieri, IC Frequency Spotter/Standard, *Popular Electronics*, Aug. 1969, p 27–32.

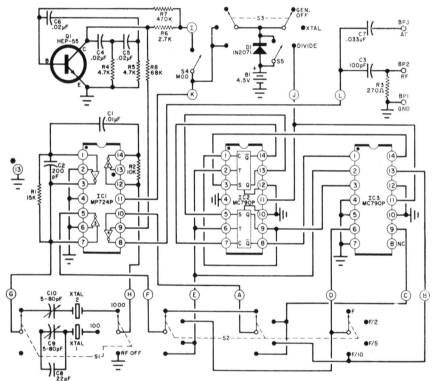

FET GRID-DIP METER—Light-weight battery-operated circuit provides greater sensitivity than comparable tube circuit, because fet is switched into operation as Q multiplier of absorption frequency meter. Covers 1.7 to 300 MHz. Developed by James Millen Mfg. Co.—Transistor Grid-Dip Frequency Meter Simplifies Portable Measurements, *IEEE Spectrum*, March 1971, p 93.

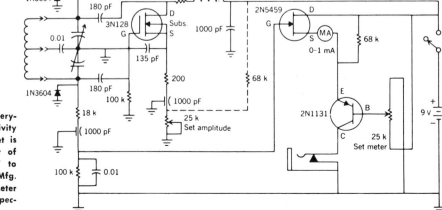

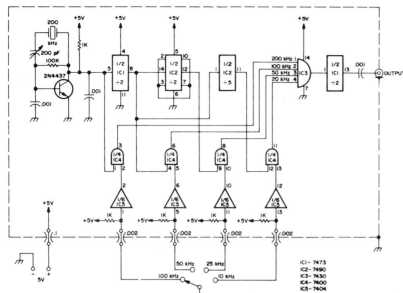

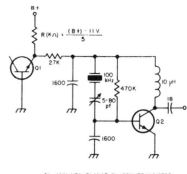

Q1 ANY NPN PLANAR SILICON TRANSISTOR
Q2 2N2925, 2N3392, 2N3565, SE4002

100-KHZ CRYSTAL CALIBRATOR—Provides usable harmonics up to about 150 MHz, has built-in voltage regulator Q1, and has padder in series with crystal for zero-beating with WWV. Regulated voltage provided by Q1 is about 11 V.—J. Fisk, "73 Useful Transistor Circuits," 73 Inc., Peterborough, NH, 1967, p 27A.

100, 50, 25, AND 10 KHZ—Switch gives choice of frequencies for symmetrical square waves derived from 200-kHz crystal oscillator with IC dividers. Used as frequency calibrator. Switch may be located remotely because lines to it have only d-c levels.—Circuits, 73, Aug. 1972, p 131.

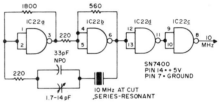

10-MHZ CRYSTAL TIME REFERENCE—Simple circuit uses four gates from IC. Frequency chosen for crystal gives excellent frequency stability with minimum of temperature compensation, and permits zero-beating against 10-MHz signal of WWV.—P. A. Stark, A Modern VHF Frequency Counter, 73, July 1972, p 5–11 and 13.

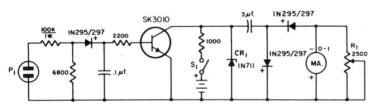

POWER-LINE FREQUENCY MONITOR—Used for checking frequency of emergency a-c generators. Calibrate by connecting P1 to regular 60-Hz power line after setting R1 for zero resistance. Readjust R1 for reading of 0.6 mA on meter. This will correspond to 60 Hz, and meter will indicate other frequencies with reasonable accuracy over upper half of scale regardless of voltage fluctuation or waveform.—"The Mobile Manual For Radio Amateurs," The American Radio Relay League, Newington, CT, 4th Ed., 1968, p 261.

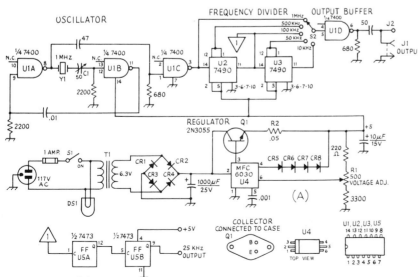

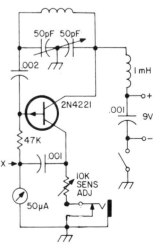

1-MHZ STANDARD—Output of crystal-controlled mvbr is divided to 500, 100, and 10 kHz by decade counters U2 and U3. For 25-kHz output, add two flip-flops as in (B). Trimmer C1 permits varying crystal frequency to align with WWV. Power supply uses Motorola IC voltage regulator U4. All diodes are HEP156 or equivalent.—D. A. Blakeslee, Double Standards, QST, April 1972, p 13–17.

1-250 MHZ OSCILLATOR WITH METER—Frequency range is covered with six plug-in coils, winding data for which is given in article. Used as grid-dip meter. 10K pot serves as sensitivity control. Jack is for headphones.—K. Brown, The Indicating Oscillator, 73, Sept. 1970, p 29–31.

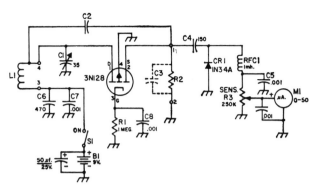

FET DIPPER—Single fet operating from 9-V battery gives portability and performance comparable to that of grid-dip meter. Article gives winding data for eight L1 plug-in coils and values of C2 and R2 for covering 1.8 to 150 MHz in eight ranges.—L. G. McCoy, A Field-Effect Transistor Dipper, QST, Feb. 1968, p 24–27.

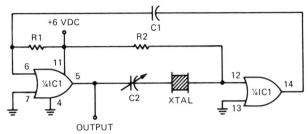

CRYSTAL CALIBRATOR—Output is square wave with high harmonic content, at frequency determined by crystal. Only half of HEP570 IC is used; for dual-frequency calibrator, use both halves. R1 and R2 are 56K, C2 is 9–35 pF, and C1 is 1,000 pF for 100-kHz crystal, 430 pF for 500 kHz, and 39 pF for 1 MHz.—"Radio Amateur's IC Projects," Motorola Semiconductors, Phoenix, AZ, HMA-36, 1971.

RECEIVER CALIBRATOR—Provides choice of 200, 100, 50, and 25 kHz calibration markers for use with general-coverage h-f receivers. 200-kHz crystal operates in series mode in Colpitts oscillator Q1. Q2 drives chain of three Hughes HRM-F/2 MOS binary dividers or equivalent, while Q3 drives Schmitt trigger Q4-Q5 when 200-kHz output is desired. Requires regulated supply. Q1, Q3, and Q5 are 2N3646 or HEP50; Q2 is NPF102 or HEP802; Q4 is 3N128. L1 is 2 mH having pot core.—H. Olson, The Ball of Wax—A Calibrator, 73, Nov. 1969, p 84–86.

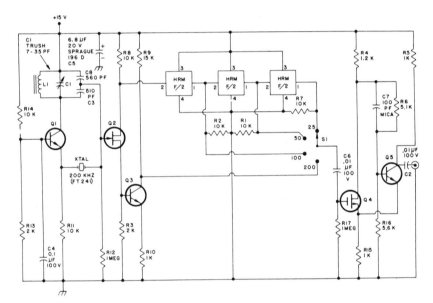

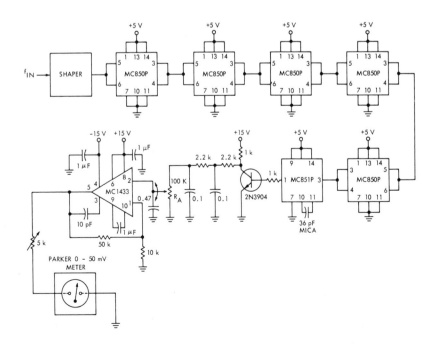

1–30 MHZ FREQUENCY METER—Uses seven IC's in linear frequency-voltage converter driving 0–50-mV meter having direct-reading frequency scale. Linearity is better than 5% over entire range. Minimum input voltage level is 2 V p-p. Report includes shaping circuit required at input, design equations, and detailed description of operation. Divide-by-32 binary uses five Motorola MC850 high-speed pulsed binary IC's feeding MC851P mono mvbr. Mono output is input to 2N3904 low-pass filter acting as averaging circuit that feeds MC1433 opamp biased to give d-c voltage gain of 5. Pot RA is adjusted to give 10-V d-c level at output of opamp for 30-MHz input, and 5K pot in series with meter is used to adjust current to 5 mA when opamp output is 10 V.—"Integrated Circuits for High Frequency to Voltage Conversion," Motorola Semiconductors, Phoenix, AZ, AN-297, 1971.

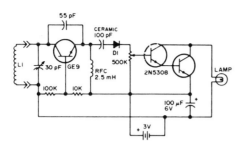

DIPPER—Simple oscillator and Darlington amplifier drive No. 48 pilot lamp that dims to indicate absorption of energy when oscillator is tuned to same frequency as circuit under test. Operates from pair of penlight cells. Oscillator will work up to 120 MHz. Vernier dial for tuning capacitor can be calibrated with any receiver having accurate tuning dial covering frequency range of input.—E. Babudro, The Dip Light, 73, March 1970, p 68–69.

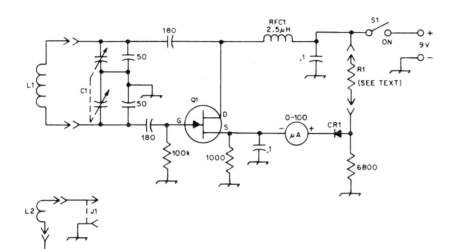

FET DIPPER—Will give indication of resonance down to 0.5 μA on meter when coupled to circuit 3 inches away. Article gives approximate winding data for eight plug-in coils covering frequency range of 1.5 to 30 MHz. With L2 added as pickup loop, jack J1 provides sampling of oscillator frequency for readout on frequency counter. Q1 is MPF102. CR1 is 1N34A. R1 is part of each plug-in coil, whose value is found experimentally by first using variable resistor and adjusting for about 95% of full-scale reading on meter; its value will be in range from 10K to 33K.—P. Lumb, High-Accuracy FET Dipper, QST, June 1972, p 46–47.

100-KHZ CRYSTAL CALIBRATOR—Provides usable signals well up into MHz range for receiver calibration. Temperature and frequency stability are excellent, especially if alkaline battery is used.—F. H. Tooker, Build FET Crystal Calibrator, Popular Electronics, March 1968, p 56–58.

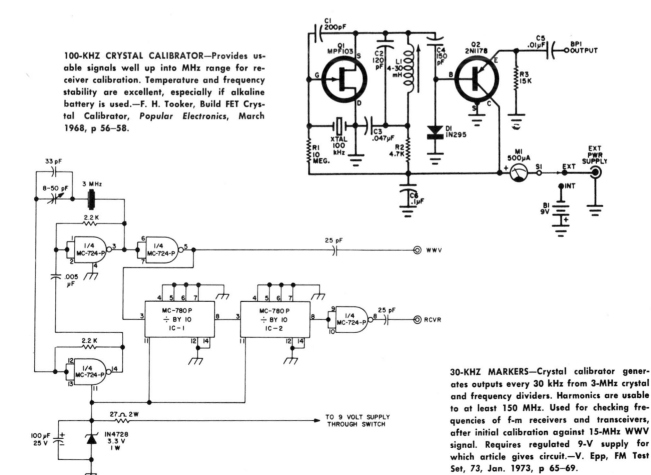

30-KHZ MARKERS—Crystal calibrator generates outputs every 30 kHz from 3-MHz crystal and frequency dividers. Harmonics are usable to at least 150 MHz. Used for checking frequencies of f-m receivers and transceivers, after initial calibration against 15-MHz WWV signal. Requires regulated 9-V supply for which article gives circuit.—V. Epp, FM Test Set, 73, Jan. 1973, p 65–69.

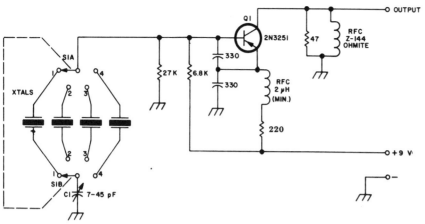

F-M TRANSMITTER OSCILLATOR—Single-transistor crystal oscillator Q1, with crystal switching, is rich in harmonics up to about 450 MHz. Used in conjunction with crystal calibrator of test set for f-m receivers and transmitters, deviation checker, and signal generator with calibrated output. C1 adjusts crystal to exact frequency desired. Output level can be adjusted with 15K linear-taper series rheostat.—V. Epp, FM Test Set, 73, Jan. 1973, p 65–69.

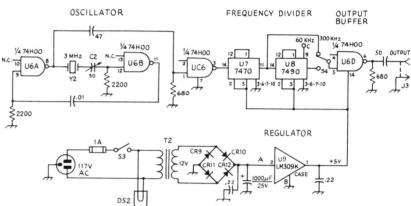

3-MHZ STANDARD—Uses high-speed gate package as oscillator and single IC as complete voltage regulator in power supply. Oscillator output is divided to 300 and 30 kHz, selected by S4, with optional 60-kHz output. All diodes are HEP156.—D. A. Blakeslee, Double Standards, QST, April 1972, p 13–17.

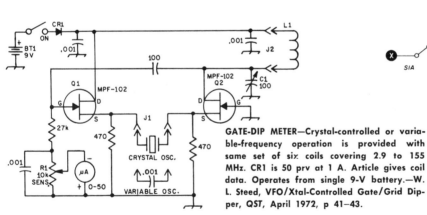

GATE-DIP METER—Crystal-controlled or variable-frequency operation is provided with same set of six coils covering 2.9 to 155 MHz. CR1 is 50 prv at 1 A. Article gives coil data. Operates from single 9-V battery.—W. L. Steed, VFO/Xtal-Controlled Gate/Grid Dipper, QST, April 1972, p 41–43.

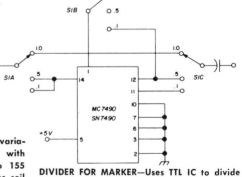

DIVIDER FOR MARKER—Uses TTL IC to divide 1-MHz crystal marker input at X by 2 or by 10 to give markers at 1, 0.5, and 0.1 MHz for locating any 100-kHz point with crystal accuracy when calibrating main tuning dial of communication receiver.—G. Tillotson, General-Coverage Receiver Frequency Calibrator, Ham Radio, Dec. 1971, p 28–29.

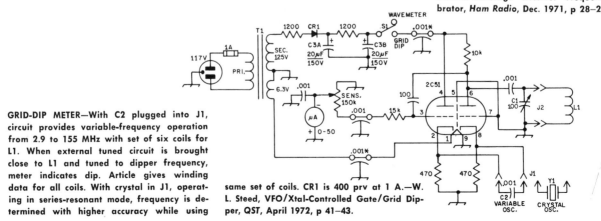

GRID-DIP METER—With C2 plugged into J1, circuit provides variable-frequency operation from 2.9 to 155 MHz with set of six coils for L1. When external tuned circuit is brought close to L1 and tuned to dipper frequency, meter indicates dip. Article gives winding data for all coils. With crystal in J1, operating in series-resonant mode, frequency is determined with higher accuracy while using same set of coils. CR1 is 400 prv at 1 A.—W. L. Steed, VFO/Xtal-Controlled Gate/Grid Dipper, QST, April 1972, p 41–43.

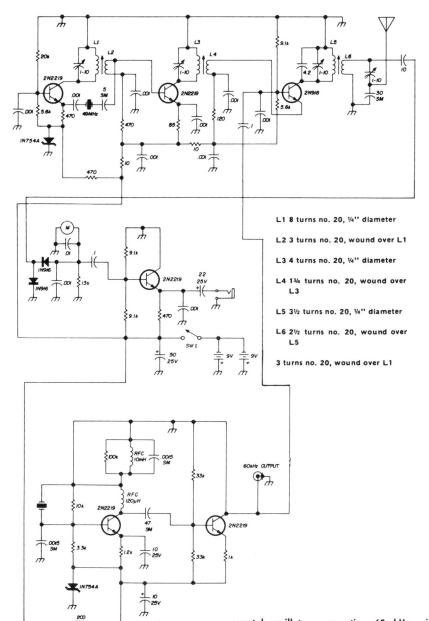

L1 8 turns no. 20, ¼" diameter

L2 3 turns no. 20, wound over L1

L3 4 turns no. 20, ¼" diameter

L4 1¾ turns no. 20, wound over L3

L5 3½ turns no. 20, ¼" diameter

L6 2½ turns no. 20, wound over L5

3 turns no. 20, wound over L1

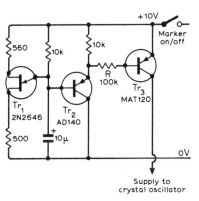

MARKER IDENTIFIER—Supply to 500-kHz crystal oscillator is switched on and off by ujt at rate of around 8 Hz to make recognition of marker signals easier in crowded bands. Creates slight warble in addition when bfo is in use. Ideal for marker recognition between 18 and 30 MHz, where harmonics of crystal are quite weak.—G. Dann, Crystal Marker Identification, *Wireless World*, Feb. 1972, p 77–78.

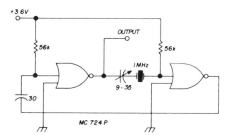

1 MHZ WITH NOR GATES—Simple dual-gate IC arrangement provides square-wave output for receiver calibration.—E. Noll, Digital IC Oscillators and Dividers, *Ham Radio*, Aug. 1972, p 62–67.

2-METER F-M FREQUENCY METER—High-accuracy heterodyne-type meter provides crystal-controlled frequency markers. Covers range of 146.94 MHz to below 145 MHz with accuracy within 15 Hz immediately after calibration and within 100 Hz long-term. 49-MHz crystal oscillator feeds buffer-tripler providing 147 MHz, coupled to 2N918 mixer. Lower crystal oscillator, generating 60 kHz with high harmonic output, drives base of mixer through buffer, to modulate 147-MHz signal. First lower sideband of mixer is 146.94 MHz, and strong signals are available at 60-kHz intervals to below 144 MHz for 19-inch whip that radiates signals for receiver calibration.—C. A. Baldwin, Two-Meter FM Frequency Meter, *Ham Radio*, Jan. 1971, p 40–43.

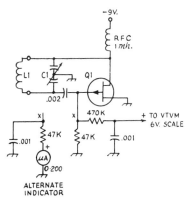

GATE DIPPER—Uses Fairchild 2N4342, 2N4360, or equivalent p-channel jfet. C1 can be transistor radio type or dual 365-pF. L1 uses plug-in coils to cover range of 3 to 200 MHz. Used same as grid-dip oscillator for determining resonant frequency.—W. Hayward, Gate-Dip Oscillator, *QST*, Sept. 1967, p 45.

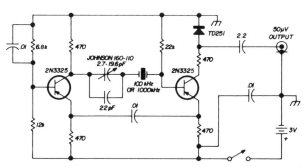

HARMONIC GENERATOR—With either 100 or 1,000 kHz crystal in reference oscillator, circuit will generate harmonics through 1,296 MHz.—C. Spurgeon, Harmonic Generator, *Ham Radio*, Oct. 1970, p 76.

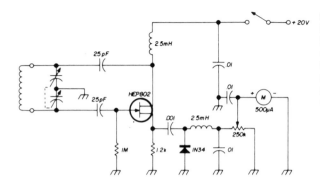

FET GRID-DIP OSCILLATOR—High input impedance of fet provides higher sensitivity for obtaining dip than is possible with tube or other transistor circuits. Supply can be two 9-V transistor batteries in series. Article covers other methods of modernizing tube-type gdo's—P. A. Lovelock, Solid-State Conversion of the GDO, *Ham Radio*, June 1970, p 20—23.

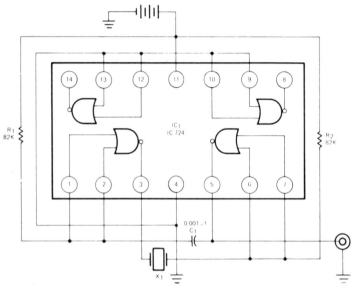

100-KHZ CRYSTAL CALIBRATOR—International Rectifier IC724 and 100-kHz crystal X1 serve with minimum of additional components as secondary frequency standard for calibrating receivers, transmitters, and other test equipment. Provides output of 3-V p-p at fundamental frequency, for use with cro. Battery voltage is critical and should be checked frequently; use two AA cells in series.—"Hobby Projects," International Rectifier, El Segundo, CA, Vol. II, p 62–63.

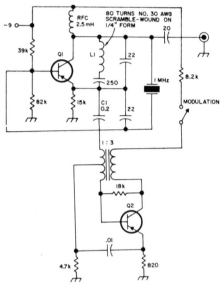

1—MHZ CRYSTAL—Provides distinctive markers up to 30 MHz, modulated at about 1,000 Hz. Particularly useful for band-edge marking. Modulation assists in identifying marker at higher frequencies where harmonics are quite weak.—J. Fisk, "73 Useful Transistor Circuits," 73 Inc., Peterborough, NH, 1967, p 28A.

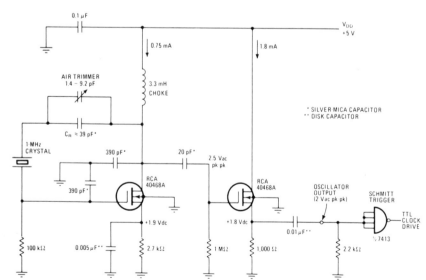

STABLE 1-MHZ CRYSTAL—Output changes less than 1 Hz over supply range of 3 to 9 V. Uses modified Pierce oscillator and isolating source-follower, with Schmitt trigger serving as TTL interface. Output is rich in r-f harmonics, for beating with 10-MHz WWV broadcast to permit adjusting oscillator easily within 0.1 Hz of nominal operating frequency. Circuit can then be used as secondary frequency standard.—T. King, Stable Crystal Oscillator Works over Wide Supply Range, *Electronics*, June 21, 1973, p 112.

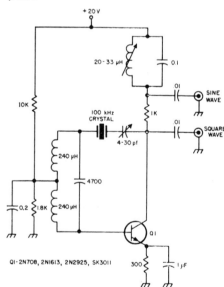

100-KHZ SINE OR SQUARE—Single-transistor Hartley crystal oscillator includes fine frequency adjustment control in series with crystal for zero-beating with WWV. Large amount of feedback drives collector of Q1 from cutoff to saturation for square-wave output, while sine-wave output is developed across tunable high-Q tank.—J. Fisk, "73 Useful Transistor Circuits," 73 Inc., Peterborough, NH, 1967, p 27A.

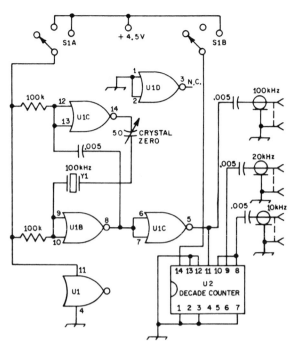

20-KHZ AND 10-KHZ MARKERS—Low-cost 100-kHz crystal mvbr provides 100, 20, and 10 kHz markers well into h-f range. U1 is quad 2-input NOR gate such as International Rectifier IC724-C, and U2 is Fairchild MS19350 decade counter.—A. C. Beresford, An Inexpensive Secondary Frequency Standard, QST, May 1972, p 37–39.

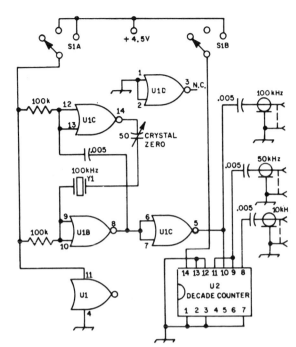

10-KHZ MARKERS—Low-cost 100-kHz crystal mvbr provides 100, 50, and 10-kHz markers well into h-f range. U1 is quad 2-input NOR gate such as International Rectifier IC724-C, and U2 is Fairchild MS19350 decade counter.—A. C. Beresford, An Inexpensive Secondary Frequency Standard, QST, May 1972, p 37–39.

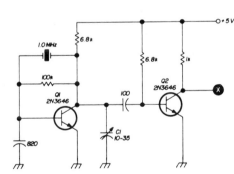

MARKER—Used to calibrate communication receiver. Output of 1-MHz crystal is rich in harmonics, to above 30 MHz.—G. Tillotson, General-Coverage Receiver Frequency Calibrator, Ham Radio, Dec. 1971, p 28–29.

SECTION 29
Measuring Circuits—General

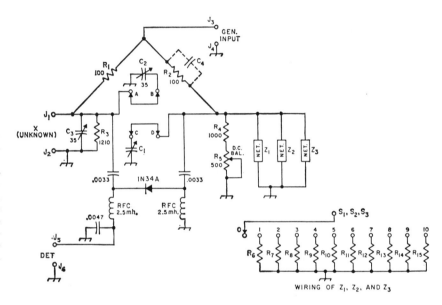

R-F ADMITTANCE BRIDGE—Can be used for measuring 10–1,000 ohm r-f resistance, 0–500 pF capacitance, 1–50 μH inductance, and complex impedances. Article covers construction and calibration. General input is transmitter or other 7,120-kHz frequency source. Meter or other detector detects null at bridge balance. C1 is about 500 pF. Table in article gives values of R6–R15 for three r-f conductance ranges.—F. Cherubini, An Admittance Bridge for R.F. Measurements, QST, Sept. 1967, p 30–33.

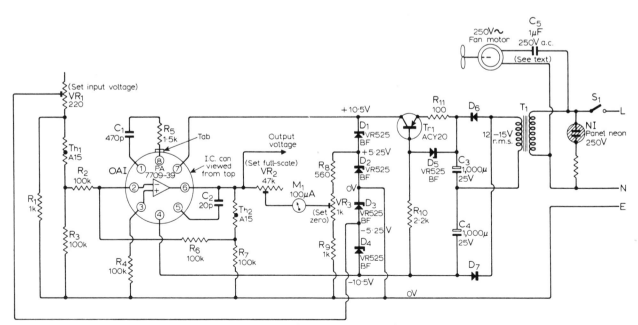

HYGROMETER—Uses Philco-Ford opamp as computing element. VR1 sets negative input voltage across type A15 wet thermistor Th1 and R3. Th2 is dry thermistor. Opamp output is capable of driving load up to 600 ohms to operate external humidity control switch. Capacitor is used in series with shaded-pole a-c motor providing continuous ventilation of wet thermistor, to reduce motor voltage and prevent overheating. Article covers construction, calibration, and use. Meter has nonlinear scale reading 0–100% humidity. Accuracy is better than 5%. D6 and D7 are A.E.I. SJ403-F diodes.—D. Bollen, A Thermistor Hygrometer, *Wireless World*, Dec. 1969, p 557–561.

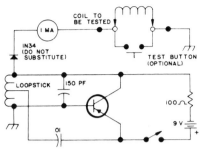

INDUCTANCE METER—After calibration with known inductance values, meter will indicate inductance values of chokes and coils accurately over range of 0.25 to 10 mH. Almost any pnp transistor may be used, but diode type is critical. To use, connect unknown coil, push test button and adjust control for 1 mA (corresponding to 0 mH since scale is like that of ohmmeter), then release button and read inductance value.—J. Clack, A Direct Reading Inductance Meter . . . The "Henryometer," 73, Dec. 1972, p 88.

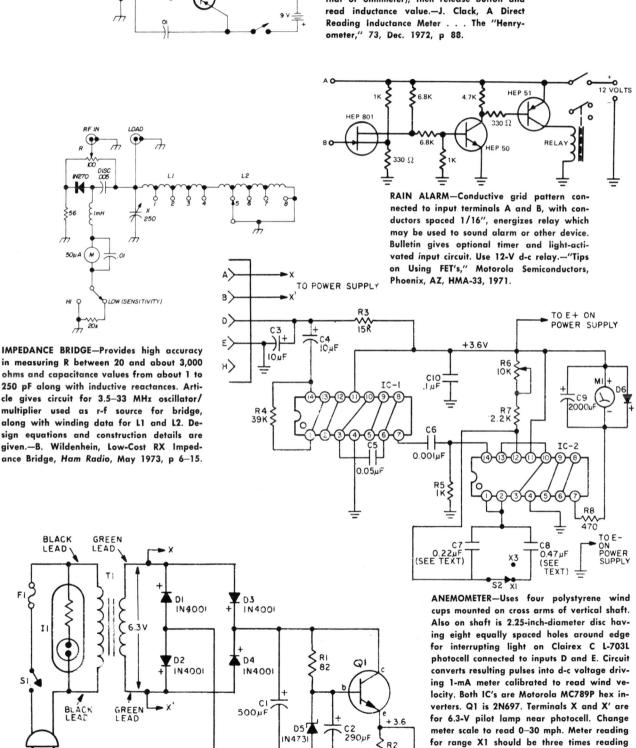

RAIN ALARM—Conductive grid pattern connected to input terminals A and B, with conductors spaced 1/16", energizes relay which may be used to sound alarm or other device. Bulletin gives optional timer and light-activated input circuit. Use 12-V d-c relay.—"Tips on Using FET's," Motorola Semiconductors, Phoenix, AZ, HMA-33, 1971.

IMPEDANCE BRIDGE—Provides high accuracy in measuring R between 20 and about 3,000 ohms and capacitance values from about 1 to 250 pF along with inductive reactances. Article gives circuit for 3.5–33 MHz oscillator/multiplier used as r-f source for bridge, along with winding data for L1 and L2. Design equations and construction details are given.—B. Wildenhein, Low-Cost RX Impedance Bridge, Ham Radio, May 1973, p 6–15.

ANEMOMETER—Uses four polystyrene wind cups mounted on cross arms of vertical shaft. Also on shaft is 2.25-inch-diameter disc having eight equally spaced holes around edge for interrupting light on Clairex C L-703L photocell connected to inputs D and E. Circuit converts resulting pulses into d-c voltage driving 1-mA meter calibrated to read wind velocity. Both IC's are Motorola MC789P hex inverters. Q1 is 2N697. Terminals X and X' are for 6.3-V pilot lamp near photocell. Change meter scale to read 0–30 mph. Meter reading for range X1 should be three times reading for range X3. Calibrate in auto having reasonably accurate speedometer.—E. A. Morris, "Windy" the Wind Gauge, Science and Electronics, April-May 1970, p 29–38 and 99–100.

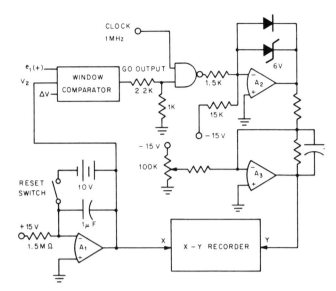

PROBABILITY DENSITY ANALYZER—Uses window comparator for which basic circuit is also given, though without values. GO output goes high each time input signal is inside window width centered on V2. If clock is also high, output of first NAND gate goes low. A2 is opamp used as comparator to generate precision voltage levels for R-C averaging filter used with remainder of circuit to drive recorder. Opamps can be Burr-Brown.—J. G. Graeme, G. E. Tobey, and L. P. Huelsman, "Operational Amplifiers," McGraw-Hill Book Co., New York, 1971, p 364—366.

ION CHAMBER—Measures polarity and amount of ionization present in air. Ions detected by ion chamber develop small voltage drop across very high resistance of R3, for amplification by Victoreen electrometer tube V1, which operates transistor bridge Q1-Q2 having polarity-indicating meter. Article covers adjustment and operation.—H. Burgess, Make Your Own Ion Chamber, *Popular Electronics*, Nov. 1969, p 31—35.

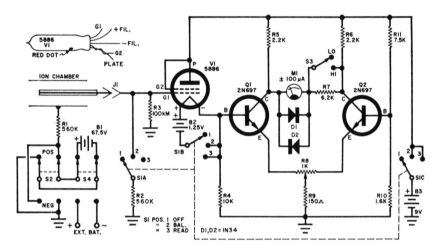

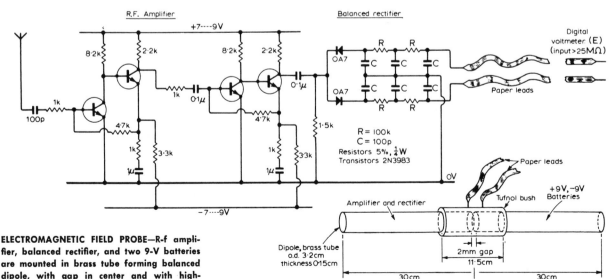

ELECTROMAGNETIC FIELD PROBE—R-f amplifier, balanced rectifier, and two 9-V batteries are mounted in brass tube forming balanced dipole, with gap in center and with high-impedance paper leads (Teledeltos pen recording paper) going from gap ends to high-impedance digital voltmeter. Used to measure near electromagnetic field in frequency range of 150 kHz to 30 MHz. Paper leads with resistance of about 25,000 ohms per foot avoid perturbation of field being measured. Has also been used to check electric field distribution in aperture of 11-dB pyramidal horn at 1,000 MHz.—J. Thickpenny, Electric Field Probe, *Wireless World*, June 1970, p 293—294.

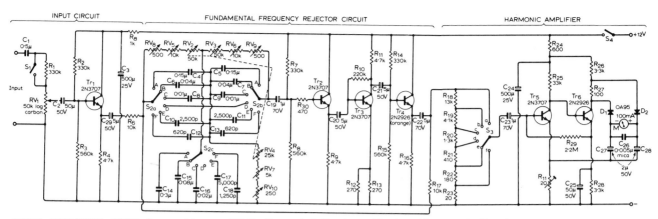

AUDIO DISTORTION METER—Covers 0.1 to 100% harmonic distortion in five ranges, with each range having frequency ratio of about 4:1 to cover audio spectrum from 20 to 20,-000 Hz. Uses twin-T filter in combination with negative feedback provided by transistor stages Tr2, Tr3, and Tr4 to give over 70 dB attenuation of fundamental frequency without significantly attenuating second and third harmonics. Article covers construction and adjustment.—L. Haigh, Distortion Factor Meter, *Wireless World*, July 1969, p 317–320.

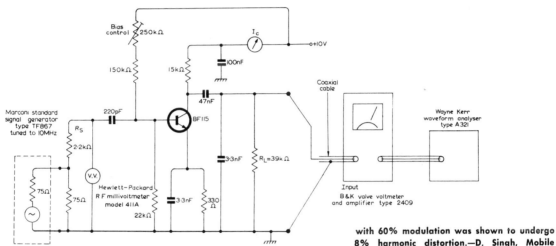

A-M DEMODULATION DISTORTION—Used in determining optimum i-f amplifier output for transistor used as detector for satisfactory demodulation with minimum percentage of harmonic distortion. With BF115 detector transistor shown in circuit, 50-mV rms signal with 60% modulation was shown to undergo 8% harmonic distortion.—D. Singh, Mobile 166 MHz A.M. Communications Receiver, *Mullard Technical Communications*, Jan. 1968, p 14–29.

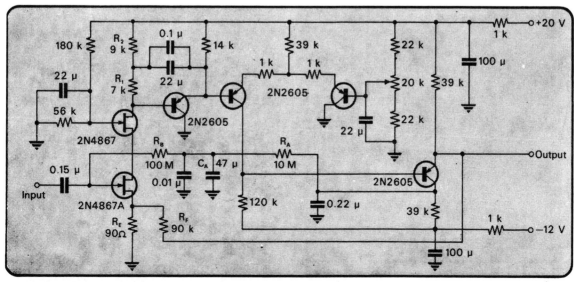

MAGNETOMETER AMPLIFIER—Low-noise fet-input amplifier is used with search coil on spacecraft to measure variations in interplanetary magnetic fields. Covers 1 Hz to 100 kHz. Power drain is 24 mW. Designed for operating temperature range of —30 to +60 C. —S. Cantarano and G. V. Pallottino, A Low-Noise FET Amplifier For A Spaceborne Magnetometer, *Electronic Engineering*, Sept. 1970, p 57–60.

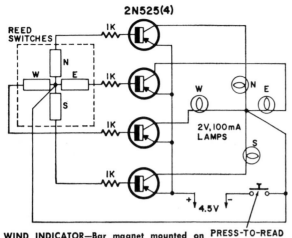

WIND INDICATOR—Bar magnet mounted on weather vane rotates over four reed switches which are normally open and close each time magnet passes over as wind fluctuates around average direction. Leads from switches are run through cable to transistors driving indicator lamps at remote point. For reading, switch is held down until lamp corresponding to N-E-S-W direction closest to average flickers on as wind varies. Hermetically sealed switches are not affected by atmospheric conditions.—R. F. Scott, Tech Topics, *Radio-Electronics*, Oct. 1969, p 63 and 68–69.

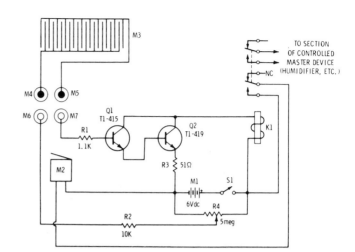

HUMIDITY CONTROL—Standard moisture-sensor plate M3 (not water sensor) completes connection between R1 and R2 to energize 6-V d-c relay K1 whenever excess of moisture builds up. Choose location for M3 at which dust accumulation is minimized. Relay can be connected to sound alarm or turn on electric heater or dehumidifier. M2 is 6-V d-c buzzer.—R. M. Brown and T. Kneitel, "49 Easy Entertainment and Science Projects," Vol. 2, Howard W. Sams, Indianapolis, IN, 1969, p 85–87.

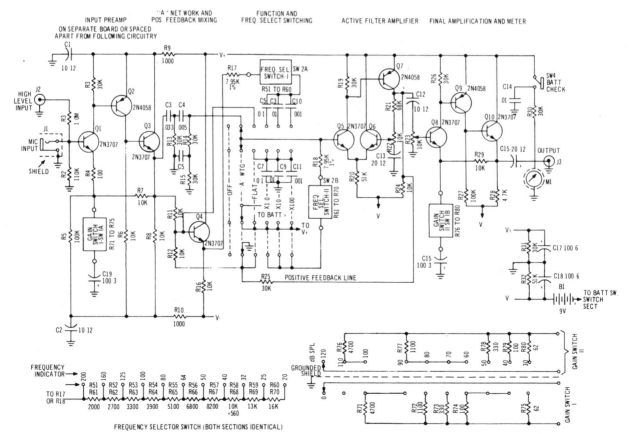

SOUND-LEVEL METER—Covers 20 to 20,000 Hz in three ranges, with frequency accuracy of 10%. Will measure sounds down to 20 dB SPL and third-octave noise down to 0 dB SPL when used with microphone sensitivity of —60 dB re 1 V/μbar. Includes bass filter to give standard "A" weightings. Ideal for measuring room acoustics. Article covers measuring techniques in detail. Preamp Q1-Q2-Q3 with overall gain adjustable from 1 to 200 feeds combination amplifier and active filter Q4. Remaining amplifier stages provide extra sensitivity for driving standard vu meter M1.—J. D. Griesinger, A Sound-Level Meter, *Audio*, Dec. 1970, p. 28, 30, 32, 34, 36, and 38.

INDUCTANCE CHECKER—Used with short-wave receiver to measure r-f inductance values from 0.3 μH to 7 mH. Consists of two-terminal oscillator to which unknown coil and known capacitor are connected in parallel. Receiver is placed near oscillator for use in determining frequency. Verify correct tuning by placing hand near tuned circuit and listening for frequency change. Oscillator range is 600 kHz to above 30 MHz, which can be covered with only two standard capacitors—10 pF and 100 pF. With 10 pF and 3 μH, frequency is 30 MHz, and 600 kHz with 7 mH. With 100 pF, 0.3 μH gives 30 MHz and 700 μH gives 600 kHz.—J. A. Rolf, Quickie Inductance Checker, *Popular Electronics*, Aug. 1973, p 98–99.

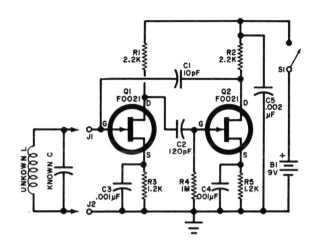

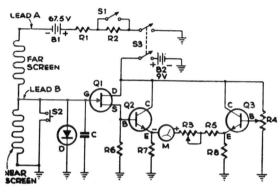

R1—10,000 ohm, ½-watt carbon resistor
R2—10 megohm, ½-watt carbon resistor
R3—3,000 ohm carbon potentiometer
R4—10,000 ohm carbon potentiometer
R5—470 ohm, ½-watt carbon resistor
R6—22,000 ohm, ½-watt carbon resistor
R7, R8—470 ohm, ½-watt carbon resistor
C—2ufd, 100-volt paper capacitor
D—Zener diode, Motorola HEP-104

BULLET TIMER—Circuit measures time required for bullet to pass between two screens placed exactly 1 ft apart. When bullet cuts first screen, B1 starts charging C through R1. When bullet cuts second screen, battery circuit is opened and charging stops; voltage across C, proportional to time of flight, is then measured with voltmeter circuit Q1-Q2-Q3. Article covers calibration of indicating meter. For bullets slower than 1,000 fps, place screens 6 inches apart. Use any 1-mA meter. B2 is 8.4-V VS146X mercury battery. Q1 is HEP 802 fet, and other transistors are 2N5172.—R. M. Benrey, Measure the Speed of a Bullet with This Electronic Stopwatch, *Popular Science*, July 1969, p 144–146 and 178.

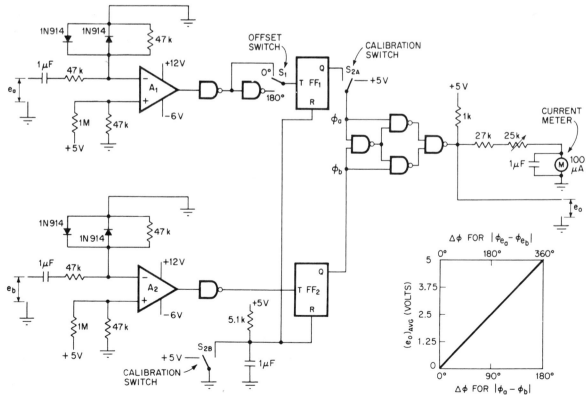

360-DEG DIGITAL PHASE METER—Detects average phase (time) difference between any two sine, square, triangle, or pulse inputs with same frequency in range from 100 Hz to above 1 MHz and amplitude range of 0.5 to 10 V p-p, using only five IC packages. A1 and A2 are type 710 comparators used for clipping and squaring input signal to give logic-level pulses that are buffered and inverted with gates from two type 7400 NAND gate packages before being applied to Texas Instruments SN7473 dual flip-flop. Flip-flops perform as binary counters, dividing input frequency by two so phase difference between inputs is also divided by two to extend measurement range to 360 deg.—C. A. Herbst, Detector Measures Phase over Full 360° Range, *Electronics*, July 19, 1971, p 73.

IMPEDANCE BRIDGE—Will measure reactive and resistive components of unknown impedance throughout high-frquency range. Applications include measurement of antenna impedance, characteristic impedance and electrical length of coaxial lines, and input impedance of r-f components and amplifiers over range of 2 to 30 MHz. Operates from single 9-V transistor radio battery. Uses diode noise generator and 3-stage amplifier to feed r-f bridge. T1 is Micro-Metals T-37-10 toroid core having 9-mm OD and 8-turn trifilar windings. Receiver having S meter is connected to DETECTOR terminal and tuned to frequency of measurement. Bridge is balanced for noise null as indicated on S meter. —G. Pappot, Noise Bridge for Impedance Measurements, *Ham Radio*, Jan. 1973, p 62–64.

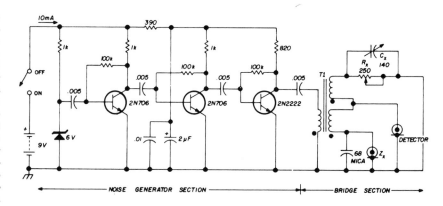

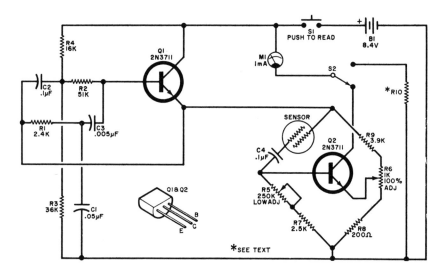

HYGROMETER—Uses 400-Hz twin-T oscillator as source for bridge having humidity sensor in one leg and transistor driving meter to indicate ambient relative humidity values up to 100%. Sensor construction details are given in article. Fiberglass cloth clamped between two pairs of brass plates is dipped in solution of lithium chloride to give sensor whose resistance varies with humidity. Article covers construction and calibration. Use 11K for R10, which should give about ¾-scale reading on meter for new battery when S2 is up for battery test.—J. Giannelli, Electro-Chemical Hygrometer, *Popular Electronics*, Oct. 1972, p 33–35.

SECTION 30
Measuring Circuits—Power

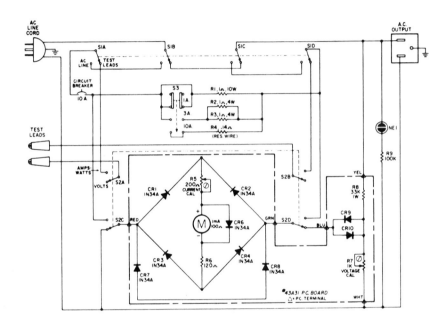

POWER DRAIN—Indicates amount of power drawn from a-c line by appliance or other device under test, on meter calibrated to read up to 345 W for line voltage of 115 V. Meter also has conventional voltage and current scales. Power conversion chart is required if line voltage differs significantly—P. Dahlen, Sencore's PM 157 Power Monitor, *Electronic Technician/Dealer*, Dec. 1970, p 57.

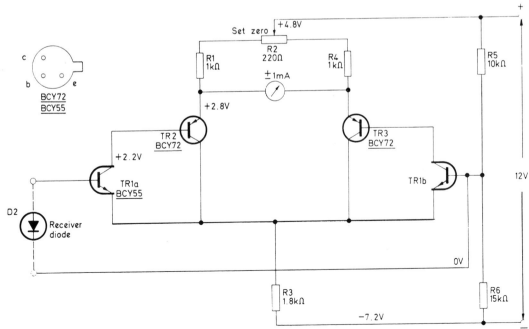

10-GHZ FIELD-STRENGTH METER—D-c amplifier used between 1N21B microwave diode D2 and indicating meter extends useful range for making field measurements at 10 GHz from Gunn-device transmitter to about 20 yards. Differential amplifier TR1 uses low-drift transistor pair.—"A Gunn Device Transmitter (10 GHz)," Educational Projects in Electronics, Mullard, London.

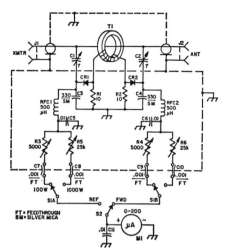

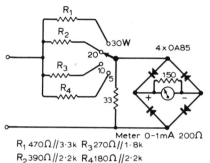

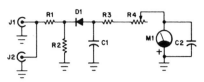

CB WATTMETER—Used in series with coax line to antenna, for continuous check of output power of transmitter or transceiver. Will also indicate jump in standing-wave ratio of antenna by showing improbably high power reading on meter. C1 and C2 are 0.001 μF, D1 is 1N60, M1 is 0–1 mA d-c meter, R1 is 3.3K, R2 4.7K, R3 10K, and R4 10K trimmer. —H. Friedman, Inline RF Wattmeter for CBers, *Electronics Illustrated*, May 1972, p 59–61 and 100.

POWER METER—Square-law characteristic of bridge diodes up to about 1.4 V makes linear scale possible for meter. Provides four ranges between 5 and 30 W at impedance of 15 ohms.—K. D. James, Linear Scale Power Meter, *Wireless World*, June 1969, p 269.

WATTS	M1	WATTS
100	200	1000
90	180	900
80	170	800
70	155	700
60	145	600
50	125	500
40	105	400
30	85	300
20	65	200
10	40	100
5	20	50

R-F WATTMETER—Provides two power ranges, 0–100 and 0–1,000 W, forward and reflected, for 3.5 to 30 MHz. Cable gives data for making calibrated meter scales. Toroidal transformer T1 has 35 turns No. 26 spread evenly to cover entire core of Amidon T-68-2 toroid. Diodes are 1N34A. Article covers construction, and gives circuits of comparable commercial versions.—D. DeMaw, In-Line RF Power Metering, *QST*, Dec. 1969, p 11–16.

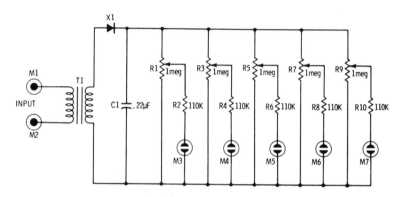

NEON WATTMETER—Pot for each neon is adjusted during calibration so bulb just lights at known a-c or a-f wattage level. Can be calibrated with various audio amplifiers having known full-volume ratings, or with variable-wattage source such as 12.6-V a-c filament transformer feeding 100-ohm pot. X1 is 100-mA 400-piv silicon diode and T1 is universal output transformer. All neons are NE-2A.—R. M. Brown and T. Kneitel, "49 Easy Entertainment & Science Projects," Howard W. Sams, Indianapolis, IN, 1971, p 29–30.

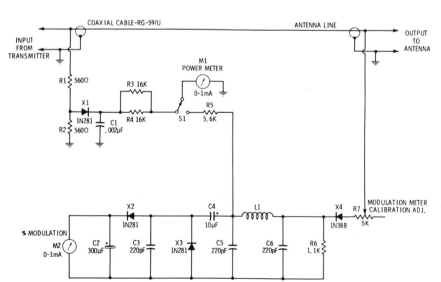

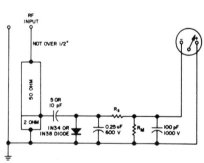

R-F WATTMETER—Reads 0–10 W on 0–1 mA meter M1. Lower portion of circuit gives percentage modulation reading for transmitter. Three-quarter swing of M2 to right corresponds to 75% modulation of a-m carrier. L1 is 24 μH.—R. M. Brown, "101 Easy CB Projects," Howard W. Sams, Indianapolis, IN, 1968, p 12–14.

500-MHZ POWER OUTPUT METER—Simple meter circuit is relatively nonreactive (vswr under 1.5:1 in operating range) and capable of dissipating 25 W for up to 30 s without overheating or 50 W for 5 s. Article covers construction of input resistor from five 10-ohm and one 2-ohm composition resistor discs. With 1-mA meter, 4K for Rs, and 180 for RM, full-scale reading is 400 W; article recommends higher-current meters and gives resistor values for them to provide more desirable lower full-scale readings.—J. A. Houser, An Improved UHF Power Output Meter, *73*, June 1973, p 37–38, 40, and 43.

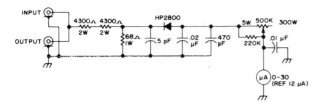

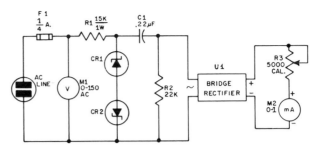

5–300 W R-F WATTMETER—Designed for use with large dummy antenna rated up to 500 MHz, for measuring transmitter output power over frequency range of 2 to 450 MHz. Article covers construction and calibration. Uses HP2800 hot carrier diode having rating of 75 piv.—F. C. Jones, RF Power Measurement with Hot Carrier Diodes, 73, Sept. 1971, p 42–44 and 46.

POWER FREQUENCY METER—Intended for use with portable a-c generator. To calibrate, connect to commercial power line and adjust R3 for reading of 0.6 mA on M2 to indicate 60 Hz. Readings are then accurate within 5% for 20 to 100 Hz. CR1 and CR2 are 1N5245 and U1 is HEP175.—J. Hall, A Field-Day AC-Power Monitor, QST, March 1971, p 40–41.

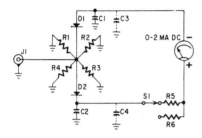

UHF WATTMETER—Provides full-scale ranges of 30 or 100 mW with acceptably flat response from 2 to 500 MHz. With separate 10-dB 50-ohm pad between transmitter and wattmeter, ranges are increased to 300 mW and 1 W. Diodes are 1N82A, R1–R4 are 200, R5 and R6 are meter multipliers selected to give desired full-scale ranges as described in text, R7 and R8 are 27, R9–R12 are 150, C1 and C2 are 1,000 pF, and C3–C4 are 0.01 μF.—H. Balyoz, An RF Wattmeter for UHF, Radio-Electronics, Sept. 1966, p 58.

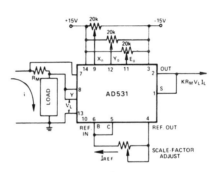

INSTANTANEOUS WATTMETER—Uses Analog Devices AD531 IC to reject common-mode interference by taking advantage of differential X input. Scale factor can be adjusted with 20K pot. Ideal for instrumentation, agc applications and for situations where two input signals must be subtracted before being subjected to further processing. NOTE: End terminals of 20K pot for Xo should be connected between +15 V and 42K resistor going to ground.—Monolithic Analog Multiplier-Dividers, Analog Dialogue, Vol. 6, No. 3, p 10.

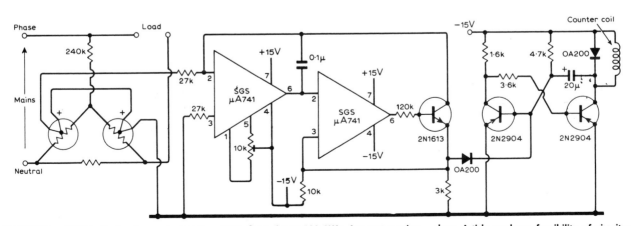

WATTHOUR METER—Electronic replacement for rotating-disc meter uses vacuum thermocouple multiplier operating on quarter-square principle to measure power consumption of loudspeaker, electric motor, or any other device drawing power from source in frequency range from d-c to 100 MHz. Integrator using opamps drives mono mvbr feeding electromagnetic counter, for converting sensed energy in watts to watthours. Heater acting on thermocouple has range of 5 mA and resistance of 400 ohms, while each couple is 2 ohms. Article analyzes feasibility of circuit.—L. A. Trinogga, Experimental Electronic Electricity Meter, Wireless World, Jan. 1972, p 33–35.

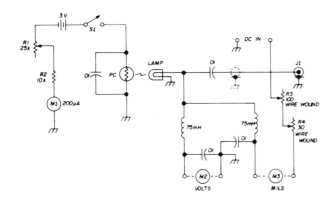

OPTOELECTRONIC R-F WATTMETER—Measures low power levels up to 148 MHz without elaborate calibration, based on fact that small lamp with short filament will reach given temperature for same amount of d-c, a-f, or r-f power. Suitable pilot lamp for transmitter power, as specified in article, is connected as transmitter load and light level is measured with cadmium sulfide photocell and meter. Direct current required through lamp to give same light reading is then measured for computing power.—H. F. Burgess, An Accurate RF Power Meter for Very Low Power Experiments, *Ham Radio*, Oct. 1972, p 58–61.

0.025–10 W R-F WATTMETER—Useful from low radio frequencies to above 450 MHz. Article covers construction and calibration. Uses HP2900 hot carrier diode having rating of 10 piv, limiting rms r-f voltage across it to 3 V for safe operation.—F. C. Jones, RF Power Measurement with Hot Carrier Diodes, *73*, Sept. 1971, p 42–44 and 46.

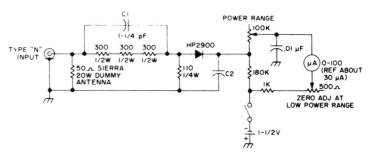

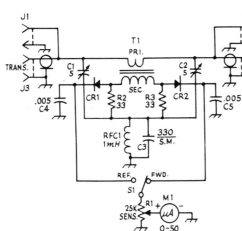

0–5 W FOR 10–80 METERS—Gives same deflection per watt on all bands in frequency range. With sensitivity control at maximum, requires only 1 W r-f energy to drive meter full-scale. Article gives construction and adjustment details. Diodes are 1N34A germanium. T1 has 60 turns No. 28 on Amidon T-68-2 toroid for secondary, with two turns over it for primary.—D. DeMaw, A QRP Man's RF Power Meter, *QST*, June 1973, p 13–15.

FIELD METER—Detector located as far as possible from antenna of CB, ham, or commercial transmitter is connected by ordinary two-wire line to indicator at operating position, for continuous check of field strength whenever transmitter is on air. Transistor can be RCA SK3009 or equivalent, and diode D is 1N34A.—J. B. White, Remote-Reading Field Meter, *Radio-Electronics*, Aug. 1967, p 60–61 and 66.

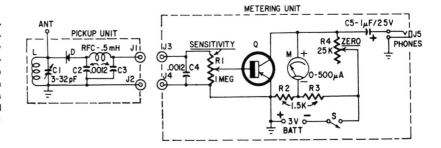

SECTION 31
Measuring Circuits—Resistance

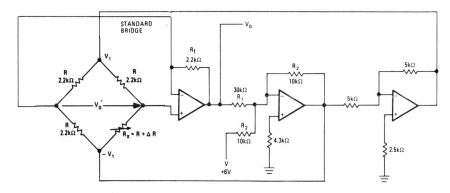

LINEAR RESISTANCE BRIDGE—Feedback circuit makes output voltage Vo vary linearly within 0.1% with value of bridge resistor Rx over range of 0 to 4.4K. Article gives design equations. Either a-c or d-c voltage can be used to excite bridge. All opamps are Burr-Brown 3024/15.—R. D. Buyton, Feedback Linearizes Resistance Bridge, *Electronics*, Oct. 23, 1972, p 102.

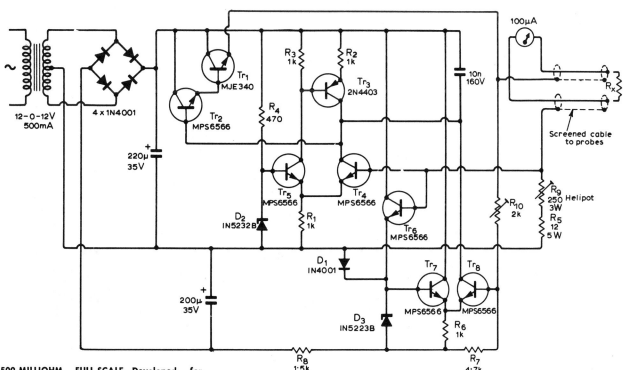

500-MILLIOHM FULL-SCALE—Developed for measuring resistance of contacts and soldered joints. R9 can be set to give either 500 milliohms or 5 ohms full-scale deflection. Output voltage is limited to under 1 V to protect meter and active devices in circuit under test. Supplies constant current to resistance under test and measures voltage drop. TR4 and TR5 form current-error amplifier controlling series pair TR1-TR2, with TR3 forming constant-current source for TR4.—J. Johnstone, Low-Range Ohmmeter, *Wireless World*, June 1971, p 294.

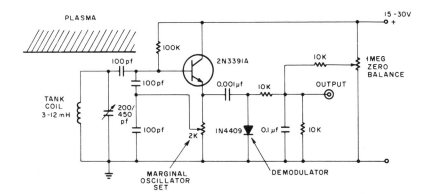

CONDUCTIVITY PROBE—Uses sensitive marginal oscillator to measure changes in Q and inductance of nonimmersive conductivity coil system because these changes are closely related to conductivity of plasma under study. Designed both for use in laboratory and under reentry flight conditions. Center-tapped capacitor divider in tank circuit is direct analog of center-tapped coil of Hartley oscillator. Basic oscillator frequency is 2 MHz.—S. Aisenberg and K. W. Chang, A Wide Range RF Coil System for the Measurement of Plasma Electrical Conductivity, *Proc. IEEE*, April 1971, p 710–712.

FOLLOWER VOLTMETER—Input impedance is 5 teraohms, as required for measuring insulation resistance of high-quality capacitors. Basic circuit accommodates input ranges of 0 to 0.5 V of either polarity for voltage measurement and 0 to +1.5 V for resistance measurement. Input range divider and function switch may be added if desired. Response is flat from 10 Hz to 100 kHz, and down 3 dB at 320 kHz for a-c voltage measurement. Voltage follower circuit with fet input gives inherently high input impedance that is further boosted by feedback. Use of balanced long-tailed pair to compare input and output voltages ensures low offset voltage.—L. E. MacHattie, Voltmeter Using F.E.T.s Measures Capacitor Insulation Resistance, *Wireless World*, Jan. 1971, p 13.

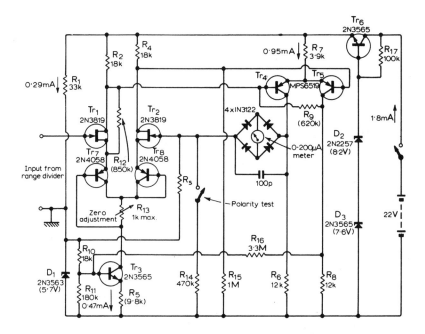

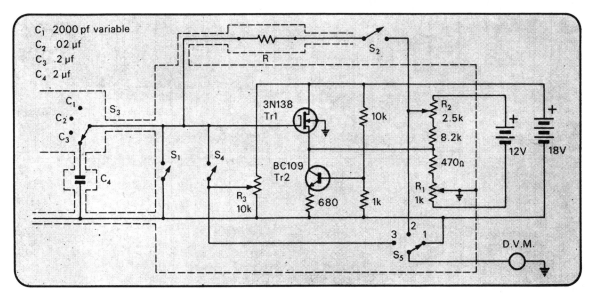

TERAOHMMETER—Designed to measure resistance values in range from 10^8 to 10^{12} ohms with precision of 1%. Based on timing the charging of an active analog integrator. Except for digital voltmeter and precision air variable capacitor, components are inexpensive. Article gives construction, operation, and performance details.—A. C. Corney, Simple, Medium-Precision High-Resistance Measuring Device, *Electronic Engineering*, Aug. 1970, p 46–47.

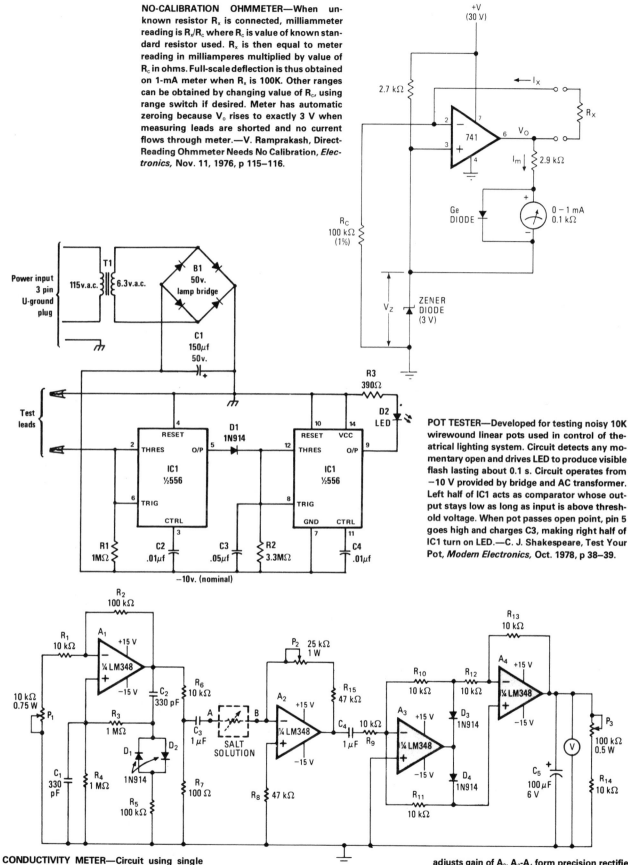

NO-CALIBRATION OHMMETER—When unknown resistor R_x is connected, milliammeter reading is R_x/R_c where R_c is value of known standard resistor used. R_x is then equal to meter reading in milliamperes multiplied by value of R_c in ohms. Full-scale deflection is thus obtained on 1-mA meter when R_x is 100K. Other ranges can be obtained by changing value of R_c, using range switch if desired. Meter has automatic zeroing because V_o rises to exactly 3 V when measuring leads are shorted and no current flows through meter.—V. Ramprakash, Direct-Reading Ohmmeter Needs No Calibration, *Electronics*, Nov. 11, 1976, p 115–116.

POT TESTER—Developed for testing noisy 10K wirewound linear pots used in control of theatrical lighting system. Circuit detects any momentary open and drives LED to produce visible flash lasting about 0.1 s. Circuit operates from −10 V provided by bridge and AC transformer. Left half of IC1 acts as comparator whose output stays low as long as input is above threshold voltage. When pot passes open point, pin 5 goes high and charges C3, making right half of IC1 turn on LED.—C. J. Shakespeare, Test Your Pot, *Modern Electronics*, Oct. 1978, p 38–39.

CONDUCTIVITY METER—Circuit using single quad opamp measures relative change in concentration of salt solution by monitoring its conductance. Use of alternating current through solution eliminates errors caused by electrolysis effect. Wien-bridge oscillator having R_4C_1 and R_2R_3 as arms of bridge generates 1-kHz signal for driving amplifier A_2 through solution. P_1 controls oscillator amplitude, and P_2 adjusts gain of A_2. A_3-A_4 form precision rectifier giving output voltage equal to absolute value of input voltage.—M. Ahmon, One-Chip Conductivity Meter Monitors Salt Concentration, *Electronics*, Sept. 15, 1978, p 132–133.

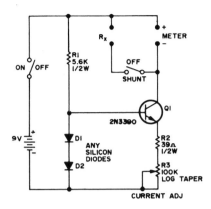

METER RESISTANCE—Practical high-accuracy circuit measures internal resistance of almost any d-c meter movement. Transistor serves as constant-current source, with R3 providing current range of about 8 μA to 13 mA. Rx is rheostat serving as shunt for meter, covering range of expected meter resistances. Connect meter and shunt, turn on circuit, turn off shunt switch, adjust R3 for full-scale meter reading, turn on shunt switch, and adjust Rx until meter reads half-scale. Rx is then exactly equal to meter resistance, and can be disconnected for measuring with ohmmeter.—Marovich, The Meter Evaluator, 73, April 1971, p 116–119.

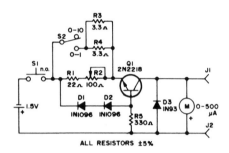

LOW-RANGE OHMMETER—Has two linear scales, 0–10 ohms and 0–1 ohm. Momentary pushbutton is used for test because of high current drain from single D cell. Q1 serves as constant-current generator providing known current through test resistance. Requires no zeroing once calibration control is set, by adjusting for full-scale reading on 10-ohm range while testing precision 10-ohm resistor.—A. Schecner, The Low-Ohm Meter, 73, Feb. 1971, p 87.

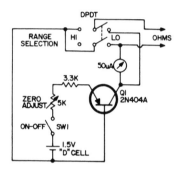

LOW-VOLTAGE OHMMETER—Can be used safely in solid-state circuits because maximum voltage across terminals is only 250 mV. Switch gives choice of 3,000-ohm low-resistance range with highest value at right end of scale, and high range with 150K near zero on microammeter scale. Article covers calibration.—I. M. Gottlieb, An Ohmmeter for Solid-State Circuits, 73, June 1973, p 91–93.

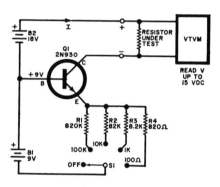

LINEARIZING VTVM SCALE—Permits reading ohmmeter scales of 100, 1K, 10K, and 100K ohms on linear scale of d-c voltage range of vtvm instead of on hard-to-read log-type resistance scale.—G. Beene, A Linear-Scale Ohmmeter, Popular Electronics, May 1972, p 45.

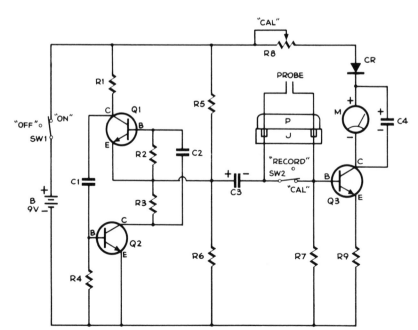

Cl, C2—0.25uf, 100v paper capacitors 2
C3, C4—25uf, 16v electrolytic capacitors 2
CR—Diode, 1N34A 1
Holder, battery— Lafayette No. 34E50053 ... 1
M—Meter, 0-1mA DC milliammeter, Lafayette No. 99E50528 ... 1
J—Connector, chassis, Lafayette No. 34E20460 ... 1
P—Connector, probe (see text), Lafayette No. 34E20015 1
Q1, Q2, Q3—Transistors 2N3704, MPS 3704, or equal 3
R1, R3, R5, R6—15,000-ohm, ½-watt carbon resistors 4
R2, R4—100,000-ohm, ½-watt carbon resistors ... 2
R7—5,100-ohm, ½-watt carbon resistor 1
R8—10,000-ohm linear-taper midget potentiometer ... 1

WATER CONDUCTIVITY GAUGE—Uses 55-Hz mvbr Q1-Q2 to convert battery voltage to about 1.5 V a-c between probes, to overcome polarization effects when measuring conductivity of water as guide to extent of pollution. Article tells how to calibrate tester with salt solution. Choose R9 between 0 and 100 ohms to make meter read 0.75 mA when 1K resistor is between probes and SW2 is open.—P. Emerson, Portable Water Tester, Popular Science, Sept. 1970, p 127–128, 132, and 140.

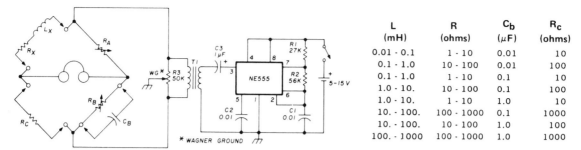

L (mH)	R (ohms)	C_b (μF)	R_c (ohms)
0.01 - 0.1	1 - 10	0.01	10
0.1 - 1.0	10 - 100	0.01	100
0.1 - 1.0	1 - 10	0.1	10
1.0 - 10.	10 - 100	0.1	100
1.0 - 10.	1 - 10	1.0	10
10. - 100.	100 - 1000	0.1	1000
10. - 100.	10 - 100	1.0	100
100. - 1000	100 - 1000	1.0	1000

RLC BRIDGE—Maxwell bridge uses only one reactive element for measuring resistance, inductance, and capacitance. Wagner ground balances stray internal capacitances to ground to obtain perfect null. Measurement ranges are shown in table. Over fixed range, R_A can be calibrated to read inductance values directly. R_A and R_B can be calibrated initially over their variable ranges by using standard resistors. Measurements are not affected by frequency of driving source. Circuit is set up as shown for measuring inductance. If standard inductance is used in place of L_x, unknown capacitor at C_B can be measured. Signetics NE555 is connected as astable oscillator running at about 1000 Hz with values shown for R and C, drawing 6.5 mA from 9-V battery. Article covers construction and calibration and gives balance equations for all measurements.—J. H. Ellison, Universal L, C, R Bridge, *Ham Radio,* April 1976, p 54–55.

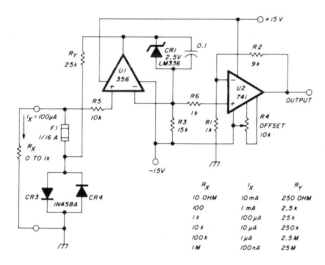

R_X	I_X	R_Y
10 OHM	10 mA	250 OHM
100	1 mA	2.5 k
1 k	100 μA	25 k
10 k	10 μA	250 k
100 k	1 μA	2.5 M
1 M	100 nA	25 M

LOW-VOLTAGE OHMMETER—Combines stable constant-current source U1-CR1-R_Y with DC amplifier U2 having gain of 10 to keep applied voltage down to 0.1 V. Output is linearly proportional to unknown resistance. Resistances well below 1 ohm can be measured accurately. U2 scales 0–100 mV unknown voltage to 0–1 V at output, so 1K resistor under test can be read as 1.000K on DVM scale. U2 should be offset-nulled to eliminate zero error, for best low-scale accuracy, by shorting input and adjusting R4 for 0.000 V out of U2. Full-scale calibration involves trimming individual range values of R_Y for correct output, while using reference value for R_X. Fuse and clamp diodes protect range resistors, and R5 protects opamp.—W. Jung, An IC Op Amp Update, *Ham Radio,* March 1978, p 62–69.

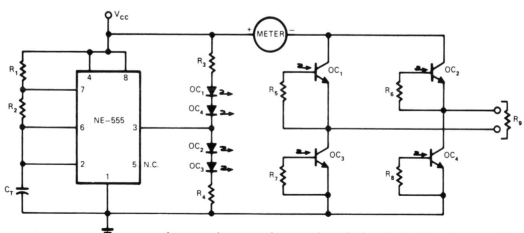

AC OHMMETER—Optoisolator circuit operating from single battery develops alternating current for measuring resistance of soils and construction materials without errors due to polarization and earth-current effects. 555 IC timer controls output at frequency determined by R_1, R_2, and C_T. R_1 is made very much less than R_2 but should not be below about 1 K. Frequency value is $1.44/(R_1 + 2R_2)C_T$. Output switching matrix is controlled by timer so OC_1 and OC_4 are on for one half-cycle and OC_2 and OC_3 are on for other half. Output current is independent of frequency and duty cycle up to 150 Hz. With Monsanto MCT-2 optoisolators, R_3 and R_4 are 330 ohms and R_5-R_8 are each 22K.—D. J. Beckwitt, AC Ohmmeter Provides Novel Use for Opto-Isolators, *EDN Magazine,* July 5, 1974, p 70.

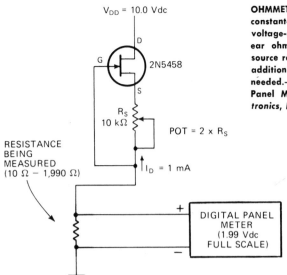

$V_{DD} = 10.0$ Vdc

D

G   2N5458

S

R_S
10 kΩ

POT = 2 x R_S

RESISTANCE
BEING
MEASURED
(10 Ω — 1,990 Ω)

$I_D = 1$ mA

+

DIGITAL PANEL
METER
(1.99 Vdc
FULL SCALE)

−

OHMMETER WITH DIGITAL VM—Simple fet constant-current source converts conventional voltage-measuring digital panel meter to linear ohmmeter having range determined by source resistance RS. Switch can be added for additional resistors if more than one range is needed.—J. L. Turino, Converting a Digital Panel Meter into a Linear Ohmmeter, *Electronics*, March 1, 1973, p 102.

$15 WHEATSTONE—Provides accuracy of 0.5% over range of 100 milliohms to 10 meg, in seven ranges. Article covers construction and calibration.—R. P. West, Jr., Inexpensive Wheatstone Bridge, *Popular Electronics*, March 1972, p 34–35.

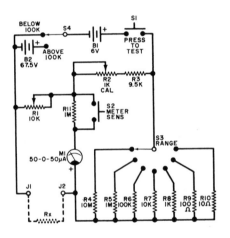

Measuring Circuits—Temperature

Convert temperature to frequency, voltage, or other parameter for driving meter or digital display that gives temperature value with desired accuracy. Includes wind-chill meter, air-velocity meter, position sensor, thermocouple multiplexer, integrator for soldering-energy pulses, and differential drive for strip-chart recorder.

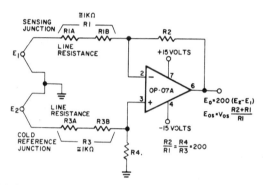

THERMOCOUPLE AMPLIFIER—Precision Monolithics OP-07A opamp has high common-mode rejection ratio and long-term accuracy required for use with thermocouples having full-scale outputs under 50 mV, frequently located in high-noise industrial environments. CMRR is 100 dB over full ±13 V range when ratios R2/R1 and R4/R3 are matched within 0.01%. Circuit is useful in many other applications where small differential signals from low-impedance sources must be accurately amplified in presence of large common-mode voltages.—D. Soderquist and G. Erdi, "The OP-07 Ultra-Low Offset Voltage Op Amp—a Bipolar Op Amp That Challenges Choppers, Eliminates Nulling," Precision Monolithics, Santa Clara, CA, 1975, AN-13, p 11.

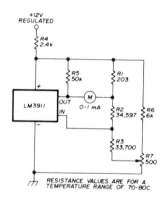

70–80°C THERMOMETER—Uses National LM3911 IC having built-in temperature sensor. If no thermometer is available for calibration, set pot R7 to its midpoint. Article gives equations for calculating resistance values for other temperature and meter ranges. Applications include monitoring of temperature in crystal oven. If permanently connected meter is not required, terminals can be provided for checking temperature with multimeter.—F. Schmidt, Precision Temperature Control for Crystal Ovens, *Ham Radio,* Feb. 1978, p 34–37.

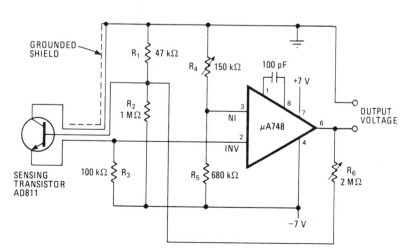

TRANSISTOR SENSOR—Use of bipolar supply for opamp makes electronic thermometer circuit fully linear even at low temperatures. Accuracy is within 0.05°C. Zero point is set by R4 and gain by R6.—C. J. Koch, Diode or Transistor Makes Fully Linear Thermometer, *Electronics,* May 13, 1976, p 110–112.

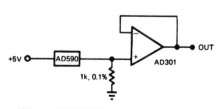

−125 to +200°C WITH 1° ACCURACY—Use of factory-trimmed AD590 IC temperature sensor gives wide temperature range with minimum number of parts. Other temperature scales can be obtained by offsetting AD301 buffer opamp.—J. Williams, Designer's Guide to: Temperature Measurement, *EDN Magazine,* May 20, 1977, p 71–77.

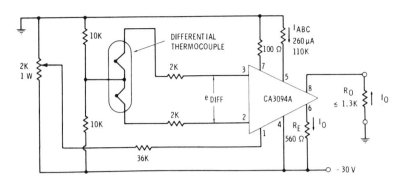

IC FOR DIFFERENTIAL THERMOCOUPLE— Amount of heat sensed by differential thermocouple is proportional to voltage between pins 2 and 3 of CA3094A programmable power switch/amplifier. Input swing of ± 26 mV gives single-ended output current range of ± 8.35 mA.—E. M. Noll, "Linear IC Principles, Experiments, and Projects," Howard W. Sams, Indianapolis, IN, 1974, p 314.

TEMPERATURE TRANSDUCER INTERFACE— Output of National LX5600 temperature-sensing transducer is inverted, level-shifted, and given extra voltage gain of 4 to give required output of 0 to +5 V for telemetry system or instrumentation recorder. Q1 furnishes constant current to thermometer, and Q2 provides inverting function. Resulting output signal is reinverted by LM201A opamp connected through zero-adjust divider to pin 3 which provides voltage reference.—P. Lefferts, "A New Interfacing Concept; the Monolithic Temperature Transducer," National Semiconductor, Santa Clara, CA, 1975, AN-132, p 3.

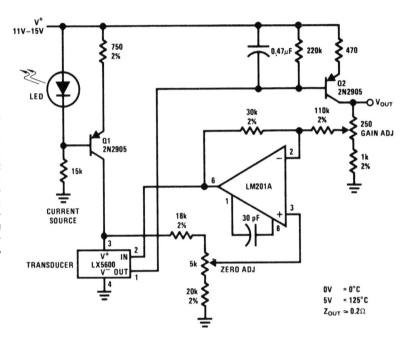

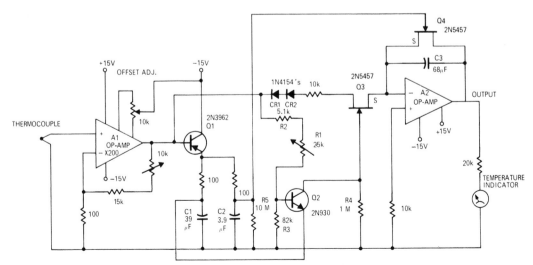

HEAT-ENERGY INTEGRATOR—Pulses of heat energy applied to solder preforms by tips of pulsed soldering machine are metered by integrate/hold-to-indicate circuit using thermocouple as input sensor. Temperature derived from area under time/temperature curve is indicated momentarily on output meter, as guide for operator when size of solder preform is changed. Article describes operation of circuit in detail and gives timing diagram.—C. Brogado, Heat-Energy Pulse Measured and Displayed, *EDN Magazine*, Sept. 15, 1970, p 61–62.

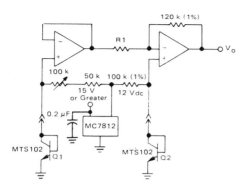

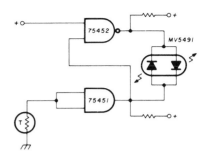

FOUR-THERMOCOUPLE MULTIPLEXING—Low power dissipation in DG306 analog switches means lower offset voltages added to thermocouple voltages by silicon in contact with aluminum in switches. Thermocouples are switched differentially to instrumentation amplifier driving meter, in order to cancel thermal offsets due to switch.—"Analog Switches and Their Applications," Siliconix, Santa Clara, CA, 1976, p 7-87.

RED/GREEN LED MONITOR—Set points are adjusted by trimming resistor shunted across thermistor, to give one color when desired temperature has been reached and other color when temperature is low. Uses Monsanto MV5491 dual red/green LED, with 220 ohms in upper lead to +5 V supply and 100 ohms in lower +5 V lead because red and green LEDs in parallel back-to-back have different voltage requirements. LED drivers are SN75452 and SN75451.—K. Powell, Novel Indicator Circuit, *Ham Radio,* April 1977, p 60–63.

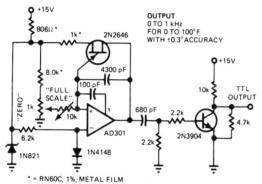

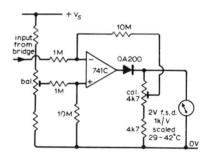

0–100°F GIVES 0–1 kHz OUTPUT—Circuit provides direct temperature-to-frequency conversion at low cost for applications where digital output is desired. Temperature sensor is 1N4148 diode having −2.2 mV/°C temperature shift, controlling AD301 opamp in relaxation oscillator circuit. Compensated 1N821 zener stabilizes against supply changes. Output network using 680 pF and 2.2K differentiates 400-ns reset edge of negative-going output ramp of opamp and drives single-transistor inverter to provide TTL output. Accuracy is within 0.3°F.—J. Williams, Designer's Guide to: Temperature Measurement, *EDN Magazine,* May 20, 1977, p 71–77.

ZERO SUPPRESSION—Opamp is used in inverting configuration at output of temperature-sensing bridge, so noninverting input of opamp can be used for suppressing meter zero when temperature range for application is 29 to 42°C. Calibration control is set for gain of about 17.2 to make meter direct-reading. Article gives operation details and methods of improving temperature stability of circuit.—R. J. Isaacs, Optimizing Op-Amps, *Wireless World,* April 1973, p 185–186.

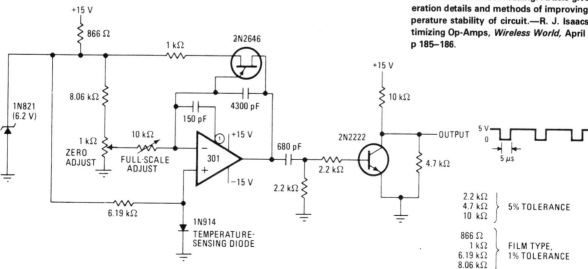

TEMPERATURE-TO-FREQUENCY CONVERTER—Frequency of relaxation oscillator varies linearly with temperature-dependent voltage across 1N914 diode sensor, with range of 0–1000 Hz for 0–100°C. Frequency meter at output shows temperature directly with accuracy of ±0.3°C. Opamp is used as integrator, with 1N821 temperature-compensated diode providing voltage reference that determines firing point of UJT. Circuit functions as voltage-to-frequency converter. Calibrate at 100°C and 0°C, repeating until adjustments cease to interact. Output frequency is then 10 times Celsius temperature.—J. Williams and T. Durgavich, Direct-Reading Converter Yields Temperature, *Electronics,* April 3, 1975, p 101 and 103; reprinted in "Circuits for Electronics Engineers," *Electronics,* 1977, p 366.

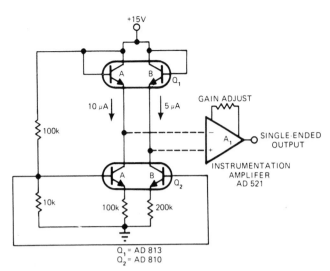

TRANSISTOR SENSOR—Current-ratio differential-pair temperature sensor uses dual transistor Q_1. Difference between base-emitter voltages of Q_{1A} and Q_{1B} varies linearly with temperature, when dual transistor Q_2 provides $10\mu A$ through Q_{1A} and $5\mu A$ through Q_{1B}. Instrumentation opamp provides single-ended output with better than 1°C accuracy over 300°C temperature range. Analog Devices AD 590 IC version of differential pair will operate over wire line thousands of feet away from instrumentation opamp, for remote sensing.—J. Williams, Designer's Guide to: Temperature Sensing, *EDN Magazine,* May 5, 1977, p 77–84.

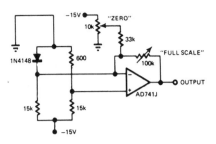

0–100°C WITH 1° ACCURACY—Low-cost diode serves as temperature sensor. To calibrate, place diode in 0°C environment and adjust zero pot for 0-V output, then place diode in 100°C environment and adjust full-scale pot for 10-V output. Repeat procedure until interaction between adjustments ceases.—J. Williams, Designer's Guide to: Temperature Measurement, *EDN Magazine,* May 20, 1977, p 71–77.

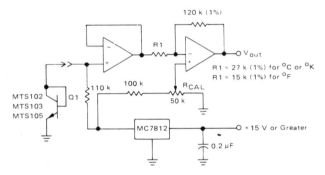

ABSOLUTE-TEMPERATURE SENSING—Silicon temperature sensor (MTS102, MTS103, or MTS105) provides precise temperature-sensing accuracy over range of −40°C to +150°C. Sensor is essentially a transistor with base and collector leads connected together externally; base-emitter voltage drop then decreases linearly with temperature over operating range. Voltage change is amplified by two opamps in series, operating from regulated output of MC7812 regulator. Opamp types are not critical. With Q1 at known temperature, adjust 50K pot to give output voltage equal to TEMP × 10 mV. Output voltage is then 10 mV per degree in desired temperature scale.—"Silicon Temperature Sensors," Motorola, Phoenix, AZ, 1978, DS 2536.

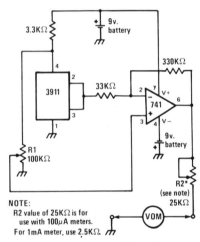

NOTE:
R2 value of 25KΩ is for use with 100μA meters.
For 1mA meter, use 2.5KΩ.

THERMOMETER—Sensor is 3911 IC whose output is 10 mV/K (kelvin temperature scale). At 0°C, output is 2.73 V. Output swing is amplified by 741 opamp to 0.1 V/°C for driving volt-ohm-milliammeter or sensitive milliammeter. R2 adjusts scaling factor, for readout in °C or °F as desired.—J. Sandler, 9 Projects under $9, *Modern Electronics,* Sept. 1978, p 35–39.

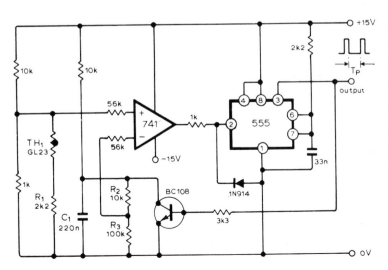

TEMPERATURE TO PULSE WIDTH—Temperature-dependent current through thermistor TH_1 develops voltage across R_1 that is compared with fraction of increasing voltage across C_1 by 741 opamp. When output of opamp goes negative, it triggers 555 IC connected as mono MVBR, to turn transistor on for about 100 μs and discharge C_1. Circuit is based on similarity between resistance-temperature curve of thermistor and inverse function of voltage across capacitor charging through resistor. For values shown, circuit gives 650-μs pulse width at 0°C, increasing 20 μs per degree with accuracy of ±1.2°C up to 60°C. If IC output is used to gate clock oscillator, number of oscillator output pulses will be directly proportional to temperature.—T. P. Y. Sander, Temperature to Pulse-Length Converter, *Wireless World,* Jan. 1977, p 76.

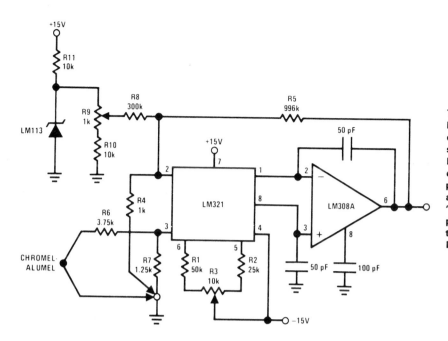

THERMOCOUPLE AMPLIFIER—Combination of LM321 preamp and LM308A opamp forms precision low-drift amplifier that includes compensation for ambient temperature variations. LM113 zener provides temperature-stable reference for offsetting output to read thermocouple temperature directly in degrees C. R4, R6, and R7 should be wirewound.—R. C. Dobkin, "Versatile IC Preamp Makes Thermocouple Amplifier with Cold Junction Compensation," National Semiconductor, Santa Clara, CA, 1973, LB-24.

TEMPERATURE-TO-FREQUENCY CONVERTER—Temperature sensor on chip of AD537 voltage-to-frequency converter IC minimizes number of external parts needed. Output frequency changes 10 Hz for each degree (kelvin or Celsius) change in temperature.—J. Williams, Designer's Guide to: Temperature Measurement, *EDN Magazine*, May 20, 1977, p 71–77.

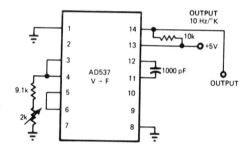

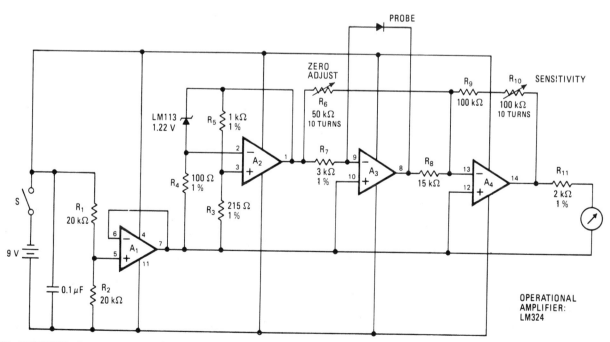

0.1°C PRECISION—Temperature sensor is LM113 diode in probe, with sections A_1 and A_2 of LM324 quad opamp maintaining constant current to diode to ensure that voltage changes across diode are direct result of temperature.

4.5-V output of A_1 is reference point for other opamps. Changes in output voltage of diode are reflected in output of A_4 through buffer A_3. Calibration involves adjusting R_6 for zero output voltage at low end of temperature range, then

adjusting R_{10} for full-scale or other convenient reading at desired upper temperature limit. Use 1-mA meter movement.—Y. Nezer, Accurate Thermometer Uses Single Quad Op Amp, *Electronics*, May 26, 1977, p 126.

SECTION 33
Metal Detector Circuits

METAL SENSOR—Simple 3-transistor design can be used for locating nails in wall studs, buried pipes, and other buried or concealed objects. L1 is ferrite-rod loopstick antenna, and L2 is standard a-m loop antenna covered with metal window screen serving as shield. Loop antenna is mounted at one end of wood rod and rest of circuit on other end. To adjust, hold loop near metal object and adjust L1 for maximum buzz or tone in phones, then adjust C6 for maximum tone. Construction details are given.—R. M. Brown and R. Kneitel, "49 Easy Entertainment & Science Projects," Howard W. Sams, Indianapolis, IN, 1971, p 21—24.

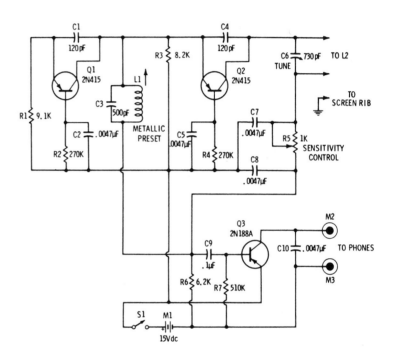

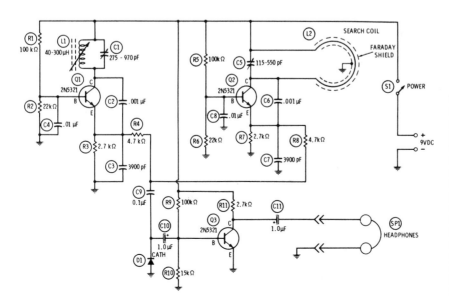

THREE - STAGE DETECTOR — Built in housing mounted on broom handle, with search coil on lower end of handle. Uses beat-frequency principle with Q2 as variable oscillator whose frequency is changed by metal near search coil and Q1 as fixed r-f oscillator tuned to get beat note when no metal is present. Q3 amplifies beat to level sufficient for headphones. Search coil is 12 turns No. 18 enamel on 12-inch diameter form, partly enclosed by Faraday shield for which construction details are given. Smaller-diameter coil will be more sensitive to coins but have less depth of penetration. L1 has adjustable ferrite core. Diode is 1N34A.—J. P. Shields, "How to Build Proximity Detectors and Metal Locators," Howard W. Sams, Indianapolis, IN, 1972, p 120—129.

TUNTED-LOOP OSCILLATOR WITH CRYSTAL FILTER—Simple, stable, and sensitive circuit for locating metal objects in ground is easy to build and operate. Will detect coins up to 8 inches from loop and larger objects up to several feet. Article gives construction and calibration details, including add-on a-f amplifier-speaker unit that can be plugged into J2 for audible indication supplementing meter reading.—C. D. Rakes, Build a Treasure Finder, *Radio-Electronics*, Nov. 1967, p 32–33.

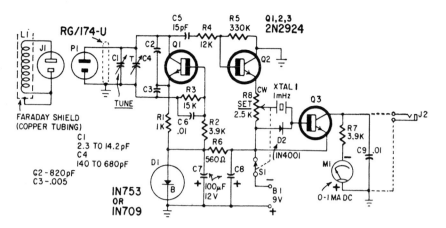

CRYSTAL-FILTER LOCATOR—Suitable for finding coins and other small objects at depths around 1 ft. Consists of Colpitts oscillator Q1 connected to 8-turn search loop mounted in 5-inch diameter Faraday shield (construction details are given), feeding buffer amplifier Q2. Both transistors are 2N2924. Oscillator is tuned to 1-MHz frequency of crystal operating in series-resonant mode as narrow-pass filter. Diodes (1N914 silicon) rectify r-f signal for driving 0-50 μA panel meter. Crystal can be any frequency from 700 kHz to 1.5 MHz as long as oscillator is tuned to it.—C. D. Rakes, "Solid State Electronic Projects," Howard W. Sams, Indianapolis, IN, 1972, p 67–73.

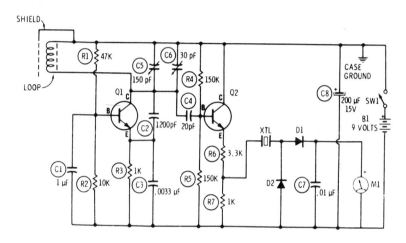

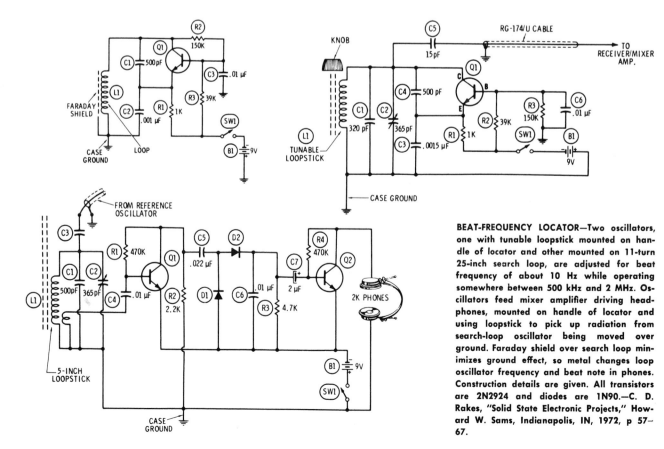

BEAT-FREQUENCY LOCATOR—Two oscillators, one with tunable loopstick mounted on handle of locator and other mounted on 11-turn 25-inch search loop, are adjusted for beat frequency of about 10 Hz while operating somewhere between 500 kHz and 2 MHz. Oscillators feed mixer amplifier driving headphones, mounted on handle of locator and using loopstick to pick up radiation from search-loop oscillator being moved over ground. Faraday shield over search loop minimizes ground effect, so metal changes loop oscillator frequency and beat note in phones. Construction details are given. All transistors are 2N2924 and diodes are 1N90.—C. D. Rakes, "Solid State Electronic Projects," Howard W. Sams, Indianapolis, IN, 1972, p 57–67.

TWO-OSCILLATOR LOCATOR—Reference oscillator mounted on handle and loop oscillator at lower end of handle are both radiation-coupled to i-f of ordinary transistor a-m radio strapped over reference oscillator. With no metal nearby, tuning capacitor of reference oscillator is adjusted to give about 1-kHz tone from radio speaker. Tune radio to spot between stations, preferably at low-frequency end of bc band. Best performance is obtained with reference oscillator initially about 1 kHz below that of loop oscillator. Ferrous or nonferrous objects near loop will then cause rise in pitch. Article covers construction and operation.—I. M. Gottlieb, New Approach for the Metal Locator, 73, Feb. 1971, p 10–14.

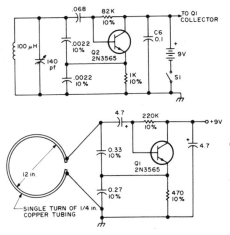

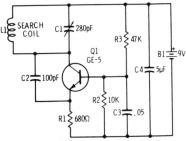

COIN FINDER—Will also locate bottle caps under beach sand. Search coil is 18 turns No. 22 enamel scramble-wound on 4-inch diameter form. Use with small transistor radio mounted on upper end of carrying handle, with search coil on lower end and its circuit just above. Tune radio to weak station, then adjust C1 for beat whistle with no metal near coil. Frequency of whistle will now change when coil is brought near buried metal.—H. Friedman, "99 Electronic Projects," Howard W. Sams, Indianapolis, IN, 1971, p 119–120.

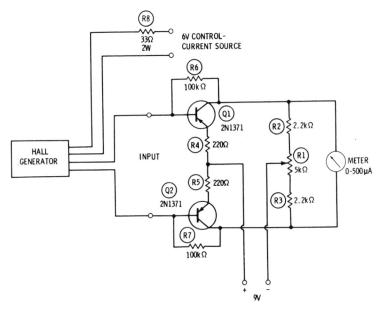

HALL-EFFECT DETECTOR—Used for detecting changes in magnetic field such as might be caused by large ferrous object. Sensor is Ohio Semitronics HR-33 or equivalent Hall generator, used with differential amplifier driving indicating meter. Construction and adjustment are covered, including use of small permanent magnet near Hall generator to null out residual Hall voltage.—J. P. Shields, "How to Build Proximity Detectors and Metal Locators," Howard W. Sams, Indianapolis, 1972, p 136–139.

PIPE FINDER—Single-coil portable unit will locate pipes or conduit inside walls or buried in ground. Operates from single 15-V battery. L2 is loop antenna of older broadcast radio, enclosed in Faraday shield to eliminate effect of external capacitance. Phones plugged into M2 should be about 2,000 ohms impedance. L1 is loopstick antenna. If no beat is heard when adjusting L1 through its range, with unit away from metal, try larger or smaller value for C8. When beat is obtained, adjust R5 for growl without metal. Tone in phones will then change in pitch when loop is brought near metal object.—R. M. Brown and T. Kneitel, "101 Easy Test Instrument Projects," Howard W. Sams, Indianapolis, IN, 1968, p 115–117.

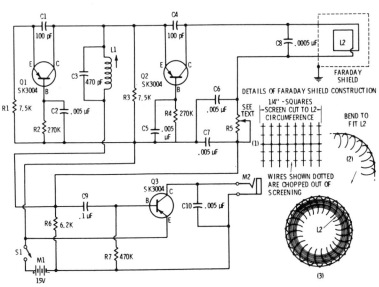

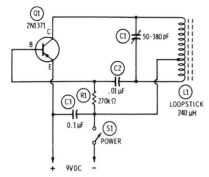

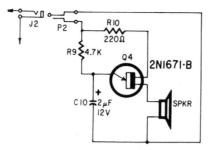

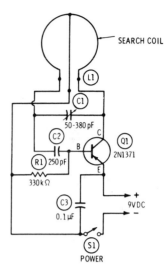

PIPE LOCATOR—Used for locating metal pipes or nails of studs concealed in walls or floors. Single-transistor oscillator using flat ferrite loop stick is mounted in flat box held against a-m transistor radio by tape or rubberbands. Radio and oscillator are adjusted until beat-frequency tone of about 1,000 Hz is heard from speaker, on side of zero beat that makes tone increase in frequency when loopstick is brought near metal object.—J. P. Shields, "How to Build Proximity Detectors and Metal Locators," Howard W. Sams, Indianapolis, IN, 1972, p 117–120.

AUDIBLE INDICATOR—Designed for use with tuned-loop oscillator type of metal detector, having crystal filter, to supplement meter reading with audible indication from speaker. Meter alone gives maximum sensitivity, so unplug ujt a-f oscillator and watch only meter when pinpointing exact location of buried metal. Article covers construction and calibration of complete locator.—C. D. Rakes, Build a Treasure Finder, *Radio-Electronics*, Nov. 1967, p 32–34.

SINGLE-TRANSISTOR USED WITH A-M RADIO—Circuit shown serves as variable-frequency oscillator for metal locator using ordinary portable a-m transistor radio as fixed-frequency oscillator. Search coil has 20 turns of No. 22 enamel on 6-inch diameter plastic hoop, tapped at tenth turn. Radio is tuned to local station between 600 and 900 kHz and C1 is adjusted until beat-frequency whistle is heard. Readjust C1 until whistle vanishes, for zero beat; metal object near search coil will bring whistle back. Construction details are given.—J. P. Shields, "How to Build Proximity Detectors and Metal Locators," Howard W. Sams, Indianapolis, IN, 1972, p 112–117.

SECTION 34

Microprocessor Circuits

Includes remote UART interfaces, code generators, and ASCII character generators.

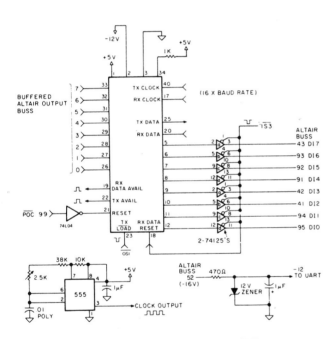

UART INTERFACE—Uses TMS-6011 UART to convert parallel data into serial data and back again for Altair 8800 microprocessor. UART mates directly to computer bus, because all outputs from UART are three-state buffers with separate enable lines provided for status bits and 8 bits of parallel output. Pin 22 is high when UART can accept another character for conversion. Pin 18 must be pulsed low to reset pin 19 so it can signal receipt of another character. Connections to pins 35–39 depend on I/O devices used, as covered in article.—W. T. Walters, Build a Universal I/O Board, *Kilobaud,* Oct. 1977, p 102–108.

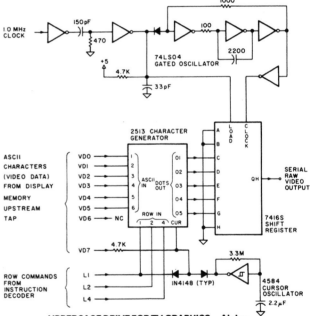

UPPERCASE DRIVE FOR TV GRAPHICS—Alphameric data-to-video converter using 2513 character generator accepts ASCII words from microprocessor memory and three line commands from instruction decoder. Five dots are outputted simultaneously, corresponding to one row on 5 × 7 dot-matrix character. 7416S eight-input one-output shift register converts dots into serial output video. Input repeats to generate all seven dot rows in row of characters. Shift register is driven by high-frequency timing circuit that delivers LOAD pulse once each microsecond along with CLOCK output running continuously at desired dot rate. Optional cursor uses 4584 5-Hz oscillator for cursor winking rate. If ASCII input bit 8 is high, cursor input goes high and output is white line on leads 01 through 05. Right diode mades this line blink, while left diode allows winking cursors only during valid character times.—D. Lancaster, TVT Hardware Design. *Kilobaud,* Jan. 1978, p 64–68.

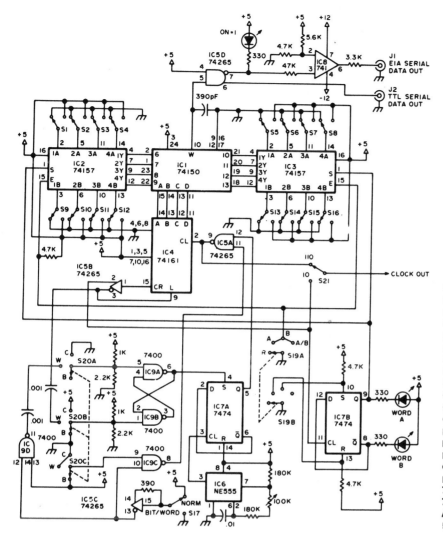

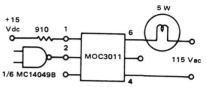

MALFUNCTION ALARM—Motorola MOC3011 optoisolator serves as interface between CMOS logic of microprocessor and 5-W 115-VAC lamp. Input logic is connected to energize infrared LED of optoisolator by providing up to 50 mA. Once triggered, indicator lamp remains on until current drops below holding value of about 100 μA.—P. O'Neil, "Applications of the MOC3011 Triac Driver," Motorola, Phoenix, AZ, 1978, AN-780, p 2.

SERIAL ASCII GENERATOR—Provides choice of two different words in standard serial ASCII asynchronous data format for troubleshooting and testing code converters and other computer peripherals. S19 gives choice of four data output patterns. R gives logic high for all 8 bits. A gives pattern determined by settings of S1-S8. B gives pattern determined by settings of S9-S16. A/B alternates words A and B. S20 gives choice of three different output modes. Mode B generates words 1 bit at a time. Mode W produces single word at rate of 110 bauds. Mode C produces continuous output of selected word pattern, for testing teleprinters and other output devices. Article covers construction and operation of circuit. IC power (+5 V) and ground pins are: 74150—24 and 12; 74157, 74161, and 74265—16 and 8; 555—8 and 1; 7474 and 7400—14 and 7.—R. J. Finger, Build a Serial ASCII Word Generator, *BYTE*, May 1976, p 50–53.

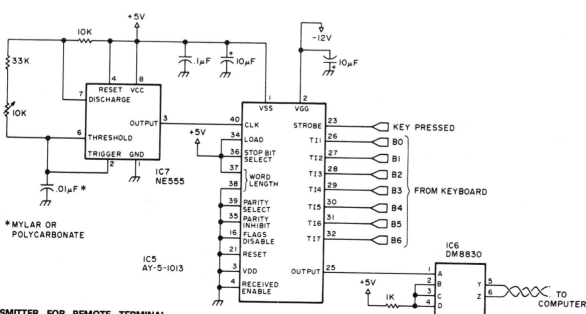

TRANSMITTER FOR REMOTE TERMINAL—Consists of AY-5-1013 UART attached to keyboard, with twisted-pair cable running to receiver unit at computer location. Coaxial extension cable for monitor is only other connection to computer system because terminal has own power supply. Transmission is in one direction only. NE555 oscillator is set at 1760 Hz ± 1% with aid of frequency counter, for 110-b/s serial rate. IC6 is 5-V National DM8830 differential line driver or equivalent. Pin 14 of IC6 goes to +5 V and pin 7 to ground.—S. Ciarcia, Come Upstairs and Be Respectable, *BYTE*, May 1977, p 50–54.

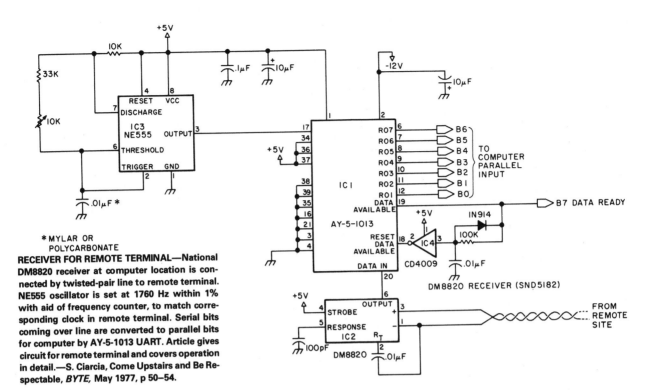

*MYLAR OR
 POLYCARBONATE

RECEIVER FOR REMOTE TERMINAL—National DM8820 receiver at computer location is connected by twisted-pair line to remote terminal. NE555 oscillator is set at 1760 Hz within 1% with aid of frequency counter, to match corresponding clock in remote terminal. Serial bits coming over line are converted to parallel bits for computer by AY-5-1013 UART. Article gives circuit for remote terminal and covers operation in detail.—S. Ciarcia, Come Upstairs and Be Respectable, *BYTE,* May 1977, p 50–54.

SECTION 35
Modem Circuits

FSK WITH SLOPE AND VOLTAGE DETECTOR —Designed for 2,125 and 2,975 Hz at data rate of about 170 Hz. Output is alternately switched between a-f inputs f1 and f2 by applying signal to each differential amplifier input pair and changing gate voltage from one extreme to other. Report describes operation.—J. Reinert and E. Renschler, "Gated Video Amplifier Applications—The MC1545," Motorola Semiconductors, Phoenix, AZ, AN 491, 1969.

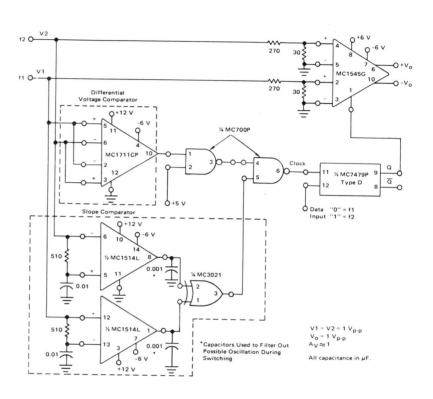

FSK FOR TWO TONES—Opamp arrangement minimizes component count in system for generating 1,070 and 1,270 Hz sine-wave audio tones representing logic 1 and logic 0 when transmitting digital data at low speed over telephone lines. First dual opamp is integrator and Schmitt trigger loop, feeding dual-op-amp second-order Butterworth active filters that provide packing to equalize signal amplitude while giving maximum suppression of second harmonics, so as not to interfere with received 2,025 and 2,225 Hz signals.—D. Kesner, Circuit Basics For FSK Modems, *Digital Design*, April 1972, p 32–34.

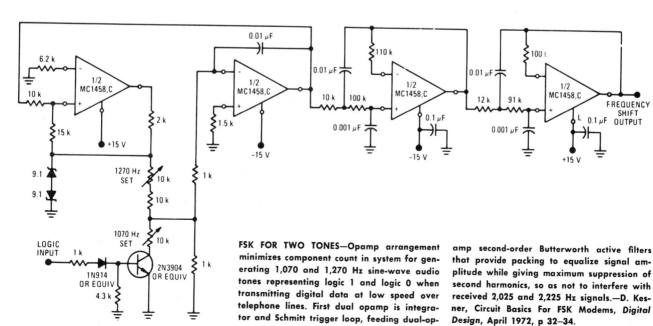

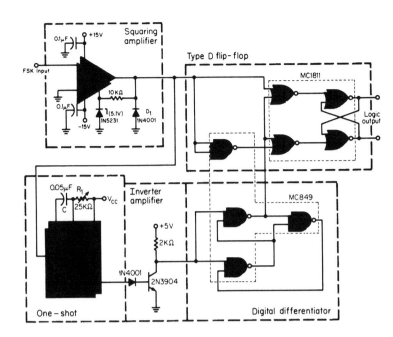

FSK RECEIVER—Converts two input audio tones to saturated logic levels. Input opamp changes sinusoidal inputs to series of square-wave pulses. D1 clips negative half. Article describes operation of circuit, which can be considered as sample-and-hold modem with one-shot providing timing signal.— D. Kesner, A Simple FSK Receiver, *The Electronic Engineer*, Dec. 1971, p 56.

FSK DECODER—Signetics 560B phase-locked loop is used as receiving converter to demodulate fsk audio tones from data communication receiver and provide shifting d-c voltage for driving printer. Will decode frequencies from near 0 to 500 kHz, giving 2-V p-p swing at up to 600-baud rate.—Frequency Shift Keying Demods with Phase-Locked Loop Devices, *Digital Design*, Feb. 1972, p 30–31.

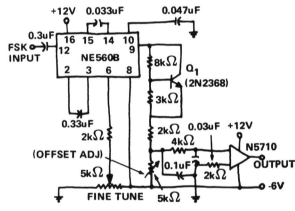

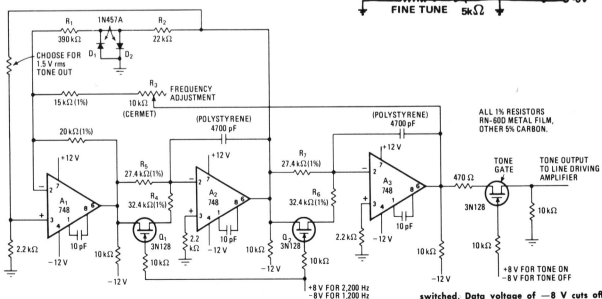

TONE OSCILLATOR—Supplies two tones, 1,-200 and 2,200 Hz, compatible with Bell type 202 modems. Uses frequency-shift tone oscillator delivering phase-continuous constant-amplitude signal that holds jitter distortion below 5% even for data rates of 1,800 bits per second, which corresponds to bit period approaching tone signal period. Circuit uses state-variable active bandpass filter A1-A2-A3 that changes its frequency when resistance is switched. Data voltage of −8 V cuts off Q1 and Q2, to give 1,200-Hz tone; +8 V unblocks these transistors to give 2,200 Hz.—B. M. Kaufman, Resistance Switching Cuts Tone Oscillator Jitter, *Electronics*, Sept. 13, 1971, p 87–88.

DEMODULATOR WITH D-C RESTORATION—Overcomes problem of d-c drift by using one LM111 opamp as accurate peak detector to provide d-c bias for one input to comparator in LM565 phase-locked loop. Circuit then acts as d-c restorer that tracks changes in drift, to make comparator self-compensating for changes in frequency or other effects of drift. Values shown are for 2,025-Hz mark and 2,225-Hz space frequencies. Will handle keying rate of 300 baud (150 Hz).—"The Phase Locked Loop IC As a Communication System Building Block," National Semiconductor, Santa Clara, CA, 1971, AN-46, p 10–11.

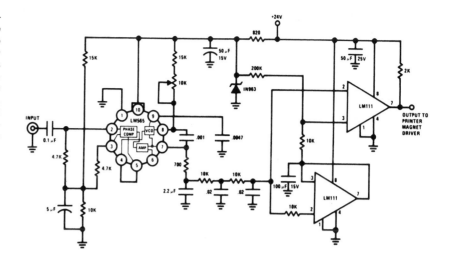

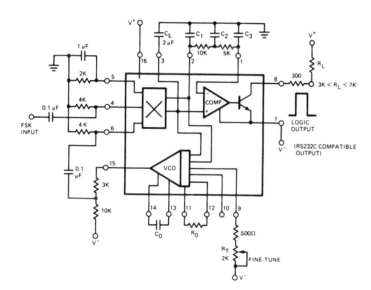

SPLIT-SUPPLY DEMODULATOR—Logic output is compatible with RS-232C because split supply is used with Exar XR-210 fsk modulator-demodulator. External components used with IC are same as for single-supply fsk demodulator, for which values are given in report both for 300-baud and 1,200-baud modem applications. Supply voltage can be 5 to 26 V.—"XR-210 FSK Modulator/Demodulator," Exar Integrated Systems Inc., Sunnyvale, CA, June 1972, p 6–7.

SELF-GENERATING FSK—Two gated oscillators combined with switching network provide self-generation of 2,125-Hz and 2,975-Hz keying signals. Negative feedback in each channel provides gain control. Switching transients are smaller than with separate oscillators because one oscillator is driven at frequency of other oscillator while first oscillator is off. Report describes operation of circuit in detail. Output voltage is about 1 V p-p.—J. Reinert and E. Renschler, "Gated Video Amplifier Applications—The MC1545," Motorola Semiconductors, Phoenix, AZ, AN-491, 1969.

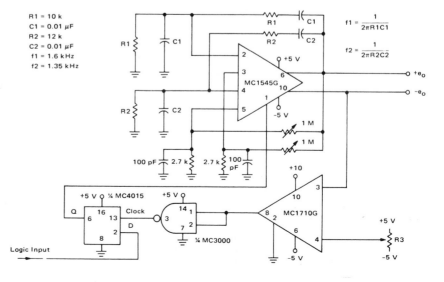

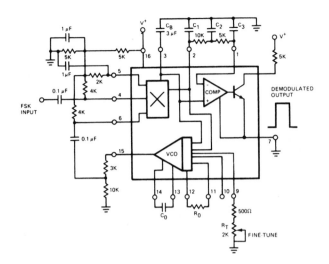

DEMODULATOR WITH SINGLE SUPPLY—Uses Exar XR-210 fsk modulator-demodulator connected as phase-locked loop system by using a-c coupling between vco output (pin 15) and pin 6 of phase detector. Fsk input is applied to pin 4. When input frequency is shifted, corresponding to data bit, polarity of d-c voltage across phase detector outputs (pins 2 and 3) is reversed. Voltage comparator and logic driver section convert level shift to binary pulse. C0 and fine-tune adjustments set vco midway between mark and space frequencies. For 1,200-baud modem, fsk values are 1,200 and 2,000 Hz, R0 is 2K, C0 0.14 μF, C1 0.033 μF, C2 0.01 μF, and C3 0.02 μF. Report also gives values for 300-baud operation. Single supply can be 5 to 26 V.—"XR-210 FSK Modulator/Demodulator," Exar Integrated Systems Inc., Sunnyvale, CA, June 1972, p 6–7.

FSK GENERATOR—Uses Exar XR-210 fsk modulator-demodulator IC for frequency-shift keying of carrier up to 100 kHz. C0 is chosen to give free-running frequency about 5% lower than space frequency. RT and RX are then set to give desired space and mark frequencies for data communication. Square-wave output is 2.5 V p-p.—"XR-210 FSK Modulator/Demodulator," Exar Integrated Systems Inc., Sunnyvale, CA, June 1972, p 7.

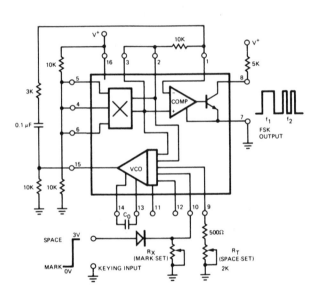

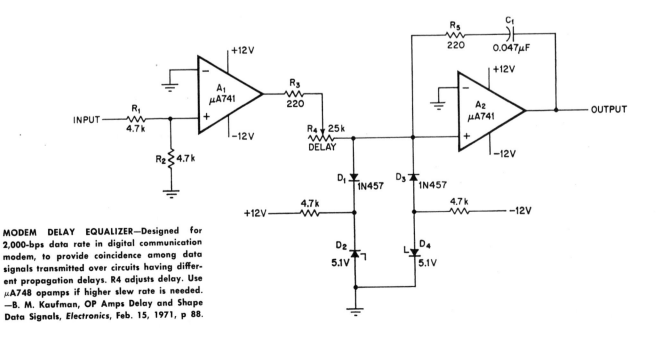

MODEM DELAY EQUALIZER—Designed for 2,000-bps data rate in digital communication modem, to provide coincidence among data signals transmitted over circuits having different propagation delays. R4 adjusts delay. Use μA748 opamps if higher slew rate is needed. —B. M. Kaufman, OP Amps Delay and Shape Data Signals, *Electronics*, Feb. 15, 1971, p 88.

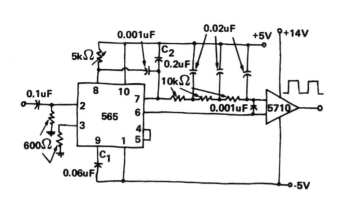

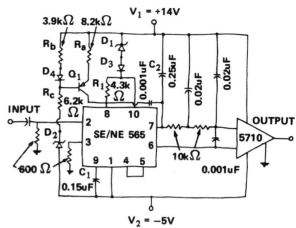

1,070 AND 1,270 HZ FSK DECODER—Three-stage R-C ladder filter removes carrier component from output of Signetics 565 phase-locked loop connected to data communication receiver, to provide shifting d-c voltage at output of opamp for driving printer. Maximum keying rate is 300 baud.—Frequency Shift Keying Demods with Phase-Locked Loop Devices, *Digital Design*, Feb. 1972, p 30–31.

FSK DECODER WITH VCO—Sophisticated fsk decoder for data communication receiver having narrow frequency deviation (1,070 and 1,270 Hz) requires adjusting free-running vco frequency in Signetics 565 phase-locked loop so output voltage swings equally above and below reference voltage on pin 6. Band edge of two-stage RC ladder network is about 800 Hz.—Frequency Shift Keying Demods with Phase-Locked Loop Devices, *Digital Design*, Feb. 1972, p 30–31.

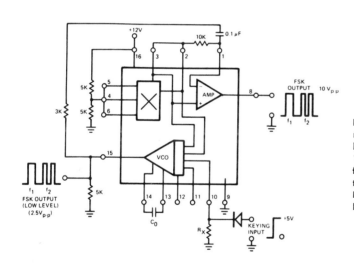

FSK GENERATOR—Utilizes digital programming capability of Exar XR-215 phase-locked loop, with control logic pulse applied to pin 10. Circuit will provide two different levels of fsk output, each with second harmonic content less than 0.3%.—"XR-215 Monolithic Phase-Locked Loop," Exar Integrated Systems Inc., Sunnyvale, CA, July 1972, p 7.

DEMODULATOR—Uses Exar XR-S200 IC with phase-locked loop connection shown, as modem suitable for Bell 103 or 202 data sets operating at up to 1,800 baud. Input frequency shift corresponding to data bit makes d-c output voltage of multiplier reverse polarity. D-c level is changed to binary output pulse by gain block connected as voltage comparator.—"XR-S200 Multi-Function Integrated Circuit," Exar Integrated Systems Inc., Sunnyvale, CA, June 1972, p 6.

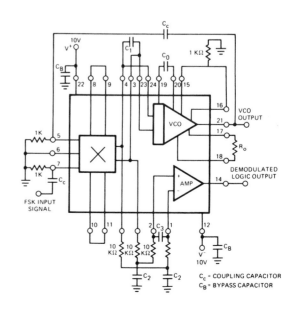

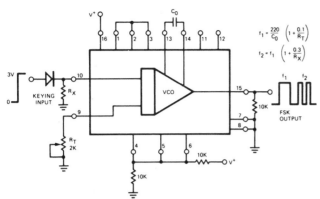

$$f_1 = \frac{220}{C_0}\left(1 + \frac{0.1}{R_T}\right)$$

$$f_2 = f_1\left(1 + \frac{0.3}{R_X}\right)$$

GENERATOR USING VCO—Generates data-communication mark and space frequencies for carrier up to 10 MHz by using vco output of Exar XR-210 fsk modulator-demodulator directly, with comparator and logic driver sections of IC removed from signal path. Supply can be anywhere between 5 and 26 V. —"XR-210 FSK Modulator/Demodulator," Exar Integrated Systems Inc., Sunnyvale, CA, June 1972, p 7–8.

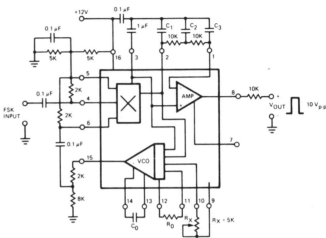

DEMODULATOR—Uses Exar XR-215 phase-locked loop IC. When input frequency is shifted for data bit, d-c voltage at phase comparator outputs 2 and 3 reverses polarity. Opamp converts d-c level shift to binary output pulse. RX sets vco frequency. Report gives typical component values both for 300-baud and 1,800-baud operation.—"XR-215 Monolithic Phase-Locked Loop," Exar Integrated Systems Inc., Sunnyvale, CA, July 1972, p 7.

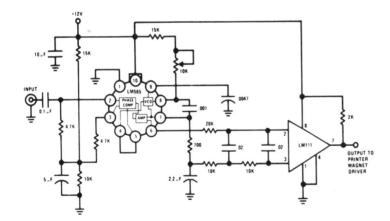

DEMODULATOR—Uses National LM565 phase-locked loop driving opamp. Values shown are for 2,025-Hz mark and 2,225-Hz space frequencies. Will handle keying rate of 300 baud (150 Hz). Chief problem is d-c drift, which may lock comparator in one state or other because demodulated output is only 150 mV; report includes circuit which overcomes drift with d-c restorer stage.—"The Phase Locked Loop IC As a Communication System Building Block," National Semiconductor, Santa Clara, CA, 1971, AN-46, p 9–11.

SECTION 36
Modulator Circuits

Covers circuits that vary amplitude or some other characteristic of carrier signal or pulse train in accordance with information contained in modulating signal.

PULSE-RATIO MODULATOR—LM111 comparator serves with single transistor to provide pulse-train output whose average value is proportional to input voltage. Frequency of output is relatively constant but pulse width varies. Pulse-ratio accuracy is 0.1%. Circuit can be used to drive power stage of high-efficiency switching amplifier, or as pulse-width/pulse-height multiplier. Article tells how circuit works.—R. C. Dobkin, Comparators Can Do More than Just Compare, *EDN Magazine,* Nov. 1, 1972, p 34—37.

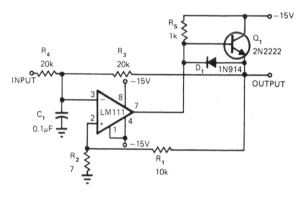

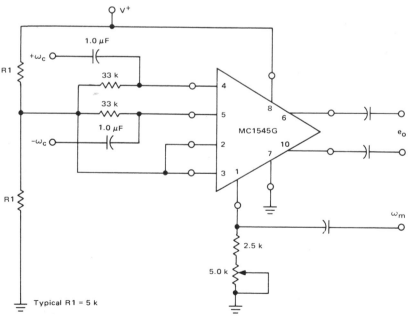

SINGLE-SUPPLY AM—Motorola MC1545 gated video amplifier is connected as amplitude modulator operating from single supply. Artificial ground is established for IC at half of supply voltage by 5K resistors R1, which should draw much more than bias current of 15 μA. All sig-nals must be AC coupled to prevent application of excessive common-mode voltage to IC.— "Gated Video Amplifier Applications—the MC1545," Motorola, Phoenix, AZ, 1976, AN-491, p 15.

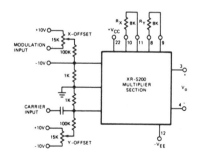

DOUBLE-SIDEBAND AM—Connection shown for multiplier section of Exar XR-S200 PLL IC gives double-sideband AM output. X-offset adjustment for modulation input sets carrier output level, and Y-offset adjustment of carrier input controls symmetry of output waveform. Modulation input can also be used as linear automatic gain control (AGC) for controlling amplification with respect to carrier input signals.—"Phase-Locked Loop Data Book," Exar Integrated Systems, Sunnyvale, CA, 1978, p 9—16.

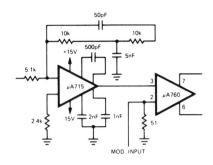

COMPARATOR FOR PWM—Ramp output of μA715 is fed into input 3 of μA760, with audio signal going to input 2. Outputs from 6 and 7 then change state each time audio input equals ramp voltage. Pulse width of output varies with instantaneous value of audio input.—P. Holtham, "The μA760—A High Speed Monolithic Voltage Comparator," Fairchild Semiconductor, Mountain View, CA, No. 311, 1972, p 7.

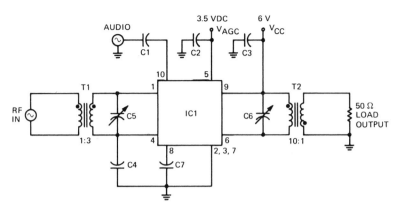

6-METER A-M—Single IC gives up to 90% modulation with very little distortion.—"Radio Amateur's IC Projects," Motorola Semiconductors, Phoenix, AZ, HMA-36, 1971.

FET BALANCED MODULATOR—Designed to operate with carrier inputs of 100–150 kHz and modulating frequencies from d-c to 10 kHz. Provides linear control by using junction fet as variable resistance. Tr4 and Tr5 form simple summing amplifier providing low output impedance.—M. E. Cook, Amplitude Modulation Using an F.E.T., *Wireless World*, Feb. 1970, p 81–82.

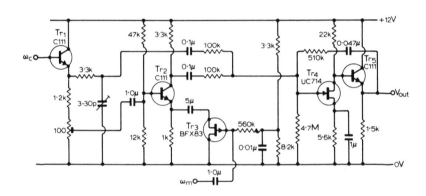

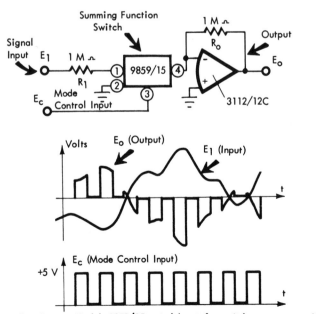

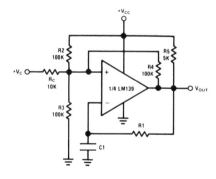

PWM DOWN TO 1 HZ—Simple connection of IC comparator gives square-wave output when input control voltage is equal to half of supply voltage. Increasing or decreasing control voltage about this value changes duty cycle. Report tells how to calculate upper and lower trip points. Center frequency depends on values of R1 and C1, with maximum being limited by value of supply voltage and output slew rate.—R. T. Smathers, T. M. Frederiksen, and W. M. Howard, "LM139/LM239/LM339 A Quad of Independently Functioning Comparators," National Semiconductor, Santa Clara, CA, AN-74, 1973, p 15.

CHOPPER—Uses Burr-Brown Model 9859/15 summing junction switch and opamp to provide chopping action for analog input signal E1. 5-V pulses are fed into Ec as mode control input for switch, so output voltage goes to zero when switch is gated OFF.—"Electronic Switches," Burr-Brown, Tucson, AZ, Sept. 1969, p 6.

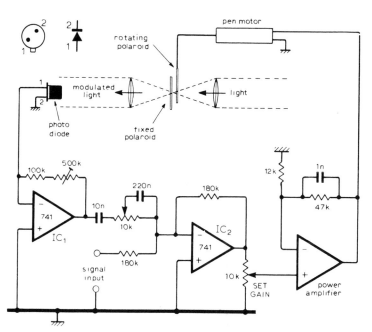

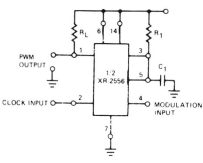

LIGHT-BEAM MODULATOR—Intensity of light beam is modulated by rotating Polaroid vane driven by small motor. Since amplitude is not constant with change in frequency between 10 and 100 Hz, compensation is provided by sampling modulated beam with silicon photodiode that is linearized by IC₁. Input and feedback signals are mixed by summing amplifier IC₂ which drives noninverting power amplifier consisting of 741 opamp driving two OC28 power transistors connected in closed feedback loop having gain of 5. Power amplifier drives pen motor of modulator.—R. F. Cartwright, Constant Amplitude Light Modulator, *Wireless World*, Sept. 1976, p 73.

PULSE-DURATION MODULATOR USES TIMER—Half of Exar XR-2556 dual timer is connected to operate in monostable mode, for triggering with continuous pulse train. Output pulses are generated at same rate as input, with pulse duration determined by R_1 and C_1. Supply voltage is 4.5–16 V.—"Timer Data Book," Exar Integrated Systems, Sunnyvale, CA, 1978, p 23–30.

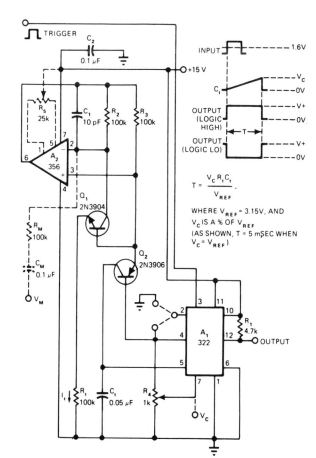

$$T = \frac{V_C R_1 C_t}{V_{REF}},$$

WHERE V_{REF} = 3.15V, AND V_C IS A % OF V_{REF} (AS SHOWN, T = 5 mSEC WHEN V_C = V_{REF})

VOLTAGE TO PULSE WIDTH—Constant-current source Q_2 produces linear timing ramp across C_t in circuit of 322 IC timer A_1, for comparison internally with 0–3.15 V applied to pin 7. Pulse is thus linearly variable function of control voltage V_C over dynamic range of more than 100:1. Circuit is highly flexible, permitting use of many other operational modes as covered in article. When AC waveform is applied to V_M, circuit operates as linear pulse-width modulator.—W. G. Jung, Take a Fresh Look at New IC Timer Applications, *EDN Magazine*, March 20, 1977, p 127–135.

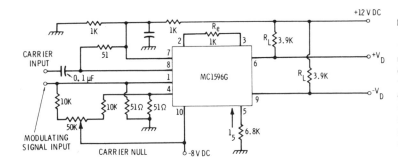

DOUBLE-SIDEBAND SUPPRESSED-CARRIER— Motorola MC1596G double-balanced modulator has carrier input between pins 8 and 7 and modulation between pins 1 and 4. Balancing carrier-null circuit, also connected between pins 1 and 4, contributes to excellent carrier rejection at output. For unbalanced output, ground one of push-pull output terminals. Requires two supplies.—E. M. Noll, "Linear IC Principles, Experiments, and Projects," Howard W. Sams, Indianapolis, IN, 1974, p 138–139.

DELTA MODULATOR—Uses LM111 comparator in basic pulse-ratio modulator circuit, with output pulse width and transition time fixed by external clock signal applied to gate of JFET switch Q_2. Average value of output is always proportional to input voltage.—R. C. Dobkin, Comparators Can Do More than Just Compare, *EDN Magazine*, Nov. 1, 1972, p 34–37.

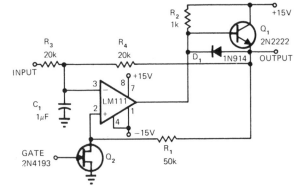

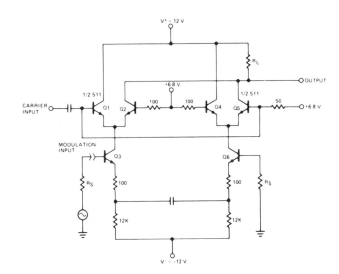

SUPPRESSED-CARRIER AM—Double-balanced modulator using Signetics 511 transistor array gives output consisting of sum and difference frequencies of carrier and modulation inputs along with related harmonics. Circuit is self-balancing, eliminating need for pots. Output includes small amounts of carrier and modulating signal. Capacitor between emitters of Q3 and Q6 is selected to have low reactance at lowest modulating frequency.—"Signetics Analog Data Manual," Signetics, Sunnyvale, CA, 1977, p 750–751.

BALANCED MODULATOR—High-performance balanced modulator for 80-meter SSB transceiver uses Motorola MC1496 IC. Adjust 50K pot for maximum carrier suppression of double-sideband output.—D. Hembling, Solid-State 80-Meter SSB Transceiver, *Ham Radio*, March 1973, p 6–17.

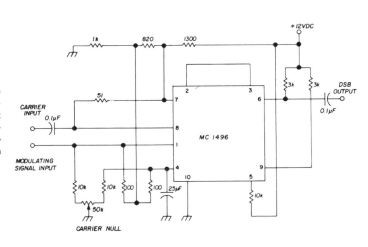

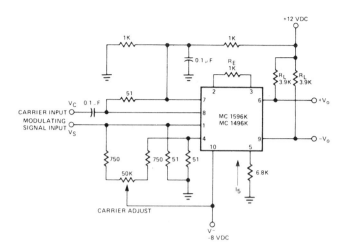

SINGLE-IC AM—Adjustable carrier offset is added to carrier differential pairs to provide carrier-frequency output that varies in amplitude with strength of modulation signal.—"Signetics Analog Data Manual," Signetics, Sunnyvale, CA, 1977, p 757.

DSB BALANCED MODULATOR—Provides excellent gain and carrier suppression by operating upper (carrier) differential amplifiers of Motorola MC1596G balanced modulator at saturated level and lower differential amplifier in linear mode. Recommended input levels are 60 mVRMS for carrier and 300 mVRMS maximum for modulating signal.—R. Hejhall, "MC1596 Balanced Modulator," Motorola, Phoenix, AZ, 1975, AN-531, p 3.

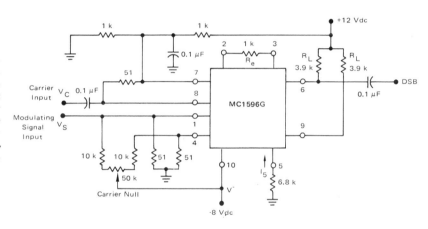

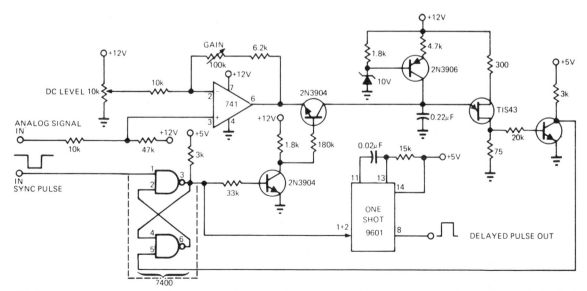

PPM WITH ANALOG CONTROL OF DELAY—Opamp, UJT, and two TTL packages generate pulse whose delay, following sync pulse, is controlled by amplitude of analog input signal at time of sync pulse. Opamp precharges timing capacitor to level depending on analog signal. Sync pulse disconnects opamp, after which timing capacitor charges up to UJT firing point. UJT output pulse then resets circuit, giving desired delayed output pulse through 9601 mono MVBR.—J. Taylor, Analog Signal Controls Pulse Delay, *EDN Magazine*, Feb. 5, 1974, p 96.

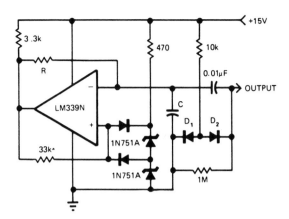

DUTY-CYCLE MODULATOR—Uses half of LM339N or LM3302N quad comparator. With no modulation signal, output is symmetrical square wave generated by one of comparators. Constant-amplitude triangle wave is generated at inverting input of second comparator, and is relatively independent of supply voltage and frequency changes. Modulating signal varies switching points to produce duty-cycle modulated wave for such applications as class D amplification for servo and audio systems.—H. F. Stearns, Voltage Comparator Makes a Duty-Cycle Modulator, *EDN Magazine,* June 5, 1975, p 76–77.

FET BALANCED MODULATOR FOR SSB—AF modulating signal is applied to gates of matched FETs in push-pull through T1 having accurately center-tapped secondary, and RF carrier is applied to sources in parallel through C3. Carrier is canceled in output circuit, leaving two sidebands. R3 is adjusted to correct for unbalance in circuit components.—R. P. Turner, "FET Circuits," Howard W. Sams, Indianapolis, IN, 1977, 2nd Ed., p 90–91.

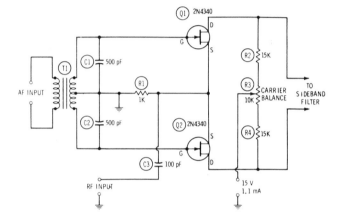

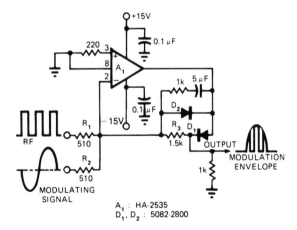

100% MODULATION OF DIGITAL SIGNALS—Developed to produce positive-going half-sine envelope modulated 100% by digital RF signal. Modulator uses loop gain of opamp to reduce diode drops very nearly to ideal zero level. D_2 prevents opamp output terminal from swinging more negative than diode drop of 0.3 V, which is not apparent at output. RF input amplitude must be sufficient to provide 100% modulation; this can be achieved by providing about 20% overdrive to give safety factor. Use hot-carrier diodes such as HP-2800 series.—D. L. Quick, Improve Amplitude Modulation of Fast Digital Signals, *EDN Magazine,* Sept. 20, 1975, p 68 and 70.

DOUBLE-SIDEBAND SUPPRESSED-CARRIER—Signetics MC1496 balanced modulator-demodulator transistor array provides carrier suppression while passing sum and difference frequencies. Gain is set by value used for emitter degeneration resistor connected between pins 2 and 3. Output filtering is used to remove unwanted harmonics.—"Signetics Analog Data Manual," Signetics, Sunnyvale, CA, 1977, p 756–757.

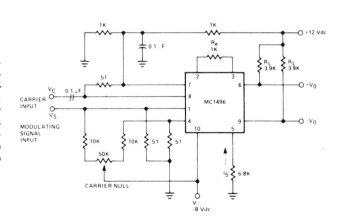

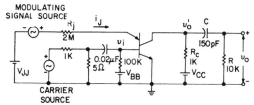

PNPN TETRODE MODULATOR—Transistor is operated in its linear mode to serve as amplitude modulator. Article gives design equations. With 100-kHz carrier and 60-Hz modulating signal, maximum percentage modulation with negligible signal distortion is about 50%.—N. C. Voulgaris and E. S. Yang, Linear Applications of a P-N-P-N Tetrode, *IEEE Journal of Solid-State Circuits*, Aug. 1970, p 146–150.

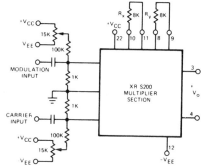

SUPPRESSED-CARRIER A-M—Multiplier section of Exar XR-S200 IC generates suppressed-carrier a-m signals, with about 60 dB carrier suppression at 500 kHz and 40 dB suppression at 10 MHz. Carrier and modulation inputs are interchangeable. 15K offset adjustments optimize carrier suppression.—"XR-S200 Multi-Function Integrated Circuit," Exar Integrated Systems Inc., Sunnyvale, CA, June 1972, p 3.

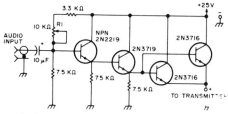

SERIES MODULATOR—Simple four-transistor circuit provides convenient means for modulating 4-W or other low-power transmitter for test purposes. Harmonic distortion is less than 1%.—D. Brubaker, A VHF AM Transmitter Using Low-Cost Transistors, *73*, Aug. 1970, p 54–59.

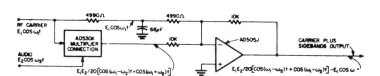

A-M USING ANALOG MULTIPLIER—Single-IC multiplier develops output proportional to product of carrier and audio input signals, shown as sum and difference sidebands. Process is known as suppressed-carrier modulation, since neither carrier nor modulation frequencies are present in output. Opamp serves as summing amplifier which reinstates carrier in output.—"AD530 Complete Monolithic MDSSR," Analog Devices, Inc., Norwood, MA, Technical Bulletin, July 1971.

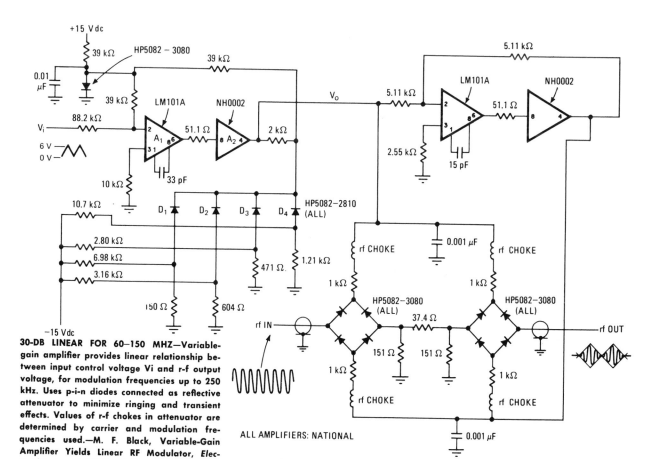

30-DB LINEAR FOR 60–150 MHZ—Variable-gain amplifier provides linear relationship between input control voltage Vi and r-f output voltage, for modulation frequencies up to 250 kHz. Uses p-i-n diodes connected as reflective attenuator to minimize ringing and transient effects. Values of r-f chokes in attenuator are determined by carrier and modulation frequencies used.—M. F. Black, Variable-Gain Amplifier Yields Linear RF Modulator, *Electronics*, Jan. 4, 1973, p 103.

ALL AMPLIFIERS: NATIONAL

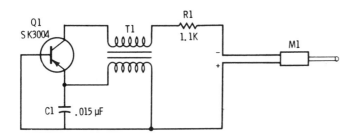

GRID-DIPPER MODULATOR—M1 is plugged into phone jack of grid-dip meter, to obtain power for feedback oscillator Q1 and add modulation to r-f generated by grid dipper. T1 can be Lafayette TR-98 transistor transformer.—R. M. Brown and T. Kneitel, "101 Easy Test Instrument Projects," Howard W. Sams, Indianapolis, IN, 1968, p 111.

AMPLITUDE MODULATOR—Uses MC1596 IC balanced modulator as amplitude modulator by unbalancing carrier null to insert proper amount of carrier into output signal. Provides excellent modulation from 0% to greater than 100%.—R. Hejhall, "MC1596 Balanced Modulator," Motorola Semiconductors, Phoenix, AZ, AN-531, 1971.

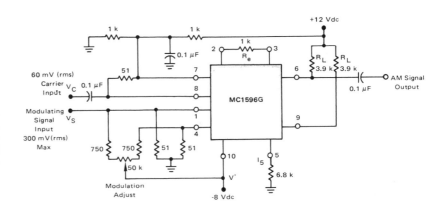

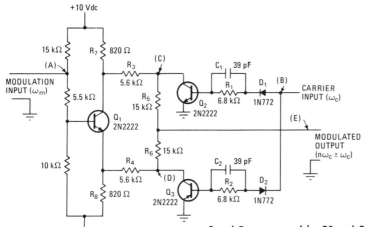

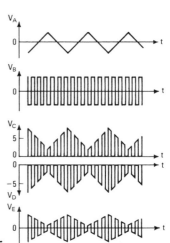

LINEAR CHOPPER—Q1 splits modulation input into equal signals having opposite polarity and phase. Q2 and Q3 pass alternate half-cycles, and chopped modulated signals at C and D are summed by R5 and R6. Provides good linearity at modulation levels up to 97.5% and modulation frequencies ranging from d-c to half of carrier frequency. Envelope shows distortion above 250 kHz and maximum linear modulation drops to 88% at 1 MHz.—D. DeKold, Amplitude Modulator Is Highly Linear, *Electronics*, June 5, 1972, p 101–102.

SECTION 37
Modulator/Demodulator Circuits

Includes detecters of various types, demodulators for quadrature encoded signals, FM detectors, and tuning indicators. See other sections on RF oscillators, amplifiers, and receivers.

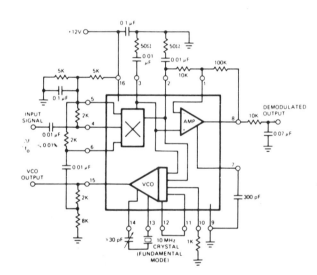

CRYSTAL FM DETECTOR—Exar XR-215 PLL IC is operated as crystal-controlled phase-locked loop by using crystal in place of conventional timing capacitor. Crystal should be operated in fundamental mode. Typical pull-in range is ± 1 kHz at 10 MHz.—"Phase-Locked Loop Data Book," Exar Integrated Systems, Sunnyvale, CA, 1978, p 21–28.

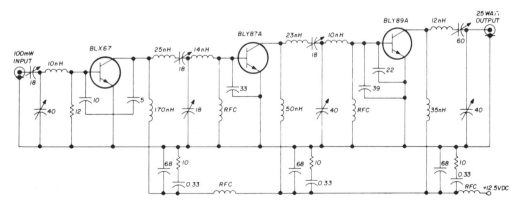

VHF POWER AMPLIFIER—Three-stage 25-W 225-MHz power amplifier module for FM applications uses three Amperex power transistors. Input and output are 50 ohms. With 100-mW input signal, output is 25 W. Four capacitive dividers serve for input, output, and interstage matching. Collectors are shunt-fed. Three decoupling networks prevent self-oscillation. Amplifier can withstand output mismatches as high as 50:1 without damage.—E. Noll, VHF/UHF Single-Frequency Conversion, *Ham Radio*, April 1975, p 62–67.

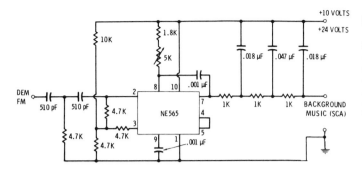

SCA DEMODULATOR—VCO of NE565 PLL is set at 67 kHz and is locked in by incoming 67-kHz subsidiary-carrier component used for transmitting uninterrupted commercial background music by FM broadcast stations. Circuit demodulates FM sidebands and applies them to audio input of commercial sound system through suitable filter. 5K pot is used to lock VCO exactly on frequency. Frequency response extends up to 7000 Hz.—E. M. Noll, "Linear IC Principles, Experiments, and Projects," Howard W. Sams, Indianapolis, IN, 1974, p 212–213.

STEREO DECODER—Single Sprague ULN-2122A IC is driven by composite signal derived at output of standard FM detector, to give original left- and right-channel audio signals for driving audio amplifiers of FM stereo receiver.—E. M. Noll, "Linear IC Principles, Experiments, and Projects," Howard W. Sams, Indianapolis, IN, 1974, p 263–266.

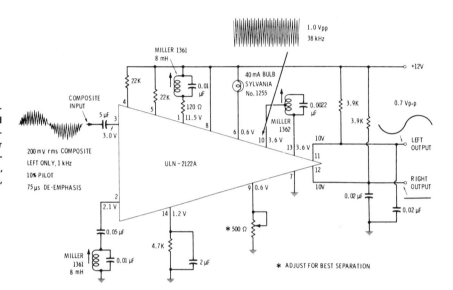

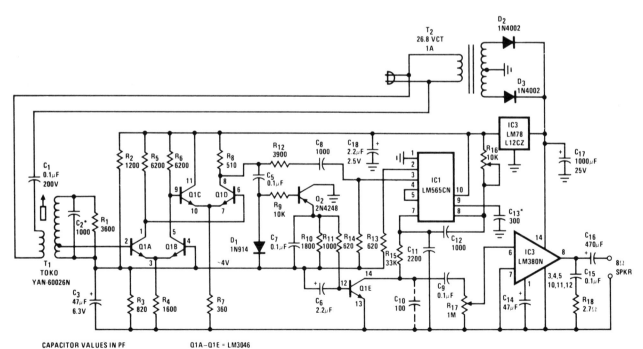

CAPACITOR VALUES IN PF
RESISTOR VALUES IN Ω
*SELECT FOR CARRIER FREQ.

Q1A–Q1E = LM3046

f_o	C_2	C_{13}
200 kHz	1000	300
100 kHz	3900	620

CARRIER-SYSTEM RECEIVER—Used to detect, amplify, limit, and demodulate FM carrier modulated with audio program, for feeding up to 2.5 W to remote loudspeaker. Can be plugged into any AC outlet on same side of distribution transformer. Carrier signal is taken from line by tuned transformer T_1. Output of two-stage limiter amplifier Q1A-Q1D is applied directly to mute peak detector D1-Q2-C7. Limiter output is reduced to 1 V P-P for driving National LM565CN PLL detector which operates as narrow-band tracking filter for input signal and provides low-distortion demodulated audio output. Mute circuit quiets receiver in absence of carrier.—J. Sherwin, N. Sevastopoulos, and T. Regan, "FM Remote Speaker System," National Semiconductor, Santa Clara, CA, 1975, AN-146.

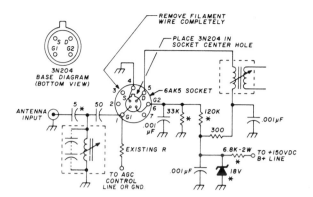

MOSFET RF STAGE—Changing 6AK5 tube to 3N204 dual-gate MOSFET improves sensitivity and lowers noise in older VHF FM communication receiver using tubes. Break off center grounding pin of tube socket and cut wires soldered to pin, then connect transistor circuit to tube socket as shown. Replace original resistor going to pin 6 with 120K and run 37K resistor from pin 6 to ground. Move antenna input lead to top of RF input coil, and remove 6-V filament wiring from socket. If tube filaments were in series, replace 6AK5 filament with 36-ohm 2-W resistor. Conversion increases sensitivity to 0.3 μV for 20-dB quieting.—H. Meyer, How to Improve Receiver Performance of Vacuum-Tube VHF-FM Equipment, *Ham Radio*, Oct. 1976, p 52–53.

TWIN-LED TUNING INDICATOR—Provides maximum sensitivity at correct tuning point and indicates direction of mistuning. Both lamps are in feedback loop of one opamp, connected to serve as highly sensitive null detector. When set is tuned correctly, output of this opamp is at midpoint of supply voltage and neither LED is lit. Circuit is used with RCA CA3089 IF chip in which AFC output is a current. Capacitor across first 741 opamp removes modulation components from this input.—M. G. Smart, F.M. Tuning Indicators, *Wireless World*, Dec. 1974, p 497.

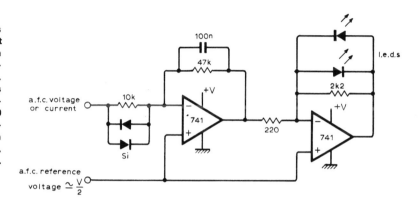

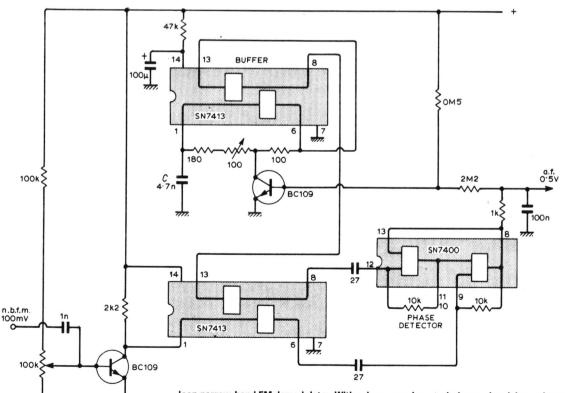

NARROW-BAND DEMODULATOR—Low-cost TTL ICs are connected to form phase-locked loop narrow-band FM demodulator. With value shown for C, circuit is suitable for IF value around 470 kHz. Article covers advantages of synchronous detection and various direct conversion techniques involving phase-locked loop.—P. Hawker, Synchronous Detection in Radio Reception, *Wireless World*, Nov. 1972, p 525–528.

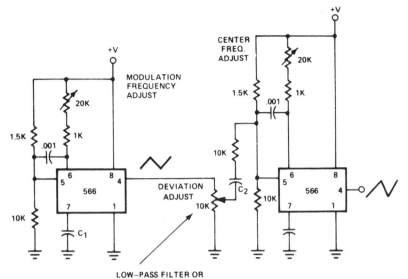

0.5 MHz WITH 20% DEVIATION—One 566 function generator serves for generating relatively low-frequency carrier (center frequency less than 0.5 MHz), and other 566 serves as modulator producing triangle output with frequency determined by C_1. Combination is suitable for deviations up to $\pm20\%$ of carrier frequency.—"Signetics Analog Data Manual," Signetics, Sunnyvale, CA, 1977, p 852–853.

FM DETECTOR—Single IC can be added to any receiver not having FM detector. Moving C_2 from pin 9 to pin 10 gives higher audio output. Receivers having less than 5 kHz IF bandwidth can be broadened by stagger-tuning IF strip slightly to improve audio clarity. Adjust tuned circuit of detector for maximum recovered audio.—I. Math, Math's Notes, *CQ*, June 1972, p 49–51 and 80.

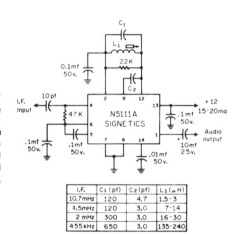

I.F.	C_1 (pf)	C_2 (pf)	L_1 (μH)
10.7 mHz	120	4.7	1.5 - 3
4.5 mHz	120	3.0	7 - 14
2 mHz	300	3.0	16 - 30
455 kHz	650	3.0	135 - 240

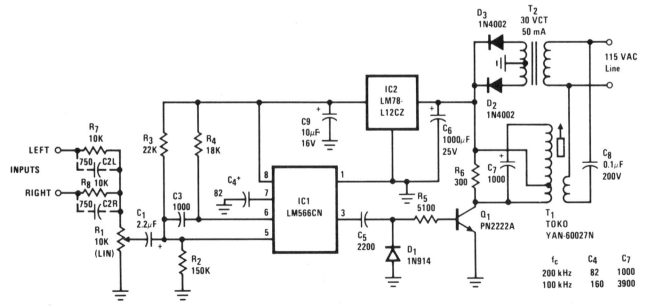

CARRIER-SYSTEM TRANSMITTER—Used to convert audio program material into FM format for coupling to standard power lines. Modulated FM signal can be detected at any other outlet on same side of distribution transformer, for demodulation and drive of loudspeaker. Input permits combining stereo signals for mono transmission to single remote loudspeaker. Uses National LM566CN VCO. Frequency response is 20–20,000 Hz, and total harmonic distortion is under 0.5% With 120/240 V power lines, system operates equally well with receiver on either side of line. Transmitter input can be taken from monitor or tape output jack of audio system.—J. Sherwin, N. Sevastopoulos, and T. Regan, "FM Remote Speaker System," National Semiconductor, Santa Clara, CA, 1975, AN-146.

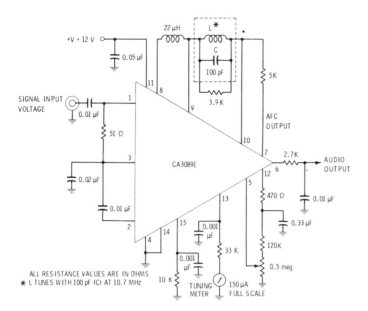

SINGLE-TUNED DETECTOR—RCA CA3089E IC serves as communication receiver subsystem providing three-stage FM IF amplifier/limiter channel, with signal-level detectors for each stage, and quadrature detector that can be used with single-tuned detector coil. Detector also supplies drive to AFC amplifier whose output can be used to hold local oscillator on correct frequency. Level-detector stages supply signal for tuning meter. Values shown are for 10.7-MHz IF.—E. M. Noll, "Linear IC Principles, Experiments, and Projects," Howard W. Sams, Indianapolis, IN, 1974, 347–349.

QUADRATURE DEMODULATOR—Quadrature coil associated with balanced-mixer demodulation system is connected to pin 6 of National LM373 IC, and output signal is taken from pin 7. Good output is obtained with only ±5 kHz deviation at either 455 kHz or 10.7 MHz. Can be operated as wideband or narrow-band circuit by choosing appropriate interstage and output LC and RC components.—E. M. Noll, "Linear IC Principles, Experiments, and Projects," Howard W. Sams, Indianapolis, IN, 1974, p 350–351.

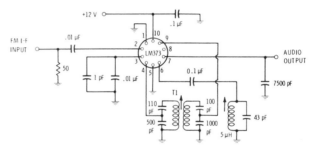

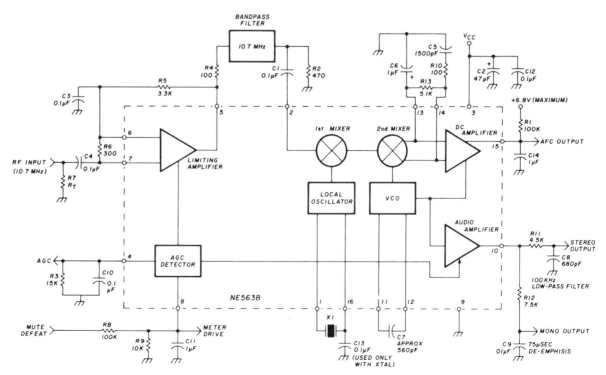

PLL IF AND DEMODULATOR—Signetics NE563B IC (in dashed lines) serves as complete IF amplifier and demodulator for FM broadcast receiver. Circuit uses downconversion from 10.7 MHz to 900 kHz, where phase detector operates. Ceramic bandpass filter provides IF selectivity at 10.7 MHz. X1 can be 9.8-MHz ceramic resonator, LC network, crystal, or capacitor.—H. Olson, FM Detectors, Ham Radio, June 1976, p 22–29.

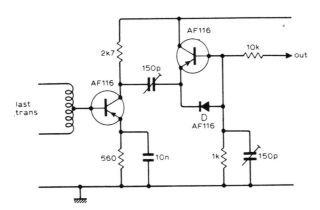

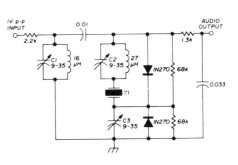

TRANSISTOR-PUMP DISCRIMINATOR—Used with 10.7-MHz IF strip of high-quality FM tuner built from discrete components. Circuit is placed between last IF stage and stereo decoder.—W. Anderson, F. M. Discriminator, *Wireless World,* April 1976, p 63.

CRYSTAL DISCRIMINATOR—Inexpensive third-overtone CB crystal used at 9-MHz fundamental serves as high-performance discriminator for VHF FM receiver. Adjust C3 for zero voltage with unmodulated carrier at or near center frequency. Adjust C1 and C2 with AF sine wave applied to FM signal generator, using CRO to check distortion of recovered sine wave. With 1 V P-P IF signal at 9 MHz and 5-kHz deviation, recovered audio will be about 1 V P-P at lower audio frequencies. Good limiter is required ahead of discriminator for AM rejection.—G. K. Shubert, Crystal Discriminator for VHF FM, *Ham Radio,* Oct. 1975, p 67–69.

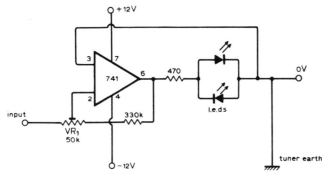

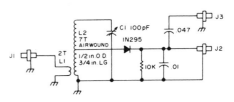

LED TUNING INDICATOR—One LED is mounted at each end of tuning scale. Tuning pointer is moved away from whichever LED is on, to dead spot at which both are off, to obtain correct tuning point. Advantages of lights-off tuning include minimum current drain and indication of even very slight mistuning by having one light come on even slightly. Adjust VR₁ to give wide enough dead spot so LEDs do not flicker on loud speech or music.—H. Hodgson, Simpler F.M. Tuning Indicator, *Wireless World,* Sept. 1975, p 413.

21–75 MHz DIODE RECEIVER—Covers 6-meter band and most 2-meter FM receiver oscillators near 45 MHz. Circuit is essentially that of crystal detector. Jack J3 gives AF output, and J2 gives DC output for meter.—B. Hoisington, Tuned Diode VHF Receivers, *73 Magazine,* Dec. 1974, p 81–84.

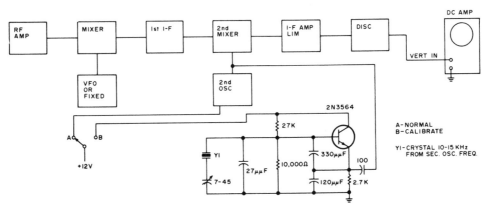

DEVIATION METER—Uses simple crystal oscillator combined with fixed or tunable FM receiver and CRO to show carrier shift on either side of center frequency. Vertical amplifier of CRO should be direct-coupled. To calibrate, tune oscillator either 10 or 15 kHz above or below second oscillator of receiver, and calibrate screen of CRO accordingly. One calibration oscillator is sufficient since transmitter usually deviates equally well both ways.—V. Epp, FM Deviation Meters, *73 Magazine,* March 1973, p 81–83.

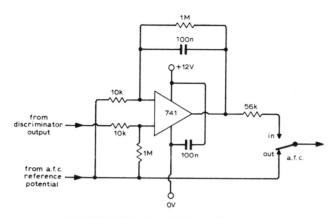

AFC AMPLIFIER—Simple DC amplifier can be added to AFC circuit of FM tuner to eliminate tuning errors over entire lock-in range.—J. S. Wilson, Improved A.F.C. for F.M. Tuners, *Wireless World,* July 1974, p 239.

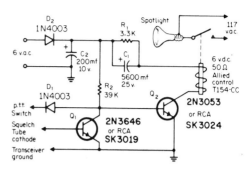

CALL ALERT—Developed to trigger relay when signal arrives at squelch tube in GE Progress Line 2-meter FM receiver. Relay is held energized about 2 s, determined by C_1-R_1, then de-energized for at least 25 s. Used for flashing red spotlight in room that is too noisy for hearing bell or buzzer. Circuit is easily adapted for any other FM receiver having squelch stage. Control circuit responds to small change in voltage at cathode of squelch tube. With no carrier present, tube conducts and places positive voltage at face of Q_1, making it conduct and turn off Q_2. When carrier arrives, Q_1 restores bias to Q_2, turning on relay. Connection to push-to-talk switch keeps lamp from flashing during transmission.—L. Waggoner, The WA0QPM "Call Alert," *CQ,* May 1971, p 48–49.

SECTION 38
Motor Control Circuits

Speed control circuits for various types and sizes of AC and DC motors, including three-phase motors.

SWITCHING-MODE CONTROLLER—Developed for driving 0.01-hp motor M at variable speeds with minimum battery drain. Circuit uses pulses with low duty cycle to set up continuous current in motor approximating almost 200 mA when average battery drain is 100 mA for output voltage of 3.5 V. Voltage comparator A_1 serves as oscillator and as duty-cycle element of controller. C_1 and R_1 provide positive feedback giving oscillation at about 20 kHz, with duty-cycle range of 10% to 70% controlled by feedback loop Q_1-R_1-C_3-R_3. D_2 is used in place of costly large capacitor for filtering.—J. C. Sinnett, Switching-Mode Controller Boosts DC Motor Efficiency, *Electronics*, May 25, 1978, p 132.

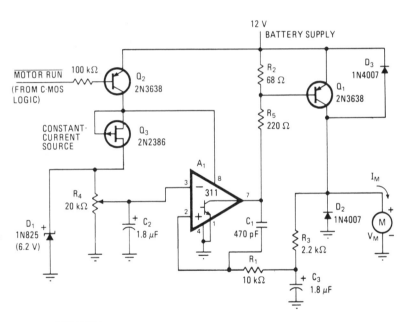

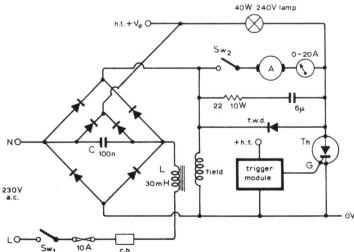

2-hp THYRISTOR CONTROL—Provides smooth variation in speed of shunt-wound DC motor from standstill to 90% of rated speed. Use thyristor rated 30 A at 600 V. Outer diodes of bridge are 35-A 600-PIV silicon power diodes, as also is thyristor diode, and inner diodes are 5-A 600-V silicon power diodes. Article gives complete circuit of trigger pulse generator used to control speed by varying duty cycle of thyristor. Larger motors can be controlled similarly by uprating thyristor and diodes. Controller will also handle other types of loads, including lamps and heaters.—F. Butler, Thyristor Control of Shunt-Wound D.C. Motors, *Wireless World*, Sept. 1974, p 325–328.

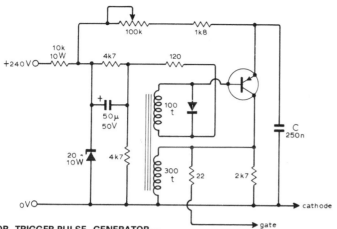

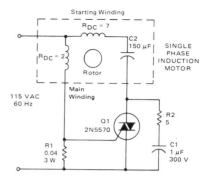

THYRISTOR TRIGGER-PULSE GENERATOR— Used with thyristor speed control for 2-hp shunt-wound DC motor. Circuit provides train of pulses with variable delay with respect to zero-crossing instants of AC supply, for feeding to cathode and gate of thyristor to vary duty cycle. Use Mullard BFX29 silicon PNP transistor or equivalent, and any small-signal silicon diode. Output pulses are suitable for triggering all types of thyristors up to largest. Article also gives motor control circuit.—F. Butler, Thyristor Control of Shunt-Wound D.C. Motors, *Wireless World,* Sept. 1974, p 325–328.

TRIAC STARTING SWITCH FOR ½-hp MOTOR— Triac replaces centrifugal switch normally used to control current through starting winding of single-phase induction motor. Value of R1 is chosen so triac turns on only when starting current exceeds 12 A. When motor approaches normal speed, running current drops to 8 A and triac blocks current through starting winding.— "Circuit Applications for the Triac," Motorola, Phoenix, AZ, 1971, AN-466, p 8.

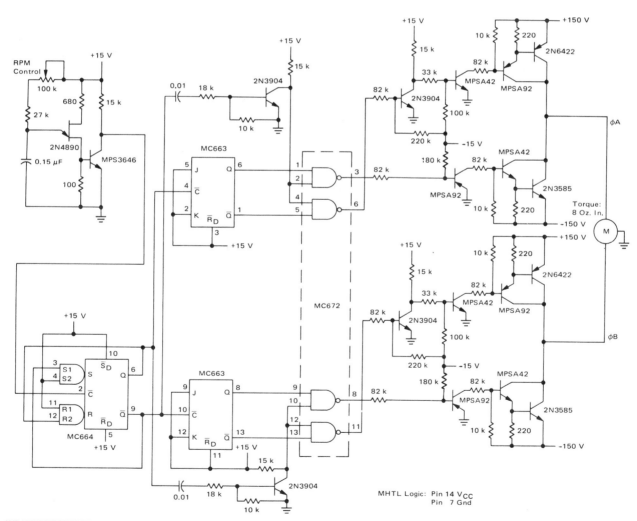

INDUCTION-MOTOR SPEED—Uses variable-frequency UJT oscillator at upper left to toggle MC664 RS flip-flop which in turn clocks MC663 JK flip-flops. Quadrature-phased JK outputs are combined with fixed-width pulses in MC672 to provide zero-voltage steps of drive signals for phase A and phase B. Outputs of RS flip-flops are differentiated and positive-going transitions amplified by pair of 2N3904 transistors, with pulse width of about 500 μs. NAND-gate outputs are then translated by small-signal amplifiers to levels suitable for driving final transistors having complementary NPN/PNP pairs. Circuit will provide speed range of 300 to 1700 rpm for permanent-split capacitor motor.—T. Mazur, "Variable Speed Control System for Induction Motors," Motorola, Phoenix, AZ, 1974, AN-575A, p 6.

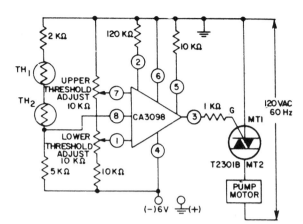

WATER-LEVEL CONTROL—Two thermistors operating in self-heating mode are mounted on sides of water tank. Thermistors change resistance when water level rises so liquid rather than air conducts heat away. Threshold adjustment pots are set so RCA CA3098 programmable Schmitt trigger turns on pump motor when water level rises above thermistor mounted near upper edge of tank, to remove water from tank and prevent overflow. Motor stays on to pump water out of tank until water level drops below location of lower thermistor inside tank.—"Linear Integrated Circuit and MOS/ FET's," RCA Solid State Division, Somerville, NJ, 1977, p 218–221.

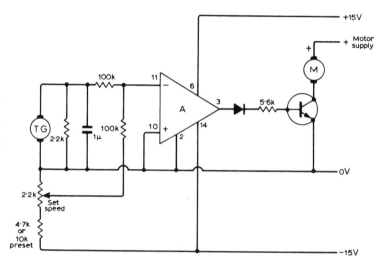

OPAMP SPEED CONTROL—Provides fine speed control of DC motor by using 0.25-W 6-V motor as tachogenerator giving about 4 V at 13,000 rpm. Opamp (RCA 3047A or equivalent) provides switching action for transistor in series with controlled motor, up to within a few volts of supply voltage. Choose transistor to meet motor current requirement.—N. G. Boreham, D.C. Motor Controller, *Wireless World*, Aug. 1971, p 386.

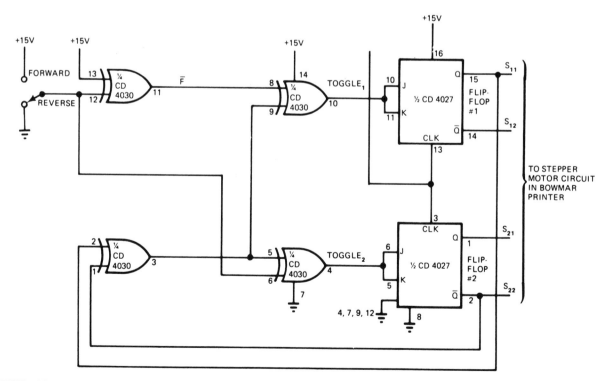

STEPPER MOTOR DRIVE—Two CMOS packages provide the four feed signals required for controlling forward/reverse drive of stepper motor for carriage drive and paper advance of Bowmar Model TP 3100 thermal printer. Outputs of flip-flops are above 10 V, enough to drive stepper motor directly. Each clock pulse to JK flip-flop advances carriage one step in direction commanded.—R. Bober, Stepper Drive Circuit Simplifies Printer Control, *EDN Magazine*, April 5, 1976, p 114.

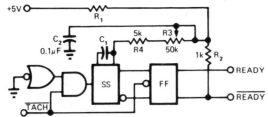

SS=MONOSTABLE MULTIVIBRATOR (SINGLE SHOT) 74122 OR 9601
OR ½ 74123 OR 9602
FF=NEGATIVE EDGE–TRIGGERED FLIP FLOP, 74H103 OR EQUIVALENT
R_1=0 TO 20% OF R_2, SELECTED FOR REQUIRED HYSTERESIS
C_1=SELECTED FOR REQUIRED TIMING RANGE.

UP-TO-SPEED LOGIC—Simple speed-sensing circuit fed by tachometer pulses makes READY output high when rotating device reaches desired minimum or threshold speed. Single-action triggering eliminates instability at decision point. Circuit also provides hysteresis, for separating pull-in and drop-out points any desired amount as determined by ratio of R_1 to R_2 in timing network. Article covers circuit operation and gives timing diagram.—W. Bleher, Circuit Indicates Logic "Ready," *EDN Magazine,* **March 5, 1974, p 72 and 74.**

CONTINUOUS-DUTY BRAKE—High or 1 bit at output port of microprocessor energizes brake solenoid of paper-tape reader through optocoupler and amplifier. When tape is to be stopped, brake solenoid is energized and tape is squeezed between top of solenoid and flat iron brake shoe that is attracted by solenoid.—D. Hogg, The Paper Taper Caper, *Kilobaud,* **March 1977, p 34–40.**

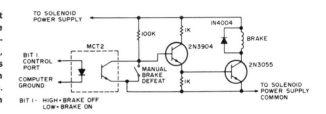

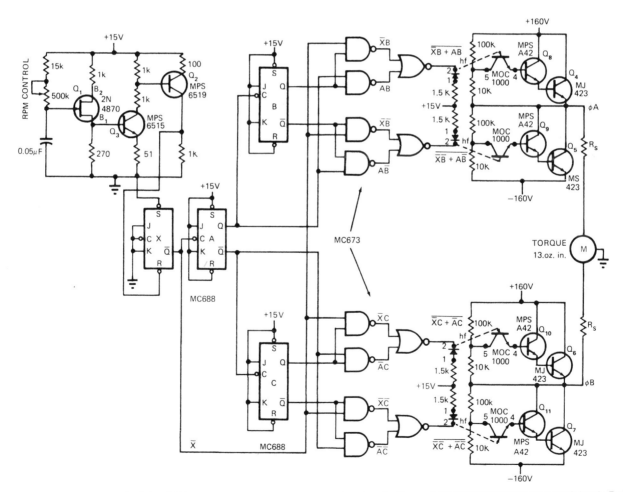

SPEED CONTROL FOR INDUCTION MOTOR—Uses UJT oscillator Q_1 to generate frequency in range from 40 to 1200 Hz for feeding to divide-by-4 configuration that gives motor source frequency range of 10 to 300 Hz. With induction motor having two pairs of poles, this gives theoretical speed range of 300 to 9000 rpm with essentially constant torque. Speed varies linearly with frequency. Circuit uses pair of flip-flops (MC673) operated in time-quadrature to perform same function as phase-shifting capacitor so motor receives two drive signals 90° apart. Article covers operation of circuit in detail. Optoisolators are used to provide bipolar drive signals from unipolar control signals. Each output drive circuit is normally off and is turned on only when its LED is on. If logic power fails, drives are disabled and motor is turned off as fail-safe feature.—T. Mazur, Unique Semiconductor Mix Controls Induction Motor Speed, *EDN Magazine,* **Nov. 1, 1972, p 28–31.**

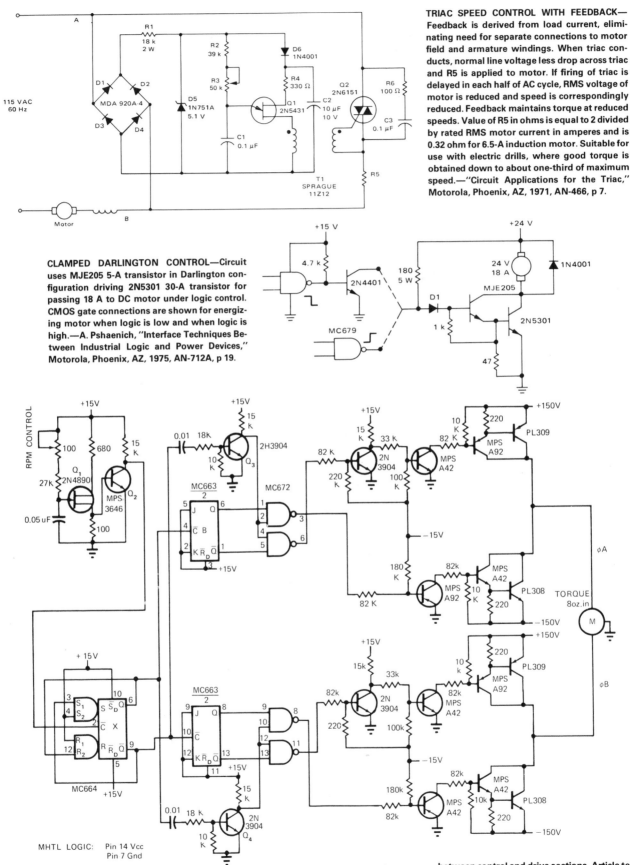

TRIAC SPEED CONTROL WITH FEEDBACK— Feedback is derived from load current, eliminating need for separate connections to motor field and armature windings. When triac conducts, normal line voltage less drop across triac and R5 is applied to motor. If firing of triac is delayed in each half of AC cycle, RMS voltage of motor is reduced and speed is correspondingly reduced. Feedback maintains torque at reduced speeds. Value of R5 in ohms is equal to 2 divided by rated RMS motor current in amperes and is 0.32 ohm for 6.5-A induction motor. Suitable for use with electric drills, where good torque is obtained down to about one-third of maximum speed.—"Circuit Applications for the Triac," Motorola, Phoenix, AZ, 1971, AN-466, p 7.

CLAMPED DARLINGTON CONTROL—Circuit uses MJE205 5-A transistor in Darlington configuration driving 2N5301 30-A transistor for passing 18 A to DC motor under logic control. CMOS gate connections are shown for energizing motor when logic is low and when logic is high.—A. Pshaenich, "Interface Techniques Between Industrial Logic and Power Devices," Motorola, Phoenix, AZ, 1975, AN-712A, p 19.

FREQUENCY CONTROLS SPEED—Circuit generates variable frequency between 10 and 300 Hz at constant voltage for changing speed of induction motor between theoretical limits of 300 and 9000 rpm without affecting maximum torque. Direct coupling between control and drive circuits is used; if motor noise affects control logic circuits, optoisolators should be used between control and drive sections. Article tells how circuit works and gives similar circuit using optical coupling.—T. Mazur, Unique Semiconductor Mix Controls Induction Motor Speed, *EDN Magazine,* Nov. 1, 1972, p 28–31.

2-hp THREE-PHASE INDUCTION—Speed is controlled by applying continuously variable DC voltage to VCO of control circuit for 750-VDC 7-A bridge inverter driving three sets of six Delco DTS-709 duolithic Darlingtons. Bridge inverter circuit for other two phases is identical to that shown for phase AA'. VCO output is converted to three-phase frequency varying from 5 Hz at 50 VDC to 60 Hz at 600 VDC for driving output Darlingtons. Optoisolators are used for base drive of three switching elements connected to high-voltage side of inverter.—"A 7A, 750 VDC Inverter for a 2 hp, 3 Phase, 480 VAC Induction Motor," Delco, Kokomo, IN, 1977, Application Note 60.

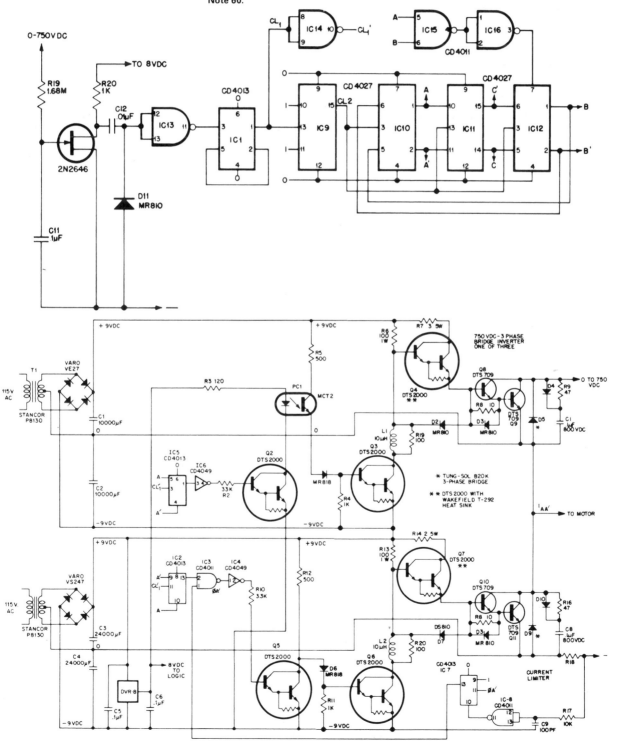

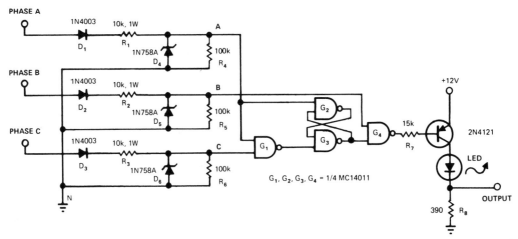

PHASE SEQUENCE DETECTOR—Circuit detects incorrect phase sequence of motor driving pump, compressor, conveyor, or other equipment that can be damaged by reverse rotation. Circuit also protects motor from phase loss that could cause rapid temperature rise and heat damage. LED is on when phasing is correct. For phase loss or incorrect sequence, output goes low and LED is dark. Diodes and zeners change sine waves for all phases to rectangular logic-level pulses that feed gates. When phases are correct, output of G_4 is train of rectangular pulses about 2.5 ns wide. Output is zero for incorrect sequences. Since leading edge of output pulse coincides with positive zero crossing of phase B, output pulses can be used to trigger SCR connected across phase B and driving relay-coil load. SCR then energizes relay only when sequence is correct.—H. Normet, Detector Protects 3-Phase-Powered Equipment, *EDN Magazine*, Aug. 5, 1978, p 78 and 80.

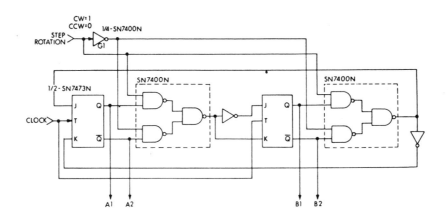

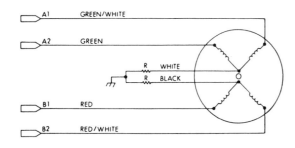

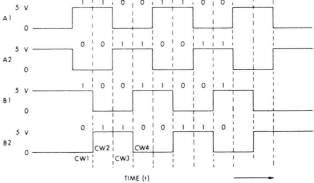

Step Sequence	A1	A2	B1	B2	Binary Code
1 CW	ON	OFF	ON	OFF	1010
2	ON	OFF	OFF	ON	1001
3	OFF	ON	OFF	ON	0101
4	OFF	ON	ON	OFF	0110
CCW 1	ON	OFF	ON	OFF	1010

STATE GENERATOR FOR STEPPER—Generates high-current square-wave pulses and provides correct switching sequence for exciting stepper motor when digital display is required to show instantaneous step angle and total revolutions traveled by shaft of stepper motor. If microprocessor is used, speed and direction of motor rotation can be controlled by programming period and level of output pulses. Clock signals trigger SN7473N JK flip-flop that changes ON/OFF states of four outputs as shown in table. Clock signal is obtained from external square-wave generator or from microprocessor such as KIM-1. Article also gives digital display circuit driven by same clock.—H. Lo, Digital Display of Stepper Motor Rotation, *Computer Design*, April 1978, p 147–148 and 150–151.

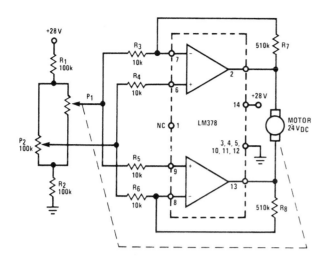

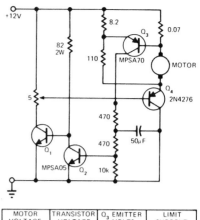

MOTOR VOLTAGE	TRANSISTOR VOLTAGE	Q_3 EMITTER VOLTS	LIMIT CURRENT
0	12	0	10A
6	6	0.42	16A
11.5	0.5	0.20	21.4A

24-VDC PROPORTIONAL SPEED CONTROL— National LM378 amplifier IC is basis for low-cost proportional speed controller capable of furnishing 700 mA continuously for such applications as antenna rotors and motor-controlled valves. Proportional control results from error signal developed across Wheatstone bridge R_1-R_2-P_1-P_2. P_1 is mechanically coupled to motor shaft as continuously variable feedback sensor. As motor turns, P_1 tracks movement and error signal becomes smaller and smaller; system stops when error voltage reaches 0 V.—"Audio Handbook," National Semiconductor, Santa Clara, CA, 1977, p 4-8–4-20.

STALLED-MOTOR PROTECTION—Modification of basic speed control circuit for small DC permanent-magnet motors provides maximum current limit under normal conditions and reduced current limit under stall conditions, to limit dissipation of series transistor Q_4 to safe value. When motor stalls, motor voltage falls, reducing voltage and motor current required to turn on Q_3 and thereby limiting stalled-motor current.—D. Zinder, Current Limit and Foldback for Small Motor Control, *EDN Magazine*, May 5, 1974, p 77 and 79.

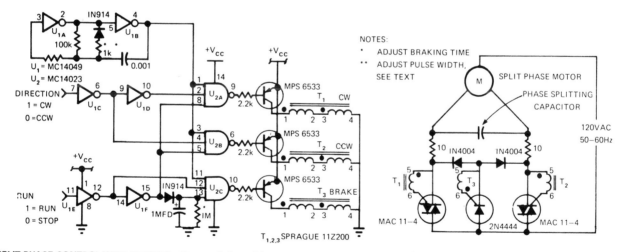

SPLIT-PHASE CONTROL WITH BRAKING—Use of CMOS logic to gate direction-controlling triacs and turn on SCR for braking provides low-cost switchless control of split-phase motor used in place of brush-type DC motor. Applications include control of ball valves and other throttling functions in process control. With shaft-position encoders, circuit generates feedback information. Overshoot and other stability problems are easily controlled by strong braking function. CMOS logic provides complete noise immunity. Oscillator pulse width is adjusted with 1K resistor in series with 1N914, and brake duration is controlled by 1-megohm resistor at input of U_{2C}. With values shown, brake is applied for about 1 s. Circuit works reliably on supply voltages of 5 to 15 V.—V. C. Gregory, Split-Phase Motor Control Accomplished with CMOS, *EDN Magazine*, Oct. 5, 1974, p 65–67.

TAPE-LOOP SPEED CONTROL—Shunt rectifier-capacitor circuit was developed for speed control of permanent split-capacitor fractional-horsepower induction motor used in some motion-picture projectors. Light-dependent resistor LDR makes Q_2 conduct when light from lamp is not blocked by tape loop. Split capacitor C_1 for motor provides both run and speed-control functions without switching. Values are: C_2 0.01 μF; D 1N4004; Q_1 2N4987; Q_2 C106B; R_1 330K; R_2 100; R_3 10.—T. A. Gross, Control the Speed of Small Induction Motors, *EDN Magazine,* Aug. 20, 1977, p 141–142.

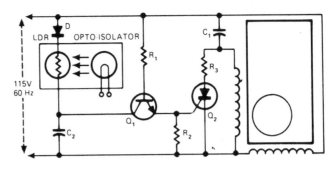

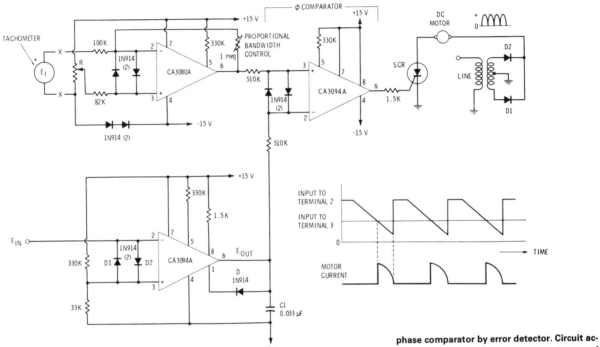

DC MOTOR SPEED CONTROLLER—Tachometer driven by motor produces output voltage proportional to speed for application to CA3080A voltage comparator after rectification and filtering. Output of CA3080A is applied to upper CA3094A phase comparator that is receiving reference voltage from another CA3094A connected as ramp generator. Output of phase comparator triggers SCR in motor circuit. Amount of motor current is set by time duration of positive signal at pin 6, which in turn is determined by DC voltage applied to pin 3 of phase comparator by error detector. Circuit action serves to maintain constant motor speed at value determined by position of pot R. Input to ramp generator is pulsating DC voltage used to control rapid charging of C1 and slower discharging to form ramp.—E. M. Noll, "Linear IC Principles, Experiments, and Projects," Howard W. Sams, Indianapolis, IN, 1974, p 321–323.

SECTION 39
Multiplexer Circuits

Includes various bit-width data multiplexers, notch filters, and multiplexer filters. See other sections on RF filters, active filters, and amplifiers.

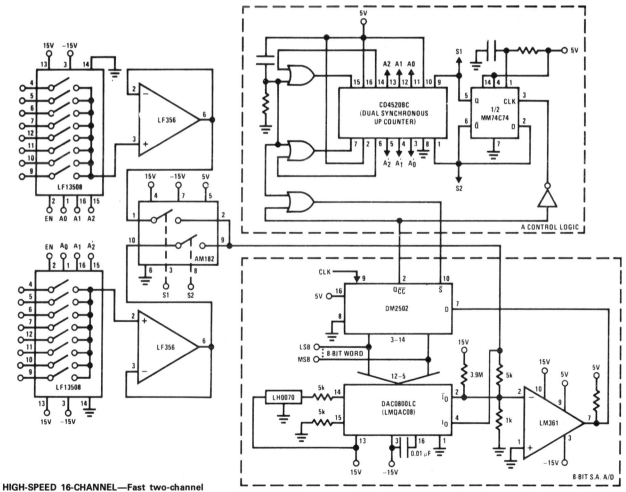

HIGH-SPEED 16-CHANNEL—Fast two-channel multiplexer using National AM182 dual analog switch provides second-level multiplexing by accepting outputs of each LF13508 eight-channel multiplexer and feeding these outputs sequentially into 8-bit successive-approximation A/D converter. Technique makes throughput rate of system independent of analog switch speed. With maximum clock frequency of 4.5 MHz, throughput rate is 31,250 samples per second per channel.—"FET Databook," National Semiconductor, Santa Clara, CA, 1977, p 5-79–5-80.

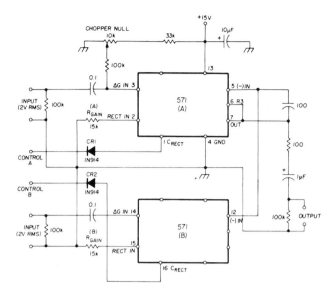

TWO-INPUT FSK MULTIPLEXER—Uses Signetics NE571 or NE570 analog compandors. Gain of each channel is unity, as determined by R_{GAIN} value for channel. When complementary control signals are provided, FSK generator switches between the two signal inputs. Outputs, when on, are summed by opamp in IC. Each channel is gated off by low control logic input. For FSK or alternate-channel use, CONTROL A and CONTROL B signals should be complementary. Control signal suppression is optimized with chopper null pot. Suppression is better than 60 dB after trimming. Circuit can also be used as summing switch, with both signals on at any given instant.—W. G. Jung, Gain Control IC for Audio Signal Processing, *Ham Radio*, July 1977, p 47–53.

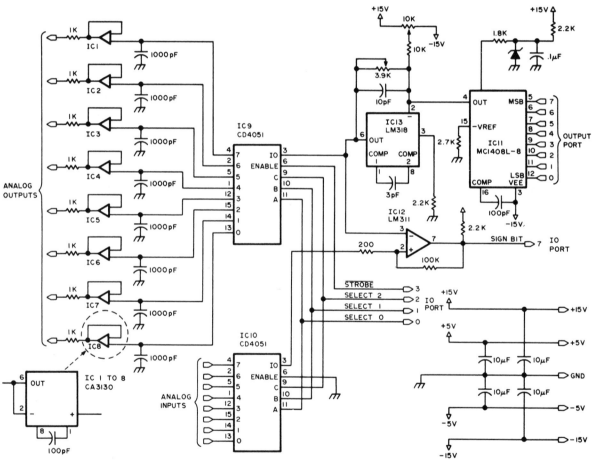

Number	Type	+5 V	GND	−15 V	+15 V	−5 V
1 to 8	CA3130	7				4
9	CD4051	16.	8		7	
10	CD4051	16	8		7	
11	MC1408L-8	13	2			
12	LM311		1	4	8	
13	LM318			4	8	

MULTIPLEXED A/D-D/A CONVERTER INTERFACE—Time-multiplexed interface minimizes hardware required for applications of personal computer system. Useful in interactive games, equipment testing, and electronic music. Optimized for 0.1-100 Hz signals. Bypass each power pin with 0.01 μF to suppress stray spikes caused by power surges. Use of LM318 opamp minimizes response time of MC1408L-8 DAC.—D. R. Kraul, Designing Multichannel Analog Interfaces, *BYTE*, June 1977, p 18–23.

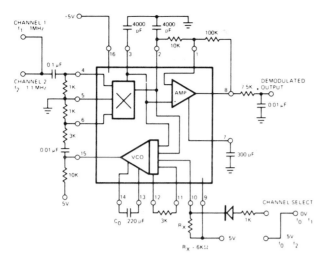

TIME-MULTIPLEXING TWO FM CHANNELS—Digital programming capability of Exar XR-215 PLL IC makes possible time-multiplexing demodulator between two FM channels, at 1.0 and 1.1 MHz. Channel-select logic signal is applied to pin 10, and both input channels are applied simultaneously to PLL input pin 4.—"Phase-Locked Loop Data Book," Exar Integrated Systems, Sunnyvale, CA, 1978, p 21–28.

THREE-CHANNEL FOR DATA—Each input channel uses CA3060 variable opamp as high-impedance voltage follower driving output MOSFET serving as buffer and power amplifier. Cascade arrangement of opamps with MOSFET provides open-loop voltage gain in excess of 100 dB.—"Circuit Ideas for RCA Linear ICs," RCA Solid State Division, Somerville, NJ, 1977, p 16.

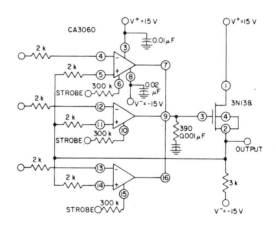

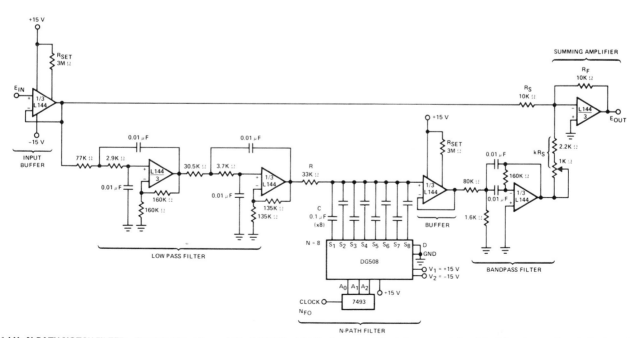

1-kHz N-PATH NOTCH FILTER—Combination of DG508 eight-channel CMOS multiplexer with low-pass and bandpass active filters provides 1-kHz notch filter having Q of 1330 and 3-dB bandwidth of 0.75 Hz at 1 kHz. Low-pass filter introduces 180° phase shift at 1 kHz. Amplifier sums original signal in phase-shifted bandpass output from N-path filter, canceling 1-kHz components in original signal to produce desired notch characteristic.—"Analog Switches and Their Applications," Siliconix, Santa Clara, CA, 1976, p 5-18–5-20.

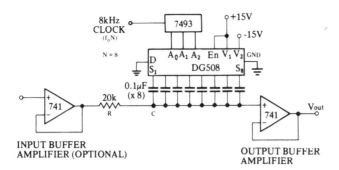

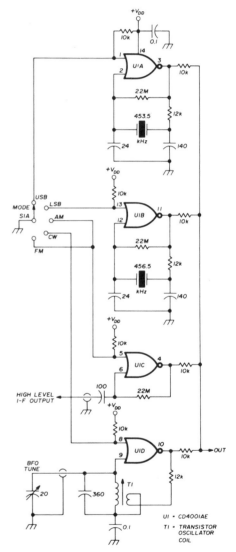

1-kHz COMB FILTER—DG508 eight-channel CMOS multiplexer is used in comb filter having fundamental frequency of 1 kHz. Sampling action provides response at each harmonic multiple except at 8 and 16 kHz (no response at Nf_0 or $2Nf_0$). Used in selective filtering of periodic signals from background of nonperiodic noise interference. 7493 TTL binary counter provides necessary 3-bit binary count sequence from 8-kHz clock. Q is 50.—"Analog Switches and Their Applications," Siliconix, Santa Clara, CA, 1976, p 5-17–5-18.

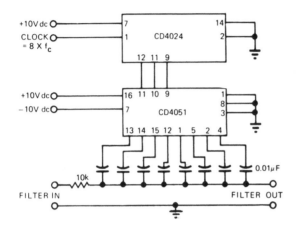

COMMUTATING BANDPASS FILTER—CD4051 analog multiplexer serves for commutation and switching of eight low-pass filter sections. Multiplexer is driven by CD4024 binary counter that is clocked at 8 times desired 100-kHz center frequency. Can be tuned by varying commutating frequency.—J. Tracy, CMOS Offers New Approach to Commutating Filters, *EDN Magazine*, Feb. 5, 1974, p 94–95.

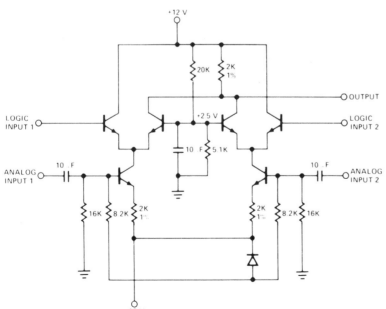

ANALOG SWITCH—Circuit using Signetics 511 transistor array provides digital selection of either of two analog signals. When logic input at left is zero, signal at analog input 1 goes to output and other analog input signal is rejected. Similarly, when logic 0 is applied to logic input 2, analog input 2 goes to output. Eight-channel analog multiplex switch can be formed by combining four 511 analog switches with Signetics 8250 binary-to-octal decoder. Analog signals up to 200 kHz are switched without amplitude degradation.—"Signetics Analog Data Manual," Signetics, Sunnyvale, CA, 1977, p 753–754.

BFO MULTIPLEXER—Signal 455-kHz multimode detection system using RCA CD4001AE quad NOR gate functions as upper-sideband or lower-sideband crystal oscillator, tunable BFO for CW, or limiter of IF signal for FM or synchronous AM reception. Desired oscillator or limiter is gated on by grounding its digital control line with S1A. Multimode reception occurs when multiplexed output of oscillators and limiter is applied to product detector.—J. Regula, BFO Multiplexer for a Multimode Detector, *Ham Radio*, Oct. 1975, p 52–55.

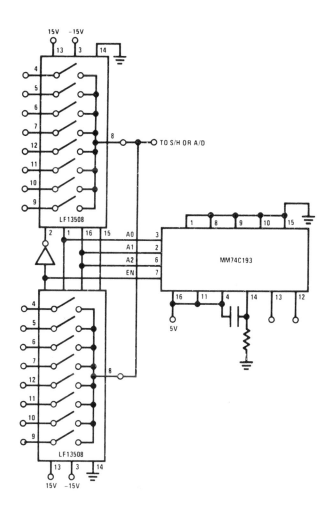

16-CHANNEL SIMPLIFIED SEQUENTIAL MULTIPLEXING—Two National LF13508 eight-channel multiplexers are connected so enable pins are used to disconnect one multiplexer while the other is sampling. Any number of eight-channel multiplexers can be connected in this way if speed is not prime system requirement.—"FET Databook," National Semiconductor, Santa Clara, CA, 1977, p 5-79–5-81.

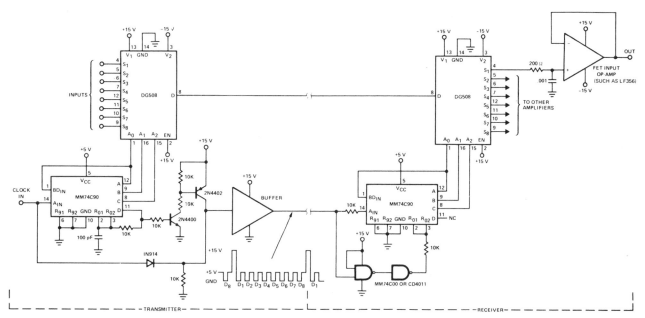

EIGHT-CHANNEL MUX/DEMUX—Provides for monitoring of all eight channels continuously at remote location instead of scanning channels at receiver. Each output of DG508 eight-channel analog multiplexer in receiver feeds its own opamp for driving readout. Similar DG508 at transmitter feeds inputs over single-wire line under control of MM74C90 presettable decade counter which also feeds pulse train over separate channel to receiver for timing and synchronization. 15-V reset pulse superimposed on 5-V clock pulses keeps channels synchronized.—"Analog Switches and Their Applications," Siliconix, Santa Clara, CA, 1976, p 7-71–7-74.

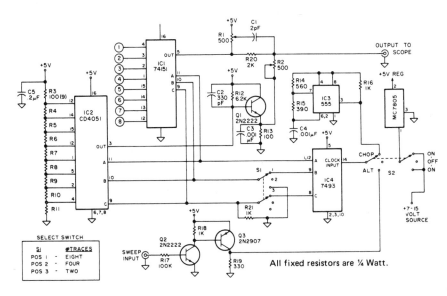

SELECT SWITCH

S1	#TRACES
POS 1	EIGHT
POS 2	FOUR
POS 3	TWO

All fixed resistors are ¼ Watt.

8-CHANNEL CRO MULTIPLEXER—Simple adapter converts any single-trace oscilloscope into professional 8-channel model by multiplexing eight input signals into one output for vertical amplifier. Discrete voltages are picked off resistor divider chain sequentially and added to digital input signal so trace is shifted fast enough to produce eight individual traces. Addressing for 74151 digital multiplexer requires 3 bits of binary code to address all eight channels in sampling sequence. As each channel is sampled, its logic level (1 or 0) appears at pin 5 of 74151 for feed to vertical input of CRO. Developed for troubleshooting in digital circuits.— W. J. Prudhomme, Build an Eight Channel Multiplexer for Your Scope, *Kilobaud,* April 1977, p 29–32.

SECTION 40
Multiplier Circuits

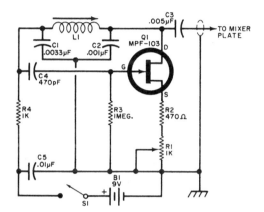

455-KHZ Q MULTIPLIER—Uses positive feed-back to narrow passband of receiver and provide Q multiplication of 20 to 30 that has effect of sharply attenuating off-resonance signal frequencies. L1 is slug-tuned bc antenna coil. R1 controls regeneration. Circuit is connected to mixer plate of communication receiver having 455-kHz i-f value.—R. N. Tellefsen, Build a FET-QM, *Popular Electronics*, Dec. 1968, p 51–53.

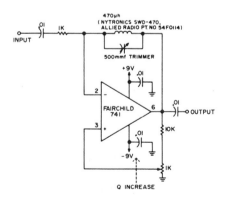

Q MULTIPLIER—Can be inserted after ssb filter in i-f amplifier of receiver or transceiver, to improve selectivity. Is particularly effective on c-w. Values shown are for 455-kHz i-f, for which this IC multiplier multiples Q of coil over 50 times. Same IC arrangement can be used to improve Q and selectivity of audio filters and tuned circuits of fsk converters.—J. J. Schultz, A Simple Integrated Circuit Q Multiplier, 73, Feb. 1970, p 134–137.

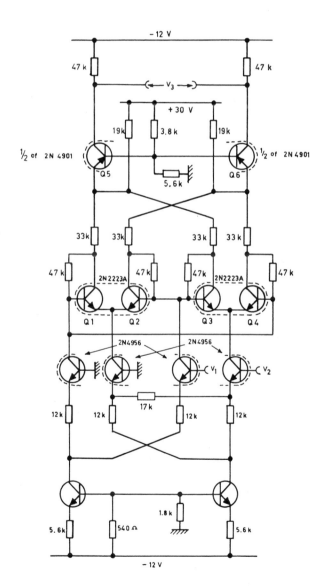

FOUR-QUADRANT MULTIPLIER—Single-stage analog multiplier provides output voltage V3 as product of input voltages V1 and V2. Circuit utilizes exact exponential characteristics of emitter-base junction of bipolar transistor to realize controlled current source. Article gives design equations. Developed for use in high-accuracy audiocorrelator.—H. Bruggemann, Feedback Stabilized Four-Quadrant Analog Multiplier, *IEEE Journal of Solid-State Circuits*, Aug. 1970, p 150–159.

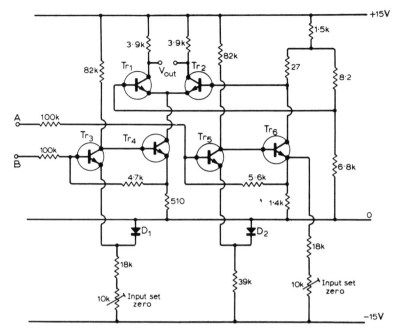

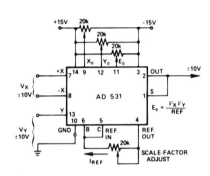

ADJUSTABLE SCALE FACTOR—Uses Analog Devices AD531 IC for computing function XY/Z, with any or all of the three input voltages variable. Scale factor (1/Z) is changed with 20K pot, or can be varied dynamically by applying externally controlled reference current. NOTE: End terminals of 20K pot for Xo should be connected between +15 V and 42K resistor going to ground.—Monolithic Analog Multiplier-Dividers, *Analog Dialogue*, Vol. 6, No. 3, p 10.

ANALOG MULTIPLIER—Linearity depends heavily on characteristic of long-tailed pair. With values shown, providing considerable negative feedback, maximum departure from linearity is about 3%. Diodes D1 and D2 compensate for base-emitter voltages of Tr3 and Tr5, so emitter current of Tr4 is zero for

zero input and emitter current of Tr6 is that required to balance long-tailed pair. Transistors are not critical, and can be BC107 or similar.—A. F. Newell, A Transistor Multiplier Circuit, *Wireless World*, June 1969, p 285–289.

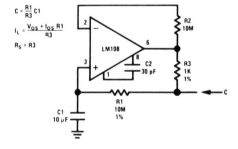

CAPACITANCE MULTIPLIER—Opamp eliminates need for large capacitance values by increasing effective capacitance of small capacitor and coupling it into low-impedance system. Circuit generates equivalent capacitance of 100,000 µF with worst-case leakage of 8 µA over −55 C to 125 C temperature range. Not suitable for tuned circuits and filters because Q is low, but satisfactory for timing circuit or servo compensation networks.—R. J. Widlar, "IC Op Amp Beats FETs on Input Current," National Semiconductor, Santa Clara, CA, 1969, AN-29, p 10–11.

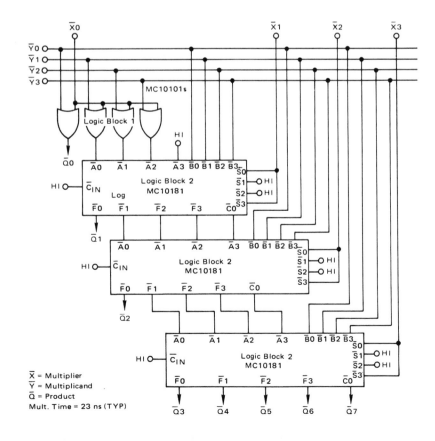

$\overline{X}$ = Multiplier
$\overline{Y}$ = Multiplicand
$\overline{Q}$ = Product
Mult. Time = 23 ns (TYP)

BINARY MULTIPLIER—Use of IC 10181 logic blocks makes ripple multiplier possible without large numbers of interconnects and parts, while giving typical multiplying time of only 23 ns for two 4-bit binary numbers. May be expanded easily to accommodate larger number of bits as covered in report. Based on use of one's complement of multiplicand and multiplier to provide one's complement of product. Product output is inverted.—T. Balph, "High Speed Binary Multiplication Using the MC10181," Motorola Semiconductors, Phoenix, AZ, AN-566, 1972.

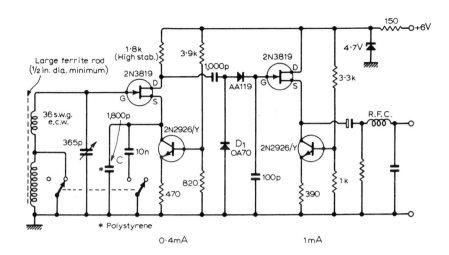

FET AS Q MULTIPLIER—Has inherent automatic bandwidth control up to 2 MHz. C determines no-signal Q; decreasing C moves stage toward oscillation. Circuit will replace two double-tuned 470-kHz i-f stages.—K. W. Mawson, High-Gain F.E.T. Tuned Amplifier, *Wireless World*, April 1970, p 182.

2-QUADRANT MULTIPLIER—Useful as gain controller for either a-c or d-c signals. Control input of 0 to 10 V at Y terminal will change gain of X-input signal in ±10-V range. Gain is unity when Y is +10 V. Linearity of control is 1% for Y input of 1 to 10 V. Bandwidth is 45 kHz at gain of 0.1 and decreases to 4.5 kHz at gain of 1.—Choosing and Using N-Channel Dual J-Fets, *Analog Dialogue*, Dec. 1970, p 4–9.

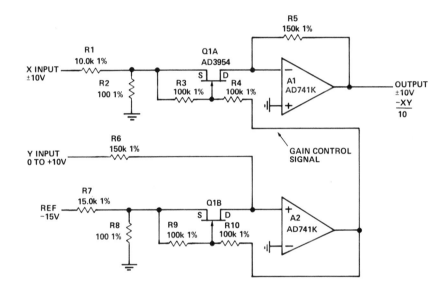

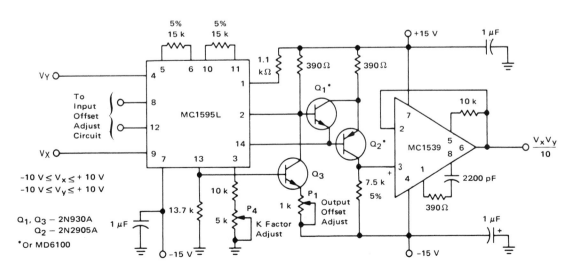

OPAMP BOOSTS CURRENT DRIVE—Uses discrete components for level-shifting of output to ground reference when multiplying two signal voltages, with opamp connected as source follower to increase current drive of single-ended output. Temperature problems are minimized by using MD60100 complementary-pair transistor package for Q1 and Q2. —E. Renschler, "Analysis and Basic Operation of the MC1595," Motorola Semiconductors, Phoenix, AZ, AN-489, 1970.

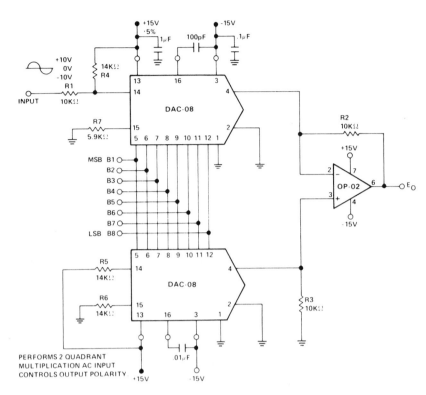

BIPOLAR ANALOG TWO-QUADRANT—Bipolar reference voltage for upper Precision Monolithics DAC-08 D/A converter modulates reference current by ±1.0 mA around quiescent current of 1.1 mA. Lower DAC-08 has same 1.1-mA reference current and effectively subtracts out quiescent 1.1 mA of upper reference current at all input codes since voltage across R3 varies between −10 V and 0 V. Output voltage E₀ is thus product of digital input word and bipolar analog reference voltage.—J. Schoeff and D. Soderquist, "Differential and Multiplying Digital to Analog Converter Applications," Precision Monolithics, Santa Clara, CA, 1976, AN-19, p 3.

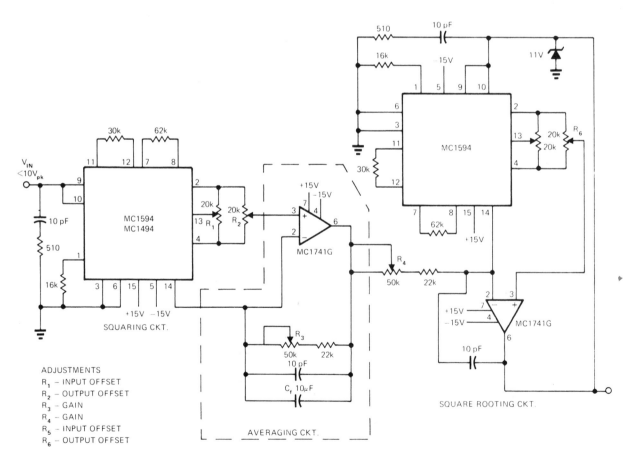

SQUARING FOR RMS—Combination of two MC1594 multipliers and two opamps gives RMS detector for squaring instantaneous input values, averaging over time interval, then taking square root to give RMS value of input waveform. First multiplier, used to square input waveform, delivers output current to first opamp for conversion to voltage and for averaging by means of capacitor in feedback path. Second opamp is used with second multiplier as feedback element for taking square root. Technique eliminates thermal response time drawback of most other RMS measuring circuits. Input voltage range for circuit is 2 to 10 V P-P; for other ranges, input scaling can be used. Since direct coupling is used, output voltage includes DC components of input. Maximum input frequency is about 600 kHz, and accuracy is about 1%.—K. Huehne and D. Aldridge, True RMS Measurements Using IC Multipliers, *EDN Magazine,* March 20, 1973, p 85—86.

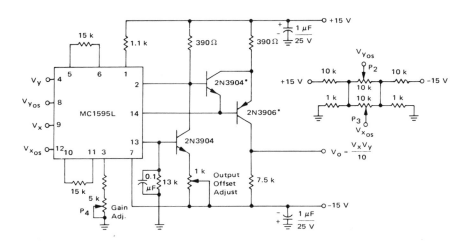

OUTPUT LEVEL SHIFTER—Transistors connected to Motorola MC1595L linear four-quadrant multiplier perform level shifting for applications requiring output having ground reference. Temperature sensitivity of circuit is minimized by using complementary transistors in same package, such as MD6100, in place of upper two transistors. If high output impedance and low current drive are drawbacks, opamp can be connected as source-follower output stage.—E. Renschler, "Analysis and Basic Operation of the MC1595," Motorola, Phoenix, AZ, 1975, AN-489, p 10.

AC-COUPLED MULTIPLICATION—Combination of Precision Monolithics REF-02 voltage reference and DAC-08 D/A converter uses compensation capacitor terminal C_C as input. With full-scale input code, output V_O is flat to above 200 kHz and 3 dB down at 1 MHz, for multiplying applications far beyond audio range. Circuit has high input impedance, as often required to avoid loading high source impedance. Dynamic range is greater than 40 dB.—J. Schoeff and D. Soderquist, "Differential and Multiplying Digital to Analog Converter Applications," Precision Monolithics, Santa Clara, CA, 1976, AN-19, p 4.

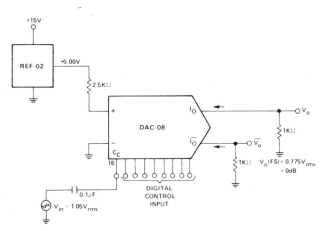

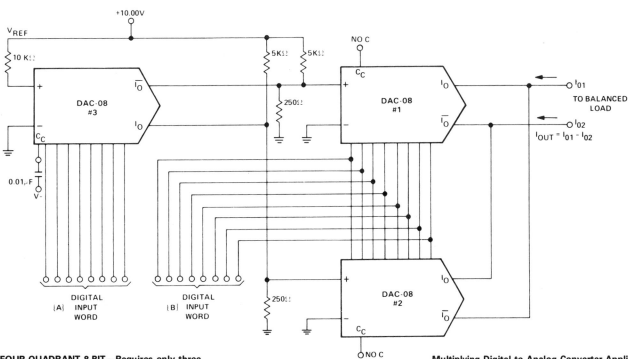

FOUR-QUADRANT 8-BIT—Requires only three Precision Monolithics DAC-08 D/A converters to provide high-speed multiplication of two 8-bit digital words and give analog output.—J. Schoeff and D. Soderquist, "Differential and Multiplying Digital to Analog Converter Applications," Precision Monolithics, Santa Clara, CA, 1976, AN-19, p 7.

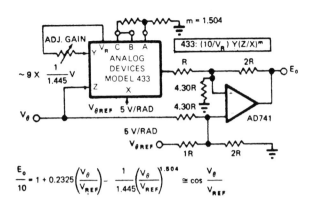

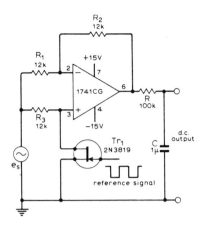

$$\frac{E_0}{10} = 1 + 0.2325\left(\frac{V_\theta}{V_{REF}}\right) - \frac{1}{1.445}\left(\frac{V_\theta}{V_{REF}}\right)^{1.504} \cong \cos\frac{V_\theta}{V_{REF}}$$

APPROXIMATING COSINES—Analog Devices 433 multiplier/divider IC approximates cosine of angle to better than 1%, by computing nonintegral exponents. Only one opamp is needed. Approximation uses arbitrary exponent as one term of cosine θ plus a linear term and a constant term, as described in article.—D. H. Sheingold, Approximate Analog Functions with a Low-Cost Multiplier/Divider, *EDN Magazine*, Feb. 5, 1973, p 50–52.

PHASE-SENSITIVE DETECTOR—Circuit using single opamp produces DC output proportional to both amplitude of AC input signal and cosine of its phase angle relative to reference signal. Can be used as synchronous rectifier in chopper-type DC amplifier or for accurate measurement of small AC signals obscured by noise. Article gives design equations.—G. B. Clayton, Experiments with Operational Amplifiers, *Wireless World*, July 1973, p 355–356.

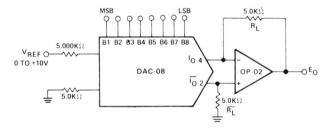

TWO-QUADRANT—Bipolar digital multiplier has output polarity controlled by offset-binary-coded digital input word. Precision Monolithics DAC-08 D/A converter drives OP-02 opamp. Output is symmetrical about ground.—J. Schoeff and D. Soderquist, "Differential and Multiplying Digital to Analog Converter Applications," Precision Monolithics, Santa Clara, CA, 1976, AN-19, p 2.

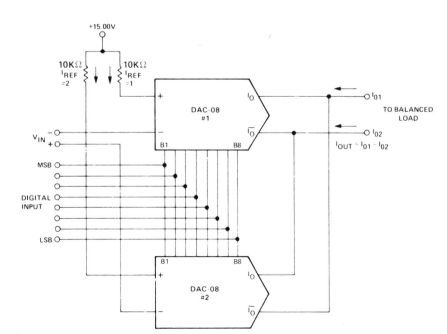

FOUR-QUADRANT MULTIPLYING DAC—Combination of two Precision Monolithics DAC-08 D/A converters accepts differential input voltage and produces differential current output. Output opamp is not normally required. Output analog polarity is controlled by analog input reference or by offset-binary digital input word. Common-mode current present at output must be accommodated by balanced load. Differential input range is 10 V.—J. Schoeff and D. Soderquist, "Differential and Multiplying Digital to Analog Converter Applications," Precision Monolithics, Santa Clara, CA, 1976, AN-19, p 3.

SECTION 41
Multivibrator Circuits

Includes monostable, astable, pulse-width variable, and multivibrators with various types of triggers. See other sections on clock circuits and oscillators, audio and RF.

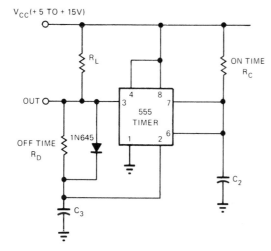

ASTABLE OSCILLATOR—Circuit for astable operation of 555 timer provides completely independent ON and OFF times. Time constant for one mode is 1.1 $R_C C_2$ and for other mode is 1.1 $R_C C_3$. Free-running period is sum of these time constants.—J. P. Carter, Astable Operation of IC Timers Can Be Improved, *EDN Magazine*, June 20, 1973, p 83.

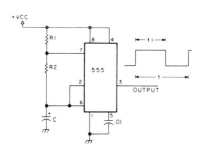

555 ASTABLE—Produces repetitive rectangular output at frequency equal to 1.443/$(R_1 + 2R_2)C$ hertz. Duty cycle is determined by values of R_1 and R_2; R_2 must be much larger than R_1 to obtain nearly a 50% duty cycle. Normal range for duty cycle is 51 to 99%. VCC is 4.5–16 V at 3–10 mA.—H. M. Berlin, IC Timer Review, *73 Magazine*, Jan. 1978, p 40–45.

LOW-POWER MONO—555 timer provides low-drain monostable operation suitable for interfacing with CMOS 4011B NAND gates. Standby drain is less than 50 μA. When mono is on, current drawn is 4.5 mA for pulse duration of T = 1.1RC.—"Signetics Analog Data Manual," Signetics, Sunnyvale, CA, 1977, p 733.

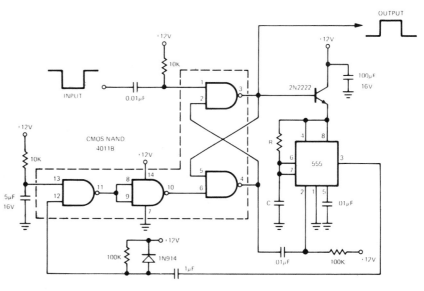

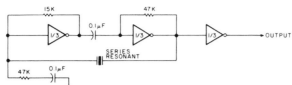

CRYSTAL MONO USING INVERTERS—Uses all three sections of CD4049 triple inverter, with series-resonant crystal connection. Supply can be in range of 3 to 15 V. Serves as compact low-power portable RF oscillator having low battery drain.—W. J. Prudhomme, CMOS Oscillators, *73 Magazine,* July 1977, p 60–63.

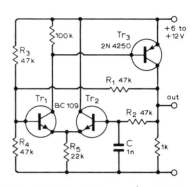

20-kHz ASTABLE—Single-capacitor circuit is reliable over wide range of temperatures, voltages, and transistor gains. Frequency varies only by 0.05% for supply voltage changes between 6 and 12 V. Timing can be changed with R_1, R_2, and C. Duty cycle depends on ratio of R_3 to R_4, and is 50% for values shown.—C. Horwitz, Tolerant Astable Circuits, *Wireless World,* Feb. 1975, p 93.

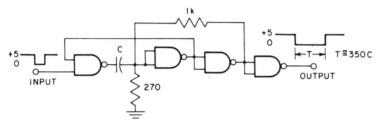

FOUR-GATE MONO—NAND-gate mono using Texas Instruments SN7400 package provides cleaner, more stable output. Feedback resistor eliminates tendency to oscillate. Output pulse width T is equal to 1.3 RC; when R is 270 ohms, T is 350 C. Input pulse widths over 30 ns can initiate output. C can be 100 pF to 100 μF.—J. E. McAlister, Single NAND Package Improves One-Shot, *EEE Magazine,* Aug. 1970, p 78.

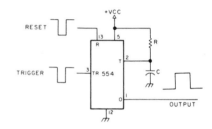

554 MONO—Uses one section of 554 quad monostable timer, connected to give output pulse for negative-going trigger pulse. Width of output pulse in seconds is equal to RC. Trigger must be narrower than output pulse. VCC is 4.5–16 V at 3–10 mA.—H. M. Berlin, IC Timer Review, *73 Magazine,* Jan. 1978, p 40–45.

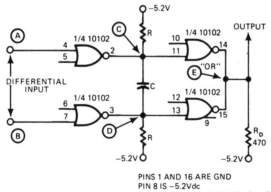

PINS 1 AND 16 ARE GND
PIN 8 IS −5.2Vdc
BYPASS PIN 8 TO GND WITH 0.1 μF

$t_w = 0.2RC$

BIDIRECTIONAL MONO—Requires only one IC, three resistors, and one capacitor. Will trigger on both positive- and negative-going transitions, as required in critical timing applications involving pulses narrower than 50 ns. Capacitor alternately discharges through one pulldown resistor to threshold, then the other. Output gates are tied together to form common output. Width of pulse is defined by values of components.—W. A. Palm, Bidirectional ECL One-Shot Uses a Single IC, *EDN Magazine,* Jan. 5, 1977, p 41–42.

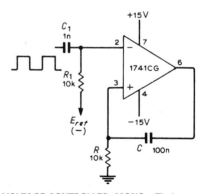

VOLTAGE-CONTROLLED MONO—Timing period of opamp operating as monostable multivibrator is controlled by magnitude of DC reference voltage. With square-wave input shown, differentiating action by C_1-R_1 gives positive pulses that cause mono to make transitions. Article gives design equation and typical waveforms.—G. B. Clayton, Experiments with Operational Amplifiers, *Wireless World,* May 1973, p 241–242.

DUAL-EDGE TRIGGERING—Although 9602 multivibrator IC can be triggered normally either on leading or falling edge of square wave, but not on both, addition of two resistors and one capacitor provides double-edge triggering. When input goes low, negative-going pulse through C_1 triggers 9602 and makes it deliver one output pulse. When input goes high again, high-going pulse is delivered directly to pin 12 of 9602, triggering it again so it produces another pulse.—J. P. Yang, Circuit Triggers One-Shot on Both Edges of Square Wave, *EDN Magazine*, Nov. 15, 1972, p 49.

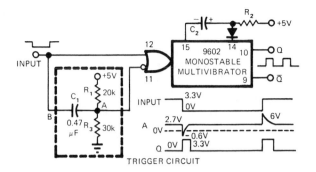

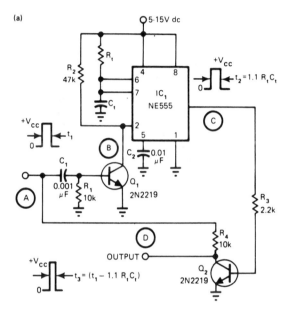

PULSE-WIDTH DETECTOR—Connections as shown for 555 timer give output only if trigger pulse width is greater than time constant ($t_2 = 1.1R_tC_t$) of mono MVBR circuit. Q_1 is normally off. Pin 2 of 555 is then high. At start of trigger pulse, output at point C is low. Positive trigger drives Q_1 on for time determined by R_1C_1, feeding negative-going pulse to trigger pin 2. Timer then acts as normal mono, driving Q_2 on for time t_2. If input pulse is still high at end of t_2, it appears at output D since Q_2 is now off. Output pulse width is thus equal to input trigger width less $1.1R_tC_t$. For greater accuracy, insert delay between point A and R_4 equal to inherent propagation delay of timer.—S. Sarpangal, Build a Pulse-Width Detector with a 555 Timer, *EDN Magazine*, Oct. 5, 1977, p 93 and 96.

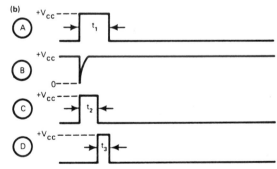

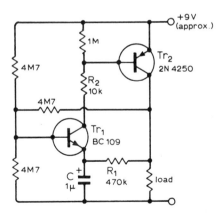

2-Hz ASTABLE PULSER—Single-capacitor circuit operates at very low duty cycles, in range of 10% to 1%. Battery drain is low because off current is about 1 μA for 50-mA on current. R_2 and C determine on time, while R_1 and C set off time. Circuit pulses about twice per second, which is suitable for animal temperature and heart-rate studies. Can be used with implanted transmitters operating from single mercury button cell for more than one year with suitable resistor values.—C. Horwitz, Tolerant Astable Circuits, *Wireless World*, Feb. 1975, p 93.

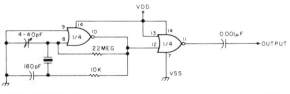

CRYSTAL WITH NOR GATES—Uses two sections of CD4001 quad NOR gate to give mono multivibrator operating in frequency range from 10 kHz up to top limit of about 10 MHz, with exact frequency depending on values used for R and C.—W. J. Prudhomme, CMOS Oscillators, *73 Magazine*, July 1977, p 60–63.

PWM MONO—Circuit provides pulse-width modulation with high duty cycles and complementary output. Strobe input to gate G_1 drives output of gate to binary 0, turning Q_1 off and letting voltage across C_1 build up until UJT Q_3 fires, discharging C_1. Output of UJT drives output of G_2 to binary 0. Article gives timing diagrams.—G. Lewis, Simple One Shot Has Complementary Outputs, *EEE Magazine,* Oct. 1970, p 78–79.

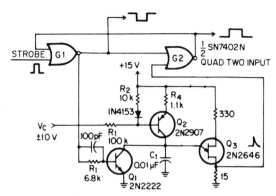

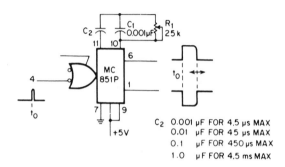

VARIABLE PULSE WIDTH—R_1 and C_2 together provide wide range of pulse widths from Motorola MC851P mono. Rise and fall times of complementary output pulses are better than 100 μs. With only four switched capacitors in combination with R_1, pulse widths can be varied between maximum of 4.5 ms and minimum well under 4.5 μs.—C. W. Stoops, Wide-Range Variable Pulse-Width Monostable, *EEE Magazine,* Dec. 1970, p 56.

LOW-POWER TTL MONO—Simple monostable circuit using DM74L03 draws only 800-μA standby current yet delivers pulses up to 1 s wide. Uses RC time control and regenerative feedback, with values of C_2 and C_3 determining frequency. Pulse width increases from 0.1 s to 0.55 s as C_2 and C_3 are increased from 10 μF to 60 μF.—C. Gilbert and C. Davis, LPTTL One-Shot Yields Wide, Clean Pulses, *EDN/EEE Magazine,* May 15, 1971, p 47–48.

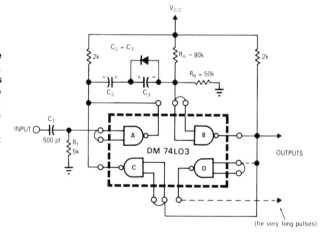

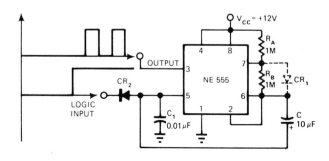

LOW OUTPUT FOR POWER-ON—Logic signal controls both turn-on and turnoff of 555 timer used as oscillator. When input signal at cathode of CR_2 goes low, oscillator remains off and output at pin 3 is low. When input goes high, oscillator starts with its first state low so there are no initial pulse errors.—K. D. Dighe, Rearranged Components Cut 555's Initial-Pulse Errors, *EDN Magazine,* Jan. 5, 1978, p 82 and 84.

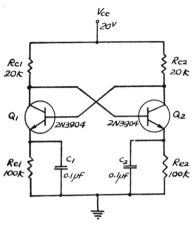

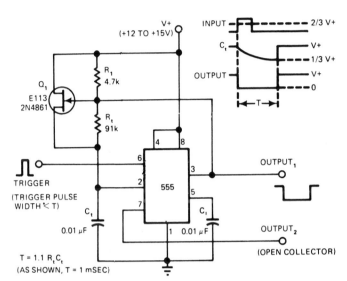

DIRECT-COUPLED ASTABLE—Collectors and bases of both emitter-biased transistors are directly coupled to each other. Switching action takes place by means of capacitor in each emitter circuit. Triangle waves are generated at emitters. Neither transistor can remain permanently cut off. Instead, circuit has two quasi-states, with switching action achieved by charging and discharging capacitor between these states. Single 0.1-μF capacitor can be used between emitters in place of C_1 and C_2.—S. Chang, Two New Direct-Coupled Astable Multivibrators, *Proceedings of the IEEE,* March 1973, p 390–391.

NEGATIVE-GOING DUAL-OUTPUT 555—Circuit triggers on positive-going pulses and delivers negative-going output timing pulses. C_1 charges when JFET switch Q_1 is held on by high output state of timer. When output of timer goes low, C_t discharges to ground through R_t. Timing accuracy is good, and duty cycles above 99% possible without jitter.—W. G. Jung, Take a Fresh Look at New IC Timer Applications, *EDN Magazine,* March 20, 1977, p 127–135.

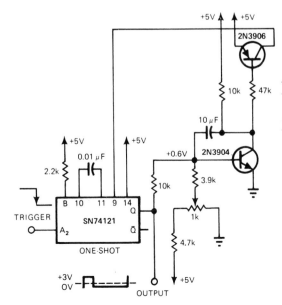

DUTY-CYCLE CONTROL—Feedback loop through two transistors automatically adjusts timing of MVBR to hold duty cycle constant over wide range of triggering rates. 2N3904 acts as integrator with time constant much longer than pulsing period. If duty cycle increases or decreases, current into integrator becomes positive or negative and DC voltage at its collector slowly decreases or increases. This collector voltage drives 2N3906 operating as current generator for adjusting automatically to give chosen duty cycle as selected by 1K pot. Range is 17% to above 50%.—J. L. Engle, Regulate Duty Cycle Automatically, *EDN Magazine,* Nov. 5, 1978, p 122.

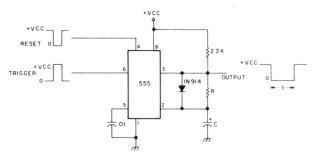

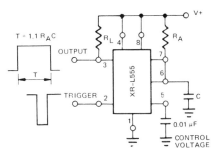

NEGATIVE-OUTPUT MONO—Timer is wired to give negative output pulse for positive-going input trigger pulse. Width of output pulse in seconds is 1.1RC. Input pulse must be narrower than desired output pulse width. When reset pin is momentarily grounded, output returns to stable state. VCC is 4.5–16 V at 3–10 mA.—H. M. Berlin, IC Timer Review, *73 Magazine,* Jan. 1978, p 40–45.

MICROPOWER MONO—Uses Exar XR-L555 having typical power dissipation of only 900 μW at 5 V, serving as direct replacement for 555 timer in micropower circuits. Time delay is controlled by one external resistor and one capacitor (R_A and C) which determine output pulse duration. Can be triggered or reset on falling waveform. Output will drive TTL circuits or source up to 50 mA.—"Timer Data Book," Exar Integrated Systems, Sunnyvale, CA, 1978, p 7–8.

SECTION 42
Music Circuits

Includes organ, piano, trombone, bell, theremin, bird-call, and other sound and music synthesizer circuits, along with circuits giving warble, fuzz, three-part harmony, reverberation, tremolo, attack, decay, rhythm, and other musical effects. Joystick control for music, active filters, contact-pickup preamp, metronomes, and tuning aids are also given.

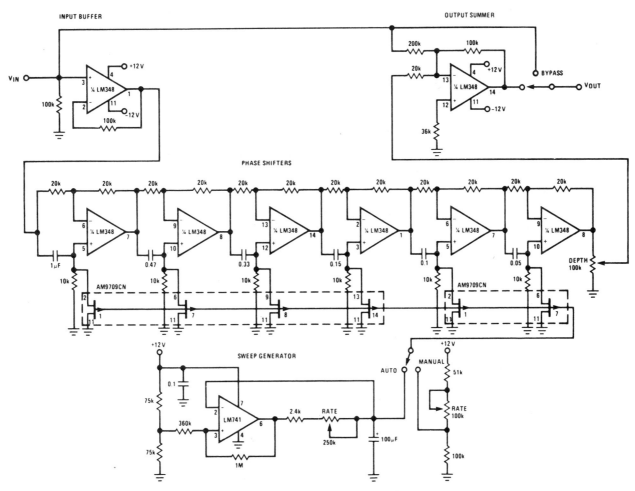

SIMULATION OF FLANGING—Sound-effect circuit sometimes called phase shifter simulates playing of two tape recorders having same material while varying speed of one by pressing on flange of tape reel. Resulting time delay causes some signals to be summed out of phase and canceled. Effect is that of rotating loudspeaker or of Doppler characteristic. Uses two LM348 quad opamps, two AM9709CN quad JFET devices, and one LM741 opamp. Phase-shift stages are spaced one octave apart from 160 to 3200 Hz in center of audio spectrum, with each stage providing 90° shift at its frequency. JFETs control phase shifters. Gate voltage of JFETs is adjusted from 5 V to 8 V either manually with foot-operated rheostat or automatically by LM741 triangle-wave generator whose rate is adjustable from 0.05 Hz to 5 Hz.—"Audio Handbook," National Semiconductor, Santa Clara, CA, 1977, p 5-10–5-11.

AUDIBLE/VISIBLE METRONOME—Produces uniformly spaced beats in synchronism with flashes of LED, at rate that can be adjusted with R1 from one beat every few seconds to ten or more beats per second. Use red Radio Shack 276-041 or similar LED. Add switch in series with battery to avoid disturbing setting of R1. R3 serves as volume control. Add 5–10 μF capacitor across loudspeaker to mellow beat sound if desired.—F. M. Mims, "Electronic Music Projects. Vol. 1," Radio Shack, Fort Worth, TX, 1977, 2nd Ed., p 55–59.

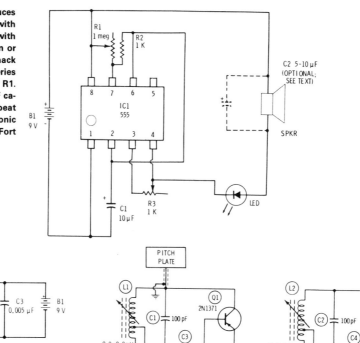

HAND-WAVING THEREMIN—Single-transistor RF oscillator is tuned to generate frequency about 455 kHz above oscillator frequency of transistor radio. With aluminum-foil antenna away from nearby objects and radio tuned between stations, R1 is adjusted until high-pitched tone is heard from radio. Now, as hand is brought toward and away from foil antenna, wailing sounds are produced. With practice, musician can produce recognizable melodies by vibrating hand. Primary controls of frequency are C1 (10–365 pF broadcast radio tuning capacitor) and adjustable antenna coil L1 (Radio Shack 270-1430). Radio can be up to 15 feet away from theremin. Rotate radio for maximum pickup from L1.—F. M. Mims, "Electronic Music Projects, Vol. 1," Radio Shack, Fort Worth, TX, 1977, 2nd Ed., p 81–89.

THEREMIN—Two transistor oscillator stages generate separate low-power RF signal in broadcast band, for pickup by AM broadcast receiver. Movement of hand toward or away from metal pitch plate varies frequency of Q1, making audio output of receiver vary correspondingly as beat frequency changes. Both circuits are Hartley oscillators, using Miller 9012 or equivalent slug-tuned coils. To adjust initially, place next to radio and set tuning slug of L1 about two-thirds out of its winding. Set slug of L2 about one-third out of its winding. Tune radio until either oscillator signal is heard. Signal can be identified by whistle if on top of broadcast station or by quieting of background noise if between stations. Adjust slugs so whistle is heard at desired location of quieting signal. Pitch of whistle should change now as hand is brought near pitch plate.—J. P. Shields, "How to Build Proximity Detectors & Metal Locators," Howard W. Sams, Indianapolis, IN, 2nd Ed., 1972, p 154–156.

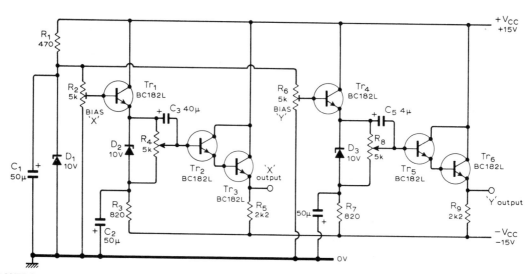

JOYSTICK CONTROL—Mechanically controlled voltage source generates two independent control voltages, proportional to stick position, to serve as one of controls for elaborate sound synthesizer used for generating wide variety of musical and other sounds. Three-part article describes circuit operation and gives all other circuits used in synthesizer.—T. Orr and D. W. Thomas, Electronic Sound Synthesizer, *Wireless World*, Part 3—Oct. 1973, p 485–490 (Part 1—Aug. 1973, p 366–372; Part 2—Sept. 1973, p 429–434).

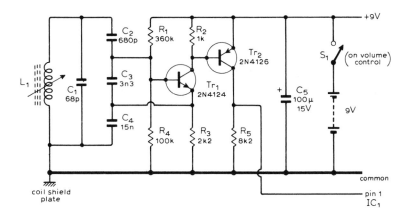

TUNING FOR EQUAL TEMPERAMENT—Instrument described enables anyone to tune such instruments as organ, piano, and harpsichord in equal temperament with accuracy approaching that of professional tuner. Only requirement is ability to hear beats between two tones sounded together. Master oscillator circuit shown generates 250.830 kHz for feeding to first of five ICs connected as programmable divider that provides 12 notes of an octave as 12 equal semitones differing from each other by factor of 1.0594. Article gives suitable power amplifier to fit along with divider connections and detailed instructions for construction, calibration, and use.—W. S. Pike, Digital Tuning Aid, *Wireless World,* July 1974, p 224–227.

TREMOLO CONTROL—National LM324 opamp connected as phase-shift oscillator operates at variable rate between 5 and 10 Hz set by speed pot. Portion of oscillator output is taken from depth pot and used to modulate ON resistance of two 1N914 diodes operating as voltage-controlled attenuators. Input should be kept below 0.6 V P-P to avoid undesirable clipping. Used for producing special musical effects.—"Audio Handbook," National Semiconductor, Santa Clara, CA, 1977, p 5-11–5-12.

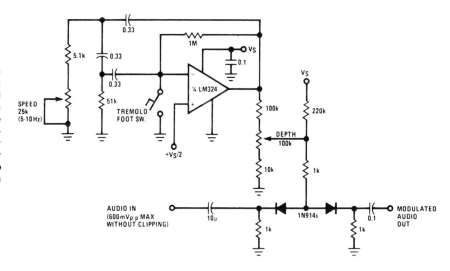

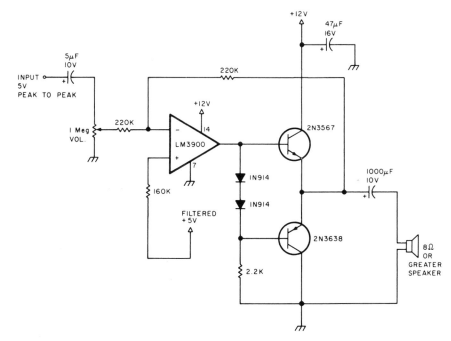

AUDIO FOR COMPUTER MUSIC—Wideband low-power audio amplifier was developed for use with DAC and low-pass active filter to create music with microprocessor.—H. Chamberlin, A Sampling of Techniques for Computer Performance of Music, *BYTE,* Sept. 1977, p 62–66, 68–70, 72, 74, 76–80, and 82–83.

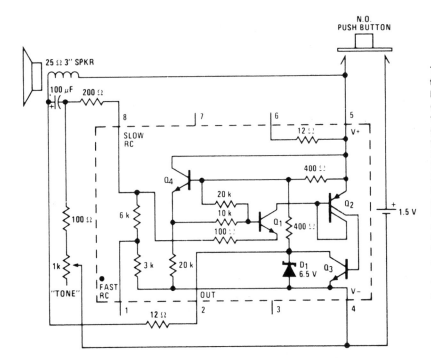

TROMBONE CIRCUIT—Unique arrangement for driving 25-ohm loudspeaker with National LM3909 IC operating from 1.5-V cell permits generation of slide tones resembling those of trombone. Operation is based on use of voltage generated by resonant motion of loudspeaker voice coil as major positive feedback for IC. Loudspeaker is mounted in roughly cubical box having volume of about 64 in³, with one end of box arranged to slide in and out like piston. Positioning of piston and operation of pushbutton permit playing reasonable semblance of simple tune. IC, loudspeaker, and battery are mounted on piston, with 2½-in length of ⁵⁄₁₆-in tubing provided to bleed air in and out as piston is moved, without affecting resonant frequency. Frequency of oscillator becomes equal to resonant frequency of enclosure.—"Linear Applications, Vol. 2," National Semiconductor, Santa Clara, CA, 1976, AN-154, p 6.

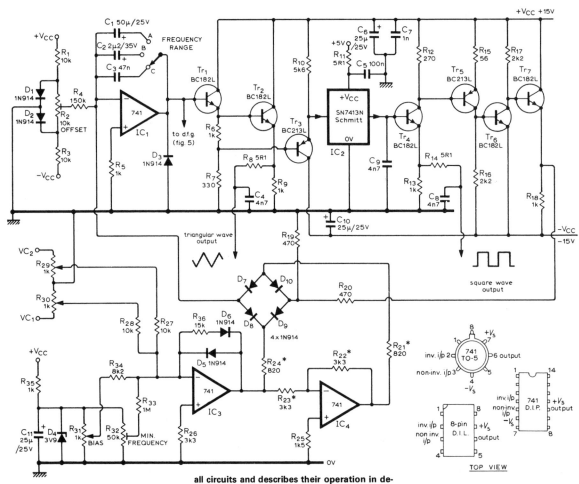

VCO SOUND SYNTHESIZER—Developed for use in instrument capable of duplicating variety of sounds ranging from bird distress calls and engine noises to spoken words and wide variety of musical instruments. Three-part article gives all circuits and describes their operation in detail. Heart of oscillator is triangle and squarewave generator built around IC Schmitt trigger. Ramp rate and operating frequency are varied by changing drive voltage or gain of integrator. Similar VCO in synthesizer also produces sine, pulse, and ramp waveforms.—T. Orr and D. W. Thomas, Electronic Sound Synthesizer, *Wireless World*, Part 1—Aug. 1973, p 366–372 (Part 2—Sept. 1973, p 429–434; Part 3—Oct. 1973, p 485–490).

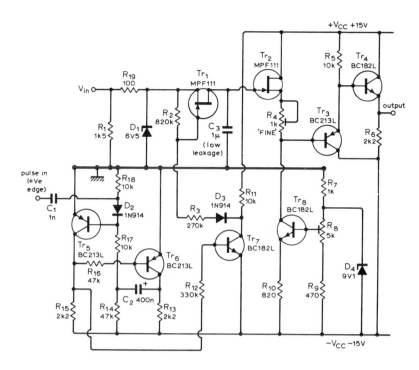

ANALOG MEMORY—Used in synthesizer for generating wide variety of musical and other sounds, to provide constant control signal for sounds requiring long fadeout. Positive input pulse initiates sampling of analog signal for preset time, with signal being held for unspecified period. Input voltage range is from about −0.5 V to +6.5 V, being deliberately limited by D_1. Three-part article describes operation in detail and gives all other circuits used in synthesizer.—T. Orr and D. W. Thomas, Electronic Sound Synthesizer, *Wireless World,* Part 3—Oct. 1973, p 485–490 (Part 1—Aug. 1973, p 366–372; Part 2—Sept. 1973, p 429–434).

TREMOLO AMPLIFIER—Provides amplitude modulation at subaudio rate (usually between 5 and 15 Hz) of audio-frequency input signal. Uses National LM389 array having three transistors along with power amplifier. Transistors form differential pair having active current-source tail to give output proportional to product of two input signals. Gain control pot is adjusted for desired tremolo depth. Interstage RC network forms 160-Hz high-pass filter, requiring that tremolo frequency be less than 160 Hz.— "Audio Handbook," National Semiconductor, Santa Clara, CA, 1977, p 4-33–4-37.

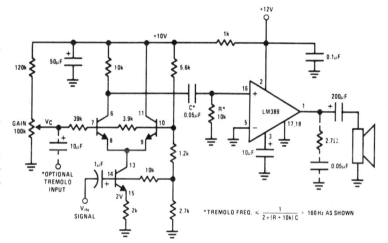

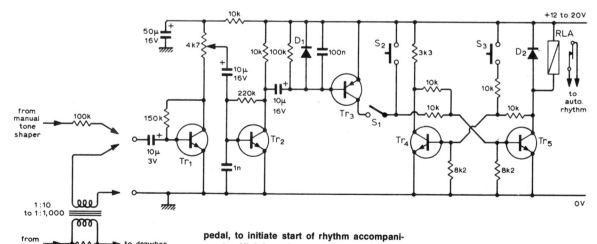

AUTOMATIC REMOTE RHYTHM CONTROL— When added to electronic organ, circuit is activated by audio signal from lower manual or pedal, to initiate start of rhythm accompaniment. High-impedance input connection through 100K is made to toneshaper output, and transformer connection is used with electromechanical Hammond organ. Transistor and diode types are not critical. If S_1 is closed, current passes through to Tr_5 and triggers bistable that pulls in relay. S_2 and S_3 are used for manual start and stop of rhythm.—K. B. Sorensen, Touch Start of Automatic Rhythm Device, *Wireless World,* Oct. 1974, p 381.

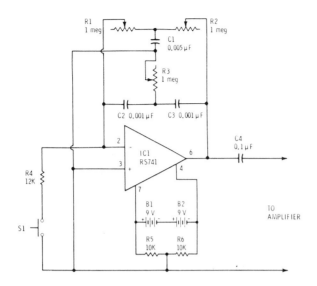

MUSICAL BELLS—Opamp connected as active filter simulates attack followed by gradual decay as produced when bell or tuning fork is struck. Filter portion of circuit uses twin-T network adjusted so active filter breaks into oscillation when slight external disturbance is introduced by closing S1 momentarily. Circuit feeds external audio amplifier and loudspeaker for converting ringing frequency into audible sound. Set R3 just below oscillation point. R1 and R2 can be adjusted to give sounds of other musical instruments, such as drums, bamboo, and triangles.—F. M. Mims, "Electronic Music Projects, Vol. 1," Radio Shack, Fort Worth, TX, 1977, 2nd Ed., p 71–80.

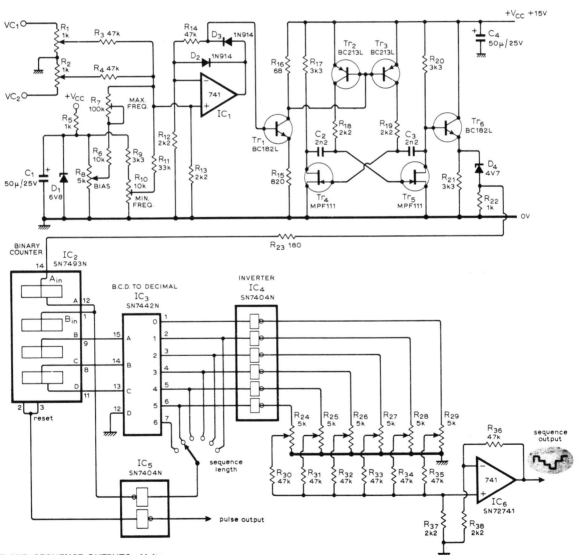

PULSE AND SEQUENCE OUTPUTS—Voltage-controlled oscillator produces sequence of steps, with amplitude of each step individually controllable up to maximum of six steps. Circuit also generates series of pulses having 1:1 mark-space ratio, each coincident with leading edge of a step. Pair of summing inputs controls os-cillator, with exponential frequency-voltage relationship extending in one range from subsonic frequencies to over 20 kHz. Used in sound synthesizer described in three-part article that gives all circuits and operating details. Applications include synthesizing sounds ranging from bird distress calls and engine noises to spoken words and wide variety of musical instruments.—T. Orr and D. W. Thomas, Electronic Sound Synthesizer, *Wireless World,* Part 2—Sept. 1973, p 429–434 (Part 1—Aug. 1973, p 366–372; Part 3—Oct. 1973, p 485–490).

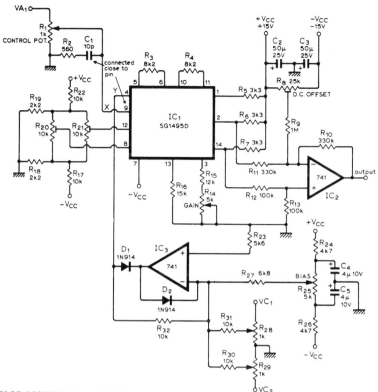

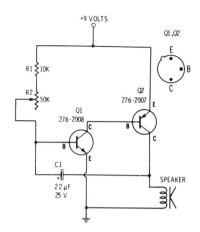

CLICKING METRONOME—Basic lamp-flashing circuit is used to produce sharp click in loudspeaker each time Q2 is turned on by RC oscillator Q1. R2 adjusts repetition rate over range of 20—280 beats per minute. Changing value of C1 varies tone of clicks.—F. M. Mims, "Transistor Projects, Vol. 1," Radio Shack, Fort Worth, TX, 1977, 2nd Ed., p 33—39.

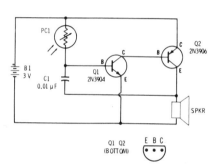

LIGHT-SENSITIVE THEREMIN—Tone of loudspeaker increases and decreases in frequency as flashlight is moved in vicinity of photocell in darkened room. Use Radio Shack 276-116 cadmium sulfide photocell. Cell resistance decreases with light, increasing frequency of audio oscillator. Continuously changing frequency resembles that produced by hand-controlled theremin.—F. M. Mims, "Electronic Music Projects, Vol. 1," Radio Shack, Fort Worth, TX, 1977, 2nd Ed., p 91—95.

VOLTAGE-CONTROLLED AMPLIFIER—Gain is linearly controlled by sum of input control voltages and a bias voltage, to provide amplitude modulation as required for synthesizer used to generate wide variety of sounds. Heart of circuit is linear four-quadrant multiplier IC. Output is taken between two load resistors, with differential amplifier IC_2 removing common-mode signal. Article describes operation in detail and gives all other circuits of synthesizer, along with procedure for aligning preset controls R_8, R_{14}, R_{20}, and R_{21}.—T. Orr and D. W. Thomas, Electronic Sound Synthesizer, Wireless World, Part 2—Sept. 1973, p 429—434 (Part 1—Aug. 1973, p 366—372; Part 3—Oct. 1973, p 485—490).

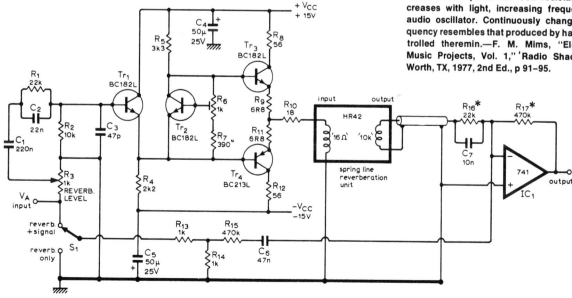

REVERBERATION—Used in sound synthesizer developed for generating wide variety of musical and other sounds. Four-transistor driver feeds spring-type reverberation unit at up to about 4 kHz, with switch giving choice of reverberation only or reverberation combined with input signal at V_A. Amount of reverberation can be controlled manually with R_3 or automatically with voltage-controlled amplifier or voltage-controlled filter of synthesizer. Three-part article gives all circuits and describes operation in detail.—T. Orr and D. W. Thomas, Electronic Sound Synthesizer, Wireless World, Part 2—Sept. 1973, p 429—434 (Part 1—Aug. 1973, p 366—372; Part 3—Oct. 1973, p 485—490).

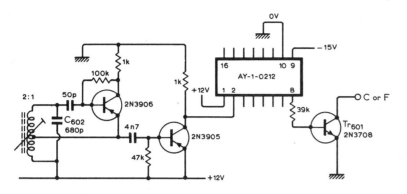

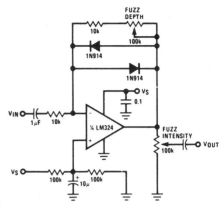

PIANO TONE GENERATOR—RF oscillator combined with General Instrument AY-1-0212 IC master tone generator replaces 12 conventional RC oscillators otherwise required in electronic piano. Frequencies generated are within 0.1% of equal-temperament scale, so piano will work well without being tuned. Three-part article gives all circuits and construction details for simple portable touch-sensitive electronic piano.—G. Cowie, Electronic Piano Design, *Wireless World,* Part 3, May 1974, p 143–145.

FUZZ CIRCUIT—Two diodes in feedback path of LM324 opamp create musical-instrument effect known as fuzz by limiting output voltage swing to ±0.7 V. Resultant square wave contains chiefly odd harmonics, resembling sounds of clarinet. Fuzz depth pot controls level at which clipping begins, and fuzz intensity pot controls output level.—"Audio Handbook," National Semiconductor, Santa Clara, CA, 1977, p 5-11.

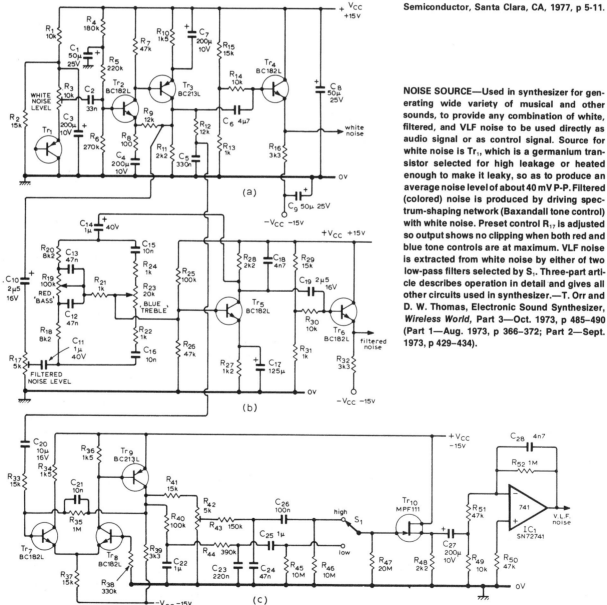

(a)

(b)

(c)

NOISE SOURCE—Used in synthesizer for generating wide variety of musical and other sounds, to provide any combination of white, filtered, and VLF noise to be used directly as audio signal or as control signal. Source for white noise is Tr_1, which is a germanium transistor selected for high leakage or heated enough to make it leaky, so as to produce an average noise level of about 40 mV P-P. Filtered (colored) noise is produced by driving spectrum-shaping network (Baxandall tone control) with white noise. Preset control R_{17} is adjusted so output shows no clipping when both red and blue tone controls are at maximum. VLF noise is extracted from white noise by either of two low-pass filters selected by S_1. Three-part article describes operation in detail and gives all other circuits used in synthesizer.—T. Orr and D. W. Thomas, Electronic Sound Synthesizer, *Wireless World,* Part 3—Oct. 1973, p 485–490 (Part 1—Aug. 1973, p 366–372; Part 2—Sept. 1973, p 429–434).

ATTACK-DECAY GENERATOR—Designed for polytonic electronic music system handling more than one note at a time. Each note to be controlled is sent through voltage-controlled amplifier (VCA) whose gain is set by charge on capacitor. Attack is changed by varying charging rate. Discharge rate sets decay of individual note. To avoid having separate adjustment pot for each VCA, duty-cycle modulation is used to change charging current through resistors. Attack pulses are generated by upper three inverters forming variable-symmetry astable MVBR. Decay pulses are generated by lower three inverters connected as half-mono MVBR. Additional half-monos can be added as needed

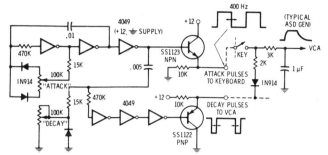

for percussion, snubbing, and other two-step decay effects.—D. Lancaster, "CMOS Cookbook," Howard W. Sams, Indianapolis, IN, 1977, p 231–232.

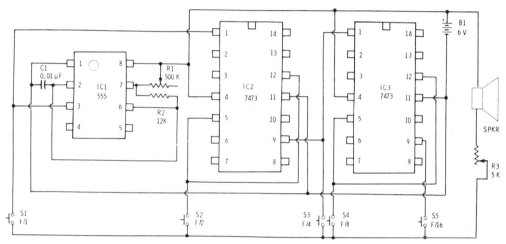

FOUR-OCTAVE ORGAN—Two 7473 dual flip-flops provide four frequency dividers for 555 timer connected as master tone generator. S1 gives fundamental frequency, and each succeeding switch gives tones precisely one octave lower. Four organ applications, pushbutton switches are added to timer circuit for switching frequency-controlling capacitors or resistors to give desired variety of notes.—F. M. Mims, "Electronic Music Projects, Vol. 1," Radio Shack, Fort Worth, TX, 1977, 2nd Ed., p 45–53.

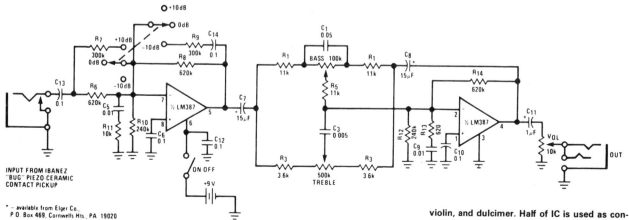

*—available from Elger Co.,
P.O. Box 469, Cornwells Hts., PA 19020

PREAMP FOR ACOUSTIC PICKUP—National LM387 dual opamp provides switchable gain choice of ±10 dB along with bass/treble tone control and volume control. Used with flat-response piezoceramic contact pickup for acoustic stringed musical instruments such as guitar, violin, and dulcimer. Half of IC is used as controllable gain stage, and other half is used as active two-band tone-control block.—"Audio Handbook," National Semiconductor, Santa Clara, CA, 1977, p 5-12.

SECTION 43
Noise Circuits

Includes many types of noise limiters, blankers, and filters for audio, IF, RF, and digital applications, along with suppression of noise from arcing contacts and motors. Circuits for white-noise and pink-noise test-signal generators are also given.

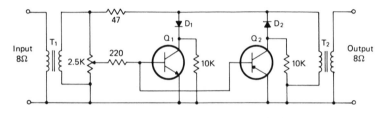

AF-POWERED CLIPPER—Designed for use just ahead of 8-ohm loudspeaker in receiver covering lower amateur phone bands (75 and 40 meters). Reduces hissing noise caused by short-wave diathermy, electric motors, and fluorescent lighting, as well as impulse noise generated by auto ignition system or atmospheric interference. T_1 and T_2 are transistor radio output transformers with 500:4 or 600:8 ohm impedance. Q_1 is 2N2222 NPN transistor. Q_2 is 2N2907 PNP transistor. D_1 and D_2 are 1N270.—C. Laster, An Audio Powered Noise Clipper, *CQ*, May 1976, p 26–27.

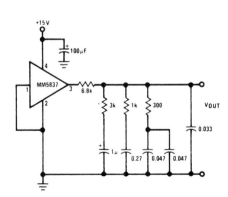

PINK-NOISE GENERATOR—Uses MM5837 broadband white-noise generator with −3 dB per octave filter from 10 Hz to 40 kHz to give pink-noise output having flat spectral distribution over entire audio band from 20 Hz to 20 kHz. Output is about 1 V P-P of pink noise riding on 8.5-VDC level. Used as controlled source of noise for adjusting octave equalizer to optimum settings for specific listening area.—"Audio Handbook," National Semiconductor, Santa Clara, CA, 1977, p 2-53–2-59.

AM NOISE SILENCER—Circuit samples mixer output (IF input) of AM receiver and, when noise pulse is detected, interrupts IF input signal for duration of noise pulse. Uses National LM372 IC having AGC loop with range of about 69 dB, for accommodating wide range of input levels. Article describes operation of circuit in detail. For frequencies above 2 MHz, use LM373 in place of LM372.—T. A. Tong, Noise Silencer for A.M. Receivers, *Wireless World*, Oct. 1972, p 483–484.

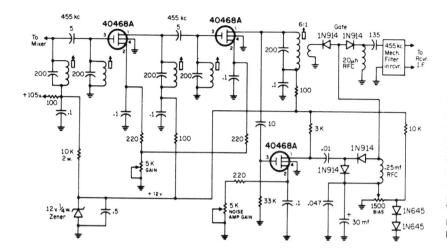

IF NOISE BLANKER—Used ahead of 455-kHz IF strip of communication receiver to provide about 40-dB attenuation of ignition and other noise pulses that can interfere with reception in 2- and 6-meter amateur bands. Two paths for noise pulses, one AC and the other DC, must be balanced for good operation. Resistor and capacitor values in noise rectifier are chosen to select sharp noise pulses in preference to signals. DC noise pulses are amplified by pulse amplifier and converted to AC noise pulses. Settings of pots are optimized for best noise blanking. Circuit requires 12-V supply, which can be obtained from receiver with appropriate dropping resistors and zener as shown for +105 V, or from separate source.—F. C. Jones, Experimental I.F. Noise Blankers, *CQ*, March 1971, p 81–83.

EXCESS-NOISE SOURCE—Develops about 18 dB of excess noise in region of 50–300 MHz for optimizing converter or receiver for best noise figure. Can also be used for noise optimizing of TV receivers and for peaking UHF TV front ends. Q1 and Q2 form cross-coupled 700-MHz MVBR. C1 is greater than C2 to favor conduction of Q2. When Q2 is on, Q3 turns on and makes current flow through broadband noise diode CR1. Diode is forward-biased because available gating voltage does not generate enough noise in reverse-bias mode. If noise output is too great, insert 2000-ohm attenuator as shown.—T. E. Hartson, A Gated Noise Source, *QST*, Jan. 1977, p 22–23.

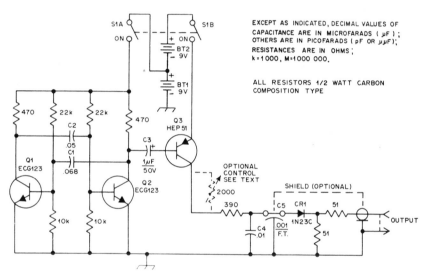

EXCEPT AS INDICATED, DECIMAL VALUES OF CAPACITANCE ARE IN MICROFARADS (μF); OTHERS ARE IN PICOFARADS (pF OR μμF); RESISTANCES ARE IN OHMS; k=1 000, M=1000 000.

ALL RESISTORS 1/2 WATT CARBON COMPOSITION TYPE

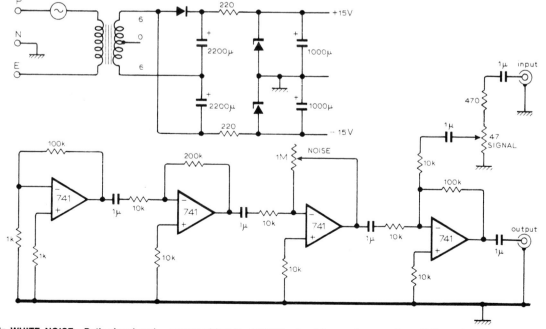

50–5000 Hz WHITE NOISE—Both signal and noise levels are continuously and independently variable from zero to maximum in simple noise generator developed to demonstrate recovery of low-level 500-Hz signal from noise. Circuit gives maximum noise output into 1500-ohm load; for lower load impedances, reduce noise level to prevent oscillation. Opamps require ±15 V supply, which can be simple voltage doubler without regulation.—J. E. Morris, Simple Noise Generator, *Wireless World*, April 1977, p 62.

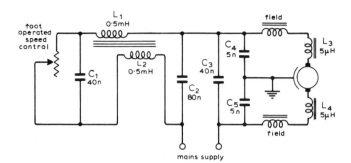

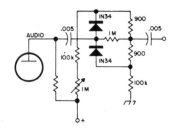

SEWING-MACHINE SUPPRESSION—Circuit is used to suppress clicks from speed control as well as interference produced by motor itself in sound and television broadcast bands.—A. S.

McLachlan, J. H. Ainley, and R. J. Harry, Radio Interference—a Review, *Wireless World,* June 1974, p 191–195.

AF NOISE LIMITER—Trough limiter eliminates background noise that is normally passed by conventional limiters, to permit use of higher volume level without annoying static when monitoring single radio channel continuously.—Circuits, *73 Magazine,* Dec. 1973, p 120.

ZENER GENERATOR—Uses National LM389 array having three transistors along with opamp. Application of reverse voltage to emitter of one grounded-base transistor breaks it down in avalanche mode to give action of zener diode. Reverse voltage characteristic, typically 7.1 V, is used as noise source for amplification by second transistor and power opamp. Third transistor (not shown) can be used to gate noise generator if desired.—"Audio Handbook," National Semiconductor, Santa Clara, CA, 1977, p 4-33–4-37.

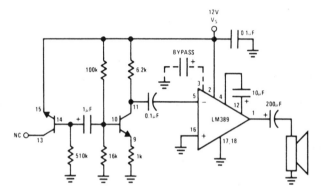

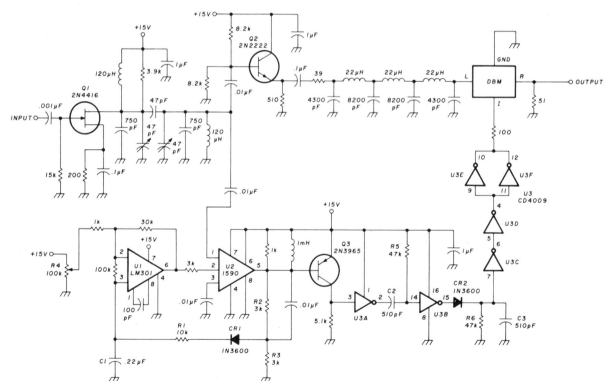

NOISE BLANKER—Minimizes effects of short-duration high-amplitude low-repetition-rate noise such as auto ignition noise, power-line arcing, and make-or-break switching. Developed for use in Collins ARR-41 receiver, where it is inserted between plate of second mixer and 500-kHz first IF amplifier. Q1 and its double-tuned drain circuit form low-gain bandpass amplifier that removes remaining local oscillator signal and sets bandwidth at about 50 kHz. Signal is then split into two channels. Q2 in main channel drives 50-ohm low-pass delay network with 700-kHz cutoff, feeding double-balanced mixer DBM operated as current-controlled attenuator. In other channel, noise amplifier U2 drives pulse detector Q3 and AGC detector CR1. Opamp U1 amplifies AGC and controls gain of U2. R4 is threshold adjustment. Gates U3B and U3C form mono used with gates of U3 to develop proper phase and current amplitude for operating blanking gate.—W. Stewart, Noise Blanker Design, *Ham Radio,* Nov. 1977, p 26–29.

LOW-PASS DIGITAL FILTER—Used to retrieve pulse train data from noisy signal line. Filtering is achieved with SN7400 quad two-input NAND gate, SN7413 dual four-input Schmitt trigger, two diodes, and two capacitors. One gate of SN7400 is used as inverter driving pulse delay operating on negative-going transition of input signal. Other Schmitt trigger, diode, and capacitor provide delay on positive-going transition. Any additional pulses occurring during delay-circuit time-out resets delay time without affecting output.—T. H. Haydon, Low-Pass Digital Filter, *EDN Magazine,* Nov. 20, 1973, p 85.

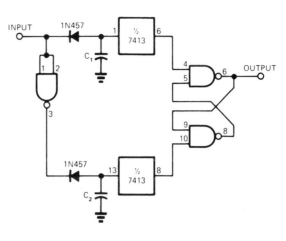

$$C_1 = C_2 = \frac{480}{F + 200} \times 10^{-6}$$
$$F = \text{CUTOFF FREQUENCY}$$

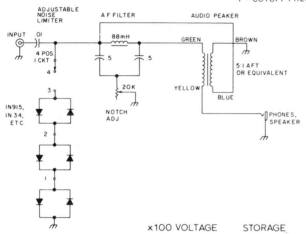

NOISE CONTROL—Circuit is plugged into headphone jack of amateur receiver. Four-position rotary switch selects desired combination of noise-limiting diodes for handling progressively more severe noise pulses. Adjustable AF T-notch filter limits passband over sufficient range for both phone and CW. Inductor is common 88-mH toroid. Audio peaker circuit overcomes insertion losses of filters.—S. T. Rappold, Noise Rejector, *73 Magazine,* Sept. 1977, p 116.

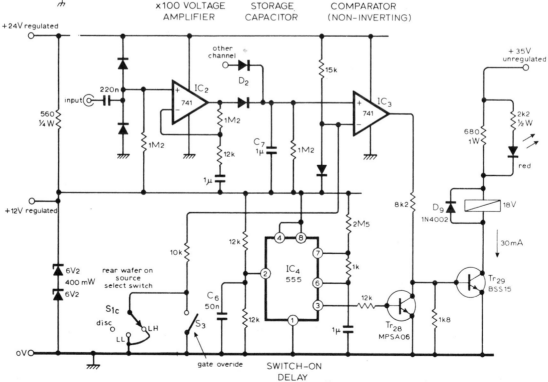

NOISE GATE FOR AF PREAMP—Used in high-performance phono preamplifier to mute output when there is no signal at phono input. Opamps each provide gains of about 100. Circuit controls muting reed relay serving both stereo channels of preamp. Delay switch-on using 555 IC overrides noise-gate opamps. Unmarked diodes are 1N914 or equivalent, and red LED is TIL209 or equivalent. Article covers circuit operation in detail and gives all other circuits of preamp.—D. Self, Advanced Preamplifier Design, *Wireless World,* Nov. 1976, p 41–46.

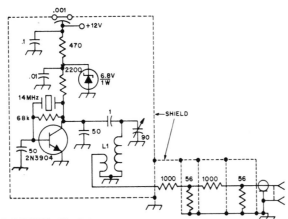

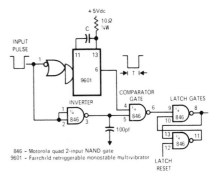

LOW-LEVEL RF SOURCE—Used to measure noise floor of receiver under test. RF source is simple, well-shielded crystal-controlled oscillator that is decoupled from battery supply. After attenuator resistors are adjusted to provide about S7 signal in receiver, oscillator housing is sealed with solder. Once calibrated, RF source is comparable to commercial signal generator, as leakage is quite low. Output is −112 dBm at 14 MHz. L1 has 24 turns enamel on Amidon T50-6 toroid, with 1 turn for output link.—W. Hayward, Defining and Measuring Receiver Dynamic Range, *QST*, July 1975, p 15–21 and 43.

NOISE-RESISTANT LATCH—False triggering of latch gates by noise spikes is prevented by generating pulse T whose width is equal to minimum width of desired input pulse. Values used for RC combination set T. If R is chosen as 10 kilohms, C should be T/3.424, where C is in picofarads and T is in nanoseconds.—S. R. Martin, Latching Circuit Provides Noise Immunity, *EDN/EEE Magazine*, Feb. 1, 1972, p 56.

AF NOISE LIMITER—Operation is similar to that of delay line. Voltage developed across voltage divider at output of 1N34 germanium diode is instantaneous, while DC voltage at output of circuit is delayed. If no pulses are present and 0.1-μF capacitor is not at ground, 1N914 silicon diode will have floating voltage. High positive pulses charge capacitor, and silicon diode shorts audio voltage. Negative pulses disable germanium diode directly. Circuit thus acts as noise blanker in both directions. Used in European communication receivers. Transistor type is not critical.—U. L. Rohde, IF Amplifier Design, *Ham Radio*, March 1977, p 10–21.

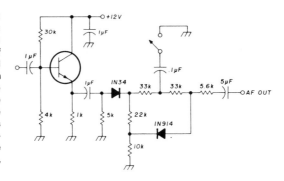

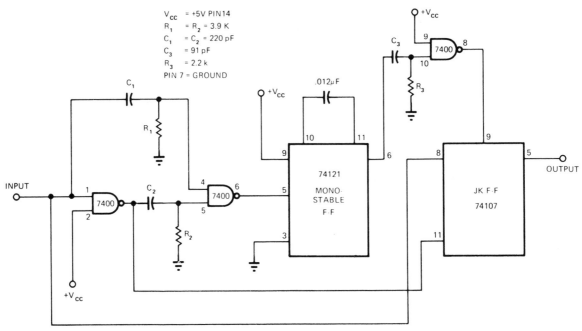

SPIKE REJECTION—Used to eliminate noise that may be present on signal line. Based on sampling input line at fixed time after each detected transition. If transition was due to noise spike, spike will no longer be present and true signal level will be sampled. If transition was caused by desired legitimate signal, sampled waveform represents true signal delayed by pulse width of mono MVBR. Mono pulse width is about 12 μs. Article gives circuit waveforms and describes operation in detail.—A. S. Bozorth, Pulse Verification Yields Good Noise Immunity, *EDN Magazine*, Nov. 5, 1973, p 75 and 77.

SECTION 44
Operational Amplifier Circuits

Versatility of modern opamps is illustrated by variety of amplification, control, signal processing, and other general-purpose functions involving frequencies ranging from DC to many megahertz. More specific applications will be found in practically all other chapters.

VARIABLE DEAD-BAND RESPONSE—Diode bridge in feedback loop of opamp provides controlled amount of dead-band response. As value of R_2 is increased from 0 ohms, voltage developed across R_2 serves to raise dead-band level at which bridge opens and circuit amplifies with normal gain of R_3/R_1. Below dead-band level, bridge is blocked and circuit gain is equal to parallel combination of R_2 and R_3 divided by R_1. Use matched diodes such as CA3019 for peaks below ±7 V; for higher peaks, use 1N914s.—W. G. Jung, "IC Op-Amp Cookbook," Howard W. Sams, Indianapolis, IN, 1974, p 207.

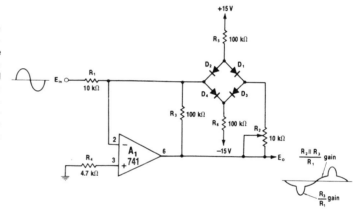

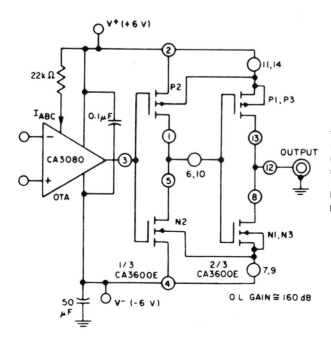

TWO-STAGE POSTAMPLIFIER—Connections shown for CA3600E CMOS transistor-pair array give total open-loop gain of about 160 dB for system. Open-loop slew rate is about 65 V/μs.— "Linear Integrated Circuits and MOS/FET's," RCA Solid State Division, Somerville, NJ, 1977, p 278–279.

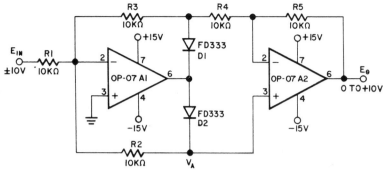

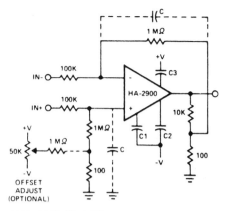

PRECISION ABSOLUTE VALUE—Circuit using two Precision Monolithics OP-07 opamps provides precise full-wave rectification by inverting negative-polarity input voltages and operating as unity-gain buffer for positive-polarity inputs. Applications include positive-peak detectors, single-quadrant multipliers, and magnitude-only measuring systems. For positive inputs, circuit simply operates as two unity-gain am-plifier stages. Negative input turns D1 off and D2 on, changing resistor currents precisely enough to give overall circuit gain of −1. Design equations are given.—D. Soderquist and G. Erdi, "The OP-07 Ultra-Low Offset Voltage Op Amp—a Bipolar Op Amp That Challenges Choppers, Eliminates Nulling," Precision Monolithics, Santa Clara, CA, 1975, AN-13, p 10.

1000 GAIN AT 2 kHz—Uses Harris HA-2900 chopper-stabilized opamp. Either input terminal may be grounded, giving choice of inverting or noninverting operation, or inputs may be driven differentially. Symmetrical input networks eliminate chopper noise, limiting total input noise to about 30 μVRMS when C is 0. Noise can be further reduced, at expense of bandwidth, by adding optional capacitors C as shown. Without these capacitors, bandwidth is 2 kHz.—"Linear & Data Acquisition Products," Harris Semiconductor, Melbourne, FL, Vol. 1, 1977, p 7-69 (Application Note 518).

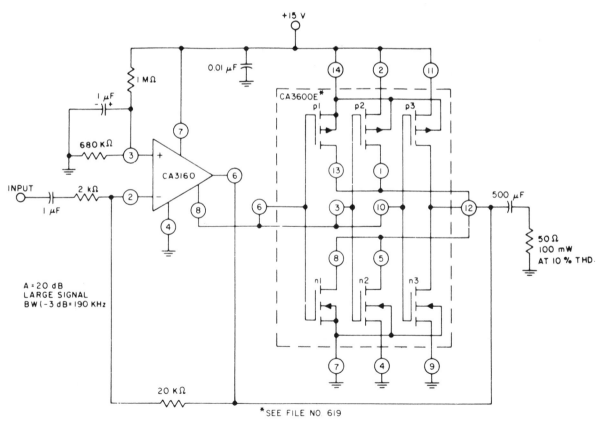

NOTE:
TRANSISTORS p1, p2, p3 AND n1, n2, n3 ARE PARALLEL-CONNECTED WITH Q8 AND Q12, RESPECTIVELY, OF THE CA3160

POWER BOOSTER—CA3600E CMOS transistor array provides parallel-connected transistors for power-boosting capability with CA3160 opamp. Feedback is used to establish closed-loop gain of 20 dB. Typical large-signal bandwidth (−3 dB) is 190 kHz.—"Linear Integrated Circuits and MOS/FET's," RCA Solid State Division, Somerville, NJ, 1977, p 271–273.

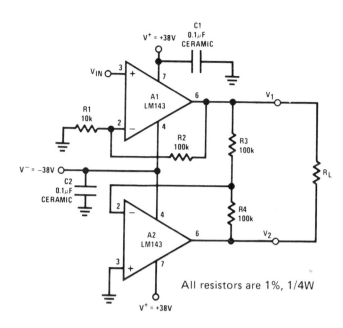

130 V P-P DRIVE—Two LM143 high-voltage opamps operating from 38-V supply can provide up to 138 V P-P unclipped into 10K floating load when connected as shown to give noninverting voltage amplifier followed by unity-gain inverter. Power supplies should be bypassed to ground with 0.1-μF capacitors.—"Linear Applications, Vol. 2," National Semiconductor, Santa Clara, CA, 1976, AN-127, p 1–3.

BUFFERED OPAMP—NPD8301 dual FET is ideal low-offset low-drift buffer for LM101A opamp. Matched sections of FET track well over entire bias range, for improved common-mode rejection.—"FET Databook," National Semiconductor, Santa Clara, CA, 1977, p 6-26–6-36.

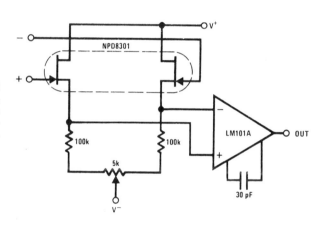

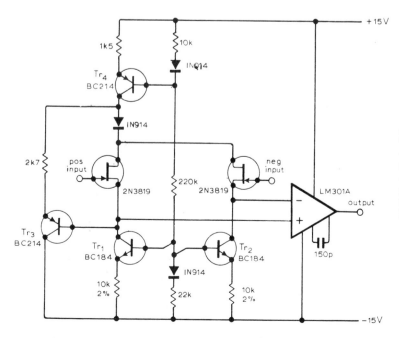

LOW-COST FET-INPUT—Uses two 2N3819 FETs as differential source-follower operated at constant source current of 200 μA provided by Tr_1 and Tr_2. Input performance is comparable to that of more expensive commercial units. Match FETs to reduce thermal drift. Trim input offset voltage to zero by adding resistor in appropriate FET source. Input impedance of circuit is greater than 10^{13} ohms.—J. Setton, F.E.T.-Input Operational Amplifier, *Wireless World*, Nov. 1976, p 61.

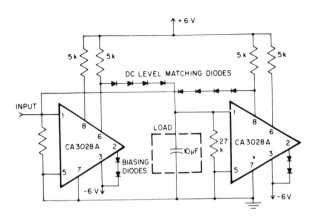

12-μH GYRATOR—Two RCA opamps in gyrator loaded with 10-μF capacitor give effective 12-μH inductor that remains constant in value over range from 10 Hz to almost 1 MHz. Q varies from 1 at 10 Hz to maximum of 500 at 10 kHz. Article gives design equations.—A. C. Caggiano, Simple Gyrator for L from C, *EEE Magazine,* Aug. 1970, p 78.

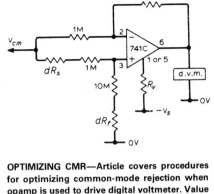

OPTIMIZING CMR—Article covers procedures for optimizing common-mode rejection when opamp is used to drive digital voltmeter. Value of R_v is determined by using resistance box connected between negative supply and pin 1 or 5 while other pin is shorted to negative supply, choosing pin which gives voltage swing in right direction on meter, then adjusting resistance box for zero output. Resistance box is similarly used at dR_s and dR_f locations.—R. J. Isaacs, Optimizing Op-Amps, *Wireless World,* April 1973, p 185–186.

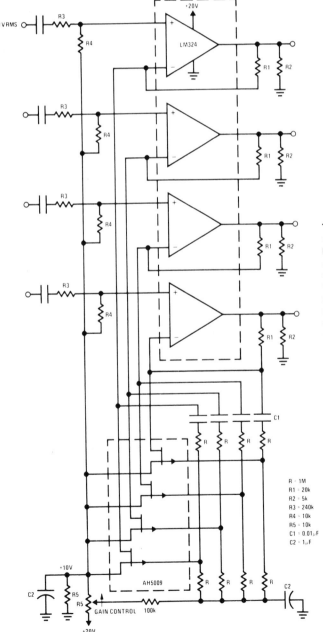

TRACKING QUAD GAIN CONTROL—Basic circuit for each channel uses section of National LM324 quad opamp with section of AH5009 quad FET in feedback path. Each channel is AC coupled and has 40-dB range (gain range of 1 to 100). Bandwidth is minimum of 10 kHz, and S/N ratio is better than 70 dB with 4.3-VRMS maximum output.—J. Sherwin, "A Linear Multiple Gain-Controlled Amplifier," National Semiconductor, Santa Clara, CA, 1975, AN-129, p 6.

POSTAMPLIFIER FOR OPAMP—High input impedance of National MM74C04 inverter makes it ideal for isolating load from output of LM4250 micropower opamp operating from single dry cell.—"Linear Applications, Vol. 2," National Semiconductor, Santa Clara, CA, 1976, AN-88, p 2.

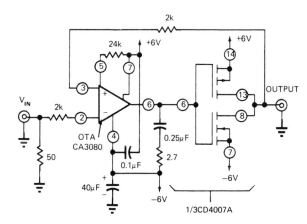

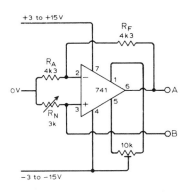

CMOS DRIVER FOR OPAMP—CMOS inverter pair (one-third of CD4007A) is used in closed-loop mode as unity-gain voltage follower for CA3080 opamp. Slew rate is 1 V/μs. Output current capability of 6 mA can be increased by paralleling two other sections of CMOS.—B. Furlow, CMOS Gates in Linear Applications: The Results Are Surprisingly Good, *EDN Magazine*, March 5, 1973, p 42–48.

NEGATIVE R—Negative-resistance connection of 741 opamp is suitable for both AC and DC applications. Requires floating power supply because 0-V terminal floats with respect to both output terminals. For DC use, adjust 10K pot to cancel offset voltage of amplifier. Value of negative resistance is varied with R_N or by adjusting ratio of R_F to R_A. Can be used to make LC circuits operate at subaudio frequencies.—D. A. Miller, Negative Resistor, *Wireless World,* June 1974, p 197.

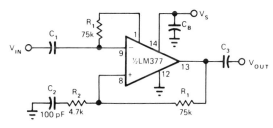

UNITY-GAIN AF CURRENT AMPLIFIER—External components are used with National LM377/378/379 family of opamps to provide stability at unity gain. Article gives design equations. At frequencies above audio band, gain rises with frequency, to well above 10 at 340 kHz for values shown.—D. Bohn, AC Unity-Gain Power Buffers Amplify Current, *EDN Magazine*, May 5, 1977, p 113–114.

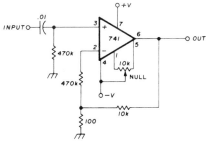

741 OPAMP—Power supply and null pot connections for TO-5 metal-can package and 8-lead DIP package are shown. Maximum rated power supply voltages are ±18 V, but lower voltages may be used. 9-V transistor battery is often used for each supply, but higher voltages will permit larger output signal swing. Pin 3 is inverting input, and pin 4 is noninverting input. With values shown, both input terminals see about same resistance, and output offset can be nulled to zero. Gain of circuit is about 100.—C. Hall, Circuit Design with the 741 Op Amp, *Ham Radio*, April 1976, p 26–29.

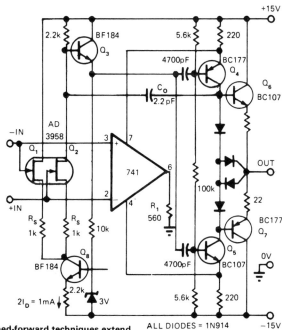

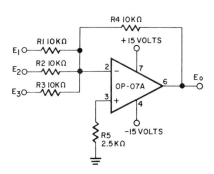

FASTER 741—Feed-forward techniques extend dynamic response of differential opamp to give unity-gain bandwidth of 18 MHz, slew rate over 200 V/μs, and DC gain above 10⁷ V/V, while preserving latchup-free operation and wide input voltage range. Composite amplifier uses fast symmetrical four-transistor output stage that is symmetrically driven by DC-coupled 741 and by AC-coupled feed-forward amplifier. Performance depends on use of nonstandard pin connections for 741, as shown. Developed for processing fast analog data in frequency domain.—J. Dostal, 741 + Feedforward = Fast-Differential Op Amp, *EDN Magazine*, Aug. 20, 1974, p 90.

SUMMING AMPLIFIER—Provides output equal to sum of all input voltages, with high precision. Use of Precision Monolithics OP-07A opamp makes circuit adjustment-free.—"Ultra-Low Offset Voltage Op Amp," Precision Monolithics, Santa Clara, CA, 1977, OP-07, p 7.

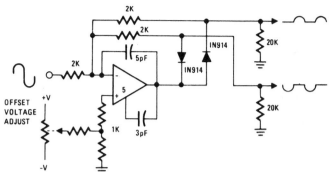

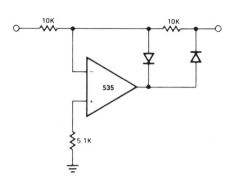

SIGNAL SEPARATOR—Circuit shown for Harris HA-2530 opamp separates input voltage into its positive and negative components. Diodes steer components to separate outputs. Applications include feeding outputs into differential amplifier to produce absolute-value circuit for multiplying or averaging functions. For bandwidth of 1 MHz, dynamic range is 100 mV to 10 V peak.—"Linear & Data Acquisition Products," Harris Semiconductor, Melbourne, FL, Vol. 1, 1977, p 7-54–7-55 (Application Note 516).

HALF-WAVE RECTIFIER—Provides accurate half-wave rectification of incoming signal. Gain is 0 for positive signals and −1 for negative signals. Diode types are not critical. Polarity can be inverted by reversing both diodes. With opamp shown, circuit will function up to 10 kHz with less than 5% distortion.—"Signetics Analog Data Manual," Signetics, Sunnyvale, CA, 1977, p 641–643.

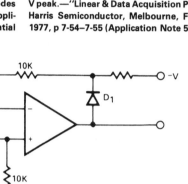

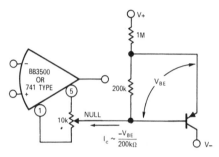

ABSOLUTE-VALUE AMPLIFIER—Generates positive output voltage for either polarity of DC input. Opamp and diode types are not critical. Accuracy is highest for input voltages greater than 1 V. Opamp is noninverting on positive signals and inverting on negative signals.—"Signetics Analog Data Manual," Signetics, Sunnyvale, CA, 1977, p 641–643.

TEMPERATURE-COMPENSATED OFFSET CONTROL—Drift effects of offset adjustment are removed by deriving correction current from emitter-base voltage of PNP signal transistor to develop appropriate temperature compensation. Correction current is divided with conventional control pot used for adjusting offset voltage. Article gives design equations.—J. Graeme, Offset Null Techniques Increase Op Amp Drift, *EDN Magazine*, April 1, 1971, p 47–48.

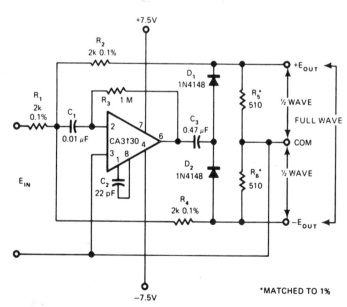

*MATCHED TO 1%

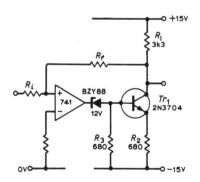

RECTIFIER WITHOUT DC OFFSET—Avoids drawback of large nonlinearity at low signal levels, by isolating AC of opamp from DC output. Circuit has wide bandwidth, as required for rectifying 20-kHz input signal with high precision. Output coupling capacitor C_3 is low-leakage Mylar; for low-frequency operation, it can be replaced with two back-to-back low-leakage tantalums. D_1 and D_2 should be matched for forward voltage at peak load current. Use Hewlett-Packard 5082-2810 hot-carrier diodes instead to improve operation at millivolt signal levels or at higher frequencies.—D. Belanger, Single Op Amp Full-Wave Rectifier Has No DC Offset, *EDN Magazine*, April 5, 1977, p 144 and 146.

FASTER SLEWING—Single transistor stage at output of opamp increases slewing rate by factor equal to gain of transistor stage. Choose R_1 to meet output impedance requirements and current rating of supply. R_2 is then made equal to R_1 divided by desired gain of transistor stage. Collector of Tr_1 should be at 0 V when output of opamp is 0 V, assuming feedback loop is not closed by R_f. Article gives design equations.—L. Short, Faster Slewing Rate with 741 Op-Amp, *Wireless World*, Jan. 1973, p 31.

SIGNAL CONDITIONER—FET-buffered opamp circuit will operate from source impedances up to 100 megohms while providing voltage gain of 5. Offset adjustment is provided for initial calibration of circuit. Developed for use with high-impedance sensors such as pH electrodes.—"Industrial Control Engineering Bulletin," Motorola, Phoenix, AZ, 1973, EB-4.

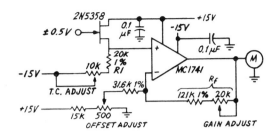

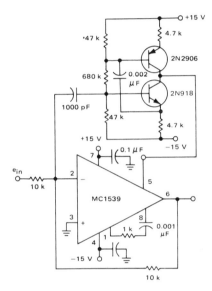

UNITY-GAIN FEED-FORWARD—Provides 10 V P-P output signal at 2 MHz when gain of feed-forward amplifier is increased to give closed-loop gain of 10. Provides fast response to step-function input, with slow settling. High-frequency circuit takes over completely when input frequency is too high for input stage to respond.—E. Renschler, "The MC1539 Operational Amplifier and Its Applications," Motorola, Phoenix, AZ, 1974, AN-439, p 20.

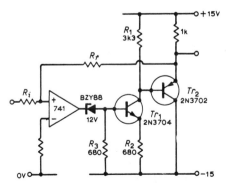

HIGH-SPEED HALF-WAVE RECTIFIER—Produces inverted half-wave replica of input signal with low error at frequencies up to 100 kHz. C_1 provides feed-forward compensation. For negative-going output, reverse connections to diodes.—W. G. Jung, "IC Op-Amp Cookbook," Howard W. Sams, Indianapolis, IN, 1974, p 191–192.

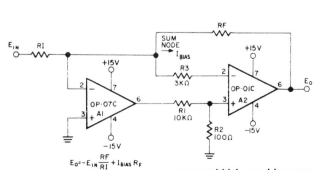

$$E_O = -E_{IN} \frac{RF}{RI} + I_{BIAS} R_F$$

SUMMING AMPLIFIER—Combination of Precision Monolithics OP-07C and OP-01C opamps gives 18 V/μs slew rate. Can be used as current-output summing amplifier for D/A converter because it requires no zero scale offset adjustments and high speed is preserved.—D. Soderquist and G. Erdi, "The OP-07 Ultra-Low Offset Voltage Op Amp—a Bipolar Op Amp That Challenges Choppers, Eliminates Nulling," Precision Monolithics, Santa Clara, CA, 1975, AN-13, p 9.

FAST SLEWING AND LOW IMPEDANCE—With values shown, Tr_1 increases slewing rate of opamp by factor of 5, and Tr_2 connected as emitter-follower reduces output impedance to meet requirements of following circuit. Feedback is taken from emitter of Tr_1 to noninverting input of opamp.—L. Short, Faster Slewing Rate with 741 Op-Amp, Wireless World, Jan. 1973, p 31.

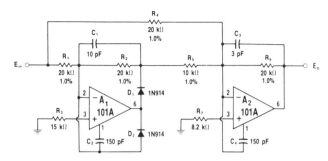

100-kHz FULL-WAVE RECTIFIER—Feed-forward connection of opamps gives high-speed full-wave rectification of signals up to 100 kHz for measurement and analysis.—W. G. Jung, "IC Op-Amp Cookbook," Howard W. Sams, Indianapolis, IN, 1974, p 193–194.

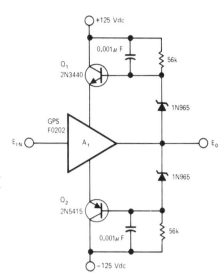

BOOSTING VOLTAGE RATING—Bootstrapping technique permits operation of low-voltage unity-gain opamp from high-voltage DC supply for handling large input signal voltage swings, while retaining gain and voltage stability of opamp. Allowable input-voltage range depends entirely on transistor rating. With 1000-V transistors, circuit can handle input signals of ±475 V. Input capability for values shown is ±100 V P-P for DC to 10 kHz. Output capability is 5 mA P at ±100 V. Input impedance is 10 teraohms.—S. A. Jensen, High-Voltage Source Follower, *EDN/EEE Magazine,* Feb. 1, 1972, p 58.

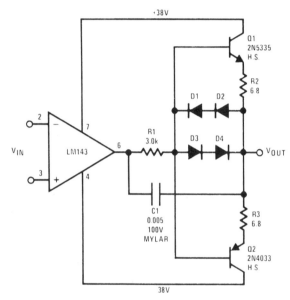

100-mA CURRENT BOOSTER—Provides short-circuit protection along with current boosting for LM143 high-voltage opamp. Diodes are 1N914. Use Thermalloy 2230-5 or equivalent heatsinks with transistors. Output is ±33 V P-P into 400-ohm load.—"Linear Applications, Vol. 2," National Semiconductor, Santa Clara, CA, 1976, AN-127, p 4.

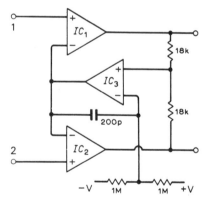

DIFFERENTIAL I/O—Arrangement shown for three 741 opamps gives amplifier having differential output as well as differential input. Circuit is designed primarily to drive meter with signal of either polarity when center-tap power supply is not available. Article covers operation and adjustment of circuit.—A. D. Monstall, Differential Input and Output with Op-Amps, *Wireless World,* Jan. 1973, p 31.

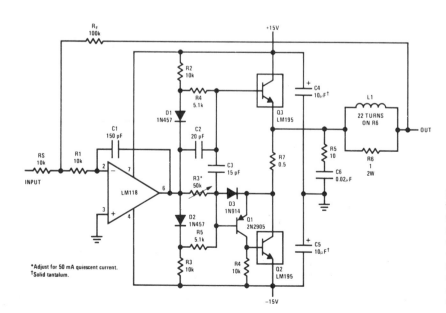

POWER OPAMP—Transistor Q1 and power transistor IC Q2 form equivalent of power PNP transistor for use with NPN LM195 power transistor IC serving as output stage for opamp. Circuit is stable for almost any load. Bandwidth can be increased to 150 kHz with full output response by decreasing C1 to 15 pF if there is no capacitive load to cause oscillation.—"Linear Applications, Vol. 2," National Semiconductor, Santa Clara, CA, 1976, AN-110, p 5–6.

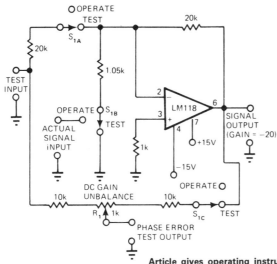

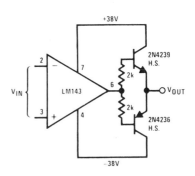

POWER BOOSTER—Simple two-transistor power stage increases power output of LM143 high-voltage opamp. Intended for loads less than 2K. Drawbacks are noticeable crossover distortion and lack of short-circuit protection. Transistors should be used with Thermalloy 2230-5 or equivalent heatsinks.—"Linear Applications, Vol. 2," National Semiconductor, Santa Clara, CA, 1976, AN-127, p 3.

PHASE-ERROR TESTER—Circuit reveals significant phase errors at relatively low frequencies, even for high-speed opamps. Technique applies to most opamps and almost any signal gain.

Article gives operating instructions based on observation of null with XY CRO connected to phase-error test output.—R. A. Pease, Technique Trims Op-Amp Amplifiers for Low Phase Shift, *EDN Magazine,* Aug. 20, 1977, p 138.

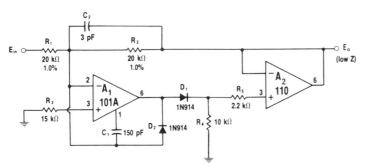

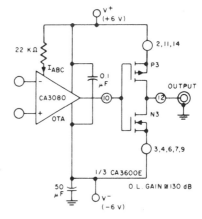

100-kHz BUFFERED RECTIFIER—High-speed 110 voltage follower is used within feedback loop of A_1 to maintain low output impedance for precision half-wave rectifier. When input signal is positive, D_1 and R_4 rectify signal and A_2 follows this signal. On opposite alternations, D_1 is off and feedback loop of A_2 is closed through D_2 so output terminal is maintained at low impedance. For opposite output polarity, reverse diode connections.—W. G. Jung, "IC Op-Amp Cookbook," Howard W. Sams, Indianapolis, IN, 1974, p 192.

POSTAMPLIFIER—CMOS transistor pair from CA3600E transistor array provides additional 30-dB gain above 100-dB gain of CA3080 opamp to give total of 130 dB. Current output is about 10 mA. Remaining transistor pairs of array can be paralleled pair shown to give greater output.—"Linear Integrated Circuits and MOS/FET's," RCA Solid State Division, Somerville, NJ, 1977, p 278–279.

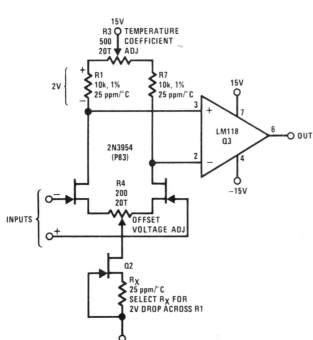

LOW TEMPERATURE COEFFICIENT—Use of National 2N3954 dual FET as input device for opamp gives fast response to thermal transients, making it possible to adjust R3 and R4 so temperature coefficient is less than 5 μV/°C from −25°C to +85°C. Common-mode rejection ratio is typically greater than 100 dB for input voltage swings of 5 V. Drain current level is set by Q2 which is 2N5457 FET.—"FET Databook," National Semiconductor, Santa Clara, CA, 1977, p 6-4–6-7.

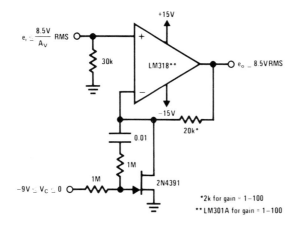

1–1000 GAIN RANGE—Control voltage of 0 to −9 V changes gain of amplifier over complete range while providing maximum output level of 8.5 VRMS and bandwidth of over 20 kHz at maximum gain. If gain range of 100 is sufficient, amplifier can be changed to LM301; 20K resistor is then changed to 2K.—"Linear Applications, Vol. 2," National Semiconductor, Santa Clara, CA, 1976, AN-129, p 5.

JFET INPUT—U401 dual JFET acting as preamp for standard bipolar opamp uses CR033 N-channel JFET as 330-μA current source. R4 is used to null initial offset. R3 is adjusted for minimum drift.—"Analog Switches and Their Applications," Siliconix, Santa Clara, CA, 1976, p 7-51.

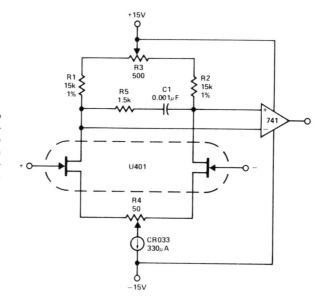

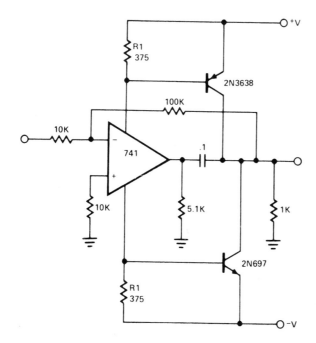

POWER BOOSTER—Opamp power booster is used after conventional opamp when greater power-handling capability is required. 741 opamp circuit shown will drive moderate loads. Other opamps may be substituted in power stage if value of R1 is appropriately changed.—"Signetics Analog Data Manual," Signetics, Sunnyvale, CA, 1977, p 640–642.

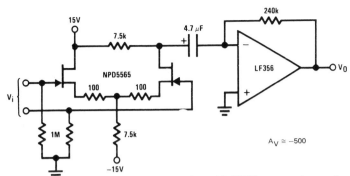

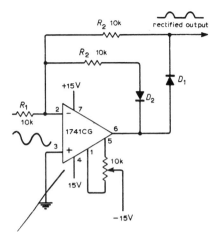

DIFFERENTIAL JFET INPUT—Differential connection of National NPD5565 dual JFET is used when balanced inputs and low distortion are main requirements for AC amplifier. Combination with LF356 opamp shown gives gain of about 500. Noise is somewhat higher than with single-ended JFET.—"FET Databook," National Semiconductor, Santa Clara, CA, 1977, p 6-17–6-19.

PRECISE RECTIFICATION—Use of opamp in combination with silicon diode overcomes non-linearity of diode at forward voltages under about 0.5 V. Offset-voltage pot is adjusted for symmetrical output waveform for small input voltages. D_1 is connected in opamp feedback path so initial forward voltage drop required to make diode conduct is supplied by amplifier output. Second feedback path through D_2 prevents output saturation on input half-cycles for which D_1 is reverse-biased.—G. B. Clayton, Experiments with Operational Amplifiers, *Wireless World*, June 1973, p 275–276.

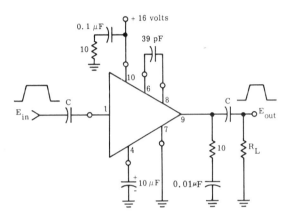

3-W PULSE AMPLIFIER—Motorola MC1554 power amplifier provides voltage gain of 18 for peak pulse power output up to 3 W. Maximum peak output current rating of 500 mA for IC should not be exceeded during peak of output pulse.—"The MC1554 One-Watt Monolithic Integrated Circuit Power Amplifier," Motorola, Phoenix, AZ, 1972, AN-401, p 3.

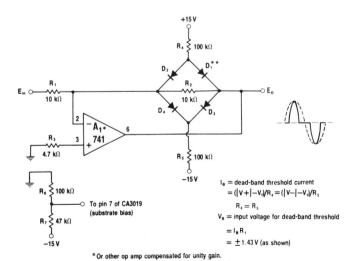

I_B = dead-band threshold current

$= (|V+|-|V_i|)/R_4 = (|V-|-|V_i|)/R_5$

$R_4 = R_5$

V_B = input voltage for dead-band threshold

$= I_B R_1$

$= \pm 1.43\,V$ (as shown)

* Or other op amp compensated for unity gain.

** $D_1 - D_4$ are matched monolithic diodes, such as the CA3019.
For peak voltage higher than ± 7 V, use 1N914s.

DEAD-BAND RESPONSE—With bridge in feedback loop of opamp, low-level input signals give essentially 100% feedback around A_1 so there is very little output voltage. When input current through R_1 rises above allowable current limit of circuit, bridge opens and output voltage jumps to new level determined by R_2. Input is then amplified by ratio of R_2/R_1 in normal linear manner. Circuit thus has dead-band property for low levels. Value of R_1 sets threshold level.—W. G. Jung, "IC Op-Amp Cookbook," Howard W. Sams, Indianapolis, IN, 1974, p 206–207.

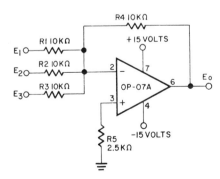

SUMMING WITHOUT ADJUSTMENTS—Single-stage opamp for analog computation provides high-precision output that is function of multiple input variables. Circuit drift is less than 2 μV per month, eliminating need for periodic calibration while ensuring long-term accuracy. Opamp is Precision Monolithics OP-07A.—D. Soderquist and G. Erdi, "The OP-07 Ultra-Low Offset Voltage Op Amp—a Bipolar Op Amp That Challenges Choppers, Eliminates Nulling," Precision Monolithics, Santa Clara, CA, 1975, AN-13, p 11.

$$GAIN = \frac{R2}{R1} = X = \frac{R3}{R1 + R2 + R3}$$

$$R3 = R1\left(\frac{X + X^2}{1 - X}\right)$$

$$FOR\ X = 0.5: \frac{5\,K\Omega}{10\,K\Omega} = \frac{R2}{R1}$$

$$R3 = 10\,k\Omega\left(\frac{0.75}{0.5}\right) = 15\,k\Omega$$

ABSOLUTE-VALUE RECTIFIER—Use of CA3140 bipolar MOS opamp in inverting gain configuration gives symmetrical full-wave output when equality of design equations is satisfied. Bandwidth for −3 dB is 290 kHz, and average DC output is 3.2 V for 20 V P-P input.—"Circuit Ideas for RCA Linear ICs," RCA Solid State Division, Somerville, NJ, 1977, p 18.

SIGN CHANGER—When switch S_1 grounds pin 3 of opamp, circuit becomes inverter providing 180° phase shift. When S_1 is at position A, input voltage acts on both inputs of A_1 and no current flows through R_1 and R_2; output voltage is then equal to input voltage. Switch permits remote programming of phase reversal. For higher input impedance, 1556 opamp can be used.— W. G. Jung, "IC Op-Amp Cookbook," Howard W. Sams, Indianapolis, IN, 1974, p 208–209.

SINGLE-ENDED JFET—Basic JFET amplifier is virtually free from popcorn noise problems of bipolar transistors and bipolar-input opamps. Combining JFET transconductance amplifier with current-to-voltage opamp adds high voltage gain and simplifies circuit applications. Gain-limiting 7.5K FET drain resistor is bypassed and removed from gain equation. Parameter variation problems are minimized by biasing FET source through 15.1K resistance to negative supply. Gain variations are minimized by leaving 100 ohms of this resistance unbypassed.—J. Maxwell, FET Amplifiers—Take Another Look at These Devices, *EDN Magazine,* Sept. 5, 1977, p 161–163.

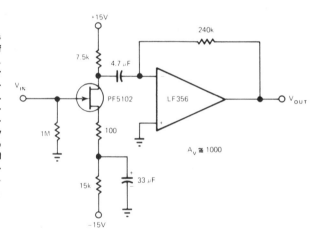

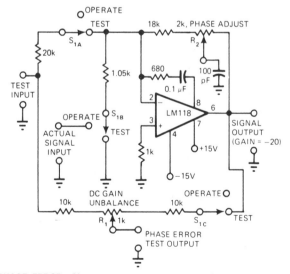

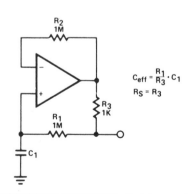

$$C_{eff} = \frac{R_1}{R_3} \cdot C_1$$

$$R_S = R_3$$

MINIMIZING PHASE ERROR—Phase compensation circuit trimmed by R_2 keeps phase error of LM118 opamp well below 1° from DC to 200 kHz. In-phase error due to gain peaking is also low. Feed-forward network connected to pin 8 improves stability, making feedback capacitor unnecessary. Step response has about 30% overshoot, and sine response has about +1 dB of peaking before going 3 dB down at about 2 MHz.—R. A. Pease, Technique Trims Op-Amp Amplifiers for Low Phase Shift, *EDN Magazine,* Aug. 20, 1977, p 138.

CAPACITANCE MULTIPLIER—Resistance ratio determines factor by which value of C_1 is multiplied when used in simple opamp circuit shown. With values shown, ratio is 1000 and 10-μF capacitor provides effective capacitance of 10,000 μF. Q of circuit is limited by effective series resistance, so R_1 should be as large as practical. Opamp type is not critical.—"Signetics Analog Data Manual," Signetics, Sunnyvale, CA, 1977, p 640–641.

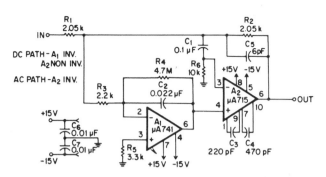

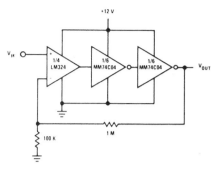

FEED-FORWARD OPAMP—DC input characteristics are determined by A_1, which is bypassed at high frequencies, while AC-coupled A_2 determines dynamic performance. Resulting composite amplifier combines such desired properties as low input current and drift, large bandwidth and slew rate, and fast settling time. Compensation network C_3-C_4-C_5 is chosen first to give desired bandwidth. Composite rolloff of 6 dB per octave is then obtained by narrow-banding A_1 with R_4 and C_2, so gain-bandwidth product is equal to ratio between unity-gain crossover frequency of A_2 and open-loop gain.—Fairchild Linear IC Contest Winners, *EEE Magazine,* Jan. 1971, p 48–49.

SINGLE-SUPPLY POSTAMPLIFIER—Use of two sections of MM74C04 as postamplifier for LM324 single-supply amplifier gives open-loop gain of about 160 dB. Additional CMOS inverter sections can be paralleled for increased power to drive higher current loads; each MM74C04 section is rated for 5-mA load.—"Linear Application, Vol. 2," National Semiconductor, Santa Clara, CA, 1976, AN-88, p 2.

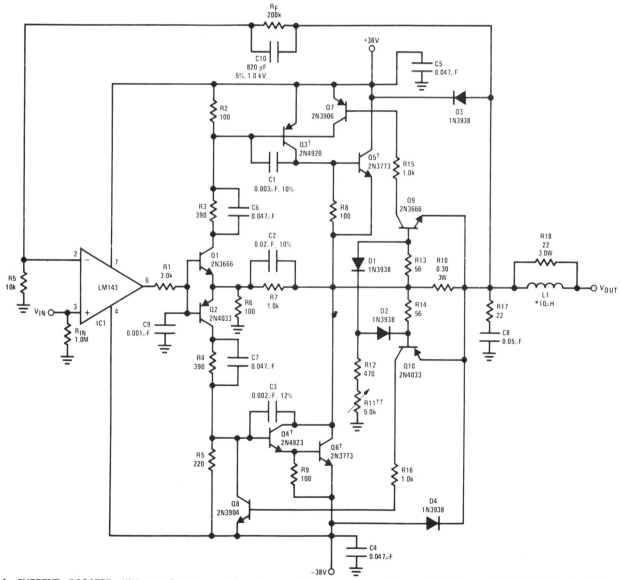

8-A CURRENT BOOSTER—High-compliance power stage for LM143 high-voltage opamp provides very high peak drive currents along with output voltage swings to within 4 V of ±38 V supply under full load. Maximum output current depends on setting of current-adjusting pot R11 and on output voltage. Limit ranges from 14 A when R11 is 0 down to about 4 A for 5K. Maximum power output is 144 WRMS, for which frequency response is 3 dB down at 10 kHz. Voltage gain is 21. Q3-Q6 should be on common Thermalloy 6006B or equivalent heatsink.—"Linear Applications, Vol. 2," National Semiconductor, Santa Clara, CA, 1976, AN-127, p 5–6.

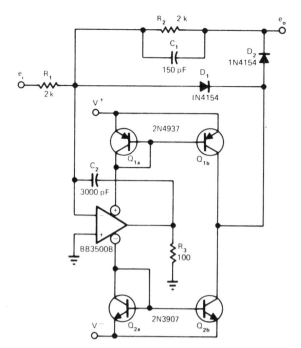

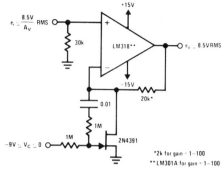

GAIN-CONTROLLED AMPLIFIER—Control voltage in range of 0 to −9 V provides gain range of 1 to 1000 for National LM318 opamp using FET in feedback path. Bandwidth is better than 20 kHz at maximum gain. Applications include remote or multichannel gain control, volume expansion, and volume compression/limiting.—J. Sherwin, "A Linear Multiple Gain-Controlled Amplifier," National Semiconductor, Santa Clara, CA, 1975, AN-129, p 5.

60-kHz PRECISION RECTIFIER—Usable full-power response of typical opamp is boosted to 60 kHz while giving 300-kHz small-signal bandwidth. Circuit uses transistors to provide speed-boosting gain during transition from one precision rectifier diode to the other in feedback loop of opamp. Added stage is driven from power-supply current drains of opamp. Article traces operation of circuit in detail.—J. Graeme, Boost Precision Rectifier BW above That of Op Amp Used, *EDN Magazine*, July 5, 1974, p 67–69.

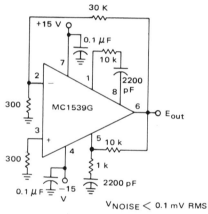

$V_{NOISE} < 0.1$ mV RMS

LOW-NOISE 5-kHz—Values shown are for operation of Motorola MC1539G opamp in closed-loop mode with noninverting gain of 100 and source impedance of about 300 ohms. Circuit bandwidth is about 5 kHz.—E. Renschler, "The MC1539 Operational Amplifier and Its Applications," Motorola, Phoenix, AZ, 1974, AN-439, p 19.

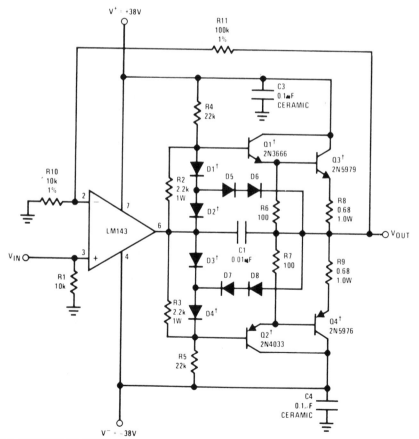

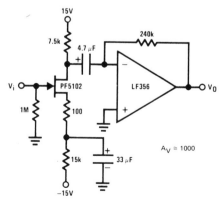

$A_V \cong 1000$

FET DRIVE—National PF5102 JFET is combined with LF356 opamp to give low noise and high gain, for use as wide-bandwidth AC amplifier. Typical gain for combination shown is about 1000. Any other opamp can be used as long as it meets slew rate and bandwidth requirements.—"FET Databook," National Semiconductor, Santa Clara, CA, 1977, p 6-17–6-19.

1-A CURRENT BOOSTER—Used with LM143 high-voltage opamp to increase output current while providing short-circuit protection and low crossover distortion. With 40-ohm load, output voltage can swing to + 29.6 V and −28 V. All four transistors should be on Thermalloy 6006B or equivalent common heatsink. All diodes are 1N3193.—"Linear Applications, Vol. 2," National Semiconductor, Santa Clara, CA, 1976, AN-127, p 4–5.

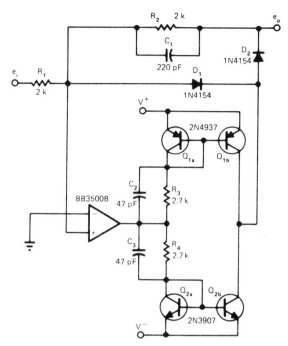

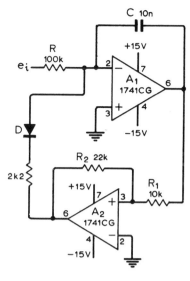

PRECISION RECTIFIER WITH GAIN—Gain is selectively added during open-loop switching transition of precision rectifier diodes D_1 and D_2 in feedback loop of opamp, to boost speed while maintaining feedback stability following switching. Q_1 and Q_2 add gain of about 250 up to 30 kHz during switching, because D_1 and D_2 are then off and do not shunt output of added stage. Following transition, one of diodes conducts heavily, shunting high output impedance of stage and dropping its gain below unity. Article covers circuit operation in detail.—J. Graeme, Boost Precision Rectifier BW above That of Op Amp Used, *EDN Magazine,* July 5, 1974, p 67–69.

VOLTAGE/FREQUENCY CONVERTER—Uses opamp A_1 as integrator and A_2 as regenerative comparator with hysteresis, to generate sequence of pulses with repetition frequency proportional to DC input voltage. Article gives design equations and typical waveforms. Input voltage range is 10 mV to 20 V for linear operation.—G. B. Clayton, Experiments with Operational Amplifiers, *Wireless World,* Dec. 1973, p 582.

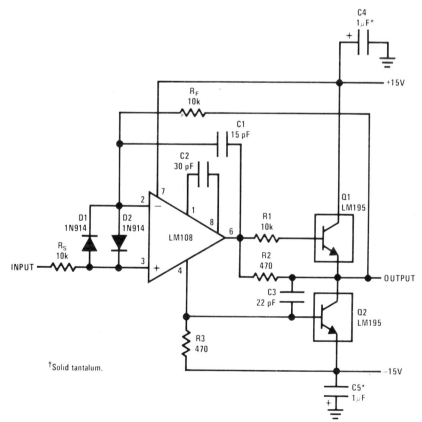

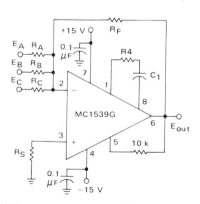

300-Hz VOLTAGE FOLLOWER—Simple LM195 power output stage provides 1-A output for voltage-follower connection of LM108 opamp.—R. Dobkin, "Fast IC Power Transistor with Thermal Protection," National Semiconductor, Santa Clara, CA, 1974, AN-110, p 6.

SUMMING OPAMP—Motorola MC1539 serves as closed-loop summing amplifier having very small loop-gain error because of high open-loop gain. R_S should equal parallel combination of R_A, R_B, R_C, and R_F.—E. Renschler, "The MC1539 Operational Amplifier and Its Applications," Motorola, Phoenix, AZ, 1974, AN-439, p 18.

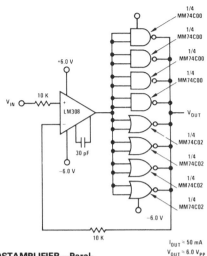

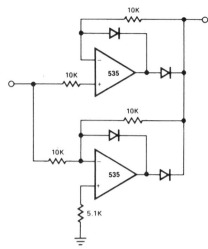

COMPLEMENTARY POSTAMPLIFIER—Paralleled NAND gates provide buffering for LM308 opamp while increasing current drive to about 50 mA for 6 V P-P output. MM74C00 NAND gates supply about 10 mA each from positive supply while MM74C02 gates supply same amount from negative supply.—"Linear Applications, Vol. 2," National Semiconductor, Santa Clara, CA, 1976, AN-88, p 2–3.

FULL-WAVE RECTIFIER—Circuit provides accurate full-wave rectification of input signal, with distortion below 5% up to 10 kHz. Reversal of all diode polarities reverses polarity of output. Output impedance is low for both input polarities, and errors are small at all signal levels.—"Signetics Analog Data Manual," Signetics, Sunnyvale, CA, 1977, p 641–643.

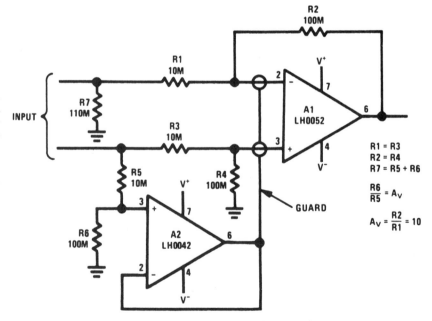

GUARDED FULL-DIFFERENTIAL—Extremely high input impedance is achieved by intercepting leakage currents with guard conductor placed in leakage path and operated at same voltage as inputs. A2 serves as guard drive amplifier, with R5 and R6 developing proper voltage for guard at their junction. R7 balances detector R5 plus R6 without degrading closed-loop common-mode rejection.—"Linear Applications, Vol. 1," National Semiconductor, Santa Clara, CA, 1973, AN-63, p 1–12.

$R1 = R3$
$R2 = R4$
$R7 = R5 + R6$

$$\frac{R6}{R5} = A_V$$

$$A_V = \frac{R2}{R1} = 10$$

SECTION 45
Optoelectronic Circuits

Basic voltage-isolating applications for optoisolators. Includes bar-code reader circuits. Other chapters may include optoisolators in circuits having specific applications.

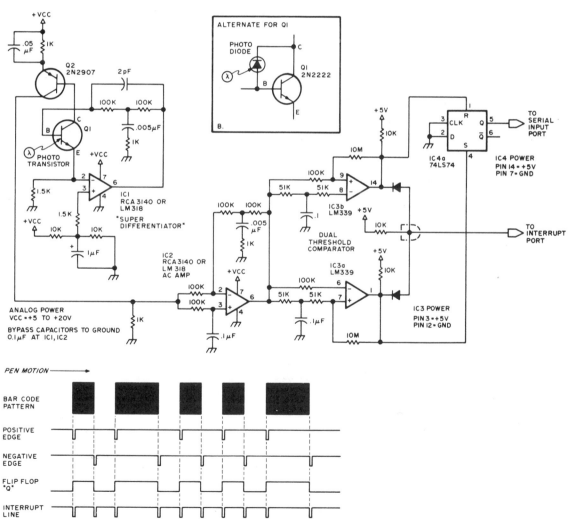

BAR-CODE READER—Edge-sensitive circuit outputs short pulses at each black-to-white or white-to-black transition. Timing diagram shows outputs corresponding to bar-code pattern indicated. Direct-current level at base of Q1 is held constant by DC servo action despite changes in temperature, ambient light, or background of pattern. Alternate sensor uses photodiode and 2N2222 transistor for increased bandwidth. Amplified differentiated signal from collector of Q2 is further amplified by IC2 and fed to dual threshold comparator. Output of comparator is short pulse for each transition, suitable for feed to microprocessor.—F. L. Merkowitz, Signal Processing for Optical Bar Code Scanning, *BYTE,* Dec. 1976, p 77–78 and 80–84.

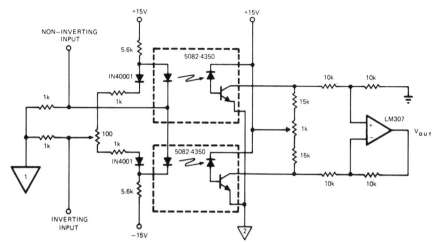

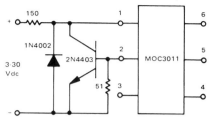

DC ISOLATOR WITH HARMONIC SUPPRESSION—Two isolators operating like push-pull amplifier minimize harmonic generation. When input signal is applied, upward change of incremental gain in one isolator is balanced by downward change in other to give harmonic cancellation. Circuit gain is about unity. Bandwidth is 2 MHz for signals below 2 V P-P. Input signals of either polarity may be applied at either inverting or noninverting input.—H. Sorensen, Opto-Isolator Developments Are Making Your Design Chores Simpler, *EDN Magazine,* Dec. 20, 1973, p 36–44.

OPTOISOLATOR INPUT PROTECTION—Combination of diode and transistor limits input current to LED of Motorola MOC3011 optoisolator to safe maximum of less than 15 mA for input voltage range of 3–30 VDC. Circuit also protects LED from accidental reversal of polarity.—P. O'Neil, "Applications of the MOC3011 Triac Driver," Motorola, Phoenix, AZ, 1978, AN-780, p 4.

SET-RESET LATCH—Provides almost complete isolation between each input and the output, as well as between inputs. Applying 2-V pulse at 14 mA momentarily to SET terminals allows up to 150 mA to flow between output terminals. This current flows until about 2 V at 15 mA is applied to RESET terminals or until load voltage is reduced enough to drop load current below 1 mA.—R. N. Dotson, Set-Reset Latch Uses Optical Couplers, *EDN Magazine,* Jan. 5, 1973, p 107.

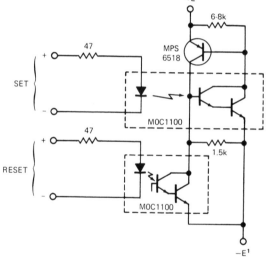

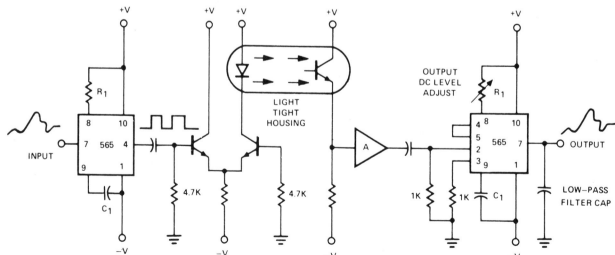

ANALOG ISOLATOR—Circuit is basically FM transmission system in which light is used as transmission medium. Transmitter uses 565 PLL as VCO for flashing LED of optoisolator at rate proportional to input voltage. Phototransistor drives amplifier having sufficient gain to apply 200 mV P-P signal to input of receiving 565 acting as FM detector for re-creating input to transmitter. Supply can be ±6 V to ±12 V.— "Signetics Analog Data Manual," Signetics, Sunnyvale, CA, 1977, p 846–847.

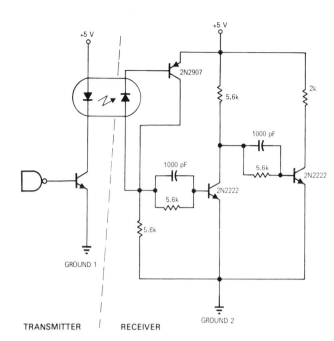

GROUND ISOLATION—Optoisolator such as HP4320 provides ground isolation up to 200 V between systems used in spacecraft. Arrangement is effective over bandwidth of DC to 1 MHz for both DTL- and TTL-driven circuits.—W. C. Milo, Simple Scheme Isolates System Grounds Optically, *EDN Magazine,* Sept. 15, 1970, p 64.

400-VDC SWITCH—Optically isolated photo-SCR serves for switching high-voltage DC. Turn-off of SCR occurs when Q_3 in MCA2 photo-Darlington shunts load current through gate, bypassing gate-cathode junction within SCR. Circuit can be operated by pulsing appropriate LEDs to turn SCR on or off. Without input signal, inverter maintains current through LED of MCA2 to keep SCR clamped off.—G. C. Riddle, Opto-Isolators Switch High-Voltage DC Current, *EDN Magazine,* Feb. 5, 1975, p 54.

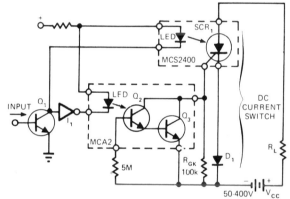

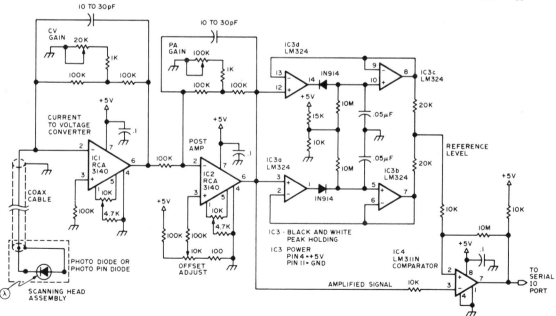

BAR-CODE SIGNAL CONDITIONER—Processes low-level signal from photodiode of bar-code scanner by converting its current output to voltage in IC1 for further amplification in IC2. Amplified signal is routed to peak holding circuits that set reference level and to comparator that outputs 0 or 1 based on reference level established. Peak values of white level and black level are held long enough to read through coded bar pattern. Difference between peak values is divided by 2 and fed to one input of comparator, while amplified signal level goes to inverting input. If signal level is greater than reference level, comparator output is 0. If signal level is less than reference level (black bar), output is 1.—F. L. Merkowitz, Signal Processing for Optical Bar Code Scanning, *BYTE,* Dec. 1976, p 77–78 and 80–84.

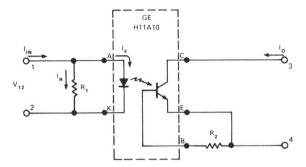

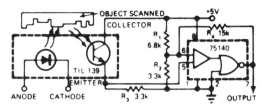

ISOLATED THRESHOLD SWITCH—Standard photocoupler programmed with 150-ohm resistor R_1 provides threshold switching function for separating high-level noise from switching-signal pulses as short as 10 μs. Current-transfer ratio of phototransistor coupler is made practically zero at some arbitrary input current, and changed rapidly back to 10% or more at slightly higher level. Programming range for threshold value extends from 60 mA for 10 ohms at R_1 to 3 mA for 400 ohms. Use of 2.7-megohm resistor R_2 across base-emitter terminals of coupler reduces low-current gain of phototransistor. Noise currents up to 5 mA on sensing line are rejected while operating currents as low as 10 mA are accepted.—J. Cook, Photocoupler Makes an Isolated Threshold Switch, *EDN Magazine*, Oct. 5, 1974, p 72, 74, and 76.

OPTOISOLATOR AS SCANNER—Consists essentially of Texas Instruments TIL 139 source/sensor assembly and common 75140 line receiver. Applications include response to reflected or interrupted light. With 5-V supply, output is at standard TTL levels. To make sensitivity adjustable, insert 500-ohm pot between R_1 and R_2. To invert output polarity, connect pin 7 of 75140 to pin 3 and take output from pin 1.— W. Grenlund, Low-Cost Photo Scanner Yields High Performance, *EDN Magazine*, Nov. 20, 1976, p 320.

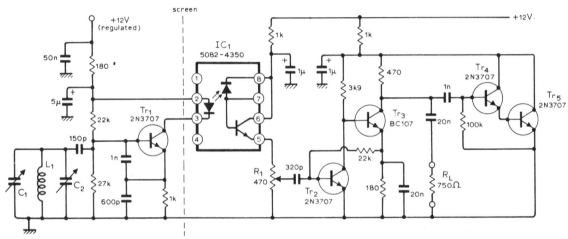

1.5–5.7 MHz OPTICALLY ISOLATED VFO—Isolation gives long-term frequency stability despite changes in ambient temperature, and eliminates effect of fluctuating load on frequency. Oscillator is emitter-coupled Colpitts using low-noise 2N3707 transistor. Article also gives circuit for output amplifier and automatic limiting control, along with alternative versions using ICs in place of transistors. Designed for use in amateur radio equipment.—A. K. Langford, Optically Coupled V.F.O., *Wireless World*, Nov. 1974, p 455–457.

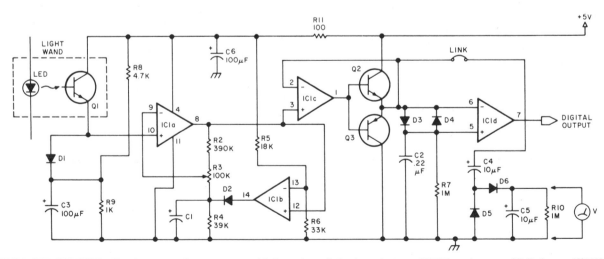

LIGHT-WAND AMPLIFIER—Signal processor is independent of most variables involved in reading printed bar data. Amplifier uses exponential forward conduction properties of silicon diode D1 to transform output of wand to logarithmically varying voltage having peak-to-peak value proportional to ratio of white and black photocurrents and independent of absolute photocurrent. White-level output of amplifier IC1a is clamped at fixed level by comparator IC1b and peak detector D2-C1. Amplified and clamped signal is converted to binary digital output required by microprocessor. Article traces operation of circuit step by step. IC1 is National LM324 quad opamp. All diodes are 1N4148 silicon or equivalent. Q2 is MPS6513 or equivalent, and Q3 is MPS6517 or equivalent. Output is TTL-compatible.—R. C. Moseley, A Low Cost Light Wand Amplifier, *BYTE*, May 1978, p 92 and 94–95.

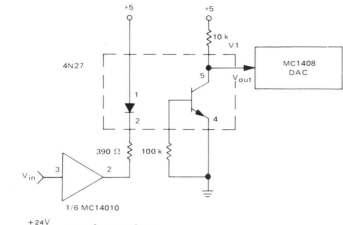

1500-V ISOLATION FOR DAC—Motorola 4N27 optoisolator provides required isolation between DAC of programmable power supply and remotely located CMOS MC14010 noninverting buffer.—D. Aldridge and N. Wellenstein, "Designing Digitally-Controlled Power Supplies," Motorola, Phoenix, AZ, 1975, AN-703, p 9.

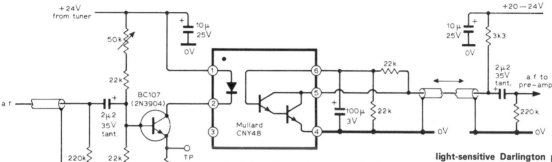

HUM-BLOCKING OPTOISOLATOR—Optoelectronic isolator for audio feed in TV set prevents circulation of ground currents at line frequency, for protection of low-level signal runs from hum interference. Used in tuner providing quality sound and video outputs, circuits for which are given in four-part article. Optoisolator uses light-sensitive Darlington pair in conjunction with infrared-emitting diode. Diode current is adjusted with 50K variable resistor to give best compromise between noise and distortion.—D. C. Read, Television Tuner Design, *Wireless World*, Jan. 1976, p 51–57.

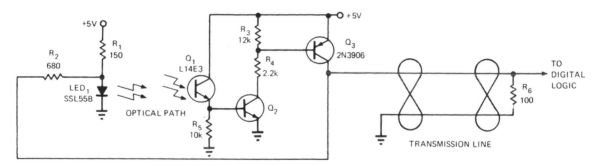

BUILT-IN HYSTERESIS—Will operate at all speeds in range from 20 kHz down to zero while still having suitable rise times for driving digital logic. When optical path is blocked, all three transistors are off and output is low. As light on Q_1 increases, Q_2 and Q_3 begin turning on; rising collector of Q_3 adds more current through R_2 to LED, giving Q_1 more light and driving Q_3 into saturation. When light dims, Q_1 begins to turn off and extra current is cut off, driving Q_3 off. With this hysteresis action, there is no constant light level at which circuit will oscillate.—D. C. Hoffman, Optical Sensor Has Built-In Hysteresis, *EDN Magazine*, June 5, 1973, p 91.

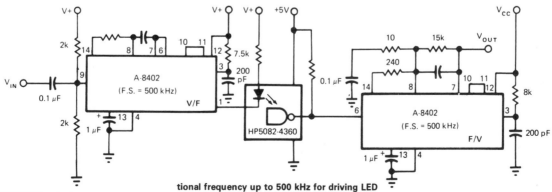

30-kHz BANDWIDTH—Isolation amplifer circuit uses Intech/Function Modules A-8402 voltage-to-frequency converter having linearity of ±0.05% to convert input voltage to proportional frequency up to 500 kHz for driving LED of optoisolator. Similar IC converts output of optoisolator back to proportional DC voltage. Supply for converters is nominally 12 V, but can be 5 to 18 V.—P. Pinter and D. Timm, Voltage-to-Frequency Converters—IC Versions Perform Accurate Data Conversion (and Much More) at Low Cost, *EDN Magazine*, Sept. 5, 1977, p 153–157.

SECTION 46
Oscillator Circuits—Audio Frequency

Includes variety of Wien-bridge, phase-shift, voltage-controlled, and multivibrator types of oscillators producing output at audio and ultrasonic frequencies. Other audio oscillators can be found in Code, Frequency Synthesizer, Function Generator, Pulse Generator, Signal Generator, Staircase Generator, Sweep, and Test chapters.

2-kHz TWO-PHASE—Dual opamp circuit uses two-pole Butterworth bandpass filter followed by phase-shifting single-pole stage that is fed back through zener voltage limiter. Circuit provides simultaneous sine and cosine outputs. Distortion is about 1.5% for sine output and about 3% for cosine. Component values shown are for 741 opamp. For higher frequencies, use 531 opamps to reduce distortion due to slew limiting.—"Signetics Analog Data Manual," Signetics, Sunnyvale, CA, 1977, p 642–644.

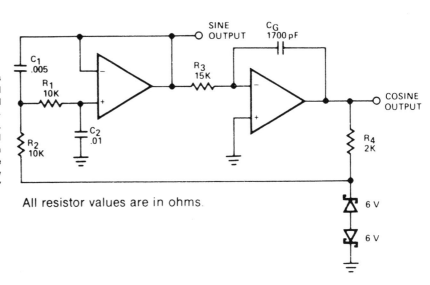

All resistor values are in ohms.

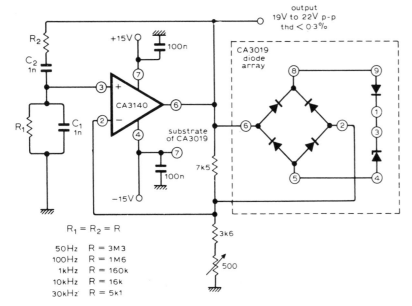

BASIC MOS OSCILLATOR—Output is 50 Hz when R_1 and R_2 are 3.3 megohms, increasing to 30 kHz as resistor values are reduced to 5100 ohms. Circuit has no inherent lower frequency limit; with 22-megohm resistors and 1-μF capacitors for C_1 and C_2, sine-wave output is 0.007 Hz. Article gives basic equations for circuit. Features include high input impedance, fast slew rate, and high output voltage capability. Combination of bridge rectifier with monolithic zener diodes in regulating system provides practically zero temperature coefficient.—M. Bailey, Op-Amp Wien Bridge Oscillator, *Wireless World*, Jan. 1977, p 77.

$R_1 = R_2 = R$

50 Hz	R = 3M3
100 Hz	R = 1M6
1 kHz	R = 160k
10 kHz	R = 16k
30 kHz	R = 5k1

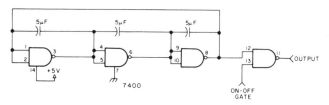

1000 Hz WITH ONE CHIP—Quad NAND gate gives sawtooth output waveform at 800 to 1000 Hz for driving other TTL circuits.—Circuits, *73 Magazine*, June 1977, p 49.

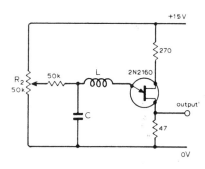

1–50 kHz SINE-WAVE—Uses unijunction transistor as negative resistance in simple RLC circuit. Maximum output with good waveform is about 200 mV. Exact frequency depends on values used for L and C.—R. P. Hart, Simple Sine-Wave Oscillator, *Wireless World*, July 1976, p 34.

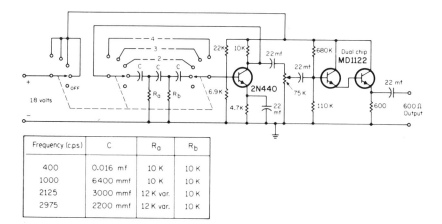

Frequency (c.ps)	C	R$_a$	R$_b$
400	0.016 mf	10 K	10 K
1000	6400 mmf	10 K	10 K
2125	3000 mmf	12 K var.	10 K
2975	2200 mmf	12 K var.	10 K

TEST TONES—Provides preset frequencies of 400, 1000, 2125, and 2975 Hz. Circuit consists of RC phase-shift oscillator driving Darlington emitter-follower that provides high-impedance load for oscillator and stable 600-ohm output impedance.—S. Kelly, A Simple Audio Test Oscillator, *CQ*, Oct. 1970, p 50 and 90.

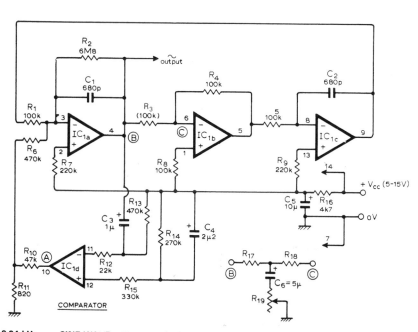

2.34-kHz SINE-WAVE—Uses low-cost LM3900N quad differential amplifier IC in low-distortion oscillator for which third harmonic distortion is typically 0.5%. Peak-to-peak amplitude of sine-wave output is typically 25% of source voltage V$_{cc}$. Frequency can be changed by altering single component, R$_3$, or by inserting between points B and C an RC network and pot connected as shown in inset. Article gives design equations for frequency and Q.—T. J. Rossiter, Sine Oscillator Uses C.D.A., *Wireless World*, April 1975, p 176.

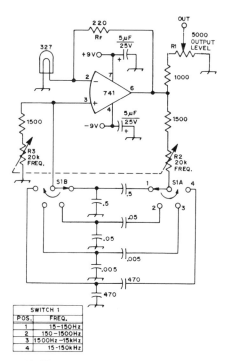

SWITCH 1	
POS.	FREQ.
1	15–150Hz
2	150–1500Hz
3	1500Hz –15kHz
4	15–150kHz

15 Hz TO 150 kHz IN FOUR RANGES—Switch gives choice of ranges, with R2 and R3 varying frequency in each. Circuit draws only 4 mA from two 9-V batteries and provides moderate output at 4–5 V. Connections shown are for TO-5 case of 741.—T. Schultz, Audio Oscillator, *QST*, Nov. 1974, p 43.

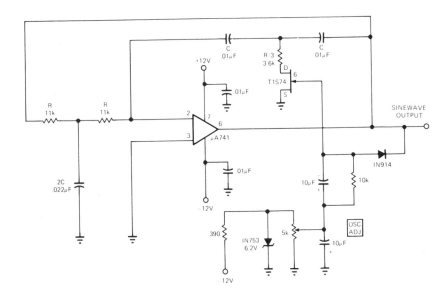

STABILIZED SINE-WAVE—Peak detector is used with FET operated in voltage variable-resistance mode, in combination with standard double-integration circuit having regenerative feedback, to give 1.46-kHz sine-wave output into 500-ohm load at 10 V P-P. Will operate at power supply voltages of 8 to 18 V without appreciable variation in output amplitude or frequency. Output varies less than 1.5% in frequency and 6% in amplitude over temperature range of 10 to 65°C. Circuit can be modified for other frequencies.—F. Macli, FET Stabilizes Sine-Wave Oscillator, *EDN Magazine*, June 5, 1973, p 87.

1-kHz LOW-DISTORTION—Total harmonic distortion is only 0.01% in amplitude-stabilized oscillator delivering 7 VRMS. Opamp A₁ has closed-loop gain of 3. Regenerative feedback through bandpass filter C_1-C_2-R_1-R_5 determines frequency of oscillation. Output is stabilized by multiplier whose control voltage is derived from integrator A_2.—R. Burwen, Ultra Low Distortion Oscillator, *EDN/EEE Magazine*, June 1, 1971, p 45.

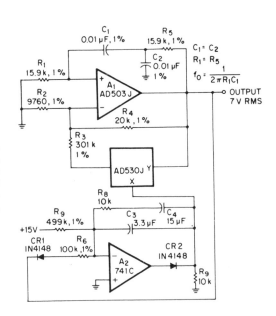

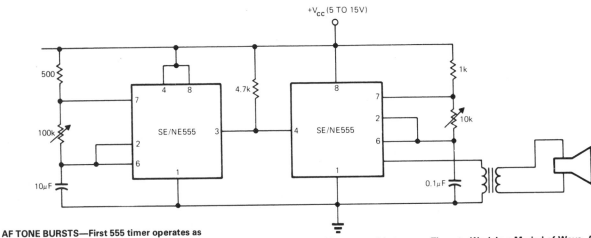

AF TONE BURSTS—First 555 timer operates as slow astable multivibrator whose output is used to gate second timer operating as AF oscillator. Arrangement provides repeatable tone-burst generation.—E. R. Hnatek, Put the IC Timer to Work in a Myriad of Ways, *EDN Magazine*, March 5, 1973, p 54–58.

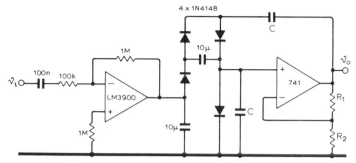

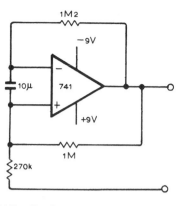

CURRENT-CONTROLLED WIEN—Small variations in input voltage to National LM3900 current-mode amplifier change frequency of four-diode current-controlled Wien bridge over range from 10 to 50,000 Hz, with frequency being proportional to control current. Value of C is 700 pF. Ratio of R_2 to sum of R_1 and R_2 should be greater than 3 to give voltage gain needed.— K. Kraus, Oscillator with Current-Controlled Frequency, *Wireless World,* Aug. 1974, p 272.

3.8 kHz—Simple opamp circuit provides convenient sine-wave AF signal.—J. S. Lucas, Unusual Sinewave Generator, *Wireless World,* May 1977, p 81.

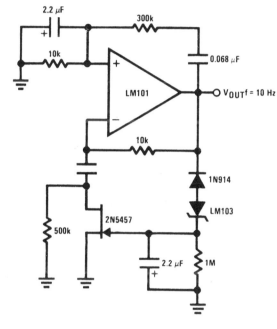

Peak output voltage

$$V_p \cong V_z + 1V$$

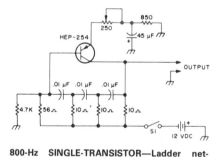

800-Hz SINGLE-TRANSISTOR—Ladder network determines frequency. For higher frequencies, decrease values of capacitors in network. Circuit also works with OC-2, SK-3004, and AT30H transistors.—Circuits, *73 Magazine,* May 1977, p 31.

10-Hz WIEN-BRIDGE—JFET serves as voltage-variable resistor in feedback loop of opamp, as required for producing low-distortion constant-amplitude sine wave. LM103 zener provides voltage reference for peak amplitude of sine wave; this voltage is rectified and fed to gate of JFET to vary its channel resistance and loop gain of opamp.—"FET Databook," National Semiconductor, Santa Clara, CA, 1977, p 6-26– 6-36.

DOT GENERATOR—Can be used by amateur radio operator to "talk" himself onto frequency while listening on downlink passband of Oscar satellite, without causing interference to other stations using satellite. Generates audio dots at rate of 12 per second. Frequency of free-running MVBR Q1-Q2 is determined by values of C1, C2, R1, and R2. Emitter-follower Q3 drives 1500-Hz audio oscillator Q4. C1 and C2 are 1-μF 16-V electrolytics. Q1-Q4 are 2N2222 or equivalent NPN general-purpose transistors.—M. Righini and G. Emiliani, Audio Dot Generator Eases OSCAR SSB Spotting, *QST,* Nov. 1977, p 45.

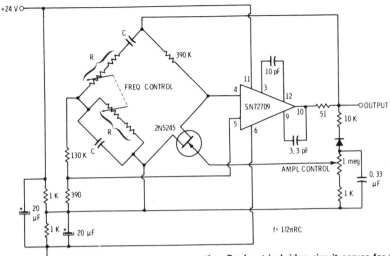

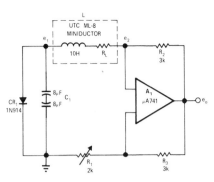

WIEN-BRIDGE AF/RF—Single JFET in basic Wien-bridge circuit drives Texas Instruments linear opamp serving as output stage. Feedback path from output of IC to base of JFET stabilizes output and provides temperature compensa-tion. Dual pot in bridge circuit serves for frequency control. Circuit performs well as either AF or RF oscillator depending on values used for R and C.—E. M. Noll, "FET Principles, Experiments, and Projects," Howard W. Sams, Indianapolis, IN, 2nd Ed., 1975, p 213–214.

25-Hz SINE-WAVE—Output voltage is 8 V P-P at about 25 Hz for values shown, with total harmonic distortion less than 0.5%. Circuit will operate from 15 Hz to 100 kHz by using other values. Set regeneration control R_1 at minimum value needed to sustain oscillation.—J. C. Freeborn, Simple Sinewave Oscillator, *EDN/EEE Magazine*, Sept. 1, 1971, p 44.

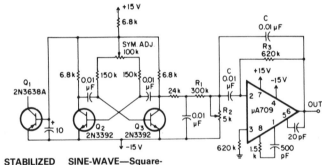

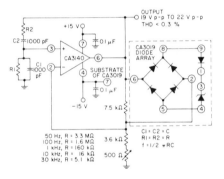

350-Hz STABILIZED SINE-WAVE—Square-wave oscillator Q_2-Q_3 stabilized by Q_1, followed by passive filter and active filter using μA709, produces amplitude-stabilized sine wave at 350 Hz, for which third harmonic is 39 dB down and other harmonics are insignificant.—E. Neugroschel and A. Paterson, Amplitude-Stabilized Audio Oscillator, *EEE Magazine*, April 1971, p 65.

50–30,000 Hz WIEN-BRIDGE—Wide-range audio oscillator utilizes high input impedance, high slew rate, and high voltage characteristics of CA3140 opamp in combination with CA3019 diode array. R1 and R2 are same value, chosen for frequency desired as given in table.—"Circuit Ideas for RCA Linear ICs," RCA Solid State Division, Somerville, NJ, 1977, p 4.

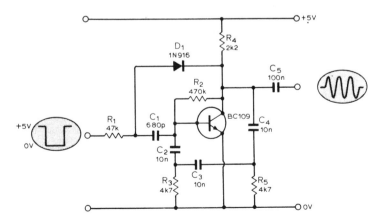

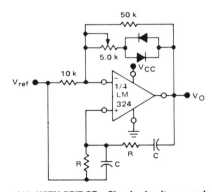

1-kHz FAST-START GATED—Circuit is conventional phase-shift oscillator in which frequency is determined by C_2, C_3, C_4, R_5, R_3, and input impedance of transistor. When input is +5 V, almost 100% negative feedback blocks oscillator. When input drops to 0 V, D_1 is reverse-biased and negative feedback is removed. At same time, edge of input pulse is applied to transistor base to kick off oscillator on its first half-cycle, which is always in phase with falling edge of input signal.—G. F. Butcher, Gated Oscillator with Rapid Start, *Wireless World*, Aug. 1974, p 272.

1-kHz WIEN-BRIDGE—Simple circuit uses only one section of LM324 quad opamp having true differential inputs. Supply voltage range is 3–32 V. Reference voltage is half of supply voltage. Values of R and C determine frequency according to equation f = 1/6.28RC. For 16K and 0.01 μF, frequency is 1 kHz. Diode types are not critical.—"Quad Low Power Operational Amplifiers," Motorola, Phoenix, AZ, 1978, DS 9339 R1.

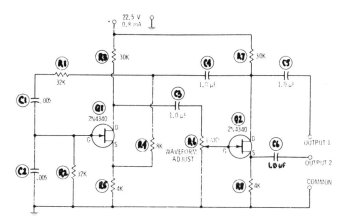

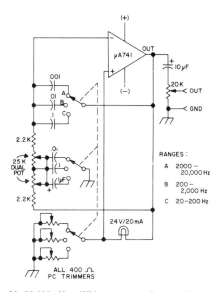

100-Hz WIEN-BRIDGE—Simple RC-tuned oscillator uses only two resistors (R1 and R2) and two capacitors (C1 and C2) to set frequency. Feedback path covers both FET stages. Set R6 for best sine-wave output. For other audio frequencies, change value of R in ohms and C in farads in equation f = 1/6.28RC where frequency is in hertz, R = R1 = R2, and C = C1 = C2.—R. P. Turner, "FET Circuits," Howard W. Sams, Indianapolis, IN, 1977, 2nd Ed., p 48–50.

20–20,000 Hz—Wide-range audio oscillator covers AF spectrum in three switch-selected ranges, with harmonic distortion as low as 0.15%, for quick checks of audio equipment. Drain is only 6 mA from two 9-V batteries. Circuit is Wien-bridge oscillator using 741 opamp. Article covers construction and calibration, including optional connection for operation from single 9-V battery with AF output reduced to 2 V.—J. J. Schultz, Wide Range IC Audio Oscillator, *73 Magazine,* Jan. 1974, p 25–28.

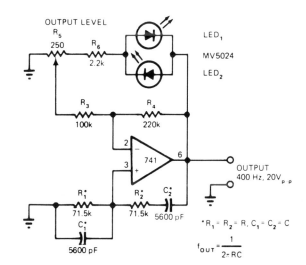

$$*R_1 = R_2 = R, \ C_1 = C_2 = C$$

$$f_{OUT} = \frac{1}{2\pi RC}$$

400-Hz LED-OPAMP SINE-WAVE—Uses LEDs as nonlinear-resistance diodes in Wien-bridge configuration with opamp operating from 15-V supply. Circuit will operate over wide range of other frequencies if values of R and C are changed. R_5 adjusts output amplitude from 10 to 20 V P-P. Total harmonic distortion is 1%.—W. G. Jung, LED's Do Dual Duty in Sine-Wave Oscillator, *EDN Magazine,* Aug. 20, 1976, p 84–85.

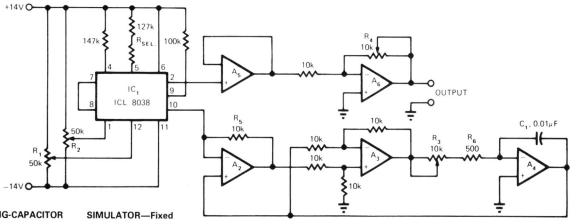

TUNING-CAPACITOR SIMULATOR—Fixed biasing network is used with Intersil 8038 variable-frequency sine-wave oscillator. Frequency is varied between 175 and 3500 Hz by circuit components forming capacitor simulator. Adjusting R_3 varies equivalent capacitor value from 500 pF to 0.01 μF. Distortion is less than 1% over frequency range. Buffer opamp A_5 provides high load impedance to IC_1 and low source impedance to variable-gain opamp A_6. All opamps are 741.—R. Gunderson, Variable-Frequency Oscillator Features Low Distortion, *EDN Magazine,* Aug. 5, 1974, p 76 and 78.

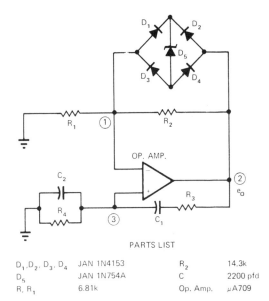

PARTS LIST

D_1, D_2, D_3, D_4	JAN 1N4153	R_2	14.3k
D_5	JAN 1N754A	C	2200 pfd
R, R_1	6.81k	Op. Amp.	μA709

ZENER CONTROLS BRIDGE—Amplitude of 10.5-kHz Wien-bridge oscillator output is maintained symmetrical above ground by using single zener with diode bridge. As output e_0 approaches soft knee threshold of conduction for zener, its impedance decreases and shunts R_2. This violates oscillator requirement that $R_2 = 2R_1$, so output begins decreasing sinusoidally. As swing decreases, gain increases until e_0 reaches negative threshold. Signal then reverses and again starts going positive.—W. B. Crittenden and E. J. Owings, Jr., Zener-Diode Controls Wien-Bridge Oscillator, *EDN Magazine*, Aug. 1, 1972, p 57–58.

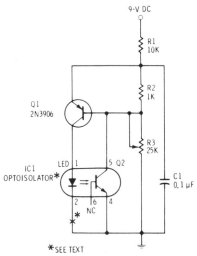

*SEE TEXT

NEGATIVE-RESISTANCE LED OSCILLATOR—Covers frequency range of about 3.2–8 kHz with values shown. Will drive loudspeaker inserted at point X. For lower frequency (range of 120–1800 Hz) and louder sound, change C1 to 1 μF. Negative-resistance portion of circuit includes Q1, Q2, LED, R2, and R3. Optoisolator can be MCT-2 or equivalent.—F. M. Mims, "Electronic Circuitbook 5: LED Projects," Howard W. Sams, Indianapolis, IN, 1976, p 26–29.

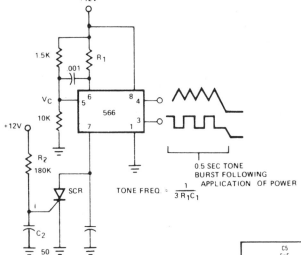

TONE FREQ. $= \dfrac{1}{3 R_1 C_1}$

0.5-s TONE BURSTS—Simple 566 function generator circuit supplies audio tone for 0.5 s after power is applied, for use as communication-network alert signal. SCR is gated on when C_2 charges up to its gate voltage, which takes 0.5 s, to shunt timing capacitor between pin 7 and ground and thereby stop tone. If SCR is replaced by NPN transistor, tone can be switched on and off manually at transistor base terminal.—"Signetics Analog Data Manual," Signetics, Sunnyvale, CA, 1977, p 852–853.

20–20,000 Hz LOW-DISTORTION—Opamp at right is driven by square-wave output of comparator at left, with feedback between opamps providing oscillation. Frequency range covered by tuning control R3 is determined by equal-value capacitors C1 and C2, which range from 0.4 μF for 18–80 Hz to 0.002 μF for 4.4–20 kHz. Distortion ranges from 0.2% to 0.4% when 20% clipping of sine wave is provided by zeners. Both positive and negative supplies should be bypassed with 0.1-μF disk ceramic capacitors.—"Easily Tuned Sine Wave Oscillators," National Semiconductor, Santa Clara, CA, 1971, LB-16.

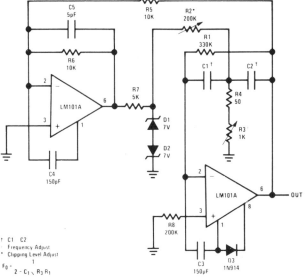

† C1 C2
  Frequency Adjust
* Clipping Level Adjust

$F_0 = \dfrac{1}{2 \cdot C_1 \cdot R_3 R_1}$

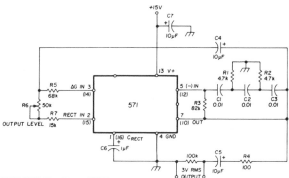

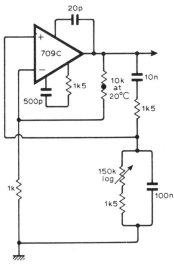

PHASE-SHIFT SINE-WAVE—Uses NE571 analog compandor as phase-shift oscillator, with internal inverting amplifier serving to sustain oscillation. Cl, C2, and C3 are timing capacitors, while R1 and R2 serve for phase-shift network. Suitable for use only as spot-frequency AF oscillator, with frequency being varied by changing values of Cl, C2, and C3. Total harmonic distortion is only 0.01% at 3-V output.—W.G. Jung, Gain Control IC for Audio Signal Processing, *Ham Radio*, July 1977, p 47–53.

SINGLE-POT WIEN—Can be tuned from 340 to 3400 Hz with single 150K logarithmic pot. Output is constant over tuning range. Opamp can also be 741. Components in the two arms of the Wien bridge have large ratio to each other, so attenuation of network is only slightly affected by change in one of resistors.—P. C. Healy, Wien Oscillator with Single Component Frequency Control, *Wireless World*, Aug. 1974, p 272.

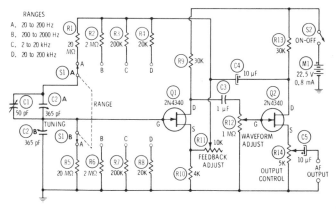

20 Hz TO 200 kHz—Variable-frequency RC-tuned oscillator uses FETs with Wien-bridge frequency-determining network. Identical resistors accurate to at least 1% are switched in pairs to change range. Dual 365-pF variable capacitor C2 is used for tuning in each range. Can be calibrated against standard audio frequency with CRO set up for Lissajous figures, or calibrated with high-precision AF meter connected to AF output terminals.—R. P. Turner, "FET Circuits," Howard W. Sams, Indianapolis, IN, 1977, 2nd Ed., p 132–134.

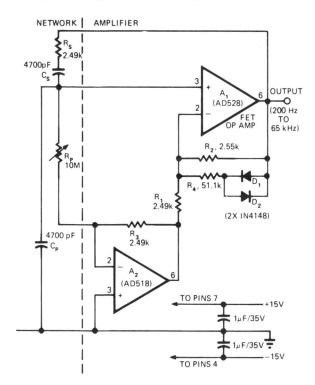

200–65,000 Hz WIEN—Adding single opamp to Wien-bridge oscillator gives wide-range oscillator having single-control tuning. R_4, D_1, and D_2 together stabilize output amplitude by providing controlled nonlinearity that reduces gain at high signal levels. AD528 opamp A_1 is FET-input complement to AD518 A_2 and has bandwidth required for wide output frequency range. R_P sweeps output from 200 Hz to 65 kHz. Since oscillation frequency is inversely proportional to square root of R_P, frequency changes rapidly near low-resistance end of pot. Use of pot with audio or log taper makes tuning more linear.—P. Brokaw, FET Op Amp Adds New Twist to an Old Circuit, *EDN Magazine*, June 5, 1974, p 75–77.

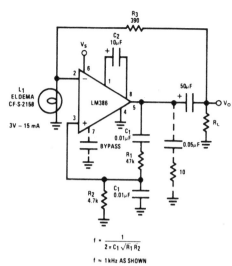

$$f = \frac{1}{2\pi C_1 \sqrt{R_1 R_2}}$$

f ≈ 1kHz AS SHOWN

1-kHz WIEN-BRIDGE—Closed-loop gain of 10, fixed by ratio of R_1 to R_2, is sufficient to avoid spurious oscillations. Frequency is easily changed by using different values for capacitors C_1. R_3 and lamp L_1 provide amplitude-stabilizing negative feedback. Supply can be 9 V.—"Audio Handbook," National Semiconductor, Santa Clara, CA, 1977, p 4-30–4-33.

WIEN SINE-WAVE—Uses NE571 analog compandor in oscillator circuit based on Wien network formed by R1-Cl and R2-C2, placed around output amplifier of section A to make it bandpass amplifier. Section B serves as inverting amplifier with nominal gain of 2. Total harmonic distortion is below 0.1%. Operating frequency is about 1.6 kHz for values shown, but can be varied from 10 Hz to 10 kHz. Frequency is $1/2\pi RC$ for R = R1 = R2 and C = Cl = C2. R should be kept between 10K and 1 megohm and C between 1000 pF and 1 μF. Useful as fixed-frequency oscillator but can be tuned if matched dual pot is used for R1-R2.—W. G. Jung, Gain Control IC for Audio Signal Processing, *Ham Radio*, July 1977, p 47–53.

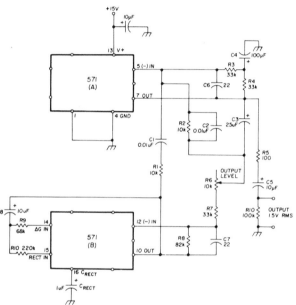

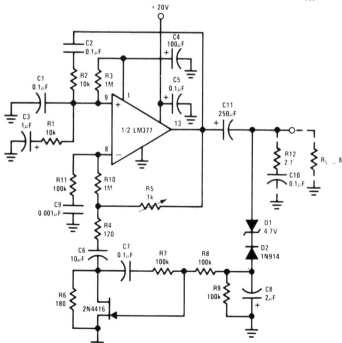

WIEN-BRIDGE 2-W—Uses half of LM377 IC connected as oscillator, with FET amplitude stabilization in negative feedback path. Total harmonic distortion is under 1% up to 10 kHz. With values shown, maximum output is 5.3 VRMS at 60 Hz. R12 and C10 are added if necessary to prevent high-frequency instability.—"Audio Handbook," National Semiconductor, Santa Clara, CA, 1977, p 4-8–4-20.

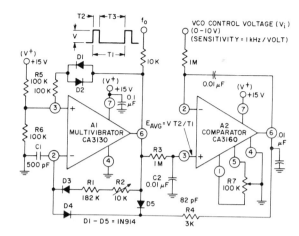

1 kHz/V FOR VCO—Voltage-controlled oscillator uses CA3130 opamp as MVBR and CA3160 opamp as comparator. Tracking error is about 0.02%, and temperature coefficient is 0.01% per degree C.—"Circuit Ideas for RCA Linear ICs," RCA Solid State Division, Somerville, NJ, 1977, p 4.

SINE-WAVE WIEN—Uses CA3140 opamp and diode array to generate low-distortion sine waves. Table gives values recommended for R and C to obtain frequencies from 50 Hz to 30 kHz. Use of zener diode clamp for amplitude control gives fast AGC.—W. Jung, An IC Op Amp Update, *Ham Radio*, March 1978, p 62–69.

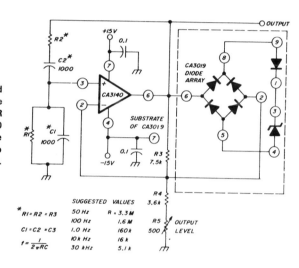

SUGGESTED VALUES	
50 Hz	R = 3.3 M
100 Hz	1.6 M
1.0 Hz	160 k
10 k Hz	16 k
30 kHz	5.1 k

* R1 = R2 = R3
  C1 = C2 = C3

$$f = \frac{1}{2\pi RC}$$

SECTION 47
Oscillator Circuits—Radio Frequency

Includes fixed and tunable Clapp, Colpitts, crystal, LC, RC, Pierce, relaxation, and wobbulator oscillators having sine or square outputs in range from AF spectrum to 200 MHz. Some can be changed in frequency by digital control or diode switching of crystals.

SECONDARY STANDARD FOR 100 AND 10 kHz—Combination of 100-kHz crystal oscillator and 10-kHz MVBR provides 100-kHz harmonics far up into high-frequency spectrum, with each 100-kHz interval subdivided by harmonics of MVBR using two FETs. Oscillator is tuned to crystal frequency with Miller 42A223CBI or equivalent slug-tuned coil L1. C1 adjusts crystal frequency over narrow range for standardizing against WWV transmissions. Synchronizing 100-kHz voltage is injected into MVBR through R5.—R. P. Turner, "FET Circuits," Howard W. Sams, Indianapolis, IN, 1977, 2nd Ed., p 127–129.

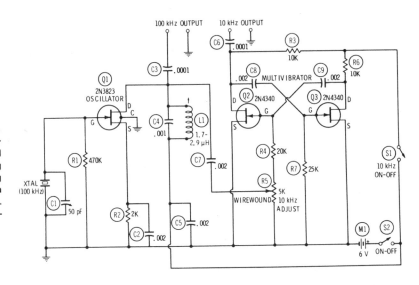

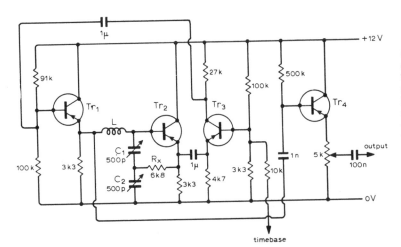

450–500 kHz WOBBULATOR—Center frequency of sweep is adjusted with C_1 and C_2. With appropriate coil, operation can be extended up to 10.7 MHz. Transistors can be BC107, BF115, BF194, or other equivalent. Choose value of R_x to give best waveform with transistor types used. Feedback for VCO is taken via Tr_3 without phase change. If control voltage for base of Tr_3 is derived from ramp output of oscilloscope time base, wobbulator output will follow variations in sweep voltage of time base.—E. C. Lay, Wobbulator, *Wireless World*, May 1975, p 226.

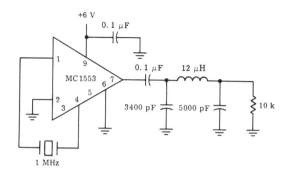

1-MHz SERIES-MODE CRYSTAL—Motorola MC1553 video amplifier provides wide bandwidth and output swing capability needed for high-frequency master clock or local oscillator in many system designs. Positive feedback is injected through crystal to input pin 1. Output is taken from pin 7 which is buffered internally from oscillator by gain and emitter-follower stages. Brute-force pi filter at output extracts desired fundamental frequency.—"A Wide Band Monolithic Video Amplifier," Motorola, Phoenix, AZ, 1973, AN-404, p 9.

9-MHz LINEAR VCO—U1A and U1C of RCA CA3046 transistor array form emitter-coupled oscillator. Portion of U1A current is diverted through U1B and L1, producing magnetic flux that reduces effective inductance of resonating coil L2. Output frequency is varied in direct proportion to voltage applied at A. L1 is 23 turns on ¾-inch Teflon form 2 inches long, with 4 turns wound between windings for L2. VR1 is 1N3828 6.2-V zener. Circuit must be well grounded and shielded to avoid hum pickup by input, which could modulate output.—D. G. Stephenson, A Second Look at Linear Tuning, *QST*, March 1977, p 40–41.

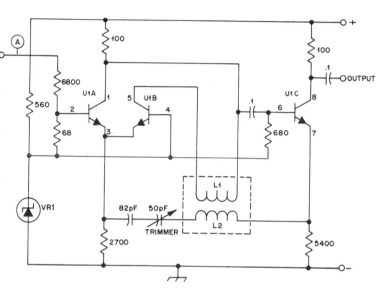

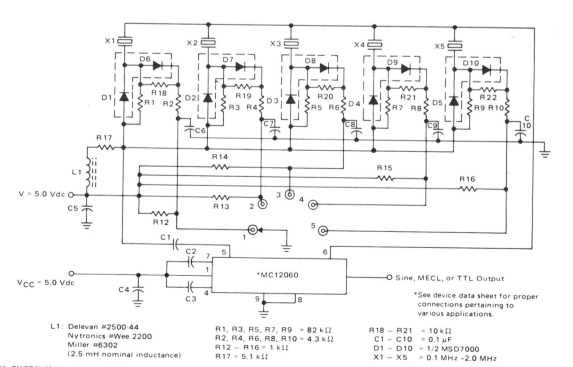

L1: Delevan #2500-44
Nytronics #Wee 2200
Miller #6302
(2.5 mH nominal inductance)

R1, R3, R5, R7, R9 = 82 kΩ
R2, R4, R6, R8, R10 = 4.3 kΩ
R12 – R16 = 1 kΩ
R17 = 5.1 kΩ

R18 – R21 = 10 kΩ
C1 – C10 = 0.1 μF
D1 – D10 = 1/2 MSD7000
X1 – X5 = 0.1 MHz –2.0 MHz

CRYSTAL-SWITCHING DIODES—Circuit for Motorola MC12060 crystal oscillator uses diodes as RF switches giving choice of five different crystal frequencies. Forward bias is applied to diode associated with desired crystal and reverse bias to diodes for other four crystals. Diode switching eliminates need to run high-frequency signals through mechanical switch, permits control of switching from remote location, and is readily adapted to electronic scanning. Requires only single 5-V supply. Frequency pulling is minimized.—J. Hatchett and R. Janikowski, "Crystal Switching Methods for MC12060/MC12061 Oscillators," Motorola, Phoenix, AZ, 1975, AN-756.

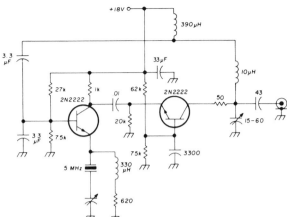

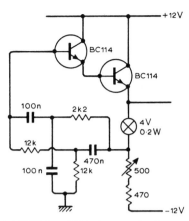

5-MHz LOW-NOISE CRYSTAL—Extremely low-noise series-mode crystal oscillator is designed for use in high-quality communication receivers. Either fundamental or overtone crystals can be used.—U. L. Rohde, Effects of Noise in Receiving Systems, *Ham Radio,* Nov. 1977, p 34–41.

RC CONTROL—Chief advantage is absence of attenuation at zero phase shift in passive RC network used to define frequency of oscillation. Output is 20 V P-P. Pilot lamp stabilizes loop gain to unity, eliminating need for thermistor.—W. R. Jackson, Oscillator Uses Passive Voltage-Gain Network, *Wireless World,* April 1975, p 175.

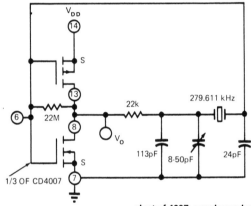

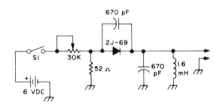

279.611-kHz CRYSTAL—DT-cut quartz crystal operating in CMOS inverter pair circuit serves as efficient timing circuit. Supply voltage can be from 5 to 15 V. With TA5987 low-voltage equivalent of 4007, supply can be 2.5 to 5 V. Stability is 4.3 PPM, not including temperature variations.—B. Furlow, CMOS Gates in Linear Applications: The Results Are Surprisingly Good, *EDN Magazine,* March 5, 1973, p 42–48.

100-kHz SINE—Tunnel-diode sine-wave oscillator uses single GE 2J-69. Frequency is stable provided there are no drastic temperature changes, but for long-term accuracy and stability a crystal oscillator is recommended.—Circuits, *73 Magazine,* May 1977, p 31.

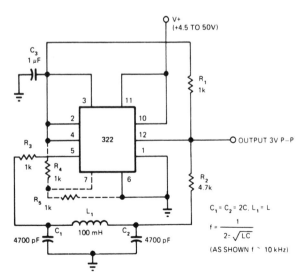

$$C_1 = C_2 = 2C, \; L_1 = L$$

$$f = \frac{1}{2 \cdot \sqrt{LC}}$$

(AS SHOWN f ≈ 10 kHz)

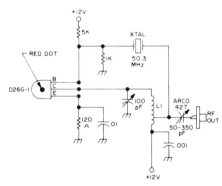

L1–25 TURNS 28 AWG TAPPED 4 TURNS FROM COLD END, WOUND ON 1/8 in DIAMETER FORM, APPROXIMATELY 1/2 in LG.

UP TO 100 kHz WITH 322 TIMER—Efficient LC oscillator uses IC timer as inverting comparator, with pi-network LC tank as resonant circuit. Output square wave is regulated to 3 V in amplitude, independently of supply voltage; upper supply limit should be 40 V instead of value shown. Sine-wave output of oscillator may also be used externally by adding single-supply opamp as buffer. Values shown give 10 kHz, but upper limit is 100 kHz.—W. G. Jung, Take a Fresh Look at New IC Timer Applications, *EDN Magazine,* March 20, 1977, p 127–135.

50-MHz CRYSTAL—Uses microtransistor as oscillator handling 100-mW input power and giving 40–50% efficiency. Article covers construction with microcomponents and gives other microtransistor circuits for low-power amateur radio use and possible bugging applications.—B. Hoisington, Introduction to "Microtransistors," *73 Magazine,* Oct. 1974, p 24–30.

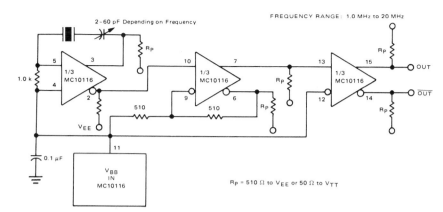

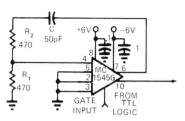

GATED 5-MHz RELAXATION—Output always starts in same phase with respect to gating signal. Frequency-selective network R_1-R_2-C provides positive feedback around MC 1545G gate-controlled wideband amplifier.—F. Macli, IC Op Amp Makes Gated Oscillator, *EDN Magazine*, Sept. 1, 1972, p 52.

1–20 MHz FUNDAMENTAL CRYSTAL—Oscillator requires no resonant tank circuit for frequencies below 20 MHz. Use of noninverting output makes oscillator section of Motorola MC10116 IC function simply as amplifier. Sec-

ond section is connected as Schmitt trigger to improve signal waveform. Third section is buffer providing complementary outputs.—B. Blood, "IC Crystal Controlled Oscillators," Motorola, Phoenix, AZ, 1977, AN-417B, p 4.

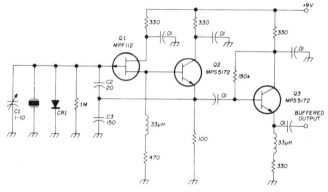

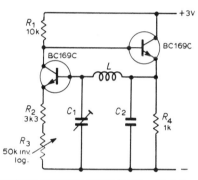

10–20 MHz CRYSTAL—Modification of basic Colpitts crystal oscillator has excellent load capacitance correlation and temperature stability. Crystal will oscillate very close to its series resonant point. Component values are optimized for 10–20 MHz. Emitter-follower Q2 provides power gain for feedback energy and gives high

crystal activity without changing phase angle of signal. Output buffer Q3 prevents loading of oscillator. Q1 is low-cost Motorola JFET, but practically any other JFET will work. CR1 is 1N914 or 1N4148.—D. L. Stoner, High-Stability Crystal Oscillator, *Ham Radio*, Oct. 1974, p 36–39.

GENERAL-PURPOSE UP TO 10 MHz—Variation of Colpitts oscillator uses negative feedback at all frequencies at which LC network does not provide phase inversion and voltage step-up. Choose values for coil and capacitors to give frequency desired. R_3 serves as regeneration control and for changing waveform of output.—G. W. Short, Good-Tempered LC Oscillator, *Wireless World*, Feb. 1973, p 84.

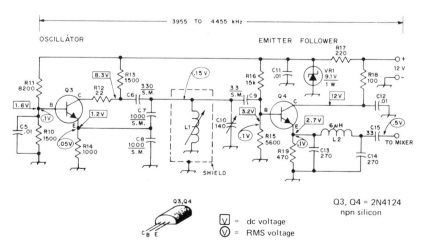

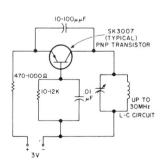

3.955–4.455 MHz VFO—Basic Colpitts LC oscillator designed for 80-meter receiver with 455-kHz IF uses zener in supply line to minimize frequency drift. Emitter-follower buffer contributes to stability by isolating oscillator from mixer. Low-pass filter C13-L2-C14 attenuates harmonic currents developed in Q3 and Q4. L1

is Miller 4503 1.7–2.7 μH variable inductor. L2 is 48 turns No. 30 enamel closewound on 1/4-inch wood dowel or polystyrene rod. Main tuning capacitor C10 can be 365-pF unit with six of rear rotor plates removed.—D. DeMaw and L. McCoy, Learning to Work with Semiconductors, *QST*, June 1974, p 18–22 and 72.

UP TO 30 MHz—Simple single-transistor RF oscillator is easily assembled from noncritical parts. Tuning capacitor and coil determine frequency.—*Circuits, 73 Magazine*, July 1977, p 35.

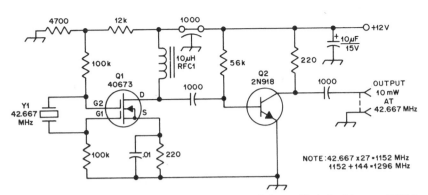

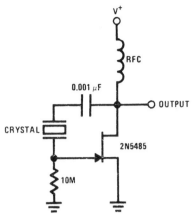

42.667-MHz MOSFET—Unusual crystal-controlled oscillator, similar to modified Pierce oscillator that uses crystal between grids 1 and 2 of tetrode tube, can be used as local oscillator in VHF and UHF converters. No trimming or tuning is required to get overtone frequency. If fundamental of crystal is desired, increase RFC1 to 100 μH or replace it with 1K resistor. Stability is excellent. Circuit works well with supply as low as 4 V.—G. Tomassetti, Dual-Gate MOSFET Offers an Unusual Crystal-Controlled Oscillator Concept, *QST*, June 1976, p 39.

JFET PIERCE CRYSTAL—Basic JFET oscillator circuit permits use of wide frequency range of crystals. High Q is maintained because JFET gate does not load crystal, thereby ensuring good frequency stability.—"FET Databook," National Semiconductor, Santa Clara, CA, 1977, p 6-26—6-36.

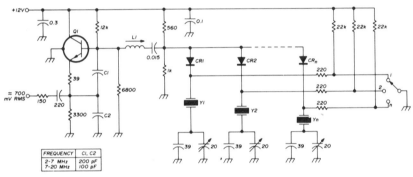

SWITCHED CRYSTALS—High stability is combined with multichannel selection by diode switching of crystals in range of 2–20 MHz, used in series-resonant mode. L1 is about 30 μH at 2 MHz and 1 μH at 20 MHz. Q1 is 2N708, HEP50, BC108, or similar NPN RF type. Diodes are switching types such as BAY67.—U. Rohde, Stable Crystal Oscillators, *Ham Radio*, June 1975, p 34–37.

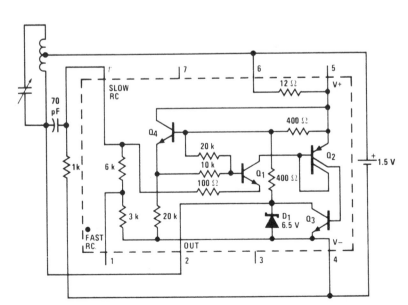

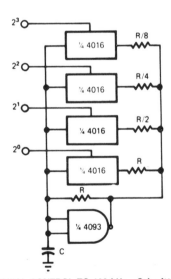

800-kHz OSCILLATOR—National LM3909 IC operating from single 1.5-V cell is used with standard AM radio ferrite antenna coil having tap 40% of turns from one end, with standard 365-pF tuning capacitor across coil. Developed for demonstrating versatility of this low-voltage IC.—"Linear Applications, Vol. 2," National Semiconductor, Santa Clara, CA, 1976, AN-154, p 8.

DIGITAL CONTROL TO 100 kHz—Schmitt trigger function of CD4093B IC gives oscillator operation over four decades of frequency without changing C. Basic frequency value is equal to k/RC, with k equal to 1.3 up to about 5 kHz and decreasing gradually to 1.0 at 100 kHz. Use of CD4016 quad transmission gate permits remote switching in of additional resistors to provide direct digital control of frequency. Arrangement shown gives choice of five unrelated frequencies, but binary selection of binary-weighted resistors will give choice of 16 unrelated frequencies.—R. Tenny, CMOS Oscillator Features Digital Frequency Control, *EDN Magazine*, June 5, 1976, p 114 and 116.

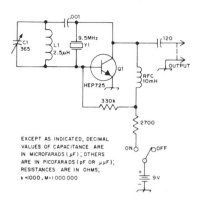

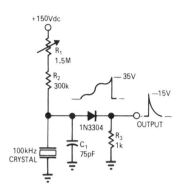

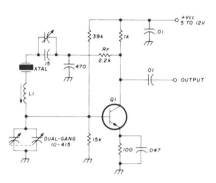

INCREASING CRYSTAL FREQUENCY—Adding parallel resonant circuit across crystal, tuned slightly above crystal frequency, makes oscillator frequency increase. Some plated crystals will work better than others in this circuit; third-overtone types operating on their fundamental generally give best results. Article covers theory of operation.—L. Lisle, The Tunable Crystal Oscillator, *QST*, Oct. 1973, p 30–32.

100-kHz CRYSTAL-DIODE RELAXATION—Crystal-controlled relaxation oscillator uses 1N3304 four-layer diode as active element. R_1 adjusts RC time constant so oscillator locks at fundamental frequency of crystal or at half this frequency.—R. D. Clement and R. L. Starliper, Crystal-Controlled Relaxation Oscillator, *EDN/EEE Magazine,* Oct. 15, 1971, p 62 and 64.

VARIABLE CRYSTAL—Maximum frequency shift is almost 10 kHz at 5 MHz. Use crystal made especially for variable operation. Frequency stability is good even at extremes of shift. Use 5–20 μH for L1 with crystals from 6–15 MHz, and 20–50 μH for 3–6 MHz. Q1 is 2N3563, 2N3564, 2N5770, BC107, BC547, BF115, BF180, SE1010, or equivalent.—R. Harrison, Survey of Crystal Oscillators, *Ham Radio*, March 1976, p 10–22.

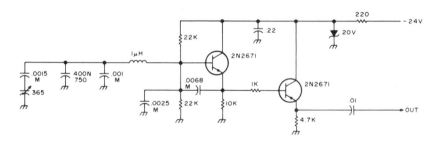

5 MHz ± 250 kHz—Simple and stable circuit using PNP transistors has tuning range of about 250 kHz in any segment of 5–9 MHz range, depending on how oscillator coil is set. Wind coil on ceramic form or use air-wound coil. Capacitors marked M should be mica for stability. Tuning capacitor is 365 pF, from AM radio. 400/N750 temperature-compensating capacitor can be replaced by 400-pF mica unless VFO is used in mobile application.—An Accessory VFO—the Easy Way, *73 Magazine,* Aug. 1975, p 103 and 106–108.

C2 — Double-bearing variable capacitor, 50 pF.
C3 — Miniature 30-pF air variable.
CR1 — High-speed switching diode, silicon type 1N914A.
L18 — 17- to 41-μH slug-tuned inductor, Q_u of 175 (J. W. Miller 43A335CBI in Miller S-74 shield can).
L19 — 10- to 18.7-μH slug-tuned pc-board inductor (J. W. Miller 23A155RPC).
RFC13, RFC14 — Miniature 1-mH rf choke (J. W. Miller 70F103AI).
VR2 — 8.6-V, 1-W Zener diode.

2.255–2.455 kHz LOCAL OSCILLATOR—Used in 1.8–2 MHz communication receiver having wide dynamic range. Oscillator has good stability, with circuit noise at least 90 dB below fundamental output. Amplifier Q14 provides required +7 dBm for injection into balanced mixer of receiver. Two-part article gives all other circuits of receiver.—D. DeMaw, His Eminence—the Receiver, *QST*, Part 1—June 1976, p 27–30 (Part 2—July 1976, p 14–17).

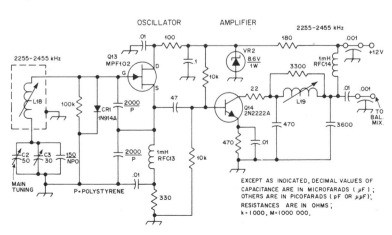

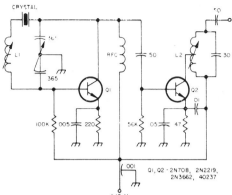

8 MHz ± 5 kHz—Tuning two-gang 365-pF variable capacitor through its range provides frequency change up to 5 kHz in output of 8-MHz crystal oscillator. L1 is 16–24 μH Miller 4507, and L2 is 40 turns No. 36 tapped at 13 turns, on ¼-inch slug-tuned form.—Circuits, *73 Magazine,* Jan. 1974, p 128.

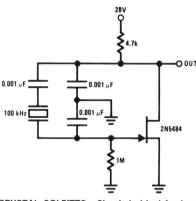

CRYSTAL COLPITTS—Circuit is ideal for low-frequency crystal oscillators because JFET circuit loading does not vary with temperature. Output frequency is determined by threshold used.—"FET Databook," National Semiconductor, Santa Clara, CA, 1977, p 6-26–6-36.

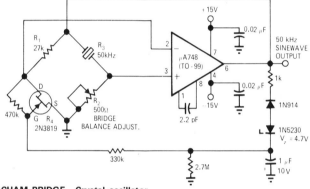

50-kHz MEACHAM BRIDGE—Crystal oscillator using Meacham bridge requires no transformers for producing low-distortion sine-wave output. Quartz crystal should be cut for operation in series-resonant mode. With minor modifications, same circuit can be used for 100- and 200-kHz crystals. By adding single-transistor stage, oscillator can be used as clock generator for TTL circuits.—K. J. Peter, Stable Low-Distortion Bridge Oscillator, *EDN/EEE Magazine,* Nov. 15, 1971, p 50–51.

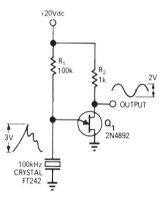

100-kHz CRYSTAL-FET RELAXATION—Adding crystal in frequency-determining circuit improves frequency stability of UJT relaxation oscillator. With charging capacitor replaced by 100-kHz quartz crystal, measured output frequency was 99.925 kHz.—R. D. Clement and R. L. Starliper, Crystal-Controlled Relaxation Oscillator, *EDN/EEE Magazine,* Oct. 15, 1971, p 62 and 64.

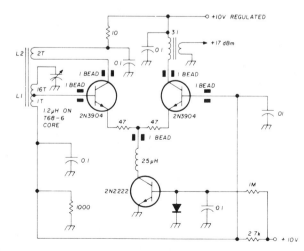

LOW-NOISE 5-MHz—Very low-noise high-Q LC oscillator operating at 5 MHz is designed for use in high-performance communication receivers. Oscillator uses two stages, one operating in class A and the other operating as limiter that also serves as feedback path.—U. L. Rohde, Effects of Noise in Receiving Systems, *Ham Radio,* Nov. 1977, p 34–41.

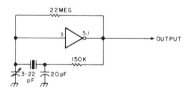

CRYSTAL WITH CMOS INVERTER—Simple mono multivibrator circuit using MC14007 or CD4007 operates in frequency range from 10 kHz up to top limit of about 10 MHz, with exact frequency depending on values used for R and C. Pin 7 of IC is VSS and pin 14 is VDD. Pins 5 and 1 must be connected together for proper operation.—W. J. Prudhomme, CMOS Oscillators, *73 Magazine,* July 1977, p 60–63.

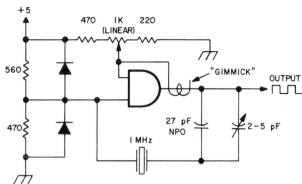

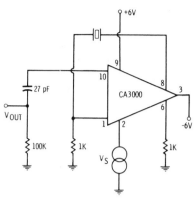

1 MHz WITH ONE GATE—Crystal oscillator uses only one section of SN7408 TTL quad AND gate. Use series-resonant crystal having 30-pF series capacitance. Adjust 1K pot for reliable start-up and symmetrical square-wave output. Diodes are 1N34A or 1N914. Gimmick is 1 or 2 turns of insulated wire wrapped around output lead.—Clyde E. Wade, Jr., An Even Simpler Clock Oscillator, *73 Magazine*, Nov./Dec. 1975, p 164.

MODULATED CRYSTAL—CA3000 differential amplifier is operated as efficient crystal-controlled oscillator. Output frequency depends on crystal. If desired, RF output can be modulated with low-frequency tone applied between pin 2 and ground.—E. M. Noll, "Linear IC Principles, Experiments, and Projects," Howard W. Sams, Indianapolis, IN, 1974, p 91.

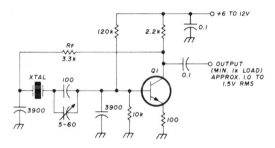

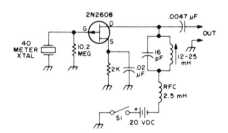

50–500 kHz CRYSTAL—Parallel-mode low-frequency oscillator makes excellent BFO for 455 kHz. If oscillator will not start, reduce value of feedback resistor R_F. Increasing R_F reduces harmonic output, but oscillator may then take up to 20 s to reach full output. For crystals with specified load capacitance of 30 or 50 pF, remove 100-pF capacitor C1 in series with crystal. Q1 is 2N2920, 2N2979, 2N3565, 2N3646, 2N5770, BC107, or BC547.—R. Harrison, Survey of Crystal Oscillators, *Ham Radio*, March 1976, p 10–22.

7 MHz—Uses single Siliconix 2N2608 FET. Keep leads short. Coil can be air-wound or permeability-tuned. If tuning capacitor is variable, coil value can be fixed. RF output level depends on circuit voltages and on activity of crystal used.—Q & A, *73 Magazine*, April 1977, p 165.

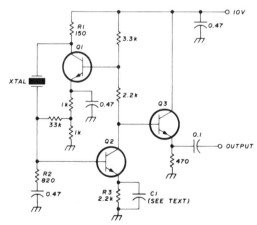

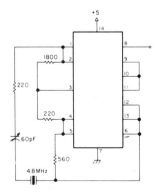

for 20-150 kHz crystals	for 150-500 kHz crystals
Q1, Q2, Q3	**Q1, Q2, Q3**
2N3565	BC107, BC547
2N2920	2N3565
2N2979	2N5770
	2N2222

20–500 kHz CRYSTAL—Series-mode oscillator requires no tuned circuit, gives choice of sine or square output, and has good frequency and mode stability. Works nicely with troublesome FT241 crystals. If any crystal fails to start reliably, increase R1 to 270 ohms and R2 to 3.3K. For square-wave operation, C1 is 1-$^\mu$F nonelectrolytic. Omit C1 for sine-wave operation; harmonic output is then quite low, with second harmonic typically −30 dB. Output is about 1.5-VRMS sine wave or 4-V square wave.—R. Harrison, Survey of Crystal Oscillators, *Ham Radio*, March 1976, p 10–22.

4.8 MHz—Uses all four sections of 7400 quad dual-input NAND gate to give 4.8 MHz output at pin 8, as harmonic-rich square wave. Can cause severe television interference during testing. Article gives five other crystal oscillator circuits using same IC.—A. MacLean, How Do You Use ICs?, *73 Magazine*, Oct. 1976, p 38–41.

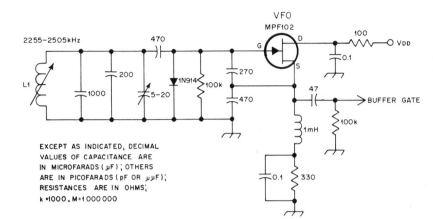

PRECISION VFO—Permeability-tuned oscillator provides stability and linearity at low cost for receivers with 160-meter tunable IF stages. L1 has 28 turns No. 36 enamel closewound on J. W. Miller form 64A022-2. Article covers construction of tuning dial, incuding contouring of L1 core to give good dial linearity. Frequency coverage is 2.255–2.505 MHz. Direct-reading dial is accurate within 1.5 kHz over entire 250-kHz tuning range.—W. A. Gregoire, Jr., A Permeability-Tuned Variable-Frequency Oscillator, *QST,* March 1978, p 26–28.

7 MHz ± 50 kHz—Requires no tuning capacitors. Collector-to-base junctions of two 2N3053 transistors perform function of varactor diodes to provide tuning over range of about 50 kHz centered on 7 MHz. Capacitors marked M should be mica.—An Accessory VFO—the Easy Way, *73 Magazine,* Aug. 1975, p 103 and 106–108.

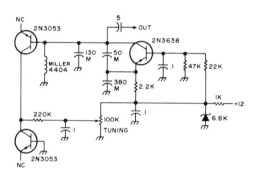

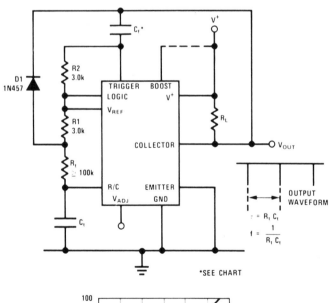

TIMER AS OSCILLATOR—Output of National LM122 timer is fed back to trigger input through capacitor to give self-starting oscillator. Frequency is $1/R_tC_t$. Output is narrow negative pulse having duration of about $2R2C_t$. Conservative value for C_t for optimum frequency stability can be chosen from graph based on size of timing capacitor C_t.—C. Nelson, "Versatile Timer Operates from Microseconds to Hours," National Semiconductor, Santa Clara, CA, 1973, AN-97, p 10.

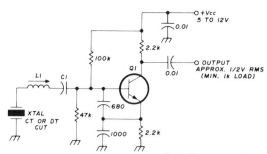

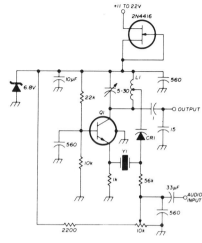

150–500 kHz CRYSTAL—Circuit is series-mode if C1 is 0.01 µF. Parallel-mode crystals can be used if C1 is equal to specified load capacitance (30, 50, or 100 pF) for crystal. Harmonic output is usually better than −30 dB. Circuit is particularly good for crystals prone to oscillate un- desirably at twice fundamental frequency. L1 is 800–2000 µH for 150–300 kHz, and 360–1000 µH for 300–500 kHz. Adjusting slug in L1 pulls crystal frequency. Q1 is 2N3563, 2N3564, 2N3693, BC107, BC547, or SE1010.—R. Harrison, Survey of Crystal Oscillators, *Ham Radio*, March 1976, p 10–22.

OSCILLATOR-DOUBLER—Overtone crystal oscillator circuit that frequency-doubles in transistor can be frequency-modulated or used as stable voltage-controlled crystal oscillator. Tuning range with 70-MHz third-overtone crystal is typically 30 kHz at crystal frequency or 60 kHz at output. L1 is resonant with C1 at desired output frequency. Tap for varactor CR1 (Motorola BB105B or BB142) is at one-fourth total number of turns. Q1 is 2N918, BF115, HEP709, or equivalent.—U. Rohde, Stable Crystal Oscillators, *Ham Radio*, June 1975, p 34–37.

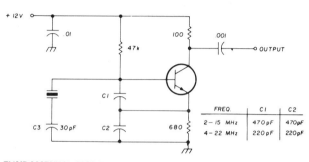

FREQ.	C1	C2
2 – 15 MHz	470 pF	470 pF
4 – 22 MHz	220 pF	220 pF

2–22 MHz FUNDAMENTAL-MODE—International Crystal OF-1 oscillator for fundamental-mode crystal has no LC tuned circuits and requires no inductors. With 28.3-MHz third-over- tone crystal, output is at fundamental of crystal or about 9.43 MHz.—C. Hall, Overtone Crystal Oscillators Without Inductors, *Ham Radio*, April 1978, p 50–51.

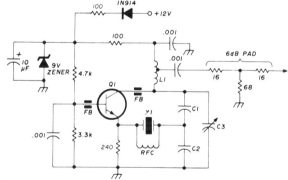

C1	10 pF mica
C2	20 to 60 pF mica. Use as high value as possible (until circuit just oscillates reliably when C3 is tuned through resonance)
C3	20 pF piston or miniature trimmer
L1	8 turns no. 24 (0.5mm) on Amidon T37-12 toroid core, tapped 3 turns from cold end
Q1	Fairchild 2N5179 recommended but 2N2857, 2N3563, 2N918 or equivalent may be substituted
RFC	0.39 µH. Resonates with crystal holder capacitance (4 to 6 pF typical) for parallel resonance at crystal frequency
Y1	90 to 125 MHz, 5th or 7th overtone, series-resonant, HC-18/U crystal. Cut leads as short as possible (¼" or 6mm maximum)

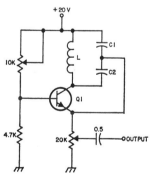

Q1 – 2N2925, 2N3392

FREQUENCY	C1	C2	L
50 kHz	3500 pf	1500 pf	10 mH
80 kHz	2200 pf	910 pf	6.2 mH
100 kHz	1800 pf	750 pf	4.7 mH
200 kHz	910 pf	390 pf	2.2 mH
455 kHz	390 pf	160 pf	1 mH
1000 kHz	180 pf	75 pf	0.47 mH

90–125 MHz CRYSTAL—Recmmended for VHF/UHF converters. Output is 5 to 15 mW. Crystal should be high-quality fifth- or seventh-overtone type. Ferrite bead FB prevents undesired oscillation above 500 MHz. For best stability, allow crystal to operate at its natural series-res- onant frequency and use regulated power sup- ply.—J. Reisert, VHF/UHF Techniques, *Ham Radio*, March 1976, p 44–48.

50–1000 kHz—Simple single-transistor circuit provides extremely stable beat-frequency oscil- lator for which frequency can be changed by using tank-circuit components listed in table.— Circuits, *73 Magazine*, Feb. 1974, p 101.

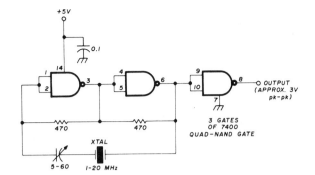

NAND-GATE TTL CRYSTAL—Overcomes problems of poor starting performance and has upper frequency limit of 20 MHz. Suitable for applications requiring high-output aperiodic oscillator. Excellent as frequency marker.—R. Harrison, Survey of Crystal Oscillators, *Ham Radio*, March 1976, p 10–22.

2–20 MHz VXO—Variable-frequency crystal oscillator plus buffer, using Signetics N7404A hex inverter or equivalent, covers 2–20 MHz. Only three inverters are used, two forming oscillator and one as output buffer. V_{CC} is +5 V. Crystals can operate at fundamental, third, or fifth overtone. Frequency-limiting capacitor C_P can be 15 pF. Only higher-frequency crystals can be moved useful amounts without creating instability problems. Article gives design equations and tables showing frequencies obtained with various crystals for various values of frequency controls C_v (0–100 pF) and L_v (0–17 μH).—B. King, Hex Inverter VXO Circuit, *Ham Radio*, April 1975, p 50–55.

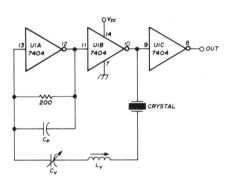

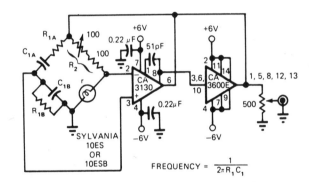

$$FREQUENCY = \frac{1}{2\pi R_1 C_1}$$

CAPACITIVELY TUNED WIEN—Output of amplifier is connected to apex of Wien bridge. Positive feedback is taken from junction of C_{1A} and C_{1B} for noninverting input of first opamp, while negative feedback is taken from other junction of bridge for inverting input. Oscillation is sustained when $R_2 = 2r$. Nonlinearity of lamp r provides stabilization of oscillator. Frequency depends on values used for bridge components.— H. D. Olson, Wien-Bridge Oscillator Is Capacitively Tuned, *EDN Magazine*, Aug. 5, 1975, p 74.

65–110 MHz OVERTONE—Uses fifth- or seventh-overtone crystals. RF choke formed by L2 is wound on low-value resistor to suppress lower-frequency resonances of crystal. Buffer is recommended. Circuit is slightly frequency-sensitive to supply voltage variations, so use well-regulated supply. Q1 is 2N3563, 2N3564, 2N5770, BF180, BF200, or SE1010.—R. Harrison, Survey of Crystal Oscillators, *Ham Radio*, March 1976, p 10–22.

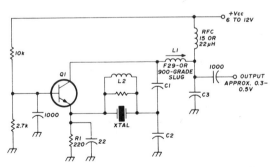

L1	65-85 MHz:	7 turns no. 22 (0.6mm) or no. 24 (0.5mm) enamelled, closewound on 3/16'' (5mm) diameter form		
	85-110 MHz:	4 turns no. 22 (0.6mm) or no. 24 (0.5mm) enamelled, on 3/16'' (5mm) diameter form, turns spaced one wire diameter		
L2	10 turns no. 34 (0.2mm) closewound on low-value ¼-watt resistor			
C1	65-85 MHz:	15 pF	85-110 MHz:	10 pF
C2	65-85 MHz:	150 pF	85-110 MHz:	100 pF
C3	65-85 MHz:	100 pF	85-110 MHz:	68 pF

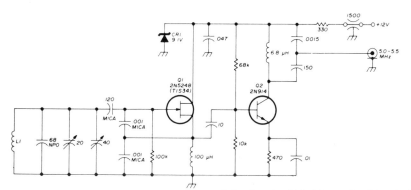

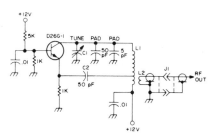

5–5.5 MHz VFO—Used in solid-state five-band communication receiver. Temperature compensation is provided by 20-pF trimmer that sets band center. L1 is 34 turns No. 24 on Amidon T50-6 toroid core.—P. Moroni, Solid-State Communications Receiver, *Ham Radio*, Oct. 1975, p 32–41.

51–55 MHz—Tunable local oscillator is padded to tune over range required for use with 1.65-MHz IF in 6-meter receiver, using Johnson type U 14-plate tuning capacitor. Can also serve as test transmitter putting out up to 20 mW. L1 is 9 turns No. 26 tapped 1 turn from low end, and L2 is 1 or 2 turns. Article covers construction in 1¼ × 1¼ × ½ inch box.—B. Hoisington, A Real Hot Front End for Six, *73 Magazine*, Nov. 1974, p 88–90 and 92–94.

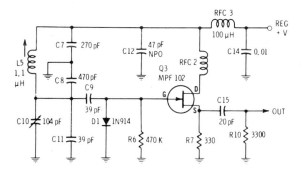

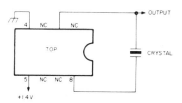

10-MHz VFO—Values shown for high-stability variable-frequency oscillator give operation in 10-MHz range. Stable supply voltage is essential. Use silver mica capacitors in gate circuit for maximum stability.—E. M. Noll, "FET Principles, Experiments, and Projects," Howard W. Sams, Indianapolis, IN , 2nd Ed., 1975, p 193–194.

465-kHz FOR IF TUNE-UP—Simple crystal oscillator using National LM3909N is adjusted to exactly desired frequency with capacitor in series with pin 8. Drain from AA cell is less than 0.5 mA at 1.2 V. Use 465-kHz crystal and couple oscillator to receiver input with 100-pF capacitor. With 100-kHz crystal, circuit will generate strong harmonics beyond 30 MHz; to zero-beat with WWV, use about 10 pF in series with crystal.—I. Queen, Simple Crystal Oscillator, *Ham Radio*, Nov. 1977, p 98.

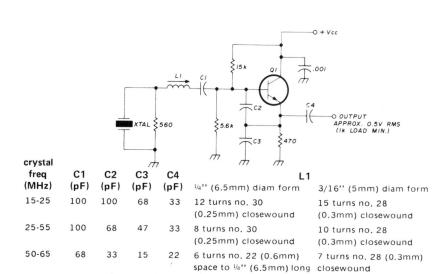

crystal freq (MHz)	C1 (pF)	C2 (pF)	C3 (pF)	C4 (pF)	L1 ¼'' (6.5mm) diam form	3/16'' (5mm) diam form
15-25	100	100	68	33	12 turns no. 30 (0.25mm) closewound	15 turns no. 28 (0.3mm) closewound
25-55	100	68	47	33	8 turns no. 30 (0.25mm) closewound	10 turns no. 28 (0.3mm) closewound
50-65	68	33	15	22	6 turns no. 22 (0.6mm) space to ¼'' (6.5mm) long	7 turns no. 28 (0.3mm) closewound

15–65 MHz IMPEDANCE-INVERTING—Uses third-overtone crystals. L1 trims crystal frequency. Resistor across crystal prevents oscillation at undesired modes. Starting is reliable and stability is good. Q1 is 2N3563, 2N3564, 2N5770, BF180, BF200, or SE1010.—R. Harrison, Survey of Crystal Oscillators, *Ham Radio*, March 1976, p 10–22.

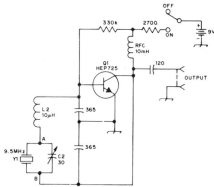

9.5-MHz TUNABLE CRYSTAL—Clapp oscillator with inductance in series with crystal can be tuned with C2 as much as 100 kHz below rated frequency of crystal. Based on making crystal act as capacitive reactance below its series-resonant frequency. Circuit can be adapted to other amateur bands by keeping reactances of various components approximately the same.—L. Lisle, The Tunable Crystal Oscillator, *QST*, Oct. 1973, p 30–32.

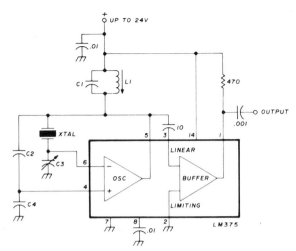

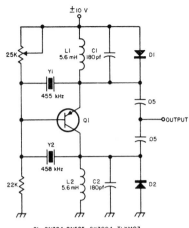

IC CRYSTAL—Uses LM375 IC with crystals from 3 to 20 MHz in parallel mode. Will oscillate with only 4-V supply, but output voltage increases with supply voltage. L1-C1 is resonant at crystal frequency. Adjust L1 only for maximum output, not for trimming frequency. If C3 is 3–30 pF, it can be used to adjust frequency of crystal.—R.

crystal freq (MHz)	C2/C3 (pF)	C4 (pF)
3-10	22	180
10-20	10	82

Harrison, Survey of Crystal Oscillators, *Ham Radio,* March 1976, p 10–22.

DUAL-FREQUENCY CRYSTAL—Uses two different crystals, with frequency being changed by reversing supply voltage. Transistor then inverts itself and gain reduces to about 2, which is adequate for oscillator operation. Provides two frequencies from single stage with minimum of switching.—Circuits, *73 Magazine,* Feb. 1974, p 101.

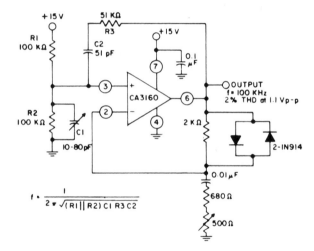

100-kHz WIEN-BRIDGE—CA3160 opamp in bridge circuit operates from single 15-V supply. Parallel-connected diodes form gain-setting network that stabilizes output voltage at about 1.1 V. 500-ohm pot is adjusted so oscillator always starts and oscillation is maintained.—"Linear Integrated Circuits and MOS/FET's," RCA Solid State Division, Somerville, NJ, 1977, p 271–272.

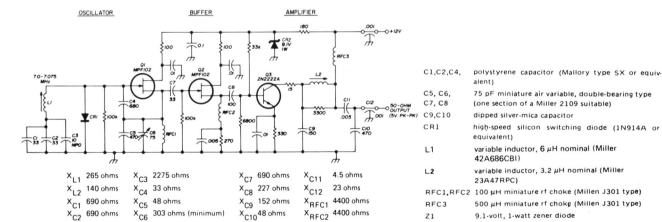

X_{L1} 265 ohms	X_{C3} 2275 ohms	X_{C7} 690 ohms	X_{C11} 4.5 ohms
X_{L2} 140 ohms	X_{C4} 33 ohms	X_{C8} 227 ohms	X_{C12} 23 ohms
X_{C1} 690 ohms	X_{C5} 48 ohms	X_{C9} 152 ohms	X_{RFC1} 4400 ohms
X_{C2} 690 ohms	X_{C6} 303 ohms (minimum)	X_{C10} 48 ohms	X_{RFC2} 4400 ohms

LOW-DRIFT 7-MHz VFO—Low-drift solid-state design for 40-meter band has maximum change of only 25 Hz from cold start to full warm-up at 25°C. After stabilization, maximum hunting is 5 Hz. Drift is minimized by paralleling two or more capacitors in critical parts of circuit. Series-tuned Colpitts oscillator is followed by two buffer stages, with second providing enough amplification for practical amateur work while further improving isolation of oscillator. Low-impedance output network minimizes oscillator pulling from load changes. Article stresses importance of choosing and using components that minimize drift.—D. DeMaw, VFO Design Techniques for Improved Stability, *Ham Radio,* June 1976, p 10–17.

Phase Control Circuits

Includes circuits for measuring, shifting, comparing, and digitally controlling phase of signal. Many use phase-locked loops. See also Lamp Control, Motor Control, Power Control, and Temperature Control chapters.

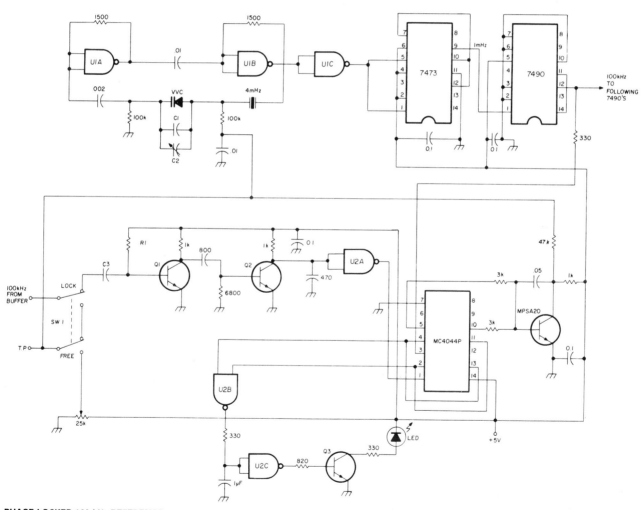

PHASE-LOCKED 100-kHz REFERENCE—Uses 4-MHz crystal in oscillator, with voltage-variable capacitor VVC in parallel with fixed and variable capacitors for setting frequency precisely. Varicap or silicon diode can also be used for VVC. Control voltage for VVC is developed by Motorola MC4044P phase-frequency detector and associated MPSA20 amplifier and filter. 7473 and 7490 ICs divide 4-MHz signal by 4 and then by 10 to give 100 kHz. Main output can be further divided with additional 7490s, down to 60 Hz for driving electric clock if desired. Adjust C3 and R1 for symmetrical square wave at pin 1 of MC4044P, with clean leading and trailing edges. Typical values are 68 pF for C3 and 300K for R1, but values will depend on transistors used. Transistor types are not critical. Gates U2B and U2C with Q3 form lock indicator circuit that turns on LED when 4-MHz oscillator is phase-locked to output of external high-stability 100-kHz frequency standard. U1 and U2 are SN7400.—C. A. Harvey, How to Improve the Accuracy of Your Frequency Counter, *Ham Radio*, Oct. 1977, p 26–28.

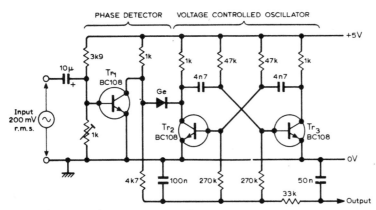

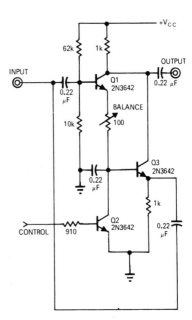

AF PLL—Addition of components to conventional two-transistor MVBR gives simple phase-locked loop. Tr_1 and diode form logic gate that conducts during alternate half-cycles of input and VCO waveforms respectively. Output of this phase detector, when filtered, is most negative when waveforms are in phase, and most positive when they are out of phase. Once phase lock has been established, it is maintained by VCO over range of 100 to 3000 Hz.—J. B. Cole, Simple Phase-Locked Loop, *Wireless World,* June 1977, p 56.

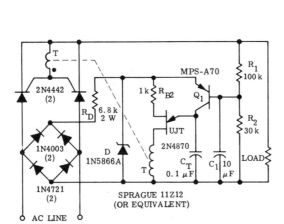

VOLTAGE-CONTROLLED PHASE SHIFTER—Circuit shifts carrier 180° by sensing polarity of modulating voltage. Operating range is 5 kHz to 10 MHz. Circuit can also be used to convert unipolar pulses to alternate bipolar pulses or vice versa when synchronized square wave is supplied to control input. With 0 V at base of Q2, Q3 will amplify RF voltage applied to input, without phase shift. To actuate switch and provide 180° phase shift, positive voltage is applied to base of Q2 so it saturates and cuts off, allowing Q1 to conduct. Output then appears across load with phase reversed.—A. H. Hargrove, Simple Circuits Control Phase-Shift, *EDN Magazine,* Jan. 1, 1971, p 39.

FULL-WAVE FEEDBACK—Used when average load voltage is desired feedback variable for full-wave phase control of load power. Circuit requires use of pulse transformer T.—D. A. Zinder, "Unijunction Trigger Circuits for Gated Thyristors," Motorola, Phoenix, AZ, 1974, AN-413, p 4.

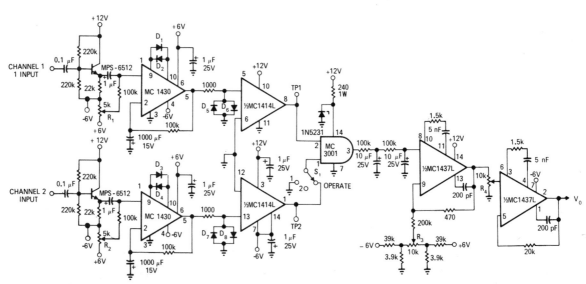

100 Hz TO 1 MHz PHASE METER—Provides better than 2% accuracy over most of frequency range, as required for making Bode plots. Based on squaring two sine waves and comparing amount of overlap to total period of an input wave. This gives directly the amount of phase difference between input wave trains, up to 180°. Instead of measuring periods, overlap is integrated over total period to give average of ON to OFF times that can be read as phase difference on voltmeter. Article gives performance specifications and describes circuit operation in detail.—D. Kesner, IC Phase Meter Beats High Costs, *EDN/EEE Magazine,* Oct. 15, 1971, p 49–52.

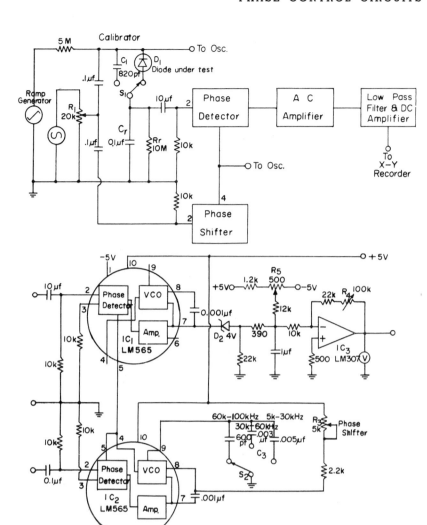

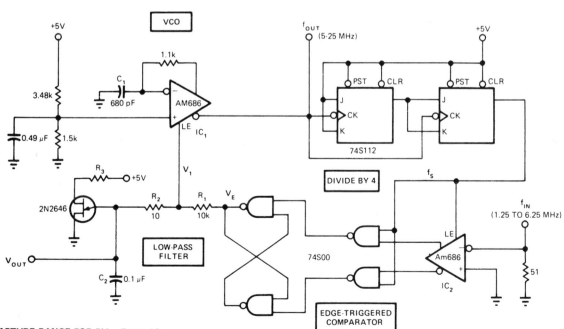

0–90° SHIFTER—Used in automatic plotter for measuring capacitance-voltage characteristics of Schottky barrier solar cells. Diode under test is connected as shown in block diagram. Phase of square-wave output from IC_2 can be shifted continuously from 0 to 90° by adjusting R_3. Article gives ramp circuit and design equations.— J. T. Lue, An Automatic C-V Plotter and Junction Parameter Measurements of MIS Schottky Barrier Diodes, *IEEE Journal of Solid-State Circuits,* Aug. 1978, p 510–514.

WIDE CAPTURE RANGE FOR PLL—Fast wideband phase-locked loop uses one Am686 latching comparator as voltage-controlled oscillator, while other is coupled with TTL latch to produce edge-triggered comparator. VCO and comparator combined with low-pass filter R_1-R_2-C_2 form PLL. When locking fails, UJT causes V_{OUT} to scan, repetitively sweeping all frequencies in VCO range until lock is restored. Capture and locking ranges are both equal at ±60% for 5-MHz input.—M. C. Hahn, PLL's Capture Range Equals Its Locking Range, *EDN Magazine,* Sept. 20, 1977, p 117 and 119–120.

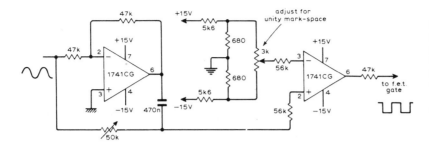

SHIFTING AND SQUARING—Circuit uses two opamps to derive phase-shifted reference square wave and DC output signal of phase-sensitive detector from same sine-wave signal source. Article gives theory of operation and waveforms for various operating conditions.—G. B. Clayton, Experiments with Operational Amplifiers, *Wireless World*, July 1973, p 355–356.

VOLTAGE FEEDBACK—Used when quantity to be sensed is isolated varying DC voltage e_s such as output of tachometer. Operating point is determined by setting of R_c. Output of voltage feedback circuit goes to thyristor in series with load.—D. A. Zinder, "Unijunction Trigger Circuits for Gated Thyristors," Motorola, Phoenix, AZ, 1974, AN-413, p 4.

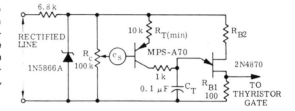

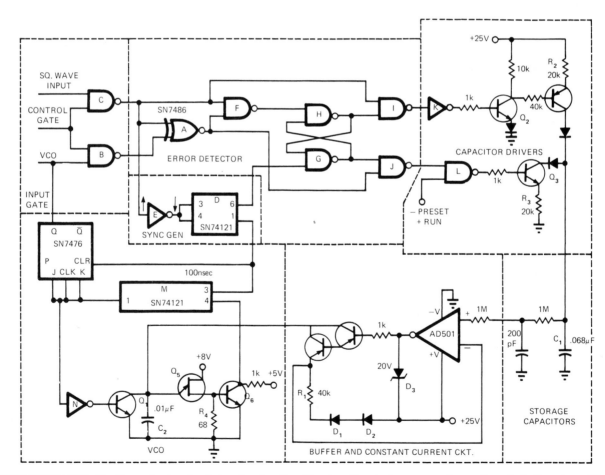

FASTER PHASE LOCK—Circuit was developed to reduce the normally long acquisition time of phase-locked loops when measuring frequency of short signal bursts. Synchronization of VCO to input phase allows correction pulses to be developed in correct polarity only, to give lockup time less than 10 cycles of input when using idling frequency of 12 kHz for VCO. Input signals are compared to those of VCO at EX-CLUSIVE-OR gate A. Gating of error pulses by gate F and flip-flop G-H allows I or J to drive current pulses of correct polarity into C_1. Voltage correction on C_1, controlled by values of R_2 and R_3, is proportional to width of error pulses. Article covers circuit operation in detail.—R. Bohlken, A Synchronized Phase Locked Loop, *EDN Magazine,* March 20, 1973, p 84–85.

SECTION 49
Photoelectric Circuits

Covers circuits involving change in light on photocell or other light-sensitive device, including punched-tape reader, transmission of voice or data signals on light beam, and solar-power oscillator. See also Fiber-Optic, Instrumentation, and Optoelectronic chapters.

LIGHT-BEAM VOICE TRANSMITTER—Opamp and transistor together provide amplitude modulation of LED in accordance with amplitude variations of microphone output signal. Requires only single 9-V supply. Other three sections of opamp are not used. Designed for dynamic microphone. Q1 is 2N2222 (Radio Shack 276-2009).—F. M. Mims, "Optoelectronic Projects, Vol. 1," Radio Shack, Fort Worth, TX, 1977, 2nd Ed., p 34–43.

RATIO OF TWO UNKNOWNS—Developed for use when two signals are time-shared on same input line, such as exists when two LEDs alternately illuminate single photocell. Measures ratio of amplitudes of unknowns with accuracy better than 1%. During time period T_1, input is sampled through S_2 and stored on C_2 for comparison with reference voltage. Result is applied through switchable amplifier network A_{FB} to gain control element which is LED-photoresistor coupled pair (CLM 6000). This closed loop adjusts signal gain to make denominator of ratio equal to reference voltage. Numerator, corresponding to time T_2, is multiplied by same gain so numerator output is proportional to desired ratio B/A of unknowns. Article describes circuit operation in detail.—R. E. Bober, Here's a Low-Cost Way to Measure Ratios, *EDN Magazine*, March 5, 1976, p 108, 110, and 112.

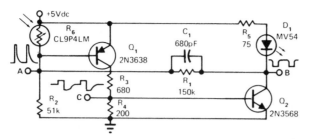

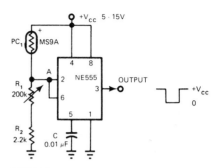

5-kHz PHOTOCELL OSCILLATOR—Provides 5-V pulses at about 5 kHz only if photocell is illuminated by its companion LED. Repetition rate varies with illumination, so interruption or attenuation of light produces easily detected frequency change that can be used as control signal. Applications include fail-safe interruption monitor and illumination transducer. Oscillation stops if beam is completely interrupted or if strong ambient light falls on photocell.—H. L. Hardy, FM Pulsed Photocell Is Foolproof, *EDN Magazine*, March 5, 1975, p 72.

PUNCHED-TAPE READER—Connection of 555 timer as Schmitt trigger produces output pulses with sharp rise and fall times that are independent of tape speed. Output is compatible with TTL or CMOS circuits. When scanning light beam hits hole in punched card or tape, resistance of light-sensitive resistor drops sharply and voltage at pins 2 and 6 rises above 0.67 V_{cc}. Voltage at output pin 3 then drops sharply from V_{cc} to 0 V. When PC_1 goes dark, circuit switches rapidly back to original state. Reverse PC_1 and R_1-R_2 for positive edge-triggered logic.—S. Sarpangal, 555 Timer Implements Tape Reader, *EDN Magazine*, Jan. 5, 1978, p 86 and 90.

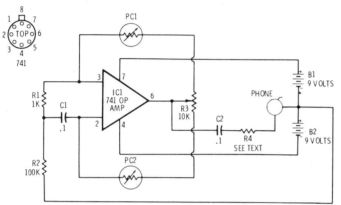

AUDIBLE LIGHT SENSOR—741 opamp is connected as audio oscillator with Radio Shack 276-677 photocells in feedback circuits. When light strikes PC1, its resistance decreases and frequency of audio tone in headphone decreases correspondingly. When light strikes PC2, which is connected to noninverting input of 741, increase in illumination serves to increase frequency. Choose R4 to reduce volume to desired level. R3 is balancing control for photocells.—F. M. Mims, "Integrated Circuit Projects, Vol. 2," Radio Shack, Fort Worth, TX, 1977, 2nd Ed., p 81—86.

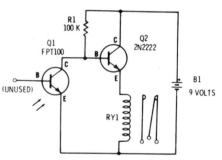

PHOTOTRANSISTOR RELAY—With phototransistor Q1 dark, R1 biases Q2 into conduction and miniature SPDT relay (Radio Shack 275-004) is energized. When light falls on Q1, Q2 is turned off and relay drops out. Battery drain is about 5 mA in darkness, dropping almost to 0 mA with light.—F. M. Mims, "Transistor Projects, Vol. 3," Radio Shack, Fort Worth, TX, 1975, p 69—74.

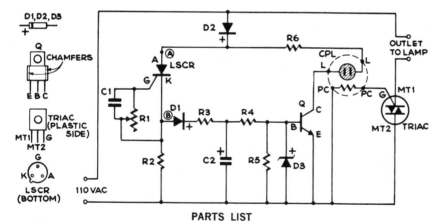

PARTS LIST

C1—0.1mfd capacitor
C2—10mfd @150V capacitor
R1—1-meg. carb. potentiometer
R2—82,000-ohm, ½w resistor
R3—390-ohm, 1w resistor
R4—2.2-megohm, ½w resistor

R5—560,000-ohm, ½w resistor
R6—22,000-ohm, ½w resistor
D1,D2—Diode (Motorola HEP 156)
D3—Zener diode, 6.2V (Motorola HEP 103 or equiv.)
Q—High-voltage transistor

(Motorola HEP S3022)
LSCR—Light-op. SCR, 200V (Radio Shack 276-1081)
Triac—Mot. HEP R1725
CPL—Light coupler Sigma 301T1-120A1 (SW Tech. Prod., 219 W. Rhapsody, San Antonio, Tex.)

GARAGE-LIGHT CONTROL—When mounted on far wall in garage, controller picks up headlight beams as car is driven in at night and turns on one or more garage lights long enough (3 min) for driver to get out of car and reach exit. Controller then flickers lights as warning and begins dimming them out. With parts specified, will handle up to 800 W of lamps. Adjust sensitivity control R1 so light in optocoupler CPL comes on when headlights strike light-operated SCR. Controller must be kept out of direct sunlight. For manual control, connect pushbutton switch between points A and B. To increase time delay, increase value of C2. With 20 μF, time will be doubled.—C. R. Lewart, Automatic Garage Light Control, *Popular Science*, July 1973, p 110.

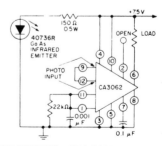

ON/OFF CONTROL—RCA CA3062 combination photodetector and power amplifier provides ON/OFF output in response to light signal. Output transistors in IC should be either saturated or blocked to avoid heat rise in silicon chip. Complementary outputs give choice of load normally on or normally off when light from infrared emitter falls on photo input of IC. Interruption of light path then produces opposite load condition.—"Linear Integrated Circuits and MOS/FET's," RCA Solid State Division, Somerville, NJ, 1977, p 156.

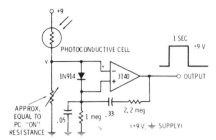

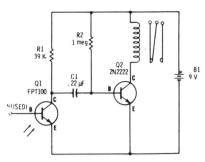

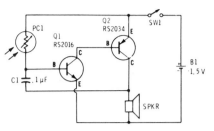

LIGHT-CHANGE DETECTOR—Combination amplifier and detector using 3140 opamp responds only to sudden changes in light on photocell while ignoring slow changes in ambient light. When beam is suddenly broken, opamp output swings positive and stays positive for delay time set by recharging of 0.05-μF capacitor on positive input. Delay locks out spurious signals until photocell resets itself to normal illumination. Values shown give time-out delay of about 1 s, with clean conditioned rectangular output pulse.—D. Lancaster, "CMOS Cookbook," Howard W. Sams, Indianapolis, IN, 1977, p 346–347.

LIGHT-CHANGE SENSOR DRIVES RELAY—Capacitive coupling between phototransistor and bipolar transistor makes circuit respond only to interruptions or rapid changes in light while ignoring normal gradual changes in ambient light as caused by clouds or at sunrise. Relay pulls in when flash of light occurs and drops out when light is removed. Use Radio Shack 275-004 miniature relay.—F. M. Mims, "Transistor Projects, Vol. 3," Radio Shack, Fort Worth, TX, 1975, p 69–74.

AUDIBLE LIGHT METER—Low light on cadmium sulfide photocell (Radio Shack 276-116) produces series of clicks in miniature 8-ohm loudspeaker. As light increases, clicks merge into audio tone that increases in frequency as light intensity increases. Can be used for classroom demonstrations or as sunrise alarm clock. Circuit is quiet in total darkness.—F. M. Mims, "Optoelectronic Projects, Vol. 1," Radio Shack, Fort Worth, TX, 1977, 2nd Ed., p 61–66.

MODULATED-LIGHT RECEIVER—Two FET stages amplify chopped or smoothly modulated output signal of silicon solar cell. With 1000-Hz modulation of 5-lm/ft² light beam, circuit will produce 1 VRMS at output when R4 is set for maximum gain. Can be used for light-beam communication and for alarm systems.—R. P. Turner, "FET Circuits," Howard W. Sams, Indianapolis, IN, 1977, 2nd Ed., p 113–114.

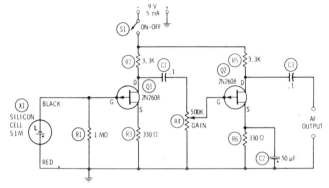

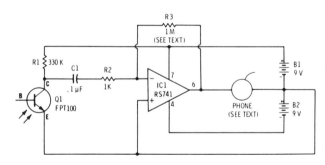

LIGHT-BEAM RECEIVER—Converts amplitude-modulated light beam back to audio signal for driving transistor radio earphone having resistance of 500–1000 ohms. Miniature 8-ohm loudspeaker can be used by adding output transformer such as Radio Shack 273-1380. Gain of opamp is controlled by R3, which can be trimmer resistor or pot. Designed for use with transmitter providing amplitude modulation of LED, for short-range voice communication.—F. M. Mims, "Optoelectronic Projects, Vol. 1," Radio Shack, Fort Worth, TX, 1977, 2nd Ed., p 44–54.

END-OF-TAPE DETECTOR—Self-compensating sensor automatically compares short-term light variations produced by beginning and end markers on digital magnetic recording tape against long-term variations of ambient light, to improve reliability of sensing marker when there are reflections from blank tape. Low-pass filter R3-C1, having time constant about 5 times expected 10-ms incoming pulse width, stores long-term light level without reacting to short signal pulse. Low-pass filter R4-C2, having 1/20 time constant of incoming pulse width, reduces spurious noise without deteriorating incoming pulses.—C. A. Herbst, Optical Tape-Marker Detector, *EEE Magazine*, March 1971, p 79.

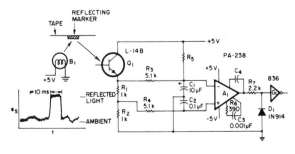

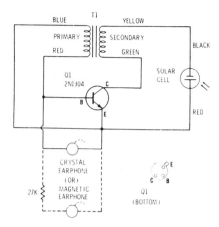

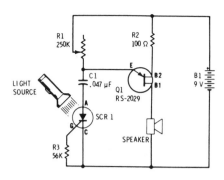

SOLAR-POWER OSCILLATOR—Supply voltage for single-transistor audio oscillator is generated by Radio Shack 276-115 selenium solar cell that produces about 0.35 V in bright sunlight. With cell 3 feet away from 75-W incandescent lamp, oscillator frequency is about 2400 Hz. Frequency drops as light increases. Transformer is 273-1378.—F. M. Mims, "Transistor Projects, Vol. 2," Radio Shack, Fort Worth, TX, 1974, p 53–58.

LASCR-CONTROLLED OSCILLATOR—UJT relaxation oscillator having loudspeaker load produces single click each time flash of light falls on light-activated SCR. Setting of R1 determines whether circuit produces series of pulses or tone burst during time light is on. Oscillator frequency increases with light intensity.—F. M. Mims, "Semiconductor Projects, Vol. 2," Radio Shack, Fort Worth, TX, 1976, p 71–77.

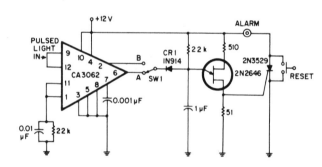

MISSING-PULSE ALARM—Developed for sensing missing light pulses or detecting absence of object on moving conveyor belt. CA3062 combination light sensor and amplifier detects light pulses synchronized to 60-Hz line. With SW1 at A, each pulse resets 20-ms timing network of 2N2646 UJT at 16.7-ms intervals, preventing UJT from firing. If light beam is interrupted by object, UJT is allowed to fire and trigger 2N3529 SCR that turns on alarm. With SW1 at B, circuit detects interruptions in steady light beam and sounds alarm only when interruption does not occur.—J. F. Kingsbury, Double Duty Photo Alarm, *EDN/EEE Magazine,* May 15, 1971, p 51.

SECTION 50
Power Control Circuits

Included are general-purpose circuits capable of handling many types of resistive or inductive loads. Although most circuits are solid-state relays, conventional relay controls are also shown. Inputs can respond to logic levels, pulses, or sensing transducers. Output devices are chiefly SCRs or triacs. Many circuits have zero-crossing action for suppressing RFI, as well as optoisolators at input or output.

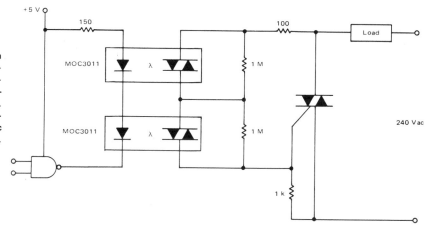

DRIVING 240-VAC TRIAC—Two Motorola MOC3011 optoisolators are used in series as interface between logic and triac controlling 240-VAC load. 1-megohm resistors across optoisolators equalize voltage drops across them. Choice of triac depends on load to be handled.— P. O'Neil, "Applications of the MOC3011 Triac Driver," Motorola, Phoenix, AZ, 1978, AN-780, p 5.

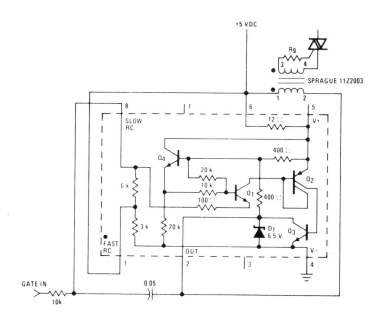

TRIAC TRIGGER—National LM3909 IC is connected as pulse-transformer driver operating from standard 5-V logic supply. IC is biased off when logic input is high. With low logic input, IC provides 10-μs pulses for transformer at about 7 kHz. Trigger is not synchronized to zero crossings but will trigger within 8 V of zero for resistive load and 115-VAC line. Triggering occurs at about 1 V, but trigger level can be changed by using other input resistors or bias dividers.—"Linear Applications, Vol. 2," National Semiconductor, Santa Clara, CA, 1976, AN-154, p 7.

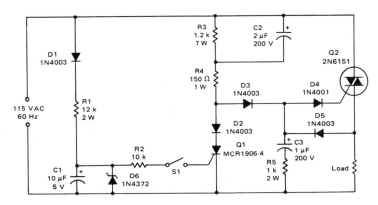

ZERO-POINT SWITCH—Used to control resistive loads. With S1 open, triac Q2 is turned on very close to zero on initial part of positive half-cycle because of large current flow into C2. Once Q2 is on, C3 charges through D5. When line voltage goes through zero and starts negative, C3 is still discharging into gate of Q2 to turn it on near zero of negative half-cycle. Load current thus flows for most of both half-cycles. When S1 is closed, Q1 is turned on and shunts gate current away from Q2 during positive half-cycles. Q2 cannot turn on during negative half-cycle because C3 cannot charge, which makes load current zero.—"Circuit Applications for the Triac," Motorola, Phoenix, AZ, 1971, AN-466, p 12.

OUTPUT CONTROL FOR CLOCK COMPARATOR—Circuit triggers 10-A triac when Q output of comparator-driven flip-flop is logic 1. LED in optoisolator is then energized, activating phototransistor pair for driving gate circuit of triac through diode bridge. Trigger voltage of triac is positive for first quadrant and negative for third quadrant, to give maximum sensitivity of triac control.—D. Aldridge and A. Mouton, "Industrial Clock/Timer Featuring Back-Up Power Supply Operation," Motorola, Phoenix, AZ, 1974, AN-718A, p 7.

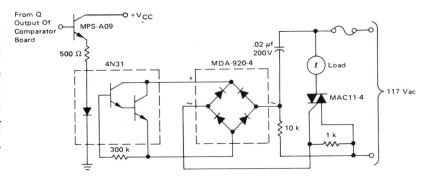

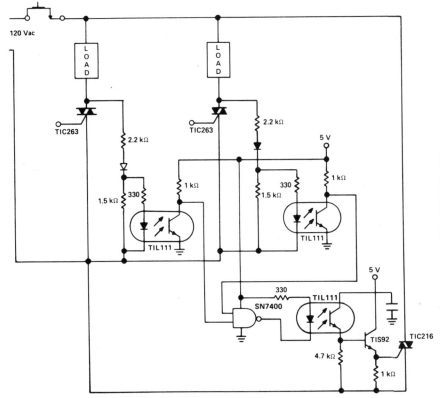

PROTECTION AGAINST SIMULTANEOUS OPERATION OF TRIACS—Optoisolators provide cross-connection between solid-state triac relay circuits to eliminate possibility that two or more triacs come on at same time due to circuit malfunction or component failure. Circuit shuts system down when this occurs.—"Thyristor Gating for µP Applications," Texas Instruments, Dallas, TX, 1977, CA-191, p 5–9.

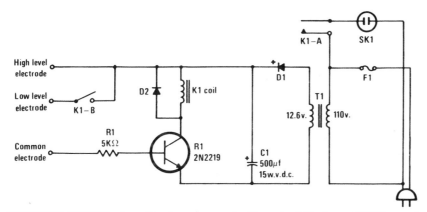

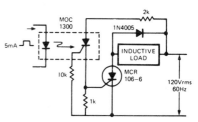

SUMP-PUMP CONTROL—Impurities in water provide conductivity for completing circuit of transistor when water reaches level of sensing electrode, energizing relay that starts pump motor. Extra set of contacts on relay keeps motor running until water drops to predeter- mined lower level. Diodes are 1N4001 or equiv- alent, rated 1 A. Fuse should be chosen to pass normal motor current. Use 12-V double-pole relay. T1 is 300-mA filament transformer.—J. H. Gilder, Automatic Turn-On, *Modern Electron- ics,* Dec. 1978, p 78.

HALF-WAVE CONTROL—Simple AC relay op- erates during positive alternations of AC source, with optoisolator providing complete isolation between control circuit and SCR han- dling inductive load. When input LED is ener- gized by control pulse, photo-SCR of optoiso- lator conducts and provides gate current for turning on power SCR. 1N4005 diode protects SCR from back EMF transients of inductive load.—T. Mazur, Solid-State Relays Offer New Solutions to Many Old Problems, *EDN Maga- zine,* Nov. 20, 1973, p 26–32.

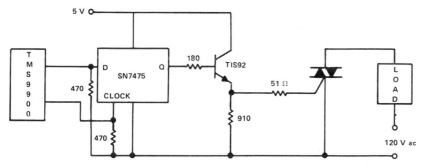

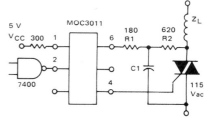

LOGIC-TRIGGERED TRIAC—Pulsed output from microprocessor controls gate drive of triac through SN7475 clock and transistor. Pulse from one output port of microprocessor is ap- plied to D input of clock simultaneously with pulse from communications register unit (CRU) going to clock input, to raise Q output of clock to logic 1. Output remains high until another pulse from CRU returns it to zero, thus giving latching action. High output turns on transistor and supplies about 100-mA gate drive to TIC263 25-A triac.—"Thyristor Gating for μP Applica- tions," Texas Instruments, Dallas, TX, 1977, CA- 191, p 4.

LOGIC DRIVE FOR INDUCTIVE LOAD—When output of NAND gate goes high and furnishes 10 mA to LED of Motorola MOC3011 optically coupled triac driver, output of optoisolator pro- vides necessary trigger for triac controlling in- ductive load. C1 is 0.22 μF for load power factor of 0.75 and 0.33 μF for 0.5 power factor. Omit C1 for resistive load. R1, R2, and C1 serve as snub- ber that limits rate of rise in voltage applied to triac.—P. O'Neil, "Applications of the MOC3011 Triac Driver," Motorola, Phoenix, AZ, 1978, AN- 780, p 2.

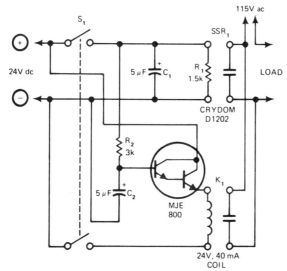

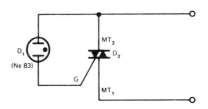

PERFECT AC SWITCH—Developed for use in computerized equipment to prevent generation of severe noise spikes if contact closure can occur at any point in AC cycle. Closing S₁ gates solid-state relay SSR₁, which noiselessly switches load at next zero crossing. During this time, C₂ charges through R₂. After time T = 3R₂C₂, MJE800 Darlington is turned on, pulling in relay K₁ to follow up SSR₁ with hard contacts. When S₁ is later opened, K₁ drops out immedi- ately but C₁ discharges through gate of SSR₁ to hold it on for about T = 6R₁C₁. Load is then switched off at next zero crossing after this delay.—E. Woodward, This Circuit Switches AC Loads the Clean Way, *EDN Magazine,* Nov. 20, 1975, p 160 and 162.

VOLTAGE-SENSITIVE SWITCH—RCA 40527 triac is triggered by small neon. After break- down occurs bidirectionally at 88 V, triac takes over as short-circuit. D₁ can be any other voltage breakdown device, such as diac or zener, and thyristor can be used in place of triac to give unilateral switching. Applications include use as power crowbar, with breakdown level set by artificial resistance-controlled zener.—L. A. Ro- senthal, Breakdown and Power Devices Form Unusual Power Switch, *EDN Magazine,* July 5, 1974, p 74–75.

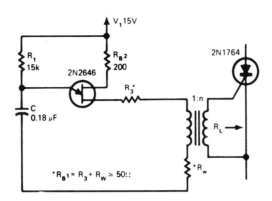

400-Hz TRIGGER FOR SCR—Simple UJT oscillator combined with pulse transformer provides pulses required for firing 2N1764 SCR. Article gives design data for pulse transformer, along with design equations.—W. Dull, A. Kusko, and T. Knutrud, Pulse and Trigger Transformers—Performance Dictates Their Specs, *EDN Magazine,* Aug. 20, 1976, p 57–62.

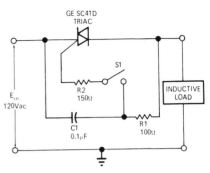

TRIAC FOR INDUCTIVE LOADS—Simple triac gating circuit applies AC power to inductive load when low-power switch S1 is closed. R1 and C1 provide dv/dt suppression.—C. A. Farel and D. M. Fickle, Triac Gating Circuit, *EDN/EEE Magazine,* Jan. 1, 1972, p 72–73.

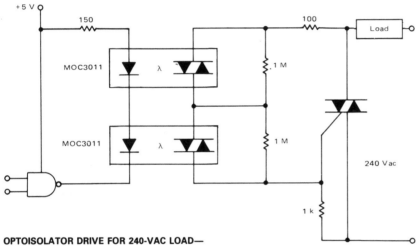

OPTOISOLATOR DRIVE FOR 240-VAC LOAD—Two Motorola MOC3011 optically coupled triac drivers are used in series to overcome voltage limitation of single coupler when triggering triac connected to control 240-VAC load. Two 1-megohm resistors equalize voltage drops across couplers.—P. O'Neil, "Applications of the MOC3011 Triac Driver," Motorola, Phoenix, AZ, 1978, AN-780, p 5.

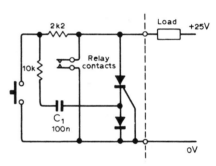

THYRISTOR SWITCH—When circuit of conventional P-gate thyristor is grounded by switch, negative-going pulse is applied to thyristor cathode, which reverse-biases the diode. When thyristor conducts, diode is forward-biased and has only about 0.7-V drop. Use low-voltage diode, rated for full load current. Opening of relay contacts makes circuit switch off.—R. V. Hartopp, Grounded Gate Thyristor, *Wireless World,* Feb. 1977, p 45.

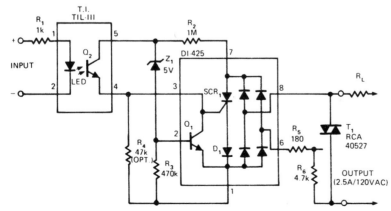

TRIAC CONTROL WITH OPTOISOLATOR—Dionics DI 425 switchable bridge circuit controls 120-VAC line in optically isolated zero-crossing solid-state relay that can be used as trigger for power triac. Small AC devices, drawing under 5 W, can be switched directly in either random or zero-crossing mode.—High-Voltage Monolithic Technology Produces 200V AC Switching Circuit, *EDN Magazine,* April 5, 1975, p 121.

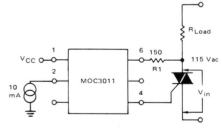

TRIAC DRIVE—Motorola MOC3011 optoisolator serves as interface between 10-mA input circuit and gate of triac controlling AC load. Choice of triac depends on load being handled. Optoisolator detector chip responds to infrared LED; once triggered on, optoisolator stays on until input current drops below holding value of about 100 μA.—P. O'Neil, "Applications of the MOC3011 Triac Driver," Motorola, Phoenix, AZ, 1978, AN-780, p 2.

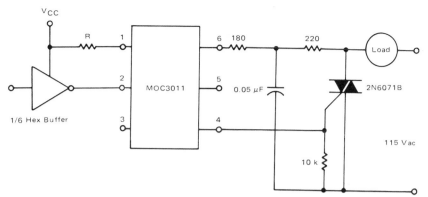

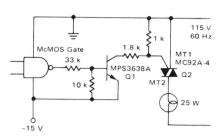

MOS DRIVE FOR TRIAC—Circuit uses one section of MC75492 hex buffer to boost 0.5-mA output of CMOS logic gate to 10 mA required for LED at input of Motorola MOC3011 optically coupled triac driver. When MOS input goes high, optoisolator provides output voltage for triggering triac that controls AC load. R is 220 ohms for 5-V supply and 600 ohms for 10-V supply. For 15 V, use MC14049B buffer and 910 ohms for R.—P. O'Neil, "Applications of the MOC3011 Triac Driver," Motorola, Phoenix, AZ, 1978, AN-780, p 4.

ACTIVE-HIGH TRIAC INTERFACE—Typical CMOS logic gate operating from negative supply triggers triac on negative gate current of 8 mA for control of 25-W AC load. High supply lines for both logic gate and interface transistor are grounded.—A. Pshaenich, "Interface Techniques Between Industrial Logic and Power Devices," Motorola, Phoenix, AZ, 1975, AN-712A, p 12.

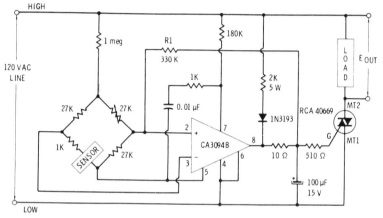

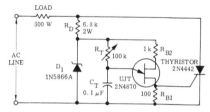

BRIDGE-TRIGGERED TRIAC—Developed for use with AC sensor in one leg of bridge. CA3094 is shut down on negative half-cycles of line. When bridge is unbalanced so as to make pin 2 more positive than pin 3, IC is off at instant that AC line swings positive; pin 8 then goes high and drives triac into conduction. Triac conduction is maintained on next negative half-cycle by energy stored in 100-μF capacitor. Bridge unbalance in opposite direction does not trigger triac.—E. M. Noll, "Linear IC Principles, Experiments, and Projects," Howard W. Sams, Indianapolis, IN, 1974, p 313–314.

600-W HALF-WAVE—UJT serves as trigger for thyristor in circuit that provides power control for load only on positive half-cycles. Thyristor acts also as rectifier, providing variable power determined by setting of R_T during positive half-cycle and no power to load during negative half-cycle.—D. A. Zinder, "Unijunction Trigger Circuits for Gated Thyristors," Motorola, Phoenix, AZ, 1974, AN-413, p 3.

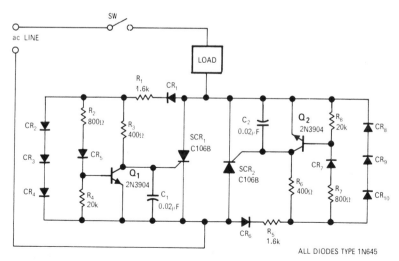

ALL DIODES TYPE 1N645

LINE-POWERED SWITCH—When AC line switch is closed, power is not applied to load until after line voltage next goes through zero. Identical circuits control each half of AC cycle. Transistor turn-on at 1.4 V prevents SCR from triggering until 0.013 ms (less than one-third electrical degree) after next zero-crossing point.—A. S. Roberts and O. W. Craig, Efficient and Simple Zero-Crossing Switch, EDN/EEE Magazine, Aug. 15, 1971, p 46–47.

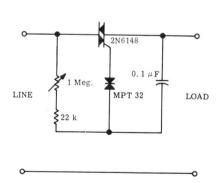

FULL-WAVE POWER CONTROL—Bidirectional three-layer trigger for triac allows triggering on both half-cycles at point determined by setting of 1-megohm pot. Triac rating determines size of load that can be handled.—"SCR Power Control Fundamentals," Motorola, Phoenix, AZ, 1971, AN-240, p 6.

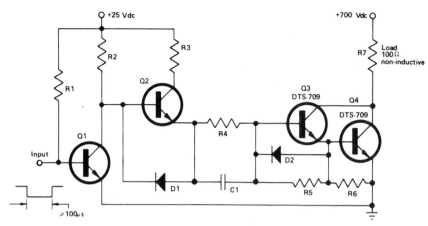

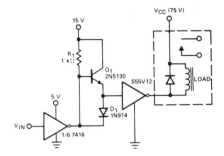

SWITCHING 4500 W AT UP TO 10 kHz—Darlington connection of Delco DTS-709 transistors will switch 7 A at 700 V with 1-μs switching time. Suitable for motor speed control, switching regulator, and inverter applications. Can be operated directly from 440-VAC line. Q1 and Q2 are 2N6100. Diodes are 1N4001. C1 is 4 μF at 15 V. R1 is 510 ohms, R2 is 100, R3 is 12, R4 is 10, R5 is 1K, R6 is 47, and R7 is 100.—"Low Cost 'Duolithic Darlington' Switches 4500 Watts at up to 10 kHz," Delco, Kokomo, IN, 1973, Application Note 54, p 2.

FAST-SWITCHING TTL INTERFACE FOR VMOS—Totem-pole TTL interface drive for S55V01 VMOS gives appreciably faster switching times (less than 30 ns). To achieve fast turn-on time without unduly small pull-up resistor, which dissipates considerable power when switch is in OFF state, emitter-follower Q₁ drives high peak currents into capacitive VMOS input.—L. Shaeffer, VMOS Peripheral Drivers Solve High Power Load Interface Problems, *Computer Design*, Dec. 1977, p 90, 94, and 96–98.

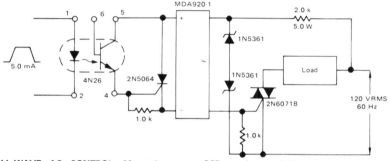

FULL-WAVE AC CONTROL—Motorola type MDA920-1 bridge rectifier provides full-wave rectification of AC line voltage for 2N5064 SCR placed across DC output of bridge. When positive logic pulse from CMOS circuit energizes optoisolator, SCR conducts and completes path for triac gate trigger current through bridge and SCR, turning on AC load. Triac rating determines size of load. Drawback of circuit is generation of EMI if logic signal occurs at other than zero crossings of AC line.—A. Pshaenich, "Interface Techniques Between Industrial Logic and Power Devices," Motorola, Phoenix, AZ, 1975, AN-712A, p 17.

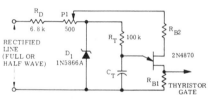

LINE-VOLTAGE COMPENSATION—Can be used with either half-wave or full-wave phase control circuit to make load voltage independent of changes in AC line voltage. P1 is adjusted to provide reasonably constant output over desired range of line voltage. As line voltage increases, P1 wiper voltage increases. This has effect of charging C_T to higher voltage so more time is taken to trigger UJT. Additional delay reduces thyristor conduction angle and thereby maintains desired average voltage.—D. A. Zinder, "Unijunction Trigger Circuits for Gated Thyristors," Motorola, Phoenix, AZ, 1974, AN-413, p 4.

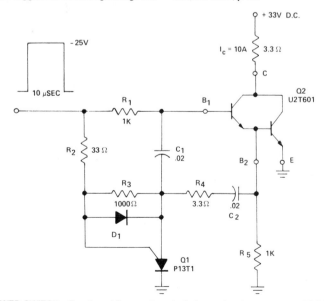

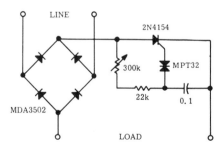

125-ns POWER SWITCH—Developed for repetitive pulse applications in which rise, fall, and storage times of pulse must be kept at absolute minimum. Circuit provides very high gain of Unitrode U2T601 Darlington and switching speeds up to 5 times greater than conventional techniques. Load power up to 10 A is typically applied within 125 ns. Applications include drive for laser diode and for radar circuits.—"Designer's Guide to Power Darlingtons as Switching Devices," Unitrode, Watertown, MA, 1975, U-70, p 19.

600-W TRIGGERED SCR—2N4154 SCR is operated from DC output of bridge rectifier and triggered by MPT32 at setting determined by position of 300K pot. Circuit provides full-wave DC control of lamp and other loads up to 600 W, using relaxation oscillator operating from DC source.—"SCR Power Control Fundamentals," Motorola, Phoenix, AZ, 1971, AN-240, p 6.

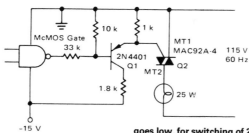

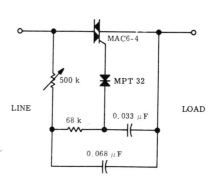

ACTIVE-LOW TRIAC INTERFACE—With connection shown for interface transistor Q1, typical CMOS gate triggers triac when gate output goes low, for switching of 25-W lamp load.—A. Pshaenich, "Interface Techniques Between Industrial Logic and Power Devices," Motorola, Phoenix, AZ, 1975, AN-712A, p 12.

FULL-RANGE CONTROL—Triggered triac is used with double phase-shift network to obtain reliable triggering at conduction angles as low as 5°, as required for control of incandescent lamps and some motors. Triac rating determines size of load.—"SCR Power Control Fundamentals," Motorola, Phoenix, AZ, 1971, AN-240, p 6.

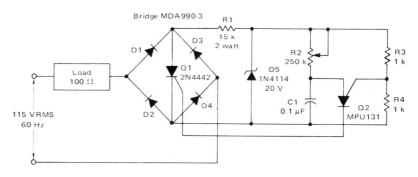

PUT CONTROLS SCR—Programmable unijunction transistor Q2 provides phase control for both halves of AC line voltage by triggering SCR connected across bridge. Relaxation oscillator formed by Q2 varies conduction interval of Q1 from 1 to 7.8 ms or from 21.6° to 168.5°, to give control over 97% of power available to load.—R. J. Haver and B. C. Shiner, "Theory, Characteristics and Applications of the Programmable Unijunction Transistor," Motorola, Phoenix, AZ, 1974, AN-527, p 10.

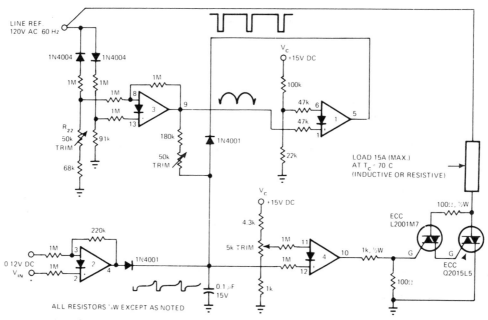

GROUND-REFERENCED RAMP-AND-PEDESTAL CONTROL—Need for transformer is eliminated by applying alternate half-cycles to inverting and noninverting inputs of section 3 of LM3900 quad opamp, so full-wave-rectified waveform is referenced to ground. Comparator opamp 1 discharges timing capacitor at zero line voltage and synchronizes circuit with line frequency. Buffer opamp 2 scales input and provides linear pedestal for capacitor. Opamp 4 is comparator serving as output driver whose output is high when capacitor is charged to level selected by high-end trimming pot. Output is sufficient for optoisolators and logic triacs.—J. C. Johnson, Ramp-And-Pedestal Phase Control Uses Quad Op Amp, *EDN Magazine,* June 5, 1977, p 208 and 211.

SECTION 51
Power Supply Circuits

Includes unregulated circuits for changing AC input voltage to variety of DC voltages ranging from 1.5 V to 3 kV. Also includes inverter circuits containing oscillator operating from DC supply and providing AC voltage at 60 Hz or 400 Hz, along with RMS AC regulator. See also Converter—DC to DC, Regulated Power Supply, Regulator, and Switching Regulator chapters.

12-V TRANSFORMERLESS PREREGULATOR—AC line voltage is converted to regulated 12 VDC by varying firing angle of 10-A SCR. Circuit provides reliable operation for AC line voltages between 50 and 140 V. Key element in triggering of SCR is programmable unijunction transistor that provides variable and accurate control of firing time. Developed for use in power supply that uses digital techniques of sample-and-hold switching to achieve high degree of isolation between power line and load without using transformer.—J. A. Dickerson, Transformerless Power Supply Achieves Line-to-Load Isolation, *EDN Magazine,* May 5, 1976, p 92–96.

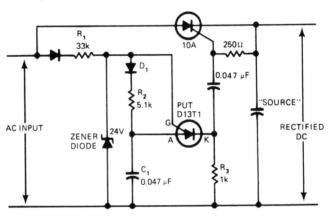

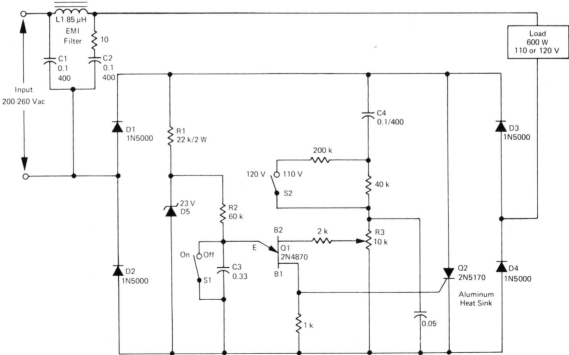

110/120 VAC ± 2.5 V AT 600 W—Simple open-loop voltage compensator for small conduction angles operates from 200–260 VAC input and provides true RMS output voltage for sensitive equipment such as photographic enlargers, oven heaters, projection lights, and certain types of AC motors. Full-wave bridge D1-D4 and SCR Q2 provide full-wave control, with UJT Q1 serving as trigger. Triggering frequency is determined by charge and discharge of C3 through R2. As input voltage increases, required trigger voltage also increases, retarding firing point of SCR to compensate for change in input.—D. Perkins, "True RMS Voltage Regulators," Motorola, Phoenix, AZ, 1975, AN-509, p 3.

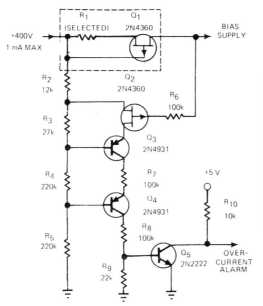

OVERCURRENT PROTECTION FOR 400-V SUP-PLY—R_1 and Q_1 form current detector for bias supply. At normal current levels, voltage drop is very small and Q_2 is reverse-biased. When current reaches 400 μA, voltage drop across R_1 forces gate of Q_1 to near pinchoff. Combined voltage drop across R_1 and Q_1 then becomes large so Q_2 is forced almost to full conduction. Q_3 and Q_4 then turn on Q_5, to provide overcurrent-alarm signal for activating logic circuit that shuts off power supply.—J. P. Thompson, Overcurrent Alarm Protects HV Supply, *EDN Magazine*, Nov. 20, 1978, p 321–322.

FULL-WAVE SYNCHRONOUS RECTIFIER—Transistors are biased on alternately by AC input voltage, to supply load current on alternate half-cycles. Silicon diodes D1 and D2 protect transistors from charging current of capacitive load when circuit is turned on. Capacitor discharge problems are minimized by use of diodes D3 and D4 in base circuits of transistors.—B. C. Shiner, "Improving the Efficiency of Low Voltage, High-Current Rectification," Motorola, Phoenix, AZ, 1973, AN-517, p 4.

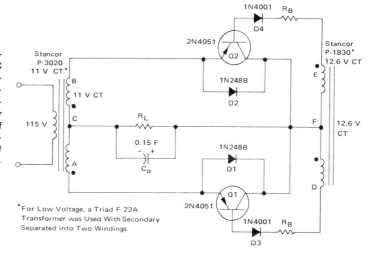

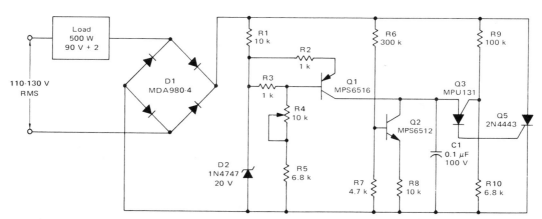

90 VRMS AT 500 W—Open-loop RMS voltage regulator acts with full-wave bridge to provide good AC voltage regulation for AC load over line voltage range of 110–130 VAC. As input voltage increases, voltage across R10 increases and serves to increase firing point of PUT Q3. This delays firing of SCR Q5 to hold output voltage fairly constant as input voltage increases. Delay network of Q1 prevents circuit from latching up at beginning of each charging cycle for C1.—R. J. Haver and B. C. Shiner, "Theory, Characteristics and Applications of the Programmable Unijunction Transistor," Motorola, Phoenix, AZ, 1974, AN-527, p 11.

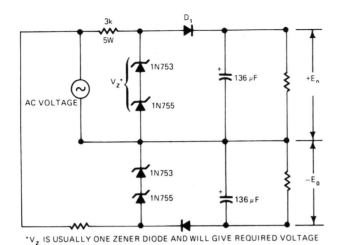

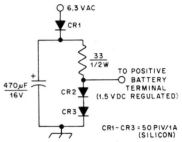

*V_z IS USUALLY ONE ZENER DIODE AND WILL GIVE REQUIRED VOLTAGE

TRANSFORMERLESS ±12 V AT 15 mA—Developed to provide bias voltage for six 741 opamps. Circuit connects directly across 120-V 60-Hz AC line. Article gives design procedure to meet performance requirements. For values shown, ripple is 1.1 V. Diode types are not critical.—C. Venditti, Build this Transformerless Low-Voltage Supply, *EDN Magazine*, Feb. 5, 1977, p 102.

1.5 V FOR VTVM—Simple rectifier circuit replaces battery in vacuum-tube voltmeter. Provides good regulation and eliminates need for frequent battery replacement. Remove battery before using supply. AC source can be 6.3-V secondary of filament transformer or terminals of 6.3-V pilot lamp in any AC equipment.—P. Alexander, Battery Replacement Circuit for VTVM, *QST*, Jan. 1976, p 42–43.

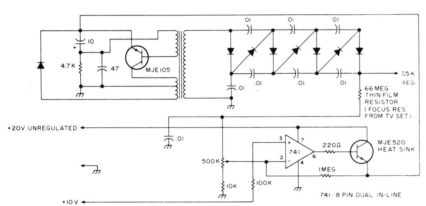

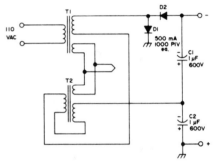

7.5-kV REGULATED SUPPLY—Power transformer is special design, but commercial unit delivering 5 to 10 kV can be used. Inverter circuit uses MJE105 transistor driving primaries of transformer. 741 opamp and transistor provide regulation for 7.5-kV output used in slow-scan TV monitor. Diodes are 1 kV, such as 1N4007. Article gives circuit of complete monitor, including low-voltage supply.—L. Pryor, Homebrew This SSTV Monitor, *73 Magazine*, June 1975, p 22–24, 26–28, and 30.

1000 V FOR CRT—Unique connection of two TV booster transformers having 125-V secondaries gives high-voltage supply for small monitor scope. T1 is connected conventionally, with its 6.3-V winding going to heater of CRT. 6.3-V winding of T2, also connected to CRT, serves as primary for second transformer. Remaining windings of T2 and high-voltage secondary of T1 are connected in series aiding to give about 367 VAC for doubling by D1-D2 and C1-C2. Since CRT drain is low, filter charges to very nearly peak voltage of 1027 VDC.—W. P. Turner, Cheap Power Supply for a CRT, *73 Magazine*, March 1974, p 53.

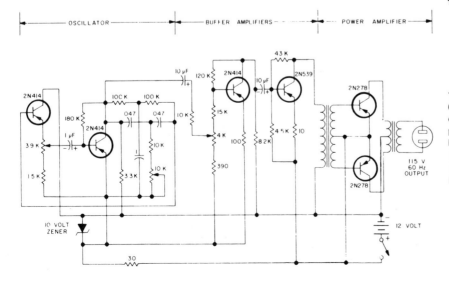

100-W SINE AT 60 Hz—Consists essentially of 60-Hz sine-wave oscillator with 10K frequency-control pot, two buffer stages, and push-pull power amplifier. Circuit eliminates noise problems of square-wave inverters when operating 115-V radio receiver or cassette player in car.—G. C. Ford, Power Inverter with Sine Wave Output, *73 Magazine*, May 1973, p 29–32.

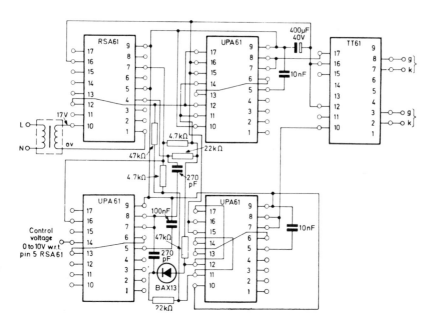

12 V TO 6 V—Permits operation of older 6-V VHF FM mobile equipment from 12-V storage battery. With transistor mounted on suitable heatsink, maximum output is 15 A. If positive and negative lines are isolated from chassis, converter may be used with either negative or positive ground.—E. Noll, Circuits and Techniques, *Ham Radio,* April 1976, p 40–43.

PARALLEL INVERTER DRIVE—Uses Mullard modules for converting DC power to AC at high power levels for such applications as driving induction motors at higher speeds than are obtainable with line frequency. DC control voltage of 0–10 V varies output frequency up to 400 Hz. UPA61 modules provide functions of level detector, pulse generator, ramp generator, capacitor discharge circuit, and bistable MVBR for parallel inverter system. RSA61 and TT61 are trigger modules, with RSA61 also providing power supplies for other modules.—"Universal Circuit Modules for Thyristor Trigger Systems (61 Series)," Mullard, London, 1978, Technical Information 66, TP1660, p 19.

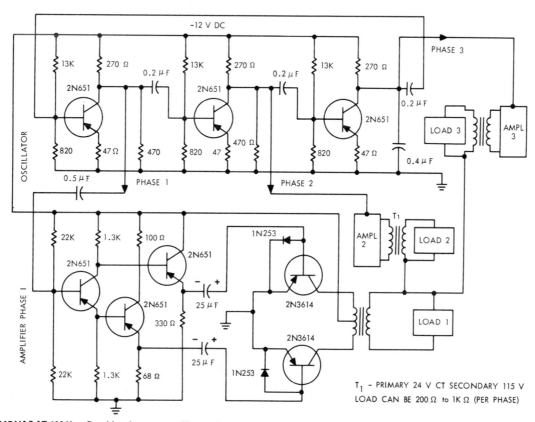

12 VDC TO 115 VAC AT 400 Hz—Provides three-phase output at 20 W by using RC coupling to oscillator in such a way that 120° phase difference exists at collectors of 2N651 transistors of oscillator. Emitter-follower amplifier driving push-pull power output transistors is shown only for phase 1; other two phases use similar amplifiers. Power transistors are operated in saturated switching mode.—R. J. Haver, "The ABC's of DC to AC Inverters," Motorola, Phoenix, AZ, 1976, AN-222, p 15.

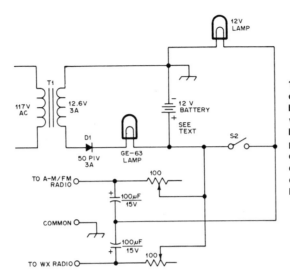

12-V EMERGENCY POWER—Trickle-charge circuit and 12-V motorcycle battery provide reliable emergency power for battery-operated weather radio, portable AM/FM receiver, or hand-held transceiver for many hours. 100K pots drop voltage to 9 V for each receiver. Lamp can be auto dome light. GE-63 pilot lamp in charging circuit acts as current limiter and charge indicator.—J. Rice, Simple Emergency Power, *QST,* March 1978, p 42.

130 AND 270 V FOR CRT—High-voltage power supply provides 270 V required for deflection plates of 2AP1-A CRT used as RTTY tuning indicator, as well as 130 V for high-voltage amplifier. Large capacitor keeps ripple voltage low.—R. R. Parry, RTTY CRT Tuning Indicator, *73 Magazine,* Sept. 1977, p 118–120.

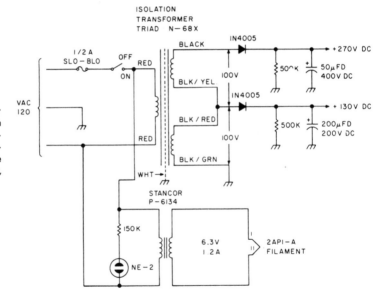

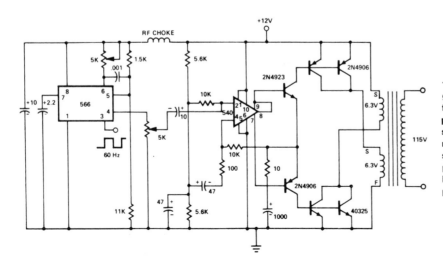

12 VDC TO 115 VAC AT 100 W—566 function generator provides triangle output at 60 Hz with frequency stability better than ±0.02%/°C. 540 power driver feeds six-transistor power output stage. Transformer load attenuates third harmonic, giving output very close to pure 60-Hz sine wave. 566 also provides square-wave output for other purposes.—"Signetics Analog Data Manual," Signetics, Sunnyvale, CA, 1977, p 853–854.

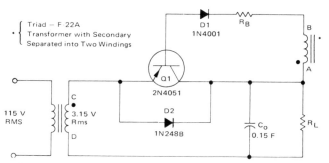

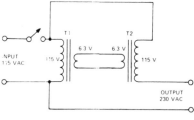

HALF-WAVE SYNCHRONOUS RECTIFIER— Transistor Q1 is synchronously biased on by AC input voltage to give efficient low-voltage regulation. When points A and C are positive with respect to points B and D, base-emitter junction of Q1 is forward-biased and collector current flows through load R_L. On negative alternations, Q1 is reverse-biased and transistor is blocked.— B. C. Shiner, "Improving the Efficiency of Low Voltage, High-Current Rectification," Motorola, Phoenix, AZ, 1973, AN-517, p 3.

230 VAC FROM 115 VAC—Connect 6.3-V filament transformers back-to-back as shown to get 230 V when step-up transformer is not available. 115-V windings must be phased properly in series; if wrong, output voltage will be zero. Output power rating at 230 V is somewhat less than twice the power (E × I) rating of smallest filament transformer. If 6.3-V 10-A transformers are used, power rating would be about 100 W (less than 2 × 6.3 × 10).—A. E. McGee, Jr., Cheap and Easy 230 Volt AC Power Supply, *73 Magazine*, Aug. 1974, p 64.

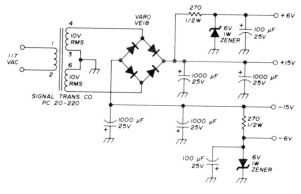

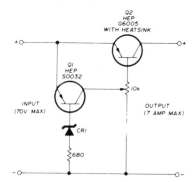

±6 V AND ±15 V—Suitable for use when frequency or some other critical parameter of load is not dependent on voltage. Developed for use in CMOS IC function generator.—R. Megirian, Inegrated-Circuit Function Generator, *Ham Radio*, June 1974, p 22–29.

TRANSIENT ELIMINATOR—Used between DC power supply and load to eliminate supply transients that might damage semiconductor devices. Zener rating should be about 10% higher than supply voltage so Q1 is normally turned off. Q2 is normally conducting. When voltage spike is present on input line, zener conducts and turns Q1 on. Q1 then places positive bias on 10K pot to turn off Q2 and protect load during transient.—J. Fisk, Circuits and Techniques, *Ham Radio*, June 1976, p 48–52.

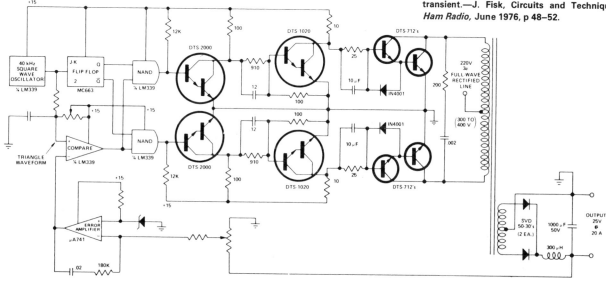

500 W AT 20 kHz—Uses four Delco DTS-712 transistors in push-pull Darlington configuration, with pulse-width modulation on push-pull inverter providing regulation. Can be operated from 220-VAC three-phase full-wave rectified line. Efficiency is up to 80%. Square-wave output of 40-kHz primary oscillator drives JK flip-flop that generates complementary square waves and divides frequency by 2 with necessary symmetry. NAND gates establish primary ON/OFF periods of power stage. Portion of output signal is compared to reference voltage, and error signal is fed to NAND gates to give regulation better than 0.1% for load range of 200–500 W or line range of 300–400 V.—"A 20 kHz, 500 W Regulating Converter Using DTS-712 Transistors," Delco, Kokomo, IN, 1974, Application Note 55.

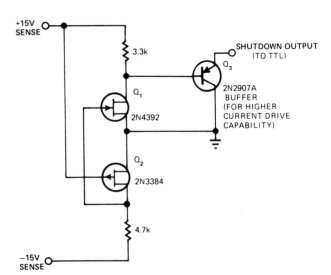

SHUTDOWN PROTECTION—Used with digital logic to prevent generation of false logic signals when power supply is turned on or turned off. FETs sense +15 V and −15 V supplies and conduct when either supply drops below pinch-off voltage, activating shutdown output. With values shown, shutdown output is disabled when supplies exceed about 4 V, to provide normal operation.—E. Burwen, Power-Supply Monitor Suppresses False Output Signals, *EDN Magazine*, Nov. 5, 1977, p 110 and 112.

117 VAC FROM 24–60 VDC—Will operate from either 24- or 32-V storage battery or from 60-VDC source. Circuit shown is set up for 60-V operation. For 24/32 V, remove F1 and F4 and insert F2 and F3, then switch S3 to 32 V. T1 is 117-V 20-A Variac with bifilar primary winding added; use 38 bifilar turns of No. 8 for 24 V and 48 turns for 32 V. Commutating capacitor C1 consists of ten 120-μF 400-VDC oil-filled capacitors (do not use electrolytics). SCRs Q1 and Q2, rated 100 A at 800 PIV (Poly Paks 92CU1167), are switched by Cornell-Dubilier 98600 60-Hz 12-V vibrator. VR1, which limits voltage across vibrator coil, consists of two 6.8-V 10-W zeners. With 60-V operation, use 1-ohm 200-W resistor in series with inverter while starting, but short it out while inverter is running. Inverter output voltage varies from 150 VAC no-load to 110 VAC with 1650-W resistive load. CR1 is 400-PIV 200-A silicon. CR2 and CR3 are 800-PIV 250-A silicon.—R. Dunaja, A High-Power SCR Inverter, *QST*, June 1974, p 36–37.

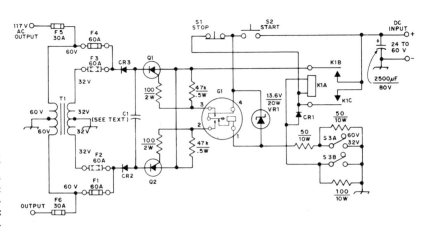

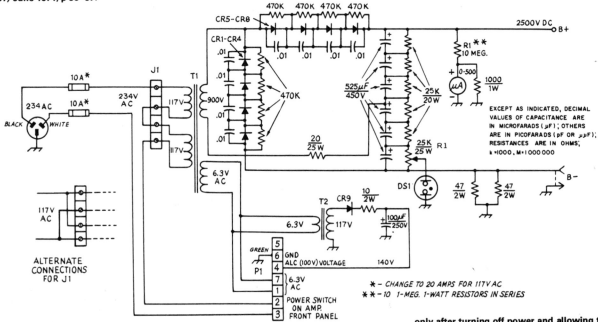

2500 V AT 500 mA—Meets power requirements for 2-kW linear amplifier using pair of 8873 conduction-cooled triodes for SSB transmitter service. Power transformer is Hammond 101165.

Diodes are 1000 PIV at 2.5 A, such as Motorola HEP170. T2 is Stancor P-8190 rated 6.3 V at 1.2 A. DS1 is 117-V neon pilot lamp. Set tap on R1 5000 ohms from B−lead. Make adjustments

only after turning off power and allowing time for capacitors to discharge; output voltages are dangerous.—R. M. Myers and G. Wilson, 8873s in a Two-Kilowatt Amplifier, *QST*, Oct. 1973, p 14–19.

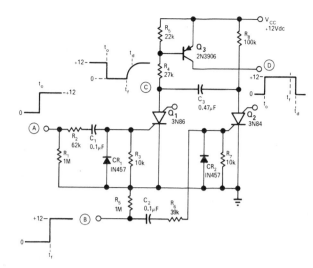

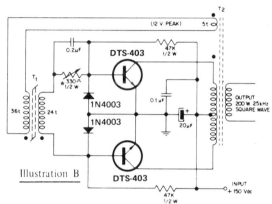

LOW STANDBY DRAIN—Positive 12-V pulse at input A triggers SCS Q_1 and turns on transistor switch Q_3. Positive pulse at input B gates SCS Q_2 on and turns off Q_3. Current drain is essentially zero (typically 3 μA). Circuit was designed to supply up to 7 mA of switched current from 12-VDC supply.—D. B. Heckman, Bistable Switch with Zero Standby Drain, *EDN/EEE Magazine*, Oct. 1, 1971, p 42.

200 W AT 25 kHz—Two Delco DTS-403 high-voltage silicon transistors are connected as push-pull oscillator operating on 150-VDC bias. Efficiency is 78% at full load. Diodes serve alternately as steering and clamp diodes.—"25 kHz High Efficiency 200 Watt Inverter," Delco, Kokomo, IN, 1971, Application Note 47.

Illustration B

T_1	Pri:	36 t #30 AWG
	Sec:	24 t #25 AWG
	Core:	Ferroxcube 266 T 125-3E2A Ferrite toroid

T_2	Pri:	126 t tapped @ 63t, 40 strands #38 AWG litz
	*Sec:	2.38 V/t
	Feedback:	5t #25 AWG
	Core:	Ferroxcube (ferrite toroid) 528T500-3C5

★ Adjust value of resistor for maximum efficiency at full load.

* To be determined by individual requirements.

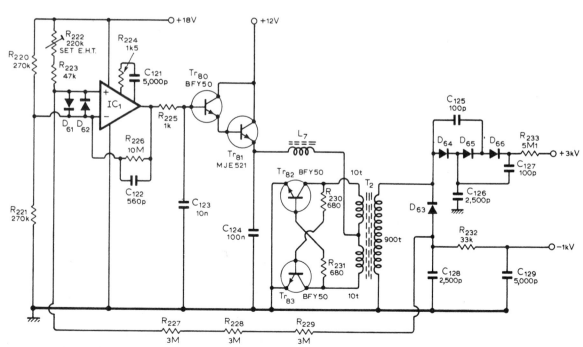

3 kV FOR CRO—Circuit also provides 1-kV negative supply at 2 mA, as required for cathode-ray tube of oscilloscope. Positive supply furnishes 50 μA at 3 kV. Design uses transistor inverter operating at about 20 kHz to simplify filtering. Tr_{82} and Tr_{83} form current-switched class D oscillator producing sine waves at high efficiency. Current multiplication is provided by Tr_{80} and Tr_{81} for 709 IC opamp.—C. M. Little, A 50 MHz Oscilloscope, *Wireless World*, July 1975, p 319–322.

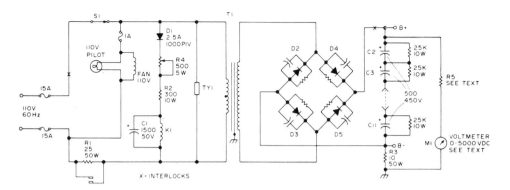

3-kV SUPPLY—Circuit uses full-wave bridge rectifier D2-D5, with each diode stack constructed from two 1000-PIV 2.5-A diodes in series. Each diode pair is shunted by 470K 1-W resistor and 0.01-μF 1000-V disk capacitor. C2-C11 are 500 μF at 450 VDC. Capacitor combination thus gives equivalent of 50 μF for filter, rated 4500 V. When using 500-μA movement for output voltmeter, R5 should be ten 1-megohm resistors in series. Thyrector TY1 is GE 6R520SP4B4. T1 has 2200-V secondary rated 500 mA. K1 is 24-V relay. Article covers construction and stresses safety precautions.—E. H. Hartz, 3000 VDC Supply, *73 Magazine*, July 1974, p 69–72.

SECTION 52

Protection Circuits

Provide protection of equipment and components from overvoltage or overcurrent conditions, ground fault, loose ground, contact arcing, and inductive transients. Also included are digital-coded or tone-coded controls for doors, auto ignition switches, and equipment ON/OFF switches, along with fail-safe interlocks and power-outage indicators. See also Burglar Alarm, Fire Alarm, Power Control, Power Supply, Regulated Power Supply, Regulator, and Siren chapters.

CROWBAR—When output of regulator for microprocessor power supply exceeds maximum safe voltage as determined by zener Q1, SCR Q2 is triggered on and conducts heavily, blowing fuse rapidly to protect equipment. Fuse rating is 125% of nominal load. Choose SCR to meet voltage and current requirements. Choose zener for desired trip voltage. Each germanium diode in series with Q1 will add 0.3 V to trip voltage, and silicon diodes add 0.6 V. To calibrate, place 1K resistor temporarily in series with Q2 and measure drop across it to see if SCR fires and produces surge on meter at desired V_{cc}.— J. Starr, Want to Buy a Little Insurance?, *Kilobaud*, Oct. 1978, p 89.

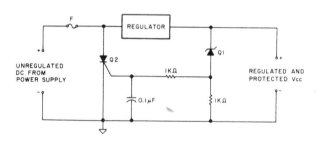

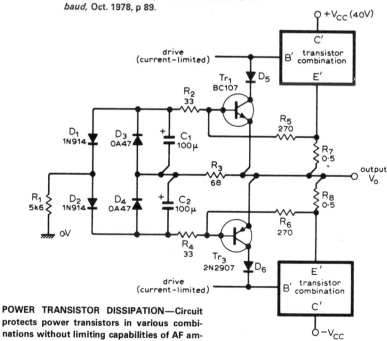

POWER TRANSISTOR DISSIPATION—Circuit protects power transistors in various combinations without limiting capabilities of AF amplifier when driving reactive loudspeaker load. With continuous signal drive into normal load, R_1 draws current from C_1 through D_1, in opposition to R_5. This gives drops of about 0.12 V across C_1 and C_2, allowing full drive. With short-circuited load, however, capacitor drops increase to about 0.55 V, thereby limiting average current in each output transistor to about 1.1 A. Diodes D_5 and D_6 are not critical, and simply prevent current flow from base to collector of transistors.—M. G. Hall, Amplifier Output Protection, *Wireless World*, Jan. 1977, p 78.

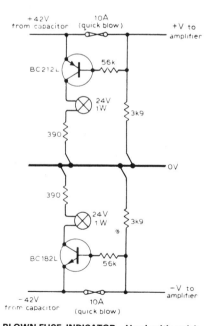

BLOWN-FUSE INDICATOR—Used with quick-blow fuses in high-power audio amplifier using split power supply. When fuse blows, transistor shunting it is turned on and passes current to corresponding indicator lamp. Maximum current in blown-fuse condition is less than 1 mA.—I. Flindell, Amplifier Blown-Fuse Indicator, *Wireless World*, Sept. 1976, p 73.

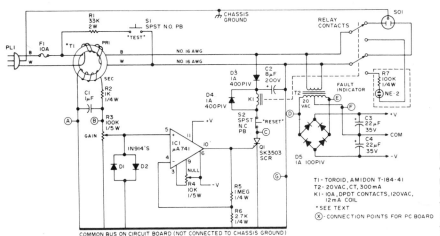

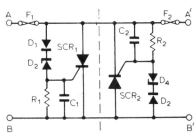

GROUND-FAULT INTERRUPTER—Compares current in ungrounded side of power line with current in neutral conductor. If currents are not equal, ground fault exists because portion of line current is taking an unintended return path through leaky electric appliance or human body. Voltage induced in toroid by unbalanced current is amplified for energizing relay K1 to break circuit. Toroid uses Amidon T-184-41 core, with 600 turns No. 30 for secondary, and 12 turns No. 16 solid twisted-pair for primary. Circuit operates on fault current of 4 mA, well below danger limit for children.—W. J. Prudhomme, The Unzapper, *73 Magazine*, Nov./Dec. 1975, p 151–156.

EQUIPMENT INTERFACE PROTECTION—When circuit shown is used to transfer signal from one piece of equipment to another, desired signal passes with very little degradation. Component values can be chosen to make thyristor SCR_1 latch at any desired voltage between A and B that is greater than 0 V, blowing fuse F_1 and giving desired protective isolation. On other side of circuit, SCR_2 will latch and blow F_2 when voltage exceeds limiting value set by diode D_2 and zener D_4. Zeners are 10-V CV7144, diodes are CV9637 small-signal silicon, resistors are 10K, capacitors are 0.047 μF, and thyristors are 2N4147.—S. G. Pinto and A. P. Bell, Thyristor Protection Circuit, *Wireless World*, Oct. 1975, p 473.

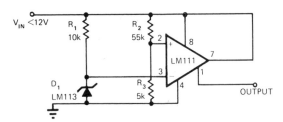

12-V OVERVOLTAGE LIMITER—Single LM111 comparator is basis for simple overvoltage protection of circuits drawing less than 50 mA. Fraction of input supply is compared to 1.2-V reference. When input exceeds reference level, power is removed from output.—R. C. Dobkin, Comparators Can Do More than Just Compare, *EDN Magazine*, Nov. 1, 1972, p 34–37.

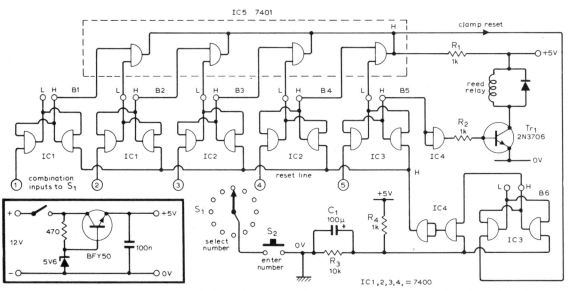

CODED LOCK—Five-digit combination lock uses five low-cost ICs operating from 5-V supply that can be derived from 12-V auto battery as shown in inset. Six set/reset bistable circuits are formed by cross-coupling pairs of dual-input NAND gates, so 0-V input is needed to change state of each. Five of bistables serve for combination, and sixth prevents operation by number in incorrect sequence. After S_1 is set to one number of code, S_2 is pushed to enter that number, with process being repeated for other four numbers of combination. Final correct number sets B5 and turns on Tr_1, to operate relay that can be used to open door.—S. Lamb, Simple Code-Operated Switch or Combination Lock, *Wireless World*, June 1974, p 196.

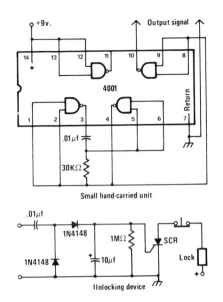

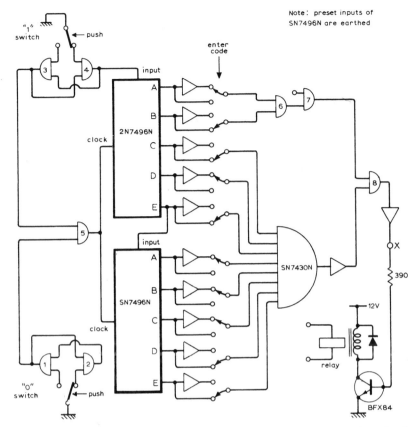

Note: preset inputs of SN7496N are earthed

3-kHz TONE LOCK—Electric door lock opens only when signal voltage of about 3 kHz is applied to two exposed terminals by holding compact single-IC AF oscillator against terminals. Will not respond to DC or 60-Hz AC. Pocket oscillator operates from 9-V transistor radio battery, with current drawn only when output prongs are held against lock terminals. SCR can be any type capable of handling current drawn by electric lock.—J. A. Sandler, 11 Projects under $11, *Modern Electronics*, June 1978, p 54–58.

10-DIGIT CODED SWITCH—Uses seven Texas Instruments positive-logic chips. NAND gates 1-4 and 5-8 are from two SN7400N packages. Two SN7404N packages each provide six of inverting opamps shown. Desired code is set up as combination of 0s and 1s by presetting ten 2-position switches. To open lock, switches at input for 0 and 1 must be pushed in sequence of code. Arrangement gives 1024 possible combinations but provides much greater protection unless intruder knows that 10 digits are required. Article describes operation of circuit. One requirement of the 2N7496N shift registers is that information be present at serial input before clocking pulse occurs.—K. E. Potter, Ten-Digit Code-Operated Switch or Combination Lock, *Wireless World*, May 1974, p 123.

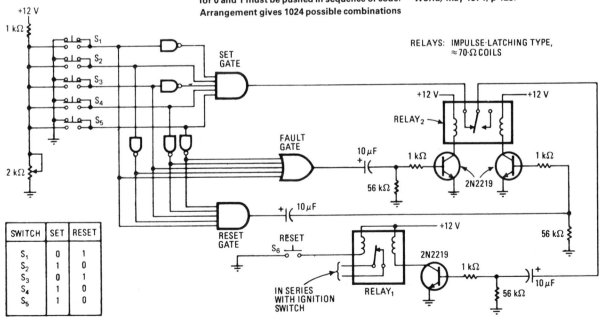

SWITCH	SET	RESET
S₁	0	1
S₂	1	0
S₃	0	1
S₄	1	0
S₅	1	0

ELECTRONIC LOCK—Correct combination of switches S_1-S_5 must be actuated to energize relay in series with ignition switch of auto or any other type of electric lock. If wrong combination is used, lock cannot be opened until resetting combination is entered. When car ignition is turned off, ignition relay should be reset (contact opened) by pressing S_6. With connections shown, switches S_2, S_4, and S_5 must be depressed simultaneously to open (set) lock. If error is made, output of fault gate goes to logic 1 and contacts of relay 2 will open. After error, S_1 and S_3 must be depressed simultaneously to reset lock before opening combination can be used again. Switches can be connected for any other desired combinations.—L. F. Caso, Electronic Combination Lock Offers Double Protection, *Electronics*, June 27, 1974, p 110; reprinted in "Circuits for Electronics Engineers," *Electronics*, 1977, p 346.

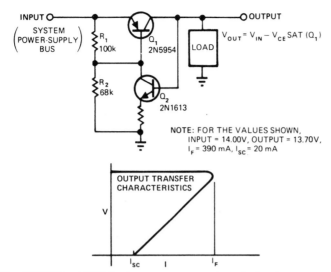

$$V_{OUT} = V_{IN} - V_{CE} \, SAT \, (Q_1)$$

NOTE: FOR THE VALUES SHOWN,
INPUT = 14.00V, OUTPUT = 13.70V,
I_F = 390 mA, I_{SC} = 20 mA.

OUTPUT TRANSFER CHARACTERISTICS

FOLDBACK CURRENT LIMITER—Provides overload and short-circuit protection for load while isolating malfunctioning circuit from other loads on common supply bus. In normal operation, Q_1 is saturated. When load attempts to draw more than this saturation value, base current of Q_1 cannot maintain saturation so voltage across unmarked resistor drops and current through Q_1 drops correspondingly. When load is shorted, Q_2 goes off and short-circuit current folds back to safe lower value. Choose value of unmarked resistor to ensure saturation of Q_1 at load current.—S. T. Venkataramanan, Simple Circuit Isolates Defective Loads, *EDN Magazine,* Jan. 20, 1978, p 114.

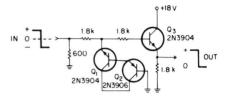

600-Hz CLAMP—Polar clamp was developed to provide overvoltage input protection for ±6 VDC teleprinter signals at 10 mA. Circuit will withstand input transients up to 120 VDC at 20 mA. When input exceeds emitter-base breakdown voltage of Q_1, Q_2 becomes forward-biased for clamping of input. With excessive negative input, Q_1 is forward-biased and emitter-base path in Q_2 completes clamping action.—R. R. Breazzano, A Polar Clamp, *EDN/EEE Magazine,* June 15, 1971, p 59.

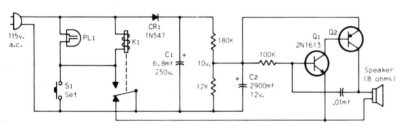

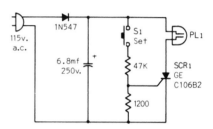

AC LINE MONITOR—Detects AC line failures of any duration and turns off neon lamp PL1 to indicate that clocks require resetting. Circuit is plugged into AC outlet, and S1 is pushed to trigger SCR on and send current through lamp.—J. R. Nelson, Some Ideas for Monitoring A.C. Power Lines, *CQ,* July 1973, p 56.

AUDIBLE LINE MONITOR—Audio oscillator coupled to simple relay circuit gives alerting tone when power fails even momentarily. C2 determines duration of tone. With 2900 µF, tone lasts about 1 s, as warning that clocks will need resetting. Q2 is any PNP audio power transistor, K1 is 115-V SPDT relay, and PL1 is neon lamp.—J. R. Nelson, Some Ideas for Monitoring A.C. Power Lines, *CQ,* July 1973, p 56.

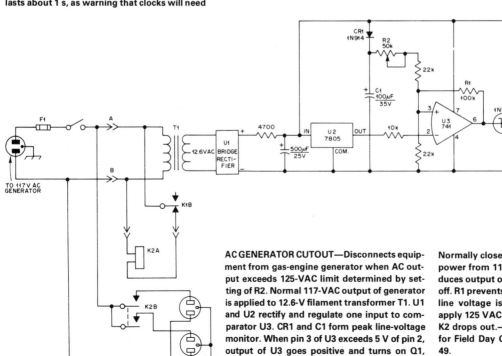

AC GENERATOR CUTOUT—Disconnects equipment from gas-engine generator when AC output exceeds 125-VAC limit determined by setting of R2. Normal 117-VAC output of generator is applied to 12.6-V filament transformer T1. U1 and U2 rectify and regulate one input to comparator U3. CR1 and C1 form peak line-voltage monitor. When pin 3 of U3 exceeds 5 V of pin 2, output of U3 goes positive and turns on Q1, which applies power to small 12-VDC relay K1. Normally closed contacts of K1 open, removing power from 115-VAC relay K2. 1N523 zener reduces output of U3 enough so Q1 can be turned off. R1 prevents relays from chattering when AC line voltage is close to threshold. To adjust, apply 125 VAC between A and B, and set R2 so K2 drops out.—P. Hansen, Overvoltage Cutout for Field Day Generators, *QST,* March 1977, p 49.

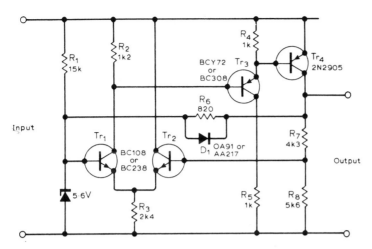

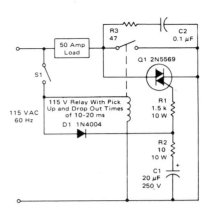

REGULATOR OVERLOAD—When output is shorted, germanium diode D₁ turns on and draws current through R₁, removing reference voltage across zener. Tr₁ is then held off and turns Tr₃ and Tr₄ off to block load current. When short is removed, circuit recovers automatically.—D. E. Waddington, Germanium Diode for Regulator Protection, *Wireless World*, March 1977, p 42.

TRIAC SUPPRESSES RELAY ARCING—Circuit prevents arcing at contacts of relay for loads up to 50 A, by turning on as soon as it is fired by gate current; this occurs after S1 is closed but before relay contacts close. Once contacts are closed, load current passes through them rather than through triac. When S1 is opened, triac limits maximum voltage across relay contacts to about 1 V. Circuit permits use of smaller relay since it does not have to interrupt full load current.—"Circuit Applications for the Triac," Motorola, Phoenix, AZ, 1971, AN-466, p 8.

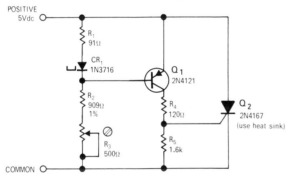

5-V CROWBAR—Simple overvoltage protection circuit for 5-V 1-A logic supply can be adjusted to trigger at 10% overvoltage or 5.5 V. Tunnel diode CR₁ senses level. At 5.5 V, diode switches slightly past its valley point, and voltage across diode biases Q₁ into saturation. Q₁ then supplies gate current to SCR Q₂, which fires and continues conducting until power supply is disconnected. Power supply must include current-limiting circuit and fuse. R₃ adjusts trip point.—L. Strahan, Logic-Supply Crowbar, *EDN/EEE Magazine*, Nov. 15, 1971, p 51.

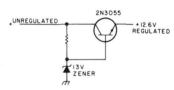

TRANSCEIVER-SAVER—Simple circuit has no effect on normal operation of CB transceiver or other solid-state equipment in auto but provides overvoltage protection if voltage regulator in auto fails. Use heatsink with transistor if transmit current is above 2 A. Choose resistance value to give output of 12.6 V during normal operation.—Circuits, *73 Magazine*, March 1977, p 152.

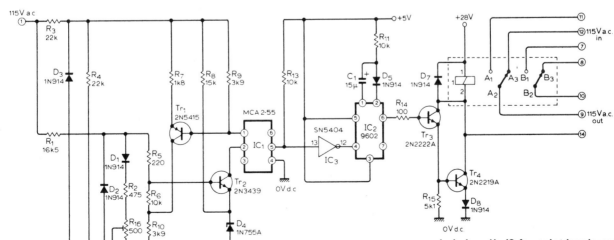

AC OVERVOLTAGE—Used to protect delicate equipment from sustained high AC line voltage, by disconnecting supply when it exceeds preset level selected by R₁₆. When base-emitter voltage of Tr₂ exceeds 7.5 V, optocoupler switches Tr₁ on to provide fast switching action. Output pulse is shaped by IC₃ for use in triggering mono IC₂. When line falls below preset level, mono reverts to stable state and switches on AC supply again.—F. E. George, A.C. Line Sensor, *Wireless World*, March 1977, p 42.

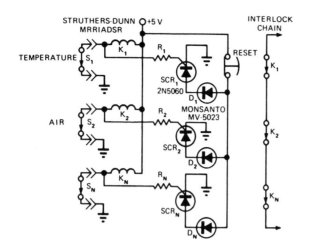

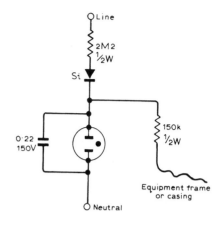

FAIL-SAFE INTERLOCK—Developed for protecting people and equipment at 40-kW RF accelerator station. All interlock switches (air flow, water flow, water pressure, temperature, etc) are normally closed, grounding one side of each relay coil. All relays are normally pulled in, to provide complete interlock chain. Failure of any component, including power supply, breaks chain and places system in safe mode.—T. W. Hardek, Interlock Protection Circuit Is Simple and Fail-Safe, *EDN Magazine*, May 20, 1975, p 74.

LOOSE-GROUND FLASHER—Uses ordinary neon lamp in series with silicon diode, with lamp normally dark. Gives warning by flashing if ground wire is accidentally or purposely disconnected from chassis of oscilloscope or other test instrument.—R. H. Troughton, Earth Warning Indicator, *Wireless World*, April 1977, p 62.

CURRENT SENSOR—Load current is sensed across base-emitter junction of output transistor. R_1 controls OFF time and R_2 controls ON time. Capacitor should be electrolytic rated above 16 V.—M. Faulkner, Two Terminal Circuit Breaker, *Wireless World*, March 1977, p 41.

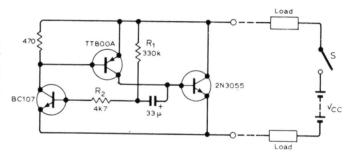

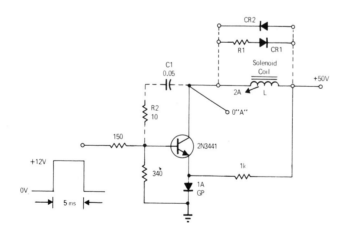

BYPASSING SOLENOID TRANSIENTS—Feedback from collector to base of power transistor through C1 and R2 protects device from destructive transients generated when inductive load such as solenoid is turned off. Alternative use of diode CR2 or CR1-R1 across coil would limit voltage transient but would increase solenoid release time.—D. Thomas, Feedback Protects High-Speed Solenoid Driver, *EDN Magazine*, Jan. 1, 1971, p 40.

SECTION 53
Pulse Generator Circuits

Generate square waves with fixed or variable width and duty cycle, at audio and radio frequencies up to 100 MHz. Includes tone-burst generators, strobe, pulse delay, PCM decoder, and single-pulse generators. See also Frequency Divider, Frequency Multiplier, Frequency Synthesizer, Function Generator, Multivibrator, Oscillator, and Signal Generator chapters.

PULSE-STRETCHING MONO—Section of CMOS MM74C04 inverter accepts positive input pulse by going low and discharging C. Capacitor is rapidly discharged, driving input of MM74C14 Schmitt trigger low. Output of Schmitt then goes positive for interval T_O which is equal to input pulse duration plus interval T that depends on values used for R, C, and supply voltage.—"CMOS Databook," National Semiconductor, Santa Clara, CA, 1977, p 5-30–5-35.

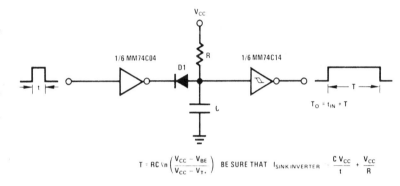

$$T = RC \ln \left(\frac{V_{CC} - V_{BE}}{V_{CC} - V_{T}} \right) \quad \text{BE SURE THAT} \quad I_{SINK\,INVERTER} = \frac{C\,V_{CC}}{t} + \frac{V_{CC}}{R}$$

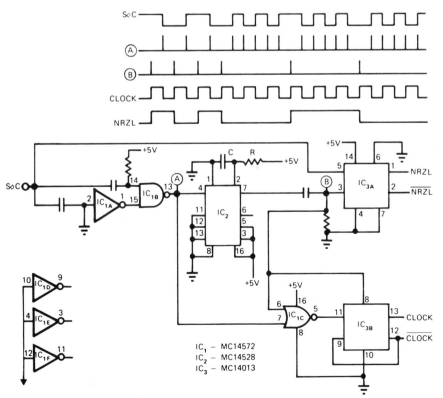

PCM DECODER—Three CMOS ICs provide decoding of Manchester (split-phase) PCM signals by generating missing mark which should occur at each change of level data to recover original clock frequency. Retriggerable mono MVBR times out at slightly longer than half of original clock frequency. Signal levels are TTL-compatible. Values of C and R depend on system frequency. Other resistors are 15K, and other capacitors are 470 pF.—M. A. Lear, M. L. Roginsky, and J. A. Tabb, PCM Signal Processor Draws Little Power, *EDN Magazine*, April 20, 1975, p 70.

IC_1 — MC14572
IC_2 — MC14528
IC_3 — MC14013

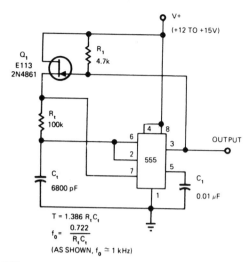

$$T = 1.386\ R_tC_t$$

$$f_0 = \frac{0.722}{R_tC_t}$$

(AS SHOWN, $f_0 \simeq 1$ kHz)

50% DUTY CYCLE WITH 555—Provides pure square-wave output without sacrificing allowable range of timing resistance. Q_1 replaces conventional timing resistor going to V+. Pull-up resistor R_1 is required to switch Q_1 fully on when it is driven by output of timer.—W. G. Jung, Take a Fresh Look at New IC Timer Applications, *EDN Magazine*, March 20, 1977, p 127–135.

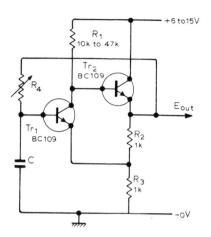

VARIABLE-FREQUENCY UP TO 0.5 MHz—Frequency is determined by choice of values for C and frequency-control potentiometer R_4. Square-wave output has almost equal mark-space ratio over wide frequency range. Regenerative action is rapid, reducing transition times. When circuit is switched on, C is uncharged and Tr_2 is on. C charges until Tr_1 begins to conduct, cutting off Tr_2 and discharging C through R_4 until Tr_1 cuts off and cycle repeats.—J. L. Linsley Hood, Square-Wave Generator with Single Frequency-Adjustment Resistor, *Wireless World*, July 1976, p 36.

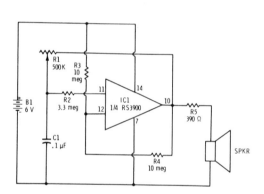

30–4000 Hz WITH OPAMP—Frequency is determined by pot R1 in feedback path. Square-wave output pulse amplitude is about 5 V. Circuit will generate almost perfect sine waves if 0.1-μF capacitor is connected between pin 12 and ground; R1 must be properly adjusted to give output of about 220 Hz.—F. M. Mims, "Integrated Circuit Projects, Vol. 6," Radio Shack, Fort Worth, TX, 1977, p 89–95.

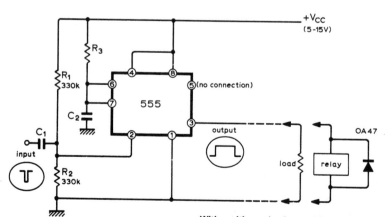

CONTROLLED-DURATION PULSES—Economical Signetics IC provides output pulse currents up to 200 mA at duration ranging from microseconds to many minutes depending on values used for R_3 and C_2. Input pulses may have duration under a microsecond, negative-going. With positive-going input pulses, output will be delayed until trailing edge occurs. Diode is required across output relay coil to suppress transients that might damage IC and cause automatic retriggering.—J. B. Dance, Simple Pulse Shaper or Relay Driver, *Wireless World*, Dec. 1973, p 605–606.

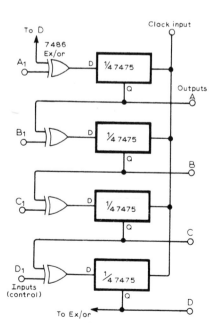

SEQUENCE GENERATOR—Uses gated shift register assembled from 7475 D-type latch, along with four EXCLUSIVE-OR gates. Clock pulse should be narrow to avoid race-around effects.—P. D. Maddison, Sequence Generator, *Wireless World*, Dec. 1977, p 80.

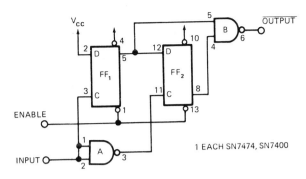

1 EACH SN7474, SN7400

SINGLE-PULSE SELECTOR—Circuit is used to select any desired single pulse from wavetrain continuously applied to input terminal. When enable pulse (not exceeding width of input pulse) is applied, flip-flop FF$_1$ clocks on leading edge of next input pulse and FF$_2$ clocks on trailing edge. Output pulse thus has same width as pulses in input wavetrain. Edge-triggering characteristics of D flip-flops prevent operation if they are enabled during input pulse; in this case, next input pulse is delivered as output.—S. J. Cormack, Pulse Catcher Uses Two ICs, *EDN Magazine,* Jan. 5, 1973, p 109.

PULSE STRETCHER WITH ISOLATION—Motorola MOC1000 optoisolator provides safe interfacing with digital logic while stretching input pulse. Circuit uses phototransistor of optoisolator as one of transistors in mono MVBR. With input pulse width of 3 μs, output pulse width is about 1.2 ms.—"Industrial Control Engineering Bulletin," Motorola, Phoenix, AZ, 1973, EB-4.

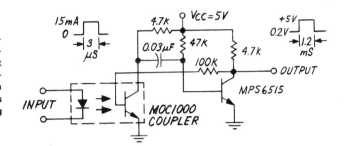

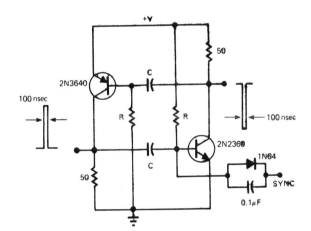

SYNCHRONIZATION TO 10 MHz—Free-running pulse generator circuit uses diode to inhibit operation until sync signal is applied. Circuit then pulses until sync signal returns to original state. Complementary outputs having pulse widths of 100 ns swing essentially from ground to power supply voltage that can be anywhere in range from 0.65 V to 15 V. Values used for R and C determine frequency. For oscillation, R must be in range of 1 kilohm to 1 megohm. For 5-V supply, frequency is 1.2/RC.—B. Shaw, Oscillator Provides Fast, Low Duty-Cycle Pulses, *EDN Magazine,* March 20, 1975, p 73.

VARIABLE-WIDTH TO 12.85 MHz—Single IC circuit uses two monostables to form pulse generator that covers over eight decades (0.054 Hz to 12.85 MHz) with only eight capacitors. Similarly, only eight capacitors cover pulse width range of over eight decades (60 ns to 18 s). Voltage control of frequency and pulse width can be obtained by connecting R$_2$ and R$_4$ to individual 1.5–4.5 V control voltage lines instead of to V$_{CC}$. Frequency will then vary almost linearly with control voltage, while pulse width will vary almost inversely with control voltage. Capacitor values range from 1 pF to over 100 μF.—M. J. Shah, Wide-Range Pulse Generator Uses Single IC, *EDN Magazine,* Jan. 5, 1973, p 107 and 109.

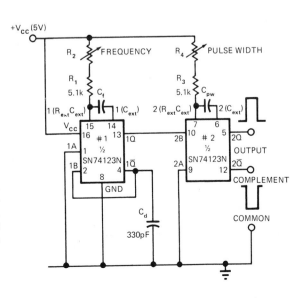

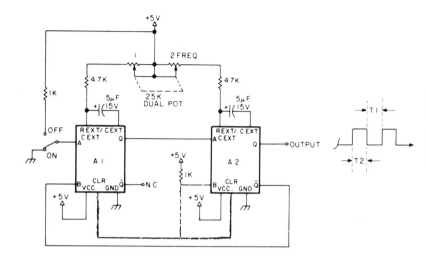

SQUARE-WAVE GENERATOR—Uses two 74122 retriggerable mono MVBRs with clear. Two single pots may be used in place of dual 25K pot if up and down times of output must be independently adjustable.—B. Voight, The TTL One Shot, *73 Magazine,* Feb. 1977, p 56–58.

120 kHz TO 4 MHz—Square-wave output of about 3.5 V can be obtained with SN7400 quad NAND gate, quartz crystal of desired frequency, and single resistor. One of unused gates may be used to gate generator output. Insertion of crystal in socket shocks crystal into oscillation at its resonant frequency, for generating square-wave output over most of frequency range. Waveform approaches clipped sine wave near 4 MHz. Output is suitable for triggering SN7490 decade counters reliably, with normal fanout.—E. G. Olson, 2 Gates Make Quartz Oscillator, *EDN Magazine,* May 5, 1973, p 74.

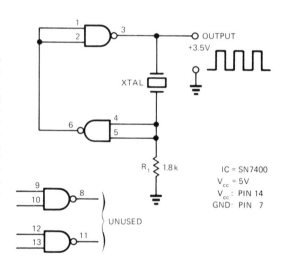

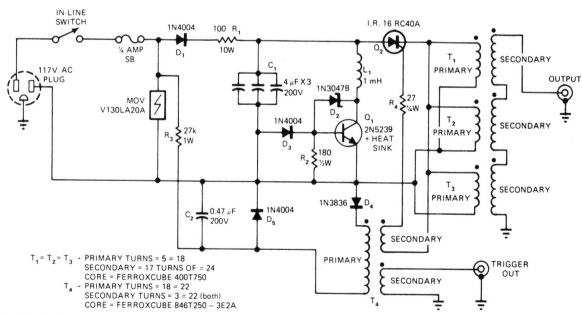

KILOVOLT PULSES—Simple circuit generates 1.5-kV pulses at fixed rate equal to line frequency. Used to drive small piezoelectric transducers for sound velocity measurements. Absence of power transformer minimizes cost, size, and weight. During half of AC cycle, C_1 charges. During other half, C_1 discharges through Q_2 into primaries T_1, T_2, and T_3 to provide output pulse. R_3, C_2, D_5, D_4, and T_4 provide trigger pulse for turning on Q_2. Shunt regulator formed by D_2, D_3, R_2, Q_1, and L_1 clamps voltage across C_1 at 130 V to ensure constant amplitude of output pulses.—S. Anderson, Portable Generator Produces Kilovolt Pulses, *EDN Magazine,* Oct. 20, 1977, p 102.

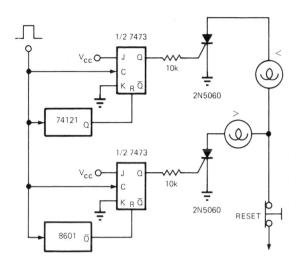

PULSE-WIDTH MONITOR—Circuit turns on upper pilot lamp when pulse width is less than predetermined minimum value, because upper 7473 JK flip-flop is clocked when pulse falls to ground before 74121 mono recovers, triggering upper SCR on. Similarly, 8601 mono is set to coincide with specified maximum pulse width; if pulse falls to ground after this mono recovers, its JK flip-flop is clocked and lower (greater than) lamp is turned on. Fault indication is held until reset button is pushed.—J. Kish, Jr., Three ICs Monitor Pulse Width, *EDN Magazine,* March 20, 1973, p 86.

60 Hz WITH 50% DUTY CYCLE—Adding single resistor R_2 to standard oscillator connection of 555 timer permits operation with 50% duty cycle independently of frequency as determined by value of C_1. For 60-Hz output, V_{CC} is 10 V, C_1 is 1 μF, R_1 is 10K, and R_2 is 75K.—R. Hofheimer, One Extra Resistor Gives 555 Timer 50% Duty Cycle, *EDN Magazine,* March 5, 1974, p 74–75.

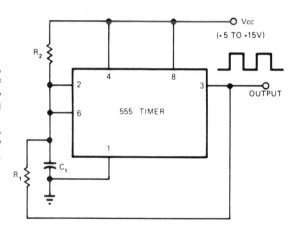

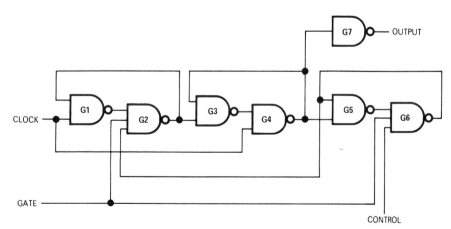

GATED PULSE TRAIN—When control is logic 0, circuit transmits train of complete clock pulses to output, beginning with first clock pulse that starts to rise after application of gate signal and ending with last clock pulse that starts before gate signal falls. When control is logic 1, circuit transmits one complete clock pulse after logic 1 gate signal rises. To send another single pulse, gate signal must be removed and reapplied. Gates are Fairchild LPDTμL9047 triple three-input NAND and 9046 quad two-input NAND; other compatible DTL or TTL NAND gates can also be used.—J. V. Sastry, Gated Clock Generates Pulse Train or Single Pulse, *EDN/EEE Magazine,* July 1, 1971, p 50.

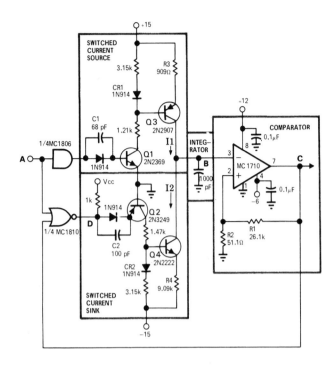

11X PULSE STRETCHER—Provides negative output pulse width equal to positive input pulse multiplied by 1 + R4/R3, which is 11 for values shown. Output pulses are TTL- or DTL-compatible. Minimum output pulse width is 70 ns, and maximum is 1/11 of pulse repetition rate. Circuit consists of switched current source, switched current sink, integrating capacitor, and comparator. Q1 and Q2 act as switches for current sources Q3 and Q4, while C1 and C2 reduce turn-on and turnoff times of switches. CR1 and CR2 provide temperature compensation for Q3 and Q4. AND gate compensates for propagation delay in NOR gate, to ensure that current sink is switched on by trailing edge of input pulse. Add inverter if output must be same polarity as input.—F. Tarico, Linear Circuit Multiplies Pulse Width, *EDN/EEE Magazine*, Dec. 1, 1971, p 45–46.

COMPLETING LAST CYCLE—Developed for applications requiring that gated oscillator must always complete its timing cycle. Circuit uses only two NAND gates and two diodes, none of which are critical as to type. With no input at A, oscillator output B is low. When A is driven high, D goes low initially and drives output B high. If input at A is removed, regenerative feedback is applied from B through diode D_2 to C until normal timing cycle is finished. Then, with B low, D becomes high and keeps output B low.—L. P. Kahhan, Gated Oscillator Completes Last Cycle, *EDN Magazine*, Jan. 5, 1977, p 43.

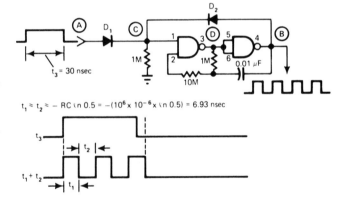

$$t_1 \approx t_2 \approx - RC \ln 0.5 = -(10^6 \times 10^{-6} \times \ln 0.5) = 6.93 \text{ nsec}$$

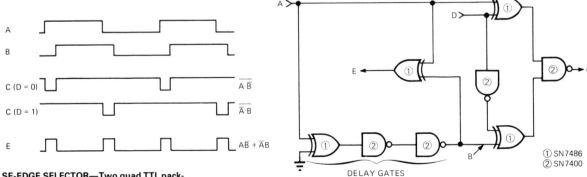

PULSE-EDGE SELECTOR—Two quad TTL packages form simple circuit that generates output pulse at C as function of either leading or trailing edge of input pulse at A, depending on logic level at terminal D. Additional output at E supplies pulses coinciding with both leading and trailing edges of input, independently of logic level at D. Maximum input frequency is 10 MHz, and edge pulses are about 35 ns wide. IC_1 is quad two-input EXCLUSIVE-OR gate, and IC_2 is quad two-input NAND gate.—C. F. Reeves, A Programmable Pulse-Edge Selector, *EDN Magazine*, April 20, 1973, p 85 and 87.

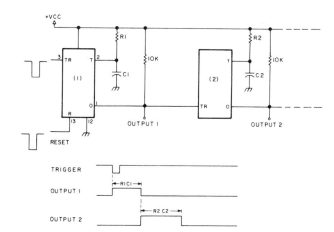

SEQUENTIAL PULSES—Any number of sections of 554 quad monostable timer can be cascaded as shown to give sequential series of output pulses of widths determined by values of R and C. No coupling capacitors are required because timer is edge-triggered. Negative reset pulse simultaneously resets all sections. Varying control voltage (in range of 4.5–16 V) affects period of all timer sections simultaneously.—H. M. Berlin, IC Timer Review, *73 Magazine,* Jan. 1978, p 40–45.

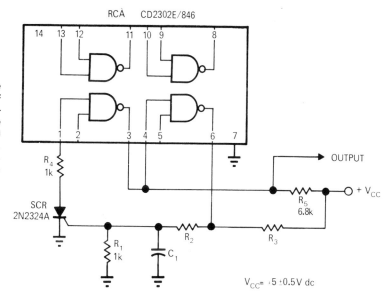

AF RECTANGULAR-WAVE—Frequency can be adjusted over wide AF range, with ON and OFF times of rectangular output signal independently varied between 35 and 60% on by choice of values for C_1 (0.05 to 40 μF), R_2 (1K or 2K), and R_3 (7.3K to 27K). Minimum value of R_3 is 6K.—D. E. Manners, Adjustable Rectangular-Wave Oscillator Interfaces with IC Logic, *EDN/EEE Magazine,* Sept. 15, 1971, p 46.

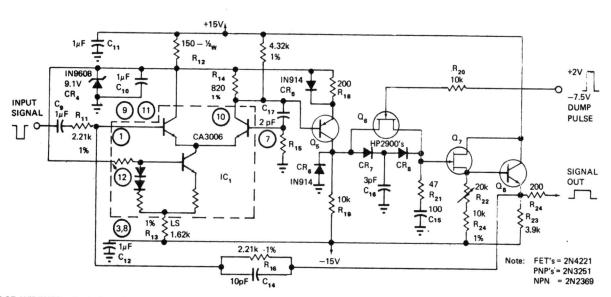

PULSE WIDENER—Peak detection diodes CR_7 and CR_8 in feedback loop of unity-gain CA3006 differential opamp form peak holder that maintains amplitude of narrow video pulses while stretching output pulses as much as 6000 times (from 50 ns to as much as 300 μs). Gain of circuit is unity. Article describes timing and control circuits required in conjunction with peak holder to achieve predictable termination times for stretched pulses. These external circuits include μA710 used as threshold limiter, 9602 dual monostable used as delay-pulse and dump-pulse timing generators, and discrete transistor stage serving as dump-pulse output stage.—B. Pearl, Peak Holder Stretches Narrow Video Pulses, *EDN Magazine,* Feb. 5, 1973, p 46–47.

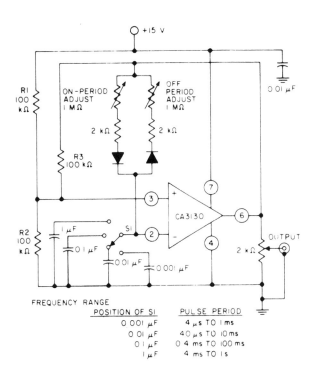

INDEPENDENT ON AND OFF PERIODS—High input resistance of CA3130 opamp permits use of high RC ratios in timing circuits, to give pulse period range of 4 μs to 1 s with switch-selected capacitors.—"Circuit Ideas for RCA Linear ICs," RCA Solid State Division, Somerville, NJ, 1977, p 5.

FREQUENCY RANGE

POSITION OF S1	PULSE PERIOD
0.001 μF	4 μs TO 1 ms
0.01 μF	40 μs TO 10 ms
0.1 μF	0.4 ms TO 100 ms
1 μF	4 ms TO 1 s

ADJUSTABLE SQUARE WAVES—Q_1 and Q_2 form flip-flop, with UJT Q_3 connected as time delay. When power is applied, one flip-flop transistor conducts and C_1 charges through one pot and diode. When C_1 reaches firing voltage of UJT, it conducts and resulting output pulse triggers flip-flop. Sequence of events now repeats, with C_1 charging through other diode. By proper selection of C_1 and pot values, circuit becomes square-wave generator with each pot controlling duration of one half-cycle. With one pot replaced by fixed resistor, circuit becomes pulse generator with other pot controlling pulse-repetition rate. If equal-value fixed resistors replace pots and R_1 is changed to pot, circuit becomes symmetrical square-wave generator with pot controlling frequency.—I. Math, Math's Notes, *CQ*, April 1974, p 64–65 and 91–92.

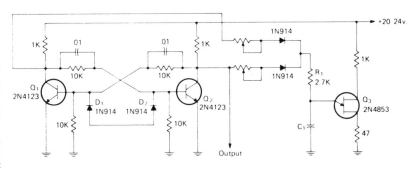

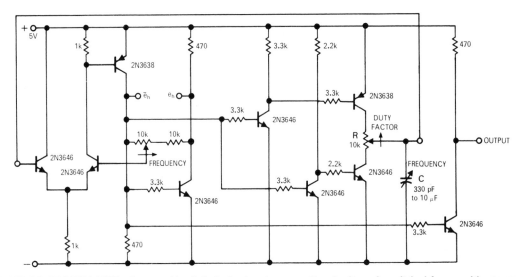

HYSTERESIS-AND-DELAY OSCILLATOR—Separate noninteracting frequency and duty-factor controls permit construction of simple telemetry oscillators having inherently linear transfer function. Absolute synchronization of independent and dependent variables is obtainable with relatively simple pulse-generating circuits. Synchronization cannot be lost. Average value of threshold voltage is maintained constant. Adjustment of hysteresis gap width moves threshold voltage limits symmetrically about average value. Resistance portion of RC delay is switched from positive to negative voltage symmetrically also. Article covers circuit operation in detail.—W. H. Swain, True Digital Synchronizer Employs Hysteresis-and-Delay Element, *EDN Magazine*, Jan. 1, 1971, p 33–35.

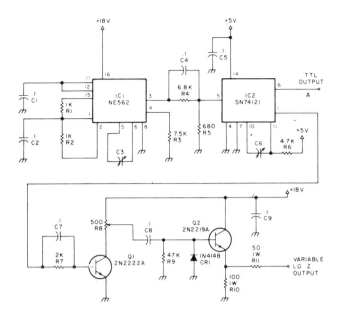

900 kHz TO 10 MHz—Pulse width is variable from about 50 ns to over 500 ms by adjusting only two components. Uses VCO portion of Signetics NE562 as pulse generator and 74121 mono MVBR to adjust pulse width. Variable capacitors C3 and C6 are broadcast-band type. VCO will operate to 30 MHz, limiting factor being stray capacitance and minimum of tuning capacitor. Low-frequency limit of VCO is about 1 Hz, obtained when C3 is 300 μF.—A. Plavcan, Pulses Galore!, *73 Magazine,* Jan. 1978, p 194–195.

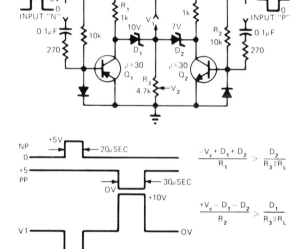

HIGH-SPEED PULSES—TTL circuit provides dual-polarity microsecond pulses. Pulse amplitude is adjusted by changing zeners D_1, D_2, or R_3. Design overcomes slew-rate problems associated with most opamps.—L. Johnson, Dual-Polarity Pulses from TTL Logic, *EDN Magazine,* April 20, 1974, p 91.

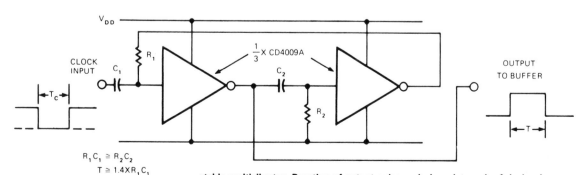

MONO PULSE-SHRINKER—Duty cycle of clock pulse is shortened by two CMOS inverters used to form negative-transition triggered mono-stable multivibrator. Duration of output pulse T is about $1.4R_1C_1$. Output pulse occurs each time input clock goes from high to low. Used with foldback current limiting for short-circuit protection in clock-driven regulated power supply. Low duty cycle of clock pulses ensures positive full-load starting of supply.—J. L. Bohan, Clocking Scheme Improves Power Supply Short-Circuit Protection, *EDN Magazine,* March 5, 1974, p 49–52.

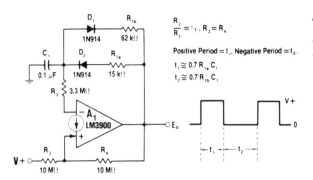

$$\frac{R_2}{R_3} = {}^1{}_3 \,,\, R_3 = R_4$$

Positive Period = t_1, Negative Period = t_2.

$t_1 \cong 0.7\,R_{1a}\,C_1$

$t_2 \cong 0.7\,R_{1b}\,C_1$

ASYMMETRICAL PULSE GENERATOR— Charge and discharge paths of timing capacitor C_1 in LM3900 IC connected as astable oscillator are individually controlled by D_1 and D_2. Value of R_{1a} controls charge rate of C_1 and period t_1, while R_{1b} controls discharge rate and period t_2. Resistors can be pots for providing variable pulse width and repetition rate. For constant frequency with variable duty cycle, R_1 can be single pot with ends going to D_1 and D_2 and tap going to output. For values shown, t_1 is 1 ms and t_2 is 4 ms.—W. G. Jung, "IC Op-Amp Cookbook," Howard W. Sams, Indianapolis, IN, 1974, p 505.

1-MHz SQUARE-WAVE FOR TDR—Fast-rise-time 1-MHz pulse generator serves with wideband CRO and T connector for time-domain reflectometry (TDR) setup used to pinpoint exact location of fault in transmission line. Will also locate multiple faults along line, measure SWR, and measure characteristic impedance of cable. With 1-MHz square-wave source having 500-ns duration for positive portions of wave, cables up to 150 feet long can be tested. R1 should equal characteristic impedance of line being tested. U1 is Signetics N7400A or equivalent quad NAND/NOR gate. Article gives instructions for use.—W. Jochem, An Inexpensive Time-Domain Reflectometer, *QST*, March 1973, p 19–21.

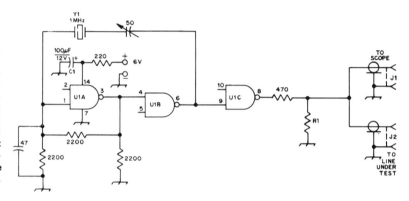

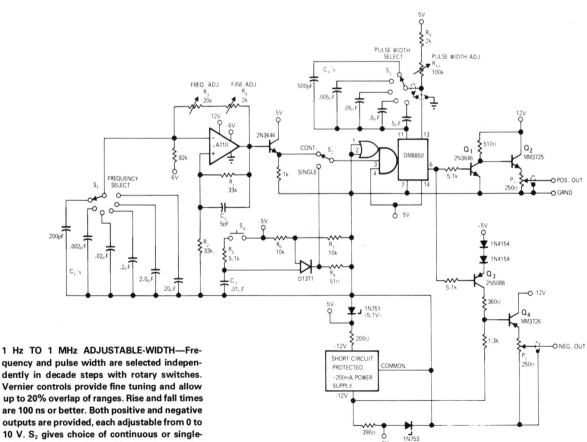

1 Hz TO 1 MHz ADJUSTABLE-WIDTH—Frequency and pulse width are selected independently in decade steps with rotary switches. Vernier controls provide fine tuning and allow up to 20% overlap of ranges. Rise and fall times are 100 ns or better. Both positive and negative outputs are provided, each adjustable from 0 to 10 V. S_2 gives choice of continuous or single-pulse operation, and pushbutton S_4 provides single-pulse outputs. μA710 comparator connected as astable MVBR provides trigger inputs for DM8850 retriggerable mono. Article gives circuit details and design equations.—C. Brogado, Versatile Inexpensive Pulse Generator, *EDN/EEE Magazine,* Oct. 1, 1971, p 37–38.

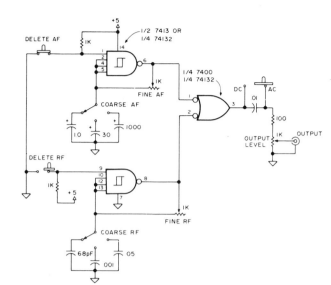

AF/RF SQUARE-WAVE—Use of feedback resistor between input and output of each gate produces oscillation in each Schmitt-trigger oscillator, one operating at audio frequencies and one operating at radio frequencies. Both AF and RF can be fed into NAND gate to give modulated RF, or outputs can be used separately as clocks for microprocessor.—B. Grater and G. Young, Build a Pulse Generator, *Kilobaud,* June 1977, p 49.

AF SQUARE WAVES—With value shown for C1, frequency of output square wave is 530 Hz. For 5300 Hz, use 0.001 μF; for 53 Hz, use 0.1 μF. Circuit will drive ordinary crystal earphone or crystal microphone used as earphone.—F. M. Mims, "Integrated Circuit Projects, Vol. 5," Radio Shack, Fort Worth, TX, 1977, 2nd Ed., p 52–56.

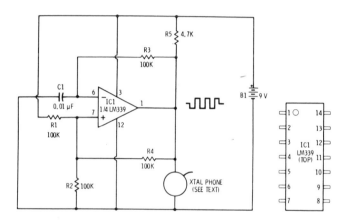

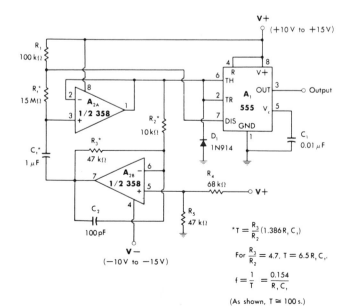

$$*T = \frac{R_3}{R_2}(1.386\,R_1\,C_1)$$

For $\frac{R_3}{R_2}$ = 4.7, T = 6.5 R_1 C_1.

$$f = \frac{1}{T} = \frac{0.154}{R_1\,C_1}$$

(As shown, T ≅ 100 s.)

EXTENDED-RANGE ASTABLE—Square-wave output is extended in frequency by combining buffer A_{2A} with opamp A_{2B} functioning as capacitance multiplier for 555 timer connected as astable MVBR. Value of 1-μF timing capacitor C_t is increased in effective value by ratio of gain of A_{2B} stage, equal to R_3/R_2. Output frequency thus corresponds to that of 4.7-μF capacitor. Negative supply should be equal and opposite to positive supply.—W. G. Jung, "IC Timer Cookbook," Howard W. Sams, Indianapolis, IN, 1977, p 118–121.

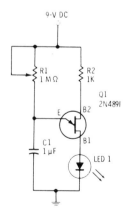

UJT/LED PULSER—Rise time of output pulse is about 200 ns and width is about 25 μs when using 1 μF for C1. Reducing value of C1 reduces pulse width. C1 charges through R1 until voltage across C1 is high enough to bias UJT into conduction. C1 then discharges through UJT and LED and cycle repeats. LED can be any common type.—F. M. Mims, "Electronic Circuitbook 5: LED Projects," Howard W. Sams, Indianapolis, IN, 1976, p 30–32.

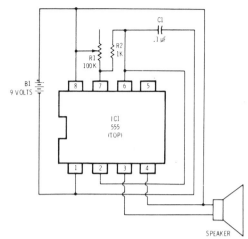

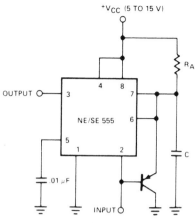

555 AS TONE GENERATOR—Connection of 555 timer as astable MVBR starts next timing cycle automatically, generating sequential square-wave output pulses in audio-frequency range with sufficient power to drive miniature 8-ohm loudspeaker. R1 controls frequency of tone.—F. M. Mims, "Integrated Circuit Projects, Vol. 2," Radio Shack, Fort Worth, TX, 1977, 2nd Ed., p 66–70.

MISSING-PULSE DETECTOR—Timing cycle of 555 timer is continuously reset by input pulse train. Change in input frequency or missing pulse allows completion of timing cycle, producing change in output level. Component values should be chosen so time delay is slightly longer than normal time between pulses.—"Signetics Analog Data Manual," Signetics, Sunnyvale, CA, 1977, p 723.

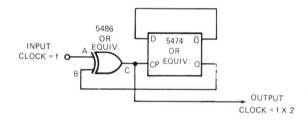

WAVEFORM-EDGE PULSER—Circuit generates square-wave output pulse for each edge of square-wave input. EXCLUSIVE-OR gate is used as programmable inverter that returns point C to quiescent low state following each transfer of data through 5474 IC. When used for frequency-doubling, input waveform should be symmetrical because output is proportional to propagation delay of flip-flop plus delay of 5486 EXCLUSIVE-OR gate.—D. Giboney, Double-Edge Pulser Uses Few Parts, *EDN Magazine*, Dec. 15, 1972, p 41.

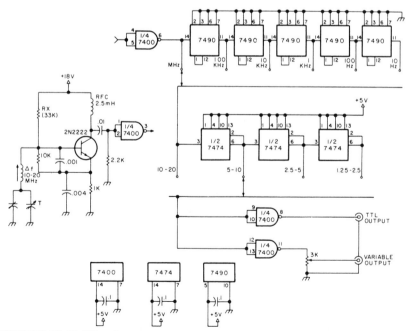

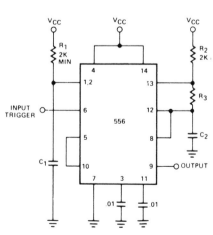

SUBAUDIO TO 20 MHz—Square-wave signal source covers wide frequency range in fully tunable decade steps, as TTL signal source for experimentation with counters, microprocessors, and other logic circuits. Uses tunable 2N2222 transistor oscillator operating at 10–20 MHz, with switchable decade dividers for range selection and switchable binary dividers for band selection. Article covers construction and calibration.—A. G. Evans, Digital Signal Source, *73 Magazine*, Dec. 1977, p 150–151.

TONE-BURST GENERATOR—One section of 556 dual timer is connected as mono MVBR and other section as oscillator. Pulse established by mono turns on oscillator, allowing generation of AF tone burst.—"Signetics Analog Data Manual," Signetics, Sunnyvale, CA, 1977, p 723–724.

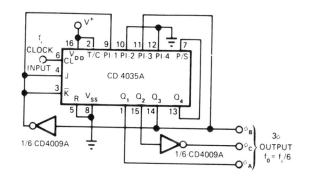

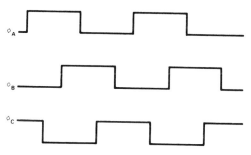

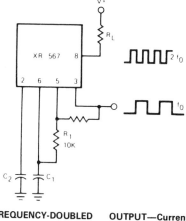

THREE-PHASE PULSE GENERATOR—Requires only CMOS 4-bit shift register and two CMOS inverters. Register is connected to operate as divide-by-6 Johnson counter giving glitch-free outputs. Circuit is driven by square-wave clock signal having frequency 6 times that of desired output frequency.—C. Rutschow, Simple CMOS Circuit Generates 3-Phase Signals, *EDN Magazine,* June 20, 1976, p 128.

FREQUENCY-DOUBLED OUTPUT—Current-controlled oscillator section of Exar XR-567 tone decoder is connected to double frequency of square-wave output by feeding portion of output at pin 5 back to input at pin 3 through resistor. Quadrature detector of IC then functions as frequency doubler to give twice output frequency at pin 8. Supply voltage range is 5–9 V.—"Phase-Locked Loop Data Book," Exar Integrated Systems, Sunnyvale, CA, 1978, p 41–48.

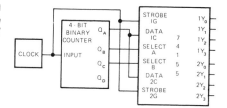

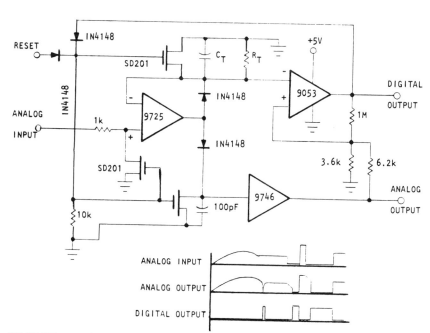

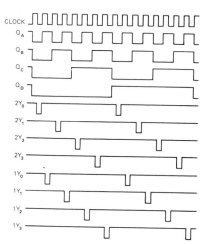

PULSE STRETCHER—Circuit also serves as analog one-shot memory and as peak sense-and-hold with automatic reset. Digital output is logic 0 until C_T and R_T have decayed by one time constant, when it goes to logic 1. Pulse duration is R_CT_C. Analog output amplitude is equal to input amplitude, and duration is same as digital pulse. Optical Electronics 9053 comparator automatically resets circuit after timing interval unless reset is performed manually. Analog output can be used as pulse stretcher with known and controllable pulse duration.—"Analog 'One-Shot'—Pulse Stretcher," Optical Electronics, Tucson, AZ, Application Tip 10292.

UNAMBIGUOUS STROBING—Combination of 74155 two-line to four-line decoder/demultiplexer with any conventional 4-bit binary counter provides family of strobe pulses staggered in such a way that pulse-edge ambiguity is impossible. Clock pulses at input serve to strobe 74155 as well as drive counter. Q_A of counter acts as data input, while Q_B and Q_C act as select lines. Action is such that edges of various 2Y pulses do not coincide with each other, with edges of 1Y pulses, or with edges of Q_B, Q_C, or Q_D pulses. Result is hazard-free strobing.—D. McLaughlin and C. Fanstini, End Edge Ambiguities with Two ICs, *EDN Magazine,* April 5, 1973, p 88.

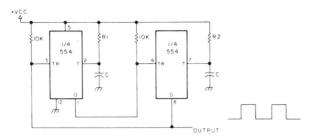

554 ASTABLE—Two sections of 554 quad monostable timer are used. Output frequency is $1/(R_1 + R_2)C$ hertz, and output duty cycle is $100R_2/(R_1 + R_2)$. When R_1 is equal to R_2, symmetrical square wave is obtained. VCC is 4.5–16 V at 3–10 mA.—H. M. Berlin, IC Timer Review, *73 Magazine,* Jan. 1978, p 40–45.

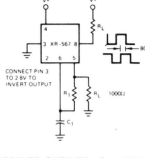

QUADRATURE OUTPUTS—Exar XR-567 tone decoder is connected as precision oscillator providing separate square-wave outputs that are very nearly in quadrature phase. Typical phase shift between outputs is 80°. Supply voltage range is 5–9 V.—"Phase-Locked Loop Data Book," Exar Integrated Systems, Sunnyvale, CA, 1978, p 41–48.

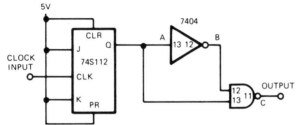

100 MHz—Developed for measuring impulse response of surface acoustic wave devices, for which pulse width had to be under 10 ns for frequency spectrum of about 100 MHz. Propagation delay time of 7404 inverter establishes output pulse width.—R. J. Lang, W. A. Porter, and B. Smilowitz, Simple Circuit Generates Nanosecond Pulses, *EDN Magazine,* Sept. 5, 1975, p 77–78.

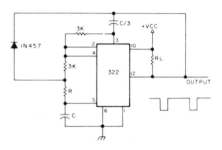

LM322 ASTABLE—National LM322 timer generates narrow negative pulse whose width is approximately 2RC seconds. VCC is 4.5–20 V. Will drive loads up to 5 mA.—H. M. Berlin, IC Timer Review, *73 Magazine,* Jan. 1978, p 40–45.

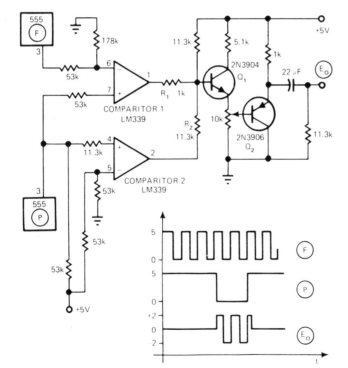

BIPOLAR PULSE TRAINS—Output of Signetics 555 timer F, consisting of unipolar waveform varying from ground to +5 V, is converted to bipolar pulse train having duration equal to that of output pulse from lower 555 timer. While P is high, comparator 2 is on, forcing R_2 to ground and placing base of Q_1 at 2.5 V (because comparator 1 is off, forcing R_1 high). Comparator 2 goes off when timer P goes low, and action of comparator 1 is turned on and off by timer F to produce bipolar pulse train at E_0.—G. L. Assard, Derive Bipolar Pulses from a Unipolar Source, *EDN Magazine,* April 5, 1977, p 144.

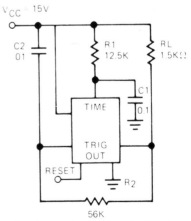

400 Hz—One section of Signetics NE558 quad timer is used as nonprecision audio oscillator providing square-wave output of about 400 Hz with values and supply voltage shown. Output frequency is affected by changes in supply voltage.—"Signetics Analog Data Manual," Signetics, Sunnyvale, CA, 1977, p 738.

SECTION 54
Receiver Circuits

Individual stages and complete circuits for entertainment, amateur radio, commercial communication, and other types of AM and FM receivers, including ultrasonic, radiotelescope, and satellite communication receivers. See also Antenna and Code.

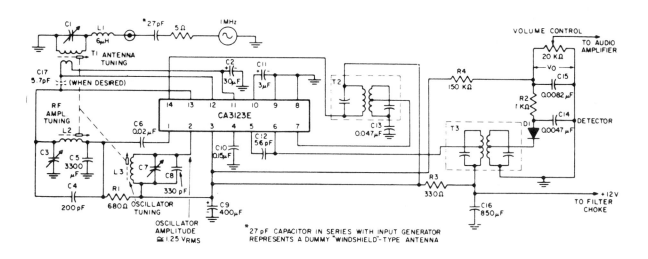

Transformer	Symbol	Frequency	Inductance μh ($\approx$)	Capacitance pF ($\approx$)	Q ($\approx$)	Total Turns To Tap Turns Ratio	Coupling
First IF:							
Primary	T_2	262 kHz	2840	130	60	none	
Secondary			2840	130	60	or 30:1 / 31:1	critical $\approx 0.017 \approx 1/Q$
Second IF:							
Primary	T_3	262 kHz	2840	130	60	8.5:1	—
Secondary			2840	130	60	8.5:1	critical $\approx 0.017 \approx 1/Q$
Antenna:							
Primary	T_1	1 MHz	195	$(C_1) - 130$	65		
Secondary		Adjusted to an impedance of 75 Ω with primary resonant at 1 MHz. Coupling should be as tight as practical. Wire should be wound around end of coil away from tuning core.					
Coils	L_1	7.9 MHz	6		50		
	L_2	1 MHz	55		50		
	L_3	1.262 MHz	41		40		

AM SUPERHET SUBSYSTEM—RCA CA3123E provides all active elements needed up to audio volume control. Table gives values of components for tuned circuits. Operates from single 12-V supply, making subsystem particularly suitable for auto radios. IF value is 262 kHz. 1-MHz signal generator shown in input circuit is used only for initial tuning.—"Linear Integrated Circuits and MOS/FET's," RCA Solid State Division, Somerville, NJ, 1977, p 361–362.

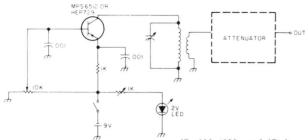

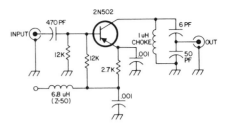

RECEIVER-CHECKING VFO—Simple variable-frequency oscillator is combined with attenuator network to generate signal of about 1 µV for checking performance of amateur radio receiver quickly. Attenuator is series arrangement of 47-, 100-, 100-, and 47-ohm resistors, with 1-ohm resistors going from each of the three junctions to ground. LC combinations are chosen for amateur band desired. Circuit will work down to at least 2 meters.—Is It the Band or My Receiver?, *73 Magazine*, Oct. 1976, p 132–133.

6-METER PREAMP—Simple transistor circuit requires no tuning, draws less than 50 mW from 9-V supply, and increases sensitivity of low-priced receiver without complicated impedance matching.—E. R. Davisson, Simple Six Pre-Amp, *73 Magazine*, Oct. 1974, p 111–112.

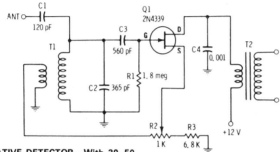

FET REGENERATIVE DETECTOR—With 30–50 foot antenna wire, circuit gives sufficient volume for driving headphones connected to secondary of Lafayette AR-104 or equivalent audio driver transformer T2, for reception of broadcast stations when tuned over AM broadcast band with C2. Feedback control R2 is backed off slightly from point of oscillation, for maximum sensitivity in removing modulation from incoming carrier. When used for CW reception, circuit is left in oscillation and audible difference frequency is produced in output corresponding to marks and spaces. T1 is Miller 2004 or equivalent antenna transformer.—E. M. Noll, "FET Principles, Experiments, and Projects," Howard W. Sams, Indianapolis, IN, 2nd Ed., 1975, p 235–237.

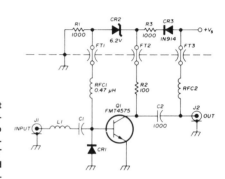

C1 100-180 pF miniature dipped mica or ceramic

C2 1000 pF miniature ceramic disc

CR1 hot-carrier diode (Hewlett-Packard 5082-2810)

CR2 6.2 volt zener diode (1N4735)

CR3 silicon diode (1N914)

FT1- feedthrough capacitors, 470-1000 pF
FT3

J1,J2 SMA-type coaxial connectors (see text)

L1 4 turns no. 24 on 0.1" (2.5mm) diameter, spaced wire diameter (approximately 30 nH)

Q1 Fairchild FMT 4575 low-noise transistor (see text)

R2 100 ohms, ¼ watt (see text)

RFC1 0.47 µH miniature rf choke (Nytronics SWD=0.47)

RFC2 0.2-0.47 µH miniature rf choke or Ohmite Z-460 (value not critical)

432-MHz LOW-NOISE PREAMP—Uses Fairchild FMT4575 transistor having 1.25-dB noise figure, equaling performance of best paramps at 432 MHz. Input matching circuit is low-loss low-Q L matching section L1-CR1. Value of blocking capacitor C1 is not critical, but should be low-loss high-Q type. Hot-carrier diode CR1 in matching section adds about 0.75 pF to circuit, and serves also as low-loss limiter that protects transistor from excessive RF. Zener-diode biasing permits direct grounding of emitter, is insensitive to transistor current gain, provides some DC protection to transistor, and requires no adjustments.—J. H. Reisert, Jr., Ultra Low-Noise UHF Preamplifier, *Ham Radio*, March 1975, p 8–19.

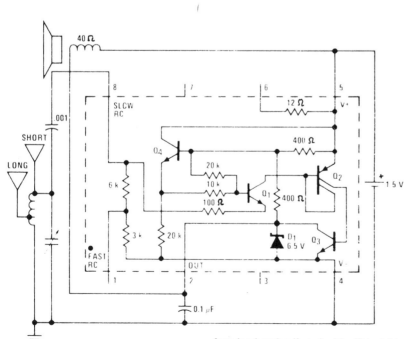

SINGLE-IC RADIO—National LM3909 IC is connected as detector-amplifier driving loudspeaker, with extremely low power gain giving continuous operation for 1 month from D cell. Tuning capability is comparable to that of simple crystal set. Provides acceptable volume from local station if used with efficient 6-inch 40-ohm loudspeaker. Coil is standard AM radio ferrite loopstick having tap 40% of turns from one end. Short antenna can be 10–20 feet, and long antenna can be 30–100 feet.—"Linear Applications, Vol. 2," National Semiconductor, Santa Clara, CA, 1976, AN-154, p 8–9.

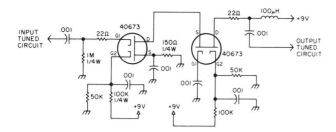

PREAMP BOOSTS GAIN 20 dB—Two RCA MOS-FETs in cascode provide extra 20 dB of gain when used ahead of older Radio Shack AX-190 shortwave receiver. Input and output tuned circuits, gang-tuned, are part of receiver preselector. Article covers construction and tune-up.—P. J. Dujmich, *Improve the AX-190 Receiver, 73 Magazine,* Jan. 1978, p 106–107.

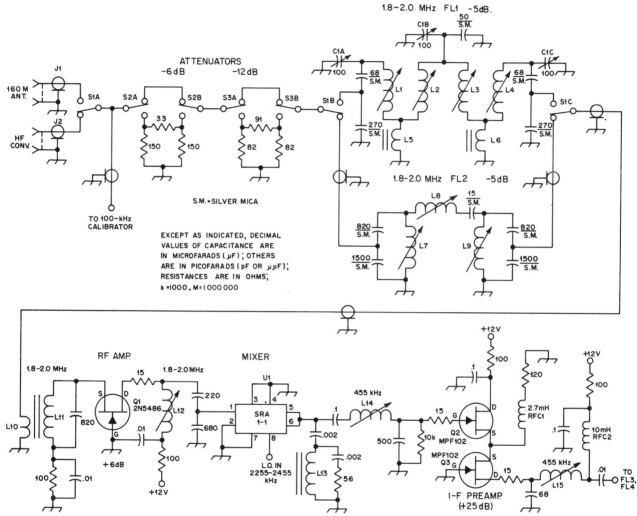

C1 — Three-section variable, 100 pF per section. Model used here obtained as surplus.

J1 — SO-239.

J2 — Phono jack.

L1, L4 — 38 to 68 μH, Q_u of 175 at 1.8 MHz, slug-tuned (J. W. Miller 43A685CBI in Miller S-74 shield can).

L2, L3 — 95 to 187 μH, Q_u of 175 at 1.8 MHz, slug tuned (J. W. Miller 43A154CBI in S-74 shield can).

L5, L6 — 1.45-μH toroid inductor, Q_u of 250 at 1.8 MHz.
15 turns No. 26 enam. wire on Amidon T-50-2 toroid.

L7, L9 — 13-μH slug-tuned inductor (J. W. Miller 9052).

L8 — 380-μH slug-tuned inductor (J. W. Miller 9057).

L10 — 16 turns No. 30 enam. wire over L11 winding.

L11 — 45 turns No. 30 enam. wire on Amidon T-50-2 toroid, 8.5 μH.

L12 — 42-μH slug-tuned inductor, Q_u of 50 at 1.8 MHz. (J. W. Miller 9054).

L13 — 8.7-μH toroidal inductor. 12 turns No. 26 enam. wire on Amidon FT-37-61 ferrite core.

L14 — 120- to 280-μH, slug-tuned inductor

(J. W. Miller 9056).

L15 — 1.3- to 3.0-mH, slug-tuned inductor (J. W. Miller 9059).

Q1, Q2, Q3 — Motorola JFET.

RFC1 — 2.7-mH miniature choke (J. W. Miller 70F273AI).

RFC2 — 10-mH miniature choke (J. W. Miller 70F102AI).

S1 — Three-pole, two-position phenolic wafer switch.

S2, S3 — Two-pole, double-throw miniature toggle.

U1 — Mini-Circuits Labs. SRA-1-1 doubly balanced diode mixer (2913 Quentin Rd., Brooklyn, NY 11229).

1.8–2 MHz FRONT END—Includes enough attenuation for comfortable listening even when nearby high-power amateur station comes on air. Used with downconverter to cover 80 meters through 10 meters. Fixed-tuned 1.8–2 MHz bandpass filter FL2 eliminates need for repeak-

ing three-pole tracking filter FL1 when tuning in band. RF amplifier Q1 compensates for filter loss by giving maximum of 6-dB gain. Double-balanced diode-ring mixer U1 handles high signal levels and has good port-to-port signal isolation. High-pass diplexer network at output of

IC mixer U1 improves noise performance without degrading 455-kHz IF. Output goes to IF filters. Two-part article gives all other circuits of receiver.—D. DeMaw, *His Eminence—the Receiver, QST,* Part 1—June 1976, p 27–30 (Part 2—July 1976, p 14–17).

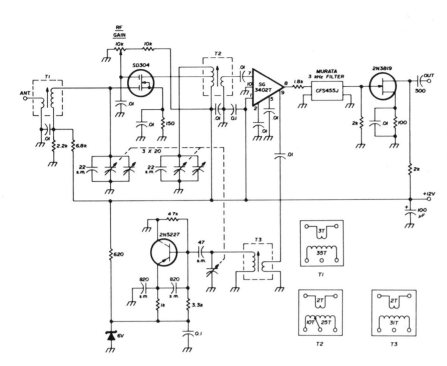

80-METER TUNER—RF stage uses dual-gate N-channel enhancement-mode Signetics SD304 operating with positive bias. With 0–6 V applied to gate 2, AGC range is about 40 dB, but circuit shown uses manual RF gain control. Extra stage of IF overcomes insertion loss of 3-kHz ceramic ladder filter. SG3402T IC is used in mixer; remove pin 6. Transformers T1, T2, and T3 are wound on standard ⅜-in IF forms.—R. Megirian, Design Ideas for Miniature Communications Receivers, *Ham Radio*, April 1976, p 18–25.

20-dB PREAMP FOR 160 METERS—Provides badly needed extra gain when using Beverage or other inefficient low-noise receiving antennas. Gate of common-source JFET is tapped down on tuned circuit by capacitive divider C3-C4 to prevent self-oscillation. Mica compression trimmer C1 provides match to antenna. L1 and L2 are J. W. Miller 43-series slug-tuned coils; L1 has tuning range of 36–57 μH, and L2 has 24–40 μH range. For 160-meter band, L1 and L2 can be peaked at 1827 kHz to provide maximum gain in 1825–1830 kHz DX window.—D. DeMaw, Build This "Quickie" Preamp, *QST*, April 1977, p 43–44.

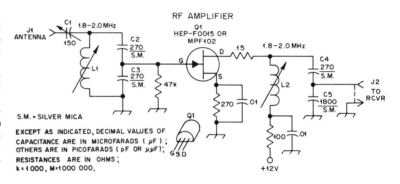

NOTE:
IC1A, B and C are sections of CD4007A IC.

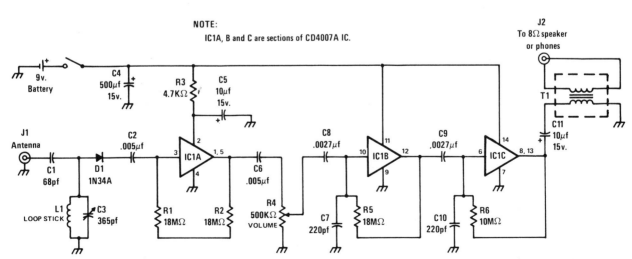

ALL-CMOS RECEIVER—Uses CD4007A IC, having complementary pair of opamps and inverter, to provide all circuits for AM broadcast radio capable of driving headphones or 8-ohm loudspeaker. Selectivity is provided by single tuned circuit and can be improved by optimizing value of C1 to adjust antenna loading. Tune with C3, adjusting L1 if necessary to get stations at low end of band.—C. Green, Easy-to-Build CMOS Radio Receiver, *Modern Electronics*, Sept. 1978, p 40–41, 46, and 59.

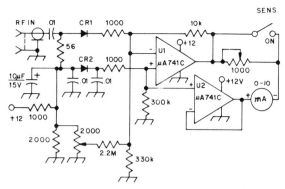

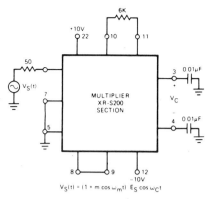

$$V_S(t) = (1 + m\cos\omega_m t) \, E_S \cos\omega_C t$$

RF METER—Simple square-law detector can detect and measure signals as low as −26 dBm, at microwatt levels. CR1 is biased with about 20 μA by opamp U1 serving as low-impedance DC source. CR2 provides temperature compensation, and U2 serves as low-impedance reference for 10-mA meter. Diodes can be hot-carrier types or 1N914s.—W. Hayward, Defining and Measuring Receiver Dynamic Range, *QST*, July 1975, p 15–21 and 43.

SYNCHRONOUS AM DETECTOR—Input signal is applied to multiplier section of Exar XR-S200 PLL IC with pins 5 and 7 grounded. Detector gain and demodulated output linearity are then determined by resistor connected between pins 10 and 11, in range of 1K to 10K for carrier amplitudes of 100 mV P-P or greater. Multiplier output can be low-pass filtered to obtain demodulated output. For typical 30% modulated input with 10-MHz carrier and 1-kHz modulation, output is clean 1-kHz sine wave.—"Phase-Locked Loop Data Book," Exar Integrated Systems, Sunnyvale, CA, 1978, p 9–16.

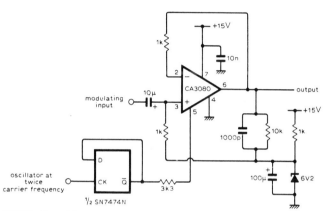

BALANCED MIXER—Uses CA3080 IC transconductance amplifier as precise low-frequency single balanced mixer with inherent carrier balance and accurately defined conversion gain. Binary divider IC halves oscillator frequency, giving carrier waveform having highly accurate unity mark-space ratio. Divided carrier is used to switch amplifier on as unity-gain voltage follower. Conversion loss is 4 dB.—R. J. Harris, Single Balanced Mixer, *Wireless World*, May 1976, p 79.

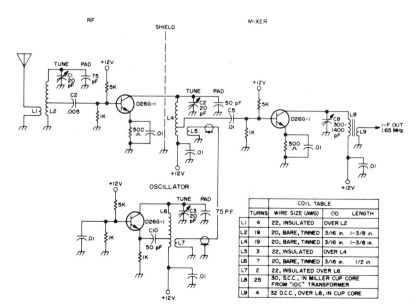

COIL TABLE			
	TURNS	WIRE SIZE (AWG)	OD	LENGTH
L1	4	22, INSULATED	OVER L2
L2	19	20, BARE, TINNED	3/16 in.	1-3/8 in.
L4	19	20, BARE, TINNED	3/16 in.	1-3/8 in.
L5	3	22, INSULATED	OVER L4
L6	7	20, BARE, TINNED	3/16 in.	1/2 in.
L7	2	22, INSULATED OVER L6
L8	25	30, S.C.C., IN MILLER CUP CORE FROM "IOC" TRANSFORMER
L9	4	32 D.C.C., OVER L8, IN CUP CORE

6-METER FRONT END—Developed for use as converter with any communication receiver having 1.65-MHz IF. Article covers construction and tune-up. Use of GE microtransistors permits miniaturization.—B. Hoisington, A Real Hot Front End for Six, *73 Magazine*, Nov. 1974, p 88–90 and 92–94.

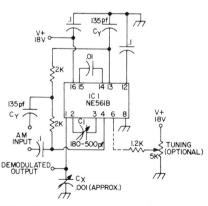

PLL AM—Phase-locked loop of Signetics NE561B is locked to AM signal carrier frequency, and output of VCO in IC is used as local oscillator signal for product detector. Tuned RF stage will generally be required, along with good antenna and ground. Simple one-transistor audio amplifier will suffice for driving loudspeaker. Circuit can be adapted for other frequencies outside of broadcast band, from 1 Hz to 15 MHz, by changing values of C_Y and C_1.—E. Kanter, PLL IC Applications for Hams, *73 Magazine*, Sept. 1973, p 47–49.

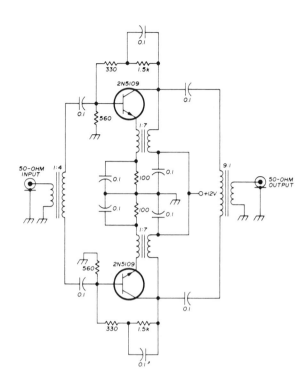

LOW-NOISE RF INPUT—Low-noise version of transistorized push-pull RF stage uses emitter feedback through transformer to give extremely high input and output impedances. Noise figure is below 2 dB. Developed for use in high-quality communication receiver.—U. L. Rohde, Optimum Design for High-Frequency Communications Receivers, *Ham Radio*, Oct. 1976, p 10–25.

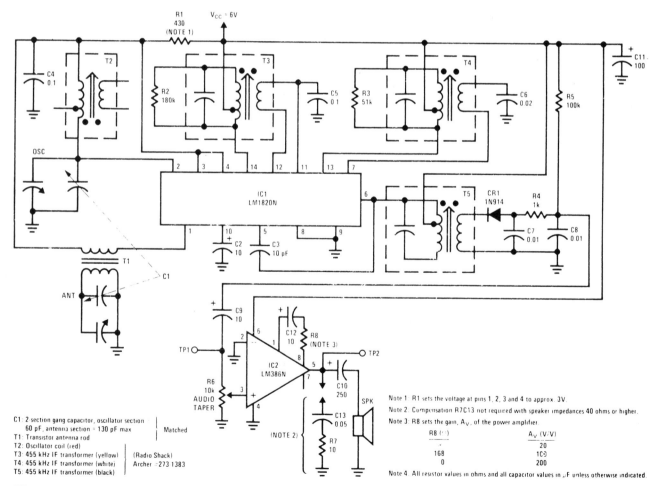

Note 1: R1 sets the voltage at pins 1, 2, 3 and 4 to approx. 3V.

Note 2: Compensation R7C13 not required with speaker impedances 40 ohms or higher.

Note 3: R8 sets the gain, A_V, of the power amplifier.

R8 (Ω)	A_V (V/V)
·	20
168	100
0	200

Note 4: All resistor values in ohms and all capacitor values in μF unless otherwise indicated.

C1: 2-section gang capacitor, oscillator section
  60 pF, antenna section = 130 pF max | Matched
T1: Transistor antenna rod
T2: Oscillator coil (red)
T3: 455 kHz IF transformer (yellow) | (Radio Shack)
T4: 455 kHz IF transformer (white) | Archer -273 1383
T5: 455 kHz IF transformer (black)

AM RADIO—National LM1820N IC provides all sections of superheterodyne broadcast-band radio up to second detector, with diode and power opamp forming rest of receiver. Output is ¼ W into 8-ohm loudspeaker when operating from 6-V supply. Total current drain is about 10 mA, making battery operation feasible.—E. S. Papanicolaou and H. H. Mortensen, "Low-Cost AM-Radio System Using LM1820 and LM386," National Semiconductor, Santa Clara, CA, 1975, LB-29.

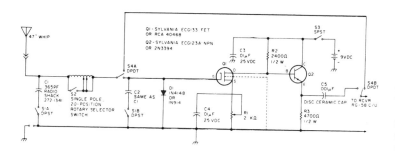

ALL-BAND PREAMP WITH WHIP—Combination of two-stage preamp and 47-inch telescoping antenna gives overall gain of over 30 dB from 160 to 10 meters, for use with communication and SWL receivers when frequent travel precludes erection of fixed antennas. Use type F, BNC, or SO-239 antenna connector. Tuning coil has 20 taps on 150 turns of No. 28 enamel wire wound on ½-inch dowel, with taps at 3, 7, 12, 18, and 25 turns and then about every 10 or 11 turns. Keep leads of Q1 shorted during handling and soldering, to avoid damage by static charges.—K. T. Thurber, Jr., Build A Vacation Special, *73 Magazine*, Aug. 1977, p 62–63.

14–30 MHz PRESELECTOR—Simple self-powered preselector using FET improves overall noise figure of shortwave receiver along with sensitivity in 14–30 MHz portion of HF band. Also helps reduce cross-modulation from strong out-of-band shortwave broadcast stations. C1 and C2 are 50–500 pF Miller 160B. L1 is 10 turns No. 22 on T50-10 Micrometals core with 1-turn link. L2 is 10 turns No. 22 with center tap and 2-turn link on T50-10 core. Q1 is MPF102, HEP-802, or HEP-F0015. D1 and D2 are 1N4002 or HEP-R0051.—H. Olson, The S38 Is Not Dead!, *73 Magazine*, Nov. 1976, p 88–89.

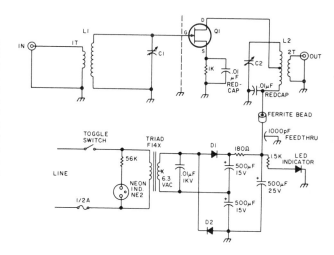

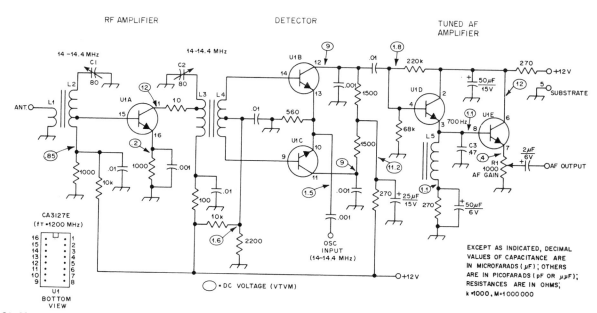

- C1, C2 — 8- to 60-pF mica or ceramic trimmer (Arco 404 or JFD DV11PS60Q suitable).
- C3 — 0.47-µF Mylar capacitor.
- L1 — Two-turn link of No. 24 enam. wire over L2.
- L2 — 25 turns No. 24 enam. wire on T50-6 powdered-iron toroid core. Tap 4 turns up from low-*Z* end. (See *QST* ads for toroid suppliers, Amidon, G. R.

Whitehouse and Palomar Eng.) Mount L1/L2 on opposite side of pc board from L3/L4. L2 = 2.5 µH.
- L3 — 25 turns No. 24 enam. wire on T50-6 toroid core. Tap 10 turns from C2 end. L3 = 2.5 µH.
- L4 — 6 turns No. 24 enam. wire, center tapped. Wind over L3.
- L5 — Pot-core inductor, 110 mH. Wind 172

turns No. 28 enam. wire on bobbin. Core kit is Amidon PC-2213-77.
- R1 — 1000-ohm linear-taper composition control, panel-mounted.
- U1 — RCA CA3127E npn transistor-array IC.

20-METER DIRECT-CONVERSION CW/SSB—Simple direct-conversion or synchrodyne receiver uses RCA CA3127E five-transistor array. Product detector follows 14-MHz RF stage. Low

drain makes receiver ideal for battery operation, but circuit has no AGC. AF output will drive headphones adequately for strong 20-meter signals, but not loudspeaker. Local-oscillator

energy at 14–14.4 MHz for product detector at 1.5–2 VRMS must be furnished by external BFO.—D. DeMaw, Understanding Linear ICs, *QST*, Jan. 1977, p 11–15.

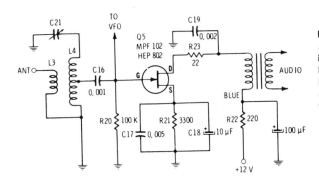

DIRECT-CONVERSION PRODUCT DETECTOR— Antenna is matched to high-impedance gate input of JFET with resonant input transformer. Demodulating carrier is applied to same gate. RC filter and audio transformer in output circuit of JFET recover demodulating audio while filtering out RF signals and undesired mixing components.—E. M. Noll, "FET Principles, Experiments, and Projects," Howard W. Sams, Indianapolis, IN, 2nd Ed., 1975, p 155.

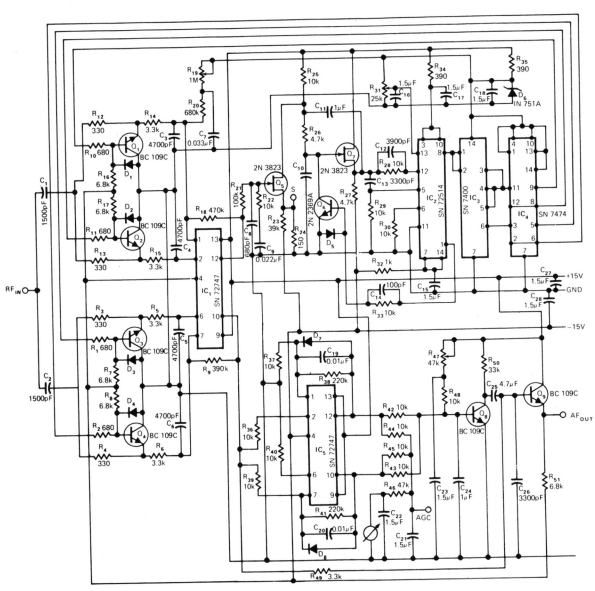

NOTE: D_1, D_2, D_3, D_4, D_5, D_7, D_8 = High speed Ge types

PLL IN AM RECEIVER—Phase-locked loops provide required stability for synchronous detection to improve reception quality of commercial double-sideband AM transmissions. Signal input and output of VCO are multiplied in phase-sensitive detector or multiplier that produces voltage proportional to phase difference between input and VCO signals. After filtering and amplifying, this voltage is used to control frequency of VCO to make it synchronize with incoming signal. Features include absence of image responses since IF is 0 Hz, almost complete immunity to selective fading, and conversion of RF to audio at very low signal levels so overall receiver gain is achieved mainly in audio amplifier. Article traces development and operation of receiver in detail.—T. Mollinga, Solve Phase Stability Problem in AM Receivers with PLL Techniques, *EDN Magazine,* Feb. 20, 1975, p 51–56.

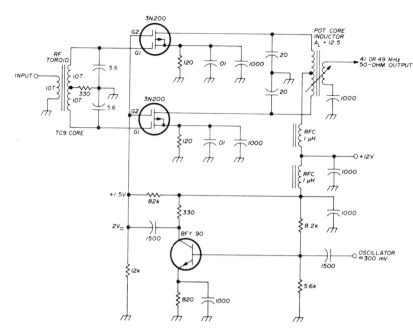

FET MIXER—Double-balanced mixer developed for use in high-quality high-fidelity communication receiver has high input impedance (about 1000 ohms). Two-tone 176-mV signal produces third-order intermodulation distortion 68 dB down.—U. L. Rohde, Optimum Design for High-Frequency Communications Receivers, *Ham Radio*, Oct. 1976, p 10–25.

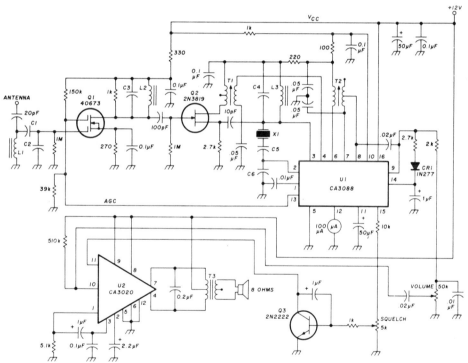

station	freq (MHz)	crystal freq (MHz)	C1 value (pF)	C2 value (pF)	C3,C4 values (pF)	C5 value (pF)	C6 value (pF)	L1&L2 turns	L1&L2 AWG	L1&L2 (mm)	L3 AWG	L3 (mm)	coil cores
WWV WWVH 2.50		2.955	300	820	220	30	150	66	32	(0.2)	32	(0.2)	T37-2
WWV WWVH 5.00		5.455	120	680	100	30	150	49	32	(0.2)	32	(0.2)	T37-2
WWV WWVH 10.00		10.455	56	330	47	30	150	40	32	(0.2)	32	(0.2)	T25-2
WWV WWVH 15.00		15.455	33	330	30	30	150	37	30	(0.25)	30	(0.25)	T25-6
WWV WWVH 20.00		20.455	30	330	27	short	10	29	32	(0.2)	32	(0.2)	T25-6
WWV WWVH 25.00		25.455	24	300	22	short	10	26	32	(0.2)	32	(0.2)	T25-6
CHU	3.33	3.785	300	820	220	30	150	50	30	(0.25)	30	(0.25)	T37-2
CHU	7.34	7.795	68	350	56	30	150	44	32	(0.2)	32	(0.2)	T37-2
CHU	14.67	15.125	33	330	30	30	150	36	32	(0.2)	32	(0.2)	T25-6

10-MHz FIXED FOR WWV—Fixed-frequency receiver has high sensitivity, portability, low power consumption, and low cost. Number of parts is minimized by using RCA CA3088 IC for converter, IF, detector, audio preamp, AGC, and tuning-meter output, along with RCA CA3020 as audio amplifier. Table gives crystal frequencies and tuned-circuit values for all nine frequencies on which frequency calibration data, propagation forecasts, geophysical alerts, time signals, and storm warnings are broadcast by American and Canadian governments. Core type numbers are for Amidon Associates cores. IF transformers come as Radio Shack set 273-1383; use only T1 (gray core) and T2 (white core). Specify load capacitance as 32 pF when ordering crystals. Use overtone crystals for 20 and 25 MHz with C5 replaced by short and C6 reduced to 10 pF.—A. M. Hudor, Jr., Fixed-Frequency Receiver for WWV, *Ham Radio*, Feb. 1977, p 28–33.

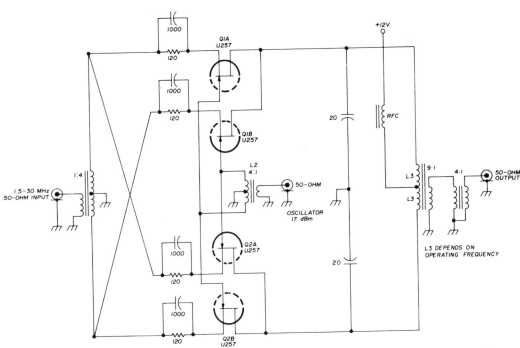

BALANCED FOUR-FET MIXER—Uses two matched FET pairs to bring third-order intermodulation distortion suppression down to 71 dB. Developed for use in high-quality communication receiver.—U. L. Rohde, Optimum Design for High-Frequency Communications Receivers, *Ham Radio,* Oct. 1976, p 10–25.

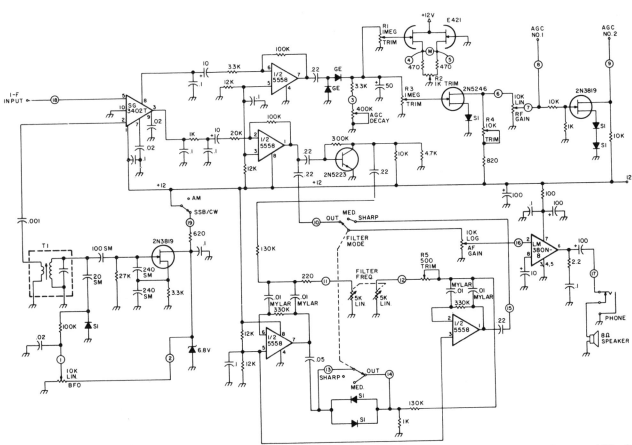

AF FOR AM/SSB/CW—Uses SG3402T as detector, with BFO disabled for AM. Pin 3 of detector output is main audio source, feeding preamp using half of dual opamp whose output goes to AF gain control except when CW filter is in use. Filter has two identical 400–1600 Hz active bandpass sections joined by threshold detector. LM380N-8 AF power amplifier is rated at 600-mW output. Audio from pin 8 of detector is amplified about 30 times in second half of dual opamp before rectification for use as AGC voltage. Circuit includes S-meter fed by AGC section. Article gives construction details of complete receiver.—R. Megirian, The Minicom Receiver, *73 Magazine,* April 1977, p 136–149.

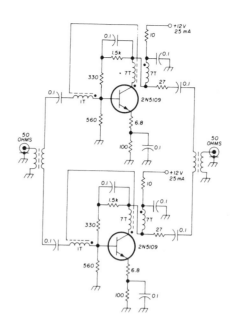

PUSH-PULL RF—Uses VHF power transistors to obtain wide dynamic range. Transformers are trifilar wound on Indiana General F625-9-TC9 toroid cores. Circuit has extremely low VSWR at both input and output, along with low noise figure. Second-order intermodulation products can be suppressed nearly 40 dB over single stage. Either RCA 2N5109 or Amperex BFR95 transistors can be used. Gain is about 11 dB. Current feedback is used through unbypassed 6.8-ohm emitter resistor, voltage feedback through unbypassed 330-ohm base-to-collector resistor, and transformer feedback through third winding on wideband transformer to stabilize input and output impedances.—U. L. Rohde, High Dynamic Range Receiver Input Stages, *Ham Radio*, Oct. 1975, p 26–31.

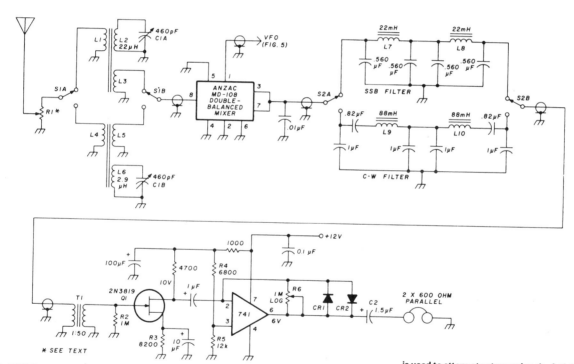

DIRECT CONVERSION—Simple direct-conversion amateur receiver uses VFO and mixer to produce AF signal directly, with no IF amplifier or second detector. For SSB reception, VFO is tuned to frequency of suppressed carrier. For CW, VFO is detuned enough to give note of desired pitch. Not suitable for AM or FM reception. Separate input tuned circuits are used for 15–40 meters and for 80–160 meters. Use ferrite or powdered iron toroid cores for coils, with turns determined experimentally. L7 and L8 are 88-mH toroids with series-connected windings. R1 is used to attenuate strong signals. Article gives circuit for VFO and buffer amplifier. Separate VFO is used for each band (160, 80, 40, 20, and 15 meters). Construction details are given, along with advantages and drawbacks of direct conversion.—D. Rollema, Direct-Conversion Receiver, *Ham Radio*, Nov. 1977, p 44–55.

SECTION 55
Regulated Power Supply Circuits

Various combinations of line-powered rectifiers and voltage regulators provide highly regulated fixed and variable positive and negative outputs ranging from 0 to ±35 V at maximum currents from 24 mA to 24 A. Dual-output supplies may have tracking. See also Power Supply, Regulator, and Switching Regulator chapters.

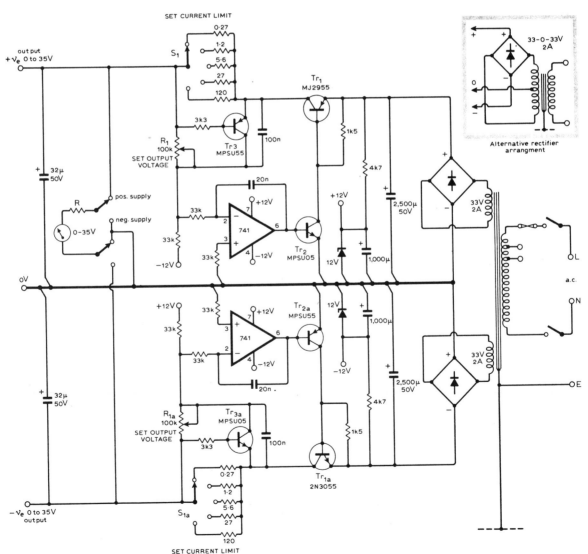

0 to ±35 V—Twin stabilized DC supply uses ganged pots R₁ and R₁a to set both positive and negative regulated outputs at any desired value up to 35 V. Input supplies from bridge rectifiers also provide ±12 V lines for 741 opamps. Load regulation is within 2 mV from no load to full 2-A maximum output. Output hum, noise, and ripple are together only 150 μV and independent of load.—J. L. Linsley Hood, Twin Voltage Stabilized Power Supply, *Wireless World,* Jan. 1975, p 43–45.

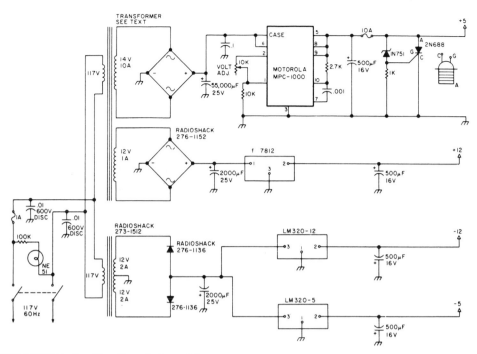

±5 AND ±12 V FOR COMPUTER—Provides all voltages required for 8080-4BD microcomputer system marketed by The Digital Group (Denver, CO). Transformer for positive supplies is 6.3-V 20-A unit with secondary replaced by two new windings giving required voltage and current. Crowbar circuit using 2N688 SCR protects ICs in memory and CPU. Use of at least 50,000 μF in filter of 5-V supply prevents noise problems in computer. MPC-1000 5-V 10-A regulator should be mounted on large heatsink at rear of computer housing in open air.—L. I. Hutton, A Ham's Computer, *73 Magazine*, Dec. 1976, p 78–79 and 82–83.

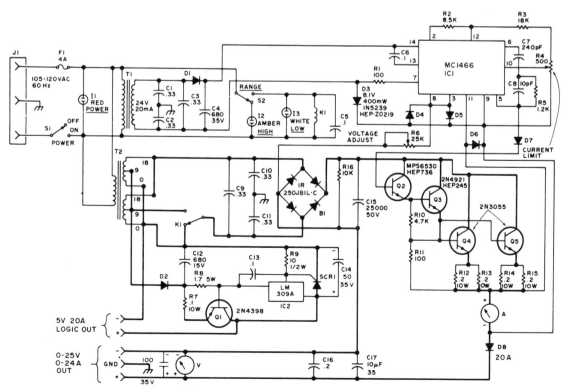

5 V AT 20 A AND 0–25 V AT 0–24 A—Developed as lab supply for experimenting with high-current TTL circuits. Motorola MC1466 monitors voltage and current requirements continuously, providing output proportional to parameters called for by front-panel controls of supply. D2 and D8 are 50-PIV 20-A diodes, and all other diodes except D3 are 1N4002 or equivalent. Article gives construction details.—J. W. Crawford, The Smart Power Supply, *73 Magazine*, March 1976, p 96–98 and 100–101.

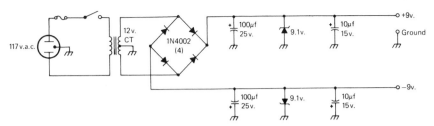

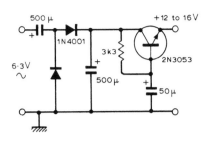

±9 V—Developed for use with demodulator of teleprinter. Regulation is provided by zeners.—I. Schwartz, An RTTY Primer, *CQ*, Feb. 1978, p 31–36.

12–16 V FROM 6.3 VAC—Designed for use with transistor or IC amplifier being fed by tube-type preamp having 6.3-V power transformer winding for filament supply.—K. D. James, Balanced Output Amplifier, *Wireless World*, Dec. 1975, p 576.

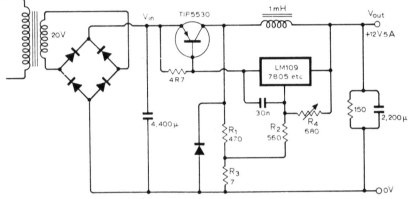

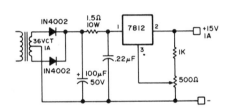

ADJUSTABLE SWITCHED REGULATOR—Circuit shows method of using LM109, 7805, or other IC voltage regulator to provide output voltage that is higher than rated output of IC. Voltage pedestal is developed across R_2 and R_3 for adding to normal regulated output of IC. R_4 adjusts amount of added voltage. Divider R_1-R_2 provides positive feedback into pedestal circuit of regulator, to allow switching of IC and transistor.—V. R. Krause, Adjustable Voltage-Switching Regulator, *Wireless World*, May 1976, p 80.

15 V AT 1 A—Developed for operating CRO from AC line. Can also be used for recharging batteries of portable CRO if pot is set to correct charging voltage for cells being used. Use good heatsink with 7812 regulator.—G. E. Friton, Eyes for Your Shack, *73 Magazine*, Jan. 1976, p 66–69.

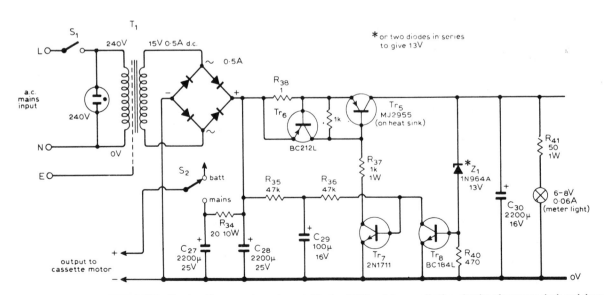

14 V AT 250 mA FOR CASSETTE DECK—Used in high-quality stereo cassette deck operating from AC line or battery. For U.S. applications, use 120-V power transformer. Power for cassette motor is taken directly from power-supply filter capacitors through 20-ohm 10-W resistor, with negative return line connecting directly to filter capacitors instead of chassis, to eliminate noise originating from pulsating current of cassette-drive motor-control circuit. Article gives all other circuits of cassette deck and describes operation in detail.—J. L. Linsley Hood, Low-Noise, Low-Cost Cassette Deck, *Wireless World*, Part 2—June 1976, p 62–66 (Part 1—May 1976, p 36–40; Part 3—Aug. 1976, p 55–56).

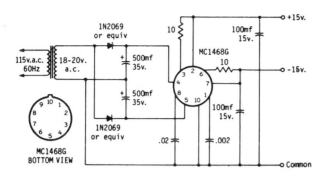

±15 V—Provides positive and negative supply voltages required by some opamps. Supply is short-circuit-proof and protects itself against overloads.—I. Math, Math's Notes, *CQ,* Jan. 1974, p 68–69.

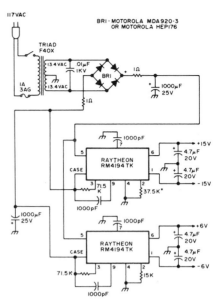

±6 AND ±15 V—Developed for use with function generator. Mount regulators on heatsinks insulated from chassis by mica wafers. Article covers construction and adjustment to give exactly desired outputs.—H. Olson, Build This Amazing Function Generator, *73 Magazine,* Aug. 1975, p 121–124.

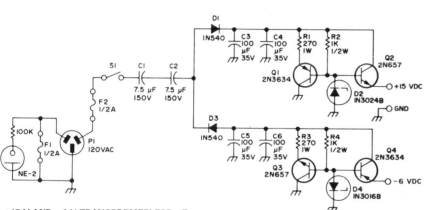

+15 V AND −6 V TRANSFORMERLESS—Transistorized regulator provides good voltage regulation with low ripple. Second ground prong is connected through fuse to grounded center conductor of AC line to guard against faulty AC wiring. If wiring is reversed, fuse will disable power supply and neon fault indicator will come on. At currents up to 55 mA, −6 V output had 0.1-V ripple and +15 V output had 0.05-V ripple.—D. Kochen, Transformerless Power Supplies, *73 Magazine,* Sept. 1971, p 14–17.

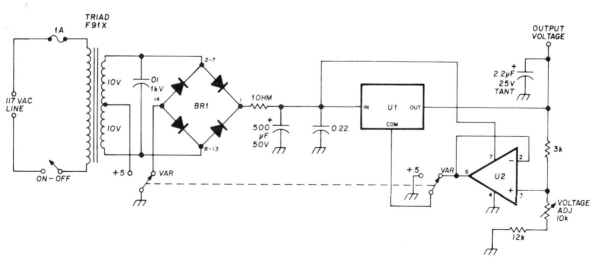

+5 V AT 200 mA OR 7–20 V AT 100 mA—Uses National LM741 opamp as noninverting follower to sample output of voltage divider and drive common terminal of National LM340-05 three-terminal voltage regulator. Heatsink tab of regulator U1 must be connected to floating heatsink. BR1 is Adva bridge.—H. Olson, Second-Generation IC Voltage Regulators, *Ham Radio,* March 1977, p 31–37.

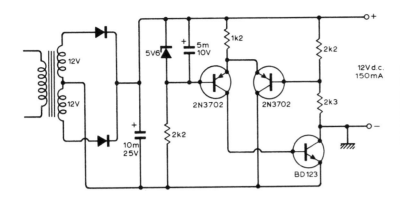

12-V LOW-RIPPLE—Three-transistor feedback circuit gives low-cost voltage stabilizer in which ripple is low and regulated output is very little less than unstabilized input voltage.—R. H. Pearson, Novel 5-Watt Class A Amplifier Uses Three-Transistor Feedback Circuit, *Wireless World*, March 1974, p 18.

+5 V WITH UNREGULATED +15 V—Developed for use with audio decoder that converts BCD output of digital display to audio tones that can be recognized by blind radio operator or experimenter.—D. R. Pacholok, Digital to Audio Decoder, *73 Magazine*, Oct. 1977, p 178–180.

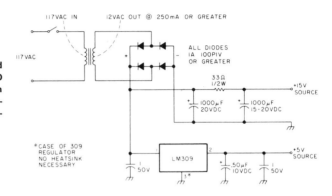

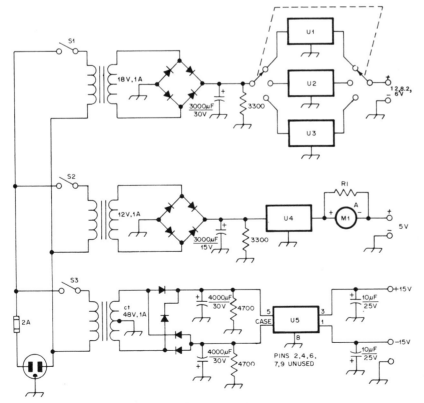

POPULAR-VOLTAGES SUPPLY—Provides most common fixed voltages required for transistor and IC projects. Eliminates cost and nuisance of replacing batteries. Provides ± 15 V at 100 mA, +5 V at 1 A, and choice of +6.0, +8.2, and +12.0 V at 1 A. Separate grounds (not chassis grounds) permit connecting supplies in series to get combination voltages. Rectifier diodes are 100 PIV at 1 A. Use meter and shunt to give full scale at 1 A; for 150-mA meter, use 0.08 ohm for R1 (six 0.5-ohm resistors in parallel). U1 is LM340K-12, U2 is LM340K-8, U3 is LM340K-6, U4 is LM340K-5, and U5 is 4195.—C. J. Appel, A Combination Fixed-Voltage Supply, *QST*, Nov. 1977, p 36–37.

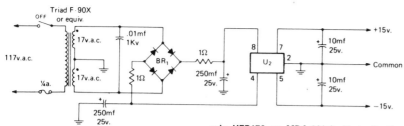

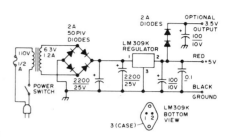

± 15 V—Developed for use in two-tone AF generator for testing SSB equipment. BR₁ is Motorola HEP176 or MDA-920-2. U₂ is Raytheon 4195DN.—H. Olson, A One-Chip, Two Tone Generator, *CQ*, April 1974, p 48–49.

5 V AT 1 A—Can handle over 30 TTL ICs in frequency counter if LM309K regulator is mounted directly on aluminum heatsink. Case of regulator is grounded, so mica insulation is not needed. Provides excellent regulation with practically no output ripple and is short-circuit-proof. Circuit also shuts itself off if temperature gets too high.—P. A. Stark, A Simple 5 V Power Supply for Digital Experiments, *73 Magazine*, Oct. 1974, p 43–44.

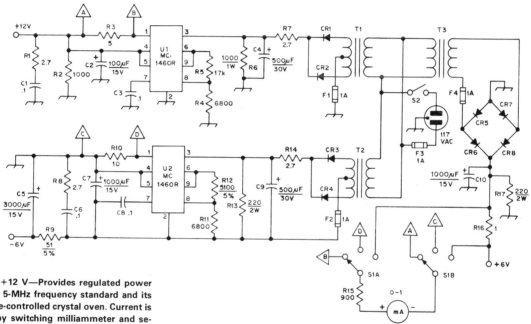

±6 V AND +12 V—Provides regulated power required by 5-MHz frequency standard and its temperature-controlled crystal oven. Current is measured by switching milliammeter and series resistor across current-limiting resistors R3 and R10, which also serve as meter shunts. DC input voltage at terminal 3 of MC1460R regulator should be at least 3 V greater than output voltage but should not exceed 20-V rating. Adjust R5 and R12 as required to get correct output voltages. Diodes are 200 PIV at 0.5 A. T1 and T2 have 117-V primary and 25.2-V secondary at 0.3 A. T3 has 117-V primary and 6.3-V secondary at 1 A.—R. Silberstein, An Experimental Frequency Standard Using ICs, *QST*, Sept. 1974, p 14–21 and 167.

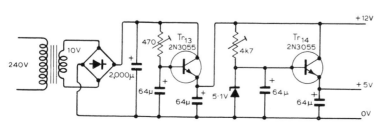

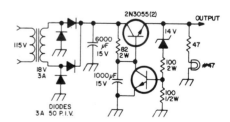

12 AND 5 V—Used with multiple photographic development timer to provide 12 V at about 5 mA and 5 V at about 300 mA for logic, control, and audible alarm circuits.—R. G. Wicker, Photographic Development Timer, *Wireless World*, April 1974, p 87–90.

12 V FOR TRANSCEIVER—Permits operation of 144-MHz transceiver from AC line at base station. If 18-V transformer at 3 A is not available, use 12-V 3-A unit and add 26 turns No. 20 teflon-coated wire to secondary in proper phase. Choose transformer having enough room for these extra turns.—W. W. Pinner, Midland 2 Meter Base or Portable, *73 Magazine*, Aug. 1974, p 61–63.

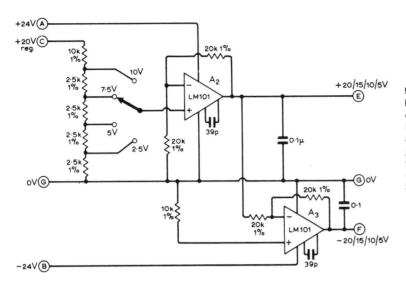

5/10/15/20 V SWITCH—Reference voltage selected by switch is applied to noninverting input of opamp having voltage gain of 2, to provide both positive and negative regulated voltages at desired value. Any standard opamp, such as μA709 or μA741, can be used in place of National LM101.—T. D. Towers, Elements of Linear Microcircuits, *Wireless World*, July 1971, p 342–346.

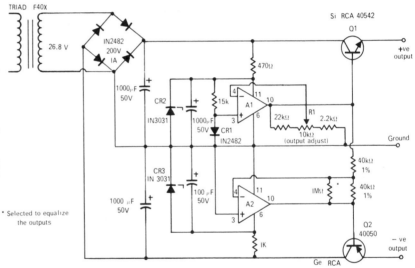

±4 V TO ±25 V—Arrangement permits varying both positive and negative regulated output voltages simultaneously with single control, with maximum load current of 400 mA for both regulators. Positive supply controls negative slave regulator to provide tracking within 0.05 V at full output. Developed for use in lab to observe effect of varying supply voltages on circuits under development.—J. A. Agnew, Dual Power Supply Delivers Tracking Voltages, *EDN Magazine*, Oct. 15, 1970, p 51.

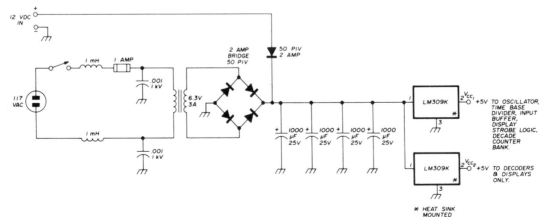

DUAL 5-V—Provides two 5-V regulated supplies for frequency counter, operating either from 9-VDC outputs of AC supply or from 12-VDC auto battery. Splitting of supply divides current demand so regulators operate well below maximum ratings and provide decoupling between sections of load.—J. Pollock, Six Digit 50-MHz Frequency Counter, *Ham Radio*, Jan. 1976, p 18–22.

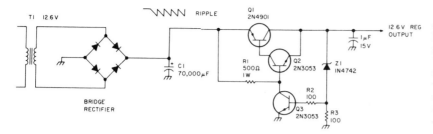

12.6 V AT 3 A—Article gives step-by-step procedure for designing simplest possible regulated supply to meet specific requirements in general service. Power transformer rated 12.6 V at 3 A delivers about 18 V P to bridge rectifier rated 50 V at 5 A. Value of C1 is chosen to keep voltage to regulator above 15-V limit at which circuit would drop out of regulation.—C. W. Andreasen, Practical P. S. Design, *73 Magazine*, June 1977, p 84–85.

+5 V AND ±6 OR ±12 V—Three power supplies for experimental use are achieved with only one transformer. LM309K regulates 5-V supply. Other two supplies are regulated by 6.2-V zeners in conventional regulator; shorting out one zener in each with gang switch reduces output to 6 V.—Design a Circuit Designer!, *73 Magazine*, Oct. 1977, p 152–153.

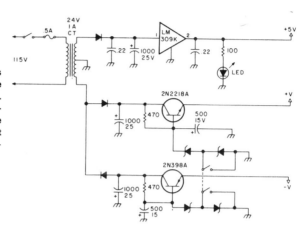

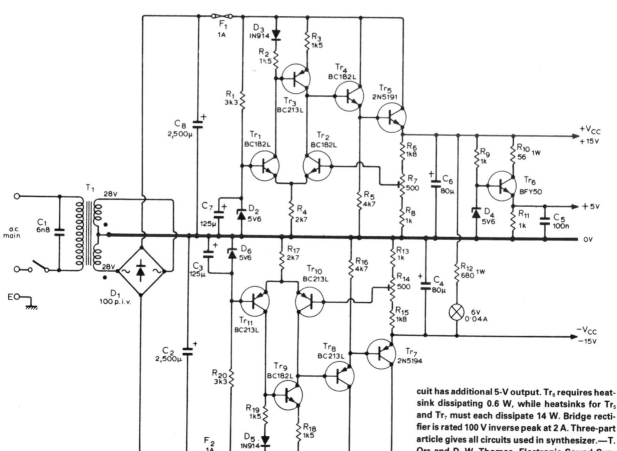

cuit has additional 5-V output. Tr6 requires heatsink dissipating 0.6 W, while heatsinks for Tr5 and Tr7 must each dissipate 14 W. Bridge rectifier is rated 100 V inverse peak at 2 A. Three-part article gives all circuits used in synthesizer.—T. Orr and D. W. Thomas, Electronic Sound Synthesizer, *Wireless World*, Part 3—Oct. 1973, p 485–490 (Part 1—Aug. 1973, p 366–372; Part 2—Sept. 1973, p 429–434).

±15 V FOR SOUND SYNTHESIZER—Provides highly stabilized voltages required by elaborate sound synthesizer developed for generating wide variety of musical and other sounds. Cir-

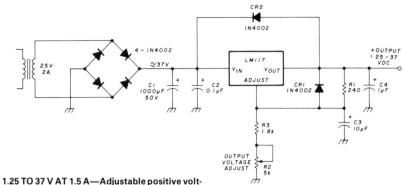

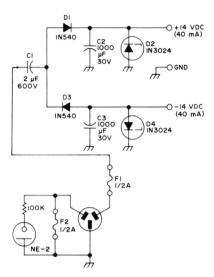

1.25 TO 37 V AT 1.5 A—Adjustable positive voltage regulator used with simple bridge rectifier and capacitor-input filter delivers wide range of regulated voltages, all with current and thermal overload protection. Load regulation is about 0.3%.—H. Berlin, A Simple Adjustable IC Power Supply, *Ham Radio*, Jan. 1978, p 95.

±14 V TRANSFORMERLESS—Simple low-current regulated supply requires no power transformer. Output current can be increased by using better filtering. Second ground prong is connected through fuse to grounded center conductor of AC line to guard against faulty AC wiring. If wiring is reversed, fuse will disable power supply and neon fault indicator will come on.—D. Kochen, Transformerless Power Supplies, *73 Magazine*, Sept. 1971, p 14–17.

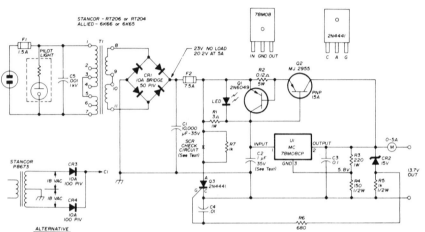

13.7 V AT 5 A—Output is constant within 0.7 V for AC line range of 98 to 128 VAC, and regulation is within tenths of a volt from 0 to 5 A. Design includes short-circuit, overcurrent, and overvoltage protection. Uses series-pass transistor to increase current-carrying capability of regulator. Transistors are mounted on but insulated from heatsink. C2 is essential to prevent oscillation under certain conditions. Use gallium arsenide phosphide LED. Article tells how to determine exact trip point of SCR crowbar.—B. Meyer, Low-Cost All-Mode-Protected Power Supply, *Ham Radio*, Oct. 1977, p 74–77.

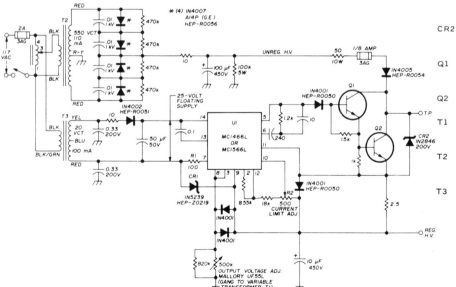

CR2	200-volt, 50-watt zener diode (heatsink to chassis)
Q1	Motorola HEP244 or MJE340 (heatsink to chassis)
Q2	Motorola HEP707 or MJ413 (heatsink to chassis)
T1	0-132 volt, 2.25 A (0.3 VA) variable auto-transformer (Superior Electric 10B)
T2	550 volts center-tapped, 110 mA (Triad R112A or R12A, filament windings not used)
T3	20 volts center-tapped, 100 mA (Triad F90X)

50–300 V VARIABLE AT 100 mA—Solid-state version of regulated high-voltage supply for tube circuit has adjustable current-limiting, instant turn-on, and long component life. Small variable autotransformer in primary circuit of high-voltage transformer is mechanically ganged to DC voltage-control pot connected to pin 8 of U1 to keep input-to-output voltage difference nearly constant. Differential voltage across Q1 never exceeds 100 V so power dissipation of Q1 is only 5 W maximum. Regulator circuit is designed around Motorola MC1466L or MC1566L floating regulator powered by 25-V supply having no common connection to ground. Use 600-V rating for 0.33 μF from T3 to ground.—H. Olson, Regulated, Variable Solid-State High-Voltage Power Supply, *Ham Radio*, Jan. 1975, p 40–44.

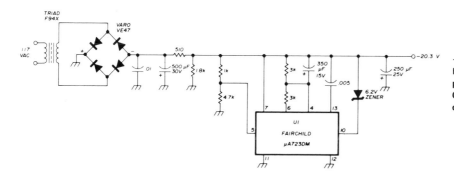

−20 V FOR VARACTORS—Precision low-ripple bias supply for varactor tuning applications can provide up to 20-mA output current.—M. A. Chapman, Multiple Band Master Frequency Oscillator, *Ham Radio*, Nov. 1975, p 50–55.

±15 V AT 10 A—Uses two Motorola positive voltage regulators, each having separate 18–24 VRMS secondary winding on power transformer T_1. Current-limiting resistor R_{SC} is in range of 0.66 to 0.066 ohm. Use copper wire about 50% longer than calculated length and shorten step by step until required pass current is obtained; thus, start with 25 ft of No. 16, 15 ft of No. 18, 10 ft of No. 20, or 6 ft of No. 22.—G. L. Tater, The MPC1000—Super Regulator, *Ham Radio*, Sept. 1976, p 52–54.

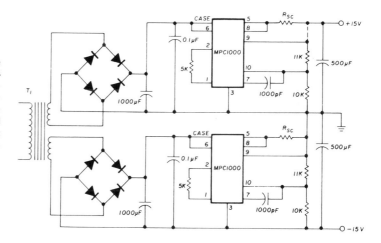

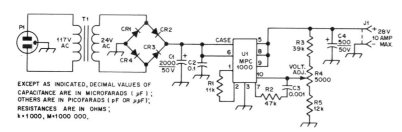

EXCEPT AS INDICATED, DECIMAL VALUES OF CAPACITANCE ARE IN MICROFARADS (µF); OTHERS ARE IN PICOFARADS (pF OR µµF); RESISTANCES ARE IN OHMS; k=1000, M=1000 000.

28 V AT 10 A—Developed for 60-W UHF linear amplifier. C1 and C4 are computer-grade electrolytics. CR1-CR4 are Motorola 1N3209 100-PIV 10-A silicon diodes. U1 is Motorola MPC100 or equivalent voltage regulator mounted on heatsink. T1 is Stancor P-8619 or equivalent 24-V 8-A transformer.—J. Buscemi, A 60-Watt Solid-State UHF Linear Amplifier, *QST*, July 1977, p 42–45.

12 V AT 10 A FOR HOUSE—Power supply is more than adequate for handling 12-V FM transceiver and even small amplifier. Series combination of three 6.3-V 10-A filament transformers drives 12-A 50-PIV bridge rectifier supplying 18 VDC to National LM305 regulator and pass transistors. Output voltage is at least 4 V less. Circuit provides foldback-current limiting for protection against load shorts. 500-ohm pot varies output from 11.2 to 14.1 V. Q1 is 2N2905, Q2 is 2N3445, and Q3 is 2N3772. T1-T3 are 6.3 V at 10 A (Essex Stancor P-6464 or equivalent).—C. Carroll, That's a Big 12 Volts, *QST*, Aug. 1976, p 26–27.

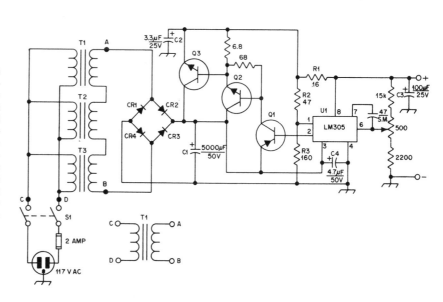

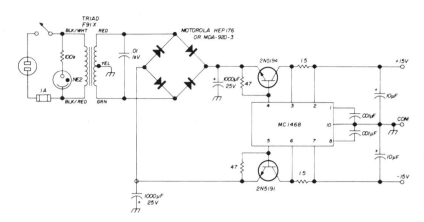

±15 V TRACKING—Uses Motorola dual-polarity regulator to provide balanced positive and negative voltages, with series-pass transistor handling major part of output current. Developed for use with audio signal generator.—H. Olson, Integrated-Circuit Audio Oscillator, *Ham Radio,* Feb. 1973, p 50–54.

±6 V AND ±15 V—Developed for use with wide-range function generator requiring these voltages for transistors and ICs. Voltage-setting 15K and 37.5K resistors are adjusted to give desired output voltages.—H. Olson, The Function Generator, *CQ,* July 1975, p 26–28 and 71–72.

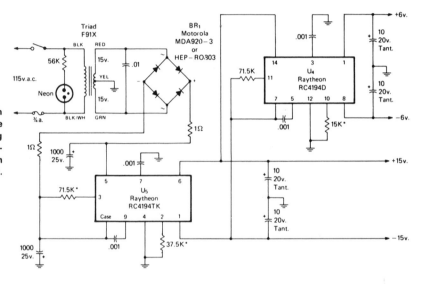

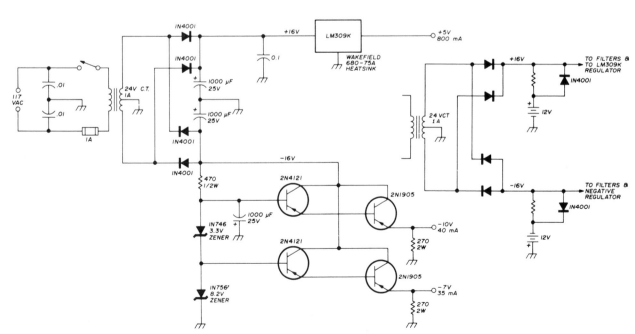

+5, −7, AND −10 V—Developed to meet power requirements of RTTY message generator having TTL and Numitrons requiring +5 V and MOS RAM requiring negative voltages. Diagram shows how to add 12-V storage batteries to prevent loss of programming if AC power fails momentarily.—B. Kelley, Random Access Memory RTTY Message Generator, *Ham Radio,* Jan. 1975, p 8–15.

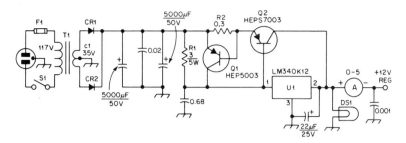

12 V AT 5 A—Uses National LM340K-12 mounted on external heatsink, with series-pass transistor Q2 boosting current rating to 5 A. Provides complete protection from load shorts; output drops suddenly to nearly zero when current exceeds 5 A. R2 is several feet of No. 22 enamel wound on phenolic form to make 0.3-ohm 60-W resistor. CR1 and CR2 are HEP R0103 or equivalent. Transformer is rated 18 V at 8 A.—C. R. Watts, A Crowbar-Proof 12-V Power Supply, *QST*, Aug. 1977, p 36–37.

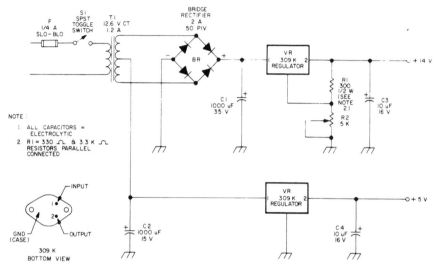

5 AND 14 V AT 1 A—Regulated dual-voltage power supply serves for experimenting with TTL, CMOS, and linear IC projects. Higher-voltage regulator must be insulated from heatsink.—A. Lorona, Dual Voltage Power Supply, *73 Magazine*, Holiday issue 1976, p 146–147.

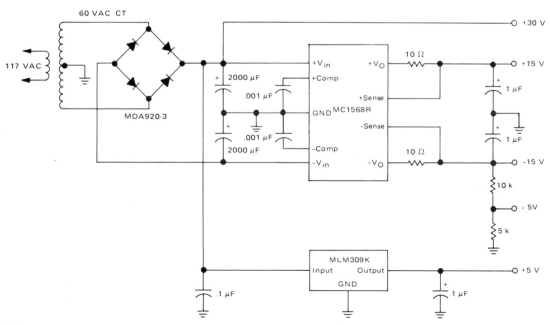

±5, ±15, AND +30 V—Provides all voltages needed for digitally controlled power supply that has voltage range from 0 to 25.5 V in 0.1-V increments. Highest positive voltage of +30 V is well above maximum output voltage that can be programmed.—D. Aldridge and N. Wellenstein, "Designing Digitally-Controlled Power Supplies," Motorola, Phoenix, AZ, 1975, AN-703, p 4.

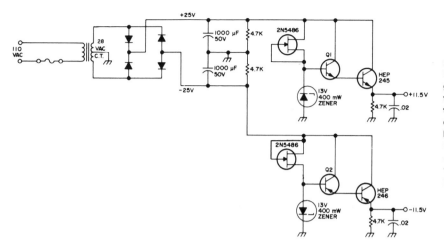

±11.5 V—Developed for experimentation with IC audio amplifiers, delivering up to 5 W, where good regulation is required to prevent oscillation caused by feedback through power supply to input stage. Q1 and Q2 are inexpensive silicon transistors, serving also as low-cost fuses because they burn out first when power supply is overloaded. Use heatsinks with silicone grease for output transistors.—D. J. Kenney, Integrated Circuit Audio Amplifiers, *73 Magazine*, Feb. 1974, p 25–30.

±0 TO 15 V AT 200 mA AND 3.8 TO 5 V AT 2 A—Developed for use with slow-scan television. Design provides equal positive and negative voltages that track each other with one manual control, adjustable from 0 to 15 V for opamps. Current limiting is provided for both positive and negative outputs. Low digital voltage can be adjusted to 3.8 V for RTL or 5 V for TTL, without current limiting. Use transformer having 35-V secondary, center-tapped, rated at 3 A.—D. Miller and R. Taggart, Popular SSTV Circuits, *73 Magazine*, March 1973, p 55–60, 62, and 64–67.

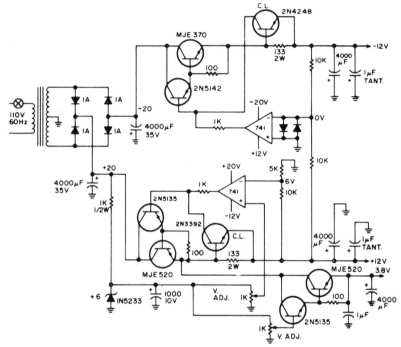

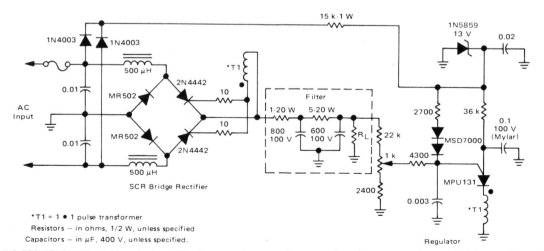

*T1 = 1 ● 1 pulse transformer
Resistors — in ohms, 1/2 W, unless specified
Capacitors — in μF, 400 V, unless specified.

80 V AT 1.5 A FOR COLOR TV—Holds output voltage across R_L within 2% over line-voltage range of 105 to 140 V. Designed for use in 19-inch color TV receiver having 700-V flyback horizontal system. Bridge rectifier has two 2N4442 SCRs that control amount of output voltage by using variable duty cycle. Regulator uses MPU131 programmable UJT, which also serves for gating SCRs. 1K pot provides control of PUT gate voltage, which in turn determines output voltage across R_L.—R. J. Valentine, "A Low-Cost 80 V–1.5 A Color TV Power Supply," Motorola, Phoenix, AZ, 1974, AN-725, p 2.

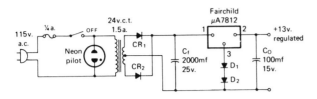

+13 V AT 1 A—Regulated output voltage can be varied upward about 0.6 V per diode by placing silicon diodes between pin 3 and ground. Two diodes boost output of regulator from 11.9 V to about 13 V. Insulate regulator from heatsink with mica washers. CR_1 and CR_2 are 50-PIV 3-A diodes. Motorola equivalent of regulator is MC7812.—A. M. Clarke, Simple, Superregulated, 12 Volt Supply, *CQ*, April 1974, p 61–62.

12 V AT 1 A—Simple supply furnishes up to 1 A with excellent regulation. Bottom of chassis can be used for heatsink. Connect additional 0.1-μF capacitor between pin 1 of VR_1 and ground. T_1 is Radio Shack 273-1505 with 12.6-V CT 1.2-A secondary, and F_1 is 0.5-A fuse.—A. Pike, Radio Shack Power Supply, *CQ*, Sept. 1977, p 66.

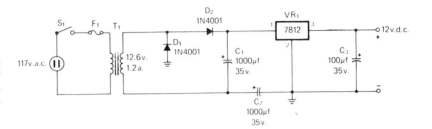

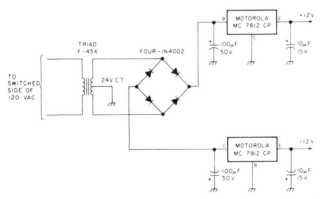

±12 V—Simple circuit provides power required for two 741 opamps used in CRT tuning indicator circuit for RTTY receiver.—R. R. Parry, RTTY CRT Tuning Indicator, *73 Magazine*, Sept. 1977, p 118–120.

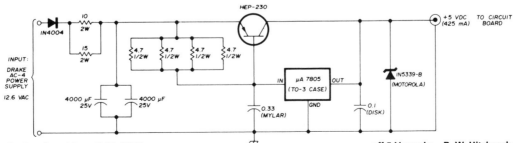

+5 V AT 425 mA—Developed for syllabic VOX system used with Drake T-4XB and R-4B transmitter and receiver. Input is taken from 12.6- VAC transformer winding of power supply for Drake, so equipment power switch also turns off 5-V supply.—R. W. Hitchcock, Syllabic VOX System for Drake Equipment, *Ham Radio*, Aug. 1976, p 24–29.

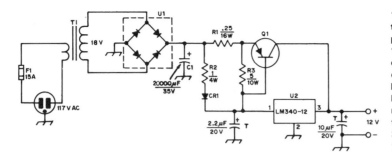

12 V AT 10 A—Permits AC operation of 12-V FM transceiver. Article tells how to rewind 12-V TV power transformer rated above 120 W with No. 12 enamel to get required 18-V secondary. If original winding has 2 turns per volt, new secondary will need 36 turns. Q1 is HEP233, HEP237, or similar transistor rated 10 A or higher, with heatsink. U1 is 25-A 100-PIV bridge rectifier, and U2 is National LM340K-12 regulator. CR1 can be any rectifier rated at least 3 A at 35 V.—L. McCoy, The Ugly Duckling, *QST*, Nov. 1976, p 29–31.

5 V AT 1 A—Simple lab supply provides voltage required for digital ICs. Rectifier is 6-A 50-PIV bridge. Power transformer has 12.6-V secondary rated 1 A, such as filament transformer.—G. McClellan, Give That Professional Look to Your Home Brew Equipment, *73 Magazine*, Feb. 1977, p 28–31.

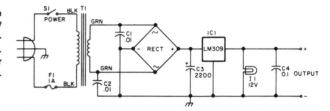

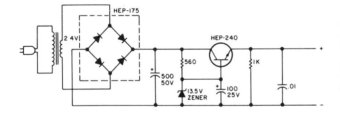

12 V FOR TRANSCEIVER—Output voltage varies only 0.2 V between transmit and receive. Transistor can be mounted directly on side of metal minibox for heatsinking. Transformer secondary is 24 V at 5 A.—Circuits, *73 Magazine*, March 1977, p 152.

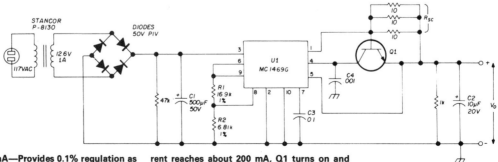

+12 V AT 50 mA—Provides 0.1% regulation as required for PLL RTTY tuning unit and other critical applications. R_{SC} and Q1 provide short-circuit protection for regulator. When output current reaches about 200 mA, Q1 turns on and limits regulator output. U1 can be Motorola MC1469G or HEP C6049G, and Q1 is any general-purpose NPN silicon transistor.—E. Lawrence, Precision Voltage Supply for Phase-Locked Terminal Unit, *Ham Radio*, July 1974, p 60–61.

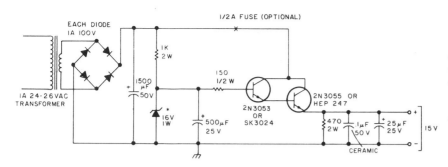

15 V AT 600 mA—Developed for 2-meter FM transceiver used as repeater. Output voltage is well filtered. Regulator allows voltage to drop only 0.1 V when repeater goes from standby to transmit. Use heatsink on 2N3055 series-regulator transistor.—H. Cone, The Minirepeater, *73 Magazine,* June 1975, p 55–57, 60–62, and 64–65.

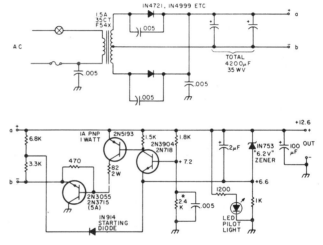

12.6 V AT 3 A—Will handle typical 15-W 2-m transceiver. Short-circuit protection is provided by 82-ohm resistor. Adjust value of resistor marked 2.4K to give desired output voltage. Transformer secondary is nominally 35 V center-tapped at 1.5 A. Output capacitor can be tantalum-slug electrolytic with any value above 10 μF.—H. H. Cross, The Chintzy 12, *73 Magazine,* Feb. 1977, p 40–41.

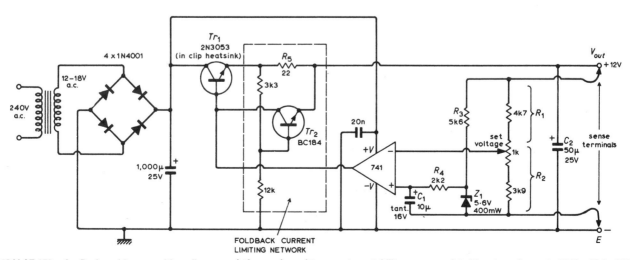

12 V AT 150 mA—Designed for use with audio preamps, FM tuners, and stereo decoders for which minimum ripple, minimum noise, good regulation, and good temperature stability are important. Uses 5.6-V reference zener that is fed from output but is inside feedback loop. Unregulated input can be up to 36 V.—M. L. Oldfield, Regulated Power Supplies, *Wireless World,* Nov. 1972, p 520–521.

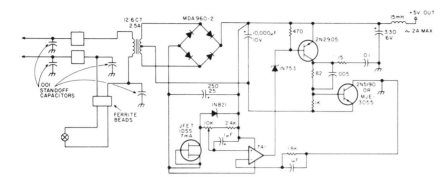

5 V AT 2 A—Developed as supply for receiver frequency counter having LED display. FET having I_{DSS} of about 7.5 mA serves in place of 1200-ohm resistor as current regulator. Power transformer is Triad F-26X with secondary rated 12.6 V center-tapped at 2.5 A.—H. H. Cross, The Chintzy 12, *73 Magazine*, Feb. 1977, p 40–41.

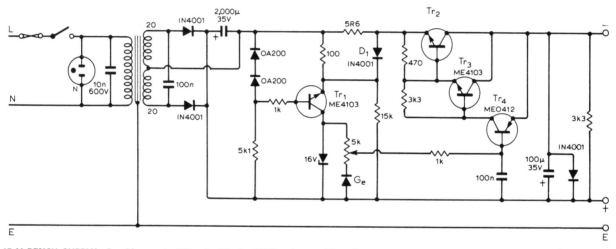

0–15 V BENCH SUPPLY—Provides up to 175 mA with ripple less than 1 mV. Choose Tr_2 to handle load current. Current limiting is provided by 5.6-ohm (5R6) resistor and D_1; when resistor drop exceeds about 1.2 V, current source Tr_1 produces less current and output voltage is reduced.—J. A. Roberts, Bench Power Supply, *Wireless World,* May 1973, p 253.

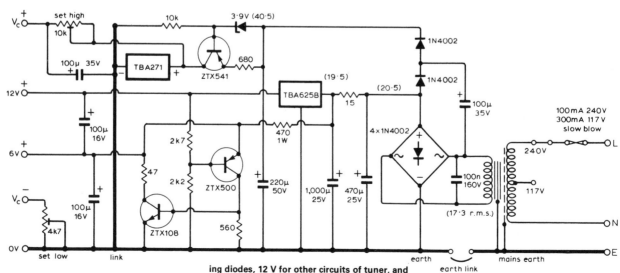

6, 12, AND 30 V FOR FM TUNER—Provides regulated 30 V for voltage-controlled varicap tuning diodes, 12 V for other circuits of tuner, and optional 6 V for stereo decoder. Uses SGS IC regulators.—L. Nelson-Jones, F.M. Tuner Design—Two Years Later, *Wireless World,* June 1973, p 271–275.

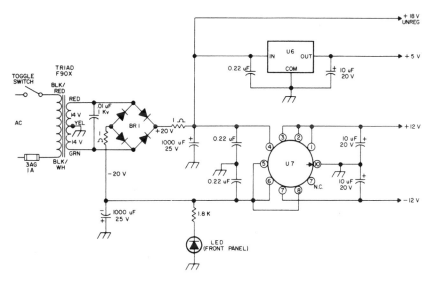

5 V AND ±12 V—Also provides 18 V unregulated for use with code regenerator driving automatic Morse-code printer. BR1 is Motorola MDA920-3 or HEP-R0802 bridge. LED is HP5082-4882 or HEP-P2000. U6 is LM341-5, MC7805, or HEP-C6110P. U7 is LM326H with TO5 finned clip-on heatsink.—H. Olson, CW Regenerator/Processor, *73 Magazine,* July 1976, p 80–82.

13 V AT 2 A WITH NPN TRANSISTORS—Q1 is reference voltage source and Q2 is series-pass regulator for basic supply suitable for running mobile FM transceiver or other 12-V portable equipment in home. Transformer secondary is 16–19 V, or can be 6-V and 12-V filament transformers in series. R1 protects rectifier diodes from surge current generated when supply is turned on. Article tells how to adapt circuit for other output voltages.—R. B. Joerger, Power Supply, *73 Magazine,* Holiday issue 1976, p 40–41.

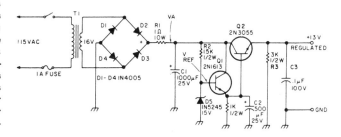

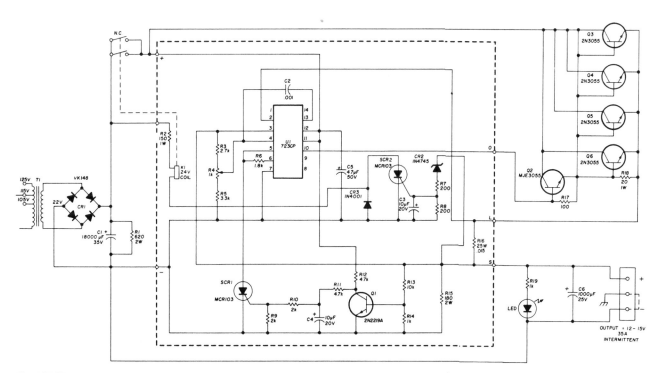

12–15 V AT 500 W—Developed to permit operation of high-power mobile solid-state amateur transmitter in home. Current sensing is done with 15-milliohm resistor R16. Short-circuit cutoff is provided by regulator along with current limiting through R16. Output voltage begins dropping as load exceeds 35 A. When voltage drops below 8 V, Q1 turns off and SCR1 turns on, cutting output power. Power supply must be turned off to unlatch SCR1. For overvoltage shutdown, CR2 starts conducting above 16 V, turning on SCR2 and activating relay K1 to cut off main DC supply. Article gives construction details. T1 has 22-V secondary.—C. C. Lo, 500-Watt Regulated Power Supply, *Ham Radio,* Dec. 1977, p 30–32.

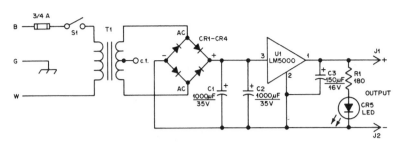

5 V AT 3 A—National LM5000 voltage regulator having built-in overload protection is basis of small bench supply for TTL work. Filament transformer rated 12.6 V at 3 A feeds full-wave bridge rectifier rated 200 PIV at 6 A, such as Radio Shack 276-1172. U1 requires heatsink insulated from chassis. Output filter C3 should be mounted directly on regulator terminals to minimize circuit oscillation. Output should read within 100 mV of 5 V. Radio Shack 276-047 LED serves as output indicator. Use 0.22-μF bypass between pins 2 and 3 of U1.—K. Powell, The 5 × 3 Power Supply, *QST*, May 1977, p 25–26.

13 V AT 2 A WITH PNP TRANSISTORS—Reference voltage source Q1 is 2N301, while series-pass regulator Q2 is 2N1523. D5 is 1N5245 15-V zener. Secondary of T1 is 16–19 V, or can be 6-V and 12-V filament transformers in series. Article tells how to adapt circuit for other output voltages.—R. B. Joerger, Power Supply, *73 Magazine*, Holiday issue 1976, p 40–41.

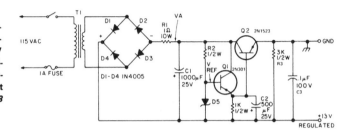

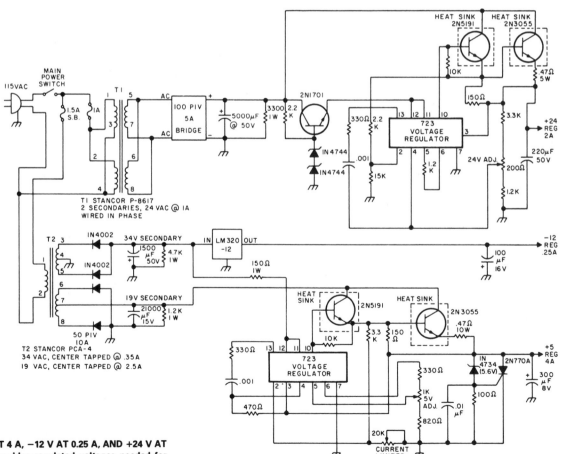

+5 V AT 4 A, −12 V AT 0.25 A, AND +24 V AT 2 A—Provides regulated voltages needed for Sykes 7158 floppy disk and its interface controller, used in Southwest Technical Products MP-68 computer system. Circuit provides adjustable current limiting and overvoltage protection on 5-V supply. Output voltage adjustments are provided for 5-V and 24-V supplies.— P. Hughes, Interfacing the Sykes OEM Floppy Disk Kit to a Personal Computer, *BYTE*, March 1978, p 178–185.

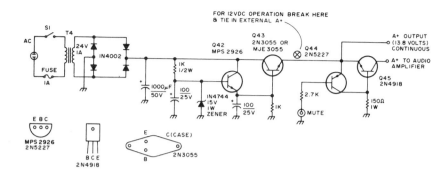

13.6 V AT 1 A—Used in all-band double-conversion superheterodyne receiver for AM, narrow-band FM, CW, and SSB operation. Simple transformer-rectifier-filter circuit is followed by zener-referenced Darlington pair. When transmitter of amateur station is on air, muting is accomplished by grounding base of Q44 through 2.7K resistor, which turns off Q45 and kills A+ to audio amplifier.—D. M. Eisenberg, Build This All-Band VHF Receiver, *73 Magazine*, Jan. 1975, p 105–112.

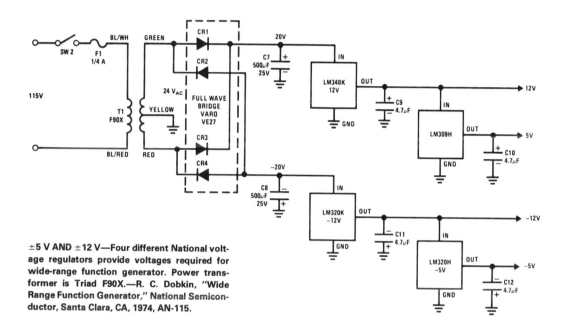

±5 V AND ±12 V—Four different National voltage regulators provide voltages required for wide-range function generator. Power transformer is Triad F90X.—R. C. Dobkin, "Wide Range Function Generator," National Semiconductor, Santa Clara, CA, 1974, AN-115.

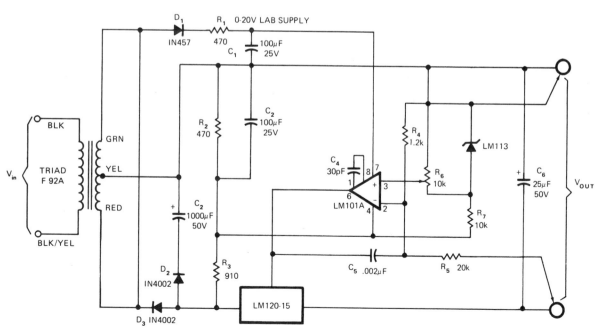

0–20 V AT 1 A—Variable-output regulated supply for lab use maintains output voltage within 2 mV of desired value for outputs up to 1 A. Arrangement uses National LM120 negative reg- ulator as pass element, LM101A opamp as error amplifier, and LM113 zener as reference. Circuit provides complete protection against load shorts. LM120 requires adequate heatsink for continuous operation.—C. T. Nelson, Power Distribution and Regulation Can Be Simple, Cheap and Rugged, *EDN Magazine*, Feb. 20, 1973, p 52–58.

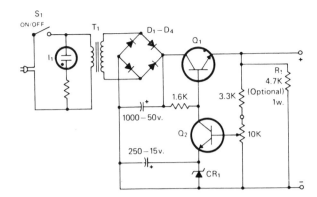

6–30 V AT 500 mA—Zener used for CR₁ should be rated 1 V less than desired minimum voltage, at 300 mW. R₁ improves regulation at low current levels. Current-limiting value is about 1 A. Diodes are 50-PIV 1-A silicon. I₁ is 117-V neon lamp. Q₁ is any 15-W NPN power transistor. Q₂ is 2N697 or equivalent. T₁ is power transformer with 24-V secondary at 0.5 A.—J. Huffman, The Li'l Zapper—a Versatile Low Voltage Supply, *CQ,* Nov. 1977, p 44.

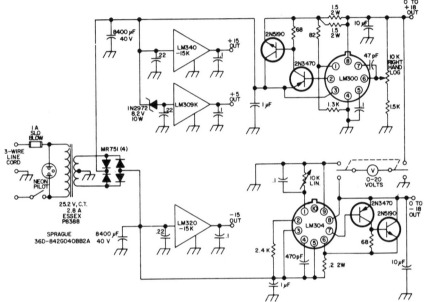

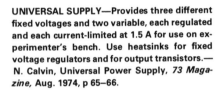

UNIVERSAL SUPPLY—Provides three different fixed voltages and two variable, each regulated and each current-limited at 1.5 A for use on experimenter's bench. Use heatsinks for fixed voltage regulators and for output transistors.— N. Calvin, Universal Power Supply, *73 Magazine,* Aug. 1974, p 65–66.

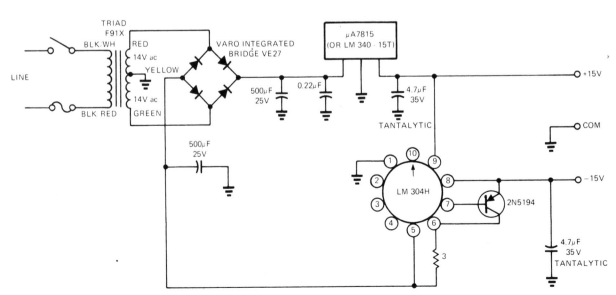

100-mA TRACKING—Circuit uses +15 V from μA7815 positive fixed-output regulator as external reference for LM304 negative regulator operating with outboard current-carrying PNP transistor. Arrangement requires only one center-tapped transformer winding yet gives required tracking of voltages. Output can be boosted to 200 mA by using larger bridge rectifier section.—H. Olson, Simple ±15V Regulated Supply Provides Tracking, *EDN Magazine,* March 20, 1973, p 87.

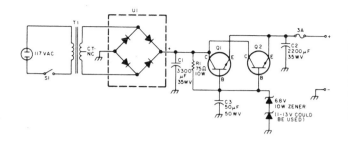

12 V AT 2.8 A—Simple supply was developed for use with 2-meter FM transceiver when operating in home. Power transistors are Radio Shack 276-592 rated 40 W. T1 is 12.6 V at 3 A, and U1 is 276-1171 rated 100 V at 6 A. Article covers construction.—M. L. Lovell, 12 Inexpensive Volts for Your Base Station, *73 Magazine*, Sept. 1976, p 60–62.

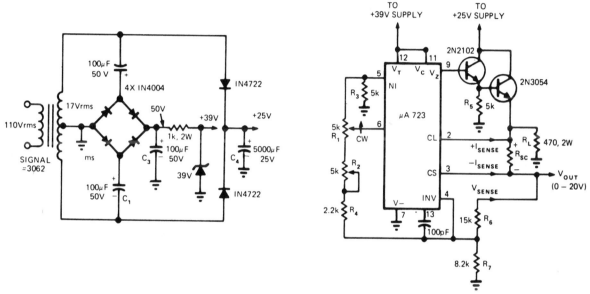

0–20 V CURRENT-LIMITING—Novel full-wave voltage doubler formed by diode bridge and C_1-C_2-C_3 provides 39 V required by μA723 regulator whose output is continuously variable with R_1.

Initially, R_2 is adjusted for minimum output voltage when R_1 is maximum counterclockwise, to balance bridge R_1-R_2-R_3-R_4 when output voltage is zero. Value used for R_{SC} determines short-cir-

cuit current. Raw DC supply provides separate 25 V for pass transistors.—L. Drake, Variable Voltage Power Supply Uses Minimum Components, *EDN Magazine*, Aug. 5, 1974, p 80 and 82.

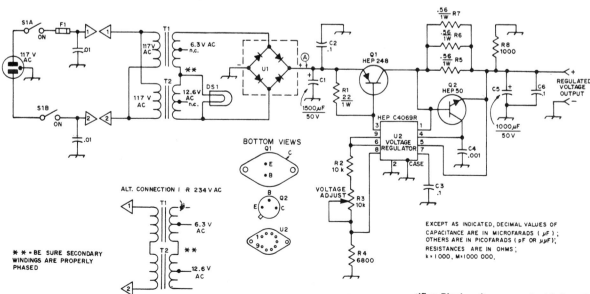

12 V AT 2 A—Will operate 10-W 220-MHz portable FM transceiver from AC line. Output voltage is adjustable from 9 to 13 V. DC voltage at point A is about 30 V. U1 is 50-V 10-A bridge rectifier. Ripple voltage on output is less than 30 mV P-P.—E. Kalin, A No-Junkbox Regulated Power Supply, *QST*, Jan. 1975, p 30–33.

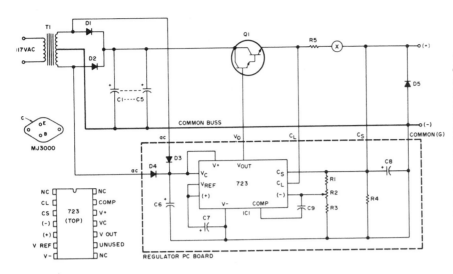

12 V AT 5 A—Uses MJ3000 Darlington power device as pass element providing gain of 1000 at 5 A. Output is set to current-limit at 6.5 A. Fuse at X is desirable. Values: R1 is 1.8K; R2 is 2.5K trimpot; R3 is 2.7K; R4 is 1.5K; R5 is 0.1 ohm at 5 W; C1-C5 are 4000 μF each at 20 V; C6 is 250 μF at 25 V; C7-C8 are 1.2 μF at 35 V; C9 is 220 pF; D1-D2 are MR1120 or equivalent rated 6 A; D3-D4 are 1N4607 or equivalent; D5 is 1N4002 or equivalent; and T1 is 24–28 V CT secondary at 4 A. Article gives design procedure for increasing regulated output to as much as 100 A.—C. Anderton, A Hefty 12 Volt Supply, *73 Magazine,* May 1975, p 85–87.

5 V FROM AC OR DC—Developed for use with secondary frequency standard to permit checking frequency of amateur radio transmitter at station or in field. Any battery capable of delivering 250 mA at 9–15 V is suitable.—T. Shankland, Build a Super Standard, *73 Magazine,* Oct. 1976, p 66–69.

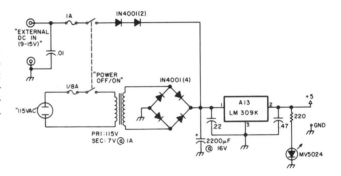

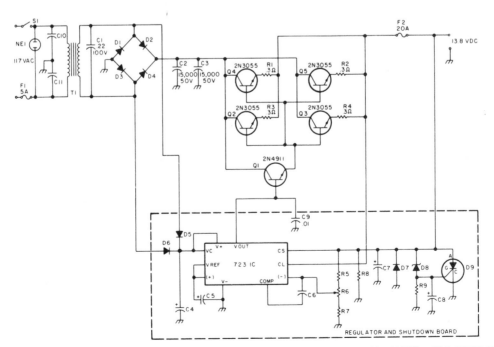

+13.8 VDC AT 18 A—Developed for use with amateur radio transceiver. Transformer secondary is rated 25 VAC at 12 A. When output voltage exceeds 15 VDC, zener D8 (1N965A or equivalent) conducts and fires 2N4441 SCR to crowbar supply and protect transceiver. Parts values are: R5 1.8K, R6 2.5K, R7 2.7K, R8 1.5K, R9 1K, C4 250 μF, C5-C6 1.2 μF, C7 220 pF, C8 100 μF, C9-C11 0.01 μF, D1-D4 1N3492 or equivalent with 100 PIV at 18 A, D5-D6 1N4607 or equivalent, and D7 1N4002 or equivalent.—T. Lawrence, Build a Brute Power Supply, *73 Magazine,* Aug. 1977, p 78–79.

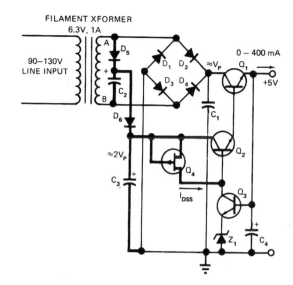

5 V WITH DOUBLER—Doubling permits use of inexpensive 6.3-V filament transformer without risking loss of regulation when line voltage drops below about 105 V. With values shown, output varied only 6 mV for line voltage range of 95 to 135 V. Doubler circuit consists of C_2, C_3, D_1, D_3, D_5, and D_6.—A. Paterson, Voltage Doubler Prevents Supply from Losing Regulation, *EDN Magazine,* Nov. 1, 1972, p 46.

COMPONENT VALUES

D_1 THRU D_4	:	1N4001
D_5, D_6	:	1N914
C_1	:	8000 μF, 15V
C_2	:	50 μF, 25V
C_3	:	250 μF, 25V
C_4	:	100 μF, 10V
Q_1	:	MJE 521 (HEAT SINK)
Q_2, Q_3	:	2N3392
Q_4	:	CONSTANT CURRENT DIODE, (e.g. 2N5033, 2-3 mA)
Z_1	:	REF. DIODE, 4.3 V @ 2-3 mA (e.g. LVA43A)

12–14 V AT 3 A—Basic circuit for operating mobile equipment off AC line uses IC voltage regulator in conjunction with series-pass transistor.—Circuits, *73 Magazine,* Holiday issue 1976, p 170.

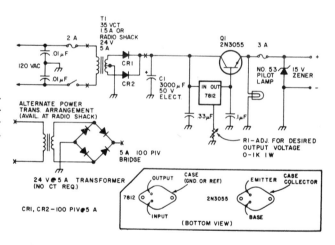

SECTION **56**
Regulator Circuits

Used at outputs of unregulated power supplies to provide highly regulated fixed and variable positive and negative output voltages ranging from 0 to ±65 V for solid-state applications and up to 1000 V at 100 W for other purposes. Maximum current ratings range from 5 mA to 20 A. Some regulators have overvoltage crowbar or foldback current limiting. Dual-output regulators may have tracking. Current regulators are included. See also Regulated Power Supply and Switching Regulator chapters.

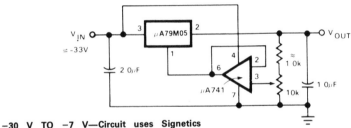

−30 V TO −7 V—Circuit uses Signetics µA79M05 adjustable voltage regulator in combination with 741 opamp to give wide negative output voltage range. Regulator includes thermal overload protection and internal short-circuit protection. Input voltage should be at least 3 V more negative than maximum output voltage desired.—"Signetics Analog Data Manual," Signetics, Sunnyvale, CA, 1977, p 670.

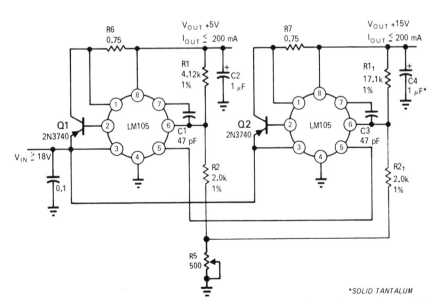

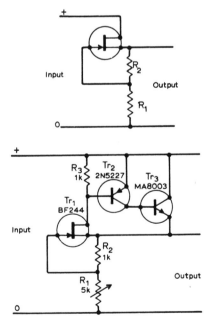

5 AND 15 V SINGLE CONTROL—Single potentiometer serves for adjusting two regulators simultaneously. Accuracy depends on output voltage differences of regulators; error decreases when output voltages are closer. Article gives design equation and covers other possible sources of error.—R. C. Dobkin, One Adjustment Controls Many Regulators, *EDN Magazine,* Nov. 1, 1970, p 33–35.

SOLID TANTALUM

5-V FET REGULATOR—Output voltage changes less than 0.1 V for load current change from 0 to 60 mA. Output voltage changes caused by change in load resistance affect gate-source voltage of FET Tr_1 via R_1 and R_2, causing compensating change in drain current. Additional transistors serve to reduce output resistance and increase output current without affecting stabilization ratio of about 1000.—C. R. Masson, F.E.T. Voltage Regulator, *Wireless World,* Aug. 1971, p 386.

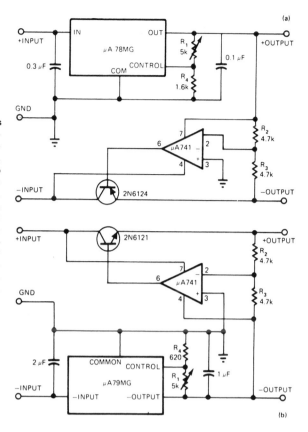

(a)

(b)

SLAVED DUAL TRACKING REGULATOR—Uses Fairchild μA78MG adjustable four-terminal regulator with opamp and power transistor for delivering output currents up to 0.5 A per side, with output voltages adjustable from ±5 V to ±20 V for component values shown. Positive side functions independently of negative side, but negative output is slave of positive output. To slave positive side, use μA79MG and 2N6121 NPN transistor as at (b). Opamp functions as inverting amplifier driving power transistor serving as series-pass element for opposite side of regulator, with R_1 adjusting both output voltages simultaneously.—A. Adamian, Dual Adjustable Tracking Regulator Delivers 0.5A/Side, *EDN Magazine*, Jan. 5, 1977, p 42.

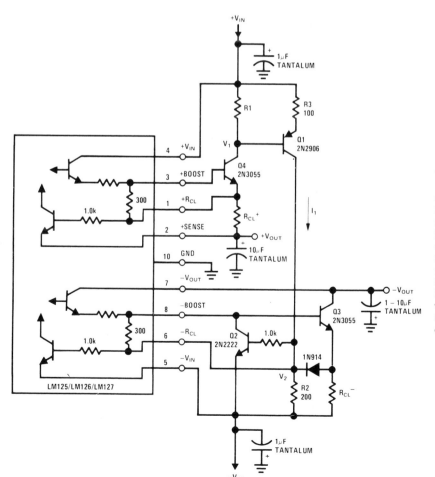

TRACKED CURRENT LIMITING—Simultaneous limiting scheme for both sections of National dual tracking regulator depends on output current of positive regulator. Voltage drop produced across R1 by positive regulator brings Q1 into conduction, with positive load current I_1 increasing until voltage drop across R2 equals negative current-limit sense voltage. Negative regulator will then current-limit, and positive side will closely follow negative output down to level of about 700 mV.—T. Smathers and N. Sevastopoulos, "LM125/LM126/LM127 Precision Dual Tracking Regulators," National Semiconductor, Santa Clara, CA, 1974, AN-82, p 13.

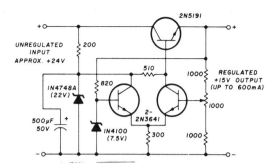

+15 V WITH DIFFERENTIAL AMPLIFIER—Series regulator uses differential amplifier as control circuit in which one side is referenced to zener and other to fraction of output voltage. Second zener provides coarse regulated voltage to differential pair.—H. Olson, Power-Supply Servicing, *Ham Radio*, Nov. 1976, p 44–50.

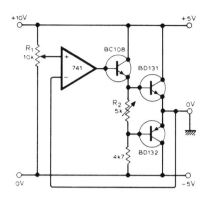

REGULATED DIVIDER FOR ±5 V—Used at output of adjustable regulated power supply providing up to 15 V, to give lower positive and negative voltages that remain steady despite changes in load current. To get +5 V and −5 V from +10 V, set R_1 at midposition and adjust R_2 for 20 mA through output transistors. Uses 741 opamp.—C. H. Banthorpe, Voltage Divider, *Wireless World*, Dec. 1976, p 41.

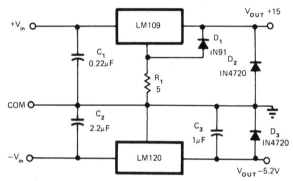

DUAL −5.2 V AND +15 V—Output voltages are equal to preset values of regulator ICs in basic arrangement shown. R_1 and D_1 ensure startup of LM109 when common load exists across supplies. D_1 should be germanium or Schottky having forward voltage drop of 0.4 V or less at 50 mA. D_2 and D_3 protect against polarity reversal of output during overloads.—C. T. Nelson, Power Distribution and Regulation Can Be Simple, Cheap and Rugged, *EDN Magazine*, Feb. 20, 1973, p 52–58.

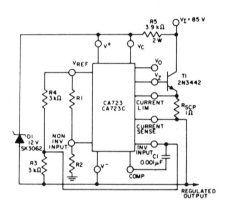

+50 V FLOATING—RCA CA723 regulator operating from 85-V supply delivers 50 V with line regulation of 15 mV for 20-V supply change and load regulation of 20 mV for 50-mA load current change.—"Linear Integrated Circuits and MOS/FET's," RCA Solid State Division, Somerville, NJ, 1977, p 61.

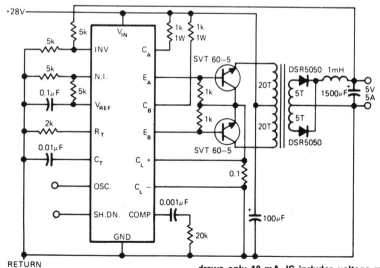

5 V AT 5 A WITH IC SWITCHER—Uses Silicon General SG1524 IC as pulse-width-modulated regulator for which operating frequency remains constant, with ON time of each pulse adjusted to maintain desired output voltage. Operating range extends above 100 kHz but device draws only 10 mA. IC includes voltage reference, oscillator, comparator, error amplifier, current limiter, pulse-steering flip-flop, and automatic shutdown for overload.—P. Franson, Today's Monolithic Switching Circuits Greatly Simplify Power-Supply Designs, *EDN Magazine*, March 20, 1977, p 47–48, 51, and 53.

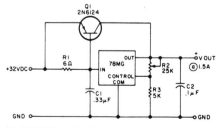

5–30 V AT 1.5 A—External series-pass transistor boosts 500-mA rated output of 78MG or 79MG regulator to 1.5 A for use as adjustable power supply in lab. Circuit has no short-circuit protection for safe-area limiting for external pass transistor, but article shows how to add protective transistor for this purpose.—J. Trulove, A New Breed of Voltage Regulators, *73 Magazine*, March 1977, p 62–64.

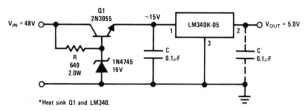

*Heat sink Q1 and LM340.

5 V FROM 48 V—Combination of zener and resistor R gives equivalent of power zener as solution to regulator protection problem when input voltage is much higher than rated maximum of regulator. Maximum load is 1 A. With optional capacitor, circuit noise is only 700 μV P-P.—"Linear Applications, Vol. 2," National Semiconductor, Santa Clara, CA, 1976, AN-103, p 10.

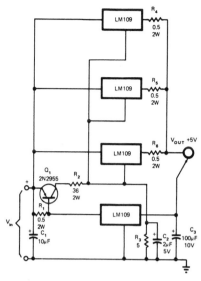

PARALLELING REGULATORS—Current-sharing problem is overcome without sacrificing ripple rejection or load regulation, by using bottom regulator as control device that supplies most of load current until current through this regulator reaches about 1.3 A. At this point Q_1 turns on and raises output voltage of other regulators to supply additional load current demands. Circuit shown will supply up to 6 A for minimum input voltage of 8 V. For optimum regulation, minimum load current should be 1 A.—C. T. Nelson, Power Distribution and Regulation Can Be Simple, Cheap and Rugged, *EDN Magazine*, Feb. 20, 1973, p 52–58.

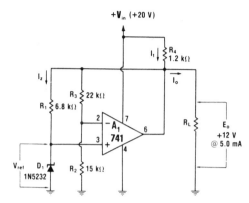

12-V SHUNT AT 5 mA—Low-power shunt regulator uses opamp to absorb excess load current. Value of R_1 is chosen to step up reference voltage of 5.6-V zener to +12 V at 5 mA. Design procedure for other output voltages is given. Output impedance is 0.01 ohm at 100 Hz, giving 120-Hz ripple-frequency filtering comparable to that of 100,000-μF capacitor.—W. G. Jung, "IC Op-Amp Cookbook," Howard W. Sams, Indianapolis, IN, 1974, p 166–168.

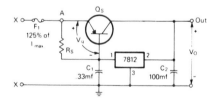

12 V AT 20 A—Regulator conducts and regulates until current demand is such that IR drop across R_S is sufficient to overcome base-emitter junction potential of switch transistor Q_S, which is two 2N174 germanium transistors in parallel. Use 2 ohms for R_S. Q_S is then turned on, with current/voltage regulation to its base controlled by regulator. Input voltage of 7812 regulator should be 2 V more than desired output voltage. Article gives three different rectifier circuits suitable for use with regulator.—A. M. Clarke, Regulated 200 Watt-12 Volt D.C. Power Supply, *CQ*, Oct. 1975, p 28–30 and 78–79.

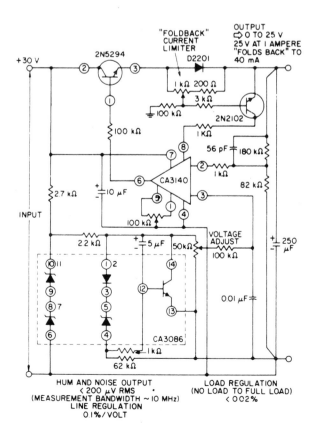

HUM AND NOISE OUTPUT < 200 μV RMS
(MEASUREMENT BANDWIDTH ~ 10 MHz)
LINE REGULATION
0.1%/VOLT

LOAD REGULATION
(NO LOAD TO FULL LOAD)
< 0.02%

0–25 V WITH FOLDBACK CURRENT LIMITING—When D2201 diode senses load current of 1 A at maximum regulated output of 25 V, 2N2102 current-sensing transistor provides foldback of output current to 40 mA. Arrangement permits use of 2N5294 transistor as series-pass element, using only small heatsink. High-impedance reference-voltage divider across 30-V supply serves CA3140 connected as noninverting power opamp.—"Linear Integrated Circuits and MOS/FET's," RCA Solid State Division, Somerville, NJ, 1977, p 248–257.

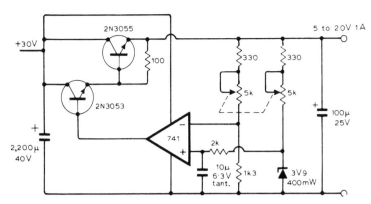

5–20 V ZENER-STABILIZED—Use of dual linear pot simplifies problem of feeding reference zener diode from variable-voltage supply.—L.

J. Baughan, Variable Power Supply with Zener Stabilization, *Wireless World,* Nov. 1975, p 520.

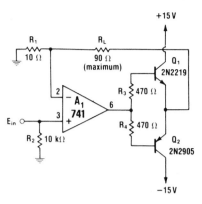

100-mA CURRENT REGULATOR—741 opamp is connected as noninverting voltage-controlled current source feeding transistors that boost output and provide bidirectional current capability in load R$_L$. If single-polarity current flow is sufficient, omit opposite-polarity transistor.—W. G. Jung, "IC Op-Amp Cookbook," Howard W. Sams, Indianapolis, IN, 1974, p 173.

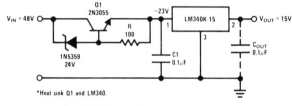

*Heat sink Q1 and LM340.

+15 V FROM HIGH INPUT VOLTAGE—Zener is used in series with resistor R to level-shift input voltage higher than rated maximum of LM340K-15 regulator. Typical load regulation is 40 mV for 0–1 A pulsed load, and line regulation is 2

mV for 1-V change in input voltage for no load. With optional output capacitor, circuit noise is only 700 µV P-P.—"Linear Applications, Vol. 2," National Semiconductor, Santa Clara, CA, 1976, AN-103, p 9–10.

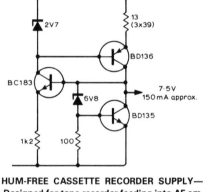

HUM-FREE CASSETTE RECORDER SUPPLY—Designed for tape recorder feeding into AF amplifier, to permit operation of recorder from power supply of amplifier without having hum due to positive feedback through shared ground connection. Circuit provides up to 150 mA at 7.5 V from supply ranging from 12 to 24 V. Transistors are connected as constant-current source in series with constant-voltage sink. Use three 39-ohm resistors in parallel as 13-ohm resistor.—G. Hibbert, Avoiding Power Supply Hum, *Wireless World,* Oct. 1973, p 515.

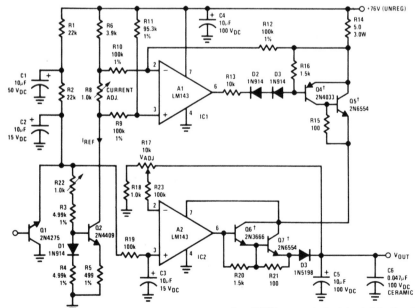

7.1–65 V AT 0–1 A—Provides continuously variable output voltage and adjustable output current range. Q1 is connected as zener to give 6.5-V reference voltage. Darlington current boosters Q4-Q7 should be on common Ther-

malloy 6006B or equivalent heatsink. Developed for use with pulsed loads. For input voltage range of 46–76 V, regulation is within 286 mV for 500-mA DC output.—"Linear Applications, Vol. 2," National Semiconductor, Santa Clara, CA, 1976, AN-127, p 8–10.

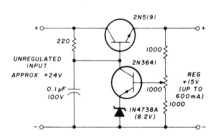

+15 V WITH FEEDBACK—Fraction of output voltage is fed back to base of 2N3641 regulator transistor. Difference between this voltage and zener diode voltage is amplified to control base of 2N5191 series transistor.—H. Olson, Power-Supply Servicing, *Ham Radio,* Nov. 1976, p 44–50.

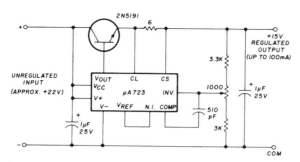

+15 V WITH μA723—Series power transistor and Fairchild IC voltage regulator provide up to 100 mA. Article covers troubleshooting and repair of all types of regulators.—H. Olson, Power-Supply Servicing, *Ham Radio*, Nov. 1976, p 44–50.

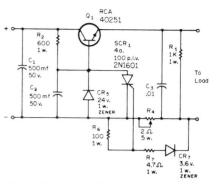

OVERLOAD PROTECTION—When critical current is exceeded, SCR_1 conducts and reduces base-ground voltage of Q_1, cutting it off. Load current then drops to very low value, and Q_1 is protected. Operation is restored by turning off current supply to power transformer after clearing short-circuit condition.—R. Phelps, Jr., Protective Circuits for Transistor Power Supplies, *CQ*, March 1973, p 44–48 and 92.

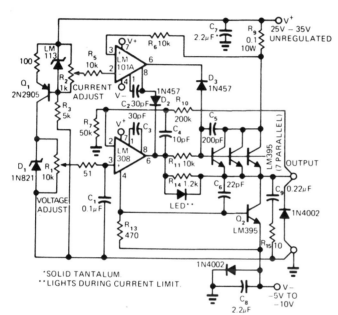

25 V AT 10 A FOR LAB—Circuit uses no large output capacitors yet has good response as constant-voltage or constant-current source. LM395 units (7 in parallel) act as current-limited thermally limited high-gain power transistor. Mount all on same heatsink for good current sharing, since 300 W will be dissipated under worst-case conditions. Only two control opamps are needed, one for voltage control and one for current control.—R. C. Dobkin, General-Purpose Power Supply Furnishes 10A and 25V, *EDN Magazine*, March 5, 1975, p 70.

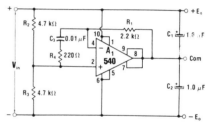

CONVERTING TO DUAL SUPPLY—With equal values for R_2 and R_3, input of 30 V is converted to ±15 V at output. If desired, R_2 and R_3 can be scaled for unequal voltage drops. Circuit uses 540 power IC having 100-mA rating for each output, for handling load imbalances up to 100 mA.—W. G. Jung, "IC Op-Amp Cookbook," Howard W. Sams, Indianapolis, IN, 1974, p 170–171.

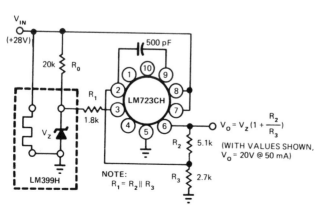

$$V_O = V_Z \left(1 + \frac{R_2}{R_3}\right)$$

(WITH VALUES SHOWN, V_O = 20V @ 50 mA)

NOTE: $R_1 = R_2 \| R_3$

10 PPM/°C—Connections shown convert LM723CH regulator into precision power reference having excellent long-term stability and temperature stability. LM399H replaces internal reference of LM723 with low-noise 6.9 V to give desired performance over temperature range from +15 to +65°C.—B. Welling, High-Stability Power Supply Uses 723 Regulator, *EDN Magazine*, Jan. 20, 1978, p 114 and 116.

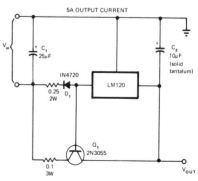

5 A AT −5 TO −15 V—Use of 2N3055 pass transistor boosts current output of LM120 regulator IC. Minimum differential between input and output voltages is typically 2.5 V, so supply voltage must be 2.5 V higher than preset output voltage of regulator chosen from National LM120 series.—C. T. Nelson, Power Distribution and Regulation Can Be Simple, Cheap and Rugged, *EDN Magazine*, Feb. 20, 1973, p 52–58.

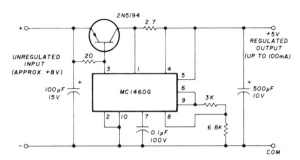

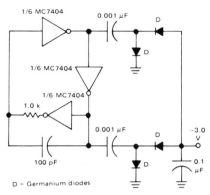

+5 V WITH MC1460G—Series power transistor and Motorola IC voltage regulator provide up to 100 mA. IC shown has been replaced by MC1469. Equivalents made by other manufacturers can also be used.—H. Olson, Power-Supply Servicing, *Ham Radio*, Nov. 1976, p 44–50.

−3 V—Circuit using three sections of Motorola MC7404 operates from +5 V supply and generates −3 V at up to 100 μA, as one of supply voltages required by Motorola MCM6570 8192-bit character generator using 7 × 9 matrix.—"A CRT Display System Using NMOS Memories," Motorola, Phoenix, AZ, 1975, AN-706A, p 5.

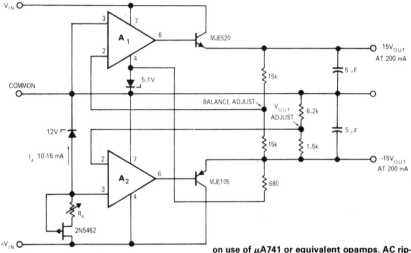

±15 V AT 200 mA—Two-opamp regulator gives dual-polarity tracking outputs that can be balanced to within millivolts of each other or can be offset as required. Negative voltage is regulated, and positive output tracks negative. Article gives step-by-step design procedure based on use of μA741 or equivalent opamps. AC ripple is less than 2 mV P-P. Conventional full-wave bridge rectifier with capacitor-input filter can be used to provide required unregulated 36 VDC for inputs.—C. Brogado, IC Op Amps Simplify Regulator Design, *EDN/EEE Magazine*, Jan. 15, 1972, p 30–34.

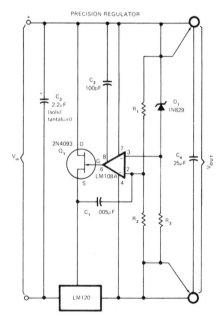

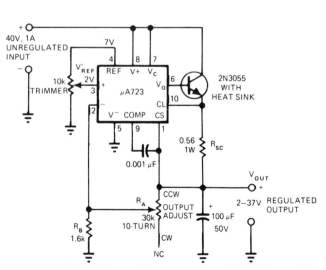

2–37 V—Simple circuit gives fine linear control with 10-turn pot over wide voltage range by first using 10K trimmer pot to divide 7-V reference down to 2 V.—G. Dressel, Regulator Circuit Provides Linear 2-37 V Adjustment Range, *EDN Magazine*, March 5, 1978, p 122.

1 A WITH 0.005% VOLTAGE ACCURACY—Use of National LM120 negative regulator with LM108A low-drift opamp and 1N829 precision reference diode gives extremely tight regulation, very low temperature drift, and full overload protection. Bridge arrangement sets output voltage and holds reference diode current constant. FET is required because 4-mA maximum ground current of regulator exceeds output current rating of opamp. R_1 and R_2 should track to 1 PPM or less. R_3 is chosen to set reference current at 7.5 mA. For output of 8 to 14 V, use LM120-5.0; for 15–17 V, use LM120-12.—C. T. Nelson, Power Distribution and Regulation Can Be Simple, Cheap and Rugged, *EDN Magazine*, Feb. 20, 1973, p 52–58.

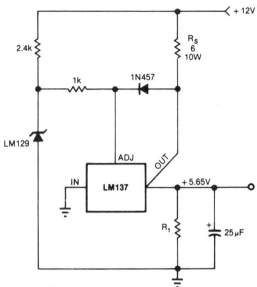

POSITIVE SHUNT REGULATION—Connection shown for LM137 negative series regulator provides high-reliability positive shunt regulation for applications having high-voltage spike on raw DC supply. Output is 5.65 V.—P. Lefferts, Series Regulators Provide Shunt Regulation, *EDN Magazine*, Sept. 5, 1978, p 158 and 160.

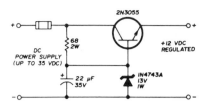

+12 V AT 2 A—Developed for unregulated 12-VDC supplies used by some amateurs with low-power VHF FM equipment, where no-load voltage may be 18 V or more. During transmit, voltage drops to about 12 V, but on receive may exceed voltage ratings of small-signal transistors in transceiver. Use heatsink with transistor, and use 2-A fuse to protect transistor from shorted load.—J. Fisk, Circuits and Techniques, *Ham Radio*, June 1976, p 48–52.

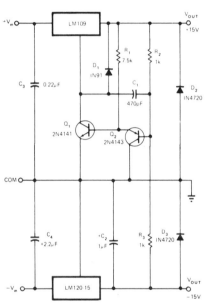

±15 V WITH TRACKING—In arrangement shown for National regulator ICs, positive output voltage tracks negative voltage to better than 1%. Ripple rejection is 80 dB for both outputs. Load regulation is 30 mV at 1 A for negative output and less than 10 mV for positive output. Circuit works well for output in range of ±6 to ±15 V. C_1 provides stability.—C. T. Nelson, Power Distribution and Regulation Can Be Simple, Cheap and Rugged, *EDN Magazine*, Feb. 20, 1973, p 52–58.

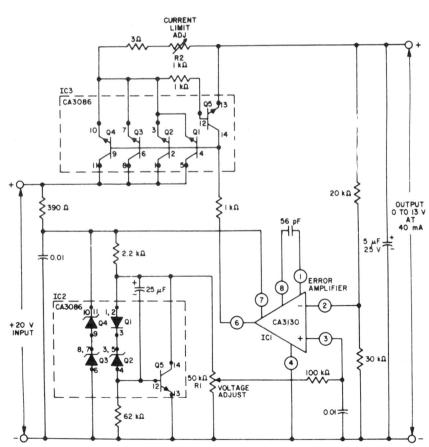

0–13 V AT 40 mA—Combination of RCA CA3130 opamp and two CA3086 NPN transistor arrays provides better than 0.01% regulation from no load to full load and input regulation of 0.02%/V. Hum and noise output is less than 25 μV up to 100 kHz.—"Linear Integrated Circuits and MOS/FET's," RCA Solid State Division, Somerville, NJ, 1977, p 236–243.

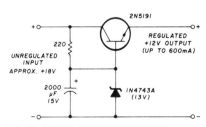

+12 V SERIES EMITTER-FOLLOWER—Base-emitter voltage is more or less constant without use of feedback because base is held at constant voltage by zener diode. Ripple at base is reduced by RC filter.—H. Olson, Power-Supply Servicing, *Ham Radio*, Nov. 1976, p 44–50.

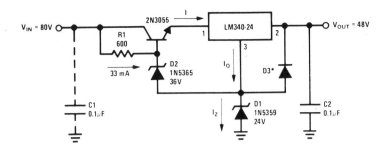

*Germanium signal diode

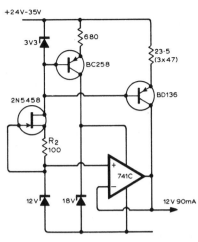

48 V FROM 80 V—Level-shifting transistor-zener combination R1-D2 is used with zener D1 to keep voltage across LM340-24 regulator below maximum rated value. Addition of zeners has drawback of increasing output noise to about 2 mV P-P. Load regulation is 60 mV for pulsed load change from 5 mA to 1 A. Line regulation is 0.01%/V of input voltage change for 500-mA load.—"Linear Applications, Vol. 2," National Semiconductor, Santa Clara, CA, 1976, AN-103, p 10–11.

HUM-FREE TUNER SUPPLY—Permits operation of high-quality FM tuner from amplifier supply without having hum due to positive feedback through shared ground connection. Circuit provides up to 90 mA at 12 V from any supply ranging from 24 to 34 V. Low output impedance eliminates all likely sources of feedback and suppresses ripple. Circuit requires careful initial adjustment to limit current sunk by 741C opamp to less than 15 mA; coarse adjustment is made by varying number of 47-ohm resistors in parallel serving as BD136 emitter resistor, and fine adjustment by changing R_2.—G. Hibbert, Avoiding Power Supply Hum, *Wireless World,* Oct. 1973, p 515.

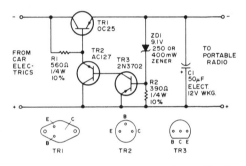

9 V FROM 12 V—Developed for economical operation of 9-V portable radio from 12-V storage battery of car.—Circuits, *73 Magazine,* March 1975, p 136.

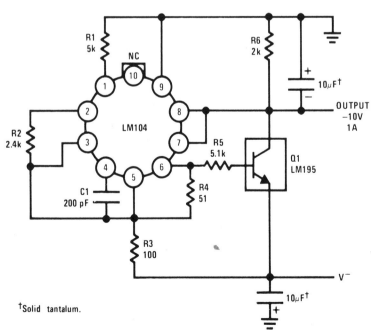

†Solid tantalum.

−10 V AT 1 A—Combination of LM195 power transistor IC and standard LM104 regulator gives negative output voltage with full overload protection and better than 2-mV load regulation. Input voltage must be only 2 V greater than output voltage.—"Linear Applications, Vol. 2," National Semiconductor, Santa Clara, CA, 1976, AN-110, p 4–5.

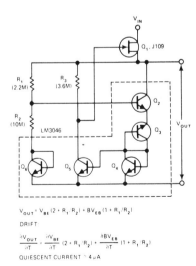

$$V_{OUT} = V_{BE}(2 + R_1/R_2) + BV_{EB}(1 + R_1/R_2)$$

DRIFT

$$\frac{\partial V_{OUT}}{\partial T} = \frac{\partial V_{BE}}{\partial T}(2 + R_1/R_2) + \frac{\partial BV_{EB}}{\partial T}(1 + R_1/R_2)$$

QUIESCENT CURRENT ≈ 4 µA

JFET SERIES-PASS—Use of JFET as series-pass element for LM3046 voltage regulator IC minimizes battery drain in microprocessor system applications. Pass element needs no preregulation because drive comes from regulated output. Gate source is isolated from line by drain and thus provides excellent line regulation.—J. Maxwell, Voltage Regulator Bridges Gap Between IC's and Zeners, *EDN Magazine,* Sept. 5, 1977, p 178–179.

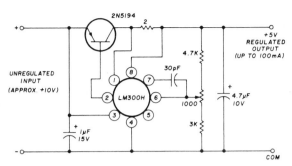

+5 V WITH LM300H—Series power transistor and National IC voltage regulator provide up to 100 mA. Improved version of regulator, LM305H, may be substituted.—H. Olson, Power-Supply Servicing, *Ham Radio*, Nov. 1976, p 44–50.

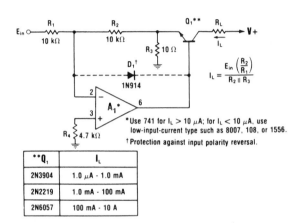

$$I_L = \frac{E_{in}\left(\frac{R_2}{R_1}\right)}{R_2 \parallel R_3}$$

* Use 741 for $I_L > 10 \mu A$; for $I_L < 10 \mu A$, use low-input-current type such as 8007, 108, or 1556.
† Protection against input polarity reversal.

Q₁	I_L
2N3904	1.0 μA - 1.0 mA
2N2219	1.0 mA - 100 mA
2N6057	100 mA - 10 A

NEGATIVE-INPUT CURRENT REGULATOR—Opamp is used as inverter starting current-boosting transistor to provide positive supply voltage. Load current range depends on transistor used. R_3 forces Q_1 to conduct much heavier current than feedback current, as required for high load current. Current gain depends on ratio of R_2 to R_3.—W. G. Jung, "IC Op-Amp Cookbook," Howard W. Sams, Indianapolis, IN, 1974, p 176–177.

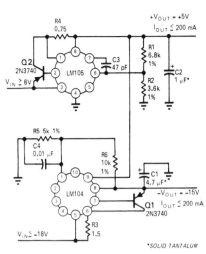

−15 V TRACKING +5 V—LM104 negative regulator is used with inverting gain to give negative output voltage that is greater than positive reference voltage. Noninverting input is tied to divider R5-R6 between negative output and ground. Positive reference determines line regulation and temperature drift, with negative output tracking.—R. C. Dobkin, One Adjustment Controls Many Regulators, *EDN Magazine*, Nov. 1, 1970, p 33–35.

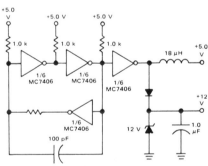

+5 AND +12 V AT 6 mA—Circuit using four sections of Motorola MC7406 provides +12 V supply required by MCM6570 8192-bit character generator using 7 × 9 matrix, along with conventional +5 V.—"A CRT Display System Using NMOS Memories," Motorola, Phoenix, AZ, 1975, AN-706A, p 5.

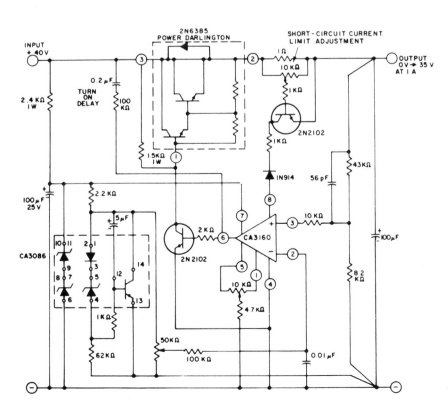

0.1–35 V AT 1 A—CA3160 serves as error amplifier in continuously adjustable regulator that functions down to vicinity of 0 V. RC network between base of 2N2102 output drive transistor and input source prevents turn-on overshoot. Input regulation is better than 0.01%/V, and regulation from no load to full load is better than 0.005%. Hum and noise output is less than 250 μVRMS.—"Linear Integrated Circuits and MOS/FET's," RCA Solid State Division, Somerville, NJ, 1977, p 267–269.

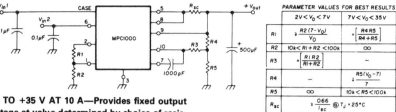

PARAMETER VALUES FOR BEST RESULTS		
	$2V < V_O < 7V$	$7V < V_O < 35V$
R1	$\geq \dfrac{R2(7-V_O)}{V_O}$	$\geq \left[\dfrac{R4\,R5}{R4+R5}\right]$
R2	$10k < R1+R2 < 100k$	∞
R3	$\approx \left[\dfrac{R1\,R2}{R1+R2}\right]$	—
R4	—	$\geq \dfrac{R5(V_O-7)}{7}$
R5	∞	$10k < R5 < 100k$
R_{sc}	$\approx \dfrac{0.66}{I_{sc}}$ @ $T_J = 25°C$	

+2 TO +35 V AT 10 A—Provides fixed output voltage at value determined by choice of resistance values, computed as given in table. Heatsink should have very low thermal resistance. For similar range of negative voltages, Motorola MPC900 regulator can be used, with circuit modified slightly as set forth in article.—H. Olson, Second-Generation IC Voltage Regulators, *Ham Radio*, March 1977, p 31–37.

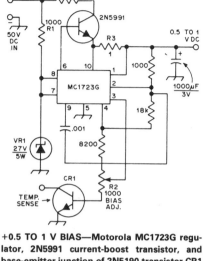

+0.5 TO 1 V BIAS—Motorola MC1723G regulator, 2N5991 current-boost transistor, and base-emitter junction of 2N5190 transistor CR1 serve as adjustable bias voltage source for 300-W solid-state power amplifier. R3 sets current limiting at about 0.65 A. Measured output-voltage variations are about ±6 mV for load changes of 0 to 600 mA.—H. O. Granberg, One KW—Solid-State Style, *QST*, April 1976, p 11–14.

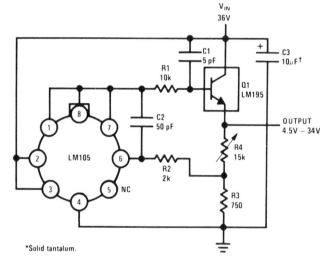

*Solid tantalum.

4.5–34 V AT 1 A—Combination of LM195 power transistor IC and standard LM105 regulator gives better than 2-mV load regulation with overload protection. Differential between input and output voltages is only 2 V.—"Linear Applications, Vol. 2," National Semiconductor, Santa Clara, CA, 1976, AN-110, p 4.

±10 V TRACKING—Fairchild 78MG and 79MG positive and negative voltage-regulator ICs provide up to 500-mA output, with protection against short-circuits and thermal overloads.—D. Schmieskors, Adjustable Voltage-Regulator ICs, *Ham Radio*, Aug. 1975, p 36–38.

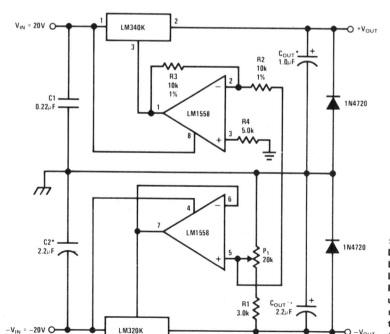

*Solid tantalum.

±5 TO ±18 V WITH TRACKING—Ground pin of LM340K-15 positive regulator is lifted by LM1558 inverter, while ground pin of negative LM320K-15 is lifted by LM1558 voltage follower. Positive regulator is made to track negative regulator within about 50 mV over entire output range. At ±15 V, typical load regulation is between 40 and 80 mV for 0–1 A pulsed load.—"Linear Applications, Vol. 2," National Semiconductor, Santa Clara, CA, 1976, AN-103, p 8–9.

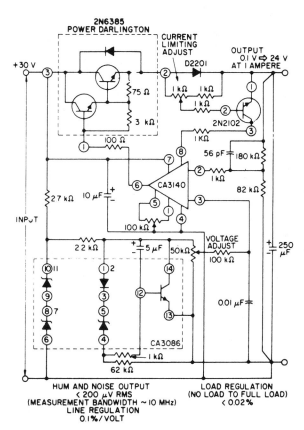

0.1–24 V AT 1 A—High-impedance reference-voltage divider across 30-V supply serves CA3140 connected as noninverting power opamp with gain of 3.2. 8-V reference input gives maximum output voltage of about 25 V. D2201 high-speed diode serves as current sensor for 2N2102 current-limit sensing amplifier. Current-limiting point can be adjusted over range of 10 mA to 1 A with single 1K pot. Power Darlington serves as series-pass element.—"Linear Integrated Circuits and MOS/FET's," RCA Solid State Division, Somerville, NJ, 1977, p 248–256.

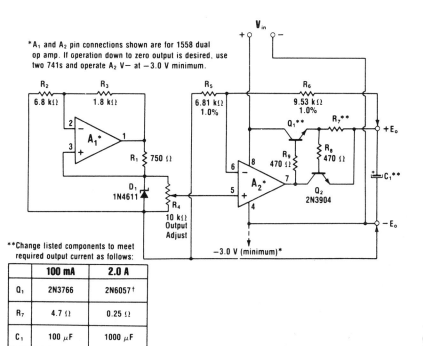

	100 mA	2.0 A
Q_1	2N3766	2N6057†
R_7	4.7 Ω	0.25 Ω
C_1	100 μF	1000 μF

†Heat sink required.

0–15 V AT 2 A—Basic zener-opamp regulator output of 6.6 V is scaled up to maximum of 15 V, adjusted with R_4, by adding buffer opamp A_2 and current-boosting transistors. Q_2 provides short-circuit protection by sensing load current through R_7. Large output capacitor C_1 maintains low output impedance at high frequencies where gain of A_2 falls off.—W. G. Jung, "IC Op-Amp Cookbook," Howard W. Sams, Indianapolis, IN, 1974, p 158–159.

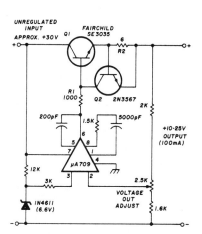

+10 TO +25 V AT 100 mA—Series regulator uses opamp as differential amplifier and extra transistor Q2 as current limiter. When 100 mA is drawn, 0.6 V is developed across R2 to make Q2 conduct, pulling Q1 base in negative direction. This action prevents excessive current from being passed by Q1.—H. Olson, Power-Supply Servicing, *Ham Radio*, Nov. 1976, p 44–50.

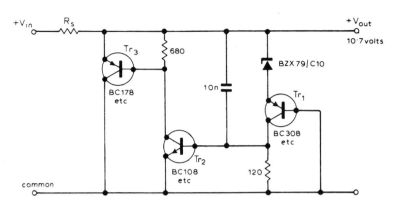

RIPPLE-PREAMP SUPPLY—Shunt regulator removes virtually all AC line ripple without using large capacitor, making it ideal for audio applications where freedom from ripple is more important than precise supply voltage level. Circuit cannot be damaged by short-circuits. Tr₃ may be power transistor or Darlington.—P. S. Bright, Ripple Eliminator, *Wireless World,* April 1977, p 62.

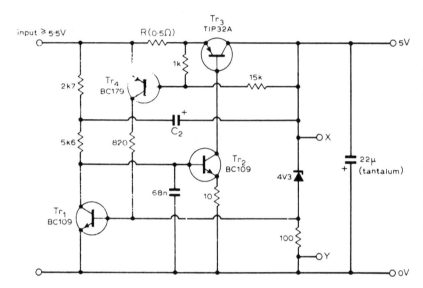

LOW COST WITH DISCRETE ELEMENTS—Performance is comparable to that of combined discrete and monolithic circuits, with load regulation of 0.01%, line regulation of 0.05%, ripple rejection of 0.1%, and output ripple and noise of 1 mV. Output is 1 A at 5 V. Foldback short-circuit protection is provided by Tr₄, with maximum current determined by value of R. C₂, which can be 100 μF, gives extra ripple rejection by introducing more AC feedback into loop. TIP32A is plastic series transistor, and is not critical; many other types will work equally well.— K. W. Mitchell, High Performance Voltage Regulator, *Wireless World,* May 1976, p 83–84.

TRIMMED DUAL SUPPLY

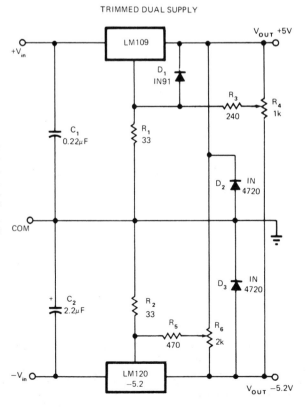

DUAL OUTPUTS WITH TRIMMING—Trimming pots connected across outputs provide positive or negative currents for producing small trimming voltages across 33-ohm ground-leg resistors of National regulators. Same components can be used for higher output voltages, but resistance values of pots should be increased if power dissipation becomes problem.—C. T. Nelson, Power Distribution and Regulation Can Be Simple, Cheap and Rugged, *EDN Magazine,* Feb. 20, 1973, p 52–58.

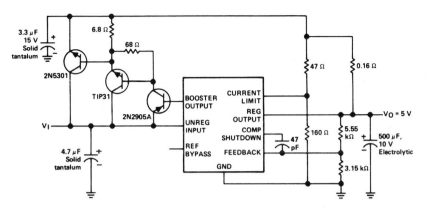

5 V AT 10 A WITH CURRENT LIMITING—Combination of three transistors and SN52105 or SN72305 regulator provides foldback current limiting for overload protection. Input voltage can be up to 40 V greater than 5-V output. Load regulation is about 0.1%, and input regulation is 0.1%/V. Regulators are interchangeable with LM105 and LM305 respectively.—"The Linear and Interface Circuits Data Book for Design Engineers," Texas Instruments, Dallas, TX, 1973, p 5-9.

0 TO ±6.6 V TRACKING AT 5 mA—Master-slave regulator combination is used to make second regulator provide mirror image of first while output of first is varied over full range from 0 to zener limit with R_4. Accuracy of tracking depends on match between R_5 and R_6, which should be 1% film or wirewound.—W. G. Jung, "IC Op-Amp Cookbook," Howard W. Sams, Indianapolis, IN, 1974, p 160–162.

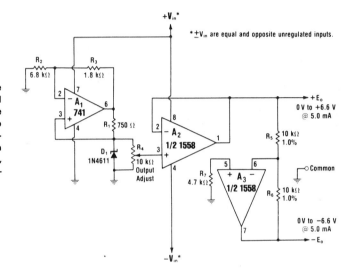

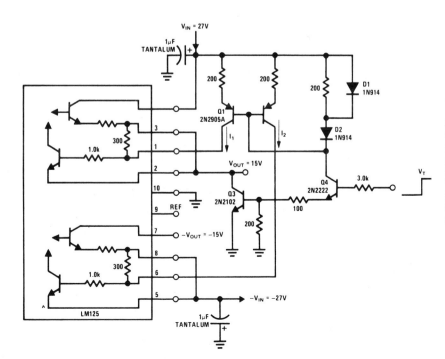

ELECTRONIC SHUTDOWN—Both sections of National LM125 dual tracking regulator are shut down by TTL-compatible control signal V_T which shorts internal reference voltage of regulator to ground. Q3 acts only as current sink.—T. Smathers and N. Sevastopoulos, "LM125/LM126/LM127 Precision Dual Tracking Regulators," National Semiconductor, Santa Clara, CA, 1974, AN-82, p 15.

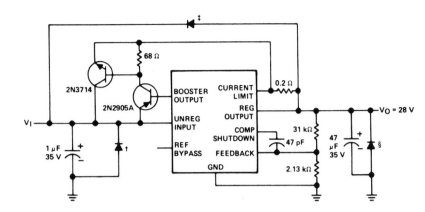

28 V AT 1 A—Circuit uses SN52105 or SN72305 regulator with three protective diodes. Feedback diode at top protects against shorted input and inductive loads on unregulated supply. Input diode protects against input voltage reversal. Output diode protects against output voltage reversal. Maximum input voltage is 50 V.—"The Linear and Interface Circuits Data Book for Design Engineers," Texas Instruments, Dallas, TX, 1973, p 5-9.

15 V AT 5 A WITH PROTECTION—External boost transistor is used with National LM340T-15 regulator to boost output current capability to 5 A without affecting such features as short-circuit current limiting and thermal shutdown. Short-circuit current is held to 5.5 A. Heatsink for Q1 should have at least 4 times capacity of heatsink for IC.—"Linear Applications, Vol. 2," National Semiconductor, Santa Clara, CA, 1976, AN-103, p 3–4.

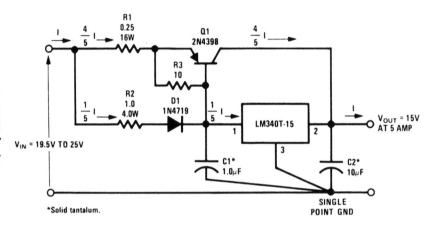

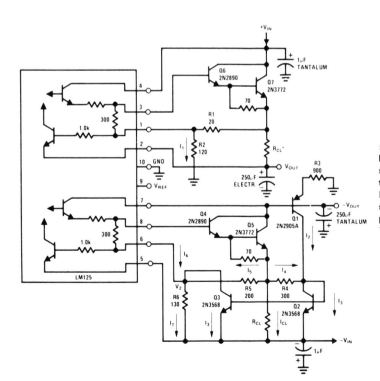

±15 V AT 10 A WITH FOLDBACK CURRENT LIMITING—Combination of Darlington pass transistors and current limiting is used with National LM125 dual tracking regulator to give high output currents with protection from short-circuits.—T. Smathers and N. Sevastopoulos, "LM125/LM126/LM127 Precision Dual Tracking Regulators," National Semiconductor, Santa Clara, CA, 1974, AN-82, p 11.

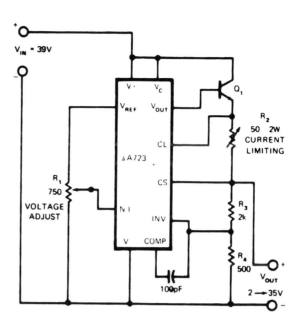

2–35 V VARIABLE—Wide voltage range is achieved by using μA723 regulator IC in simple feedback arrangement requiring only single pot to vary output voltage continuously and linearly from 2 to 35 V. Resistors R_3 and R_4 divide output voltage by 5, so inverting input of regulator sees one-fifth of output voltage. R_1 is connected between 7-V reference of IC and ground to present any intermediate voltage to noninverting input. IC acts to keep these two voltages equal. Maximum input voltage limit is 40 V; if possibility of higher voltages exists in lab applications, protect IC with 40-V zener across it.—J. Gangi, Continuously Variable Voltage Regulator, *EDN Magazine,* Feb. 20, 1973, p 91.

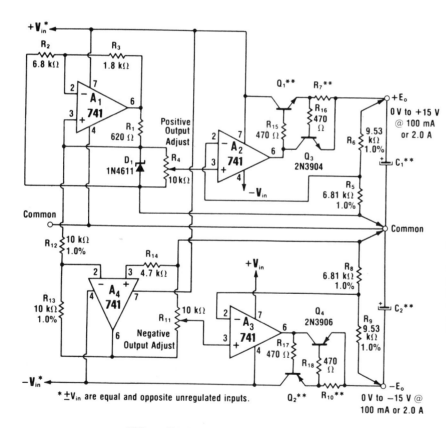

0 TO ±15 V INDEPENDENTLY VARIABLE—Common zener reference serves for both regulators. Buffer A_3 uses negative reference voltage developed from 6.6-V positive voltage across D_1 by inverter A_4. Both regulators provide 100 mA or 2 A depending on transistors used.—W. G. Jung, "IC Op-Amp Cookbook," Howard W. Sams, Indianapolis, IN, 1974, p 162–164.

* ±V_{in} are equal and opposite unregulated inputs.

**Change listed components to meet required output current as follows:

	100 mA	**2.0 A**
Q_1	2N3766	2N6057†
Q_2	2N3740	2N6050†
R_7, R_{10}	4.7 Ω	0.25 Ω
C_1, C_2	100 μF	1000 μF

†Heat sink required.

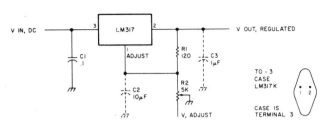

1.2–37 V AT 1.5 A—Uses National LM317 adjustable three-terminal positive voltage regulator. Output voltage is determined by ratio of R1 and R2. Output can be adjusted from 37 V down to 1.2 V with R2. If DC input is 40 V, regulation is about 0.1% at all settings when going from no load to full load. Regulator includes overload and thermal protection. If current limit is exceeded, regulator shuts down. C2 and C3 are optional; C2 improves ripple rejection, and C3 prevents instability when load capacitance is between 500 and 5000 pF.—Adjustable Bench Supply, *73 Magazine*, Dec. 1977, p 192–193.

+5 V AT 3 A—Uses Motorola MPC1000 positive voltage regulator to provide high current required for large TTL project. Current-limiting resistor R_SC is in range of 0.66 to 0.066 ohm. Use copper wire about 50% longer than calculated length and shorten step by step until required pass current is obtained; thus, start with 25 ft of No. 16, 15 ft of No. 18, 10 ft of No. 20, or 6 ft of No. 22.—G. L. Tater, The MPC1000—Super Regulator, *Ham Radio*, Sept. 1976, p 52–54.

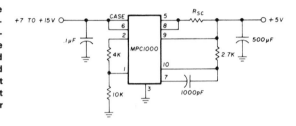

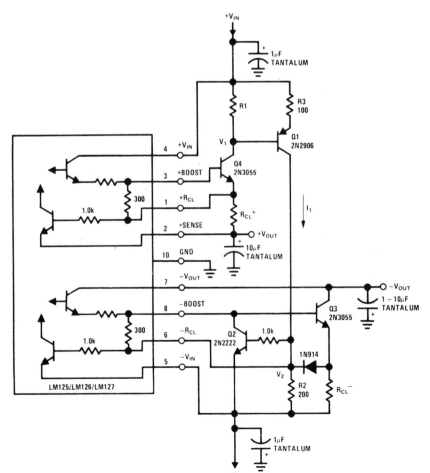

SIMULTANEOUS CURRENT LIMITING—Limiting action of circuit depends on output current of positive regulator but acts simultaneously on both positive and negative outputs of National LM125 dual tracking regulator. Positive output current produces voltage drop across R1 that makes Q1 conduct. When increase in current makes voltage drop across R2 equal negative current limit sense voltage, negative regulator will current-limit. Positive regulator closely follows negative output down to level of about 700 mV. Q2 turns off negative pass transistor during simultaneous current limiting. Output voltages are ±15 V.—"Linear Applications, Vol. 2," National Semiconductor, Santa Clara, CA, 1976, AN-82, p 12–13.

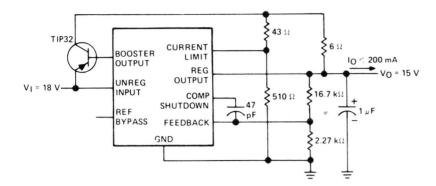

15 V AT 200 mA—Linear regulator using Texas Instruments SN52105, SN72305, or SN72376 is connected for foldback current limiting. Regulators are interchangeable with LM105, LM305, and LM376 respectively. Load regulation is 0.1%, and input regulation is 0.1%/V.—"The Linear and Interface Circuits Data Book for Design Engineers," Texas Instruments, Dallas, TX, 1973, p 5-9.

5 V AT 1 A—Use of Darlington at output boosts power rating of standard opamp voltage regulator circuit. Article gives step-by-step design procedure. With μA741 opamp, circuit gives good regulation along with short-circuit protection. AC ripple is less than 2 mV P-P. Required input of 30 V is obtained from conventional full-wave bridge rectifier with capacitor-input filter.—C. Brogado, IC Op Amps Simplify Regulator Design, *EDN/EEE Magazine*, Jan. 15, 1972, p 30–34.

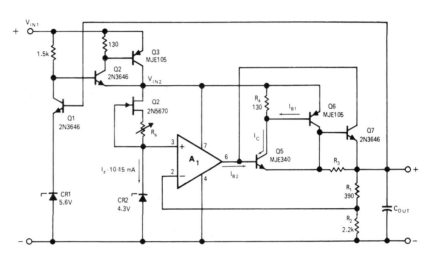

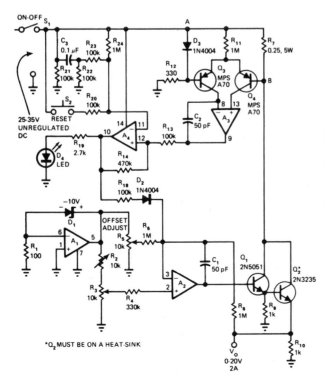

0–20 V AT 2 A—R_3 provides control of output voltage for regulator built around LM3900 quad Norton opamp. Output is well regulated against both line and load variations and is free of ripple. Opamp sections A_3 and A_4 provide overcurrent sensing and shutdown functions; after output fault is cleared, S_2 is closed momentarily to restore output power. Article describes circuit operation and initial setup in detail.—J. C. Hanisko and W. Wiseman, Variable Supply Built Around Quad Amp Outputs 2A, *EDN Magazine*, June 20, 1976, p 128 and 130.

*Q₂ MUST BE ON A HEAT-SINK

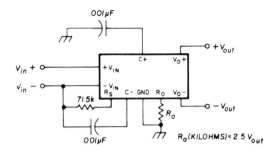

VARIABLE DUAL-POLARITY—External resistor R_o determines values of positive and negative regulated output voltages provided by Silicon General SG3501 dual regulator.—H. Olson, Second-Generation IC Voltage Regulators, *Ham Radio*, March 1977, p 31–37.

R_o (KILOHMS)= 2.5 V_{out}

5 V AT 200 mA—Article gives step-by-step design procedure for developing special opamp regulator when commercial unit meeting desired specifications is not available. Opamp is μA741. Circuit gives good regulation along with short-circuit protection, with less than 2 mV P-P AC ripple. Required input of 20 V is obtained from conventional full-wave bridge rectifier with capacitor-input filter.—C. Brogado, IC Op Amps Simplify Regulator Design, *EDN/EEE Magazine*, Jan. 15, 1972, p 30–34.

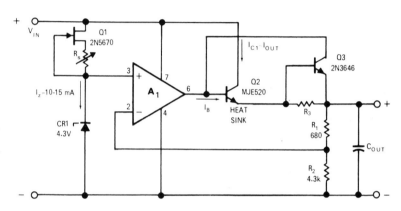

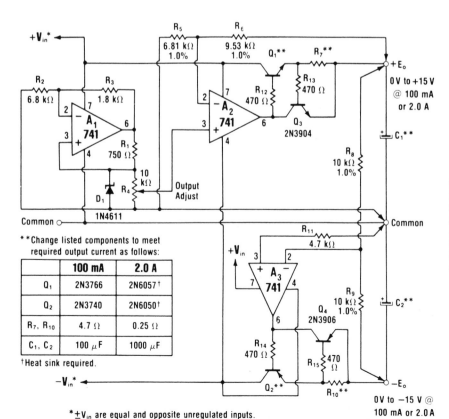

0 TO ±15 V TRACKING AT 100 mA OR 2 A—Basic tracking regulator is combined with transistors to extend output to voltages higher than zener reference and provide higher output currents. Choice of transistors for Q_1 and Q_2 determines maximum load current.—W. G. Jung, "IC Op-Amp Cookbook," Howard W. Sams, Indianapolis, IN, 1974, p 161–163.

**Change listed components to meet required output current as follows:

	100 mA	**2.0 A**
Q_1	2N3766	2N6057 †
Q_2	2N3740	2N6050 †
R_7, R_{10}	4.7 Ω	0.25 Ω
C_1, C_2	100 μF	1000 μF

†Heat sink required.

*±V_{in} are equal and opposite unregulated inputs.

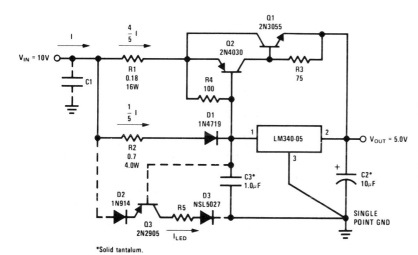

5 V AT 5 A FOR TTL—Typical load regulation is 1.8% from no load to full load. Q1 and Q2 serve in place of single higher-cost power PNP boost transistor. Dotted lines show how to add overload indicator using National NSL5027 LED and R2 as overload sensor. When load current exceeds 5 A, Q3 turns on and D3 lights. Circuit includes thermal shutdown and short-circuit protection.—"Linear Applications, Vol. 2," National Semiconductor, Santa Clara, CA, 1976, AN-103, p 5.

NEGATIVE SHUNT REGULATION—Connection shown for LM117 positive series regulator provides spike-suppressing negative shunt regulation of −5 V output. With capacitor shown, regulator will withstand 75-V spikes on raw DC supply. For larger spikes, increase capacitor value.—P. Lefferts, Series Regulators Provide Shunt Regulation, *EDN Magazine*, Sept. 5, 1978, p 158 and 160.

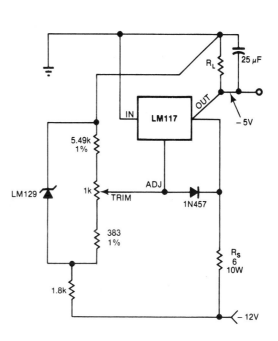

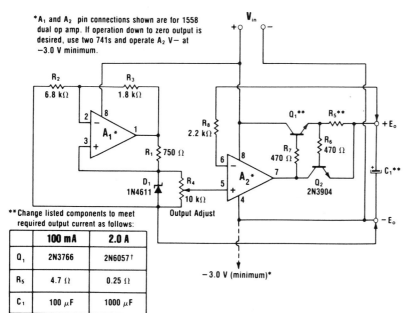

0–6.6 V AT 2 A—High-power circuit is suitable for low-voltage logic devices that require high current at supply voltages between 3 and 6 V. Maximum output of 2 A is obtained with 2N6057 Darlington pair for Q_1. Single 2N3766 can be used if load is only 100 mA. Q_2 provides short-circuit protection for Q_1. Since supply does not have to be adjusted down to 0 V, negative supply for A_2 can go to common negative of circuit. Optional connection to −3 V is used only when voltage range must go down to 0 V.—W. G. Jung, "IC Op-Amp Cookbook," Howard W. Sams, Indianapolis, IN, 1974, p 157–158.

*A_1 and A_2 pin connections shown are for 1558 dual op amp. If operation down to zero output is desired, use two 741s and operate A_2 V− at −3.0 V minimum.

**Change listed components to meet required output current as follows:

	100 mA	**2.0 A**
Q_1	2N3766	2N6057†
R_5	4.7 Ω	0.25 Ω
C_1	100 μF	1000 μF

†Heat sink required.

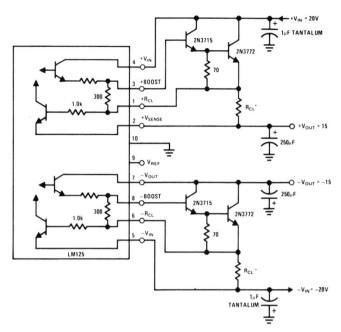

±15 V AT 7 A—External Darlington stages boost output currents of LM125 dual tracking regulator and increase minimum input/output voltage differential to 4.5 V. Maximum output current is limited by power dissipation of 2N3772. Typical load regulation is 40 mV from no load to full load.—T. Smathers and N. Sevastopoulos, "LM125/LM126/LM127 Precision Dual Tracking Regulators," National Semiconductor, Santa Clara, CA, 1974, AN-82, p 6.

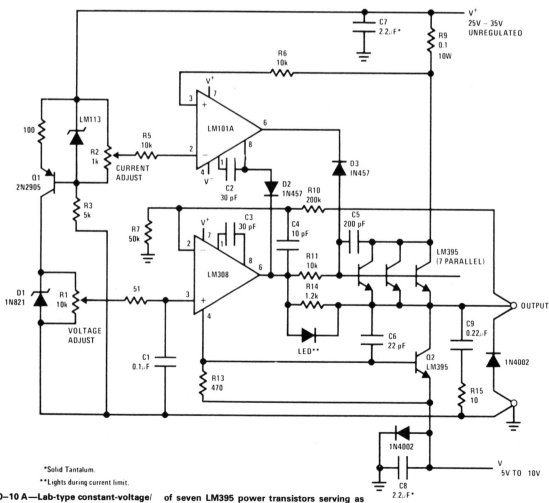

*Solid Tantalum.

**Lights during current limit.

0–25 V AT 0–10 A—Lab-type constant-voltage/ constant-current power supply using standard ICs achieves high current output by paralleling of seven LM395 power transistors serving as pass element. Current limiting is provided on LM395 chip for complete overload protection.— "Linear Applications, Vol. 2," National Semiconductor, Santa Clara, CA, 1976, LB-28.

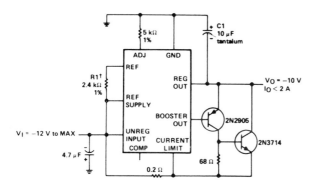

−10 V AT 2 A—Negative-voltage regulator using SN52104 or SN72304 accepts input voltage of −12 V to −40 V and uses only single external resistor to provide regulated output of −10 V with typical load regulation of 1 mV and input regulation of 0.06%. ICs are interchangeable with LM104 and LM304 respectively.—"The Linear and Interface Circuits Data Book for Design Engineers," Texas Instruments, Dallas, TX, 1973, p 5-5.

±15 V TRACKING—Single NE/SE5554 dual tracking regulator is used with pass transistors to give higher output current than 200-mA limit for each section of regulator, with close-tolerance tracking.—"Signetics Analog Data Manual," Signetics, Sunnyvale, CA, 1977, p 672–673.

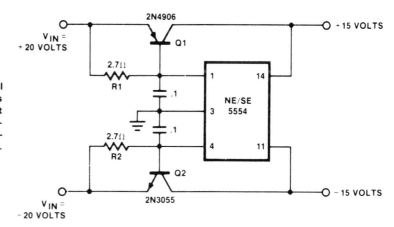

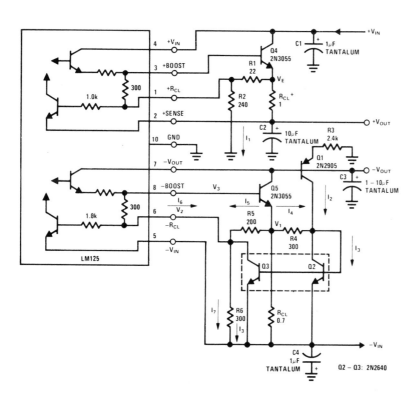

FOLDBACK CURRENT LIMITING—Reduces short-circuit output current of National LM125 dual tracking regulator sections to fraction of full-load output current, avoiding need for larger heatsink. Programmable current source is used to give constant voltage drop across R5 for negative regulator. Simple resistor divider serves same purpose for positive regulator. Design examples are given.—T. Smathers and N. Sevastopoulos, "LM125/LM126/LM127 Precision Dual Tracking Regulators," National Semiconductor, Santa Clara, CA, 1974, AN-82, p 7.

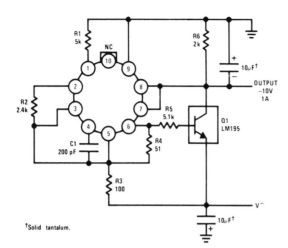

†Solid tantalum.

−10 V AT 1 A—National LM195 power transistor, used with LM105 regulator, provides full overload protection. Load regulation is better than 2 mV. Circuit requires only 2-V differential between input and output voltages.—R. Dobkin, "Fast IC Power Transistor with Thermal Protection," National Semiconductor, Santa Clara, CA, 1974, AN-110, p 5.

15 V AT 1 A WITH LOGIC SHUTDOWN—Arrangement shown provides practical method of shutting down LM340T-15 or similar regulator under control of TTL or DTL gate. Pass transistor Q1 operates as saturated transistor when logic input is high (2.4 V minimum for TTL) and Q2 is turned on. When logic input is low (below 0.4 V for TTL), Q2 and Q1 are off and regulator is in effect shut down.—"Linear Applications, Vol. 2," National Semiconductor, Santa Clara, CA, 1976, AN-103, p 11.

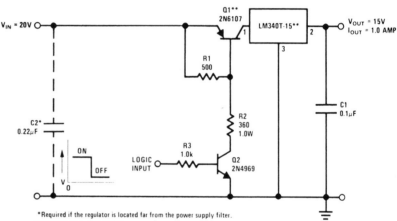

*Required if the regulator is located far from the power supply filter.
**Head sink Q1 and the LM340.

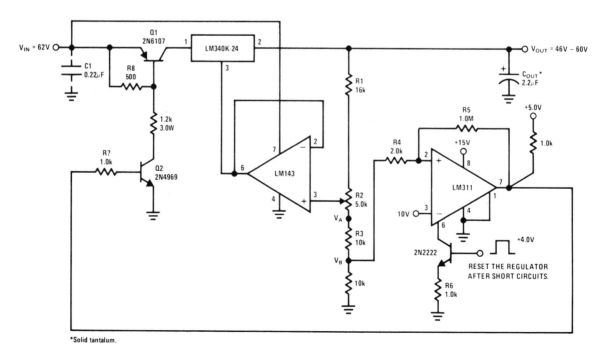

*Solid tantalum.

46–60 V FROM 62 V—Variable-output high-voltage regulator includes short-circuit and overvoltage protection. When LM340K-24 regulator has been shut down by shorted load, LM311 must be activated by applying 4-V strobe pulse to 2N2222 transistor to make Q1 close again and start regulator.—"Linear Applications, Vol. 2," National Semiconductor, Santa Clara, CA, 1976, AN-103, p 11–12.

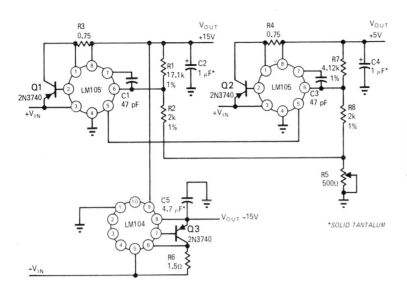

+15, +5, AND −15 V—Single potentiometer R5 serves for adjusting all three regulated output voltages simultaneously. Accuracy of adjustment is within 2%.—R. C. Dobkin, One Adjustment Controls Many Regulators, *EDN Magazine*, Nov. 1, 1970, p 33–35.

*SOLID TANTALUM

0–15 V—Addition of 307 or 301A opamp and three inexpensive components to standard three-terminal voltage regulator provides programming capability from maximum terminal voltage down to zero. With adequate heatsink, output current can be up to 1 A. Opamp A_2 provides floating reference voltage to normally grounded common terminal of A_1, with pot allowing ground to be positioned anywhere along voltage drop of 15 V across pot. Unregulated negative supply is not critical, and drain is 10 mA.—W. G. Jung, Three Components Program Regulator from Maximum to Zero, *EDN Magazine*, May 20, 1977, p 126 and 128.

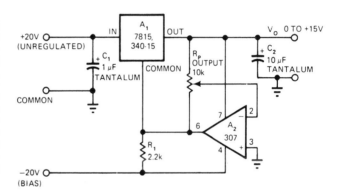

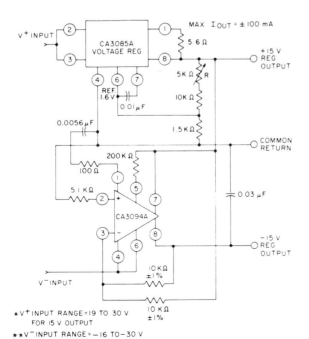

±15 V TRACKING AT 100 mA—Provides line and load regulation of 0.075% by using CA3094A programmable opamp and CA3085A series voltage regulator. V+ input range is 19 to 30 V for 15-V output, while V− input range is −16 to −30 V for −15 V output.—"Circuit Ideas for RCA Linear ICs," RCA Solid State Division, Somerville, NJ, 1977, p 18.

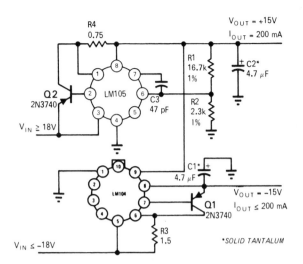

±15 V TRACKING—Arrangement uses LM104 negative regulator to track positive regulator, with both regulators adjusted simultaneously by changing R1. Inverting opamp can be added to provide negative output voltage while using positive voltage as reference.—R. C. Dobkin, One Adjustment Controls Many Regulators, *EDN Magazine,* Nov. 1, 1970, p 33–35.

BOOSTING OUTPUT CURRENT—External NPN pass transistor is added to each section of LM125 precision dual tracking regulator to increase maximum output current by factor equal to beta of transistor. To prevent overheating and destruction of pass transistors and resultant damage to regulator, series resistor R_{CL} is used to sense load current. When voltage drop across R_{CL} equals current-limit sense voltage in range of about 0.3 to 0.8 V (related to junction temperature), regulator will current-limit. Maximum load current is about 1 A for 25°C junction and 0.6 ohm for R_{CL}. LM125 provides ±15 V, LM126 provides ±12 V, and LM127 provides +5 V and −12 V.—T. Smathers and N. Sevastopoulos, "LM125/LM126/LM127 Precision Dual Tracking Regulators," National Semiconductor, Santa Clara, CA, 1974, AN-82, p 5.

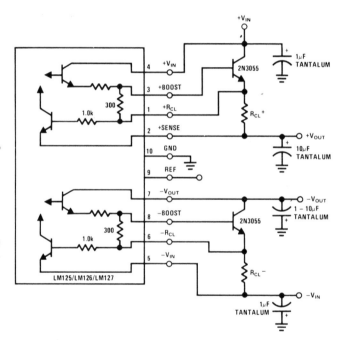

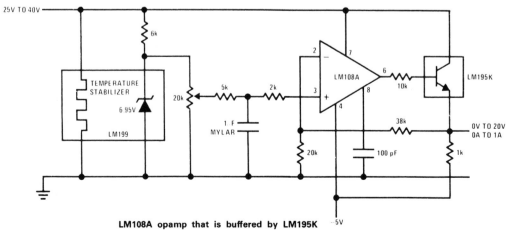

0–20 V HIGH-PRECISION—National LM199 temperature-stabilized 6.95-V reference feeds LM108A opamp that is buffered by LM195K power transistor IC which provides full overload protection.—"Linear Applications, Vol. 2," National Semiconductor, Santa Clara, CA, 1976, AN-161, p 6.

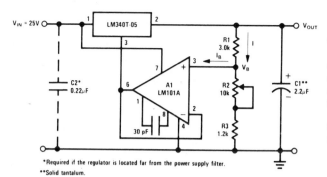

7–23 V AT 1.2–2 A—Ground terminal of LM340T-05 regulator is raised by amount equal to voltage applied to noninverting (+) input of opamp, to give output voltage set by R2 in resistive divider. Short-circuit protection and thermal shutdown are provided over full output range.—"Linear Applications, Vol. 2," National Semiconductor, Santa Clara, CA, 1976, AN-103, p 6–7.

0 TO −6.6 V AT 5 mA—Voltage follower A₂ buffers output that can be adjusted over full range from 0 V to zener limit with R₄. Positive supply of A₂ must go to voltage slightly more positive than +3 V common if linear output operation is required over full range.—W. G. Jung, "IC Op-Amp Cookbook," Howard W. Sams, Indianapolis, IN, 1974, p 159–160.

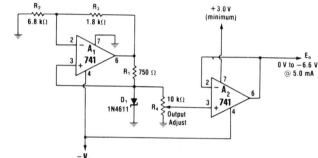

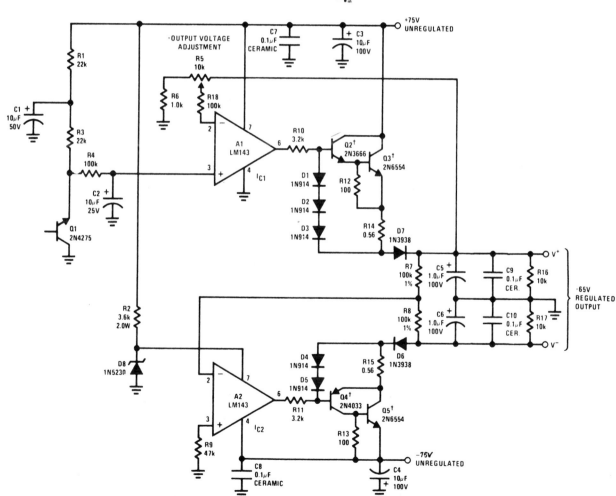

±65 V TRACKING AT 1 A—Circuit uses two LM143 high-voltage opamps in combination with zener reference and discrete power-transistor pass elements. Q1 is transistor used as stable 6.5-V zener voltage reference. Opamp A1 amplifies reference voltage from 1 to about 10 times for application through R10 to Darlington-connected transistors Q2 and Q3. Feedback resistor R5 is made variable so positive output voltage can be varied from 6.5 V to about 65 V. This output is applied to unity-gain inverting power opamp A2 to generate negative output voltage. Q2-Q5 should be on common Thermalloy 6006B or equivalent heatsink. Supply includes short-circuit protection. Maximum shorted load current is about 1.25 A.—"Linear Applications, Vol. 2," National Semiconductor, Santa Clara, CA, 1976, AN-127, p 8–9.

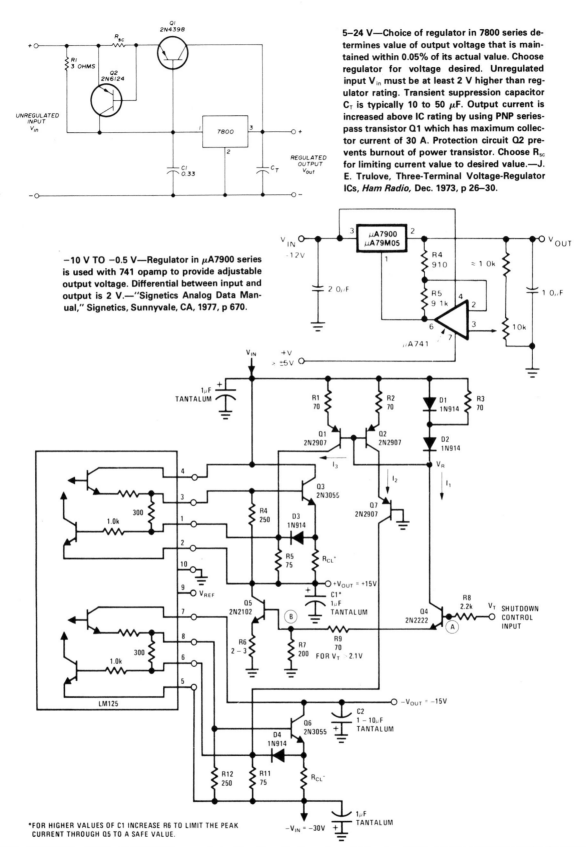

5–24 V—Choice of regulator in 7800 series determines value of output voltage that is maintained within 0.05% of its actual value. Choose regulator for voltage desired. Unregulated input V_{in} must be at least 2 V higher than regulator rating. Transient suppression capacitor C_T is typically 10 to 50 μF. Output current is increased above IC rating by using PNP series-pass transistor Q1 which has maximum collector current of 30 A. Protection circuit Q2 prevents burnout of power transistor. Choose R_{sc} for limiting current value to desired value.—J. E. Trulove, Three-Terminal Voltage-Regulator ICs, *Ham Radio*, Dec. 1973, p 26–30.

−10 V TO −0.5 V—Regulator in μA7900 series is used with 741 opamp to provide adjustable output voltage. Differential between input and output is 2 V.—"Signetics Analog Data Manual," Signetics, Sunnyvale, CA, 1977, p 670.

*FOR HIGHER VALUES OF C1 INCREASE R6 TO LIMIT THE PEAK CURRENT THROUGH Q5 TO A SAFE VALUE.

CURRENT BOOSTING WITH ELECTRONIC SHUTDOWN—Circuit provides complete shutdown for both sections of National LM125 dual tracking regulator without affecting unregulated inputs that may be powering additional equipment. Shutdown control signal is TTL-compatible, but regulator may be shut down at any desired level by adjusting values of R8 and R9. Control signal is used to short internal reference voltage of regulator to ground, thereby forcing positive and negative outputs to about +700 mV and +300 mV respectively. When shutdown signal is applied, Q4 draws current through R3 and D2, establishing voltage V_R that starts current sources Q1 and Q2. Currents I_1, I_2, and I_3 are then equal so both sides of regulator are shut down simultaneously.—T. Smathers and N. Sevastopoulos, "LM125/LM126/LM127 Precision Dual Tracking Regulators," National Semiconductor, Santa Clara, CA, 1974, AN-82, p 14.

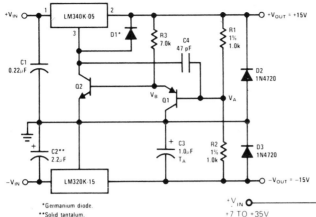

±15 V TRACKING AT 1 A—Positive regulator tracks negative. If Q1 and Q2 are perfectly matched, tracking action is unchanged over full operating temperature range, with tracking better than 100 mV. Regulation from no load to full load is 10 mV for positive side and 45 mV for negative side.—"Linear Applications, Vol. 2," National Semiconductor, Santa Clara, CA, 1976, AN-103, p 8–9.

*Germanium diode.
**Solid tantalum.

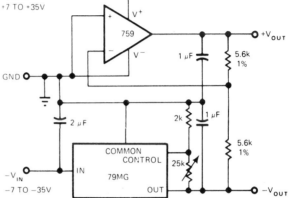

DUAL TRACKING—Uses one 759 power opamp for positive output, connected to track with 79MG negative voltage regulator having adjustable output. Common-mode range of 79MG includes ground, permitting operation from single supply. Circuit can also be built with two power opamps, one inverting and the other noninverting.—R. J. Apfel, Power Op Amps—Their Innovative Circuits and Packaging Provide Designers with More Options, *EDN Magazine*, Sept. 5, 1977, p 141–144.

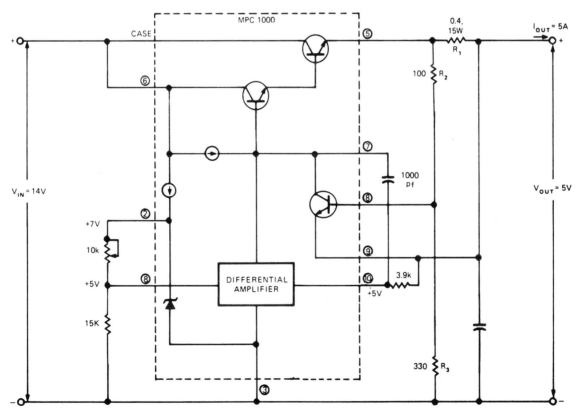

CURRENT-FOLDBACK PROTECTION— MPC1000 hybrid regulator provides regulated output of 5 V at 5 A from 14-V input. Values of components are based on foldback current of 6 A and short-circuit current of 2 A; this ensures that dissipation of regulator on short-circuit is less than dissipation at rated load. Short-circuit current is controlled by diode drop across R_1 and foldback current by drop across R_2. Article gives design equations and procedure for obtaining other output voltages. Circuit also serves to limit starting surges into capacitive load, and reduces heatsink size and transistor ratings. Returning R_3 to pin 2 of MPC1000 instead of pin 3 gives lower short-circuit current, improves efficiency, and reduces heat generation. Foldback protection is not suitable for variable-output supplies because foldback current is proportional to output voltage.—R. L. Haver, Use Current Foldback to Protect Your Voltage Regulator, *EDN Magazine*, Aug. 20, 1974, p 69–72.

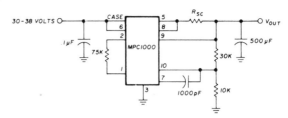

28 V AT 7 A—Uses Motorola MPC1000 positive voltage regulator to provide regulated voltage for aircraft radio equipment being used at ground station. Current-limiting resistor R_{SC} is in range of 0.66 to 0.066 ohm. Use copper wire about 50% longer than calculated length and shorten step by step until required pass current is obtained; thus, start with 25 ft of No. 16, 15 ft of No. 18, 10 ft of No. 20, or 6 ft of No. 22. Input voltage is obtained from 30-V transformer and bridge rectifier.—G. L. Tater, The MPC1000— Super Regulator, *Ham Radio,* Sept. 1976, p 52–54.

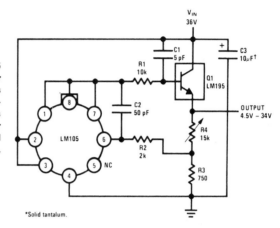

4.5–34 V VARIABLE AT 1 A—National LM195 power transistor is used with LM105 regulator to give fully adjustable range of output voltages with overload protection and only 2-V input-to-output voltage differential. Load regulation is better than 2 mV.—R. Dobkin, "Fast IC Power Transistor with Thermal Protection," National Semiconductor, Santa Clara, CA, 1974, AN-110, p 4.

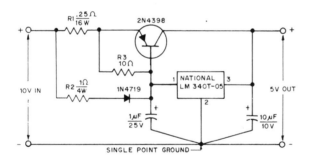

5 V AT 5 A—Current-sharing design provides short-circuit protection, safe-operating-area protection, and thermal shutdown. Typical load regulation is 1.4%.—W. R. Calbo, A High-Current, Low-Voltage Regulator for TTL Circuits, *QST,* Sept. 1975, p 44.

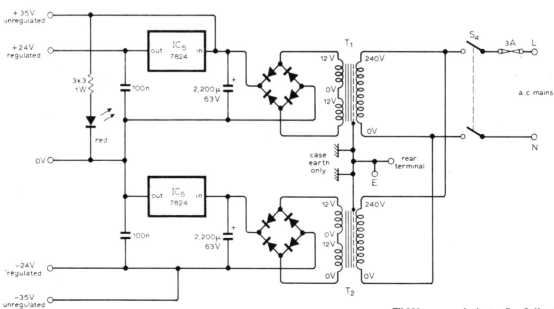

±24 V REGULATED AND ±35 V UNREGULATED—Developed for use with high-performance stereo preamp. Each IC regulator requires about 7 cm² of heatsink area. Red LED is TIL209 or equivalent.—D. Self, Advanced Preamplifier Design, *Wireless World,* Nov. 1976, p 41–46.

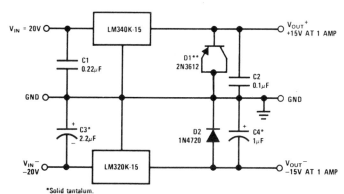

±15 V SYMMETRICAL AT 1 A—Connection shown gives same line and load regulation characteristics as for individual regulators. D1 ensures start-up of LM340K-15 under worst-case conditions of common load and 1-A load current over full temperature range.—"Linear Applications, Vol. 2," National Semiconductor, Santa Clara, CA, 1976, AN-103, p 8.

*Solid tantalum.

**Germanium diode (using a PNP germanium transistor with the collector shorted to the emitter).

Note: C1 and C2 required if regulators are located far from power supply filter.

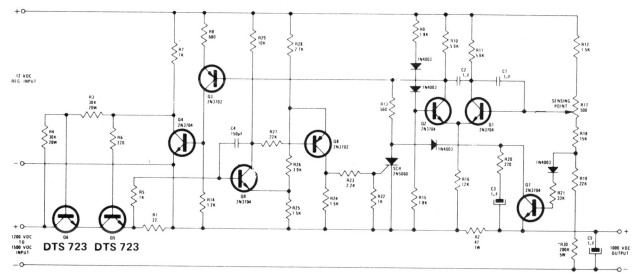

1000 V AT 100 W—Two Delco DTS-723 transistors in series function as pass element of regulator in which differential amplifier Q1-Q2 senses output voltage and compares it with reference voltage at base of Q2. Difference signal is amplified by Q3-Q4 for feed to Q5. 12-V regulated supply is referenced to high side of output voltage through R2. R1 is chosen so regulator shuts down when load current reaches 120 mA and triggers Schmitt trigger Q8-Q9 which fires SCR. When overload is removed, circuit returns to normal operation. Input voltage range of 1200–1500 V gives 0.1% regulation at full load.—"1000-Volt Linear DC Regulator," Delco, Kokomo, IN, 1974, Application Note 45.

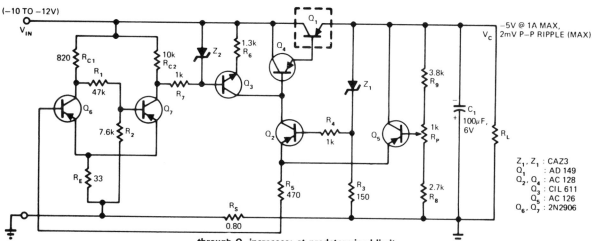

−5 V WITH PROTECTION—Switching-type short-circuit protection network uses R_7 connected to Schmitt trigger Q_6-Q_7. Ground is provided by Q_7 which is normally conducting. If output of regulator is short-circuited, current through Q_1 increases; at predetermined limit, Q_6 conducts and cuts off Q_7, breaking ground connection of R_7 and thus cutting off Q_3. Power transistor Q_1 is also cut off, and output current begins to decrease. When load current drops below another predetermined level, Q_6 again goes off and Q_7 turns on to begin another ON/OFF cycle, with switching process continuing until short-circuit is removed.—H. S. Raina and R. K. Misra, Novel Circuit Provides Short Circuit Protection, *EDN Magazine*, June 5, 1974, p 84.

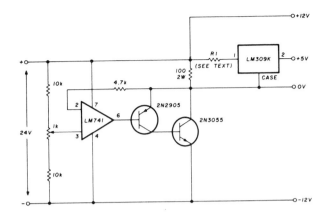

VOLTAGE ADAPTER—Bench power supply provides ± 12 V and +5 V from single regulated 24-V source, for use with many ICs. Both 12-V supplies can be adjusted in same direction by varying 24-V source or in opposite directions by adjusting 1K pot. R1 is used to decrease power dissipated in LM309K voltage regulator and is normally 2.2 ohms.—J. A. Piat, Voltage Adapter for MSI/LSI Circuits, *Ham Radio,* March 1978, p 115.

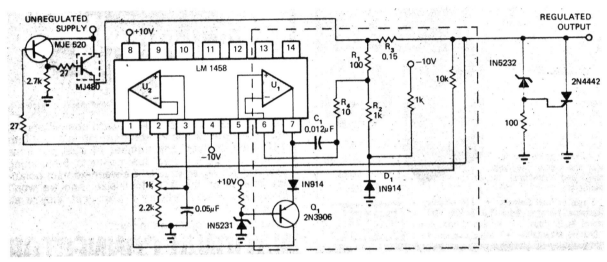

OVERVOLTAGE CROWBAR—Components within dashed lines protect regulator IC from overcurrent condition frequently encountered when zener-SCR crowbar is used across output.

Regulated output is 5 V with IC shown. Article gives operating details of circuit and equation for shutdown time, which is about 1 s.—S. J.

Pirkle, Circuit Protects Power Supply Regulator from Overcurrent, *EDN Magazine,* Feb. 5, 1973, p 89.

Servo Circuits

Includes logic-controlled preamps and power amplifiers for driving two-phase, stepper, and other types of 60-Hz and 400-Hz servomotors in either direction for correct time at correct speed for bringing servo shaft exactly to desired new position.

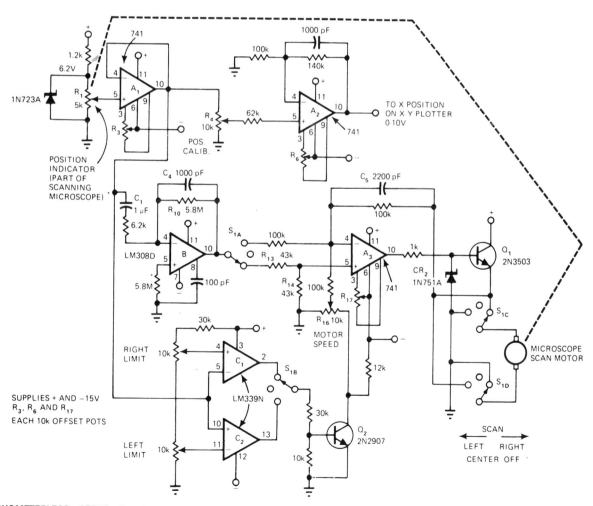

TACHOMETERLESS SERVO—Developed to provide speed control for motor enclosed in such a way that tachometer cannot be used for feedback. Position pot R_1 and differentiator B substitute for tachometer in controlling rate of scanning-microscope eyepiece used for measuring CRT line width. Buffer A_1 feeds X input of XY plotter through opamp A_2, and also feeds differentiator B and limit-detector voltage comparator C_1-C_2. S_1 switches A_3 between inverting and noninverting operation each time scanning direction changes, to keep feedback negative.— H. F. Stearns, Differentiator and Position Pot Sub for Tachometer, *EDN Magazine*, Aug. 5, 1977, p 50–52.

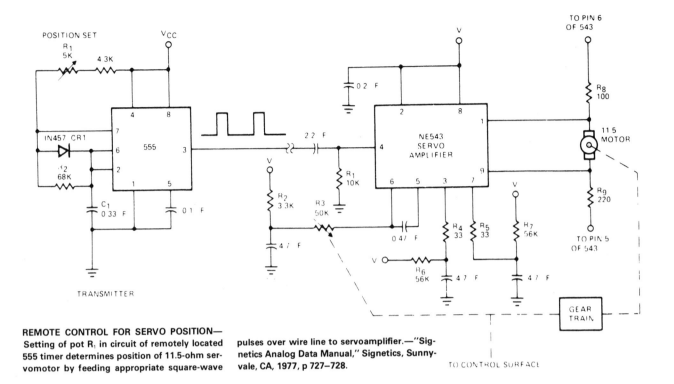

REMOTE CONTROL FOR SERVO POSITION—
Setting of pot R_1 in circuit of remotely located 555 timer determines position of 11.5-ohm servomotor by feeding appropriate square-wave pulses over wire line to servoamplifier.—"Signetics Analog Data Manual," Signetics, Sunnyvale, CA, 1977, p 727–728.

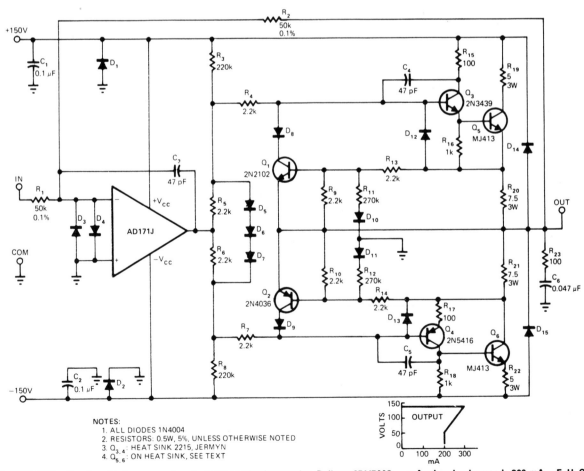

NOTES:
1. ALL DIODES 1N4004
2. RESISTORS: 0.5W, 5%, UNLESS OTHERWISE NOTED
3. $Q_{3,4}$: HEAT SINK 2215, JERMYN
4. $Q_{5,6}$: ON HEAT SINK, SEE TEXT

400-Hz AMPLIFIER—Developed to increase output power of digital-to-synchro converter systems while providing stable and accurate output and overall gain even with reactive loads. Includes overload protection. Delivers 95 VRMS at 400 Hz continuously into 500-ohm load. Power bandwidth is about 20 kHz. Foldback current limiting drops short-circuit current to 200 mA when load exceeds 300 mA.—F. H. Cattermolen and J. A. Pieterse, Digital/Synchro Amplifier Features Overload Protection, *EDN Magazine,* Nov. 5, 1977, p 107–108.

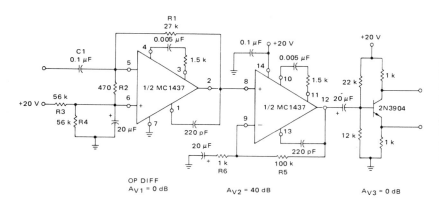

DUAL-OPAMP PREAMP—First section of Motorola MC1437 dual opamp is connected as operational differentiator driving direct-coupled noninverting opamp. Single-ended output is converted to push-pull by following phase-splitting amplifier for driving power amplifier of 115-V 60-Hz servomotor.—A. Pshaenich, "Servo Motor Drive Amplifiers," Motorola, Phoenix, AZ, 1972, AN-590.

SERVO DRIVE—Combination of Fairchild μA795 multiplier and μA741 opamp generates AC error signal for driving two-phase servomotor. Phase-shifted signal from R_1-C_1 is applied to input pin 4 of multiplier, DC signal input is applied to pin 9, and servo position signal goes to pin 12. Multiplier takes difference between signals on 9 and 12, multiplies this by signal on pin 4, and feeds resulting sine wave from pin 14 to opamp for amplification and transfer to servo driver. When servomotor action makes voltages on 9 and 12 equal, system is nulled.—Fairchild Linear IC Contest Winners, *EEE Magazine*, Jan. 1971, p 48–49.

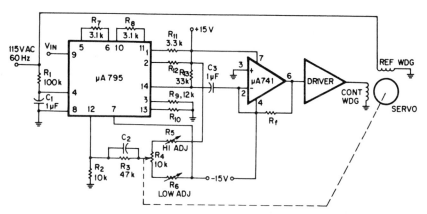

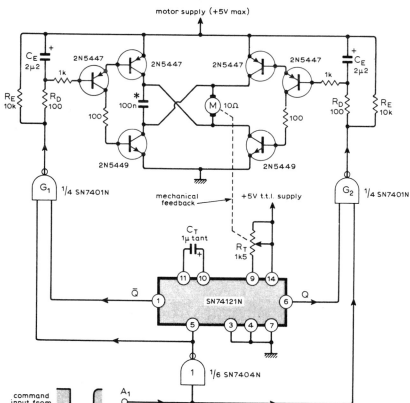

TTL SERVO CONTROL—Used in nine-channel remote control system having nine identical servos fed by decoder at receiving end of data link. Variable-width pulse command from decoder is fed into TTL IC pulse-width comparator that feeds bridge-type motor drive. Command pulse controls both direction and duration of motor rotation. Article describes operation in detail and gives associated coder and decoder circuits.—M. F. Bessant, Multi-Channel Proportional Remote Control, *Wireless World*, Oct. 1973, p 479–482.

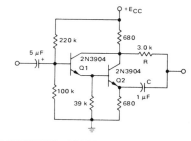

DARLINGTON PHASE SHIFTER—Basic 90° push-pull RC phase shifter using discrete transistors is connected as phase-splitting amplifier. Used at output of follow-up pot in servoamplifier driving 115-V 60-Hz servomotor. Supply is 28 V. Motorola MPSA13 Darlington IC can be used if 39K resistor is omitted.—A. Pshaenich, "Servo Motor Drive Amplifiers," Motorola, Phoenix, AZ, 1972, AN-590.

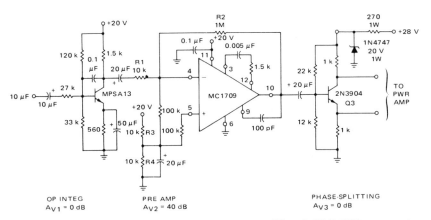

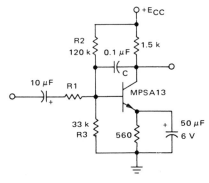

PHASE-SPLITTING PREAMP—Uses Motorola MPSA13 operational integrator to provide 90° phase shift for MC1709 inverting opamp, with single-ended output complemented by phase-splitting amplifier to provide push-pull drive for power amplifier of 115-V 60-Hz servomotor. Voltage gain is about 40 dB.—A. Pshaenich, "Servo Motor Drive Amplifiers," Motorola, Phoenix, AZ, 1972, AN-590.

OPERATIONAL-INTEGRATOR PHASE SHIFTER—Motorola MPSA13 Darlington IC provides 90° phase shift required in servoamplifier for 115-V 60-Hz servomotor. Two cascaded 2N3904 discrete Darlingtons can be used in place of IC.—A. Pshaenich, "Servo Motor Drive Amplifiers," Motorola, Phoenix, AZ, 1972, AN-590.

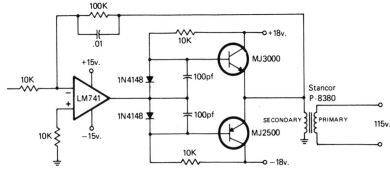

20 W AT 60 Hz—Adding high-current complementary transistors to opamp gives servoamplifier with 115-V output. Opamp drives low impedance of 10-V filament transformer connected in reverse to boost output to 115 V for driving servo. Use heatsink for transistors.

Bringing opamp feedback resistor to actual output point makes nonlinearities and crossover point between transistors insignificant by placing them in feedback loop.—I. Math, Math's Notes, *CQ*, Jan. 1978, p 53–54 and 70.

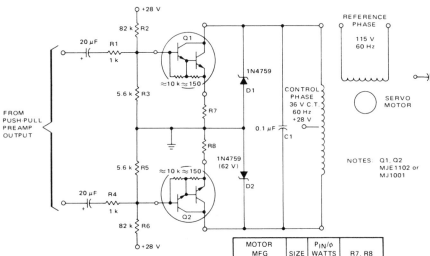

MOTOR MFG	SIZE	P$_{IN}$/ø WATTS	R7, R8
WESTON 11 MA2 U-211663	11	4	6.8 Ω 1/2W
DAYSTROM U-207263	18	10	3.3 Ω 1W

28-V PUSH-PULL POWER AMPLIFIER—Power Darlingtons are used in common-emitter configuration to give high current gain for driving control phase of 60-Hz servo while providing high input impedance for preamp. No transformers are required. Darlingtons require heatsinks. Suitable for driving size 11 servo at 4 W and size 18 at 10 W if emitter resistors R7 and R8 are changed as in table.—A. Pshaenich, "Servo Motor Drive Amplifiers," Motorola, Phoenix, AZ, 1972, AN-590.

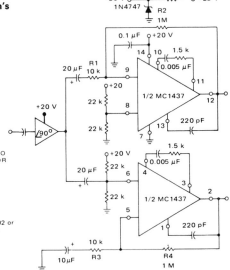

PARALLEL-OPAMP PREAMP—Provides differential output required for driving power amplifier of 115-V 60-Hz servomotor. One opamp section is connected inverting and the other noninverting to give required complementary outputs. Voltage gain is 40 dB, operating from single 20-V zener-regulated supply. High DC feedback gives excellent DC stability. Bandwidth is about 6 kHz. Input is driven by 90° phase shifter.—A. Pshaenich, "Servo Motor Drive Amplifiers," Motorola, Phoenix, AZ, 1972, AN-590.

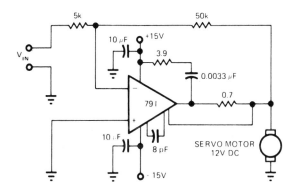

12-VDC DRIVE—Circuit uses 791 power opamp in inverting configuration with gain of 10 for driving size 8 12-VDC servomotor in either direction. Article tells how to calculate heatsink requirements for opamp.—R. J. Apfel, Power Op Amps—Their Innovative Circuits and Packaging Provide Designers with More Options, *EDN Magazine*, Sept. 5, 1977, p 141–144.

DIFFERENTIAL INPUT AND OUTPUT—Preamplifier for servosystem uses 90° operational integrator to drive MC1420 opamp having differential input and differential output connected in inverting configuration. With values shown, voltage gain is about 38 dB. Bandwidth is about 4 kHz, giving stability when using 510-pF compensating capacitors. Zener provides 12 V required for opamp operation from single supply.—A. Pshaenich, "Servo Motor Drive Amplifiers," Motorola, Phoenix, AZ, 1972, AN-590.

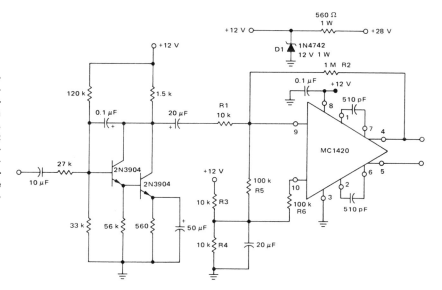

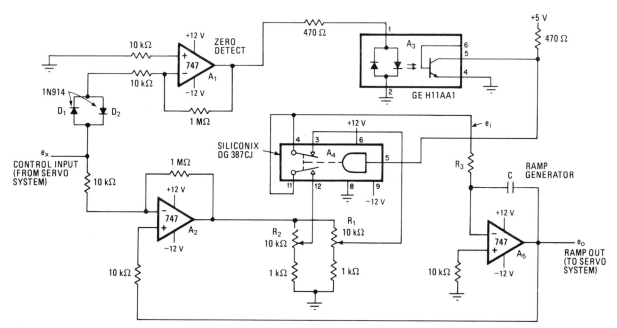

UP/DOWN RAMP CONTROL—Siliconix DG387CJ solid-state relay A_4 provides switching from up ramp to down ramp for decelerating servo when it zeroes in on correct new position. Slopes are determined by settings of R_1 and R_2. Arrangement ensures optimum servo system response at low cost. A_1 detects that input is other than 0 V and energizes optoisolator A_3 for switching A_4. Resulting positive-going ramp from A_5 moves system load toward desired position, making feedback voltage of servo reduce control-input voltage. When this drops to within 0.7 V of ground, A_1 goes low and A_3 turns off. A_4 now initiates down-ramp waveform to decelerate system to stop. For ramp rate of 20 V/s, C can be 0.33 μF and R_3 1.8 megohms.—R. E. Kelly, Up-Down Ramp Quickens Servo System Response, *Electronics*, July 20, 1978, p 121 and 123.

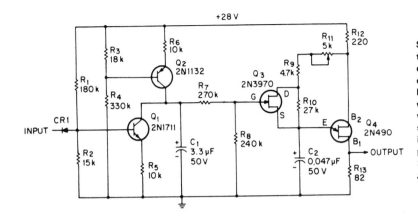

STEP-SERVO CONTROL—Variable UJT oscillator generates train of pulses under control of digital input logic levels, at 1000 pulses per second for logic 1 or 4400 pulses per second for logic 0, with smooth transitions between rates when logic changes, for driving stepping servomotor. Q_1 and Q_2 are constant-current sources. JFET Q_3 acts as voltage-controlled variable resistor in parallel with R_{10}, controlling pulse rate of UJT oscillator Q_4.—C. R. Forbes, Step-Servo Motor Slew Generator, *EEE Magazine*, Oct. 1970, p 76–77.

TWO-PHASE SERVO DRIVE—Both sections of National LM377 power amplifier are connected to provide up to 3 W per phase for driving small 60-Hz two-phase servomotor. Power is sufficient for phonograph turntable drive. Lamp is used in simple amplitude stabilization loop. Motor windings are 8 ohms, tuned to 60 Hz with shunt capacitors.—"Audio Handbook," National Semiconductor, Santa Clara, CA, 1977, p 4-8–4-20.

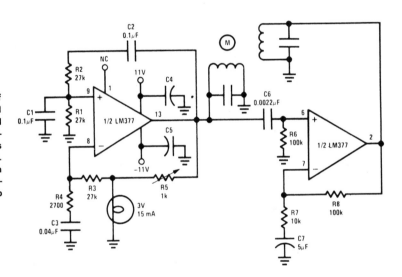

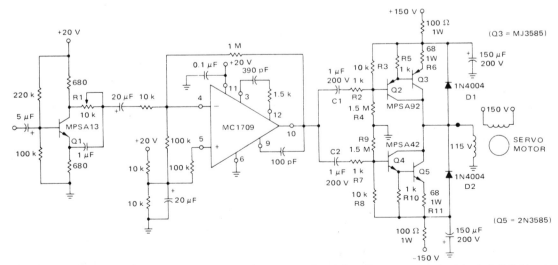

LINE-OPERATED AMPLIFIER—Push-pull RC phase shifter, single-ended preamp, and push-pull class B power amplifier all obtain supply voltages from AC supply that can either use power transformer or operate directly from line with diode rectifiers. Power output is enough to drive size 18 servomotor at 10 W. Larger servomotors can be used if reduced supply voltages can be tolerated. Suitable power supply circuits are given.—A. Pshaenich, "Servo Motor Drive Amplifiers," Motorola, Phoenix, AZ, 1972, AN-590.

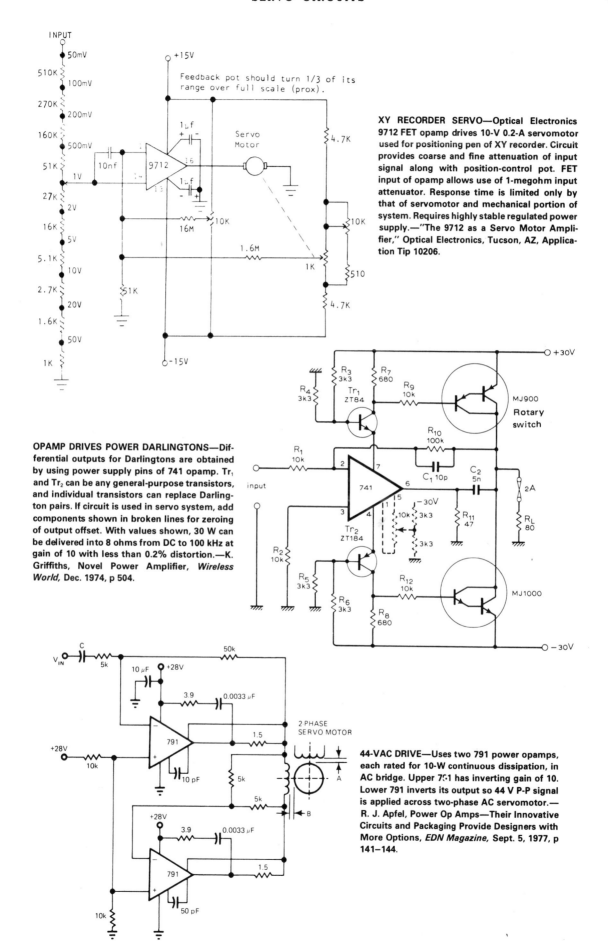

XY RECORDER SERVO—Optical Electronics 9712 FET opamp drives 10-V 0.2-A servomotor used for positioning pen of XY recorder. Circuit provides coarse and fine attenuation of input signal along with position-control pot. FET input of opamp allows use of 1-megohm input attenuator. Response time is limited only by that of servomotor and mechanical portion of system. Requires highly stable regulated power supply.—"The 9712 as a Servo Motor Amplifier," Optical Electronics, Tucson, AZ, Application Tip 10206.

OPAMP DRIVES POWER DARLINGTONS—Differential outputs for Darlingtons are obtained by using power supply pins of 741 opamp. Tr_1 and Tr_2 can be any general-purpose transistors, and individual transistors can replace Darlington pairs. If circuit is used in servo system, add components shown in broken lines for zeroing of output offset. With values shown, 30 W can be delivered into 8 ohms from DC to 100 kHz at gain of 10 with less than 0.2% distortion.—K. Griffiths, Novel Power Amplifier, *Wireless World,* Dec. 1974, p 504.

44-VAC DRIVE—Uses two 791 power opamps, each rated for 10-W continuous dissipation, in AC bridge. Upper 791 has inverting gain of 10. Lower 791 inverts its output so 44 V P-P signal is applied across two-phase AC servomotor.—R. J. Apfel, Power Op Amps—Their Innovative Circuits and Packaging Provide Designers with More Options, *EDN Magazine,* Sept. 5, 1977, p 141–144.

SECTION 58
Switching Regulator Circuits

Covers regulators in which DC input voltage is converted to pulse-width-modulated frequency in range of 9–100 kHz, with duty cycle or frequency being varied automatically to maintain essentially constant output voltage at desired value. Circuit may use discrete components or switching-regulator IC.

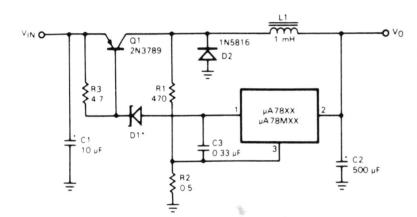

5–24 V SWITCHING—Choice of regulator in μA7800 series determines fixed output voltage. Devices are available for rated outputs of 5, 6, 8, 12, 15, 18, and 24 V, positive or negative, with output current ratings of 100 mA, 500 mA, or 1 A. If input voltage is greater than maximum input rating of regulator used, add voltage-dropping zener D1 to bring voltage between pins 1 and 3 down to acceptable level.—"Signetics Analog Data Manual," Signetics, Sunnyvale, CA, 1977, p 668.

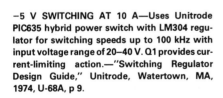

–5 V SWITCHING AT 10 A—Uses Unitrode PIC635 hybrid power switch with LM304 regulator for switching speeds up to 100 kHz with input voltage range of 20–40 V. Q1 provides current-limiting action.—"Switching Regulator Design Guide," Unitrode, Watertown, MA, 1974, U-68A, p 9.

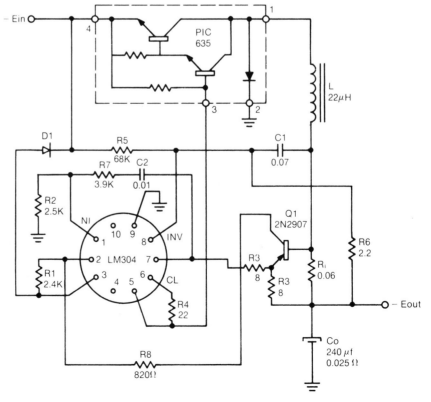

460

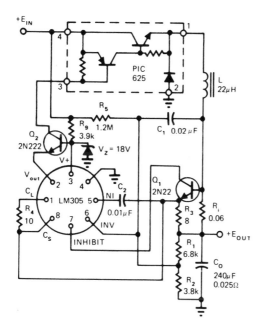

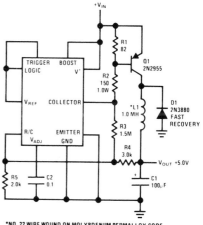

5-V FIXED OFF-TIME SWITCHING—Uses LM305 regulator and Unitrode hybrid power switch in PIC600 series. Operates in fixed OFF-time mode. Output ripple of 100 mV P-P is independent of input voltage range of 20 to 40 V for output of +5 V ± 1%. Switching speed is nominally 50 kHz but can go up to 100 kHz. Article covers theory of operation in detail.—L. Dixon and R. Patel, Designers' Guide to: Switching Regulators, *EDN Magazine*, Oct. 20, 1974, p 53–59.

*NO. 22 WIRE WOUND ON MOLYBDENUM PERMALLOY CORE

5 V AT 1 A—National LM122 timer is connected as switching regulator by using internal reference and comparator to drive switching transistor Q1. Minimum input voltage is 5.5 V. Line and load regulation are less than 0.5%, and output ripple at switching frequency is only 30 mV. Output voltage can be adjusted between 1 V and 30 V by using appropriate values for R2-R5.—C. Nelson, "Versatile Timer Operates from Microseconds to Hours," National Semiconductor, Santa Clara, CA, 1973, AN-97, p 9.

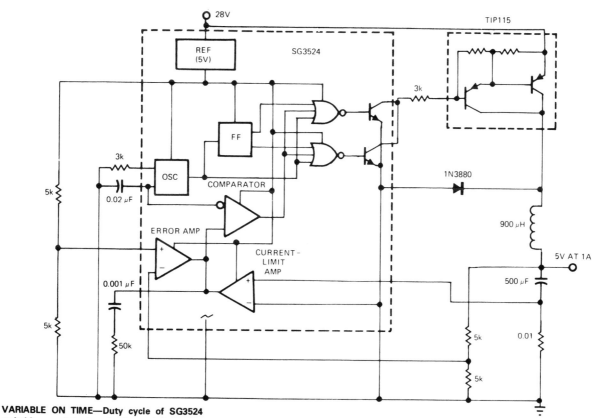

VARIABLE ON TIME—Duty cycle of SG3524 switching regulator is varied by modulating ON time while maintaining constant switching fre- quency, using pulse-duration-modulation con- trol circuit.—J. Spencer, Monolithic Switching Regulators—They Fit Today's Power-Supply Needs, *EDN Magazine*, Sept. 5, 1977, p 117–121.

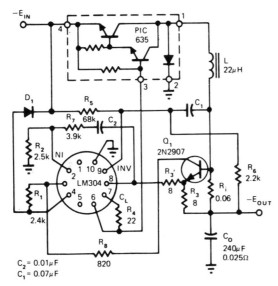

–10 V SWITCHING—Uses LM304 regulator and Unitrode hybrid power switch in PIC600 series to provide output of 10 A. R_1 and R_2 determine reference voltage. Current limiting is achieved by reducing reference voltage to ground instead of turning off base drive to power output switch. Article covers operating theory.—L. Dixon and R. Patel, Designers' Guide to: Switching Regulators, *EDN Magazine*, Oct. 20, 1974, p 53–59.

STEPPING 5 V UP TO 15 V—Fairchild μA78S40 switching regulator transforms 5 V to 15 V at efficiency of 80% for 150-mA load. Average input current is only 550 mA. Article gives design equations.—R. J. Apfel and D. B. Jones, Universal Switching Regulator Diversifies Power Subsystem Applications, *Computer Design*, March 1978, p 103–112.

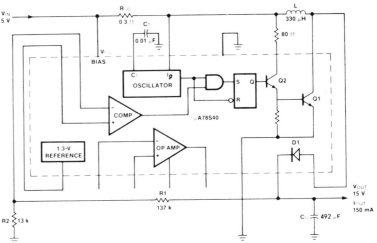

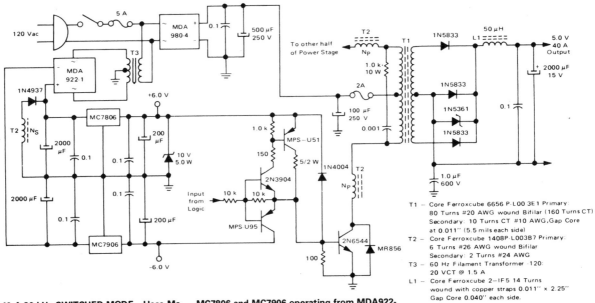

5-V 40-A 20-kHz SWITCHED-MODE—Uses Motorola 2N6544 power transistors operating with 3-A collector current (other half of power stage is identical). Bridge rectifier and capacitive filter connected directly to AC line form 150-VDC supply for inverter operating at 20 kHz. Regulators MC7806 and MC7906 operating from MDA922-1 bridge rectifier of 15-W filament transformer T3 provide ±6 V for logic circuits that provide pulse-width modulation for inverter. When logic signal is high, MPS-U51 saturates and supplies 1 A to base of 2N6544 inverter power transistor. When logic is low, MPS-U95 Darlington holds inverter transistor off.—R. J. Haver, "Switched Mode Power Supplies—Highlighting a 5-V, 40-A Inverter Design," Motorola, Phoenix, AZ, 1977, AN-737A, p 10.

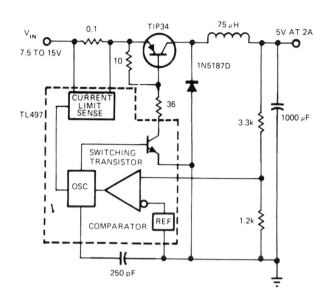

VARIABLE SWITCHING FREQUENCY—TL497 switching regulator operates at maximum frequency under maximum load conditions. For smaller loads, duty cycle is varied automatically by maintaining fixed ON time and varying switching frequency. Circuit optimizes efficiency at about 75% by reducing switching losses as load decreases.—J. Spencer, Monolithic Switching Regulators—They Fit Today's Power-Supply Needs, *EDN Magazine*, Sept. 5, 1977, p 117–121.

**−5 V FLYBACK SWITCHING—Uses Unitrode PIC625 regulator operating at 25 kHz and TL497 control circuit operating in current-limiting mode to give line and load regulation of 0.2% for input voltage of 12 V ±25%. Efficiency is 75%. Short-circuit current is automatically limited to 3 A.—"Flyback and Boost Switching Regulator Design Guide," Unitrode, Watertown, MA, 1978, U-76, p 5.

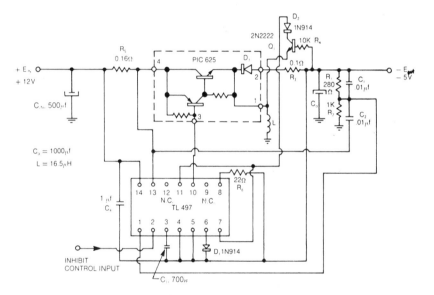

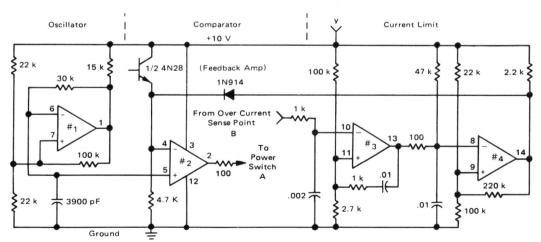

CONTROL FOR SWITCHING REGULATOR— Uses all four sections of Motorola MC3302 quad comparator. First section is connected as 20-kHz oscillator that supplies sawtooth output sweeping between voltage limits set by 100K positive feedback resistor and 15-V supply. Section 2 compares sawtooth output to feedback signal, to produce variable-duty-cycle output pulse for power switch of switching regulator. Sections 3 and 4 initiate current-limiting action; section 3 senses overcurrent and triggers section 4 connected as mono MVBR. Limiting occurs at about 4 A. When load short is removed, regulator resets automatically. Point A goes to push-pull drive for power switch of regulator, and point B goes to current-sensing resistor in output circuit of regulator. Point y goes to 10-V supply.—R. J. Haver, "A New Approach to Switching Regulators," Motorola, Phoenix, AZ, 1975, AN-719, p 7.

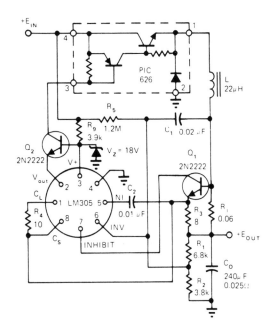

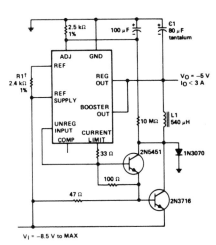

HIGH-VOLTAGE POSITIVE SWITCHING—Uses 18-V zener in series with 3.9K resistor to provide power for LM305 IC regulator. Q₂ provides base drive for PIC626 hybrid power switch and isolates output of LM305 from switch.—L. Dixon and R. Patel, Designers' Guide to: Switching Regulators, *EDN Magazine*, Oct. 20, 1974, p 53–59.

−5 V AT 3 A SWITCHING—Negative-voltage regulator using SN52104 or SN72304 accepts input voltage range of −8.5 V to −40 V and provides regulated output of −5 V with typical load regulation of 1 mV and input regulation of 0.06%. ICs are interchangeable with LM104 and LM304 respectively. L1 is 60 turns No. 20 on Arnold Engineering A930157-2 molybdenum permalloy core or equivalent.—"The Linear and Interface Circuits Data Book for Design Engineers," Texas Instruments, Dallas, TX, 1973, p 5-5.

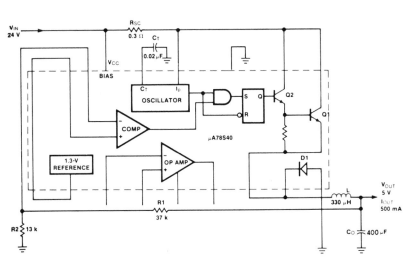

REDUCING 24 V TO 5 V—Uses Fairchild μA78S40 switching regulator having variety of internal functions that can provide differing voltage step-up, step-down, and inverter modes by appropriately connecting external components. Connections shown provide step-down from 24 V to 5 V at 500 mA with 83% efficiency. Applications include running TTL from 24-V battery. Output ripple is less than 25 V. Article gives design equations.—R. J. Apfel and D. B. Jones, Universal Switching Regulator Diversifies Power Subsystem Applications, *Computer Design*, March 1978, p 103–112.

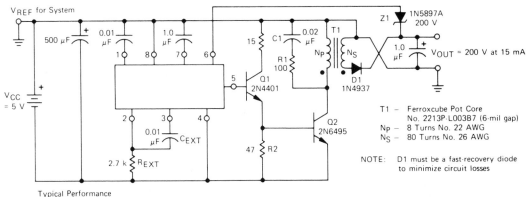

Typical Performance

Line Reg. (4 V < V$_{IN}$ < 6 V) = .3%	Overall Efficiency = 66%
Load Reg. (P$_{OUT}$ ≤ 3 W) = .2%	20-kHz Ripple = .1 V p-p

5 V TO 200 V WITH SWITCHING REGULATOR—Converts standard logic supply voltage to high voltage required by gas-discharge displays, using Motorola MC3380 astable MVBR as control element in switching regulator. Will drive up to 15 digits. Operating frequency is about 20 kHz.—H. Wurzburg, "Control Your Switching Regulator with the MC3380 Astable Multivibrator," Motorola, Phoenix, AZ, 1975, EB-52.

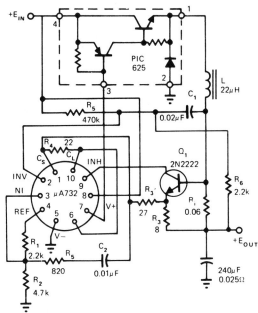

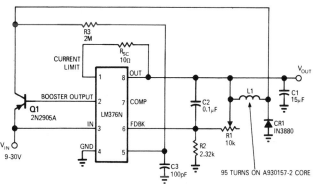

+10 V SWITCHING—Positive switching regulator circuit uses μA732 with Unitrode PIC625 hybrid power switch and single transistor, operating in fixed OFF-time mode. Article covers regulator theory of operation in detail.—L. Dixon and R. Patel, Designers' Guide to: Switching Regulators, *EDN Magazine,* Oct. 20, 1974, p 53–59.

BATTERY REGULATOR—Uses LM376N positive voltage regulator in switching mode to compensate for voltage changes of battery supply during discharge cycle, without adjusting series rheostat. Load regulation is 0.3% for unregulated input of 9 to 30 V, with R1 and R2 setting output voltage anywhere between 5 and 27 V. Maximum output current is 25 mA. Switching frequency of regulator is 33 kHz.—E. R. Hnatek and L. Goldstein, Switching Regulator Designed for Portable Eqiupment, *EDN/EEE Magazine,* Sept. 15, 1971, p 39–41.

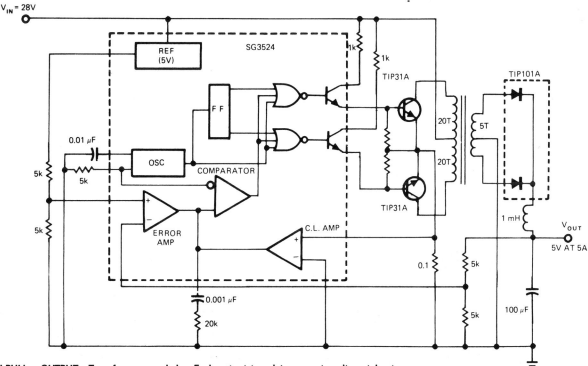

PUSH-PULL OUTPUT—Transformer-coupled push-pull output for SG3524 fixed-frequency pulse-duration-modulated switching regulator gives output flexibility, allowing for multiple outputs and wide range of output voltages. Each output transistor operates alternately at half of switching frequency. Switching regulator applies voltage alternately to opposite ends of transformer primary, making transformer perform as if it had AC input. TIP101A rectifier then provides desired 5-VDC output at 5A.—J. Spencer, Monolithic Switching Regulators—They Fit Today's Power-Supply Needs, *EDN Magazine,* Sept. 5, 1977, p 117–121.

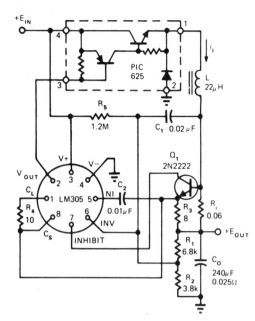

5-V SWITCHING—Fixed OFF-time mode of operation is used in switching regulator design to provide 5-V output that is constant within 100 mV P-P for input range of 20–40 V, for loads ranging from 10 A maximum to 2 A minimum. Switching frequency can be in range of 1–50 kHz. Operation above 20 kHz eliminates possibility of audio noise but with some drop in efficiency. Values shown are for 50 kHz. Article gives design equations and design procedure.—L. Dixon and R. Patel, Designers' Guide to: Switching Regulators, Part 2, *EDN Magazine,* Nov. 5, 1974, p 37–40.

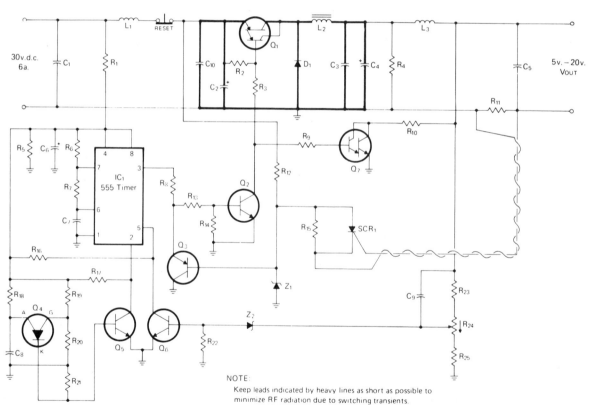

NOTE:
Keep leads indicated by heavy lines as short as possible to minimize RF radiation due to switching transients.

C_1, C_3, C_5, C_{10}—1.0μF, Polycarb	R_8, R_{13}, R_{23}—1.2K, ½ W
C_2, C_6—100μF, 50V	R_9—15K, ½ W
C_4—1000μF, 50V	R_{10}—20Ω, 10W
C_7—0.0082μF	R_{11}—0.075Ω, 6 watts
C_8—390pF	R_{12}—1.5K, 1W
C_9—0.002μF	R_{14}—330Ω, ½ W
D_1—1N3890	R_{15}, R_{19}—680Ω, ½ W
L_1, L_3—10μhy, 10 amps	R_{16}—22K, ½ W
L_2—180μhy	R_{17}—4.7K, ½ W
Q_1—D45E2 (General Electric)—	R_{18}—120K, ½ W
Q_2, Q_5—D33D25	R_{20}—1K, ½ W
Q_3—D29E25	R_{21}—100Ω, ½ W
Q_4—2N6027	R_{22}—18K, ½ W
Q_6—D32S4	R_{24}—1K, 1W Pot.
Q_7—D40K2—Use Thermalloy	R_{25}—390Ω, ½ W
6063B heatsink	SCR-1—C103B
R_1, R_3, R_4, R_5—1.2K, ½ W	Z_1—1N5233B
R_2, R_7—110Ω, ½ W	Z_2—1N5226B
R_6—4.7K, ½ W	IC-1—555 Timer

150-W SWITCH-MODE—Unregulated DC voltage is applied to power Darlington Q_1 serving as switch that chops voltage so rectangular waveform is applied to RLC output filter. Average voltage to filter depends on duty cycle of switch. 555 timer operates in mono MVBR mode as pulse generator and pulse-duration modulator. R_{24} applies varying voltage to pin 5 to modulate pulse duration linearly with respect to applied voltage. Actions of Q_1, Q_2, and Q_6 maintain constant 3.6 V at arm of control pot.

Q_4 and Q_5 provide 20-kHz clock pulse, above audible range. Overcurrent protection of transistors is provided by R_{11}, SCR, and Q_3. Adjust R_{11} so SCR turns on and shuts down circuit when current through R_{11} reaches 8 A. Circuit must be reset manually after overload. Q_7 and R_{10} load circuit to prevent oscillation at low output voltage and light load.—R. J. Walker, A 150 Watt Switch-Mode Regulator, *CQ,* March 1977, p 40–43 and 74–75.

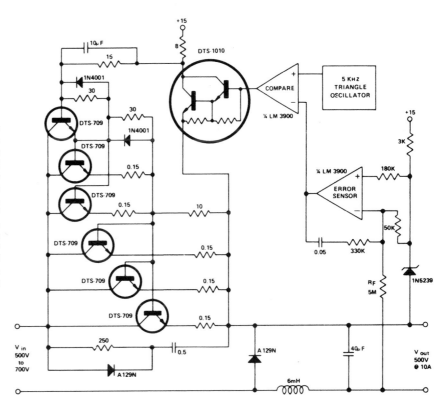

5-kW SWITCHING—Six Delco DTS-709 transistors are connected in progressive Darlington configuration to provide stable and efficient switching at high voltages. Can be operated from 480-V three-phase full-wave rectified line to minimize filter cost. Control circuit uses one LM3900 IC operating from isolated 15-V supply, along with 5-kHz triangle oscillator and error sensor feeding into comparator. In power stage, one DTS-709 drives two DTS-709s which drive three DTS-709s. Efficiency is better than 90% for all loads above 500 W.—"Economical 5 kW Switching Regulator Using DTS-709 Transistors," Delco, Kokomo, IN, 1974, Application Note 56.

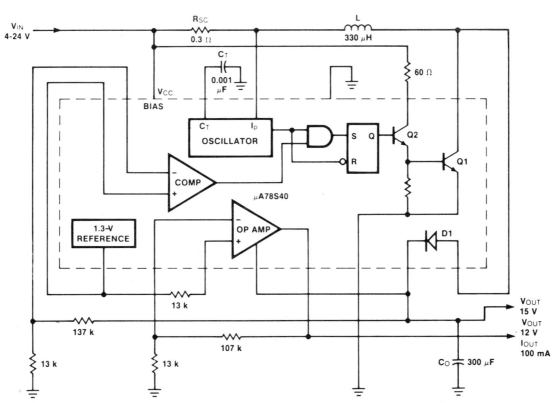

+12 V AND +15 V FROM 4–24 V—Connections shown for Fairchild μA78S40 switching regulator give universal regulator providing either step-up or step-down, for loads up to 100 mA. Efficiency is about 50% for input extremes of 4 and 24 V, increasing to maximum of 75% for other input voltages. Output ripple is essentially eliminated at 12-V output.—R. J. Apfel and D. B. Jones, Universal Switching Regulator Diversifies Power Subsystem Applications, *Computer Design*, March 1978, p 103–112.

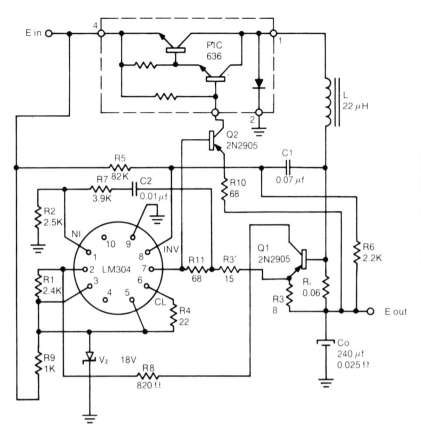

HIGH-VOLTAGE NEGATIVE-SWITCHING—De-signed for operation from supply voltages above maximum of −40 V for LM304 regulator. Output is −5 V at up to 10 A. Q2 provides voltage isolation between regulator and Unitrode PIC636 hybrid power switch. R9 limits current through zener under steady-state and start-up conditions.—"Switching Regulator Design Guide," Unitrode, Watertown, MA, 1974, U-68A, p 9.

6 V FOR CALCULATOR—Can be mounted in housing of calculator or small transistor radio, for operation from AC line. D_1 and D_2 produce 15 VDC across filter capacitor C_2 as supply for inverter Tr_1 operating at 13 kHz. Transformer is wound with No. 37 wire on small core such as Phillips P14/8 337 pot core. Primary windings are bifilar. Use grounded shield to reduce radiated switching noise.—M. Faulkner, Miniature Switch Mode Power Supply, *Wireless World,* Oct. 1977, p 65.

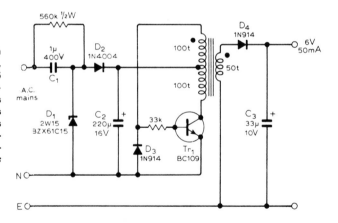

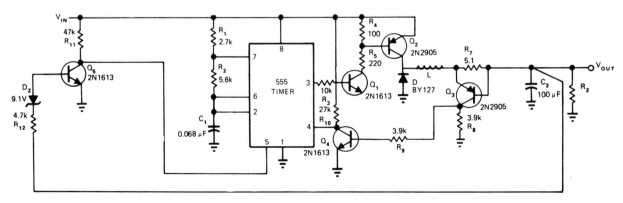

10-V SWITCHING AT 100 mA—Use of 555 timer as pulse-width-modulated regulator gives line regulation of 0.5% and load regulation of 1%.   Circuit includes current foldback. With 15-V input, output is 10 V.—P. R. K. Chetty, Put a 555 Timer in Your Next Switching Regulator Design, *EDN Magazine,* Jan. 5, 1976, p 72.

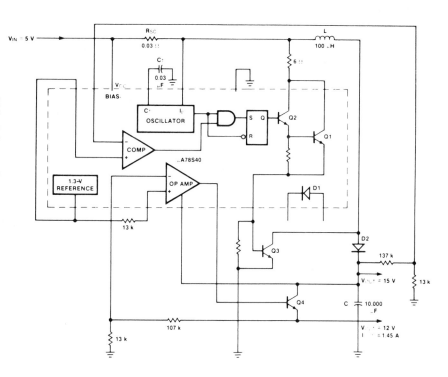

+12 V AND +15 V FROM 5 V—Uses Fairchild μA78S40 switching regulator having variety of internal functions that can provide differing voltage step-up, step-down, and inverter modes by appropriately connecting external components. External NPN transistor Q3 boosts step-up regulator, and NPN transistor Q4 increases series-pass regulator output well above 1 A. Total of 1.5 A is available from two outputs. Transistor and diode types are not critical. Efficiency is 80% for 15-V output and 64% for 12-V output.—R. J. Apfel and D. B. Jones, *Universal Switching Regulator Diversifies Power Subsystem Applications, Computer Design,* March 1978, p 103–112.

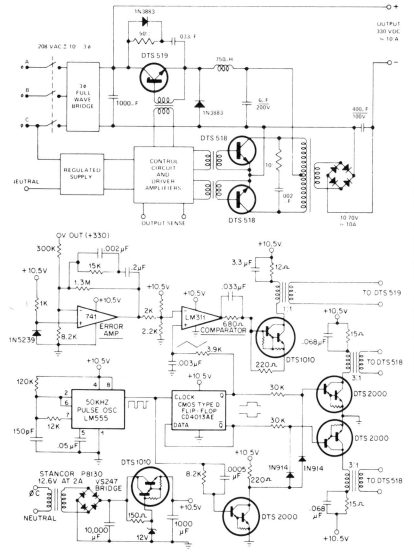

3.3-kW SWITCHING—Delco DTS-518 and DTS-519 power transistors in high-efficiency stacked supply are operated at 25-kHz switching rate to provide 330 VDC at 10 A. Control circuit operates at primary 50-kHz pulse frequency, with negative-going pulses having 2-μs duration. Flip-flop converts this to 25-kHz complementary square-wave signal driving Darlington DTS-2000s. Transformer cores are Magnetics EE No. 42510 each having 15-turn primary and 5-turn secondary for driving DTS-518s. Error amplifier compares portion of total output voltage to zener reference for control of DTS-519 power transistor switching at 25 kHz. Efficiency is 95% at full load.—"3.3kW High Efficiency Switch Mode Regulator," Delco, Kokomo, IN, 1977, Application Note 59.

4.5–30 V SWITCHING AT 6 A—LM105 positive regulator serves as amplifier-reference for LM195 power transistor IC in switching regulator. Duty cycle of switching action adjusts automatically to give constant output. Q2 consists of four LM195s in parallel since each is rated at only about 2 A. R8 serves as output voltage control.—"Linear Applications, Vol. 2," National Semiconductor, Santa Clara, CA, 1976, AN-110, p 4.

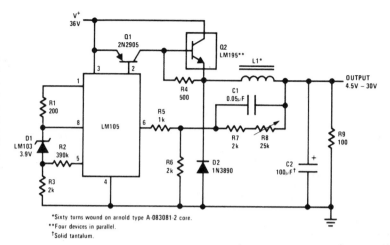

*Sixty turns wound on arnold type A-083081-2 core.
**Four devices in parallel.
†Solid tantalum.

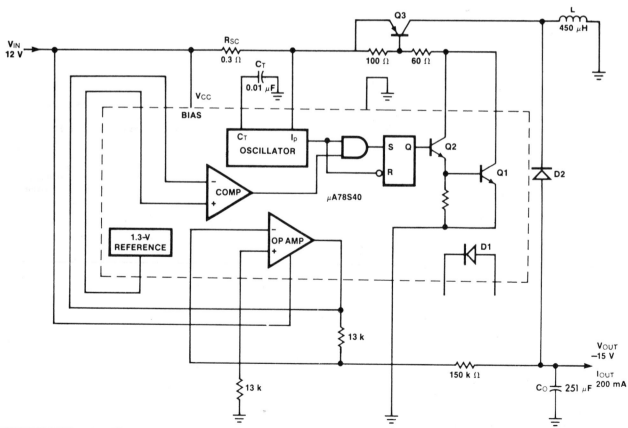

TRANSFORMING +12 V TO −15 V—External PNP transistor Q3 and catch diode D2 (types not critical) are used with Fairchild μA78S40 switching regulator so no pin of IC substrate has voltage more negative than substrate, which is grounded. Efficiency is 84% with 200-mA load. Output voltage ripple is 50 mV but can be reduced by increasing value of C_0.—R. J. Apfel and D. B. Jones, Universal Switching Regulator Diversifies Power Subsystem Applications, *Computer Design,* March 1978, p 103–112.

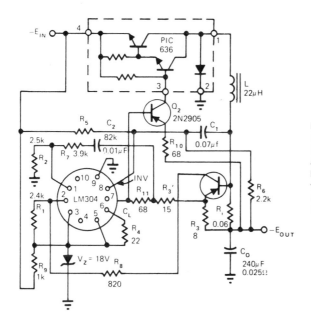

HIGH-VOLTAGE NEGATIVE SWITCHING—Uses zener to reduce supply voltage to acceptable level for LM304 IC regulator. Base drive and voltage isolation are provided by Q_2, R_{10}, and R_{11} for PIC636 hybrid power switch. Circuit operates in fixed OFF-time mode.—L. Dixon and R. Patel, Designers' Guide to: Switching Regulators, *EDN Magazine,* Oct. 20, 1974, p 53–59.

−12 V AT 300 mA FROM −48 V—Uses Fairchild μA78S40 switching regulator having variety of internal functions that can provide differing voltage step-up, step-down, and inverter modes by appropriately connecting external components. Efficiency is 86%, and output ripple is 300 mV. Extra opamp on chip is used to derive required reference voltage of −2.6 V from internal 1.3-V reference.—R. J. Apfel and D. B. Jones, Universal Switching Regulator Diversifies Power Subsystem Applications, *Computer Design,* March 1978, p 103–112.

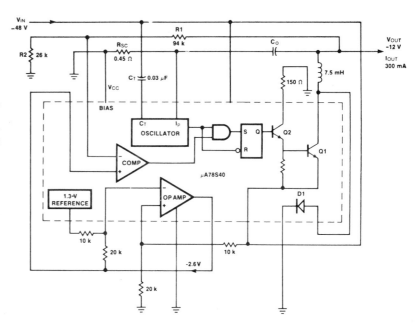

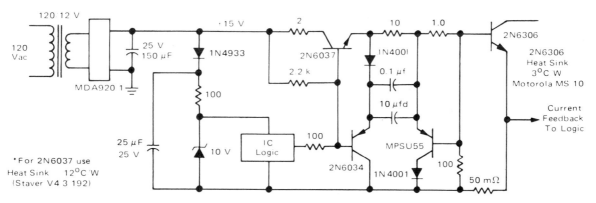

POWER SWITCH FOR SWITCHING REGULA-TOR—Circuit operating from 12-V step-down transformer includes push-pull driver providing interface between logic drive signal and 2N6306 high-voltage power transistor. Switching is provided at 3 A and 20 kHz, with artificial negative bias supply created from single positive supply to improve fall time. Current limiting is added to base current to limit overdrive and reduce storage time. Power switch is turned off by forcing IC to logic low. Used in 24-V 3-A switching-mode power supply operating from AC line.—R. J. Haver, "A New Approach to Switching Regulators," Motorola, Phoenix, AZ, 1975, AN-719, p 5.

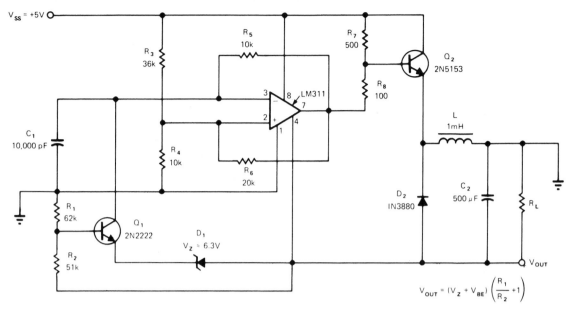

+5 V TO −15 V—Use of switching regulator for voltage conversion permits generation of higher output voltage along with polarity reversal. LM311 operates as free-running MVBR with low duty cycle. Frequency is determined by C_1 and R_5 and duty cycle by divider R_3-R_4. Extra loop function performed by Q_1 and zener operating in conjunction with resistor network modifies oscillator duty cycle until desired output level is obtained. Nominal frequency is 6 kHz, duty cycle is 20% for −15 V output, and maximum load current is 200 mA. Design equations are given.—H. Mortensen, IC Comparator Converts +5 to −15V DC, *EDN Magazine,* Dec. 20, 1973, p 78–79.

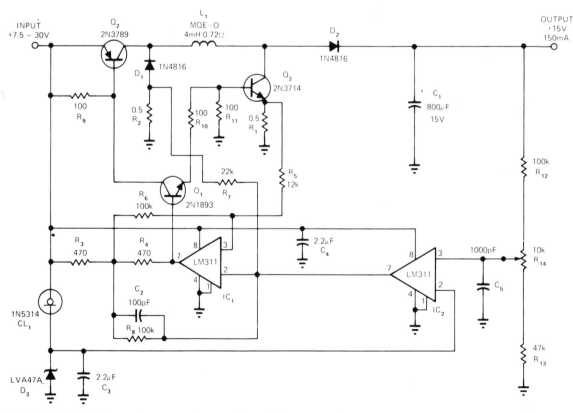

15 V FROM 7.5–30 V—Switching regulator operation is independent of input voltage level. When power is applied, Q_1 conducts and turns on Q_2 and Q_3. When linear rising current of Q_1 exceeds upper threshold as sensed by R_1, IC_1 switches to low output state and turns off all three transistors. Voltage across L_1 reverses, and current flows into C_1 through D_1 and D_2. When this current as sensed by R_2 falls below lower threshold, IC_1 switches back to its high output state. This oscillating action continues until output voltage as sensed by IC_2 rises above desired level, when IC_2 switches to its low output state and holds IC_1 low until output drops back below preset level to complete one cycle of oscillation.—A. Delagrange, Voltage Regulator Can Have Same Input and Output Level, *EDN Magazine,* Aug. 5, 1973, p 87 and 89.

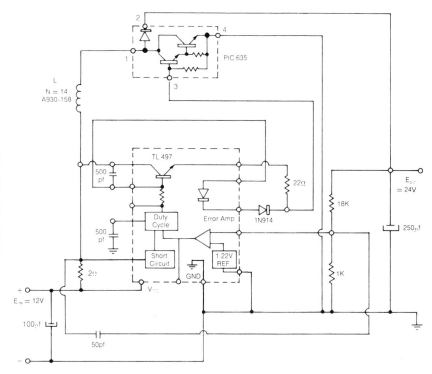

+24 V FROM +12 V AT 2 A—Combination of PIC635 boost switching regulator and TL497 control circuit accepts DC input voltage and provides regulated output voltage that must be greater than input voltage. When transistor switch is turned on, input voltage is applied across L. When transistor is turned off, energy stored in L is transferred through diode to load where it adds to energy transferred directly from input to output during diode conduction time. Output voltage is regulated by controlling duty cycle.—"Flyback and Boost Switching Regulator Design Guide," Unitrode, Watertown, MA, 1978, U-76, p 9.

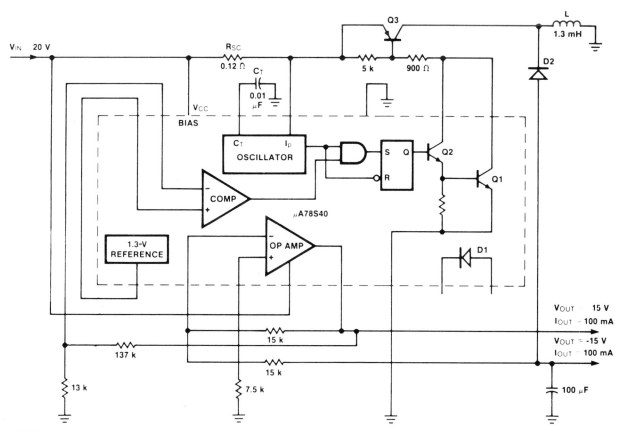

±15 V TRACKING—Dual-tracking connection for Fairchild μA78S40 switching regulator operates from single 20-V input. Efficiency is 75% for +15 V and 85% for −15 V, both at 100 mA. Output ripple is 30 mV.—R. J. Apfel and D. B. Jones, Universal Switching Regulator Diversifies Power Subsystem Applications, *Computer Design,* March 1978, p 103–112.

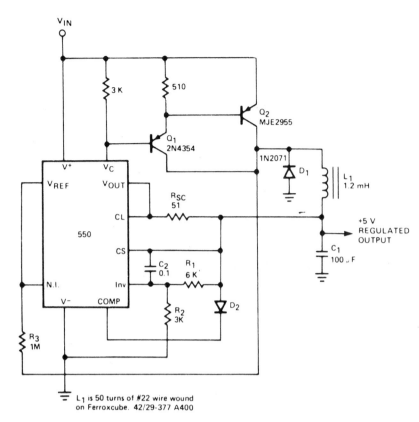

5-V SWITCHING—Darlington pair Q_1-Q_2 serves as switch for regulator using Signetics 550 as threshold detector. Design equations are given. Exact frequency of self-oscillating switching regulator depends primarily on parasitic components. If frequency is important, as in applications requiring EMI suppression, regulator may be locked to external square-wave drive signal fed to reference terminal.—"Signetics Analog Data Manual," Signetics, Sunnyvale, CA, 1977, p 661–662.

L_1 is 50 turns of #22 wire wound on Ferroxcube. 42/29-377 A400

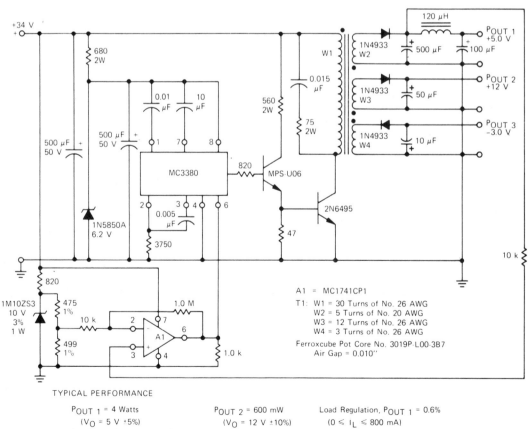

TYPICAL PERFORMANCE

$P_{OUT\ 1}$ = 4 Watts
(V_O = 5 V ±5%)
5-V Ripple Component = 50 mV
(120 Hz + 20 kHz)

$P_{OUT\ 2}$ = 600 mW
(V_O = 12 V ±10%)
$P_{OUT\ 3}$ = 3 mW
(V_O = 3 V ±10%)

Load Regulation, $P_{OUT\ 1}$ = 0.6%
(0 ≤ I_L ≤ 800 mA)

MULTIPLE-OUTPUT SWITCHING REGULATOR—Additional outputs are obtained from switching regulator by adding secondary windings to power transformer. Motorola MC3380 astable MVBR serves as control element. Feedback is achieved by amplifying output error with opamp A1 and applying this voltage to pin 6. Report covers design of transformer and power circuit.—H. Wurzburg, "Control Your Switching Regulator with the MC3380 Astable Multivibrator," Motorola, Phoenix, AZ, 1975, EB-52.

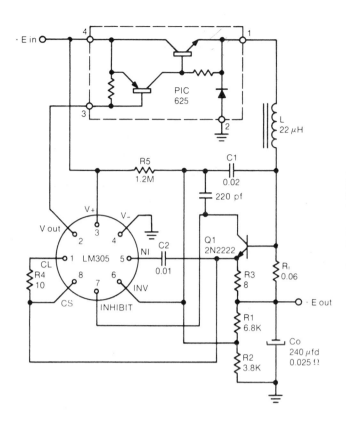

+5 V SWITCHING AT 10 A—Unitrode PIC625 hybrid power switch provides switching action for LM305 regulator at switching speeds up to 100 kHz for input voltage range of 20–40 V. Circuit operates in fixed OFF-time mode that makes output ripple independent of input voltage. Q1 provides current-limiting action.— "Switching Regulator Design Guide," Unitrode, Watertown, MA, 1974, U-68A, p 7.

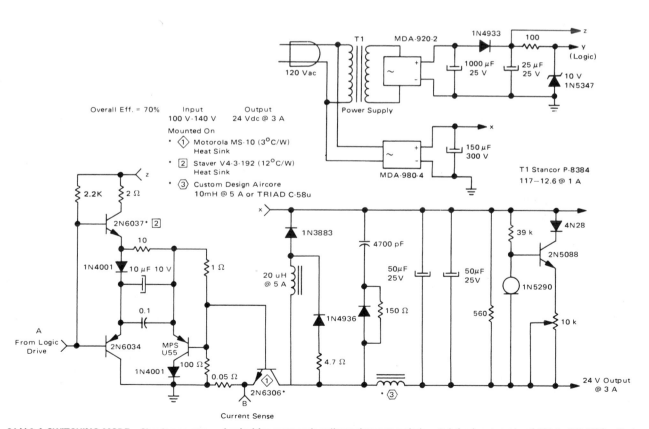

24-V 3-A SWITCHING-MODE—Circuit operates at 20 kHz from AC line with 70% efficiency. Control portion uses quad comparator and optoisolator and provides short-circuit protection. Logic drive uses push-pull transistors to switch 2N6306 power transistor at 20-kHz rate. Load regulation is 0.8% over output range of 1.5 to 3 A with 120-VAC input. Line regulation is 3% at 3 A for input range of 100 to 140 VAC.—R. J. Haver, "A New Approach to Switching Regulators," Motorola, Phoenix, AZ, 1975, AN-719, p 11.

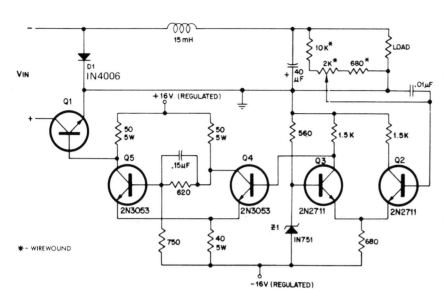

250 V AT 3 A—Single high-voltage silicon power transistor Q1 serves as series element in switching regulator, with regulation obtained by pulse-width modulation. Delco DTS-431 provides output of 250 V for maximum input of 325 V; other Delco transistors in same series give different combinations of output voltage and current in range of 300–750 W maximum output power. Efficiency is 92% at full load. Differential amplifier Q2-Q3 senses output voltage of regulator and feeds Schmitt trigger Q4-Q5 for turning series transistor Q1 on and off. Resulting square wave of voltage is smoothed by LC filter between Q1 and load.—"Pulse Width Modulated Switching Regulator," Delco, Kokomo, IN, 1972, Application Note 39, p 3.

–5 V SWITCHING—Unitrode U2T201 Darlington serves as switching element for LM304 step-down switching regulator operating from input of –25 V. Operating frequency can be about 25 kHz. Darlington will handle peak currents up to 10 A.—"Designer's Guide to Power Darlingtons as Switching Devices," Unitrode, Watertown, MA, 1975, U-70, p 10.

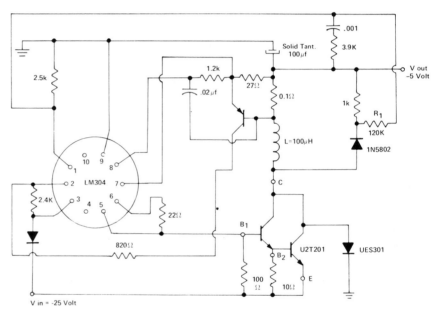

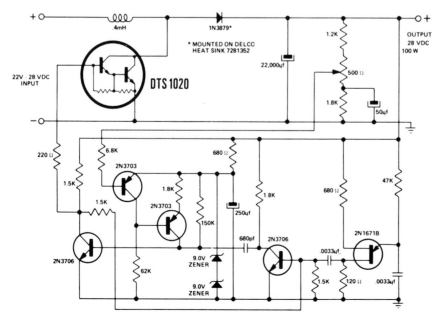

28 V AT 100 W—Circuit using Delco DTS-1020 Darlington silicon power transistor operates over input range of 22–28 V. Switching rate is 9 kHz. Efficiency is about 85% at full load. Output voltage is sensed to control pulse width of mono MVBR which is triggered at 9 kHz by oscillator.—"28 Volt Darlington Switching Regulator," Delco, Kokomo, IN, 1971, Application Note 49, p 4.

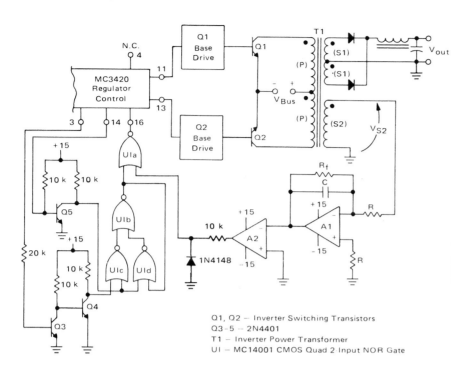

Q1, Q2 — Inverter Switching Transistors
Q3-5 — 2N4401
T1 — Inverter Power Transformer
UI — MC14001 CMOS Quad 2-Input NOR Gate

SYMMETRY CORRECTION—Low-cost external correction circuit for MC3420 switching-mode regulator ensures balanced operation of power transformer in push-pull inverter configuration. Circuit senses voltage impressed on primary of T1 through sensing secondary S2, for integration by opamp A1 so voltage on C represents volt-second product applied to T1. During conduction period of Q1, voltage on C ramps up to some positive value and output of A2 is low. Conduction period for output 2 then begins, Q2 turns on, and C ramps down to 0 V. A2 output then goes high, inhibiting output 2 and Q2. Times for C2 to charge and discharge are equal so conduction periods are equal.—H. Wurzburg, "A Symmetry Correcting Circuit for Use with the MC3420," Motorola, Phoenix, AZ, 1977, EB-66.

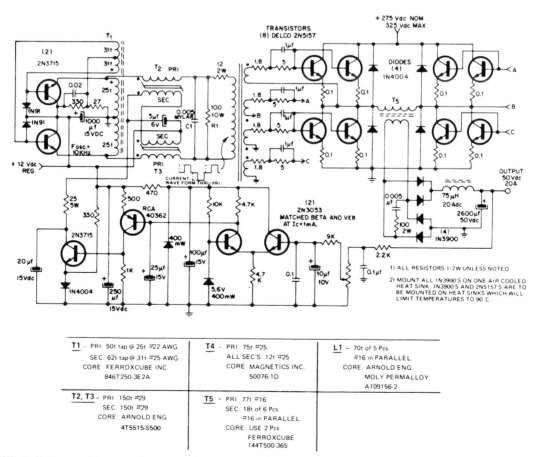

T1 — PRI 50t tap @ 25t #22 AWG.	T4 — PRI 75t #25	L1 — 70t of 5 Pcs.
SEC 62t tap @ 31t #25 AWG.	ALL SEC'S 12t #25	#16 in PARALLEL
CORE FERROXCUBE INC.	CORE MAGNETICS INC.	CORE ARNOLD ENG.
846T250-3E2A	50076-1D	MOLY PERMALLOY
		A109156-2
T2, T3 — PRI 150t #29	**T5** — PRI 77t #16	
SEC 150t #29	SEC 18t of 6 Pcs.	
CORE ARNOLD ENG	#16 in PARALLEL	
4T5515-S500	CORE USE 2 Pcs.	
	FERROXCUBE	
	144T500-365	

50 V AT 1 kW—Switching regulator operating at 10 kHz uses pulse-width modulation to give 87% efficiency at full load. Input voltage is 275 VDC. Inverter output drives combination of eight Delco 2N5157 power transistors con- nected in paralleled pairs in each leg of bridge circuit. Clamp diodes in each bridge leg prevent reverse conduction through collector-base diodes of transistors. Regulator consists of dif- ferential amplifier and two-stage DC amplifier controlling direct current through windings of magnetic amplifier.—"One Kilowatt Regulated Power Converter with the 2N5157 Silicon Power Transistor," Delco, Kokomo, IN, 1972, Application Note 44, p 3.

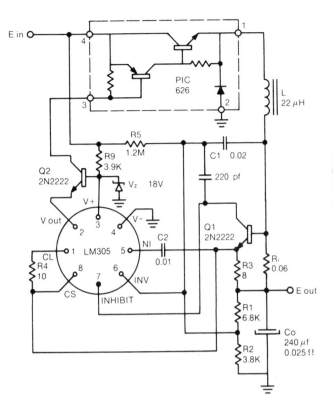

HIGH-VOLTAGE POSITIVE-SWITCHING—Designed for operation from supply voltages above 40-V maximum rating of LM305 regulator. Output is +5 V at up to 10 A. Circuit uses fraction of input voltage as determined by R9 and zener, with Q2 providing voltage isolation between regulator and Unitrode PIC626 hybrid power switch.—"Switching Regulator Design Guide," Unitrode, Watertown, MA, 1974, U-68A, p 9.

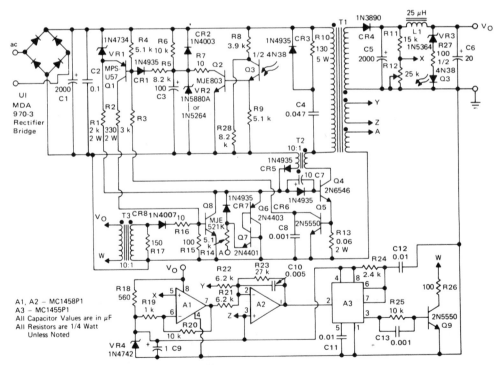

24 V AT 3 A FOR CATV—Switching regulator design meets requirements for cable television systems where small size, low weight, and high efficiency are prime considerations. Circuit operates above 18 kHz either from 40–60 V 60-Hz square-wave source (CATV power line from ferroresonant transformer) or from DC standby source. Control circuit consists of dual opamp and linear IC timer used to vary ON time of 2N6546 power transistor. At start-up, Q4 is saturated and full input voltage is applied to primary of power transformer T1. Current then ramps up linearly until Q4 is switched off by opamps A1 and A2 and timer A3. Power transistor is operated between saturation and OFF state at above 18 kHz, with ON time varied while OFF time is fixed, to maintain constant output voltage as sensed by A1.—J. Nappe and N. Wellenstein, "An 80-Watt Switching Regulator for CATV and Industrial Applications," Motorola, Phoenix, AZ, 1975, AN-752, p 5.

SECTION 59
Telephone Circuits

Includes coders and decoders for standard Touch-Tone pairs of frequencies and for single-tone remote ON/OFF control, along with repeater autopatch circuits, Touch-Tone to dial converter, phone-call counter, ring detector, ring simulator, and busy-signal generator.

DIAL-TONE GENERATOR—Simultaneous pairs of Touch-Tone frequencies used by telephone company are generated by adjustment-free circuit using Motorola Touch-Tone dialer with external 1-MHz crystal. Internal circuits of IC select proper division rates and convert outputs to synthesized sine waves of correct frequencies. Grounding one of row inputs by pressing key gives lower-frequency tone, while grounding one of column inputs gives higher-frequency tone. Special Touch-Tone keyboard provides this grounding action automatically when single key is pressed.—D. Lancaster, "CMOS Cookbook," Howard W. Sams, Indianapolis, IN, 1977, p 239–240.

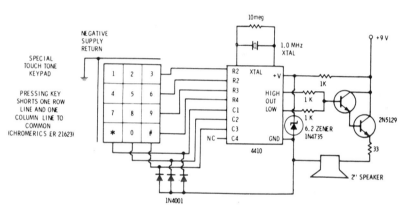

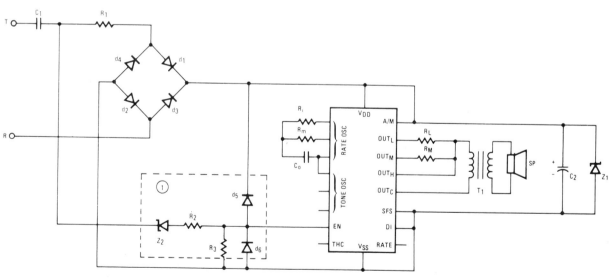

TYPICAL VALUES							
C_1	1µF/200V	R_m	750kΩ	T_1	2000Ω/8Ω TRANSFORMER	Z_2	27V ZENER
R_1	2kΩ	C_0	330pF	SP	8Ω SPEAKER	R_2	150kΩ
d_1 d_4	1N 4004	R_L	18kΩ	C_2	47µF/25V	R_3	300kΩ
R_i	200kΩ	R_M	3.3kΩ	Z_1	12V ZENER 1N4742	d_5, d_6	1N914

BELL SIMULATOR—Uses AMI S2561 CMOS IC to simulate effects of telephone bell by producing tone signal that shifts between two predetermined frequencies at about 16 Hz. In applications where dial pulse rejection is not necessary, network inside dashed lines can be omitted and pins EN and DI connected directly to V_{DD}, which is typically 10 V. Power is derived from telephone lines by diode-bridge supply. Values shown give tone frequencies of 512 and 640 Hz. Power output to 8-ohm loudspeaker is at least 50 mW, fed through 200:8 ohm transformer.—"Tone Ringer," American Microsystems, Santa Clara, CA, 1977, S2561, p 7.

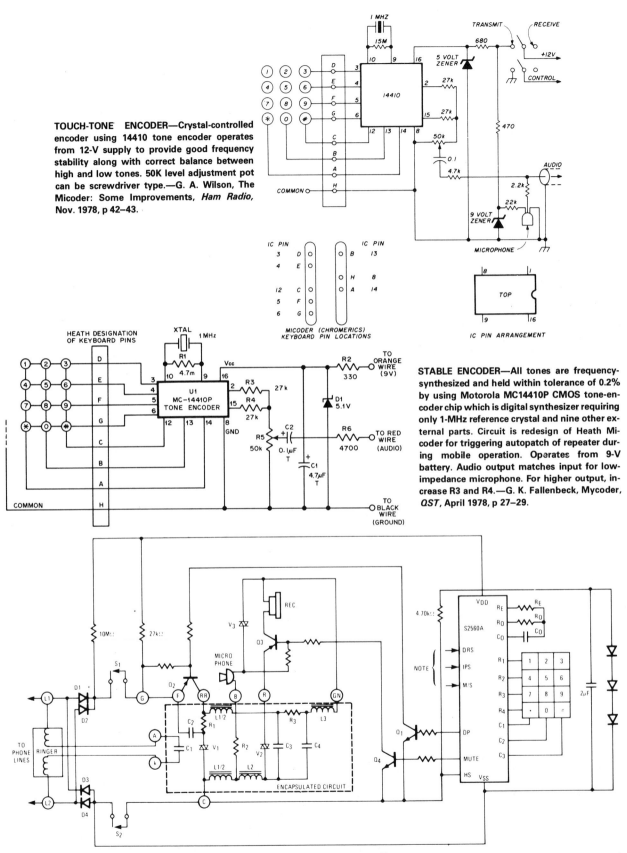

TOUCH-TONE ENCODER—Crystal-controlled encoder using 14410 tone encoder operates from 12-V supply to provide good frequency stability along with correct balance between high and low tones. 50K level adjustment pot can be screwdriver type.—G. A. Wilson, The Micoder: Some Improvements, *Ham Radio*, Nov. 1978, p 42–43.

MICODER (CHROMERICS) KEYBOARD PIN LOCATIONS

IC PIN ARRANGEMENT

HEATH DESIGNATION OF KEYBOARD PINS

STABLE ENCODER—All tones are frequency-synthesized and held within tolerance of 0.2% by using Motorola MC14410P CMOS tone-encoder chip which is digital synthesizer requiring only 1-MHz reference crystal and nine other external parts. Circuit is redesign of Heath Micoder for triggering autopatch of repeater during mobile operation. Operates from 9-V battery. Audio output matches input for low-impedance microphone. For higher output, increase R3 and R4.—G. K. Fallenbeck, Mycoder, *QST*, April 1978, p 27–29.

KEY PULSER—American Microsystems S2560A CMOS IC pulser converts pushbutton inputs to series of pulses suitable for telephone dialing, as replacement for mechanical telephone dial. Circuit shows typical connection to dial telephone set using 500-type encapsulated circuit. Dialing rate can be varied by changing dial rate oscillator frequency. IC includes 20-digit memory that makes last dialed number available for redialing until new number is entered. Entered digits are stored sequentially in internal memory, with dial pulsing starting as soon as first digit is entered. Arrangement permits entering digits much faster than output rate. Last number is redialed by going off hook and pressing # key.—"Key Pulser," American Microsystems, Santa Clara, CA, 1977, S2560A/S2560B, p 8.

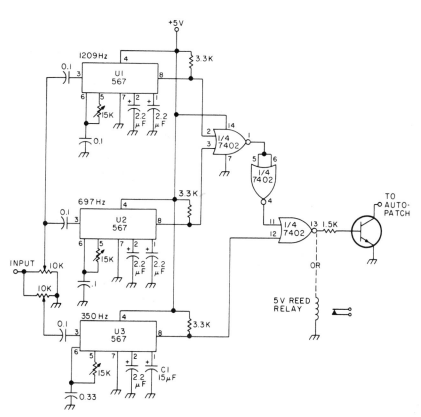

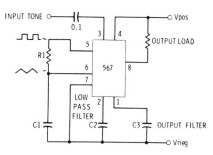

PLL SINGLE-TONE DECODER—Can be used for Touch-Tone decoding as well as for telephone-line and wireless control applications using single audio frequency. Operating center frequency depends on R1 and C1. R1 should be between 2K and 20K. C1 in microfarads is computed from $f = 1/R1C1$, where R is in megohms and f is in hertz. C2 is low-pass filter in range of 1–22 μF; the larger its value, the narrower its bandwidth. C3 is not critical and can be about twice C2.—C. D. Rakes, "Integrated Circuit Projects," Howard W. Sams, Indianapolis, IN, 1975, p 68–73.

TOLL-CALL KILLER—Prevents unauthorized direct long-distance dialing through repeater autopatch from areas where "1" must be dialed ahead of desired out-of-town phone number. Based on simultaneous detection of 350-Hz component of dial tone and 1209- and 697-Hz tones assigned to "1" in Touch-Tone system.

Circuit requires only three 567 tone decoders, 7402 quad gate, and either transistor or relay for controlling autopatch. Article covers installation and operation.—W. J. Hosking, Long Distance Call Eliminator, *73 Magazine,* April 1976, p 44–45.

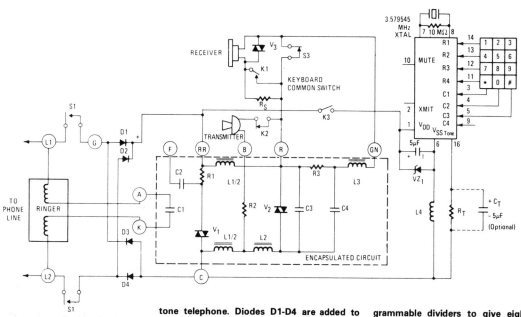

DUAL-TONE SIGNALING—American Microsystems S2559 digital tone generator IC at upper right interfaces directly with encapsulated 500-type telephone set to give pushbutton dual-tone telephone. Diodes D1-D4 are added to telephone set to ensure that polarity of direct voltage across device is unchanged even if connections to phone terminals are reversed. Generator IC requires external crystal feeding programmable dividers to give eight standard audio frequencies with high accuracy for combining in pairs as required for dual-tone signaling.—"Digital Tone Generator," American Microsystems, Santa Clara, CA, 1977, S2559, p 11.

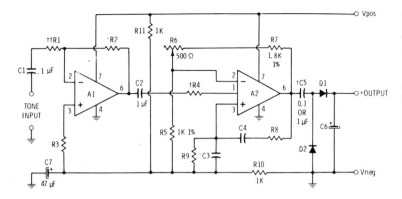

SINGLE-TONE DECODER—Used at receiving end of leased telephone line in which single tone frequency serves for alarm and other purposes. A1 is 741 opamp connected as inverting amplifier, with R1 and R3 chosen to match input impedance and R2 chosen to give gain required for available input signal level. For 10K input impedance, R1 and R3 are 10K and R2 in kilohms is 10 times required gain (500K for gain of 50). Actual tone decoding is performed by A2, which is also 741; here C3, C4, R8, and R9 are frequency-determining components and R6 is gain control. R4 is chosen to give desired bandwidth; use 470K for 5–10%, 1 megohm for 3–5%, and 2.2 megohms for 1–3%. R8 is same as R2, and R9 equals R3. Diodes are 1N914.—C. D. Rakes, "Integrated Circuit Projects," Howard W. Sams, Indianapolis, IN, 1975, p 60–66.

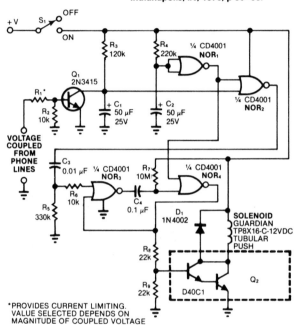

PHONE-CALL COUNTER—Circuit actuates solenoid that depresses R/S counting key of SR-56 calculator for each interrogation event consisting of sequence of pulse bursts each corresponding to ring of phone. Bursts are separated by 4-s pauses, so circuit includes time delay that prevents actuation of solenoid until line has remained quiescent for more than 5 s after burst. Article includes program that is inserted in calculator to total number of times R/S key is depressed. Applications include counting number of telephone calls received while away.—M. Bram, Hardware + Program Makes SR-56 Event Counter, *EDN Magazine*, Aug. 5, 1978, p 84 and 86.

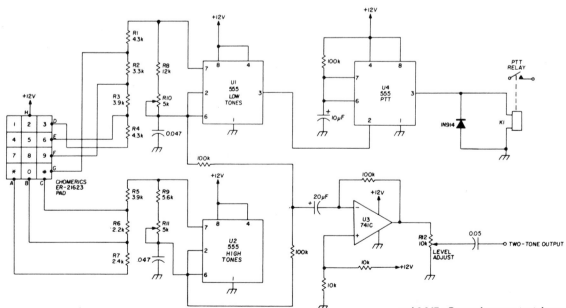

TOUCH-TONE ENCODER—Uses 555 timers to generate Touch-Tone frequencies in pairs using two of seven possible frequencies, under control of standard 12-button pad. Adjust R10 so low-group oscillator reads 941 Hz at pin 3 of U1 when * key is pressed. Frequencies of 852, 770, and 697 Hz will then be correct within 2% when 7, 4, and 1 are pressed, if 1% resistors are used and 0.047-μF capacitors are tantalum or Mylar. Automatic push-to-talk control uses U4 connected as 1-s mono MVBR driving relay K1.—H. M. Berlin, Homebrew Touch-Tone Encoder, *Ham Radio*, Aug. 1977, p 41–43.

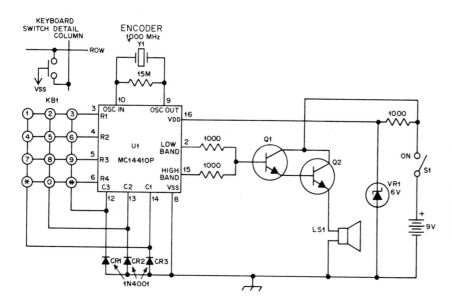

TOUCH-TONE DRIVE FOR LOUDSPEAKER—Encoder is held in front of microphone to access and use autopatch of repeater. Acoustic coupling eliminates need for opening new transceiver to make wire connections, which would void guarantee. Uses Motorola MC14410P IC with KB1 keyboard (Polypaks 92CU3149). Q1 and Q2 are 2N3643 or equivalent. Y1 is 1.000 MHz crystal (Mariann Labs ML18P or Sherold Crystal HC-6). Transistors Q1 and Q2 boost output enough to drive 8-ohm loudspeaker. Total current drain is 35 mA idling and 100 mA with full drive.—C. Gorski, A Low-Cost Touch-Tone Encoder, *QST*, Oct. 1976, p 36–37.

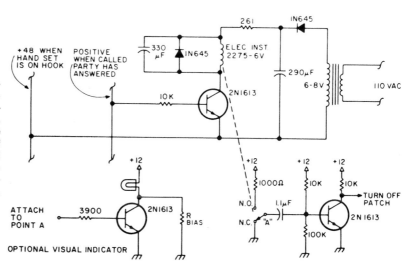

AUTOPATCH RELEASE—Control circuit automatically releases telephone autopatch at receiver when called party hangs up, by generating disconnect signal for patch control logic. Action is based on reversal of polarity of phone line when called party answers, and return of polarity to preanswered condition when called party hangs up. Article describes circuit operation and use.—T. R. Yocom, Automatic Autopatch Release, *73 Magazine*, April 1977, p 52.

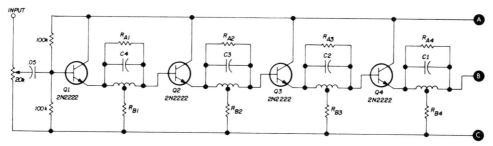

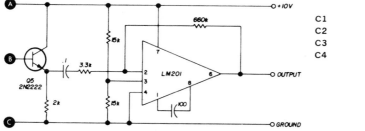

	low group	high group
C1	.68	.22
C2	.5	.18
C3	.39	.15
C4	.33	.1

TOUCH-TONE BAND-REJECT FILTER—Cascaded notch filters with active limiter at output provide 20-dB attenuation of either low (697–941 Hz) or high (1209–1633 Hz) groups of tones, as aid to decoding for repeater control functions. All coils are 88-mH toroid. R_A is between 5600 and 22,000 ohms, and R_B is 1000 to 3000 ohms. Article gives tuning procedure for selecting resistor values and adjusting toroids so each stage rejects different tone in its band.—B. Bretz, Multi-Function FM Repeater Decoder, *Ham Radio*, Jan. 1973, p 24–32.

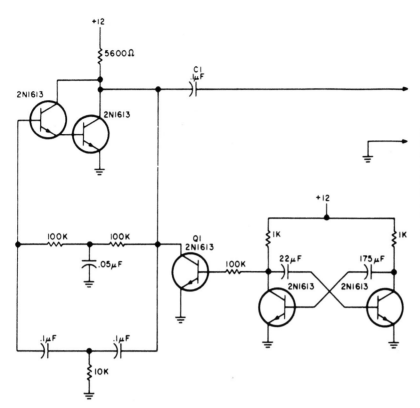

BUSY-SIGNAL GENERATOR—Conventional Bell System busy signal is provided by turning twin-T oscillator at left on and off with low-frequency asymmetrical square wave generated by transistor pair at right. Q1 acts as switch for turning oscillator on and off. Developed for use at repeater in home when autopatch connects to family telephone, to inhibit use of autopatch by mobile station when phone is in use. Article also covers connections to phone line and to repeater.—T. Yocom, An Autopatch Busy Signal, *73 Magazine,* Holiday issue 1976, p 148 and 150.

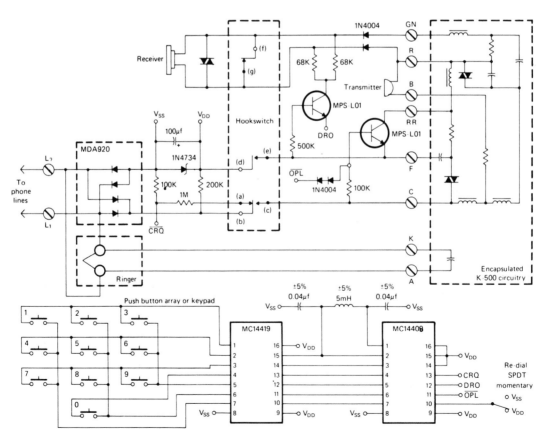

PUSHBUTTON-TO-DIAL CONVERTER—Combination of Motorola MC14419 keypad-to-binary converter and MC14408 BCD-to-dial telephone pulse converter is used with 10-switch pushbutton array to provide correct chain of pulses for dialing number on conventional dial-telephone system. Eleventh SPDT button is used for redial feature; if line called is busy, one press of redial button dials number over again. Number is stored for repeated use until new number is dialed. Check local telephone company regulations before making connections to telephone lines.—I. Math, A Push-Button to Dial Telephone Converter, *CQ,* Sept. 1976, p 36–37.

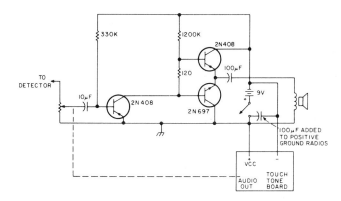

TOUCH-TONE ENCODER—Consists of SME Touch-Tone generator and keyboard made by Data Signal (Albany, GA) mounted on any small transistor radio. Only audio section is used, with output tones from loudspeaker being fed acoustically to microphone of FM amateur station. Article gives construction details.—D. Ingram, The Shirt Pocket Touch-Tone, *73 Magazine,* Nov. 1976, p 58–59.

45-kHz LOW-PASS STATE-VARIABLE FILTER—Used in precision telephone-network active equalizer. Damping value is 0.082, which requires 1% components. For high pass, take output from first opamp; for bandpass, take output from second opamp.—D. Lancaster, "Active-Filter Cookbook," Howard W. Sams, Indianapolis, IN, 1975, p 147.

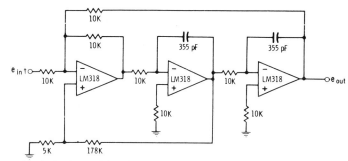

† must return to ground via low-impedance dc path.

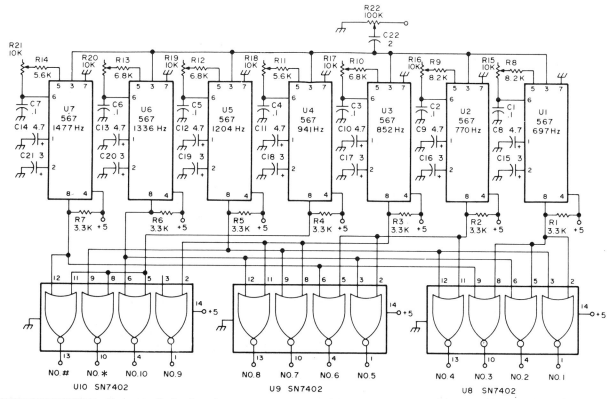

TOUCH-TONE DECODER—Uses seven National LM567 phase-locked loop decoders (U1–U7) having high noise rejection, immunity to false signals, and stable center frequency. Each 567 activates proper gate of SN7402, making output of gate go to high or 1 state for driving NPN transistor that can turn on LED labeled with corresponding Touch-Tone number. Alternatively, gate outputs can drive 12 relays, with relay contacts going to LEDs and/or to keyboard switches of ordinary calculator used as digital display. Article tells how to adjust 10K pot for each 567 for detection of desired frequency.—W. MacDowell, Touch-Tone Decoder, *73 Magazine,* June 1976, p 26–27.

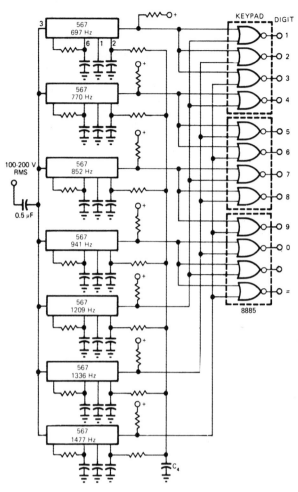

PLL TOUCH-TONE DECODER—Seven 567 PLLs sense presence of selected tones from common 100–200 VRMS input line, while 8885 NOR gates perform necessary decoding logic to generate decimal outputs. Circuit takes advantage of good frequency selectivity provided by lock-and-capture ranges of PLLs, as required for discriminating against many tones.—E. Murthi, Monolithic Phase-Locked Loops—Analogs Do All the Work of Digitals, and Much More, *EDN Magazine,* Sept. 5, 1977, p 59–64.

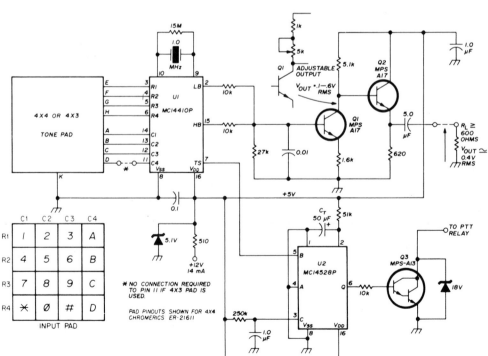

TONE ENCODER—Motorola MC14410 CMOS IC is basis of accurate low-power Touch-Tone encoder system providing full 2-of-8 encoding from basic 1-MHz crystal oscillator. Can be used with 2-of-7 or 2-of-8 keypad switch matrix such as Chromerics ER-21623 or ER-21611. Q1-Q2 form tone-amplifier/emitter-follower line driver. U2 is push-to-talk mono 1-s timer. Supply can be any voltage from 5 to 12 V if zener is used to supply 5 V to ICs. Article covers circuit operation in detail and gives tone-encoder frequency table.—J. DeLaune, Digital Touch-Tone Encoder for VHF FM, *Ham Radio,* April 1975, p 28–31.

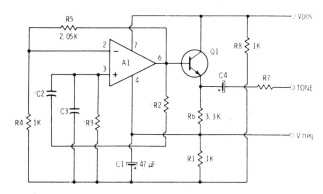

SINGLE-TONE SIGNALING—Wien-bridge oscillator using 741 opamp drives 2N2924 or equivalent NPN transistor to generate stable audio tone for signaling over telephone lines. Tuning capacitor (C2 and C3 are equal) and resistor (R2 and R3 are equal) values range from 0.1 μF and 15.9K for 100 Hz to 0.005 μF and 6.3K for 5000 Hz. For other frequencies, use f = 0.159/R2C2. With 12-V supply, tone output is about 7 V P-P. Select R7 to match impedance of driven circuit.—C. D. Rakes, "Integrated Circuit Projects," Howard W. Sams, Indianapolis, IN, 1975, p 55—60.

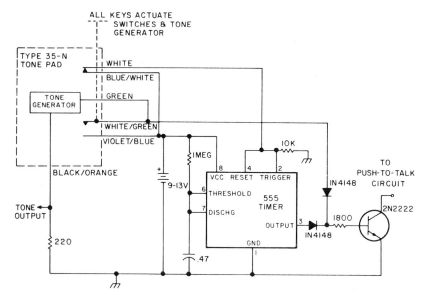

MOBILE AUTOPATCH—Circuit operates push-to-talk of mobile station automatically when any button on Touch-Tone pad is pushed for dialing telephone number after making autopatch, eliminating need for engaging microphone before dialing. Circuit remains active for about 2 s after Touch-Tone button is released.—Circuits, *73 Magazine,* May 1977, p 19.

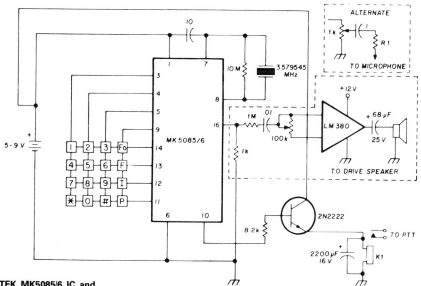

TOUCH-TONE IC—MOSTEK MK5085/6 IC and keyboard together form inexpensive Touch-Tone generator producing tones within 0.75% of required values. Uses 3.579545-MHz TV color-burst crystal. Pin 15 is grounded to provide dual tones only. Pin 10 provides output when keyboard entry has been made, for keying push-to-talk (PTT). Loudspeaker can be eliminated if output is fed directly into microphone input of transmitter. Choice of IC depends on type of keyboard used.—T. Ahrens, Integrated-Circuit Tone Generator, *Ham Radio,* Feb. 1977, p 70.

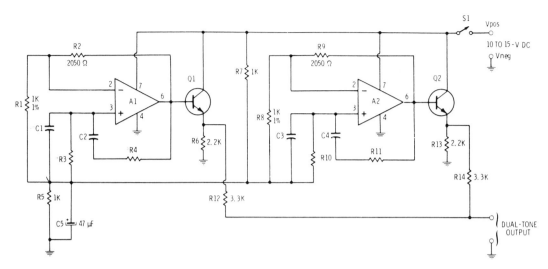

TWO-TONE ENCODER—741 opamps and 2N2924 transistors are connected as single-tone encoders producing different audio frequencies, with outputs connected together. For 2000-Hz tone, use 0.01 μF for C1 and C2 and use 8K for R3 and R4. Formula for frequency of each encoder is f = 0.159/RC where f is in hertz, R in megohms, and C in microfarads; R = R3 = R4 and C = C1 = C2. Frequencies can be chosen for Touch-Tone signaling.—C. D. Rakes, "Integrated Circuit Projects," Howard W. Sams, Indianapolis, IN, 1975, p 95–97.

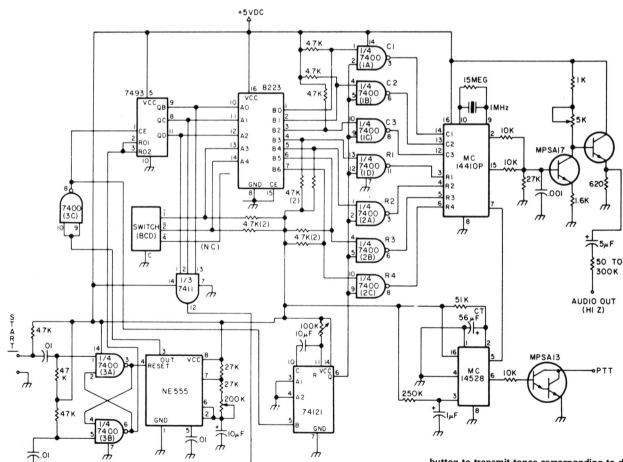

FOUR-NUMBER CALLER—Motorola MC14410 CMOS Touch-Tone generator chip forms basis for automatic dialer using BCD thumbwheel switch to choose telephone number desired. Numbers are stored in 256-bit PROM by conventional programming. Article shows how autopatch access and disconnect switches are added. To make telephone call from car through repeater, select number desired, push access button and, when dial tone is heard, push start button to transmit tones corresponding to desired number. Article covers circuit operation, programming, and coding, and gives additional circuit using 512-bit PROM to provide eight telephone numbers.—W. J. Hosking, Drive More Safely with a Mobile Dialer, *73 Magazine*, Feb. 1977, p 102–104.

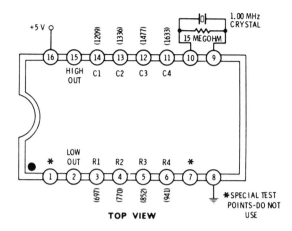

TOUCH-TONE DIALER—Single Motorola 4410 chip requires only two external components and 2-of-8 keyboard to generate two sine waves simultaneously for Touch-Tone dialing and telephone modem communication. Each key on keyboard grounds one of C inputs and one of R inputs. As example, when 6 key is pressed, R2 and C3 are grounded to give 770-Hz sine wave on pin 2 and 1477-Hz sine wave on pin 15. Designed for driving 1K load. Output voltage is about 600 mV P-P for low output and 800 mV P-P for high output.—D. Lancaster, "CMOS Cookbook," Howard W. Sams, Indianapolis, IN, 1977, p 132.

RING DETECTOR—Optoisolator using neon lamp and light-dependent resistor serves as interface between telephone line and line-operated remote bell. Neon fires reliably from nominal 100-VAC ring signal, while capacitor C_1 provides isolation required to prevent latch-up by sustaining voltages within range of phone-line quiet battery. If optional protective varistor R_{IC1} is added, rating of capacitor can be reduced to 400 V. Triac Q_1 in series with primary of line transformer provides synchronization to 20-Hz ringing frequency of phone system.—W. D. Kraengel, Jr., Ring Detector Optically Interfaces Phone, *EDN Magazine*, Aug. 5, 1978, p 80 and 82.

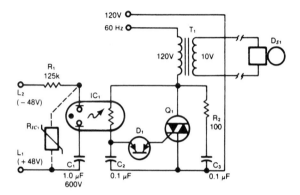

SECTION 60
Voltage-Level Detector Circuits

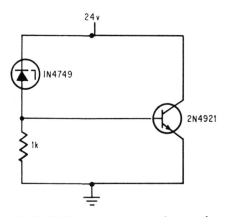

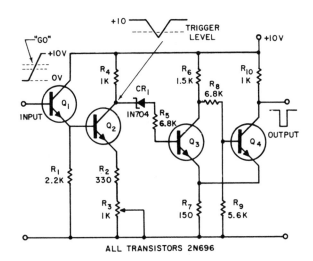

POWER ZENER—Zener conducts when supply voltage reaches 24 V, developing voltage across 1K resistor that turns on transistor. Resulting increased power is obtained at fraction of cost of power zener otherwise required.—J. O. Schroeder, Transistor Increases Zener's Power Capability, *Electronics*, Sept. 16, 1968, p 99—100.

5.6—5.8 V GO-NO-GO—Provides voltage discrimination within 0.2-V passband for automatic testing. Only voltages within passband cause switching of Schmitt trigger. R3 controls width of passband. For other GO bands, input voltages should be applied to Q1 through suitable voltage dividers or zener diodes.—R. Vokoun, Voltage Level Discriminator Provides Go-No-Go Indication, *Electronic Design*, Nov. 23, 1964, p 67—69.

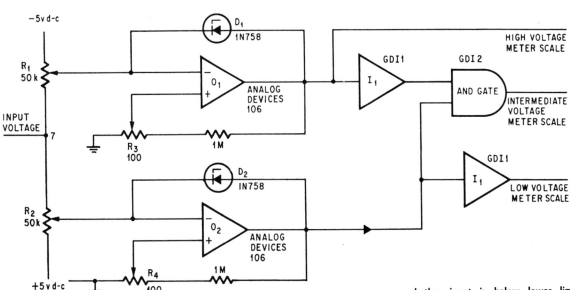

AUTOMATIC VOLTMETER—IC opamps and other standard logic circuits connected as shown give automatic three-range switching for voltage comparator. Voltage limits are independently adjustable. One output will indicate whether input is above upper limit, another will indicate whether input is between limits, and third output will indicate whether input is below lower limit. Any limit may be either positive or negative.—W. Ellermeyer, Voltage Comparator Is Made With Op Amps and Logic Gates, *Electronics*, July 8, 1968, p 91—92.

VOLTAGE-LEVEL DETECTOR—Differential amplifier has two stable states as in flip-flop, and changes state whenever d-c input level goes above preset reference level. With components shown, differential of only 100 mV with respect to reference will change state. —G. Richwell, Adjustable Level-Detector, *EEE*, Oct. 1966, p 139.

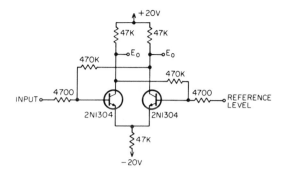

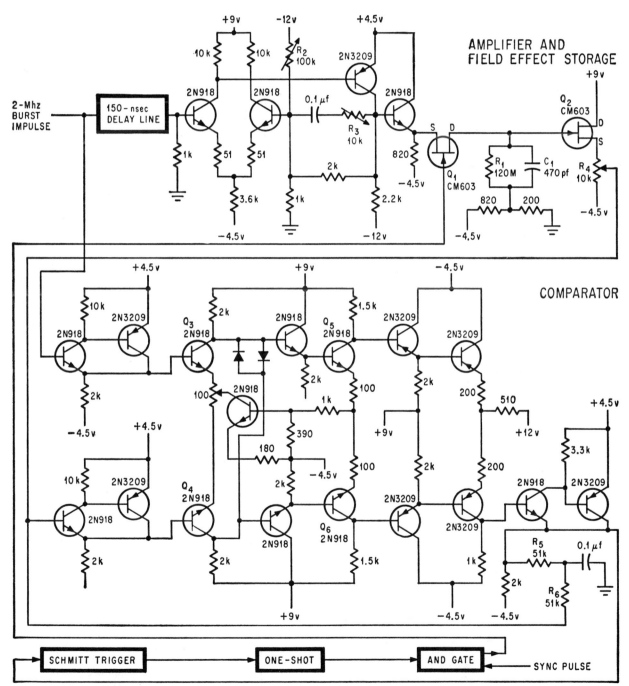

VIDEO BURST PEAK DETECTOR—Useful where position of peak in video signal is required, as for sampling time delay signals and for video, radar, and sonar equipment. Comparator circuit can sense highest peak of 2-MHz a-m burst to within 5 mV for peaks up to 1 V.—D. S. Greenstein, Detector Stores Peaks of Video Bursts, *Electronics*, Oct. 3, 1966, p 104–106.

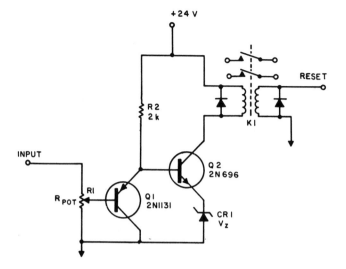

FAIL-SAFE VOLTAGE MONITOR—Operates latching relay when input voltage exceeds threshold level set by R1. For fail-safe application, relay contacts are connected to remove power from system or perform other required function. Transistors provide temperature compensation to improve accuracy. —M. Furukawa, Voltage-Sensing Circuit Is Temperature-Compensated, *Electronic Design*, May 10, 1967, p 90—91.

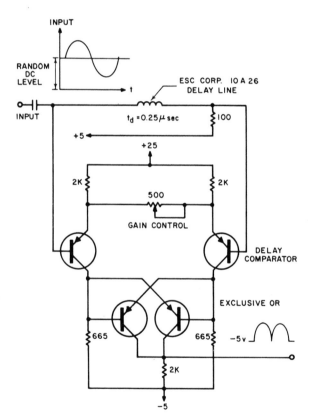

CORE PULSE-HEIGHT DETECTOR—Self-tracking detector is virtually independent of duty cycle, recovers core-memory outputs having random d-c component, and can detect signals in presence of noise frequencies that are one order of magnitude lower than signal. Will detect 200-ns pulse with rise time of 100 ns. Uses comparator method, by making comparison between each half-cycle preceding and following d-c level crossing. All transistors are 2N711.—R. T. Shevlin, Pulse-Height Detector Operates Independently of D-c Input Level, *Electronic Design*, March 29, 1965, p 32—33.

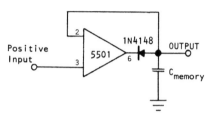

PEAK SENSE AND HOLD—IC analog comparator serves for comparing input voltage with that across memory capacitor. If input is greater than that of capacitor, output moves in positive direction to charge capacitor until it equals input voltage. 5501 then changes state and reverse-biases diode, isolating capacitor to hold value of input voltage.— Applying the Model 5501 Monolithic Analog Comparator, Optical Electronics, Tucson, Ariz., No. 10147.

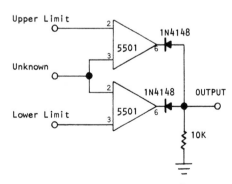

WINDOW DETECTOR—Uses two analog comparators to give zero output when unknown input voltage is within predetermined acceptable range or "window." Output is positive when unknown is beyond limits that are independently set.—Applying the Model 5501 Monolithic Analog Comparator, Optical Electronics, Tucson, Ariz., No. 10147.

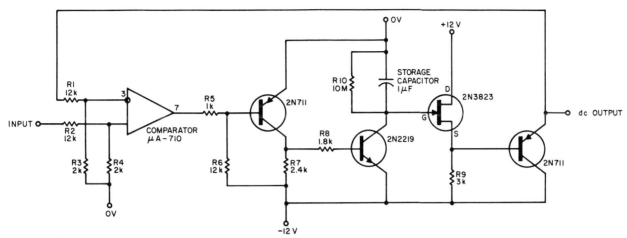

NARROW-PULSE PEAK SENSOR—Provides exact proportionality of d-c output to input signal peaks. Has low threshold, for response to very slight increases in peak amplitudes. Will handle pulses as narrow as 0.2 μs and duty cycle below 1% for repetitive signals. IC differential comparator samples peak and compares it with existing d-c output voltage, then drives level shifter controlling transistor switch that supplies current to storage capacitor that governs d-c output voltage.—W. C. Dillon, An Operational Peak Detector Captures Very Narrow Pulses, *Electronic Design*, Oct. 25, 1967, p 138 and 140.

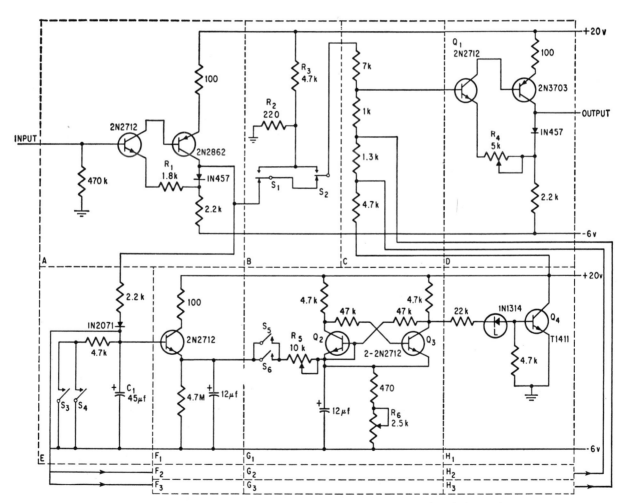

PEN RECORDER SCALE CHANGER—Although developed to operate with spectrometer that repetitively scans frequency band of light emissions in upper atmosphere, this low-cost scale changer can be used with other meters and recorders as well. Three ranges are provided, with automatic changing at end of scan when input during scan rises to 1 V, 2 V, and 4 V. Article gives operating details. Peak-level detector stage E drives three identical chains of F, G, and H stages, one for each scale factor. F stages provide impedance matching. G stages are flip-flops adjusted to switch at different input voltage levels. H stages operate as ground switches.—B. E. Bourne and R. L. Gattinger, Automatic Scale Changer Shifts Recorder Range, *Electronics*, May 15, 1967, p 95—97.

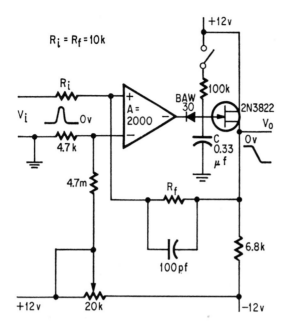

$R_i = R_f = 10k$

PEAK DETECTOR—Basic diode-capacitor peak detector accuracy is improved **1,000 times** by adding fet and IC opamp. With nulling switch open and positive pulse at input, circuit acts as inverting amplifier. When downward transition of pulse begins, output is held constant at peak value because detector and source follower cannot follow change in polarity. To adjust output for zero offset with zero input voltage, close nulling switch and adjust 20K pot.—A. E. Vinatzer, High Accuracy Obtained from Peak Detector Using Op Amp, *Electronics*, March 17, 1969, p 94.

DETECTOR AND COUNTER—Circuit shown will detect envelope of bursts of high-frequency pulses in range of 1 kHz to 1 MHz if suitable component values are used. For those shown, upper limit for input prf is 1 kHz. If D1 is connected to point A instead of to +5 V, circuit becomes pulse counter that delivers output pulse after arrival of specified number of input pulses. After each circuit lock-up, 0.01-μF capacitor must be discharged manually or given time to discharge before input is reapplied.—K. Sheth, Pulse-Train Detector and Counter, *EEE*, Nov. 1966, p 156.

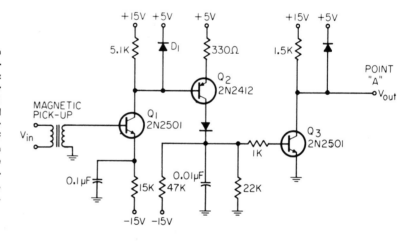

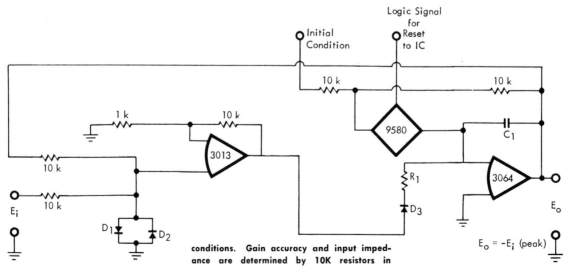

INVERTING PEAK DETECTOR—Uses Burr-Brown 9580 electronic switch and two opamps to provide high-speed reset to initial conditions. Gain accuracy and input impedance are determined by 10K resistors in summing network. Can reset to either negative or positive initial conditions, permitting use as peak detector even if some maxima are negative and some positive.—Operational Amplifiers, Burr-Brown Research, Tucson, Ariz., LI-327, 1969, p 43.

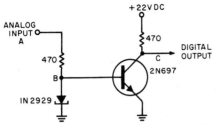

ANALOG VOLTAGE SENSOR—Action of positive-going analog input voltage on tunnel diode makes transistor saturate, at predetermined input point, so as to switch transistor on for positive halves of input signal cycles.—Tunnel Diode-Transistor Level Sensor (Circuit Digest), *Electronic Design*, March 2, 1964, p 58.

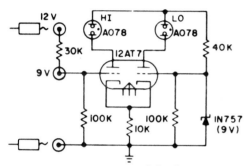

EQUAL VOLTAGE TEST—For balancing two voltage-divider points in production testing. Only 200-mV difference in either direction causes variation in illumination level of lamps. Any reference from zero to dissipation level of tube can be used instead of 9 V shown.—E. Bauman, "Applications of Neon Lamps and Gas Discharge Tubes," Signalite, Neptune, N.J., p 145.

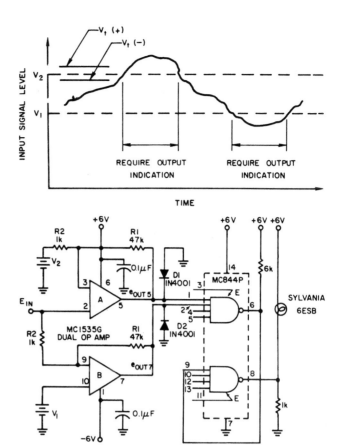

ENVELOPE DETECTOR—Pilot lamp is turned on when input signal voltage is above upper limit V2 or below lower limit V1. Amount of hysteresis about each limit is function of R1 and R2, and is 69 mV for limits of 2.5 V and 3.5 V.—K. Wolf, Signal-Level Envelope Detector Uses Dual Operational Amplifier, *Electronic Design*, Feb. 1, 1969, p 78 and 80.

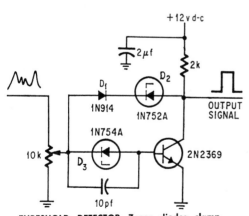

THRESHOLD DETECTOR—Zener diodes clamp collector of transistor at 0.5 V, just outside saturation region, to eliminate turnoff delay and thereby give faster rise and fall times, of order of 30 ns. Threshold value is independent of temperature if temperature coefficient of D1 is matched to D2 and that of D3 is matched to base-emitter junction of Q1. Potentiometer sets threshold level.—B. Fugit, Collector Clamping Improves Threshold Detector, *Electronics*, Sept. 30, 1968, p 81.

5 MV—10 V LEVEL DETECTOR—High-sensitivity voltage level detector, using only tunnel diode and two transistor switches, requires no diode limiters to protect from overload. To sample input, Q2 is turned off by sample pulse. Q1 will switch only if input signal is positive during sampling time. With minor changes, circuit can be used as transient detector or level comparator.—A. J. Welty, Tunnel-Diode Level Detector Is Ultra-Sensitive, *Electronic Design*, Jan. 4, 1965, p 73–74.

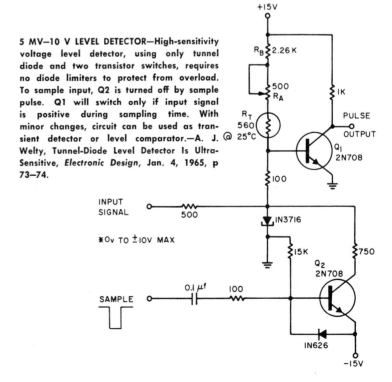

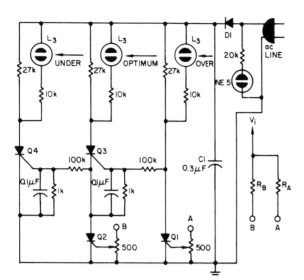

AMBIGUITY — Three-level voltage monitor automatically turns off lower-level indicator lamps when lamp at higher level is triggered, to eliminate ambiguity in display. Uses C106B1 scr's and any suitable neon lamps. If circuit is to monitor 100 V d-c to tolerance of 2 V, RA and RB are chosen to make Q1 and Q2 fire when Vi is 102 V and 98 V. Technique can be extended to more levels by adding more sections like middle one (Q2-Q3-L3).—A. Prokop, SCR Threshold Detector Eliminates Ambiguities, *Electronic Design*, Sept. 1, 1969, p 104 and 106.

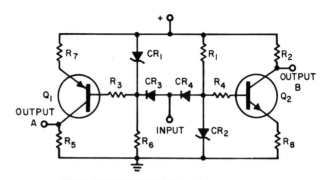

PULSE AMPLITUDE-POLARITY SORTER—Q1 detects positive input pulses and amplifies them sufficiently to drive several output circuits, while Q2 similarly detects and amplifies negative input pulses. Transistor types are not critical, as long as Q1 is pnp and Q2 is npn. Zeners CR1 and CR2 are chosen to maintain associated transistors back-biased until incoming pulse makes transistors conduct. Other two diodes simply provide isolation. Signal-level adjustments are best made by using variable resistors for R1 and R6. Pulses of different levels can be sorted by using several circuits in parallel and adjusting each for different level. Other resistor values are not critical.—F. W. Kear, Pulse-Sorter Network Detects, Amplifies Bi-Polar Signals, *Electronic Design*, Aug. 30, 1965, p 48—49.

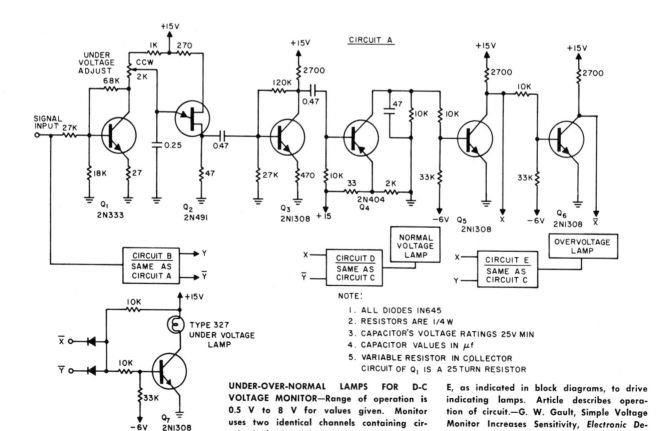

NOTE:
1. ALL DIODES 1N645
2. RESISTORS ARE 1/4 W
3. CAPACITOR'S VOLTAGE RATINGS 25V MIN
4. CAPACITOR VALUES IN μf
5. VARIABLE RESISTOR IN COLLECTOR CIRCUIT OF Q_1 IS A 25 TURN RESISTOR

UNDER-OVER-NORMAL LAMPS FOR D-C VOLTAGE MONITOR—Range of operation is 0.5 V to 8 V for values given. Monitor uses two identical channels containing circuit A having six transistors. These two channels feed circuit C (lower left), D, and E, as indicated in block diagrams, to drive indicating lamps. Article describes operation of circuit.—G. W. Gault, Simple Voltage Monitor Increases Sensitivity, *Electronic Design*, Sept. 14, 1964, p 70.

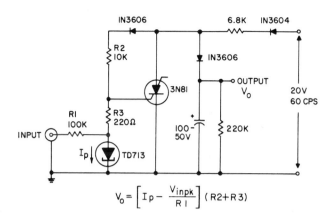

$$V_o = \left[I_p - \frac{V_{inpk}}{R1} \right] (R2+R3)$$

PEAK SENSING WITH TUNNEL DIODE—Used to measure peak voltage of waveform that repeats well above 60 times per second. 100-μF capacitor is charged by half-wave rectified 60-Hz voltage that also is applied to anode of scs 3N81. When input waveform brings tunnel diode to its peak point current, diode switches and turns on scs, terminating charging of capacitor. In negative half-cycle, capacitor discharges slightly but is recharged on next positive half-cycle. Output voltage is thus proportional to peak of input waveform.—W. R. Spofford, Jr., Applications For The New Low Cost TD700 Series Tunnel Diodes, General Electric, Syracuse, N.Y., No. 90.66, 3/67, p 10.

ADJUSTABLE HYSTERESIS—Differential comparator and zener provide independently adjustable voltage trip point and hysteresis. Q1 and Q3 can be 2N3904 or equivalent, Q2 2N3096 or equivalent, and D1 1N708. R7 varies reference level, for changing trip point from 0.7 V to 5 V, while R5 changes hysteresis.—R. Billon, Level Detector Has Independently Adjustable Hysteresis and Trip Point, Electronic Design, Oct. 11, 1967, p 98.

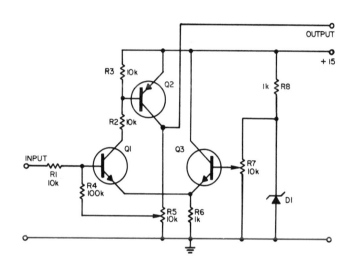

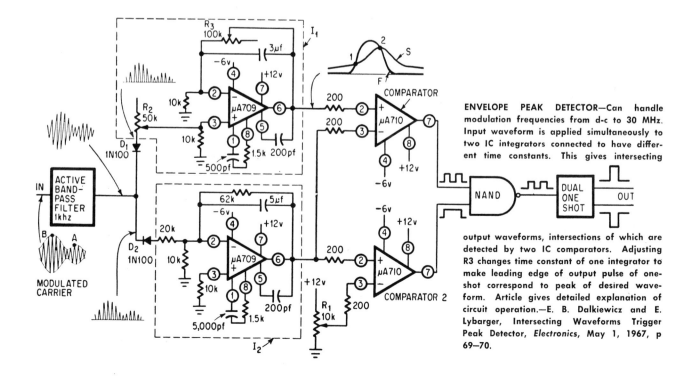

ENVELOPE PEAK DETECTOR—Can handle modulation frequencies from d-c to 30 MHz. Input waveform is applied simultaneously to two IC integrators connected to have different time constants. This gives intersecting output waveforms, intersections of which are detected by two IC comparators. Adjusting R3 changes time constant of one integrator to make leading edge of output pulse of one-shot correspond to peak of desired waveform. Article gives detailed explanation of circuit operation.—E. B. Dalkiewicz and E. Lybarger, Intersecting Waveforms Trigger Peak Detector, Electronics, May 1, 1967, p 69—70.

MULTIPLE-VOLTAGE MONITOR—Simple comparison amplifier and summing junction are used to compare junction voltage with reference voltage that is actually system ground or zero. Used for monitoring system having many power supplies with voltage levels of both polarities. Any change in one of monitored voltage levels will unbalance comparison amplifier and make warning lamp glow. Voltage measurements must then be made to identify defective supply. Sensing accuracy is 5%.—R. C. Gerdes, Voltage Sensor Monitors Multiple Power Supply Outputs, "400 Ideas for Design Selected from Electronic Design," Hayden Book Co., N.Y., 1964, p 216.

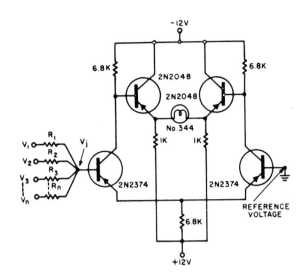

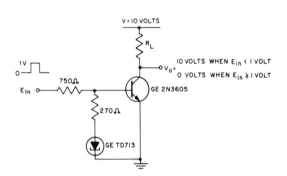

TUNNEL-DIODE LEVEL DETECTOR—Adjusting value of input resistor will make tunnel diode switch at any desired threshold voltage, and turn back off when input voltage is dropped to zero.—W. R. Spofford, Jr., Applications For The New Low Cost TD 700 Series Tunnel Diodes, General Electric, Syracuse, N.Y., No. 90.66, 1967, p 8.

LOGIC-LEVEL LAMP—Lamp lights when probe is touched to terminal having pulse amplitude or d-c level above 0.7 V. Speeds troubleshooting of digital circuits by eliminating need to look at cro or meter. Lamp is mounted directly on probe. Light-emitting diodes such as HP5082-4400 in series with 470-ohm resistor may be used in place of lamp for faster response. Level detector Q1 triggers 1-ms mono mvbr Q2-Q3.—J. M. Firth, Go/No Go Circuit Gives Visual Indication of RTL Logic Level, Electronic Design, Feb. 15, 1970, p 88.

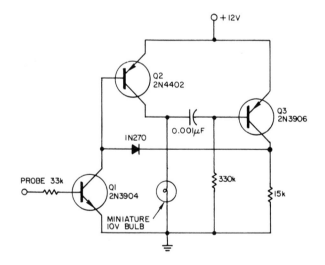

HOLDING PEAK VOLTAGE—Will hold peak voltage of short-duration analog signal for any required period up to several hundred ms. Permits use of pen recorders having slow response. Value of C1 determines holding period. First three transistors form combined Schmitt trigger and one-shot.—P. P. Tong, Peak-Hold Circuit, EEE, March 1967, p 164 and 166.

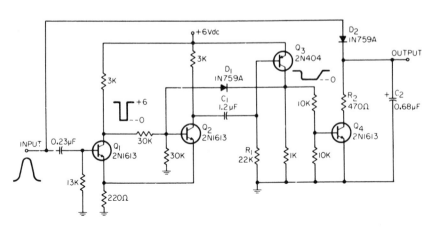

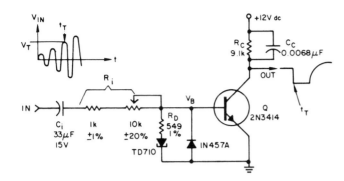

10-KHZ SINE-WAVE PEAK DETECTOR—Simple tunnel-diode threshold detector senses peak values exceeding preset level. Output voltage is near supply voltage while input is low, and drops to about zero when input exceeds threshold value controlled by setting of Ri.—D. B. Heckman, Sense Signal Levels with a Tunnel Diode, *Electronic Design*, Feb. 1, 1969, p 48—51.

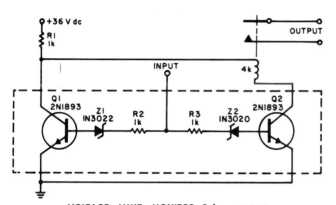

VOLTAGE LIMIT MONITOR—Relay contacts are closed as long as d-c input voltage is within specified limits. If voltage rises above limit, Q1 conducts and shorts out relay coil. If voltage drops below limit, both transistors cut off and relay drops out. Zeners determine voltage limits.—O. Tedenstig, Monitor DC Levels with a Simple Circuit, *Electronic Design*, June 6, 1968, p 110 and 112.

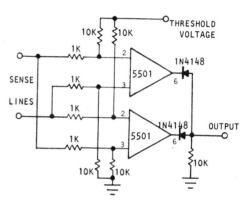

DUAL SENSE AMPLIFIER—Uses two 5501 analog comparators to indicate when any unbalance or voltage difference exists on two input lines. Output is positive regardless of which input line is more positive. Noise on lines does not cause false output indication.—Applying the Model 5501 Monolithic Analog Comparator, Optical Electronics, Tucson, Ariz., No. 10147.

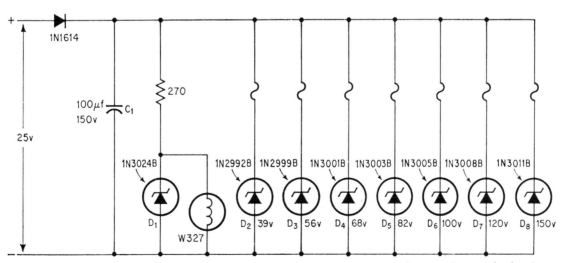

MONITORING POWER-LINE TRANSIENTS—Provides continuous surveillance of 28-V d-c supply line, storing approximate voltage magnitude of largest transient spike. Each zener is in series with sensitive 10-ohm 0.066-A instrument fuse, and each zener-fuse combination corresponds to detected voltage level about 5 V above zener breakdown voltage. Detection levels for zeners shown are 44, 61, 73, 87, 105, 125, and 155 V. 37-V spike on 28-V line means total peak of 65 V, which blows fuses for D2 and D3 because they avalanche into conduction. Operator then knows that spike was greater than 61 V but less than 73 V required to blow fuse for D4.—O. Pitzalis, Jr., Zener Circuit Detects Transients in Power Lines, *Electronics*, May 12, 1969, p 111.

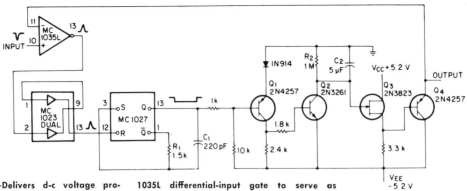

PEAK DETECTOR—Delivers d-c voltage proportional to peak amplitude of pulses as narrow as 10 ns. Negative output voltage, proportional to voltage across C2, is fed back to inverting input of Motorola MC-1035L differential-input gate to serve as arbitrary reference against which negative-going input pulse is compared. Reference rises with each input pulse until equilibrium is reached and C2 stops charging. Output is independent of pulse width.—M. J. Prickett, Peak Detector for Very Narrow Pulses, *EEE*, June 1970, p 92 and 94.

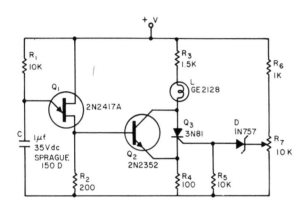

CURE FOR VOLTAGE-DROPPING HYSTERESIS —Use of ujt oscillator Q1 in voltage indicator overcomes hysteresis effect. Circuit will trigger off within 2% of desired level for triggering on, over wide temperature range. Supply voltage is 28 V. Lamp glows when voltage reaches desired level. If two circuits are used, with relays in place of lamps, they can be used to remove power whenever voltage goes above or below 2% of desired trigger level determined by R7.—J. V. Crowling, Solid-State Voltage Indicator Overcomes Hysteresis Problem, *Electronic Design*, Dec. 21, 1964, p 51—52.

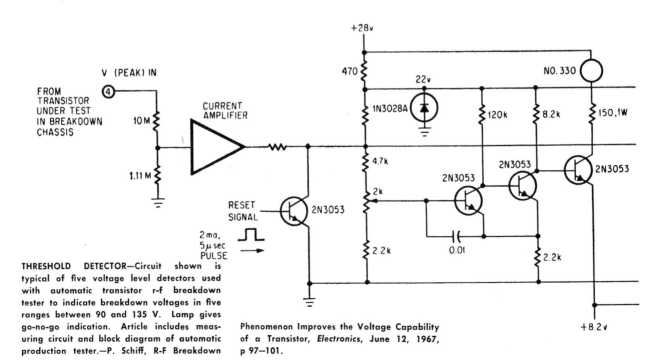

THRESHOLD DETECTOR—Circuit shown is typical of five voltage level detectors used with automatic transistor r-f breakdown tester to indicate breakdown voltages in five ranges between 90 and 135 V. Lamp gives go-no-go indication. Article includes measuring circuit and block diagram of automatic production tester.—P. Schiff, R-F Breakdown Phenomenon Improves the Voltage Capability of a Transistor, *Electronics*, June 12, 1967, p 97—101.

HYSTERESIS CIRCUIT DESIGN—Article gives mathematical procedure requiring calculation of only four resistor values to obtain desired upper and lower threshold points on hysteresis curve. Design approach assumes opamp input impedance is much larger than source impedance, and output impedance much smaller than load impedance, with amplifier switching state when differential input voltage is 0 V. Values shown are for 6.2 V upper threshold and —0.7 V lower threshold, with 3 V input required for switching amplifier from low to high state and 0.5 V for high to low state.—W. A. Cooke, A Simplified Design Approach for Hysteresis Circuits, *Electronics*, Nov. 10, 1969, p 106.

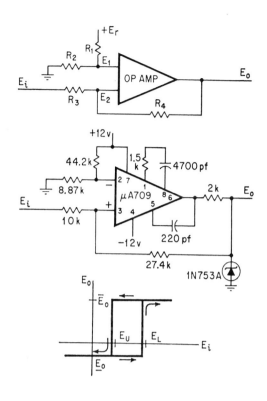

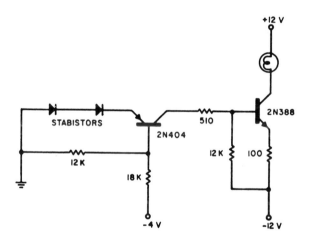

THREE-VOLTAGE MONITOR—Lamp glows only when all three supply voltages are present in typical pnp digital system having negative collector and clamp supplies, and positive bias supply. For npn system, use complementary version of circuit. Absence of any one voltage extinguishes lamp, giving true AND function. Stabistors provide tight control over error level of clamp voltage.— D. Chin, DC Presence Indicator Checks Three Voltages, "400 Ideas for Design Selected from Electronic Design," Hayden Book Co., N.Y., 1964, p 155.

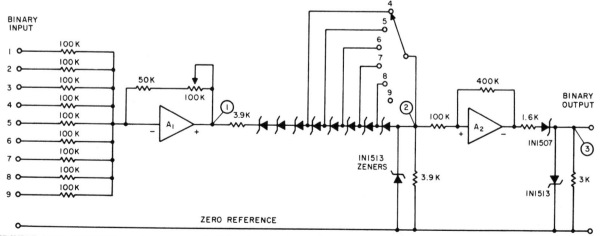

NINE-INPUT LEVEL DETECTOR—Simple opamp circuit with diode limiting serves in place of expensive binary logic. Binary output depends on whether each binary input is off (0 V for logical zero) or on (—11 V d-c for logical one). Output is binary one only when more than preselected number of inputs is one. Developed in connection with design of radio telescope.—W. Leroy Gahm, Binary Level Detector for Input Step-Voltages, *Electronic Design*, Oct. 26, 1964, p 50.

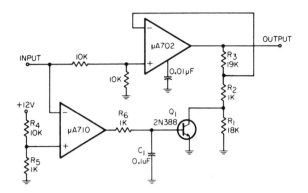

AUTOMATIC RANGE SELECTOR—Electronic range-changing switch for voltmeter uses two opamps to compare input voltage with fixed reference derived from divider R4-R5. With values shown, ranges provided are 0–1 V and 0–10 V full-scale. When input voltage exceeds first range, output of opamp µA710 goes negative and shuts off Q1, decreasing gain by factor of 10 for upper range. Isolation diode can be added to Q1 for driving scale indicator showing which range is in use.—C. Becklein, Automatic Scaling Circuit Uses ICs, *EEE*, Dec. 1966, p 118.

VOLTAGE SENSING—When input voltage to be sensed exceeds reference voltage applied by external source, ujt fires and produces output pulse that can be used to fire scr or other pulse-sensitive device. R1 adjusts trigger level. Trigger current required is less than 5 µA. Long-term voltage stability is 10 mV, and short-term is 1 mV.—T. P. Sylvan, The Unijunction Transistor Characteristics and Applications, General Electric, Syracuse, N.Y., No. 90.10, 1965, p 79.

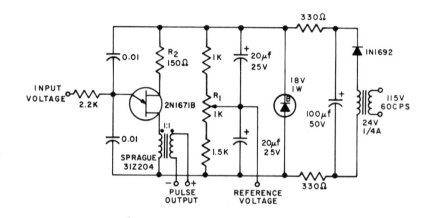

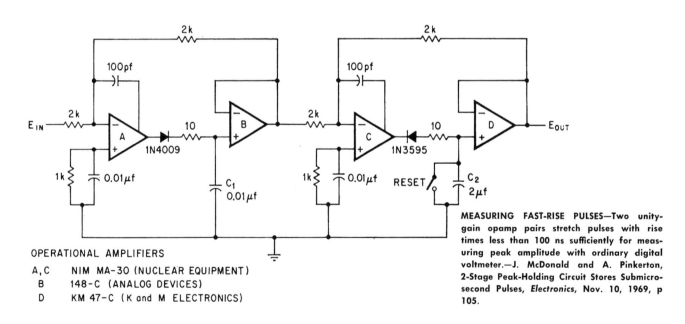

OPERATIONAL AMPLIFIERS

A,C   NIM MA-30 (NUCLEAR EQUIPMENT)
B   148-C (ANALOG DEVICES)
D   KM 47-C (K and M ELECTRONICS)

MEASURING FAST-RISE PULSES—Two unity-gain opamp pairs stretch pulses with rise times less than 100 ns sufficiently for measuring peak amplitude with ordinary digital voltmeter.—J. McDonald and A. Pinkerton, 2-Stage Peak-Holding Circuit Stores Submicrosecond Pulses, *Electronics*, Nov. 10, 1969, p 105.

NOTE: To locate additional circuits in the category of this chapter, use the index at the back of this book. Check also the author's "Sourcebook of Electronic Circuits," published by McGraw-Hill in 1968.

SECTION **61**
Voltage Reference Circuits

5-V REFERENCE SUPPLY—Circuit actually gives highly stable 6.7 V, but this can be adjusted to exactly 5 V by R5 in precision-type output control circuit. Transistor Q4 serves as temperature-compensating voltage-regulator diode. Article gives construction and adjustment details.—C. D. Todd, Stable, Low-Cost Reference Power Supplies, *Electronics World*, Dec. 1967, p 39—41 and 79.

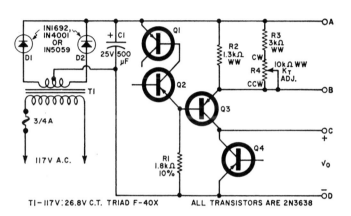

TI—117V:26.8V C.T. TRIAD F-40X   ALL TRANSISTORS ARE 2N3638

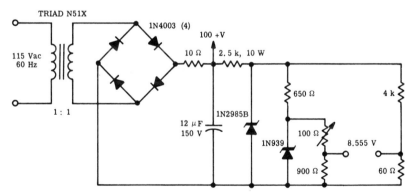

CASCADED ZENER REFERENCE—Provides reference voltage supply of 8.555 V, which changes only 0.0001% for 10% input voltage change. 1N2985B 22-V diode absorbs large changes in input, permitting operation of second diode over linear range.—"Zener Diode Handbook," Motorola, Phoenix, Ariz., 1967, p 6—19.

—4 KV BRIDGE REFERENCE—Diode bridge coupled to astable mvbr clamps random-frequency random-width digital input pulses to d-c level without distortion. Used to unblank crt where grid is biased at —4 kV below ground.—W. E. Peterson, D-c Restorer Clamps Random Pulses to a Reference, *Electronics*, July 7, 1969, p 106.

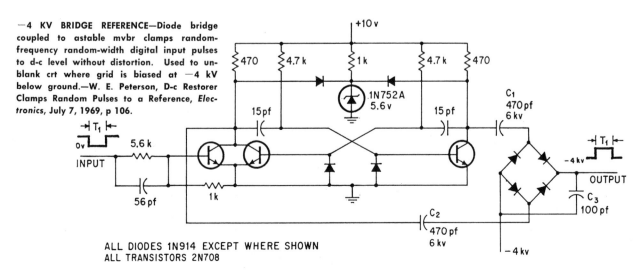

ALL DIODES 1N914 EXCEPT WHERE SHOWN
ALL TRANSISTORS 2N708

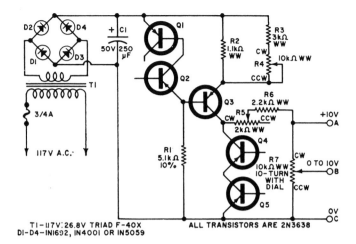

10-V REFERENCE SUPPLY—Provides constant voltage, independent of temperature, for calibrating lab or shop instruments and for monitoring very small voltage changes. Stability is achieved by using low-cost transistors as temperature-compensated voltage regulator diodes. Article gives construction and adjustment details, and chopper circuits for making a-c measurements.—C. D. Todd, *Stable, Low-Cost Reference Power Supplies, Electronics World,* Dec. 1967, p 39—41 and 79.

T1—117V:26.8V TRIAD F-40X
D1-D4—1N1692, 1N4001 OR 1N5059
ALL TRANSISTORS ARE 2N3638

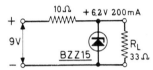

6.2 V D-C WITH POWER ZENER—Converts nonstabilized 9-V d-c supply to 6.2 V with stabilization factor of 50:1. Diode must be mounted on heat sink. Hum and ripple from source are attenuated in same ratio.— "Zener Diodes and Their Applications," Philips, Pub. Dept., Elcoma Div., Eindhoven, The Netherlands, No. 17, 1966.

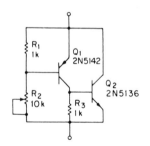

ADJUSTABLE ZENER—Performs functions of zener diode, with R2 determining operating voltage at any value above about 0.8 V. Impedance is maximum of 3 ohms below 5 V, where conventional zeners have considerably higher impedance and are not always readily available.—K. Karash, *Adjustable, Low-Impedance Zener, EEE,* March 1970, p 130.

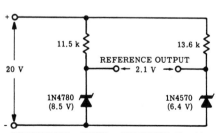

DIFFERENTIAL ZENER REFERENCE—Provides 2.1-V reference by utilizing difference between 8.5-V and 6.4-V zener references. Technique is particularly useful for temperature-compensated reference voltage sources. —"Zener Diode Handbook," Motorola, Phoenix, Ariz., 1967, p 6—19.

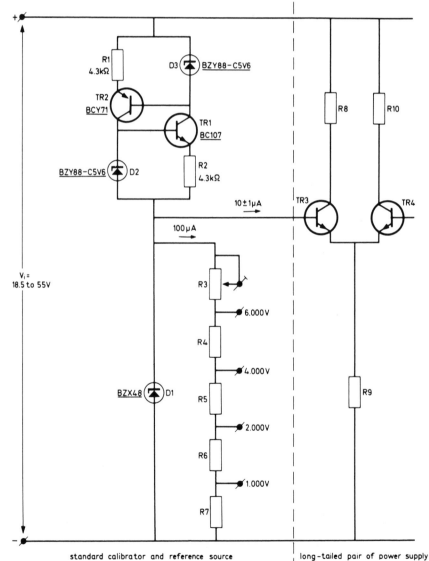

standard calibrator and reference source   long-tailed pair of power supply

MULTIPLE-VOLTAGE REFERENCE SOURCE—Stabilizing circuit at left of dashed line, using zener D1 in series with two-transistor constant-current source, can be incorporated between differential amplifier and any supply between 18.5 and 55 V d-c, for excellent voltage and temperature stability. When D1 is changed, R3 can be adjusted to give same reference voltages and 6.5 V d-c across entire resistor string R3-R7.—High Stability Reference Diodes BZX47 Family, Philips, Pub. Dept., Elcoma Div., Eindhoven, The Netherlands, No. 331, 1968.

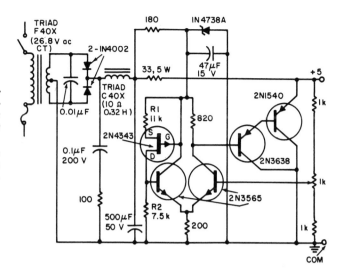

5-V FET REFERENCE—Use of p-channel 2N-4343 fet as constant-current source in combination with resistor gives much better temperature stability than convenional zener reference. Ouput impedance is below 0.1 ohm up to full load of 180 mA.—H. Olson, Stabilize Voltage Regulator by Replacing Zener with a FET, *Electronic Design*, Sept. 27, 1967, p 72 and 74.

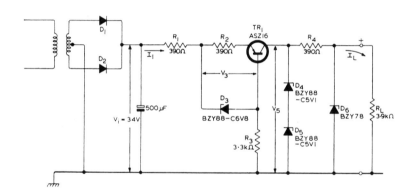

DIFFERENTIAL-AMPLIFIER SUPPLY—Provides highly stabilized output of about 5.3 V at 1.5 mA, as required by many differential amplifiers. Change of 20% in input voltage changes output less than 0.011%.—"Voltage Regulator (Zener) Diodes," Philips, Pub. Dept., Elcoma Div., Eindhoven, The Netherlands, Application Book, p 46.

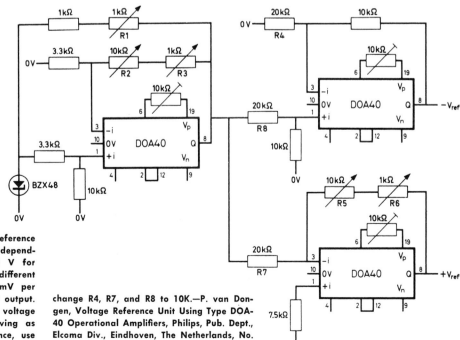

POSITIVE AND NEGATIVE 5 V—Reference voltages of opposite polarity are independently adjustable to same value, 5 V for resistor values shown, or 10 V with different values. Voltage drift is only 0.2 mV per deg C. Can deliver up to 5 mA per output. Uses three Philips DOA40 opamps as voltage comparators, with BZX48 zener serving as reference voltage. For 10 V reference, use 15-V positive and negative supplies and change R4, R7, and R8 to 10K.—P. van Dongen, Voltage Reference Unit Using Type DOA-40 Operational Amplifiers, Philips, Pub. Dept., Elcoma Div., Eindhoven, The Netherlands, No. 65.

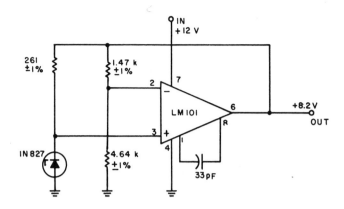

8.2 V D-C REFERENCE—Opamp and five components provide 0.01 mV per V d-c regulation for low-current applications (10 mA) such as for vco's in feedback loops and for precision digital-analog converters. Other output voltages are obtained by changing zener and resistor divider values.—F. R. Shirley and L. Vanderlosk, A Stable Voltage Reference Uses Only Six Components, *Electronic Design*, Jan. 18, 1968, p 128.

CONSTANT CURRENT FOR ZENER—Precision voltage obtained by sending constant current through temperature-compensated reference diode D1 is compared with voltage drop across R2 by differential amplifier A1. Amplifier adjusts its output to produce constant voltage drop across R2 for driving constant current through D2. Output voltage varies less than 2 mV for load currents of 0 to 300 mA and supply voltages of 15 to 40 V.—S. Miller, Precision Voltage Reference Combined with Voltage Regulator, *Electronic Design*, Sept. 1, 1969, p 102 and 104.

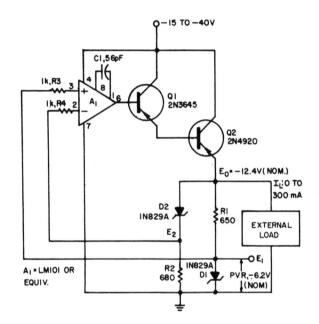

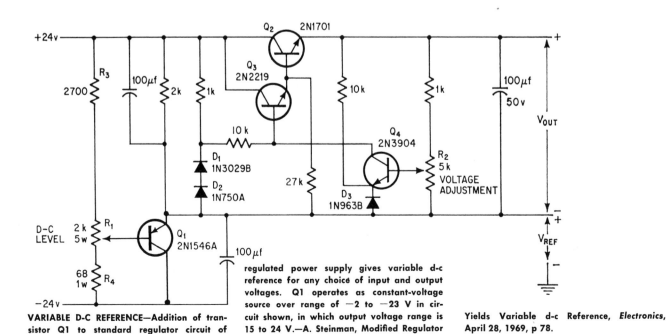

VARIABLE D-C REFERENCE—Addition of transistor Q1 to standard regulator circuit of regulated power supply gives variable d-c reference for any choice of input and output voltages. Q1 operates as constant-voltage source over range of −2 to −23 V in circuit shown, in which output voltage range is 15 to 24 V.—A. Steinman, Modified Regulator Yields Variable d-c Reference, *Electronics*, April 28, 1969, p 78.

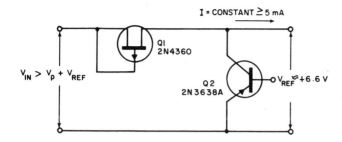

6 V AT 5 MA—Utilizes collector-emitter breakdown characteristic of bipolar transistor and constant-current property of junction fet to give low-cost reference source having output variation of only 3%.—E. J. Kennedy, Inexpensive 6-V Reference Is Also Temperature-Stable, *Electronic Design*, Nov. 8, 1967, p 112 and 114.

STANDARD-CELL REPLACEMENT — Provides constant output of about 4.5 V over temperature range of 30 to 60 C, is undamaged by intermittent short-circuits, and much more shock-resistant than Weston standard cell. D1-D6 are BZY88-C6V2 zeners and D7 is BZY88-C4V3. Values of R3 and R5 are determined by experiment as described in book. —"Voltage Regulator (Zener) Diodes," Philips, Pub. Dept., Elcoma Div., Eindhoven, The Netherlands, Application Book, p 26.

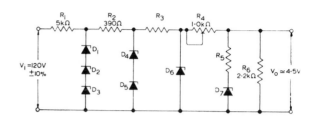

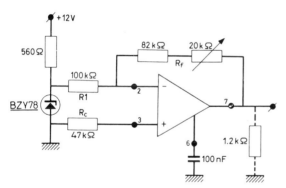

ADJUSTABLE REFERENCE—Uses Philips TAA-241 difference amplifier in circuit with 5.3 V zener. Can be adjusted to steady value close to −5 V with Rf. Rc provides compensation for bias current drift. Temperature drift of circuit is 0.5 mV per degree C. —J. Cohen and J. Oosterling, Applications of a Practical D. C. Difference Amplifier, Philips, Pub. Dept., Elcoma Div., Eindhoven, The Netherlands, No. 321, 1968.

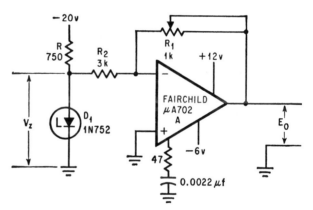

ADJUSTABLE REFERENCE VOLTAGE—IC op-amp and resistor divider translate zener diode reference to any desired precise voltage in range from −5 to +5 V, at low cost. Useful when required voltage reference is low, as for powering thermistor bridge. R1 adjusts reference voltage.—J. Althouse, IC Amplifier Provides Variable Reference Voltage, *Electronics*, Oct. 17, 1966, p 88.

8.2 V—Used as reference voltage supply for 100-V series-regulated d-c power supply.— "Silicon Power Circuits Manual," RCA, Harrison, N.J., SP-51, p 216.

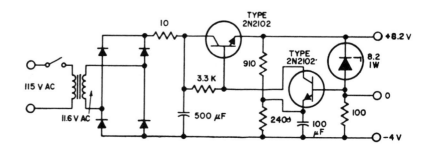

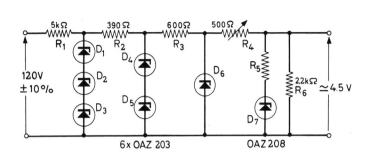

4.5-V ZENER SUBSTITUTE FOR WESTON CELL —Provides reference voltage between 4 and 6 V with better constancy than standard cell, despite a-c line voltage fluctuations up to 10% and temperature variations between 30 and 60 C. Supply voltage is first stabilized by shunt-regulated circuits connected in cascade. Final zener has opposite temperature coefficient to that of preceding diode, to eliminate effect of temperature.—S. T. Ho, Reference-Voltage Unit Equipped with Zener Diodes, *Electronic Applications*, Philips, Pub. Dept., Elcoma Div., Eindhoven, The Netherlands, Vol. 24, No. 4, 1963–1964, p 153–161.

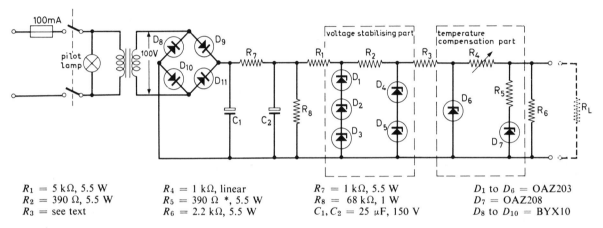

R_1 = 5 kΩ, 5.5 W	R_4 = 1 kΩ, linear	R_7 = 1 kΩ, 5.5 W	D_1 to D_6 = OAZ203
R_2 = 390 Ω, 5.5 W	R_5 = 390 Ω *, 5.5 W	R_8 = 68 kΩ, 1 W	D_7 = OAZ208
R_3 = see text	R_6 = 2.2 kΩ, 5.5 W	C_1, C_2 = 25 μF, 150 V	D_8 to D_{10} = BYX10

4–6 V SECONDARY STANDARD—Bridge rectifier across secondary of power-line transformer energizes zeners providing output voltage between 4 and 6 V, depending on components, having greater constancy than standard cells. Requires calibration against standard cell or equivalent-accuracy measuring instrument. Value of R5 is based on assumed permissible output voltage variation of 0.1 mV for 10% line voltage variation.—S. T. Ho, Reference-Voltage Unit Equipped with Zener Diodes, *Electronic Applications*, Philips, Pub. Dept., Elcoma Div., Eindhoven, The Netherlands, Vol. 24, No. 4, 1963–1964, p 153–161.

NOTE: To locate additional circuits in the category of this chapter, use the index at the back of this book. Check also the author's "Sourcebook of Electronic Circuits," published by McGraw-Hill in 1968.

SECTION 62
Zero-Voltage Detector Circuits

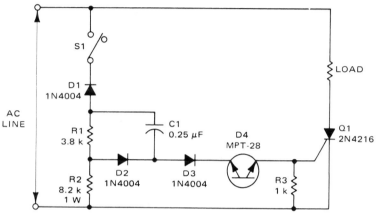

SENSITIVE-GATE SCR SWITCH—Zero-point switch ensures that control scr turns on at start of each positive alternation. If turned on later in cycle, voltage and current spikes produced could cause electromagnetic interference. Circuit actually oscillates near zero-crossing point and provides series of pulses to assure zero-point switching.—"Semiconductor Power Circuits Handbook," Motorola, Phoenix, Ariz., 1968, p 6–5.

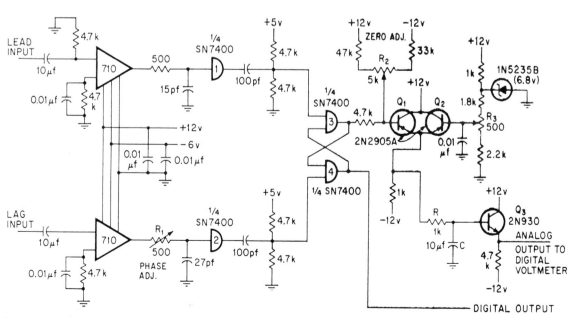

ZERO-CROSSING DETECTOR—Two voltage comparators and logic gate feed digital voltmeter to form digital-output phase comparator giving 1% accuracy at 1 MHz. Voltage comparators detect zero crossings of each input signal, for amplification by SN7400 TTL gate, differentiation, and setting of latch gates 3 and 4 for driving analog switch at duty cycle determined by phase relationship of input signals. D-c complement of resulting pulse train is applied to digital voltmeter.—R. H. Gruner, Phase Comparator Yields Digital Output, *Electronics*, Sept. 15, 1969, p 118.

RFI-FREE A-C SWITCH—Triacs close circuit only when line voltage is near zero and open it only when current is near zero, to minimize rfi. With S1 open, T1 fires at beginning of each a-c half-cycle and prevents T2 from firing. When S1 is closed, T1 commutates off at next zero crossing and R2 supplies gate current to T2 so it fires at beginning of each following half-cycle until S1 is opened.—W. B. Miles, Switch Your AC Loads at Zero Voltage or Current, *Electronic Design*, Sept. 13, 1967, p 128, 130, and 132.

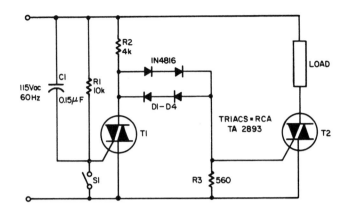

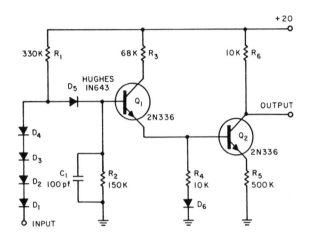

ZERO CROSSOVER—Gives output when positive-going signal is within 0.4 V of zero crossover, over temperature range of −55 to 80 C, for input signal peaks of up to 100 V in both directions and frequencies from d-c to 3,000 Hz. To reduce detection range to within 0.2 V of zero, diodes should be selected to have same forward voltage characteristics.—F. E. Olson, Zero Crossover Detector Shows High Sensitivity, *Electronic Design*, Dec. 7, 1964, p 67—68.

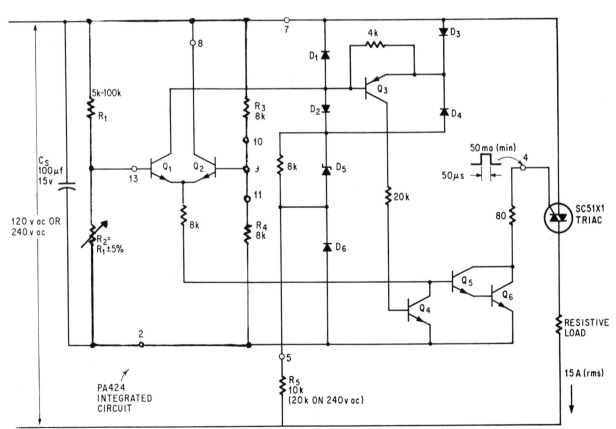

IC WITH TRIAC CONTROLS 1 KW—Uses GE PA424 IC with only five external components, by taking advantage of zero-voltage switching. Triac is fired only at zero-crossing points between half-cycles of a-c voltage. Requires no d-c power supply, and can control up to 15 A through resistive load. Article gives other power control applications for this IC. For temperature control, thermistor is used for R1.—F. W. Gutzwiller and J. H. Galloway, Power Grab By Linear IC's, *Electronics*, Aug. 21, 1967, p 81—86.

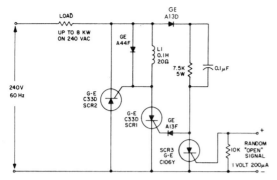

ZERO-VOLTAGE 8-KW SWITCHING—Combination of scr slaving circuit and synchronous switching insures that load voltage is always applied in essentially complete cycles, so voltage is zero at instant of switching and r-f interference is minimum. SCR2 fires at beginning of next half-cycle after SCR1 fires, to deliver even number of half-cycles and thereby reduce magnetic saturation effects in inductive load.—"SCR Manual," 4th Edition, General Electric, 1967, p 149.

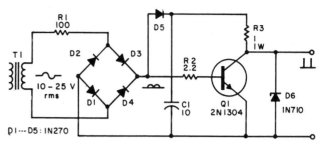

SYNC FOR FIRING SCR—Simple, inexpensive circuit requires no supply voltage or center-tapped transformer and can be used with any waveform. Used when precise synchronization with a-c line frequency is required, as in scr firing circuits. Output pulse is about 200 μs wide, with rise time of 20 μs. Q1 is normally ON; when signal level drops below that for conduction, in vicinity of zero crossing, Q1 cuts off and positive-going pulse appears at collector and output.—R. Billon, Zero-Crossing Detector Needs No Supply Voltage, *Electronic Design*, Jan. 4, 1968, p 144—145.

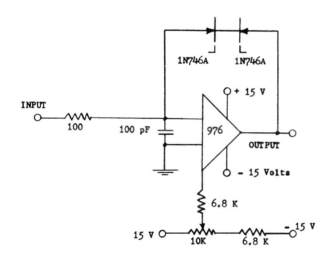

HIGH-SPEED ZERO-CROSSING DETECTOR—Uses OEI Model 976 opamp having slewing rate of 250 V per μs and minimum gain of 50 dB at 1 MHz. Requires negative input signal up to 1 V maximum and is inverting, with output of approximately 3.3 V when input is negative and with output transition occurring within nanoseconds of input zero crossing.—A High Speed Zero Crossing Detector Using an Operational Amplifier, Optical Electronics, Tucson, Ariz., No. 10041.

SCR A-C STATIC SWITCH—Used for zero-point switching of resistive load on a-c line. Application of control signal makes scr Q1 turn on during first positive half-cycle after signal is applied. Q1 provides gate current for turning Q2 on. Q2 charges C1 through D2 and D3 to peak line voltage. When D2 becomes reverse-biased by decaying line voltage, C1 discharges and turns on Q4 for desired switching action. Will not handle large inductive load.—"Semiconductor Power Circuits Handbook," Motorola, Phoenix, Ariz., 1968, p 4—8.

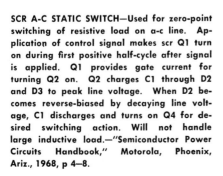

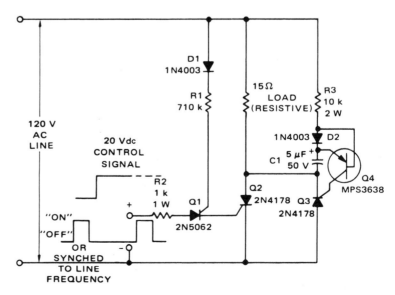

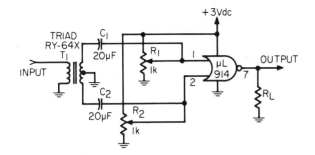

200—10,000 HZ ZERO-VOLTAGE DETECTOR— Delivers output pulse whenever a-c input waveform goes through zero. Applications include counters, phase-control circuits, and some types of analog-digital converters. Provides 1.2 V peak output for 1 V rms input.—T. Polaneczky and A. Brand, IC NOR Gate Detects Zero-Axis Crossing, *EEE*, April 1968, p 107.

PRECISION ZERO-CROSSING DETECTOR— Uses OEI Model 9186 opamp with feedback diodes that limit output to 0.5 V with either polarity, depending on polarity of input before crossing zero. Output lags actual zero crossing by 90 deg because of amplifier phase shift, which means that detector output will be zero 95 ns after input is zero for 10-MHz signal.—A Precision Zero Crossing Detector, Optical Electronics, Tucson, Ariz., No. 10114.

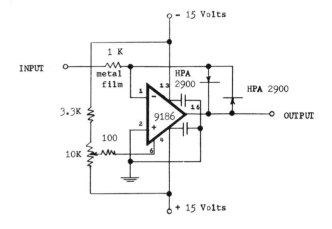

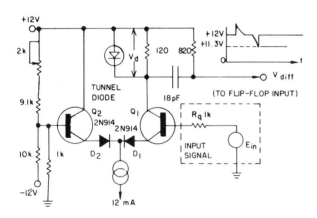

ZERO-CROSSING DISCRIMINATOR — High-speed circuit has broad dynamic range, high input resistance, and high input sensitivity. Collector currents of Q1 and Q2 are equal only when input signal is zero. When negative input signal is fed to Q1, collector current and voltage drop Vd decrease. When drop goes below tunnel diode voltage, it switches from high to low state, corresponding to low-level discrimination. With tunnel diode also in Q2 collector circuit, zero crossings of sine-wave input can be discriminated. —M. Sampaleanu, High-Speed Zero Crossing Detector Uses Tunnel Diodes, *Electronic Design*, March 15, 1970, p 214 and 216.

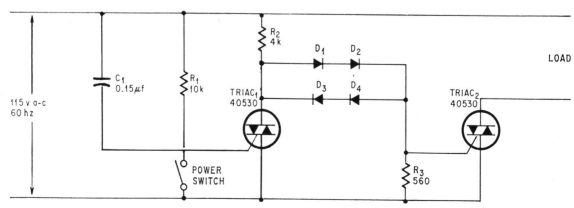

NOISE-FREE TRIAC A-C SWITCH— Permits switching of a-c voltages and currents only at zero points, to prevent generation of rfi that could destroy transistors.—W. B. Miles, Gated Semiconductors Clean A-C Switching, *Electronics*, May 27, 1968, p 105—106.

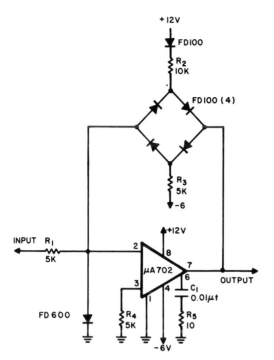

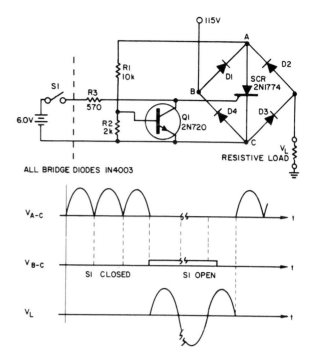

ZERO-CROSSING DETECTOR—Uses Fairchild IC opamp to perform power spectrum analysis of analog signal waveform, within specified passband, that is extracted from broadband signal and noise. Amplifier is driven to maximum output of 1.5 V for input voltage excursions as small as 500 μV to give uniform clipping of random analog waveform with negligible hysteresis and negligible loss of average reference level.—L. A. Watts, The Integrated Operational Amplifier: A Versatile and Economical Circuit, "Microelectronic Design," Hayden Book Co., N.Y., 1966, p 189–194.

ZERO-VOLTAGE SWITCHING—Regardless of time S1 is closed, scr is turned on or off only at zero-voltage crossover point of a-c line voltage, for minimizing power-line rfi.—A. J. Marek, Simple Zero Crossing Detector Minimizes Power-Line RFI, *Electronic Design*, March 1, 1970, p 81.

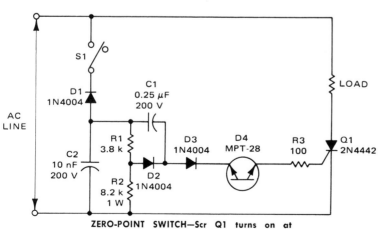

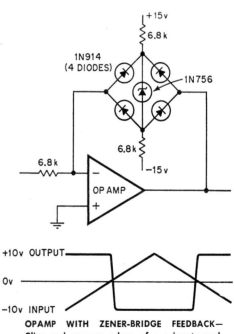

ZERO-POINT SWITCH—Scr Q1 turns on at start of first positive half-cycle following closing of S1, to prevent generation of electromagnetic interference. Book describes capacitor-charging action of circuit.—"Semiconductor Power Circuits Handbook," Motorola, Phoenix, Ariz., 1968, p 6–5.

OPAMP WITH ZENER-BRIDGE FEEDBACK—Clips and squares edges of a-c inputs each time zero crossover is detected. Useful in phase-sensitive demodulation networks. Positive amplitudes of input generate negative output pulse, while negative inputs generate positive voltage level.—R. Liu, Zener Diode in Op Amp's Loop Enables Symmetrical Clipping, *Electronics*, Feb. 16, 1970, p 105.

Index

Absolute-temperature sensing, 227
Absolute value circuits:
  amplifier, 304
  opamp, 300
  rectifier, 310
AC control, full-wave, 356
AC-coupled multiplier, 277
AC-generator cutout, 370
AC-line monitors, 201, 370, 499
AC noise-free switch, 512
AC ohmmeter, 223
AC overvoltage, 371
AC switch:
  perfect, 353
  RFI-free, 510
  static, 511
AC voltage, variable, 176
Acoustic pickup, preamp for, 293
Active filters:
  CW, 121, 122, 138
  DC level shifter, 142
  with emphasis, 123
  four-function, tunable, 300–3000 Hz, 141
  gyrator, 10-kHz, 119
  line-frequency, 116
  octave audio equalizer, 129
  speaker crossovers, 120
  for speech, 122
  three-function, 1-kHz, 126, 132
  tracking: 1-MHz, 129
    10-MHz, 141
  variable-Q, 117, 123, 139
  variable-Q and frequency, 137
  voltage-tuned: 1 Hz to 500 kHz, 128
    10-kHz, 121
  wideband, 300–3000 Hz, 123
  (See also Bandpass filters; High-pass filters;
    Low-pass filters; Notch
    filters; State-variable filters)
Active-high triac interface, 355
Active load for lighting control, 174
Active-low triac interface, 357
A/D converter circuits:
  0–10 V input, 66
  4-μs, 65
  with 10-bit accuracy, 64
  autoranging, 63, 65
  BCD output for, 69
  buffers for, 66, 67, 71
  compressing, 70
  cyclic converter for, 66
  latch comparators for, 49
  ratiometric, 68
  repetitive-mode, 68
  serial data output, 62
  software-controlled, 63
  successive approximation, 64, 69, 70

A/D converter circuits (cont.)
  tracking, 67, 70
  video compressor, 64
  for voice digitizer, 71
A/D-D/A converter interface, 268
Adjustable-Q notch filter, 122
Adjustable-width pulse generator, 382
AF circuits:
  amplifiers, 6
  for AM/SSB/CW, 396
  attenuator, voltage-controlled, 7
  bandpass filter, variable-Q, 139
  clamping of, with opamps, 7
  clipper powered by, 294
  compander, 9
  compressor, 7
  compressor/limiter, 5
  constant level, 9
  current amplifier, 303
  DC level shifter for, 142
  differential-amp clipper, 6
  distortion meter for, 211
  equalizers, 7, 129
  expander, 8
  exponential control, 6
  filters: pi-section, 115
    switchable, 115
    variable-Q, 139
  function generator, 149
  low-pass filter, 128, 138
  measuring circuits: distortion meter, 182
    frequency meter, 181
    level detector, 181
    peak program meter tester, 180
    S-meter, 181
    voltmeter, 180
  mixer pot, panning, 9
  noise limiter, 296, 298
  notch filter, 130, 136
  oscillator circuits (see AF oscillator circuits)
  PLL, 344
  preamplifier, noise gate for, 297
  preemphasis, 10
  rectangular-wave generator, 379
  relay-operated, 6, 7
  speech processor, 8
  square-wave, 383
  suppressor, switching-click, 8
  switching gate, 9
AF oscillator circuits:
  10-Hz, Wien-bridge, 323
  15 Hz to 150 kHz, 321
  20–20,000 Hz, 325, 326
  20-Hz to 200-kHz, 327
  25-Hz, sine-wave, 324
  50–30,000 Hz, Wien-bridge, 324
  60-Hz, Wien-bridge, 328

AF oscillator circuits (cont.)
  100-Hz, Wien-bridge, 325
  200–65,000 Hz, Wien-bridge, 327
  340–3400 Hz, single-pot, 327
  350-Hz, sine-wave, stabilized, 324
  400-Hz, sine-wave, LED-opamp 325
  400 Hz, 1000 Hz, 2125 Hz, and 2975 Hz,
    321
  800 Hz, single-transistor, 323
  1-kHz: fast-start gated, 324
    low-distortion, 322
    one-IC, 321
    for VCO, 329
    Wien-bridge, 324, 328
  1–50 kHz, sine-wave, 321
  1.46-kHz, sine-wave, 322
  1.6-kHz, sine-wave, 328
  2-kHz, two-phase, 320
  2.34-kHz, sine-wave, 321
  3.8 kHz, 323
  10.5-kHz, Wien-bridge, zener-controlled,
    326
  for CW monitoring, 44
  dot generator, 323
  MOS, 320
  negative-resistance LED, 326
  tone bursts, 0.5-s, 322, 326
  tuning capacitor simulator, 325
  (See also Sine waves and sine-wave
    circuits, oscillators; Wien-bridge
    oscillators)
AFC amplifier, 257
AGC circuits:
  50-dB, 154
  high-accuracy, 155
  limited, for oscillator, 154
  with multiplier, 157
  multiplier module for, 153
  optimizing, with FET, 156
  pulse, boxcar for, 157
  for SSB, 158
  voltage-controlled, 158
Alarms:
  fire, infrared, 165
  headlights-on, 12, 13
  intrusion, 159, 160
  malfunction, microprocessor, 235
  missing-pulse, 350
  rain, 209
  RPM-limit, 15
  touch, 35
All-band preamplifier, 393
Alternating current (see AC)
Alternator, regulator for, 16
AM circuits:
  AF for, 396
  broadcast rejection filter, 115

AM circuits (*cont.*)
  demodulation distortion meter, 211
  detector, synchronous, 391
  modulators, 250
    6-meter, 244
    with analog multiplier, 249
    double-sideband, 243
    single-IC, 247
    single supply, 243
    suppressed-carrier, 246, 249
  noise silencer, 294
  radio, 392
    with metal detector circuit, 233
  receiver, 391, 394
  superhet subsystem, 387
Ambiguity eliminator, voltage monitor with, 496
Ammeters:
  AC, 5-A, 198
  DC, 196
  microammeter, 194, 195, 198
    amplifier for, 195
  nanoammeter, 170, 193
Amplifiers:
  absolute value, 304
  acoustic pickup, 293
  AF, 6
  AFC, 257
  camera pulse, 160
  capacitance-meter, 184
  current, AF unity-gain, 303
  DC, 38, 198
  differential-input, 168, 170
  dual sense, 499
  erasing, 157
  gain-controlled, 312
  galvanometer, 193, 194
  high-gain, 169
  hodoscope, 172
  lamp, 165
  light-wand, 318
  line-operated, for servo, 458
  magnetometer, 211
  meter, 167, 184, 194, 196
  microammeter, 195
  power, 251, 456
  pulse, 160, 309
  rezeroing, 101
  for servo, 400-Hz, 454
  solid-state-lamp, 165
  squelchable, 6
  strain-gage, 167
  summing, 303, 305, 313
    without adjustments, 309
  thermocoupler, 225, 229
  tremolo, 289
  voltage-controlled, 291
  for weak signals, 169
  (*See also* Postamplifiers; Preamplifiers)
Amplitude modulation (*see* AM)
Analog circuits:
  isolator, 316
  memory for music circuit, 289
  multipliers, 249, 274
  switch, 270
  voltage sensor, 495
Analog-to-digital converters (*see* A/D converter circuits)
Analyzer, probability density, 210
Anemometer, 209
Answer filter, bandpass, 105
Answerer, modem, 96
Antenna circuits:
  attenuator, 3
  bridges: RF, 3
    SWR, 4
    VSWR, 2

Antenna circuits (*cont.*)
  far-field transmitter, 1
  loop preamp, 3
  meters: field-strength, 2
    VSWR, 1
  modulator monitor, 2
  Q multiplier, 2
Arcing, relay, triac suppressor for, 371
ASBS-triac for lighting control, 179
ASCII generator, serial, 235
Astable multivibrators, 279
  20-kHz, 280
  direct-coupled, 283
Astable pulse generators, 281, 386
  extended-range, 383
Asymmetric pulse generator, 383
Attack-decay generator, 293
Attenuators:
  5-step, 3
  Darlington, 155
  voltage-controlled, 7
Audible circuits:
  capacitor tester, 183
  light sensor, 348
  line monitor, 370
  metal detector, 233
  metronome, 286
Audio frequency (*see* AF circuits)
Automatic circuits:
  calibrator for light circuits, 169
  range selector, 502
  remote rhythm control, 289
  shutoff for battery charger, 21, 22
  voltmeter, 490
Automatic gain control (*see* AGC circuits)
Automobile circuits:
  battery, capacitor as, 15
  battery charger, 20
  battery monitor, 11
  breaker points, transistorized, 13
  electric vehicle control, 12
  headlights alarm, 12, 13
  ignition: CD, 12, 14
    cold-weather, 11
  light interface for, 16
  regulators, 11, 13, 16
  rpm limit alarm, 15
  tach/dwell meter, 12
  tachometer, 16
  wiper-delay control, 15
Autopatch circuits, 483, 487
Autoranging, A/D converters with, 63, 65
Autoreference in D/A converter, 94
A/X function generator, 146

Balance detector, bridge, 30
Balanced chopper, 36
  modulator for, 37
Balanced-line transmission, 99
Balanced mixers, 391, 396
Balanced modulators, 244, 246
  chopper, 37
  DSB, 247
  SSB, FET, 248
Ballast, rapid-start, 173
Band-reject filter, touch-tone, 483
Bandpass filters:
  1-Hz, 127, 131
  1 Hz to 500 kHz, 128
  15–3500 Hz, 117
  20–2000 Hz, 135
  100-Hz, 127
  200–400 Hz, 136
  700–2000 Hz, 141
  800-Hz, 139

Bandpass filters (*cont.*)
  1-kHz, 131, 140
    cascaded-opamp, 132
    high-Q, 141
    multiple-feedback, 126
    state-variable, 136
  1.4-kHz, twin-T, 130
  5-kHz, series-switched, 118
  10–20 kHz, 118
  20-kHz, 126
  1.8–1.9 MHz, 139
  50.5 MHz, 112
  160-meter, 112
  AF: variable-bandwidth, 135
    variable-Q, 139
  answer, for phones, 105
  commutating, 270
  narrow, for speech, 122
  originate, for modems, 97
  RTTY, 225-Hz, 115
  Sallen-key, 127
  voice, 114
  wideband, 123
Bar-code circuits:
  reader, 315
  signal conditioner, 317
Batteries:
  ignition, capacitor as, 15
  monitor for, 11
  switching regulator circuit for, 465
Battery chargers:
  9.6 V at 20 mA, 17
  12-volt, 18–23
  14-volt, 21
  18-volt monitor for, 22
  adjustable finish-charge, 17
  auto, 20
  automatic shutoff, 21, 22
  constant-current, 23
  gelled-electrolyte, 23, 24
  monitors for, 19, 22
  nickel-cadmium, 18–24
  overcharge protection for, 21
  silver-zinc, 18
  solar-power, 23
  solar-power backup for, 19
  third electrode, 19
  trickle, 20
  UJT, 22
  voltage indicator for, 19, 20
BCD:
  for A/D converter, 69
  comparator for, 53
  conversion of binary to, 85
  conversion of excess-three Gray code to, 79
  conversion of 7-segment display to, 84, 88
  D/A converter with, 90, 92, 93
Beat-frequency locator in metal detector circuit, 231
Bell:
  musical, 290
  phone, simulator for, 479
Bell-system, D/A converter for, 95
Bench regulated power supply, 0–15 V, 414
BFOs:
  20-meter, 46
  multiplexer with, 270
Bias, op amp, 29
Binary, conversion of:
  to BCD, 85
  from Gray code, 82
  to process current, 89, 94
Binary-coded decimal (*see* BCD)
Binary multiplier, 274
Bipolar power supply, 10-volt, 27

Blanker, noise, 295, 296
Blown-fuse indicator, 367
Body-capacitance control, 32
Boosters:
    current: 100-mA, 306
        1-A, 312
        8-A, 311
        regulators for, 446, 448
    power, 307, 308
    for voltage rating, opamp, 306
Bounceless circuits (see Contact-bounce
        suppression circuits)
Boxcar pulse for gain control, 157
Brakes, motor control, 261, 265
Breaker points, transistorized, 13
Break-in, fast, 47
Bridge circuits:
    10-volt bipolar power supply from, 27
    balance detector, 30
    comparator-controlled FET, 25
    diamond gate, 30
    exclusive-OR, 27
    impedance, 209, 214
    with lamps, 26
    Meacham, 336
    null meter, logarithmic, 28
    opamp bias, 29
    phase-measuring, 26
    photocell, 171
    potentiometer, 27
    quadrupler, 27
    reference, −4000 V, 503
    resistance, linear, 219
    RF, for coax, 3
    RF admittance, 208
    RLC, 223
    SCR control, three-phase, 25
    series-gate, 30
    shunt gate, 29
    solar-cell isolation, 28
    SWR, self-excited, 4
    thermistor thermometer, 29
    triac trigger, 355
    tube warmup control, 27
    VSWR, 2
    for weak DC signals, 30
    Wheatstone, 224
    (See also Wien-bridge oscillators)
Buffered opamp, 301
Buffered rectifier, 100-kHz, 307
Buffers for A/D converter, 66, 67, 71
Bullet timer, 213
Burglar detector, 33
Burst peak detector, 491
Busy-signal generator, 484
Button, no-push elevator, 31

Calculator, 6 V for, 468
Calibrators:
    automatic, for light circuits, 169
    crystal, 202
        100-kHz, 201, 203, 206
        1-MHz, 206
    ham-band, 199
    receiver, 202
Call alert, 257
Call counter, 482
Call synthesizer, CQ, 44
Caller, four-number, 488
Camera pulse amplifier, 160
Camera time base, 163
Capacitance:
    body, control with, 32
    multipliers of, 274, 310
    oscillator tuned by, Wien-bridge, 340
    (See also Capacitor control circuits; Capaci-
        tors)

Capacitance measuring circuits:
    leakage, of electrolytics, 183, 187
    meters: 1 pF to 1 μF, 191
        4 pF to 0.1 μF, 183
        5 pF to 1 μF, 188
        15 pF to 10 μF, 185
        0.001–0.1 μF, 186
        0.1 μF, up to, 184
        adapter for counter for, 190
        amplifier for, 184
        counter as, 186
        digital, 189, 191
        five-range, 189, 190
        by frequency shift, 187
        oscillator for, 184
        seven-range, 185
        VOM, 185, 191
        VTVM, 186
    testers, 188
        audio-tone, 183
        electrolytic, 187
        neon-lamp, 187
        by substitution, 189
Capacitor control circuits:
    elevator button, no-push, 31
    intrusion detector, 33
    liquid detectors, 31, 32
    proximity detector, 34
    proximity switch, 32, 33
    touch alarm, 35
    touch switch, 34, 35
    variable capacitor, 34
Capacitors:
    as ignition battery, 15
    liquid-sensing, 32
    tuning, simulator for, 325
    variable, 34
Carrier detect, FSK demodulator with,
        102
Carrier-system circuits:
    receiver, 252
    transmitter, 254
Cascaded zener reference, 503
Cassette recorders:
    regulated power supply for, 400
    regulator for, 7.5 V, 426
CATV, regulator for, 478
CB wattmeter, 216
CD ignition, 12
    low-emission, 14
Cell replacement, 507, 508
Chopper circuits:
    amplifier, low-level DC, 38
    balanced, 36, 37
    D/A converter, 38
    driver, 37
    fast switch, 39
    FET, 38
    full-wave photoelectric, 38
    gain control, 155
    lamp driver, 38
    modulators, 244
        DC, 37
        linear, 250
        noise canceller, 40
    opamp, 40
    photochoppers, 38, 39
    series-shunt, 36
    for servo-driven pen, 39
    transformerless, 39, 40
Clamp, 600-Hz, 370
Clamped comparator, 50, 52
Clamped Darlington for motor control circuit,
        262
Clamping, with op amps, 7
Clicking metronome, 291
Clinical thermometer, 162

Clippers:
    AF-powered, 294
    differential-amplifier, 6
Clock comparator, power control for, 352
CMOS circuits:
    A/D converter, 70
    driver for opamp, 303
    inverter, crystal oscillator, 336
    keyer, 45
    postamplifier, 307
    pseudorandom code, 105
    receiver, 390
CMR, optimizing, 302
Coax:
    line drivers for, 97, 99, 100, 105
    RF bridge for, 3
Code circuits:
    BFO, 20-meter, 46
    CQ, 42, 44
    CW, 43–45, 48
    direct-conversion, 40-meter, 41
    fast break-in, 47
    filter, CW, 48
    identifier, CW, 48
    keyers, 42–45, 47
    monitors, 45, 47
    Morse-code set, 42
    product detector, 46
    timer, 43
    tone detector, 46
Coded infrared source, 163
Coded-light detector, 162
Coded lock, 368
Coded switch, 10-digit, 369
Coder-decoder, 101
Coin finder, 232
Cold-weather ignition, 11
Color TV, power supply for, 410
Colpitts oscillators, 333, 336
Comb filter, 1-kHz, 270
Communication detector, 164
Communication source, 164
Commutating bandpass filter, 270
Compander, 100-dB range, 9
Comparators:
    BCD, 4-bit, 53
    clamped, 5-volt, 50
    clock, power control for, 352
    to drive lamps, 55
    to drive LED, 56
    dual limit, 57
    FET bridge controlled by, 25
    frequency, 52, 54
    hysteresis, 53, 54
    for independent signals, 53
    internally gated window, 51
    latch, for A/D converter, 49
    level-crossing display, 52
    measuring threshold of, 56
    microvolt, 50
    oscillator-suppressing, 50
    phase, 50, 52
    for PWM, 244
    slew rate, 54
    staircase window, 51
    strobed, 50, 55
    three-level, 55
    variable clamping, 52
    variable window, 50
    voltage, 53
    voltage-window, 57
    zener reference, 54
Complementary postamplifier, 314
Compressing A/D converter, 70
Compressors:
    AF, 7
    with limiter, 5

Compessors (*cont.*)
  video, 64
Computer music, audio for, 287
Computers, power supply for, 399
Conductivity gauge, 222
Conductivity meter, 221
Conductivity probe, 220
Constant AF level, 9
Constant-brightness LED, 179
Constant-current circuits:
  battery charger, 23
  for zener, 506
Constant-output limiter, 156
Contact-bounce suppression circuits:
  counter, 58, 61
  delayed start, 59
  gate-formed switch, 59
  interlock, 61
  isolator, 60
  keyboard, 59
  latching gates, 61
  make/break switch, 61
  one-pulse-per-push, 60
  rotary switch, 60
  square wave, 58
Continuous-duty motor brake, 261
Controlled-duration pulses, 374
Controls:
  body-capacitance, 32
  duty-cycle, 283
  electric-vehicle, 12
  exponential, two-quadrant, 6
  garage-light, 348
  humidity, 212
  nicad battery charge, 24
  noise, 297
  on-off, photoelectric, 348
  SCR, 25
  soft-touch, 34
  sump-pump, 353
  touch, 35
  tracking quad gain, 302
  tremolo, 287
  tube warmup, 27
  wiper-delay, 15
  (*See also* Lighting control circuits; Motor control circuits)
Converter circuits:
  12 VDC to 115 VAC: at 100 W, 362
    at 400 Hz, 361
  24–60 VDC to 117 VAC, 364
  A/D-D/A interface, 268
  binary to BCD, 85
  current to voltage, 81, 195, 197, 198
  cyclic, for A/D converter, 66
  digital-to-analog, 38
  digital-to-frequency, 85
  Gray code to binary, 79, 82
  phase, 87
  pulse height-to-time, 85
  pushbutton-to-dial, 484
  seven-segment display to BCD, 84, 88
  square-wave to DC, 81
  square-wave to sine wave, 86, 147
  temperature-to-frequency, 227, 229
  temperature-to-pulse width, 228
  time-to-voltage,84
  triangle-wave to sine wave, 152
  voltage-to-current, 86
  voltage-to-frequency, 78–84, 87, 88, 313
  voltage-to-pulse width, 82, 86
  voltage-to-time, 80
  (*See also* A/D converter circuits; D/A converter circuits; DC-to-DC converters)
Core pulse-height detector, 492

Cosines, approximating, 278
Counters:
  adapter for, in capacitance meter, 190
  as capacitance meter, 186
  debouncing with, 61
  and detector, 494
  Geiger, 192, 197
  phone-call, 482
  programmable, 144
  switch-closure, 58
CQ circuits:
  call synthesizer, 44
  recorder, 42
CROs:
  3000 V for, 365
  multiplexer for, 272
Crossover, speaker, active filter for, 120
Crowbar, 367, 371, 452,
CRTs:
  130 V and 270 V for, 362
  1000 V for, 360
Crystal calibrators, 202
  100-kHz, 201, 203, 206
  1-MHz, 206
Crystal detector, FM, 251
Crystal discriminator, 256
Crystal filters:
  diode-switched, 113, 114
  in metal detector circuit, 231
Crystal multivibrators, 280, 281
Crystal oscillators:
  20–500 kHz, 337
  50–500 kHz, 337
  100-kHz, relaxation, 335, 336
  150–500 kHz, 339
  279.611-kHz, 332
  465-kHz, 341
  1-MHz, series-mode, 331
  1–20 MHz, 333
  2–20 MHz, 340
  2–22 MHz, 339
  3–20 MHz, 342
  4.8-MHz, 337
  5-MHz, low-noise, 332
  7-MHz, 337
  9.5-MHz, tunable, 341
  10–20 MHz, 333
  15–65 MHz, impedance inverting, 341
  42.667-MHz, 334
  50-MHz, 332
  65–110 MHz, overtone, 340
  90–125 MHz, 339
  with CMOS inverter, 336
  Colpitts, 336
  dual, 342
  modulated, 337
  NAND-gate TTL, 340
  Pierce, 334
  with switching, 331, 334
  variable, 335
Current:
  amplifier of, unity-gain, 303
  conversion of, to voltage, 81
  conversion of voltage to, 86
  process, conversion of binary to, 89, 94
Current boosters:
  100-mA, 306
  1-A, 312
  8-A, 311
  regulator for, 446, 448
Current-controlled circuits:
  function generator, 150
  Wien oscillator, 323
Current limiting, regulators with, 425, 436, 443
  0–20 V, 419
  5 V at 5 A, negative-input, 431

Current limiting (*cont.*)
  5 V at 10 A, 435
  ±15 V at 10 A, 436
  simultaneous, 438
  tracked, 423
Current measuring circuits:
  ammeters: AC, 5-A, 198
    DC, 196
  current-to-voltage converters, 81, 195, 197, 198
  DC amplifier, 198
  DC transformer, 196
  electrometers, 193–197
  galvanometer amplifier, 193, 194
  Geiger counter, 192, 197
  with high-gain preamp, 192
  meter amplifiers, 194, 196
  microammeters, 194, 195, 198
    amplifier for, 195
  nanoammeter, 193
  preamp for, 192
  supply current, 193
Current regulators:
  100-mA, 426
  negative input, 431
Current sensor, 372
Current-to-voltage converters, 81, 195, 197, 198
Cutout, AC generator, 370
CVSD encoder for secure radio, 99
CW:
  AF for, 396
  filters for, 48, 121
    bandpass, 112
    low-pass, 138
    two-stage, 122
    variable-Q, 122
  signals for, 43
    identifier for, 48
    oscillator to monitor, 44
    regenerated, 45
Cyclic converter for A/D converter, 66

D/A converter circuits:
  autoreference, 94
  BCD, 90, 92, 93
  Bell-system, 95
  binary-to-process current, 89, 94
  chopper, 38
  using digital switches, 91
  for digital-to-frequency conversion, 85
  four-quadrant algebraic, 95
  four-quadrant multiplying, 278
  high-speed, 90
  isolation of, 1500-V, 319
  ratiometric, 68
  for speech, 91
  sum of digital numbers, 92, 95
  temperature compensation in, 93
Darlington circuits:
  attenuator, 155
  clamped, for motor control, 262
  FET-bipolar, 170
  power, in servo driver, 459
Data coupler with fiber optics, 106
Data demodulator, 0–20 Hz, 169
Data links:
  10-megabit, 108
  24-megabit, 106
  optically coupled, 98
Data transmission circuits:
  balanced-line, 99
  coax drivers, 97, 99, 100, 105
  interface, 100-ohm line, 97
  line drivers: differential, 103, 104

Data transmission circuits (*cont.*)
    polarity-reversing, 98, 101, 102
    single-ended, 103
  optically coupled, 98
  PLL for, 104
  twisted-pair, 104
  wired-OR terminal, 105
  (*See also* Modem circuits)
DC amplifiers, 38, 198
DC isolator, 316
DC level shifter for AF, 142
DC modulator, 37
DC motor speed control, 266
DC multipliers, 76
DC signals, weak, amplifying, 30
DC-to-DC converters:
  2 V to 20 V and 150 V, 73
  5 V to −7 V, 76
  5 V to 8 V, 76
  5 V to ±12 V, 74
  5 V to 400 V, 75
  6 V to ±15 V, 73, 74
  12 V to −12 V, 72
  12 V to −8 V, 74
  12 V to 6 V, 361
  12 V to ±15 V, 77
  15 V to −12 V, 77
  15 V to ±15 V, 75
  15 V to 27 V, 77
  280 V to 600 V, 75
  multipliers, 76
  polarity reverser, 76
DC transformer, 96
DC voltage conversion:
  from square-wave, 81
  from time, 84
  to time, 80
Dead-band response, 299, 309
Debouncing (*see* Contact-bounce suppression
    circuits)
Decoders:
  FSK, 238
    1070 Hz and 1270 Hz, 241
    with VCO, 241
  Manchester, 109
  PCM, 373
  stereo, 252
  tone, 46
  touch-tone, 481, 482, 485, 486
Delay equalizer for modem circuit, 240
Delayed start for contact bounce, 59
Delayed turn-off for lighting control, 177, 179
Delta modulator, 246
Demodulation distortion meter, AM, 211
Demodulators:
  data, 169
  FSK, 241, 242
    with carrier detect, 102
    with DC restoration, 239
    PLL in, 104
    single-supply, 240
    split-supply, 239
  Manchester-code, 109
  narrow-band, 253
  PLL, 255
  quadrature, 255
  SCA, 252
  (*See also* Modulator/demodulator circuits)
Detectors:
  AM, synchronous, 391
  bridge balance, 30
  coded-light, 162
  communication, 164
  end-of-tape, 349
  fire, 160
  FM, 251, 254

Detectors (*cont.*)
  FSK, 102
  Hall-effect, 232
  indium antimonide, 161
  infrared, 109, 163
  intrusion, 33
  lead sulfide, 164
  level, AF, 181
  light-change, 349
  liquid level, 31
  missing-pulse, 384
  phase-sensitive, 278
  phase sequence, 264
  preamp for, 192
  product, 46, 394
  proximity, 34
  pulse-width, 281
  regenerative, FET, 388
  ring, 489
  single-tuned, 255
  window, 56, 492
  (*See also* Metal detector circuits; Voltage
     level detector circuits;
     Zero-voltage detector circuits)
Deviation meter, 256
Dialer, touch-tone, 489
Dial-tone generator, 479
Diamond gate, 30
Differential circuits:
  amplifiers, 6, 168, 170, 424
  I/O opamps, 67, 306
  input and output for servo, 457
  JFET input opamp, 301, 309
  line driver, 103, 104
  power supply, 5.3 V at 1.5 mA, 505
  thermocoupler, 226
  zener reference, 504
Digital control for RF oscillator, 334
Digital signals, modulator for, 248
Digital switches, D/A converter using, 91
Digital-to-analog converters (*see* D/A convert-
    er circuits)
Digital-to-frequency conversion, 85
Digitizer, voice, A/D converter for, 71
Dimmers, triac:
  400-W, 175
  800-W, 174
  with brightening, 175
  time-dependent, 178
Diode receiver, 21–75 MHz, 256
Diodes:
  for crystal switching, 114, 331
  for IF filter switching, 113
  tunnel, peak sensing with, 497
Dipper, 202, 203
Direct conversion:
  40-meter, 41
  CW/SSB, 20-meter, 393
  product detector, 394
  receiver, 397
Direct current (*see* DC)
Discriminators:
  crystal, 256
  transistor-pump, 256
  zero-crossing, 512
Displays:
  level-crossing, 52
  seven-segment, 84, 87
Distortion meters:
  AF, 182, 211
  AM demodulation, 211
Dividers:
  for marker, 204
  regulated, for ±5 V, 424
  (*See also* Frequency, dividers of)
Dot generator, 323

Double-sideband modulators:
  AM, 243
    suppressed-carrier, 246, 248
Doubler, frequency, 339
DSB balanced modulator, 247
Dual-edge triggering multivibrator, 281
Dual-limit comparator, 57
Dual-output multivibrator, 283
Dual-sense amplifier, 499
Dual-tone dialing, 481
Duty-cycle control for multivibrator, 283
Duty-cycle modulator, 248

Electric-vehicle control, 12
Electrolytic capacitor testers, 187
  leakage, 183, 187
Electromagnetic field probe, 210
Electrometers:
  fast-response, 195
  high-sensitivity, 197
  MOSFET, 194, 196
  seven-range, 196
  wide-range, 193
Electronic lock, 369
Elevator button, no-push, 31
Elliptic high-pass/low-pass filter, 111
Emergency power, 12-V, 362
Emphasis, filter with, 123
Encoders:
  CVSD, for secure radio, 99
  touch-tone, 480, 482, 485–488
End-of-tape detector, 349
Envelope detector, 495
  peak, 497
Equal voltage test, 495
Equalizers, 7, 129, 240
Equipment interface protection, 368
Erasing amplifier, 157
Excess-noise source, 295
Excess-three Gray code-to-BCD converter, 78
Exclusive-OR, 27, 103
Expander, 8
Exponential control, two-quadrant, 6

Fail-safe interlock, 372
Fail-safe voltage monitor, 492
Far-field transmitter, 1
Fast break-in, 47
Fast switch, 39
Feed-forward opamp, 311
FET circuits:
  AGC, 156
  balanced mixer, 396
  balanced modulators, 244, 248
  bridge, computer-controlled, 25
  buffer for A/D converter, 67
  choppers, 38
  grid-dip meter, 200
  mixer, 395
  opamp, 312
  as Q multiplier, 275
  reference, 5 V, 505
  regulator, 5 V, 422
Fiber optic circuits:
  data coupler, 106
  data link, 106
  infrared detector, 109
  LED pulsers, 107, 110
  light transmission checker, 107
  Manchester-code demodulator, 109
  Manchester decoder, 109
  receiver, 108, 109
  transmitter, 107, 108
  TTL link for microprocessors, 110
  (*See also* Optoelectronic circuits; Optoisola-
    tor circuits)

Field meter, 218
Field-strength meters, 2, 215
Filter circuits:
  band-reject, touch-tone, 483
  bandpass answer, 105
  bandpass originate, for modems, 97
  comb, 270
  CW, 48
  receive, for transmissions, 98
  (See also Active filters; Bandpass filters;
    High-pass filters; Low-pass
    filters; Notch filters; Passive filters)
Fire alarm, 165
Fire detector, 160
Flanging, simulation of, 285
Flasher, loose-ground, 372
Flash-on suppression for lighting control, 179
Flattest response high-pass filter, 138
Floating regulator, +50 V, 424
Flyback switching regulator, 463
FM circuits:
  detectors, 251, 254
  oscillator, transmitter, 204
  transmitter, optical, 108
  with time-multiplexing, 269
  tuner, power supply for, 414
Foldback current limiting (see Current limiting)
Follower voltmeter, 220
Fourier function generator, 151
Four-quadrant D/A converters:
  algebraic, 95
  ratiometric, 68
Four-quadrant multipliers, 273
  8-bit, 277
  DAC, 278
Frequency:
  comparators of, 52, 54
  converters of, 52
    0.1 Hz to 100 kHz, 79
    10 Hz to 10 kHz, 78
    from digital, 85
    from temperature, 227, 229
    from voltage, 78–84, 87, 88, 313
    to voltage, 84
  crystal, increasing, 335
  dividers of: by 1, 2, 5, or 10, 144
    by 3 to 29, 143
    by 5, 144
    by 15, 143
    odd-modulo, 143
    programmable counter, 144
    square-wave, 144
    UHF prescaler, 145
  doubler of, 339
  measuring (see Frequency measuring cir-
    cuits)
  motor-speed control by, 262
  power-line, meter for, 217
  pulse generator to double, 385
  quadrupler for, 27
  shifting of, capacitance measuring by, 187
  sweep function generator, 150
Frequency measuring circuits:
  1-MHz with NOR gates, 205
  calibrators, crystal, 202
    100-kHz, 201, 203, 206
    1-MHz, 206
  dipper, 202, 203
  gate-dip meter, 204, 205
  grid-dip meter, 200, 204
  grid-dip oscillator, 206
  ham-band calibrator, 199
  harmonic generator, 205
  marker divider, 204
  marker identifier, 205
  marker pips, 10-kHz to 30-MHz, 200

Frequency measuring circuits (cont.)
  markers: 1-kHz, 199
    10-kHz, 207
    20-kHz, 207
    30-kHz, 203
    50-kHz, 199
  meters: 1–30 MHz, 202
    2-meter, 205
    AF, 181
  oscillators: FM transmitter, 204
    with meter, 1–250 MHz, 201
  power-line monitor, 201
  references: 1-MHz, 200, 201
    3-MHz, 204
  sine-square wave, 100-kHz, 206
  time reference, 10-MHz, 201
Front end circuits:
  1.8–2 MHz, 389
  6-meter, 391
FSK circuits:
  decoders, 238
    1070 Hz and 1270 Hz, 241
    with VCO, 241
  demodulators: with carrier detect, 102
    PLL in, 104
  detector, 102
  function generator, 146
  generator, 240, 241
  multiplexer, two-input, 268
  receiver, 238
  self-generating, 239
  with slope and voltage detector, 237
  for two tones, 237
Full-range power control, 357
Full-wave lighting control, 179
Full-wave photoelectric chopper, 38
Full-wave power control, 350, 355
Full-wave rectifiers:
  100-kHz, 306, 314
  synchronous, 359
Function generator circuits:
  10-Hz to 2-MHz, 148
  1,000,000:1 frequency range, 150, 152
  A/X, hyperbolic, 146
  Fourier, digital, 151
  frequency sweep, 150
  pulse/sawtooth, 152
  sine-square-triangle: 0.5-Hz to 1-MHz,
    147
    FSK, 146
  with single control, 152
  square-to-sine, 147
  square-triangle, 149
    AF, 149
    current-controlled, 150
    variable, 151
  triangle-to-sine, 152
  voltage-controlled: nonlinear, 147
    two-phase, 148
  (See also Multivibrator circuits; Oscillators;
    Pulse generating circuits)
Fuzz circuit, 292

Gain control circuits:
  60-dB range, 154
  1000:1, 153
  amplifier, 312
  attenuator, Darlington, 155
  chopper-controlled, 155
  erasing, 157
  limiter, constant-output, 156
  optoelectronic, 158
  quad, 302
  variable resistor, 154
  (See also AGC circuits)
Galvanometer amplifiers, 193, 194

Garage-light control, 348
Gate-dip meter, 204, 205
Gate-formed switch, 59
Gated pulse train, 377
Gated relaxation oscillator, 5-MHz, 333
Gates, latching, 61
Gauge, water conductivity, 222
Geiger counter, 192, 197
Gelled-electrolyte battery charger, 23, 24
Generators:
  AC, cutout for, 370
  ASCII, serial, 235
  attack-decay, 293
  busy-signal, 484
  dial-tone, 479
  dot, audio, 323
  FSK, 240, 241
  harmonic, 205
  for modem circuit, 242
  piano tone, 292
  pink-noise, 294
  sequence, 374
  tone, 384
  tone-burst, 384
  zener noise, 296
  (See also Function generator circuits; Pulse
    generator circuits)
Go/No-go indicators, 80, 490
Gray code converters:
  to BCD, 79
  to binary, 82
Grid-dip meters, 200, 204
Grid-dip oscillator, 206
Grid-dipper modulator, 250
Ground, loose, flasher for, 372
Ground-fault interrupter, 368
Ground isolation, 317
Guarded full-differential opamp, 314
Gyrators:
  12-μH, 302
  in active filter, 119

Half-wave lighting-control circuit, 176
Half-wave power control, 353, 355
Half-wave rectifiers, 304
  high-speed, 305
  synchronous, 363
Half-wave trigger for thyristor, 355
Hall-effect detector, 232
Ham-band calibrator, 199
Hand-waving theremin, 286
Harmonic generator, 205
Headlights-on reminder, 12, 13
Heat-energy integrator, 226
High-current infrared LED pulser, 107
High-gain circuits:
  amplifier, 169
  for lighting control, 177
  preamp, 192
High-impedance buffer, 66
High-input Z, 171
High-intensity lamp, control for, 179
High-pass filters:
  10-Hz, 121
  100-Hz, 122
  1-kHz: flattest response, 138
    fourth-order, 139
    third-order, 135
  2.4-kHz, 138
  2.955-MHz, 114
  elliptic, 111
High-Q filters:
  bandpass, 1-kHz, 141
  notch, 131
High-speed circuits:
  A/D converter, 69

High-speed circuits (*cont.*)
  D/A converter, 90
  half-wave rectifier, 305
  multiplier, 267
  pulse generator, 381
  zero-voltage detector, 511
High-voltage regulator circuits:
  negative switching, 468, 471
  positive switching, 464, 478
High-voltage supply, 161
Hodoscope amplifier, 172
Hum-blocking optoisolator, 319
Hum-free supplies:
  for cassette recorder, 426
  for tuner, 430
Humidity control, 212
Hygrometers, 208, 214
Hyperbolic A/X function generator, 146
Hysteresis:
  adjustable, 497
  built-in, 319
  design of, 501
  independent adjustment of, 54
  voltage-controlled, 53
  voltage-dropping, 500
Hysteresis-and-delay oscillator, 380

Identifiers:
  CW, 48
  marker, 205
IF circuits:
  filter, diode-switched, 113
  noise blanker, 295
  PLL, 255
  tune-up, 465-kHz oscillator for, 341
Ignition:
  CD, 12, 14
  cold weather, 11
Impedance bridge, 209, 214
Impedance inverting oscillator, 341
Indicators:
  audible: light sensor, 348
    line monitor, 370
    metal detector, 233
    metronome, 286
  blown-fuse, 367
  Go/No-go, 80, 490
  power drain, 215
  tuning, LED, 253, 256
  voltage, LED, 19, 20
  wind direction, 212
Indium antimonide detector, 161
Inductance checker, 213
Inductance meter, 209
Induction motor speed control, 259, 261, 263
Inductive loads, 353, 354
Infrared circuits:
  in 12 kV supply, 161
  camera pulse amplifier, 160
  camera time base, 163
  clinical thermometer, 162
  coded source, 163
  coded-light detector, 162
  communication detector, 164
  communication source, 164
  detector, 109, 163
  fire alarm, 165
  fire detector, 160
  indium antimonide detector, 161
  intrusion alarm, 159, 160
  lamp amplifier, 165
  lead sulfide detector, 164
  LED pulser, 107
  modulator, 165
  nuvistor preamp, 162
  radiation thermometer, 161

Infrared circuits (*cont.*)
  temperature sensor, 164
  thermal microscope, 159
Input protection, optoisolator, 316
Instantaneous wattmeter, 217
Instrumentation circuits:
  ±34 V common-mode range, 168
  calibrator, automatic, 169
  Darlington, 170
  data demodulator, 169
  differential I/O amplifiers, 168, 170
  high-gain amplifier, 169
  high-input Z, 171
  hodoscope amplifier, 172
  light meters, 171, 172
  meter amplifier, 167
  metal detectors, 167, 168, 172
  nanoammeter, 170
  pH meter, 170
  photocell bridge, 171
  strain-gage amplifier, 167
  wind speed, 166
  (*See also* Measuring circuits)
Integrator, heat-energy, 226
Interference, CW filter for, 112
Interlocks:
  bounceless, 61
  fail-safe, 372
Internally gated window comparator, 51
Interrupter, ground-fault, 368
Intrusion alarms, 159, 160
Intrusion detector, 33
Inverters:
  in monostable multivibrator, 280
  parallel driver for, 361
Inverting peak detector, 494
Ion chamber, 210
Isolation and isolators:
  bounceless, 60
  data coupler, 98, 106
  solar-cell, 28
  (*See also* Optoisolator circuits)

JFET circuits:
  opamp, 301, 308, 309
  Pierce oscillator, 334
  series-pass regulator, 430
Joystick control for music circuit, 286

Key pulser for telephone, 480
Keyboard bounce eliminator, 59
Keyers, 43
  CMOS, 45
  with memory, 42, 47
  sensor, 44

Lamps:
  amplifier for, 165
  comparator-driven, 55
  driver using, for chopper, 38
  high-intensity, control of, 179
  logic-level, 498
  regulator for, 176
  for voltage regulation, 26
Laser-controlled oscillator, 350
Latch comparators, 49
Latches:
  noise-resistant, 298
  set-reset, 316
Latching gates, 61
Lead sulfide detector, 164
Leakage, capacitor, 183, 187
LEDs:
  comparator-driven, 56
  constant-brightness, 179

LEDs (*cont.*)
  pulse modulation using, 110
  pulser using, 107
  trickle battery charger using, 20
  voltage indicator using, 19, 20
Level-crossing display, 52
Level detector, AF, 181
Level shifters, DC, 142, 277
Light-beam circuits:
  modulator, 245
  receiver, 349
  voice transmitter, 347
Light-change circuits, 349
Light meters:
  audible, 349
  with LED readout, 171
  linear, 172
  six-range, 171
Light-sensitive theremin, 291
Light sensor, audible, 348
Light transmission checker, 107
Light-wand amplifier, 318
Lighting control circuits:
  active load for, 174
  delayed turn-off, 177, 179
  with flash-on suppression, 179
  full-wave, 179
  half-wave feedback for, 176
  high-gain control for, 177
  for high-intensity lamp, 179
  lamp regulator, 176
  with LED, constant-brightness, 179
  limited range, 900-W, 177
  line compensation, 175
  load control interface for, 174
  low-cost, 179
  low-loss control, 178
  mood-lighting control, 175
  projection-lamp voltage regulator, 174
  rapid-start ballast, 173
  soft-start, 500-W, 177
  standby light, 6 V, 176
  with thyristor, 176
  timer-dimmer, 178
  with triac, 173
  triac dimmers: 400-W, 175
    800-W, 174
    with brightening, 175
    time-dependent, 178
  with UJT-triac, 1.2-kW, 178
  with variable AC voltage, 176
  voltage reducer, 176
Lights, auto-trailer interface for, 16
Limiters, 5
  AF noise, 296, 298
  constant-output, 156
  foldback current, 370
  overvoltage, 12-V, 368
Line drivers (*see* Data transmission circuits)
Line-frequency filter, 116
Line monitors 201, 370, 499
Line-operated amplifier for servo, 458
Line-powered switch, 355
Line voltage, compensation for, 356
  for lighting control, 175
Linear circuits:
  chopper modulator, 250
  light meter, 172
  modulator, 249
  resistance bridge, 219
  VCO, 9-MHz, 331
  for VTVM scale, 222
Liquid level detector, 31
Liquid-sensing capacitor, 32
Load control interface for lighting control, 174
Local oscillator, 335

Locks:
    coded, 368
    electronic, 369
    tone, 3-kHz, 369
Logarithmic null voltmeter, 28
Logic drive for inductive load, 353
Logic-level lamp, 498
Logic shutdown, regulator with, 444
Logic-triggered triac, 353
Loop antennas:
    160-meter preamp for, 3
    Q multiplier for, 2
Loose-ground flasher, 372
Low-cost opamp, 301
Low-distortion oscillators:
    20–20,000 Hz, 326
    1-kHz, 322
    AF, 326
    Wien-bridge, 29
Low-drift oscillator, 7-MHz, 342
Low-emission CD ignition, 14
Low-loss lighting control, 178
Low-noise circuits:
    AC switch, 512
    opamp, 312
    oscillator, 336
    preamp, 432-MHz, 388
    RF input, 392
Low-pass filters:
    250-Hz, third-order, 136
    500-Hz, unity-gain, 129
    600-Hz, third-order, 140
    1-kHz: second-order, 140
        fifth-order, 114
    2-kHz, 129
    2.125-kHz, 113
    2.4-kHz, 138
    10-kHz, 127, 131
    100-kHz, unity-gain, 137
    480-kHz, 123
    42.5-MHz, 114
    AF, 128, 138
    elliptic, 111
    for noise, 297
    pi-section, 115
    Sallen-key, 127
    state-variable, 45-kHz, 485
    tunable fourth-order, 138
Low-ripple regulated power supply, 12 V, 402
Low-standby drain power supply circuit, 365
Low-voltage indicator, LED, 20
Low-voltage ohmmeters, 222, 223

Magnetometer amplifier, 211
Make/break switch, bounceless, 61
Malfunction alarm, for microprocessor circuit, 235
Manchester-code demodulator, 109
Manchester decoder, 109
Markers:
    1-kHz, 199
    10-kHz, 207
    20-kHz, 207
    30-kHz, 203
    50-kHz, 199
    divider for, 204
    identifier of, 205
    pips, 10-kHz to 30-MHz, 200
Meacham bridge oscillator, 50-kHz, 336
Measuring circuits:
    anemometer, 209
    bullet timer, 213
    distortion meters, 211
    electromagnetic field probe, 210
    humidity control, 212

Measuring circuits (cont.)
    hygrometers, 208, 214
    impedance bridge, 209, 214
    inductance checker, 213
    inductance meter, 209
    ion chamber, 210
    magnetometer amplifier, 211
    phase meter, 213
    probability density analyzer, 210
    rain alarm, 209
    resistance, RF, 208
    sound-level meter, 212
    wind-direction indicator, 212
    (See also specific measuring circuits, e.g.,
        Audio; Capacitance; Current; Fre-
        quency; Meters; Power; Resistance;
        Temperature)
Memory:
    analog, for music circuit, 289
    keyer with, 42
Metal detector circuits, 168
    with AM radio, 233
    audible indicator for, 233
    beat-frequency locator, 231
    coin finder, 232
    crystal-filter locator, 231
    Hall-effect detector, 232
    metal sensor, 230
    pipe finder, 232, 233
    PLL in, 172
    three-stage, 230
    tuned-loop oscillator, 231
    twin-oscillator, 167
    two-oscillator locator, 232
Meters:
    amplifiers for, 167, 184, 194, 196
    capacitance: 1 pF to 1 μF, 191
        4 pF to 0.1 μF, 183
        5 pF to 1 μF, 188
        15 pF to 10 μF, 185
        0.001–10 μF, 186
        0.1 μF, up to, 184
    adapter for counter for, 190
    counter as, 186
    digital, 189, 191
    direct-reading, 189
    five-range, 189, 190
    frequency shift, 187
    seven-range, 185
    VOM, 185, 191
    VTVM, 186
    conductivity, 221
    deviation, 256
    distortion, 211
    field, 218
    field-strength, 2, 215
    frequency: 1–30 MHz, 202
        2-meter, 205
        AF, 181
        power-line, 217
    gate-dip, 204, 205
    grid-dip, 200, 204
    inductance, 209
    light: audible, 349
        with LED readout, 171
        linear, 172
        six-range, 171
    with oscillator, 1–250 MHz, 201
    pH, 170
    phase, 213, 344
    power, 216
    resistance of, measuring, 222
    RF, 391
    sound-level, 181, 212
    tach/dwell, 12
    VSWR, 1

Meters (cont.)
    watthour, 217
    (See also Ammeters; Ohmmeters;
        Voltmeters; Wattmeters)
Metronomes:
    audible/visible, 286
    clicking, 291
Microammeters, 194, 195, 198
    amplifier for, 195
Micropower comparator, 55
Microprocessor circuits:
    malfunction alarm, 235
    receiver, remote terminal, 236
    serial ASCII generator, 235
    transmitter, remote terminal, 235
    TTL link for, 110
    UART interface, 234
    uppercase driver, 234
Microscope, thermal, 159
Microvolt comparator, 50
Missing-pulse alarm, 350
Missing-pulse detector, 384
Mixer pot, panning, 9
Mixers, 391, 395, 396
Mobile autopatch, 487
Mobile equipment, 12 V to 6 V for, 361
Modem circuits, 100
    answerer, 96
    bandpass originate filter, 97
    delay equalizer for, 240
    FSK: decoders, 238, 241
        demodulators, 239–242
        generator, 240, 241
        receiver, 238
        self-generating, 239
        with slope and voltage detector, 237
        for two tones, 237
    tone oscillator, 238
Modulated-light receiver, 349
Modulation, pulse, using LEDs, 110
Modulation monitor, 2
Modulator circuits:
    0.5-MHz, 254
    6-meter, 244
    AM, 250
        with analog multiplier, 249
        double-sideband, 243
        single-IC, 247
        single-supply, 243
        suppressed-carrier, 246, 249
    balanced, 246
        chopper, 37
        DSB, 247
        FET, 244, 248
    chopper, 244
        balanced, 37
        linear, 250
    comparator for PWM, 244
    DC, 37
    delta, 246
    for digital signals, 248
    double-sideband suppressed-carrier, 246, 248
    duty-cycle, 248
    FET balanced, 244, 248
    grid-dipper, 250
    infrared, 165
    light-beam, 245
    linear, 30-dB for 60–150 MHz, 249
    PNPN tetrode, 249
    PPM, with analog control, 247
    pulse-duration, 245
    pulse-ratio, 243
    PWM, 244, 282
    series, 249
    SSB, balanced, FET, 248
    voltage to pulse width, 245

Modulator/demodulator circuits:
  AFC amplifier, 257
  call alert, 257
  demodulators: narrow-band, 253
    PLL, 255
    quadrature, 255
    SCA, 252
  detectors: FM, 251, 254
    single-tuned, 255
  deviation meter, 256
  discriminators, 256
  power amplifier, VHF, 251
  receivers: carrier-system, 252
    diode, 21–75 MHz, 256
  RF stage, MOSFET, 253
  stereo decoder, 252
  transmitter, carrier-system, 254
  tuning indicator, LED, 252, 256
Monitors:
  18-volt, 22
  AC line, 370
  with ambiguity eliminator, 496
  battery, 11
  battery charger, 19
  code, 47
  CW, 44
  modulation, 2
  multiple-voltage, 498
  nicad charging, 21
  power-line, 201, 370
  pulse-width, 377
  sidetone, 45
  supply current, 193
  temperature, LED, 227
  three-voltage, 501
  transients, power-line, 499
  voltage, 492, 496, 499
Mono multivibrators:
  bidirectional, 280
  crystal, 280
  four-gate, 280
  low-power, 279
  micropower, 284
  negative output, 284
  pulse-shrinking, 382
  pulse-stretching, 373
  PWM, 282
  voltage-controlled, 280
Mood-lighting control, 175
Morse-code set, 42
MOS circuits:
  A/D buffer, 71
  AF oscillator, 320
  driver, for triac, 355
  electrometers, 194, 196
  RF oscillator, 334
  RF stage, 253
  variable resistor, 154
Motor control circuits:
  brake, continuous-duty, 261
  clamped Darlington, 262
  DC, speed, 266
  frequency-controlled, 262
  induction speed, 259, 261
    three-phase, 263
  opamp, 260
  phase sequence detector, 264
  proportional speed, 265
  split-phase, with braking, 265
  stalled-motor protection, 265
  stepper motors: state generator for, 264
    drive for, 260
  switching-mode, 258
  tape-loop speed, 266
  thyristor, 258, 259
  triac, with feedback, 262

Motor control circuits (cont.)
  triac-starting switch, 259
  up-to-speed logic, 261
  water-level control, 260
Multiple-feedback bandpass filter,
  126
Multiplexer circuits:
  3-channel, 269
  8-channel, 271, 272
  16-channel: high-speed, 267
    sequential, 271
  A/D-D/A converter interface, 268
  analog switch, 270
  BFO, 270
  comb filter, 1-kHz, 270
  commutating bandpass filter, 270
  FSK, two-input, 268
  N-path notch filter, 269
  thermocoupler, 227
  time-multiplexing, 269
Multiplier circuits:
  AC-coupled, 277
  adjustable scale factor, 274
  AGC, 157
  analog, 274
    in AM modulator, 249
  binary, 274
  capacitance, 274
    opamp, 310
  cosines, approximating, 278
  DC, 76
  four-quadrant, 273
    8-bit, 277
    DAC, 278
  with opamp, 275
  output level shifter, 277
  phase-sensitive detector, 278
  Q: 455-kHz, 273
    FET as, 275
    for loop, 2
  squaring, for RMS, 276
  two-quadrant, 275, 278
    bipolar/analog, 276
Multiplier module for again control, 153
Multivibrator circuits:
  astable, 279
    20-kHz, 280
    direct-coupled, 283
    pulser, 2-Hz, 281
  crystal: using inverters, 280
    with NOR gates, 281
  dual-edge triggering, 281
  duty-cycle control, 283
  low output, for power on, 282
  negative-going, dual output, 283
  pulse-width, variable, 282
  pulse-width detector, 281
  (See also Mono multivibrator circuits;
    Oscillators)
Music circuits:
  acoustic pickup preamp, 293
  analog memory for, 289
  attack-decay generator, 293
  bells, musical, 290
  for computers, 287
  equal temperament, tuning for, 287
  flanging, simulation of, 285
  fuzz circuit, 292
  joystick control, 286
  metronomes: audible/visible, 286
    clicking, 291
  noise source, 292
  organ, four-octave, 293
  piano tone generator, 292
  pulse and sequence outputs, 290
  reverberation, 291

Music circuits (cont.)
  rhythm control, automatic remote, 289
  synthesizer, 288
  theremins: hand-waving, 286
    light-sensitive, 291
  tremolo: amplifier for, 289
    control for, 287
  trombone, 288
  VCO for, 291

NAND-gate crystal oscillator, 340
Nanoammeters, 170, 193
Narrow-band demodulator, 253
Narrow-pulse peak detector, 493
Negative-output multivibrators, 283, 284
Negative-resistance circuits:
  LED oscillator, 326
  opamp, 303
Negative shunt regulation, 441
Neon-lamp capacitor tester, 187
Neon wattmeter, 216
Nickel-cadmium battery chargers, 18–24
No-calibration ohmmeter, 221
No-push elevator button, 31
Noise circuits:
  AF limiter, 298
  blanker, 296
  canceller, 40
  clipper, AF-powered, 294
  control, 297
  excess-noise source, 295
  gate, for AF preamplifier, 297
  IF, blanker, 295
  latch, noise-resistant, 298
  limiter, AF-noise, 296
  low, opamp, 312
  low-pass filter, 297
  pink-noise generator, 294
  RF source, low-level, 298
  sewing-machine suppression, 296
  silencer, AM, 294
  source for, 292
  spike rejection, 298
  white-noise, 50–5000 Hz, 295
  zener generator, 296
Noncontacting temperature sensor, 164
NOR gates:
  1-MHz with, 205
  with crystal multivibrator, 281
Notch filters:
  50-Hz Wien-bridge, 126
  60-Hz, 120
    adjustable-Q, 122
    high-Q, 131
    tunable, 112
  100 Hz–10 kHz, tunable, 142
  225.8 Hz, 137
  600-Hz, 130
  693–2079 Hz, 115
  1-kHz, 140
    N-path, 269
  1.5-kHz, 119
  4.22-kHz, 127
  19-kHz, 132
  AF, 130, 136
N-path notch filter, 269
Null voltmeter, logarithmic, 28
Nuvistor preamp for infrared, 162

Octave audio equalizer, 129
Odd-modulo frequency divider, 143
Offset control, opamp, 304
Ohmmeters:
  500-milliohm full scale, 219
  AC, 223
  digital, 224

Ohmmeters (cont.)
  low-voltage, 222, 223
  no-calibration, 221
  teraohmeter, 220
On-off control, photoelectric, 348
One-pulse-per-push switch, 60
Opamp circuits:
  130 V P-P drive, 301
  1000 gain at 2 kHz, 300
  741 type, 303
  absolute value, 300, 304
    rectifier for, 310
  bandpass filter, 132
  bias, 29
  buffered, 67, 301
  capacitance multiplier, 310
  chopper, 40
  clamping with, 7
  CMOS driver for, 303
  CMR, optimizing, 302
  converter, voltage to frequency, 313
  current boosters: 100-mA, 306
    1-A, 312
    8-A, 311
  dead-band response, 299, 309
  differential I/O, 306
    JFET input, 301, 308–310
  faster slewing, 304, 305
  feed-forward, 311
  FET drive, 312
  gain: 1–1000, 308
    2000 at 2-kHz, 300
  gain-controlled, 312
  guarded full-differential, 314
  gyrator, 12-μH, 302
  JFET input, 301, 308–310
  low-cost, FET-input, 301
  low-impedance, 305
  low-noise, 312
  motor control with, 260
  for multiplier, 275
  negative resistance, 303
  offset control, 304
  phase error: minimizing, 310
    tester for, 307
  postamplifiers, 302
    CMOS, 307
    complementary, 314
    single-supply, 311
    two-stage, 299
  power, 306
  power booster, 300, 307, 308
  pulse amplifier, 3-W, 309
  pulse generator, 374
  rectifiers: buffered, 100-kHz, 307
    without DC offset, 304
    full-wave, 306, 314
    half-wave, 304, 305
    precise, 309, 312, 313
  sign changer, 310
  signal conditioner, 305
  signal separator, 304
  summing, 303, 305, 313
    without adjustments, 309
  temperature coefficient, low, 307
  tracking quad-gain control, 302
  unity-gain: AF current amplifier, 303
    feed-forward, 305
  voltage follower, 313
  voltage rating, boosting of, 306
Operational-integrator phase shifter, 456
Optoelectronic circuits:
  automatic calibrator, 169
  bar-code circuits: reader, 315
    signal conditioner, 317
  gain control, 158

Optoelectronic circuits (cont.)
  hysteresis, built-in, 319
  input protection, 316
  isolation circuits: 1500-V, for DAC, 319
    30-kHz bandwidth, 319
    analog, 316
    DC, with harmonic suppression, 316
    ground, 317
    hum-blocking, 319
    threshold switch, isolated, 318
  latch, set-reset, 316
  light-wand amplifier, 318
  scanner, 318
  switch, 400-VDC, 317
  VFO, 1.5–5.7 MHz, 318
  wattmeter, RF, 218
  (See also Fiber optic circuits)
Optoisolator circuits:
  analog, 316
  bandpass, 30-kHz, 319
  data link, 98
  driver, 354
  ground, 317
  triac control, 354
OR terminals, wired, 105
Organ, four-octave, 293
Oscillation, suppressor for, 50
Oscillators:
  1–250 MHz, with meter, 201
  astable, 279
  capacitance-meter, 184
  CW-monitoring, 44
  FM transmitter, 204
  grid-dip, 206
  hysteresis-and-delay, 380
  laser-controlled, 350
  limited gain control for, 154
  for metal detectors, 167, 231, 232
  photocell, 348
  sine-square, 100-kHz, 206
  solar-power, 350
  tone, for modem circuit, 238
  (See also AF oscillators; Multivibrators; RF
    oscillators; Sine waves and sine-
    wave circuits, oscillators; Wien-bridge
    oscillators)
Output level shifter, 277
Overcharge protection, battery charger, 21
Overcurrent protection, for 400-V supply, 359
Overload protection, regulator for, 371, 427
Overtone oscillator, 65–110 MHz, 340
Overvoltage:
  AC protection, 371
  crowbars for, 367, 371, 452
  limiter for, 12-V, 368

Panning mixer pot, 9
Parallel inverter driver, 361
Parallel-opamp preamp, 456
Paralleling of regulators, 425
Passive filters:
  AF, 115
  AM broadcast rejection, 115
  bandpass, 112, 115
  CW, 112
  using diode-switched crystals, 113, 114
  elliptic high-pass/low-pass, 111
  high-pass, 2.955-MHz, 114
  IF, 113
  low-pass: 1-kHz fifth-order, 114
    2125-Hz, 113
    42.5 MHz, 114
    pi-section, 115
  notch: 60-Hz, tunable, 112
    693–2079 Hz, tunable, 115
  voice bandpass, 114

PCM decoder, 373
PDM telemetry, synchronous sawtooth for,
  97
Peak detectors, 494, 500
  10-kHz sine-wave, 499
  envelope, 497
  holding, 498
  inverting, 494
  narrow-pulse, 493
  sense and hold, 492
Peak program meter tester, 180
Peak sensing with tunnel diode, 497
Pen, servo-driven, 39
Pen recorder scale changer, 493
PH meter, 170
Phase:
  comparator for, 50, 52
  converter of, 87
  measuring of, at 100 MHz, 26
  meter for, 213, 344
  sequence detector for motor control, 264
Phase control circuits:
  full-wave feedback, 344
  phase meter, 100 Hz to 1 MHZ, 344
  phase shifters: 0–90 degree, 345
    Darlington, for servo, 455
    operational-integrator, for servo, 456
    voltage-controlled, 344
  PLLs: 100-kHz reference, 343
    AF, 344
    fast, 346
    wide-capture range for, 345
  shifting and squaring, 346
  voltage feedback, 346
Phase error, opamp, 310
Phase-error tester, 307
Phase-locked loops (see PLL circuits)
Phase-sensitive detector, 278
Phase-shift sine-wave oscillator, 327
Phase-splitting preamp, 456
Phone-call counter, 482
Photocells:
  bridge using, 171
  for gain control, 158
  oscillator, 5-kHz, 348
Photochopper, 39
Photoelectric circuits:
  alarm, missing-pulse, 350
  chopper, 38
  detectors: end-of-tape, 349
    light-change, 349
  garage-light control, 348
  light sensor, audible, 348
  meter, audible light, 349
  on-off control, 348
  oscillators: 5-kHz, photocell, 348
    laser-controlled, 350
    solar-power, 350
  punched-tape reader, 348
  for ratio of two unknowns, 347
  receivers: light-beam, 349
    modulated-light, 349
  relays: light-change driven, 349
    phototransistor, 348
  transmitter, voice, 347
  (See also Fiber optic circuits; Optoelectronic
    circuits)
Pi-section filter, 115
Piano tone generator, 292
Pierce oscillators, 334
Pink-noise generator, 294
Pipe finder, 232, 233
PLL circuits:
  0.01 Hz to 100 kHz, 104
  AM receivers, 391, 394
  FSK demodulators, 104, 241, 242

PLL circuits (*cont.*)
  function generator, 147
  for IF and demodulator, 255
  metal detector, 172
  phase-control, 343, 344, 346
  touch-tone decoders, 481, 486
Points, breaker, transistorized, 13
Polarity-reversing circuits:
  line driver, 98, 101, 102
  power-supply, 76
Positive shunt regulator, 429
Postamplifiers:
  CMOS, 307
  complementary, 314
  single-supply, 311
  two-stage, 299
Potentiometers:
  bridge with, 27
  panning mixer, 9
  tester for, 221
Power amplifiers:
  28-V, push-pull, 456
  VHF, 251
Power boosters, 300, 307, 308
Power control circuits:
  AC, full-wave, 356
  for clock comparator, 352
  full-wave, 355, 356
  half-wave, 353, 355
  inductive loads: driver for, 353
    triac for, 354
  line voltage compensation, 356
  optoisolator drive for 240-VAC load, 354
  ramp-and-pedestal control, 357
  SCRs: PUT control for, 357
    trigger for, 600-W, 356
  sump-pump control, 353
  switches: 125-ns, 356
    AC, 353
    line-powered, 355
    thyristor, 354
    voltage-sensitive, 353
    zero-point, 352
  switching, of 4500 W, 356
  thyristors in, 354, 355
  VMOS, interface for, 356
  (*See also* Triac circuits)
Power drain indicator, 215
Power frequency meter, 217
Power-line monitors, 201, 370
  for transients, 499
Power measuring circuits:
  field meter, 218
  field-strength meters, 2, 215
  power drain, 215
  power frequency meter, 217
  power meters, 216
  watthour meter, 217
  (*See also* Wattmeters)
Power opamp, 306
Power output meter, 500-MHz, 216
Power supply circuits:
  1.5 V for TVTM, 360
  6 V from 12 V, 361
  ±6 V and ±15 V, 363
  10-volt, bipolar, 27
  12 V emergency power, 362
  ±12 V at 15 mA, 360
  90 VRMS at 500 W, 359
  110/120 VAC, ±2.5 V at 600 W, 358
  115 VAC from 12 VDC: at 100 W, 362
    at 400 Hz, 361
  117 VAC from 24–60 VDC, 364
  130 V and 270 V for CRT, 362
  230 VAC from 115 VAC, 363
  1000 V, for CRT, 360

Power supply circuits (*cont.*)
  2500 V at 500 mA, 364
  3000 V, 366
  7500 V, 360
  12 kV, 161
  100 W at 60 Hz, sine-wave, 360
  200 W at 25 kHz, 365
  500 W at 20 kHz, 363
  bipolar, 10-volt, 27
  inverter driver, parallel, 361
  low-standby drain, 365
  overcurrent protection, 359
  preregulator, 358
  rectifiers, synchronous: full-wave, 359
    half-wave, 363
  shutdown protection, 364
  transformerless, 358, 360
  transient eliminator, 363
  (*See also* DC-to-DC converters; Protection
    circuits; Regulated power
    supply circuits; Regulator circuits;
    Switching regulator circuits)
Power switch for regulator circuit, 471
Power transistor dissipation, 367
Power zener, 490
  voltage regulator with, 504
PPM modulator, 247
Preamplifiers:
  20 dB, 389, 390
  432-MHz, low-noise, 388
  6-meter, 388
  acoustic-pickup, 293
  all-band, with whip, 393
  high-gain, 192
  loop, 3
  noise gate for, 297
  nuvistor, 165
  parallel-opamp, 456
  phase-splitting, 456
  regulator-ripple, 434
  for servo, 455
Preemphasis, 1500 Hz, 10
Preregulator, 358
Prescaler, UHF, 145
Preselector, 14–30 MHz, 393
Probability density analyzer, 210
Probes:
  conductivity, 220
  electromagnetic field, 210
Process current, conversion of binary to, 89, 94
Processor, speech, 8
Product detector, 46
  direct-conversion, 394
Programmable counter, 144
Projection-lamp voltage regulator, 174
PROM CW identifier, 48
Protection circuits:
  AC generator cutout, 370
  blown-fuse indicator, 367
  clamp, 600-Hz, 370
  crowbars, 367, 371, 452
  current sensor, 372
  equipment interface, 368
  flasher, loose-ground, 372
  ground-fault interrupter, 368
  interlock, fail-safe, 372
  limiters: foldback current, 370
    overvoltage, 12-V, 368
  locks: coded, 368
    electronic, 369
    tone, 3-kHz, 369
  monitors, ac line, 370
  overcurrent, 359
  overvoltage, AC, 371
  power transistor dissipation, 367
  regulator overload, 371

Protection circuits (*cont.*)
  relay arcing, triac suppressor for, 371
  shutdown, 364
  switch, coded, 369
  transceiver-saver, 371
  transients, solenoid, bypassing, 372
  triac, 352
Proximity switches, 32–34
Pseudorandom code, 105
Pulse amplitude-polarity sorter, 496
Pulse and sequence outputs, 290
Pulse-duration modulator, 245
Pulse generator circuits, 152
  1 Hz to 1 MHz, adjustable-width, 382
  2-Hz, astable, 281
  30–4000 Hz, with opamp, 374
  60-Hz, with 50% duty cycle, 377
  400 Hz, 386
  120 kHz to 4 MHz, 376
  900 kHz to 10 MHz, 381
  1-MHz square-wave, 382
  20-MHz, up to, 384
  100-MHz, 386
  50% duty cycle, 374, 377
  astable, 383, 386
  asymmetric, 382
  for completing last cycle, 378
  controlled-duration, 374
  detector, missing-pulse, 384
  frequency-doubled, 385
  gated pulse train, 377
  high-speed, 381
  with independent on-off periods, 380
  kilovolt, 376
  monitor, pulse-width, 377
  oscillator, hysteresis-and-delay, 380
  PCM decoder, 373
  pulse stretchers, 385
    11X, 378
    with isolation, 375
    mono for, 373
  pulse trains, bipolar, 386
  with quadrature outputs, 386
  rectangular-wave, AF, 379
  selectors: pulse-edge, 378
    single-pulse, 375
  sequence generator, 374
  sequential, 379
  shrinker, 381
  square-waves, 376
    1-MHz, 382
    adjustable, 380
    AF, 383
    AF/RF, 383
  stretcher, 385
  strobing, unambiguous, 385
  synchronization, to 10-MHz, 375
  three-phase, 385
  as tone generator, 384
  tone-burst, 384
  UJT/LED, 383
  variable frequency, 374
  variable-width, 375
  waveform-edge, 384
  widener, 379
Pulse-ratio modulator, 243
Pulse-width detector, 281, 384
Pulse-width modulation circuits:
  1 Hz, down to, 244
  comparator for, 244
  multivibrator, 282
Pulse-width multivibrator, 282
Pulsers, 107, 281
Pulses:
  amplifiers of, 160, 309
  boxcar, for gain control, 157

Pulses (*cont.*)
 fast-rise, measuring, 502
 height of, conversion of, to time, 85
 modulation of, LED, 110
 width-conversion to: from temperature, 228
  from voltage, 82, 86
  (*See also* Pulse-width modulation circuits)
Punched-tape reader, 348
Push-pull circuits:
 power amplifier, 28-V, 456
 RF, 397
 for switching regulator circuit, 465
Pushbutton-to-dial converter, 484
PUT control for SCR, 357
PWM circuits (*see* Pulse-width modulation
  circuits)

Q multipliers:
 455-kHz, 273
 FET as, 275
 for loop, 2
 (*See also* Variable-Q filter circuits)
Quadrature demodulator, 255
Quadrupler, frequency, 27

Radiation thermometer, 161
Radio frequency (*see* RF)
Radios:
 AM, 392
 secure, CVSD encoder for, 99
 single-IC, 388
Rain alarm, 209
Ramp control, up/down, 457
Ramp-and-pedestal control, 357
Random code generation, 105
Range selector, automatic, 502
Rapid-start ballast, 173
Ratio of two unknowns, circuit for, 347
Ratiometric D/A converter, 68
Readers:
 bar-code, 315
 punched-tape, 348
Receive filter for transmissions, 98
Receiver circuits:
 40-meter, 41
 AF, for AM/SSB/CW, 396
 AM, 392
  PLL in, 394
  superhet subsystem, 387
 calibrator for, 202
 carrier-system, 252
 CMOS, 390
 coax, 100
 detectors: AM, synchronous, 391
  diode, 256
  product, direct-conversion, 394
  regenerative, FET, 388
 direct-conversion, 393, 394, 397
 fiber-optic, 108, 109
 front-end: 1.8–2 MHz, 389
  6-meter, 391
 FSK, 238
 light-beam, 349
 meter, RF, 391
 mixers, 391, 395, 396
 modulated-light, 349
 PLL, AM, 391, 394
 preamplifiers: 20 dB, 389, 390
  6-meter, 388
  432-MHz, low noise, 388
  all-band, with whip, 393
 preselector, 14–30 MHz, 393
 radios: AM, 392
  single-IC, 388
 remote terminal, 236
 RF: low-noise, 392
  push-pull, 397

Receiver circuits (*cont.*)
 tuner, 80-meter, 389
 VFO to check, 388
 for WWV, 10-MHz fixed, 395
 (*See also* Antenna circuits; Code circuits)
Recorder, CQ, 42
Rectangular-wave generator, AF, 379
Rectifiers:
 absolute-value, 310
 buffered, 307
 without DC offset, 304
 full-wave: 100-kHz, 306, 314
  synchronous, 359
 half-wave, 304
  high speed, 305
  synchronous, 363
 precise, 309
 60-Hz, 312
 with gain, 313
References:
 frequency: 100-kHz, 343
  1-MHz, 200, 201
  3-MHz, 204
 time, 10-MHz, 201
 (*See also* Voltage reference circuits)
Regenerated CW, 45
 filter for, 48
Regenerative detector, 388
Regulated power supply circuits:
 −20 V, for varactors, 407
 ±0 V to 15 V at 200 mA and 3.8 V to 5 V at
  2 A, 410
 0–15 V bench supply, 414
 0–20 V: at 1 A, 417
  current-limiting, 419
 0 to ±35 V, 398
 1.25 V to 37 V at 1.5 A, 406
 ±4 V to ±25 V, 404
 5 V: at 425 mA, 411
  at 1 A, 403, 412
  at 2 A, 414
  at 3 A, 416
  from AC or DC, 420
  with doubler, 421
  dual, 404
 5 V at 4 A, −12 V at 0.25 A, and +24 V at
  2 A, 416
 5 V, −7 V, and −10 V, 408
 5 V at 20 A and 0–25 V at 0–24 A, 399
 5 V and ±6 V or ±12 V, 405
 5 V at 200 mA or 7–20 V at 100 mA, 401
 5 V, 10 V, 15 V, 20 V, switched, 404
 5 V and 12 V, 403
 5 V and ±12 V, 415
 ±5 V and ±12 V, 417
  for computer, 399
 5 V and 14 V at 1 A, 409
 5 V with unregulated +15 V, 402
 5 V at 1 A, ±15 V at 100 mA, and choice of
  other, 402
 ±5 V, ±15 V, and ±30 V, 409
 ±6 V and +12 V, 403
 6 V, 12 V, and 30 V for FM tuner, 414
 ±6 V and ±15 V, 401, 408
 6–30 V at 500 mA, 418
 6.3 VAC to 12–16 VDC, 400
 ±9 V, 400
 ±11.5 V, 410
 12 V: at 50 mA, 412
  at 150 mA, 413
  at 1 A, 411
  at 2 A, 419
  at 2.8 A, 419
  at 5 A, 409, 420
  at 10 A, 407, 412
  low-ripple, 402

Regulated power supply circuits (*cont.*)
 for transceiver, 403, 412
 ±12 V, 411
 12–14 V at 3 A, 421
 12–15 V at 500 W, 415
 12–16 V from 6.3 VAC, 400
 12.6 V at 3 A, 405, 413
 13 V: at 1 A, 411
  at 2 A, 415, 416
 13.6 V at 1 A, 417
 13.7 V at 5 A, 406
 13.8 V at 18 A, 420
 14 V at 250 mA, for cassette deck, 400
 ±14 V, transformerless, 406
 15 V: at 600 mA, 413
  at 1 A, 400
 15 V and −6 V, transformerless, 401
 ±15 V, 401, 403
  100 mA tracking, 418
  at 10 A, 407
  for sound synthesizer, 405
  tracking, 408
 28 V at 10 A, 407
 50–300 V at 100 mA, 406
 80 V at 1.5 A, for color TV, 410
 7.5-kV, 360
 universal supply, 418
 (*See also* Power supply circuits; Regulator
  circuits; Switching regulator circuits)
Regulator circuits:
 −30 V to −7 V, 422
 −15 V to −5 V at 5 A, 427
 −15 V tracking +5 V, 431
 −10 V: at 1 A, 430, 444
  at 2 A, 443
 −10 V to −0.5 V, 448
 −5.2 V and +15 V, 424
 −5 V, with protection, 451
 −3 V, 428
 0 V to −6.6 V at 5 mA, 447
 0–6.6 V at 2 A, 441
 0 V to ±6.6 V at 5 mA with tracking, 435
 0 V to ±15 V: at 100 mA or 2 A with tracking, 440
  independently variable, 437
 0–13 V at 40 mA, 429
 0–15 V, 445
  at 2 A, 433
 0–20 V: at 2 A, 439
  high-precision, 446
 0–25 V: 0–10 A, 442
  with foldback current limiting, 425
 0.1–24 V at 1 A, 433
 0.1–35 V at 1 A, 431
 0.5 V to 1 V bias, 432
 1.2–37 V at 1.5 A, 438
 2 V to 35 V at 10 A, 432
 2–35 V, 432, 437
 2–37 V, 428
 4.5–34 V at 1 A, 432, 450
 5 V: at 200 mA, 440
  at 1 A, 439
  at 3 A, 438
  at 5 A, 424, 441, 450
  at 10 A, with current limiting, 435
  from 48 V, 425
  FET, 422
  with LM300H, 431
  with MC1460G, 428
 5 V and +12 V at 6 mA, 431
 5 V and ±12 V, from 24 V, 452
 5 V, +15 V, and −15 V, 445
 5 V and 15 V, single control, 422
 ±5 V, divider for, 424
 ±5 V to ±18 V, with tracking, 432
 5–20 V, zener stabilized, 424

Regulator circuits (*cont.*)
  5–24 V, 448
  5–30 V at 1.5 A, 424
  7–23 V at 1.2–2 A, 447
  7.1–65 V at 0–1 A, 426
  7.5 V, for cassette recorder, 426
  9 V from 12 V, 430
  ±10 V, with tracking, 432
  10–25 V at 100 mA, 433
  12 V: at 5 mA, shunt, 425
    at 2 A, 429
    at 20 A, 425
    series emitter-follower, 429
  15 V: at 200 mA, 439
    at 1 A, with logic shutdown, 444
    at 5 A, with protection, 436
    with differential amplifier, 424
    with feedback, 426
    from high input voltage, 426
    with μA723, 427
  ±15 V: at 100 mA, with tracking, 445
    at 200 mA, 428
    at 1 A, symmetrical, 451
    at 1 A, with tracking, 449
    at 7 A, 442
    at 10 A, with foldback current limiting, 436
    with tracking, 429, 443, 446
  ±24 V, with ±35 V unregulated, 450
  25 V at 10 A, for lab, 427
  28 V at 1 A, 436
  28 V at 7 A, 450
  46–60 V from 62 V, 444
  48 V from 80 V, 430
  50 V, floating, 424
  ±65 V at 1 A, with tracking, 447
  110/120 VAC, ±2.5 V at 600-W, 358
  1000 V at 100 W, 451
  adjustable switched, 400
  auto, 11, 13, 16
  current: 100-mA, 426
    boosting, 446, 448
    negative-input, 431
  with discrete elements, 434
  divider, for ±5 V, 424
  dual outputs, with trimming, 434
  dual supply, conversion to, 427
  high-accuracy, 1 A, 428
  with lamps, 26
  low-cost, 434
  negative shunt, 441
  overload protection, 371, 427
  overvoltage, crowbar for, 452
  paralleling, 425
  positive shunt, 429
  ripple-preamp, 434
  series pass, JFET, 430
  with shutdown, 435, 444
  with temperature stability, 427
  tracking, slaved dual, 423
  tuner supply, hum-free, 430
  variable, dual-polarity, 440
  (*see also* Current limiting; Power supply circuits; Regulated power supply circuits; Switching regulator circuits)
Relaxation oscillators:
  100-kHz, 335, 336
  5-MHz, gated, 333
Relays:
  arcing in, suppressor for, 371
  audio-operated, 6
  light-change driven, 349
  phototransistor, 348
  tone-driven, 7
Reminder, headlight, 12, 13
Remote control for servo, 454
Remote rhythm control, 289

Remote terminals:
  receiver for, 236
  transmitter for, 235
Repetitive-mode A/D converter, 68
Resistance, negative:
  LED oscillator, 326
  opamp, 303
Resistance measuring circuits:
  bridges: RLC, 223
    Wheatstone, 224
  conductivity meter, 221
  conductivity probe, 220
  follower voltmeter, 220
  meter resistance, 222
  pot tester, 221
  resistance bridge, 219
  RF, 208
  VTVM scale, linearizing, 222
  water conductivity gauge, 222
  (*See also* Ohmmeters)
Resistor, variable, MOS transistor as, 154
Reverberation, 291
Rezeroing amplifier, 101
RF circuits:
  bridge, for coax, 3
  input, low-noise, 392
  meter, 391
  noise source, 298
  oscillator circuits (*see* RF oscillator circuits)
  push-pull, 397
  resistance, measuring of, 208
  RF stage, MOSFET, 253
  wattmeters: 0.025–10 W, 218
    0–5 W, 218
    0–10 W, 216
    0–100 and 1000 W, 216
    5–300 W, 217
    optoelectronic, 218
RF oscillator circuits:
  2.225–2.455 kHz, 335
  10-kHz and 100-kHz, 330
  20–500 kHz, crystal, 337
  50-kHz, Meacham bridge, 336
  50–500 kHz, crystal, 337
  50–1000 kHz, 339
  100-kHz: digital control to, 334
    relaxation, 335
    sine-wave, 332
    with timer, 332
    Wien-bridge, 342
  150–500 kHz, crystal, 339
  279.611-kHz, crystal, 332
  450–500 kHz, wobbulator, 330
  465-kHz, 341
  800-kHz, 334
  1-MHz: with one gate, 337
    series-mode crystal, 331
  1–20 MHz, crystal, 333
  2–20 MHz, crystal, 340
  2–22 MHz, fundamental mode, 339
  2.225–2.505 MHz, 338
  3.955–4.455 MHz, Colpitts, 333
  4.8 MHz, 337
  5-MHz, 335
    crystal, low-noise, 332
    low-noise, 336
    relaxation, 333
  5–5.5 MHz, 341
  7-MHz, 337, 338
    low-drift, 342
  8-MHz, crystal, 336
  9-MHz, linear VCO, 331
  9.5-MHz, crystal, tunable, 341
  10-MHz, 341
    Colpitts, 333

RF oscillator circuits (*cont.*)
  10–20 MHz, crystal, 333
  15–65 MHz, impedance inverting, 341
  30-MHz, 333
  42.667-MHz, Pierce, 334
  50-MHz, crystal, 332
  51–55 MHz, 341
  65–110 MHz, overtone, 340
  90–125 MHz, crystal, 339
  crystal-switching diodes, 331
  doubler, 339
  precision, 338
  RC, 332
  timer as, 332, 338
  Wien-bridge, capacitively-tuned, 340
  (*See also* Crystal oscillators)
RFI-free AC switch, 510
Rhythm control, automatic remote, 289
Ring detector, 489
Ripple-preamp regulator, 434
RLC bridge, 223
RMS, squaring for, 276
Rotary switch, debouncing, 60
RPM-limit alarm, 15
RTTY, bandpass filter for, 225-Hz, 115

Sallen-key filters, 127
Sawtooth generators, 97, 152
SCA demodulator, 252
Scale changer, pen recorder, 493
Scale factor, adjustable, 274
Scanner, optoisolator as, 318
SCRs:
  PUT control for, 357
  as switches: AC-static, 511
    for zero-voltage detector, 509
  sync for firing, 511
  three-phase control with, 25
  trigger for, 600-W, 356
Secure radio, CVSD encoder for, 99
Selectors:
  pulse-edge, 378
  range, automatic, 502
  single-pulse, 375
Self-controlled autoranging, 63
Self-excited SWR bridge, 4
Self-generating FSK, 239
Sensors:
  analog voltage, 495
  current, 372
  for keyers, 44
  light, audible, 348
  metal, 230
  transistor, 225, 228
Separator, signal, 304
Sequence generator, 374
Sequential pulses, 379
Serial ASCII generator, 235
Serial data output, 62
Series gate, 30
Series-mode crystal oscillator, 1-MHz, 331
Series modulator, 249
Series pass regulator, JFET, 430
Series-shunt chopper, 36
Series-switched bandpass filter, 118
Servo circuits:
  12-VDC drive for, 457
  20 W at 60 Hz, 456
  amplifiers for: 400-Hz, 454
    line-operated, 458
  differential input and output, 457
  drives for, 455
    44-VAC, 459
    two-phase, 458
  pen driven by, 39

Servo circuits (*cont.*)
  phase shifters: Darlington, 455
    operational-integrator, 456
  power amplifier, 456
  power Darlingtons in, 459
  preamps for: dual opamp, 455
    parallel-opamp, 456
    phase-splitting, 456
  ramp control, up/down, 457
  remote control for, 454
  step-servo, 458
  tachometerless, 453
  TTL, 455
  XY recorder, 459
Set-reset latch, 316
Seven-segment display, conversion of, to
  BCD, 84, 88
Sewing-machine suppression, 296
Shrinker, pulse, 381
Shunt gate, 29
Shunt regulators:
  12-V at 5 mA, 425
  negative, 441
  positive, 429
Shutdown:
  for power supply circuit, 364
  regulator, 435, 444, 448
Sidetone monitor, 45
Sign changer, 310
Signal conditioners, 305, 317
Signal separator, 304
Signaling:
  dual-tone, 481
  single-tone, 487
Silencer, AM noise, 294
Silver-zinc battery charger, 18
Sine waves and sine-wave circuits:
  conversion of: from square-wave, 86,
    147
    from triangle-wave function generator,
    152
  generators: 0.5-Hz to 1-MHz, 147
    AF, 149
    FSK, 146
  oscillators: 25-Hz, 324
    350-Hz, stabilized, 324
    400-Hz, 325
    1–50 kHz, 321
    1.46-kHz, stabilized, 322
    2.34-kHz, 321
    3.8-kHz, 323
    100-kHz, 206, 332
    phase-shift, 324, 327
    tuning-capacitor simulator, 325
    Wien-bridge, 328, 329
  peak detector for, 10-kHz, 499
Single-ended circuits:
  JFET opamp, 310
  line driver, 103
Single-slope voltage-to-frequency converter,
  82
Single-tone signaling, 487
Single-tuned detector, 255
Slew rate:
  measuring, 54
  opamp with fast, 304, 305
S-meter, 181
Smoke, locating fire in, 160
Soft-start lighting control, 177
Soft-touch control, 34
Software controlled A/D converter, 63
Solar-cell isolation, 28
Solar-power circuits:
  backup for battery charger, 19
  battery charger, 23
  oscillator, 350

Solar-power circuits (*cont.*)
  overcharge protection, 21
Solenoid transient bypassing, 372
Solid-state-lamp amplifier, 165
Sound-level meter, 212
Sound synthesizer, 288
  regulated power supply for, 405
Speakers:
  crossovers for, active filter for, 120
  touch-tone drive for, 482
Speech:
  D/A converter for, 91
  narrow bandpass filter for, 122
  processor for, 8
Speed control, motor (*see* Motor control circuits)
Spike rejection, 298
Split-phase circuits:
  line driver, 98
  motor control, 265
Square-waves and square-wave circuits:
  1-MHz, for TDR, 382
  adjustable, 380
  AF/RF, 383
  bounceless, 58
  conversion of: to DC, 81
    to sine waves, 86, 147
  frequency divider, 144
  function generator for: 0.5-Hz to 1-MHz, 147
    AF, 149
    current controlled, 150
    FSK, 146
    variable, 151
  oscillator, 100-kHz, 206
  (*See also* Pulse generator circuits)
Squaring for RMS, 276
Squelchable amplifier, 6
SSB:
  AF for, 396
  gain control for, 158
  modulator for, 248
Staircase window comparator, 51
Stalled-motor protection, 265
Standby light, 6 V, 176
State generator for stepper motor, 264
State-variable filters, 118, 119
  700-Hz, 119
  1-Hz, 126
  1-kHz, 128
    bandpass, 136
    with Q of 10, 118
    variable gain, 127
  45-kHz, 485
Step-servo control, 458
Stepper motors:
  drive control for, 260
  state generator for, 264
Stereo decoder, 252
Strain-gage amplifier, 167
Stretcher, pulse, 385
  11X, 378
  with isolation, 375
  mono for, 373
Strobed comparators, 50, 55
Strobing, unambiguous, 385
Successive approximation A/D converter, 64
  CMOS-compatible, 70
  high-speed, 69
Sum of digital numbers, D/A converter for, 92, 95
Summing amplifiers, 303, 305, 313
  without adjustments, 309
Sump-pump control, 353
Superhet subsystem, AM, 387
Supply current monitor, 193
Suppressed-carrier AM modulator, 246, 249
Suppressors:

Suppressors (*cont.*)
  oscillation, 50
  switching-click, 8
Switch-closure counter, 58
Switched-crystals oscillator, 334
Switched regulator, adjustable, 400
Switches:
  400-VDC, 317
  AC, 353
    RFI-free, 510
  analog, 270
  bounceless make/break, 61
  closure counter for, 58
  coded, 10-digit, 369
  digital, 91
  fast, 39
  gate-formed, 59
  line-powered, 355
  one-pulse-per-push, 60
  power, 356, 471
  proximity, 32–34
  for regulated power supply, 404
  rotary, debouncing, 60
  SCR, AC static, 511
  threshold, isolated, 318
  thyristor, 354
  touch, 34, 35
  triac, AC noise-free, 512
  triac-starting, 262
  voltage-sensitive, 353
  zero-point, 352, 513
Switching:
  of 4500 W up to 10 kHz, 356
  zero-voltage, 513
Switching-click suppressor, 8
Switching gate, audio, 9
Switching mode motor control, 258
Switching regulator circuits:
  −15 V: from +5 V, 472
    from +12 V, 470
  −12 V at 300 mA, from −48 V, 471
  −10 V, 462
  −5 V, 476
    at 3 A, 464
    at 10 A, 460
    flyback switching, 463
  4.5–30 V at 6 A, 470
  5 V, 466, 474
    at 1 A, 461
    at 10 A, 475
    at 40 A, 20-kHz, 462
    from 24 V, 464
    off-time, fixed, 461
  5–24 V, 460
  6 V, for calculator, 468
  10 V, 465
    at 100 mA, 468
  12 V and 15 V: from 4–24 V, 467
    from 5 V, 469
  15 V: from 5 V, 462
    from 7.5–30 V, 472
  ±15 V, with tracking, 473
  24 V: at 3 A, 475, 478
    from 12 V at 2 A, 473
  28 V at 100 W, 476
  50 V at 1000 W, 477
  200 V from 5 V, 464
  250 V at 3 A, 476
  150-W, 466
  3300-W, 469
  5000-W, 467
  battery regulator, 465
  control for, 463
  high-voltage: negative switching, 468, 471
    positive switching, 464, 478
  motor controller, 258

Switching regulator circuits (*cont.*)
  multiple-output, 474
  on-time, variable, 461
  power switch for, 471
  with push-pull output, 465
  with symmetry correction, 477
  (*See also* Power supply circuits; Regulated
    power supply circuits; Regulator cir-
    cuits)
SWR bridge, self-excited, 4
Symmetry correction, regulator with, 477
Synchronization to 10-MHz, 375
Synchronous detector, AM, 391
Synchronous sawtooth, 97
Synthesizers:
  CQ call, 44
  sound, 288
    regulated power supply for, 405

Tach/dwell meter, 12
Tachometer, digital, 16
Tachometerless servo, 453
Tape-loop motor speed control, 266
TDR, square-wave for, 1-MHz, 382
Telephone circuits:
  autopatch, 483, 487
  bell simulator, 479
  busy-signal generator, 484
  call counter, 482
  caller, four-number, 488
  converter, pushbutton-to-dial, 484
  decoder, touch-tone, 481, 482, 485, 486
  dial-tone generator, 479
  dialer, touch-tone, 489
  encoder, touch-tone, 480, 482, 485–488
  key pulser, 480
  low-pass filter, state-variable, 45-kHz, 485
  mobile autopatch, 487
  ring detector, 489
  signaling: dual-tone, 481
    single-tone, 487
  toll-call killer, 481
  touch-tone band-reject filter, 483
  touch-tone drive, for speaker, 482
Temperature coefficient, opamp, low, 307
Temperature compensation in D/A converter,
  93
Temperature measuring circuits:
  absolute-temperature sensing, 228
  converters: temperature-to-frequency, 227,
    229
    temperature-to-pulse width, 228
  heat-energy integrator, 226
  monitor, LED, 227
  noncontacting, 164
  thermocouplers: amplifiers for, 225, 229
    differential, 226
    multiplexing with, 227
  transducer interface, 226
  transistor sensor, 225, 228
  zero suppression, 227
  (*See also* Thermometers)
Temperature stability, regulator with, 427
Teraohmmeter, 220
Terminals, wired-OR, 105
Terminations, twisted-pair, 104
Test tones, 321
Testers:
  capacitor, 188
    audio-tone, 183
    electrolytic, 187
    neon-lamp, 187
    with substitution, 189
  peak program meter, 180
  phase-error, 307
  pot, 221

Theremins:
  hand-waving, 286
  light-sensitive, 291
Thermal microscope, 159
Thermistor thermometer, 29
Thermocouplers:
  amplifiers for, 225, 229
  differential, 226
  multiplexing with, 227
Thermometers:
  −125 to 200°C, 225
  0 to 100°C, 228
  0 to 100°F, 227
  70–80°C, 225
  0.1°C precision, 229
  clinical, 162
  radiation, 161
  thermistor, 29
Third-electrode battery charger, 19
Three-function filters, 126
  1-kHz, 132
Three-phase control, SCR, 25
Three-phase induction motor, speed control
  for, 263
Threshold detector, 495, 500
Threshold switch, isolated, 318
Thresholds, measuring, 56
Thyristors:
  half-wave trigger for, 600-Hz, 355
  in lighting control, 176
  in motor control, 258, 259
  as switch, 354
Time, conversion of:
  to DC voltage, 84
  from DC voltage, 80
Time base, camera, 163
Time-dependent dimmer, 178
Time reference, 10-MHz, 201
Timer-dimmer, 178
Timers:
  bullet, 213
  for code practice, 43
  as oscillator, 332, 338
  with pulse-duration modulator, 245
  for pulses, 384, 386
  for voltage-to-frequency conversion, 88
Toll-call killer, 481
Tone circuits:
  bursts, 322, 326, 384
  decoder, 46
  generator, 384
    piano, 292
  lock, 3-kHz, 369
  oscillator, 238
  relay driver, 7
Touch circuits:
  alarm, 35
  control, 34, 35
  switch, 34, 35
Touch-tone circuits:
  band-reject filter, 483
  decoders, 481, 482, 485, 486
  dialer, 489
  drive, for speaker, 482
  encoders, 480, 482, 485–488
Tracked current-limiting regulator, 423
Tracking A/D converters, 67, 70
Tracking filters:
  1-MHz, 129
  10-MHz, 141
  line-frequency, 116
Tracking quad gain control, 302
Tracking regulators:
  −15 V tracking +5 V, 431
  0 V to ±6.6 V at 5 mA, 435

Tracking regulators (*cont.*)
  0 V to ±15 V at 100 mA or 2 A, 440
  ±5 V to ±18 V, 432
  ±10 V, 432
  ±15 V, 408, 429, 443, 446, 473
    at 100 mA, 445
    at 1 A, 449
  ±65 V at 1 A, 447
  100 mA, 418
  dual, 423
Trailer lights, interface for, 16
Transceivers:
  power supplies for, 403, 412
  saver circuit for, 371
Transducer interface, temperature, 226
Transformer, DC, 196
Transformerless circuits:
  chopper, 39, 40
  power supply, ±12 V at 15 mA, 360
  preregulator, 12-V, 358
  regulated power supplies: ±14 V, 406
    15 V and −6 V, 401
Transients:
  eliminator of, 363
  power-line, monitor for, 499
  solenoid, bypassing, 372
Transistor-pump discriminator, 256
Transistor temperature sensor, 225, 228
Transmitters:
  carrier-system, 254
  far-field, 1
  fiber-optic, 107, 108
  FM, oscillator for, 204
  remote terminal, 235
  voice, light-beam, 347
  (*See also* Data transmission circuits)
Tremolo circuits:
  amplifier for, 289
  control for, 287
Triac circuits:
  active-high interface for, 355
  active-low interface for, 357
  bridge-triggered, 355
  for clock comparator, 352
  control for, 1 kW, 510
  dimmers: 400-W, 175
    800-W, 174
    with brightening, 175
    time-dependent, 178
  driver for, 351, 354
  full-wave AC control, 356
  for inductive load, 354
  lighting control with, 173
  logic-triggered, 353
  MOS driver for, 355
  for motor control, 262
  optoisolator control for, 354
  protection circuit for, 352
  for suppressing relay arcing, 371
  switch, AC noise-free, 512
  trigger for, 351, 354
  voltage-sensitive switch, 353
  zero-point switch, 352
Triac-starting switch for motor control,
  259
Triangle-wave generators:
  0.5-Hz to 1-MHz, 147
  AF, 149
  current-controlled, 150
  FSK, 146
  to sine wave, 152
  variable, 151
Trickle battery charger, 20
Triggers:
  for SCR, 600-W, 356
  for triac, 351, 354

Trombone circuit, 288
TTL link for microprocessor, 110
Tube warmup control, 27
Tuned-loop oscillator in metal detector circuit, 231
Tuners:
  80-meter, 389
  FM, regulated power supply for, 414
  power supply for, hum free, 430
Tuning circuits:
  for equal temperament, 287
  indicators, LED, 253, 256
  tuning-capacitor simulator, 325
Tunnel diodes:
  level detector with, 498
  peak sensing with, 497
TVs:
  graphics on, uppercase driver for, 234
  regulated power supply for, 410
TVTMs:
  1.5 V for, 360
  linearizing scale of, 222
Twin-T bandpass filter, 1.4-kHz, 130
Twisted-pair terminations, 104
Two-phase circuits:
  function generator, 148
  oscillator, 2-kHz, 320
  servo drive, 458
Two-quadrant circuits:
  exponential control, 6
  multipliers, 275, 276, 278
Two-tone touch-tone encoder, 488

UART interface, 234
UHF circuits:
  prescaler, 145
  wattmeter, 217
UJT circuits:
  battery charger, 22
  for lighting control, 178
  pulser, 383
Under-over-normal lamps for voltage monitor, 496
Unity-gain circuits:
  AF current amplifier, 303
  feed-forward opamp, 305
  high-pass filter, 129
  low-pass filters, 129, 137
Universal regulated power supply, 418
Uppercase driver for TV graphics, 234
Up-to-speed logic for motor control, 261

Varactors, power supply for, 407
Variable AC voltage for lighting control, 176
Variable-bandwidth bandpass filter, 135
Variable comparator, 52
Variable dead-band response, 299
Variable frequency circuits:
  filter, 137
  oscillators (see VFOs)
  switching regulator circuit, 463
Variable-gain circuits:
  differential-input amplifier, 170
  state-variable filter, 127
Variable-Q filter circuits, 137
  10-kHz, 117
  AF bandpass, 139
  for CW, 122
Variable resistor, MOS transistor as, 154
VCOs:
  AF oscillator for, 1-kHz, 329
  in FSK decoder, 241
  linear, 9-MHz, 331
  in modem circuit generator, 242
  in sound synthesizer, 288

VFOs:
  1.5–5.7 MHz, optically-coupled, 318
  3.955–4.455 MHz, 333
  to check receiver, 388
VHF power amplifier, 251
Video circuits:
  burst peak detector, 491
  compressor, 64
VMOS, interface for, 356
Voice circuits:
  bandpass filter, 114
  digitizer, A/D converter for, 71
  transmitter, light-beam, 347
Voltage-controlled circuits:
  attenuator, 7
  function generator, 147
    two-phase, 148
  gain, 158
  hysteresis, 53
  monostable multivibrator, 280
  oscillators (See VCOs)
  phase-shifter, 344
Voltage follower, 313
Voltage level detector circuits, 491
  5 mV to 10 V, 495
  5.6–5.8 V, Go-No-Go, 490
  amplifier, dual sense, 499
  core pulse-height, 492
  counter and detector, 494
  envelope detector, 495
  envelope peak, 497
  equal voltage, 495
  hysteresis: adjustable, 497
    design of, 501
    voltage-dropping, 500
  lamp, logic-level, 498
  monitors: ambiguity eliminator, 496
    multiple-voltage, 498
    three-voltage, 501
    transients, power-line, 499
    under-over-normal lamps for, 496
    voltage, fail-safe, 492
    voltage limit, 499
  nine-input, 501
  peak, 494, 500
    10-kHz sine wave, 499
    holding, 498
    inverting, 494
    narrow-pulse, 493
    sense and hold, 492
    sensing, with tunnel diode, 497
  pen recorder scale changer, 493
  pulses: fast-rise, 502
    amplitude-polarity sorter, 496
  range selector, automatic, 502
  sensor, analog voltage, 495
  threshold, 495
  tunnel diode, 498
  video burst peak, 491
  voltage sensing, 502
  voltmeter, automatic, 490
  window detector, 492
  zener, power, 490
  (See also Regulator circuits)
Voltage reference circuits:
  −4000 V bridge, 503
  4–6 V, secondary standard, 508
  4.5 V, cell replacement, 507, 508
  5 V, 503
    FET, 505
  ±5 V, 505
  5.3 V at 1.5 A, differential, 505
  6 V at 5 mA, 507
  6.2 V, with power zener, 504
  8.2 V, 506, 507
  10 V, 504

Voltage reference circuits (cont.)
  auto, 13
  constant current, for zener, 506
  lamps for, 26
  multiple-voltage, 504
  projection-lamp, 174
  variable, 506, 507
  zener: 4.5 V, 508
    adjustable, 504
    cascaded, 503
    constant current for, 506
    differential, 504
    power, 504
Voltage-sensitive switch, 353
Voltage-tuned filters:
  1 Hz–500 kHz, bandpass, 128
  10-kHz, 121
Voltage-window comparator, 57
Voltages:
  comparator of, 53
  conversion of: from current, 81, 195, 197, 198
    to current, 86
    to frequency, 78–84, 87, 88, 313
      Go/No-go indicator using, 80
    to pulse width, 82, 86
    to pulse width modulator, 245
  indicator of, LED, 19, 20
  level detector (see Voltage level detector circuits)
  reducer of, for lighting control, 176
Voltmeters:
  AF, 180
  automatic, 490
  for capacitance meter, 185, 191
  follower, 220
  logarithmic null, 28
VSWR circuits, 1, 2
VTVMs:
  capacitance meter, 186
  linearizing scale of, 222

Warmup control, tube, 27
Water-conductivity gauge, 222
Water-level motor control, 260
Watthour meter, 217
Wattmeters:
  CB, 216
  instantaneous, 217
  neon, 216
  RF, 216
    0.025–10 W, 218
    0–5 W, 218
    0–100 and 1000 W, 216
    5–300 W, 217
    optoelectronic, 218
  tester for peak program, 180
  UHF, 217
Waveform-edge pulser, 384
Weak signals, amplifier for, 169
Wheatstone bridge, 224
White-noise, 50–5000 Hz, 295
Wideband chopper, 39
Wideband filter, 300–3000 Hz, 123
Widener, pulse, 379
Wien-bridge notch filter, 50-Hz, 126
Wien-bridge oscillators, 26
  10-Hz, 323
  10 Hz–5 MHz, 28
  20–20,000 Hz, 325
  50–30,000 Hz, 324
  100-Hz, 325
  200–65,000 Hz, 327
  340–3400 Hz, single-pot, 327
  1-kHz, 324, 328
  1.6-kHz, sine-wave, 328

Wien-bridge oscillators (*cont.*)
  10.5-kHz, zener-controlled, 326
  100-kHz, 342
  2-W, 328
  AF/RF, 324
  capacitively-tuned, 340
  current-controlled, 323
  low-distortion, 29
  sine-wave, 329
  single-pot, 327
Wind-direction indicator, 212
Wind speed, 166, 209
Windows:
  comparators for: 1.9–2.1 V, 51
    internally gated, 51
    staircase, 51
  detectors, 56, 492
  variable, 50
Wiper-delay control, 15
Wired-OR terminals, 105

Wobbulator oscillator, 450–500 kHz, 330
WWV, 10-MHz fixed for, 395

XY recorder servo, 459

Zapper, nicad, 20
Zener circuits:
  adjustable, 504
  bridge feedback in zero-voltage detector,
    513
  constant current for, 506
  noise generator, 296
  power, 490, 504
  references, 54
    4.5 V, 508
    cascaded, 503
    differential, 504
  regulator stabilized by, 426

Zener circuits (*cont.*)
  Wien-bridge oscillator, 326
Zero-point switches, 352, 513
Zero suppression in temperature measuring
  circuit, 227
Zero-voltage detector circuits:
  200–10,000 Hz, 512
  discriminator, 512
  high-speed, 511
  precision, 512
  SCRs: switches for, 509, 511
    sync for firing, 511
  switches, 513
    AC, RFI-free, 510
    SCR, AC static, 511
    triac, AC noise-free, 512
  switching, 8-kW, 511
  triac control, 1 kW, 510
  zener-bridge feedback, 513
  zero crossover, 509, 510

MARKUS

ABOUT THE EDITORS

JOHN MARKUS (deceased) was a professional writer and served as a special consultant to McGraw-Hill Book Company, an organization he was associated with for 27 years before striking out on his own as a writer and consultant.

He held many positions at McGraw-Hill, including that of feature editor on *Electronics* magazine. In this capacity he was responsible for many state-of-the-art reports in the field of electronics. One of these reports earned him the Jesse H. Neal Editorial Award for outstanding journalism.

He also served as technical director on the Technical Research Staff of the McGraw-Hill Book Company, applying electronic techniques to the mechanization of information publishing systems.

Mr. Markus was a senior member of the Institute of Electrical and Electronics Engineers. He authored, coauthored, and edited over 15 books for McGraw-Hill, including *Electronics Dictionary*, Fourth Edition, and *Sourcebook of Electronic Circuits*. He was also a consulting editor and contributing editor to the *McGraw-Hill Dictionary of Scientific and Technical Terms*, and contributed over 30 articles to the *McGraw-Hill Encyclopedia of Science and Technology*.

CHARLES D. WESTON has accumulated over twenty years of experience in the fields of computer design engineering, oceanography, audio-visual product design, and manufacturing engineering. He has been a technical writer for a number of years, and is currently employed as a technical editor at *Portable Computer Review* magazine.